Topics in Astroparticle and Underground Physics 2009

(TAUP 2009)

Journal of Physics: Conference Series Volume 203

Rome, Italy
1-5 July 2009

ISBN: 978-1-61738-274-1

Printed from e-media with permission by:

Curran Associates, Inc.
57 Morehouse Lane
Red Hook, NY 12571

Some format issues inherent in the e-media version may also appear in this print version.

Printed by Curran Associates, Inc. (2010)

For permission requests, please contact the Institute of Physics
at the address below.

Institute of Physics
Dirac House, Temple Back
Bristol BS1 6BE UK

Phone: 44 1 17 929 7481
Fax: 44 1 17 920 0979

techtracking@iop.org

Additional copies of this publication are available from:

Curran Associates, Inc.
57 Morehouse Lane
Red Hook, NY 12571 USA
Phone: 845-758-0400
Fax: 845-758-2634
Email: curran@proceedings.com
Web: www.proceedings.com

TABLE OF CONTENTS

WORKSHOP SESSION: COSMOLOGICAL ISSUES

WORKSHOP SESSION: DOUBLE BETA DECAY

WORKSHOP SESSION: GRAVITATIONAL WAVES AND NEUTRINOS

WORKSHOP SESSION: SOLAR, REACTOR AND LOW-ENERGY NEUTRINOS

WORKSHOP SESSION: ATMOSPHERIC NEUTRINOS AND HIGH ENERGY NEUTRINO BEAMS

WORKSHOP SESSION: EDUCATION AND OUTREACH

WORKSHOP SESSION: HIGH ENERGY ASTROPHYSICS

POSTER SESSION

This volume is dedicated to the memory of Julio Morales who passed away on Saturday July 4, 2009 at the age of 61 after a strenuous battle with cancer.

He had devoted most of his intense professional career to the field of underground physics, and contributed in an essential way to the foundation and development of the Canfranc Laboratory. The TAUP Conferences owes him a great debt for a long collaboration with, for the remarkable organization of the TAUP session in 2005 in Zaragoza and his activity as a member of the Steering Committee. In Julio Morales, our community has lost a highly appreciated colleague, a charming person and a great friend. He will be missed by all of his colleagues.

The eleventh meeting of the TAUP Workshop series, TAUP 2009, was held in Rome on 1–5 July 2009, organized by the INFN Gran Sasso National Laboratory (LNGS).

The workshop venue was originally L'Aquila. The earthquake of 6 April changed our plans. The earthquake killed about 300 people, wounded about 15,000, made homeless about 60,000 and destroyed a large part of the built environment, as well as essential infrastructures. The LNGS buildings were not seriously affected by the event, and the underground site not at all. But after the event most of the LNGS staff were homeless and the local hotels were damaged.

So we were forced to move the Workshop to Rome, and found a good venue in the Angelicum, the Pontifical University of Saint Thomas Aquinas, in the very center of Rome. This University, founded in 1580, has its historical roots in the medieval House of Studies of the Dominican Order in Rome, and we think that all the TAUP participants enjoyed the spiritual atmosphere which emanated from the ancient walls of the building and the cloister.

All the various aspects of our interdisciplinary field of research at the intersection of Particle Physics, Astrophysics and Cosmology, named Astroparticle Physics, have been covered: Dark Components of the Universe, Gravitational Waves, Neutrino Physics and Astrophysics, High Energy Astrophysics, Cosmic Rays and Gamma-Rays Astronomy. Exciting scientific results were presented and the main issues of the field debated in the plenary and parallel sessions. The program also included a Session on Outreach and Education, with contributions from major Laboratories, Experimental Collaborations and International Organizations.

Unhappily, the sad news from Zaragoza of the passing of our dear colleague and friend Julio Morales, to whom these Proceedings are dedicated, marked the last day of the Workshop.

On behalf of the Organizing Committee of TAUP 2009, we thank all the Speakers and Conveners, who made possible a detailed vision of the field, and the members of the International Advisory Committee and of the TAUP Steering Committee for assistance and advice on the scientific program. The support received by IUPAP and by the National Institute for Nuclear Physics (INFN) is gratefully acknowledged.

Eugenio Coccia
Luciano Pandola
Nicolao Fornengo
Roberto Aloisio

A personal special thank you to my fellow organizers, to the Secretariat and to the LNGS staff. They worked successfully for TAUP 2009 in a difficult moment for our Laboratory, and even more, for several of them, in a difficult moment of their lives.

Eugenio Coccia
Chairman of the TAUP 2009 Organizing Committee

TAUP 2009 has been conducted in accordance with IUPAP principles as stated in the ICSU-Document 'Universality of Science' (sixth edition, 1989) regarding the free circulation of scientists for international purposes. In particular, no bona fide scientist have been excluded from participation on the grounds of national origin, nationality, or political considerations unrelated to science.

TAUP STEERING COMMITTEE

F T Avignone, University of South Carolina
B C Barish, CALTECH
E Bellotti, University of Milano, INFN
J Bernabeu, University of Valencia
A Bottino (Chair), University of Torino, INFN
N Fornengo, University of Torino, INFN
T Kajita, ICRR University of Tokyo
C W Kim, Johns Hopkins University, KIAS
V Matveev, INR Moscow
J Morales, University of Zaragoza
G Raffelt, MPI Munchen
D Sinclair, University of Carleton
M Spiro, IN2P3

TAUP 2009 INTERNATIONAL ADVISORY COMMITTEE

J J Aubert, CNRS Marseille
M Baldo-Ceolin, University of Padova, INFN
G Bellini, University of Milano, INFN
L Bergstrom, University of Stockholm
R Bernabei, University of Roma Tor Vergata, INFN
A Bettini, University of Padova, INFN, LSC
S Bilenky, JINR Dubna
D O Caldwell, UCSB
J Cronin, University of Chicago
A Dar, Technion Haifa
G Domogatsky, INR Moscow
J Ellis, CERN
E Fernandez, IFAE Barcelona
E Fiorini, University of Milano, INFN
T Gaisser, University of Delaware
G Gelmini, UCLA
G Gerbier, CEA Saclay
A Giazotto, INFN Pisa
F Halzen, University of Wisconsin
W Haxton, University of Washington
T Kirsten MPI Heidelberg
L Maiani, University of Roma La Sapienza, INFN
A McDonald, Queen's University
K Nakamura, KEK
R Petronzio, INFN, University of Roma Tor Vergata
L Resvanis, University of Athens
F Ronga INFN, LNF
C Rubbia INFN, LNGS

A Smirnov, ICTP Trieste
C Spiering, DESY
N Spooner, University of Sheffield
A Suzuki, KEK
S Ting MIT, CERN
M S Turner, FNAL, University of Chicago
J W F Valle, IFIC Valencia
D Vignaud, APC Paris
G Zatsepin, INR Moscow

TAUP 2009 ORGANIZING COMMITTEE

R Aloisio, LNGS
R Antolini, LNGS
F Arneodo, LNGS
Z Berezhiani, University of L'Aquila, INFN
V Berezinsky, LNGS
R Cerulli, LNGS
E Coccia [Chair], LNGS/INFN, U of Roma Tor Vergata
N D'Ambrosio, LNGS
N Fornengo, University of Torino, INFN
M Laubenstein, LNGS
O Palamara, LNGS
L Pandola [Scientific Secretary], LNGS

Journal of Physics: Conference Series **203** (2010) 011004 doi:10.1088/1742-6596/203/1/011004

Conference Program

Eleventh International Conference on Topics in Astroparticle and Underground Physics

TAUP 2009

1-5 July 2009, Rome, Italy

PLENARY SESSIONS

COSMOLOGY

(Chair: *Daniel Vignaud*)

PROGRESS AND PROBLEMS IN COSMOLOGY — *Alexander Vilenkin (Tufts University, Medford, MA, USA)*
THE PHYSICS OF THE COSMIC MICROWAVE BACKGROUND — *Andrew Jaffe (Imperial College, UK)*
DARK ENERGY PROBES — *Enrique Gaztanaga (Instituto de Ciencias del Espacio, CSIC, Spain)*
THE DARK MATTER STRUCTURE OF THE UNIVERSE — *Ravi Sheth (University of Pennsylvania, USA)*

GRAVITATIONAL WAVES

(Chair: *Lucia Votano*)

THE PRESENT GRAVITATIONAL WAVE DETECTION EFFORT — *Keith Riles (University of Michigan, USA)*
ADVANCED GRAVITATIONAL WAVE DETECTORS — *Giovanni Losurdo (INFN/Firenze, Italy)*

DIRECT DARK MATTER SEARCHES

(Chair: *Manfred Lindner, Milla Baldo Ceolin*)

RESULTS FROM DAMA/LIBRA EXPERIMENT — *Pierluigi Belli (INFN/Roma Tor Vergata, Italy)*
RESULTS FROM THE CDMS EXPERIMENT — *Jody Cooley-Sekula (Stanford University, USA)*
RESULTS FROM THE XENON EXPERIMENT — *Elena Aprile (Columbia University, USA)*
THE WARP EXPERIMENT — *Claudio Montanari (INFN/Pavia, Italy)*
PARTICLE DARK MATTER INTERPRETATIONS OF DIRECT SEARCHES — *Alessandro Bottino (INFN/Torino, Italy)*
AXION SEARCHES — *Leslie Rosenberg (University of Washington, USA)*

NEUTRINOS

(Chair: *David Sinclair, Henry Sobel, Vladimir Gavrin*)

PROGRESS IN THE UNDERSTANDING OF NEUTRINO PROPERTIES — *Jose W.F. Valle (Universidad de Valencia/IFIC, Spain)*
FRONTIERS OF LOW ENERGY NEUTRINO PHYSICS AND ASTROPHYSICS — *Gioacchino Ranucci (INFN Milano, Italy)*
STATUS AND PERSPECTIVES OF SHORT BASELINE STUDIES — *Mark Dierckxsens (University of Chicago, USA)*
STATUS AND PERSPECTIVES OF ATMOSPHERIC NEUTRINOS AND LONG BASELINE STUDIES — *Takaaki Kajita (ICRR and University of Tokyo, Japan)*
NEUTRINOS FROM CERN TO GRAN SASSO: THE CNGS PROJECT — *Marcos Dracos (IPHC Strasbourg, France)*
NEUTRINOLESS DOUBLE BETA DECAY — *Stefan Schoenert (Max-Planck Institute for Nuclear Physics, Heidelberg, Germany)*
STATUS OF THE SUDBURY NEUTRINO OBSERVATORY — *Joshua Klein (University of Pennsylvania, USA)*
SUPERNOVA NEUTRINO OSCILLATIONS: WHAT DO WE UNDERSTAND? — *Amol Dighe (Tata Institute, Mumbai, India)*

Cosmic Rays and High Energy Neutrinos
(Chair: *Venyamin Berezinsky*)

OBSERVATIONS OF UHE COSMIC RAYS — *Giorgio Matthiae (Universita' di Roma Tor Vergata and INFN, Italy)*
THE HIGH ENERGY UNIVERSE: COSMIC RAYS — *Pasquale Blasi (INAF/Arcetri, Italy and Fermilab, USA)*
UHE COSMIC NEUTRINOS: THEORETICAL ASPECTS — *Michael Kachelriess (Institute for Physics, Trondheim, Norway)*
DETECTION OF HIGH ENERGY NEUTRINOS — *Gregory Sullivan (University of Maryland, USA)*

Gamma Rays and Antimatter in Space
(Chair: *Jose Bernabeu, Alessandro Bettini*)

ASTROPHYSICAL SOURCES OF GAMMA-RAYS — *Isabelle Grenier (CEA and AIM, France)*
GROUND BASED OBSERVATIONS OF GAMMA RAYS — *Mosè Mariotti (INFN/Padova)*
SPACE BASED OBSERVATIONS OF GAMMA RAYS — *Philippe Bruel (LLR - CNRS/IN2P3)*
DETECTION OF ANTIMATTER IN OUR GALAXY — *Piergiorgio Picozza (INFN/Roma Tor Vergata, Italy)*
DARK MATTER SIGNALS IN SPACE — *Pierre Salati (LAPTH/Annecy, France)*

Underground Laboratories

OVERVIEW OF UNDERGROUND LABORATORIES AND FACILITIES — *Eugenio Coccia (INFN/LNGS and INFN/Tor Vergata)*

PARALLEL SESSIONS

Parallel Session on Cosmology
(Conveners: *A. Diaferio, T. Moroi, K. Olive, M. Shaposhnikov*)

Gravitational reheating in quintessential inflation — *Stefano Scopel (Seoul National University)*
Relic density of neutrinos with primordial asymmetries — *Sergio Pastor IFIC - U. Valencia)*
Interplay between the ``Low'' and ``High'' Energy CP-violation in Leptogenesis — *Serguey Petcov (SISSA/INFN Trieste)*
Constraining Super-Critical String/Brane Cosmologies with Astrophysical Data — *Vasiliki Mitsou (IFIC - U. Valencia)*
Phantom dark energy and cosmological solutions without the Big Bang singularity — *Anton Baushev (JINR, Dubna)*
Dirac-Born-Infeld and k-inflation: the CMB anisotropies from string theory — *Christophe Ringeval (Louvain University)*

Parallel Session on Gravitational Waves
(Conveners: *A. Giazotto, E. Katsavounidis, K. Kuroda*)

Status of Virgo — *Viviana Fafone (INFN, Sezione di Roma Tor Vergata)*
Underground Cryogenic Interferometer CLIO — *Shinji Miyoki (ICRR, The University of Tokyo)*
Cosmic rays detection with Explorer and Nautilus — *Carlo Ligi (INFN Frascati)*
ET design study: science case and perspectives — *Fulvio Ricci (University of Rome La Sapienza & INFN)*
ET site selection: seismic and gravity gradient noise — *Jo van den Brand (Nikhef)*
Recent results from LIGO, GEO and Virgo searches — *Carlo Nicola Colacino (INFN Pisa)*
Search for bursts with LIGO, GEO and Virgo — *Nicolas Leroy (University of Paris-Orsay)*
Status of neutrino detectors — *Walter Fulgione (INAF - IFSI Torino)*
Search for neutrinos from core-collapse supernova from the global network of detectors — *Alex Habig (University of Minnesota)*
The future of neutrino detectors *Kate* — *Scholberg (Duke University)*
Supernovae rates, models and waveforms — *Christian Ott (CalTech)*
Using neutrinos to search for GW signals — *Giulia Pagliaroli (INFN, Gran Sasso)*
Multiwavelength observations with LIGO — *Laura Cadonati (University of Massachusetts-Amherst)*
Gravitational wave-neutrino joint search for core-collapse supernovae — *Erik Katsavounidis (MIT)*

Parallel Session on Dark Matter searches

(Conveners: *F. Donato, J. Morales, C. Munoz, N. Spooner*)

The ArDM experiment (preliminary)	*Christian Regenfus (University of Zürich)*
Development of the MiniCLEAN dark matter detector	*James Nikkel (Yale University)*
Results from the first science run of ZEPLIN-III	*Isabel Lopes (LIP-Coimbra)*
Improved understanding of liquid xenon's response in dark matter searches	*Aaron Manalaysay (University of Zurich)*
The XMASS experiment at the Kamioka Observatory	*Kai Martens (IPMU, The University of Tokyo)*
The LUX Dark Matter Search	*Dan McKinsey (Yale University)*
Investigation of ambient and muon-induced neutron background in LSM	*Klaus Eitel (Forschungszentrum Karlsruhe)*
Status and Prospects of a Deep Underground Laboratory in China	*Qian Yue (Tsinghua University)*
Results from DMTPC 10-liter prototype	*Denis Dujmic (MIT)*
Underground Low Flux Neutron Background Measurements in LSM	*Ilias Savvidis (Aristotle University of Thessaloniki)*
DRIFT update	*Mark Pipe (University of Sheffield)*
Dark Matter Searches with the PICASSO experiment at SNOLab	*Berta Beltran (University of Alberta)*
COUPP	*Andrew Sonnenschein (Fermilab)*
Dark Matter Search with CCDs (DAMIC)	*Juan Estrada (Fermilab)*
Limits on Low-Mass WIMP Dark Matter with an Ultra-Low-Energy Ge at 220 eV Threshold	*Henry Wong (Academia Sinica, Taiwan)*
LIPSS Experiment: Free-Electron Laser in Search for Dark Matter Candidates	*Andrei Afanasev (Hampton University/Jefferson Lab)*
Axion Dark matter Revisited	*Luca Visinelli (University of Utah)*
Latests results from the CERN Axion Solar Experiment	*Igor Irastorza (University of Zaragoza)*
Dark Matter Searches with the experiment EDELWEISS-II	*Véronique Sanglard (Institut de Physique Nucléaire de Lyon)*
BGO scintillating bolometer: its application in dark matter experiments	*Ysrael Richard Ortigoza Paredes (University of Zaragoza)*
The expected background spectrum in NaI dark matter detectors and the DAMA result	*Vitaly Kudryavtsev (University of Sheffield)*
Technical aspects and dark matter searches	*Francesco Nozzoli (Università e INFN Roma Tor Vergata)*
Inert Doublet Model and DAMA: elastic or inelastic dark matter candidates	*Chiara Arina (Universite Libre de Bruxelles)*
Relevance of ion channeling for direct dark matter detection	*Graciela Gelmini (Physics Dept. UCLA)*
Elastic and Inelastic LSP-Nucleus Scattering on Medium-Heavy Nuclei	*Jouni Suhonen (University of Jyväskylä)*
ANAIS collaboration. Status Report on the ANAIS experiment	*Carlos Pobes (University of Zaragoza)*
Tidal effect in the Radon-due neutron flux from the Earth's crust	*Yuri Stenkin (Inst. for Nucl. Research of Russian Academy of Sciences)*
Dark matter search by means of segmented scintillator (PICO-LON)	*Ken-Ichi Fushimi (The Univ. of Tokushima)*
(In)visible Z' and dark matter	*Yann Mambrini (LPT, Orsay)*
Dark matter annihilation at cosmological redshifts: possible relic signal from WIMPs annihilation	*Anton Baushev (JNR, Dubna)*
Right-handed sneutrino as WIMP dark matter in the NMSSM	*David G. Cerdeno (Universidad Autonoma de Madrid)*
Sneutrino Dark Matter	*Bhaskar Dutta (Texas A&M University)*
Sneutrino as DM candidate and dynamical generation of neutrino masses	*Federica Bazzocchi (Vrije Universitet)*
Light neutralinos and neutrino signal	*Viviana Niro (MPIK, Heidelberg)*
Pamela and Fermi data: A new background for future dark matter searches?	*Carlos Yaguna (Universidad Autonoma de Madrid)*
Results and Prospects of Dark Matter Searches with IceCube	*Gustav Wikström (Stockholm University)*
Astrophysical boost factor and Dark Matter indirect detection	*Timur Delahaye (Torino University / LAPTh Annecy)*
Galactic positrons and electrons from dark matter and astrophysical sources	*Roberto Lineros (Torino University)*

PARALLEL SESSION ON DOUBLE BETA DECAY
(Conveners: *F. Avignone, C. Cattadori, A. Faessler, F. Piquemal*)

Brief Overview of Experimental Decay, and the Majorana Project	*Steve Elliott (Los Alamos National Lab)*
Theoretical status of the double beta decay	*Amand Faessler (Univ. Tuebingen)*
Development and installation of the GERDA experiment	*Anatoly Smolnikov (INR/JINR Dubna)*
Probing the Majorana Nature of Neutrinos by the Neutrinoless Double Beta decay	*Stefano Morisi (Univ. of Valencia)*
Variances and Covariances of the Nuclear Matrix Elements for the Neutrinoless Double Beta Decay	*Anna Maria Rotunno (Univ. of Bari)*
New Concepts for the Gaseous Xenon Detector for Double-Beta Decay	*David Sinclair (Carleton University)*
Test of the Single State Dominance Hypothesis for the Two-Neutrino Double Beta Decay	*Oscar Moreno (Univ. Complutense, Madrid)*
MOON for Double-Beta Decays and Neutrino Nuclear Responses	*Ken-Ichi Fushimi (Osaka University)*
The NEMO-III and SuperNemo	*Francois Mauger (LPC, Caen)*
Neutrino mass, neutrinoless double EC and rare beta decays	*Jouni Suhonen (Univ. Jyväskylä)*
Searching for Double-Beta Decay with the Enriched Xenon Observatory	*Lisa Kaufman (University of Maryland)*
The effective neutrio mass at the milli-eV-frontier	*Serguey Petcov (SISSA, Trieste)*
NEXT: Searching for the Neutrinoless Double Beta Decay with a Gas-Xenon TPC	*Pau Novella (CIEMAT)*
Effective Operators for the pionic Contributions to the 0ν Double Beta Decay derived from Quarks	*John Vergados (Univ. of Ioannina)*
From CUORICINO to CUORE: Investigating Neutrino Properties with 2β Decay	*Monica Sisti (INFN Milano Bicocca and INFN)*
The Lepton Number Violating Double Electron Capture	*Alexander Merle (Max Planck Inst., Heidelberg)*
Status of the COBRA Experiment	*Kai Zuber (Technische Universitaet Dresden)*
The DCBA Experiment for Studying Neutrino less Double-Beta Decay	*Nubuhiro Ishihara (KEK)*
Double Beta-Decay of ^{109}Cd with the TGV-2 Experiment	*Nikolay Rukhadze (JINR Dubna)*
Candles for the Study of Neutrinoless Double-Beta Decay	*Izumi Ogawa (Osaka University)*
SNO+	*Carsten Krauss (Alberta U.)*

PARALLEL SESSION ON SOLAR, REACTOR AND OTHER LOW-ENERGY NEUTRINO PHYSICS
(Conveners: ***M. Chen, K.T. Inoue, V. Matveev, G. Raffelt*****)**

Results of the Borexino experiment — *Lothar Oberauer (Technical University Munich)*
Low energy neutrino physics at Super-Kamiokande — *Michael Smy (University of California, Irvine)*
InP solid state detector for a measurement of low energy solar neutrinos — *Yoshiyuki Fukuda (Miyagi University of Education)*
The neutrino source experiment on SAGE two-zone gallium target — *Valery Gorbachev (INR, Moscow)*
Linear solar models: the role of opacity and metals in the sun — *Francesco Villante (University L'Aquila)*
Probing non-standard neutrino interactions with solar and reactor neutrinos — *Mariam Tortola (University Hamburg)*
Remaining inconsistencies with solar neutrinos: Can spin flavour precession provide a clue? — *Joao Pulido (CFTP and IST, Lisbon)*
Solar flare in underground megaton detectors — *Daniele Fargion (University and INFN Rome 1)*
Geo-neutrinos — *Stephen Dye (University of Hawaii)*
A search for supernova relic neutrino at Super-Kamiokande — *Takashi Iida (ICRR, University of Tokyo)*
Diffuse neutrino flux from failed supernovae — *Cecilia Lunardini (Arizona State University)*
Neutrino and antineutrino spectral splits from collective effects in supernovae — *Irene Tamborra (University and INFN Bari)*
The nuclear response of the COBRA detector to supernova neutrino energy spectra — *Theocharis Kosmas (University of Ioannina)*
Long-term study of low energy counting rate with the Large Volume Detector — *Gianmarco Bruno (INFN-LNGS and University L'Aquila)*
Reactor neutrino detection for non proliferation with the Nucifer experiment — *Amanda Porta (CEA-Saclay)*
Investigating the nuclear response of Te isotopes to supernova neutrinos — *Vaitsa Tsakstara (University of Ioannina)*
Lepton-nucleon interactions at the next-to-leading order — *Aleksandrs Aleksejevs (Memorial University Newfoundland)*
The Double Chooz Experiment - toward the high precision measurement in the quest for theta13 — *Jelena Maricic (Drexel University)*
A High-Precision Measurement of theta13 with the Daya Bay Reactor Neutrino Experiment — *Karsten Heeger (Univ. of Wisconsin)*
Moessbauer experiment with neutrinos and neutrino oscillations — *Evgeny Akhmedov (MPI Nuclear Physics, Heidelberg)*
Magnus approximation for neutrino oscillations with three flavors in matter — *A. Aguilar-Arevalo (Inst. de Ciencias Nucleares, UNAM)*
The KATRIN Experiment — *Marcus Beck (Univ. Muenster)*
Measuring neutrino mass with radioactive ions in a storage ring — *Christopher Orme (IPPP, Durham University, UK)*
Measurements of neutrino-electron scattering cross section and the EW Parameters at the Kuo-Sheng reactor neutrino laboratory — *Muhammed Deniz (Academia Sinica, Taiwan)*
Electromagnetic properties of neutrinos — *Alexander Studenikin (Moscow State University)*

PARALLEL SESSION ON ATMOSPHERIC NEUTRINOS AND HIGH ENERGY NEUTRINO BEAMS
(Conveners: ***J. Bernabeu, A. Guglielmi, E. Lisi, Y. Suzuki*****)**

Current MINOS neutrino oscillation results — *Alec Habig (Minnesota Duluth U.)*
Probing theta_13 with global neutrino data analysis — *Antonio Palazzo (IFIC, Valencia)*
Status of the T2K long baseline neutrino oscillation experiment — *Atsuko Kondo-Ichikawa (Kyoto U.)*
The ND280m detectors contribution to the T2K neutrino measurements — *Magali Besnier (Lab. Leprince Ringuet)*
CP violation studies via Beta and EC-decay channels of the same ion — *Maria-Catalina Espinoza-Hernandez (IFIC, Valencia)*
Emulsion analysis in the OPERA experiment — *Naotaka Naganawa (Nagoya U.)*
Status report of ICARUS T600 at the LNGS — *Alessandro Menegolli (INFN, Pavia)*
ArgoNeuT, a Liquid Argon TPC in a low energy neutrino beam — *Joshua Spitz (Yale U.)*
The Long Baseline Neutrino Oscillation Experiment at DUSEL — *Jelena Maricic (Drexel U.)*
New Results from the MiniBooNE Experiment at Fermilab — *Michael Shaevitz (Columbia U.)*
The Double LAr project — *Paola Sala (University of Milano and INFN)*
Non-standard neutrino interactions from low-scale seesaw models — *He Zhang (KTH, Stockholm)*
Results of nucleon decay search in Super-Kamiokande-I+II — *Kenji Kaneyuki (ICRR, Tokyo)*
Final results of a search for neutron-antineutron oscillation in SK-I — *Jun Kameda (ICRR, Tokyo)*
Oscillations of very low energy atmospheric neutrinos — *Orlando Peres (Unicamp)*
Atmospheric Neutrinos in Super-Kamiokande — *Roger Wendell (Duke U.)*
Sensitivity on Earth Core and Mantle densities using Atmospheric Nu's — *Ofelia Pisanti (U.Federico II, Naples)*
Cosmic Neutrinos: Flavor Composition at Source and Mixing Parameters — *Arman Esmaili Taklimi (IPM, Tehran)*

PARALLEL SESSION ON HIGH ENERGY ASTROPHYSICS
(Conveners: ***R. Aloisio, J. Buckley, T. Montaruli*****)**

M87 and CenA: laboratories for VHE physics of jets and near super-massive black holes — *Martin Raue (Max-Planck-Institut fuer Kernphysik, Heidelberg)*
VERITAS — *David Hanna (McGill University)*
Gamma Ray Astronomy with the ARGO-YBJ experiment — *Tristano Di Girolamo (Universita' di Napoli Federico II and INFN Napoli)*
Have Cherenkov telescopes detected a new light boson? — *Marco Roncadelli (INFN Pavia)*
CTA (Cherenkov Telescope Array) — *Robert Wagner (Max-Planck-Institute for Physics, Munich)*
Search for neutrinos of astrophysical origin with IceCube — *Andreas Gross (University of Canterbury)*
First results of the ANTARES neutrino telescope — *Juan Dios Zornoza (IFIC and Valencia U.)*
Lake Baikal: from Megaton to Gigaton — *Bair Shaybonov (JINR, Dubna)*

KM3NeT: a cubic kilometre-scale deep sea neutrino telescope in the Mediterranean Sea *Anna Rita Margiotta (Bologna Univ. and INFN)*
Limit on the diffuse flux of ultra-high energy neutrinos using the Pierre Auger Observatory *Pierre Billoir (LPHE, Paris)*
Interpretations of the positrons flux observed by PAMELA *Paolo Lipari (INFN, Roma)*
Dark Matter Searches with gamma-ray measurements *Jennifer Siegal-Gaskins (CCAPP, Ohio State University)*
Evolution of the cosmic ray anisotropy above 100 TeV *Piera Luisa Ghia (IFSI/INAF, Torino, LNGS/INFN, and IPN/CNRS, Orsay)*
Hybrid detection of UHECR with the Pierre Auger Observatory *Mariangela Settimo (INFN Lecce)*
Study of Acoustic Ultra-high energy Neutrino Detection (SAUND) phase II *Naoko Kurahashi (Stanford University)*
Neutrino oscillations in MHD supernova explosions *Shio Kawagoe (National Astronomical Observatory of Japan)*
Cosmic Ray Studies with the OPERA detector at Gran Sasso *Nicoletta Mauri (Bologna University and INFN)*

Parallel Session on Education and Outreach
(Conveners: *R. Antolini, C. Bradaschia, J. Jackson*)

How CERN deals with high visibility *Antonella Del Rosso (CERN)*
Education and Public Outreach of the Pierre Auger Observatory *Greg Snow (University of Nebraska)*
Astronomy's New Messengers: A LIGO traveling exhibit to reach out to a young adult audience
Marco Cavaglià (University of Mississippi)
European Week of Astroparticle Physics *Arnaud Marsollier (ASPERA)*
Outreach activities of KamLAND *Tadao Mitsui (Research Center for Neutrino Science, Tohoku University)*
ESOF 2010 exactly one year ahead *Enrico Predazzi (Università di Torino)*
Science in the Neighborhood *Judy Jackson (Fermilab)*
The "100 Euro" interferometer *Carlo Bradaschia (INFN - Pisa)*
Gran Sasso National Laboratory: Outreach and Communication activities *Adriano Di Giovanni (LNGS - INFN)*

POSTER SESSIONS

Poster Session on Dark Matter Searches
(Conveners: *F. Donato, J. Morales, C. Munoz, N. Spooner*)

Dark matter annihilation in the gravitational field of a black hole *Anton Baushev (JNR, Dubna)*
Dark matter search with DMTPC 10-liter prototype *Denis Dujmic (MIT)*
Development of the light readout for the ArDM experiment *William Creus (University of Zurich - CERN)*
Technique to measure the cosmic muon fluxes at the Old Canfranc Underground Laboratory
Asun Rodríguez (Universidad de Zaragoza)
Detection of fast neutrons with LiF and Al203 scintillating bolometers *Jorge Puimedón (Universidad de Zaragoza)*

Poster Session on Double Beta Decay
(Conveners: *F. Avignone, C. Cattadori, A. Faessler, F. Piquemal*)

New observation of the 2nu-2beta decay of Mo-100 to the 0_1^+ level of Ru-100 in the ARMONIA experiment
Vladimir Tretyak (Inst. for Nuclear Research, Kyiv)
Investigation of Double Beta Decay in Nd-150 and Nd-148 to Excited State of Daughter Nuclei
Alexander Barabash (Inst. for Theoretical and Experimental Physics, Moscow)
Background Level Estimations Using Pulse Shape Analysis in the New Generation Ge Experiments
Hector Gomez Maluenda (University of Zaragoza)
Development ov Segmented Germanium Detectors for the Neutrinoless Double Beta Decay Experiment GERDA
Jing Liu (Max-Planck Institute for Physics, Munich)
The BiPo detector *Susana Cebrian (University of Zaragoza)*
The GERDA Cherenkov Muon Veto *Markus Knapp (Kepler Center for Astro and Particle Physics)*
Production of enriched Stable Isotopes for Large-Scale Experiments in Underground Physics
Elena Nikitina (JCS Production Association, Electrochemical Plant, Russia)

Poster Session on Solar, Reactor and other Low Energy Neutrinos
(Conveners: *M. Chen, K.T. Inoue, V. Matveev, G. Raffelt*)

Hanohano: deep underwater neutrino observatory *Jelena Maricic (Drexel University)*
Uncertainty in the low energy stopping power and anomalous enhancement in the measured D+D fusion rates
Massimo Coraddu (INFN Cagliari)

Poster Session on High Energy Astrophysics

Journal of Physics: Conference Series **203** (2010) 011004 doi:10.1088/1742-6596/203/1/011004

(Conveners: *R. Aloisio, J. Buckley, T. Montaruli*)

Results of the radio-detection experiment HASRA (Hawaii Askaryan in Salt Radio Array) *Radovan Milincic (Drexel University)*
Multimessenger point source search: UHECR and neutrinos *Jelena Petrovic (NIKHEF)*
R-process and Neutrinos *Fernando Rossi Torres (DRCC, Unicamp, Brasil)*
Alignment, positioning and absolute pointing of the undersea ANTARES telescope
Felix Fehr (Erlangen Centre for Astroparticle Physics)
PAMELA sub-detectors capability in light nuclei identification: status and perspectives *Lucia Consiglio (INFN Napoli)*
PAMELA mission: constraints to cosmic rays propagation from nuclei *Laura Marcelli (INFN)*
Limit on the diffuse flux of ultra-high energy neutrinos using the Pierre Auger Observatory
Evgeny Akhmedov (Max Planck Institute for Nuclear Physics, Heidelberg)

FURTHER INFORMATION ON THE TAUP CONFERENCES AT: HTTP://WWW.TAUP-CONFERENCE.TO.INFN.IT/

All papers published in this volume of *Journal of Physics: Conference Series* have been peer reviewed through processes administered by the proceedings Editors. Reviews were conducted by expert referees to the professional and scientific standards expected of a proceedings journal published by IOP Publishing.

Topics in Astroparticle and Underground Physics (TAUP 2009) IOP Publishing
Journal of Physics: Conference Series **203** (2010) 012001 doi:10.1088/1742-6596/203/1/012001

Perspectives in cosmology

Alexander Vilenkin

Institute of Cosmology, Department of Physics and Astronomy,
Tufts University, Medford, MA 02155, USA

E-mail: vilenkin@cosmos.phy.tufts.edu

Abstract.
The "new standard cosmology", based on the theory of inflation, has very impressive observational support. I review some outstanding problems of the new cosmology and the global view of the universe – the multiverse – that it suggests. I focus in particular on prospects for further observational tests of inflation and of the multiverse.

1. Introduction

We live at a very interesting juncture in cosmology, when a new paradigm appears to be shaping up. Its true significance will only be clear from the perspective of the future. In this talk I will try to outline where I think we stand and where we might be going.[1] This is, of course, a tricky business. When speaking about perspectives, it's good to remember the quote from George Orwell, who wrote that even "to see what is in front of one's nose needs a constant struggle."

There are three basic questions in cosmology that we would like to answer. The first is: What is out there?, or What is the content of the universe? The second is: How did it get there?, or How did the universe evolve to its present state? And finally, What does it all mean? What is the big picture and how is it related to fundamental physics? There have been remarkable developments in all three of these areas over the last few decades.

The picture of the universe which has emerged from these developments can be called "the new standard cosmology." According to this picture, the universe has a nearly flat geometry, and its main three ingredients are dark energy (72%), dark matter (23%), and baryonic matter (less than 5%). The key role in the new cosmology is played by inflation – a period of super-fast, accelerated expansion in the early universe. Density perturbations that led to structure formation originated as quantum fluctuations during this inflationary period. Baryonic matter was formed in non-equilibrium, baryon-number-violating processes after inflation. After that, the evolution is along the lines of the hot big bang model. The new cosmology is still under construction, with details remaining largely uncertain. I will discuss some of these details in the rest of the talk.

2. What is out there?

With some luck, we may be on the verge of discovering what dark matter is. Some of the dark matter candidates are well motivated from the particle physics perspective. For example, weakly

[1] Limitations of space do not allow me to provide adequate references to the literature, so I chose to give no references at all.

interacting massive particles (WIMPs) could be produced as thermal relics in the early universe and are well suited for direct and indirect searches and for collider experiments. Axions could be produced by vacuum misalignment during inflation. There is no shortage of other candidates: supermassive particles, gravitinos, Q-balls, vortons, etc. I will stop at that, since dark matter will be discussed extensively at this meeting.

Turning now to dark energy, in Frank Wilczek's words, it is "the most mysterious fact in all of physical science." We have several dark energy candidates, but all of them have serious problems. First, the dark energy may simply be the vacuum energy, or the cosmological constant, ρ_v. The vacuum should then have pressure $P_v = -\rho_v$, with both P_v and ρ_v constant in time. A widely discussed alternative is a dynamical dark energy. Phenomenologically, it can be characterized by an equation of state,

$$P = w\rho, \tag{1}$$

where $w \neq -1$ and may be time-dependent. In quintessence models, the dark energy is attributed to a scalar field with a slowly declining potential, e.g.,

$$V(\phi) \propto \phi^{-\alpha}. \tag{2}$$

As the field rolls down the potential, its energy density gradually gets smaller. The dark energy is so small today because the universe is so old, and the field ϕ had time to roll to very large values. Finally, there are modified gravity models, which explain the observed accelerated expansion by deviations from Einstein's gravity. Of all three alternatives, I think the front runner is the cosmological constant, for the reasons that I will now explain.

As I have mentioned, the cosmological constant has its problems. The first of these is the old cosmological constant problem, which has been with us for nearly half a century. All quantum fields, like the electromagnetic field, contribute to the vacuum energy. This contribution is given by the sum of zero-point oscillation energies over modes with all possible frequencies. This is a divergent sum, and we should cut it off at some mass scale M, which may be set, for example, by the supersymmetry breaking scale,

$$\rho_v = \frac{1}{V}\sum \frac{1}{2}\omega \sim M^4. \tag{3}$$

(I use the system of units where $\hbar = c = 1$.) With $M \sim 1$ TeV, this gives $\rho_v \sim 10^{60}\rho_m$, where ρ_m is the average density of matter, while observations indicate that $\rho_v \sim 3\rho_m$. Fermions and bosons contribute to ρ_v with opposite signs, but we would need cancellation up to the 60th decimal point in order to explain the observed value.

Another problem is related to the fact that the vacuum energy density remains constant in time, while the matter density is diluted by the expansion. Why then do we happen to live at the very special epoch when the two densities are comparable? This is the so-called coincidence problem.

As of now, the only viable resolution of these problems is provided by the anthropic approach. It assumes that ρ_v takes a wide range of values in different parts of the universe. Of course, this can work only if the underlying particle physics admits many vacua with different values of ρ_v. This picture has some support from string theory, and I will get back to it toward the end of the talk. For now, let us assume that ρ_v is indeed variable from one region to another.

The vacuum energy acts as a repulsive force, and it prevents galaxies from forming once it comes to dominate the universe. So, a large ρ_v means no galaxies. We can calculate the probability distribution for an observer randomly picked in the universe to measure a given value of ρ_v. In regions with large values of ρ_v there are no galaxies and therefore no observers, so the probability is strongly suppressed. Very small values, where ρ_v is much smaller than it needs to be for galaxy formation, are also rather unlikely. One finds that the observed value of

ρ_v is well within the 2σ range of the resulting distribution. It is worth pointing out that the anthropic prediction of $\rho_v \sim \rho_m$ was made well before the observation.[2]

The main drawback of the alternative dynamical dark energy models is that they do not address the cosmological constant problems. For example, quintessence models assume that the dark energy density approaches zero as $\phi \to \infty$, while in general it should approach a nonzero vacuum energy density ρ_v. These models also offer no solution to the coincidence problem. The same criticisms apply to the modified gravity approach. Moreover, the equation of state parameter w derived from the WMAP data combined with supernova observations is

$$w = -0.97 \pm 0.07, \tag{4}$$

which is very close to the cosmological constant value $w = -1$.

Apart from the three main components of the universe, what else may be out there? We know there are ultrahigh-energy cosmic rays. Until recently their origin appeared mysterious, but now it seems that they may have a rather mundane astrophysical origin.

There is much to be learned about cosmic magnetic fields. We have a rather sketchy information about the field distribution on the largest scales, and the origin of the magnetic fields remains a mystery.

We believe there should be gravitational waves, and we expect to detect them in the coming decade.

There are also some more exotic things which may or may not be there. For example, topological defects, such as cosmic strings.

Of course, there may also be some surprises – things we have not thought about which will change the direction of current research in new and unexpected ways.

I would like to discuss cosmic strings in a bit more detail. Strings are linear defects which could be formed at a phase transition in the early universe. They are predicted in a wide class of elementary particle models. Some superstring-inspired models suggest that fundamental strings may also have astronomical dimensions and play the role of cosmic strings.

Once they are formed, strings evolve in a scale-invariant manner. A typical horizon volume at any time has some long strings stretching across it and a large number of small closed loops. The average number of long strings per horizon is[3]

$$N \sim 10/p, \tag{5}$$

where p is the probability for two intersecting strings to reconnect. Field-theory, solitonic strings always reconnect, so $p = 1$, while for superstrings $p \ll 1$. This opens an exciting possibility of directly testing superstring theory. If we detect $N \gg 1$ long strings in the sky, we will know that we are looking at superstrings.

Strings can be detected through a variety of observational effects. Oscillating loops of string emit gravitational waves – both bursts and a stochastic background. They can also be sources of ultrahigh-energy cosmic rays. Long strings can act as gravitational lenses and can produce characteristic signatures in the CMB: discontinuous temperature jumps along the strings and

[2] I would like to emphasize that this prediction is not based on the trivial statement, often called "the anthropic principle", that observers can live only in regions where the constants of nature are consistent with their existence. Rather, it is based on "the principle of mediocrity", which asserts that we are typical representatives in a class of similar observers in the universe. Although not always explicitly stated, this principle is always assumed in statistical predictions, such as the prediction of the multipoles of the CMB.

[3] Both the normalization and the p-dependence in this equation are somewhat uncertain. The dynamics of a cosmic string network involves highly non-linear processes on a wide range of length scales; as a result, a quantitative description of string evolution is an extremely difficult problem. This problem is now successfully being addressed in new, high-resolution numerical simulations of cosmic strings.

a B-mode polarization pattern due to rotational perturbations induced by strings. If cosmic strings are discovered, this would open a window into physics of very high energies, which are not likely ever to be accessed in accelerators.

3. Cosmic history

There is a great deal yet to be learned about the structure formation in the universe, from the dark ages, through star and galaxy formation, and the formation of supermassive black holes. We are now in the era of large surveys and massive computer simulations, and we can expect a lot of progress in this area in the coming decade. Here, I will focus on the theory of inflation and its upcoming tests.

A period of inflation in the early universe explains why the universe is large, expanding, homogeneous, isotropic, and flat. It is remarkable that no viable alternatives to inflation have been suggested, despite some heroic attempts. This is paralleled only by Darwin's theory in biology. And like Darwin's theory, inflation is not a specific model, but rather a paradigm encompassing a wide class of models.

Simple models of inflation make very definite predictions: (i) a flat universe, with the total density very close to the critical density,

$$\Omega_{total} = 1 \pm 10^{-5}; \tag{6}$$

(ii) a nearly flat spectrum of adiabatic, Gaussian density fluctuations,

$$P_s(k) \propto k^{-n_s}, \tag{7}$$

with the spectral index in the range

$$0.95 < n_s < 0.98; \tag{8}$$

(iii) a stochastic background of gravitational waves with a flat spectrum. The amplitude of these gravitational waves is model-dependent, but in the simplest model with a quadratic inflaton potential,

$$V(\phi) = \frac{1}{2}m^2\phi^2, \tag{9}$$

the ratio of the power in tensor and scalar perturbations is

$$r \equiv P_t/P_s \approx 0.14. \tag{10}$$

A gravitational wave background of this magnitude may be within the reach of Planck satellite. Current observations yield the values

$$\Omega_{total} = 1 \pm 0.01, \qquad n_s = 0.96 \pm 0.01, \qquad r < 0.22, \tag{11}$$

and are consistent with the simplest model.

If the simplest quadratic model (9) is supported by subsequent tests, this would be a spectacular triumph of inflationary cosmology. On the other hand, the CMB data show some tantalizing hints of non-Gaussianity. The parameter f_{NL}, which is related to the three-point correlation function and should vanish for Gaussian fluctuations, is

$$f_{NL} = 30 \pm 15. \tag{12}$$

This is consistent with the Gaussian value $f_{NL} = 0$ at the 2σ level, but barely. If indeed $f_{NL} \sim 30$, then Planck will detect it at 6σ level. That would be a very dramatic development, as it would rule out a wide class of models of inflation, including most single-field models.

Moreover, a number of anomalies have been reported in the CMB data: (i) the CMB quadrupole is much smaller than expected; (ii) a strange alignment between the low multipoles; (iii) an asymmetry between the correlation functions in the Northern and Southern hemispheres; (iv) a statistically unlikely cold spot in the WMAP microwave map of the sky. Are these anomalies due to chance? If you look hard enough at any stochastic pattern, you will start seeing things. For example, you can see my initials "AV" in the left upper part of the map.[4] But it is hard to believe that all anomalies are due to chance, so some new physics may well be forthcoming from all this.

4. The big picture

Inflationary cosmology has profound implications for the global structure of the universe, particularly for what the universe is like beyond the horizon, and for the beginning and the end of the universe.

The key point is that inflation is generically eternal. Even though it has ended in our local neighborhood, it still continues in remote parts of the universe. The reason is that the end of inflation involves quantum, probabilistic processes and does not occur everywhere at once. The details are somewhat model-dependent. For example, in models where inflation ends through bubble nucleation, we have an inflating high-energy false vacuum. Low-energy bubbles nucleate and expand in this inflating background. We live in one of the bubbles and can see only a small part of it. Even though the bubbles expand at speeds close to the speed of light, they rarely collide, because the false vacuum that separates them is expanding even faster. This makes room for more bubbles to form, and the process continues forever.

The physical properties of all low-energy bubbles do not have to be the same. In fact, string theory, which is at present our best candidate for the fundamental theory, appears to have a multitude of solutions describing vacua with different values of the low-energy constants of nature. These solutions are characterized by different compactifications of extra dimensions, by branes wrapped around extra dimensions is different ways, by different values of the fluxes, etc. The number of possibilities is combinatorial and can be as high as 10^{1000}. Bubbles of all possible vacua will nucleate in the course of eternal inflation. The resulting multiverse, with bubbles within bubbles within bubbles, provides a natural arena for anthropic applications, such as the one I described for the cosmological constant.

The multiverse picture makes many people rather unhappy. There have been complaints that it cannot be tested, even in principle. If only a small part of one bubble is accessible to observation, how can we ever verify that other bubbles really exist?

I think such a test may in fact be possible. We can look for observational signatures of bubble collisions. Even though collisions are rare, our bubble will collide with an infinite number of other bubbles in the course of its expansion. A bubble collision could produce an imprint in the CMB; its detection would provide direct evidence for eternal inflation. There is, however, no guarantee that a bubble collision has occurred within our cosmic horizon.

Another approach is to use theoretical models of the multiverse to make statistical predictions for what we can expect to see in our observable part of the universe. One such prediction, for the cosmological constant, has already been confirmed. This may be our first evidence that there is indeed a huge multiverse out there. We can also derive predictions for other parameters, such as the density parameter Ω, the ratio of baryonic and dark matter densities, the neutrino masses, etc. With more successful predictions, we may be able to prove the case for the multiverse beyond reasonable doubt.[5]

[4] I am grateful to Andrei Linde for pointing this out.

[5] A long-standing problem with making predictions in the multiverse stems from the fact that any event having a nonzero probability will happen in the course of eternal inflation, and it will happen an infinite number of times. Statistical predictions are based on relative frequencies of events in the limit of $t \to \infty$. One finds however that

If inflation has no end, could it also have no beginning? This would allow us to avoid many perplexing questions associated with the beginning of the universe. However, a theorem that I proved with Arvind Borde and Alan Guth states that, under very general conditions, inflationary spacetimes are incomplete to the past. This indicates that inflation must have had some sort of a beginning.

What could have happened before inflation? In my (biased) view, the most attractive possibility is the spontaneous nucleation of a small closed universe out of nothing. The nucleation process can be elegantly described in terms of a Euclidean instanton solution of Einstein's equations. Going beyond this semiclassical description will require further progress in quantum gravity.

I would like to conclude with some important news about the end of the universe. It is often said that if the dark energy is a cosmological constant, then the universe will expand forever. This is true for the universe as a whole, but not necessarily for our local region. In the multiverse picture, there must be a large number of negative-energy vacua, and bubbles of such vacua will inevitably form within our (nearly zero-energy) vacuum. At some point, probably in a very distant future, our neighborhood will be engulfed by a negative-energy bubble. The expansion will then locally turn into contraction, and our region will collapse to a big crunch. The good news is that the bubble will arrive at nearly the speed of light, so we will have no time to worry about it.

the outcome sensitively depends on the limiting procedure. In the last few years there has been a significant progress towards resolution of this problem (which is known as "the measure problem").

The present gravitational wave detection effort

Keith Riles for the LIGO and Virgo Scientific Collaborations
Randall Lab, University of Michigan, Ann Arbor, MI 48109-1040, USA
E-mail: kriles@umich.edu

Abstract. Gravitational radiation offers a new non-electromagnetic window through which to observe the universe. The LIGO and Virgo Collaborations have completed a first joint data run with unprecedented sensitivities to gravitational waves. Results from searches in the data for a variety of astrophysical sources are presented. A second joint data run with improved detector sensitivities is underway, and soon major upgrades will be carried out to build Advanced LIGO and Advanced Virgo with expected improvements in event rates of more than 1000. In parallel there is a vigorous effort in the radio pulsar community to detect nHz gravitational waves via the timing residuals in an array of pulsars at different locations in the sky.

1. Introduction

Einstein's General Theory of Relativity predicts the existence of gravitational waves, disturbances of space-time that propagate at the speed of light and have two transverse quadrupolar polarizations [1]. Their amplitude is usually parametrized by a dimensionless strain h representing the induced fractional change in distance between two test masses.

The search for gravitational waves has many motivations. First, is fundamental scientific curiosity about new phenomena. More prosaically, one can use graviational radiation to test general relativity. For example, one can test the tranverse and quadrupolar nature of the radiation, and one can test whether or not the radiation travels at the speed of light (massless graviton). One can also directly probe highly relativistic phenomena, such as black-hole formation. Perhaps more intriguing is the new view one gains of the universe. Gravitational waves cannot be appreciably absorbed by dust or stellar envelopes, and most detectable sources are some of the most interesting objects in the universe. More generally, gravitational wave astronomy opens up an entirely new non-electromagnetic spectrum. Astronomy has found surprises since the mid 20th-century, as non-optical light bands have been explored, from the radio to gamma rays. New surprises likely await in the exploration of the gravitational spectrum.

The radiation arises at lowest order from quadrupolar motions [1]. Specifically, the radiation can be described by a metric tensor disturbance $h_{\mu\nu}$ that depends on the time derivative of the source quadrupole:

$$h_{\mu\nu} \quad = \quad \frac{2G}{rc^4}\frac{d^2}{dt^2}\Big[I_{\mu\nu}\Big], \tag{1}$$

where G is Newton's gravitational constant, r is the distance to the source, c is the speed of light, and $I_{\mu\nu}$ is the source's reduced quadrupole tensor [1]. Given the tiny multiplying constant, the source quadrupole's 2nd time derivative must be enormous to give detectable effects far from the source, implying enormous masses ($\sim M_\odot$) with relativistic motions, such as two neutron stars in a coalescing binary system, as discussed below.

Strong indirect evidence already exists for gravitational wave emission. The famous Hulse-Taylor binary system, consisting of an observed pulsar with 17-Hz radio emission in an 8-hour orbit with an unseen neutron star companion, has shown a small but unmistakable quadratic decrease in orbital period (~40 seconds over 30 years), in remarkably good agreement with expectation from gravitational wave energy loss [2]. The gravitational wave emission frequency (~70 μHz) is far too low to be observed directly by present gravitational wave detectors, but if we were to wait about 300 million years, the system would eventually spiral into a spectacular coalescence easily visible with present gravitational wave detectors.

Perhaps our best hope for gravitational wave discovery lies with corresponding binary systems in the numerous galaxies far away from us, but there are large uncertainties in estimated coalescence rates for compact binary coalescence of systems containing neutron stars (NS) and/or black holes (BH). For example, a recent paper [3] estimates a 90% CL range of 2.9×10^{-2}–0.46 per year for initial LIGO detection of a NS-NS, NS-BH, or BH-BH coalescence.

Other candidate transient sources of gravitational waves include supernovae and gamma ray bursts (some of which may well be coalescing binary systems). If we are fortunate, electromagnetic transients will be seen simultaneously by other astronomers, allowing more confident gravitational wave detection with lower signal-to-noise ratio (SNR). Potential non-transient gravitational wave sources include rapidly spinning neutron stars in our own galaxy, emitting long-lived continuous waves, or a cosmological background of stochastic gravitational waves, analogous to the cosmic microwave background radiation. Results from searches for both transient and long-lived gravitational-wave sources will be discussed below.

2. LIGO and Virgo Detectors

2.1. Principles of Operation

The principles of gravitational wave interferometry are described in detail by Saulson [1]; only a brief summary is provided here. A Michelson interferometer with orthogonal arms is a natural gravitational wave detector in that a wave incident along the normal to the plane of the arms can induce a quadrupole distortion (differential arm length change) to which the interferometer is designed to be most sensitive. More generally, the sensitivity of the interferometer to a gravitational wave depends on the wave's direction of incidence and polarization. The antenna pattern of sensitivity is broad, giving good sky coverage, but at the expense of directional precision for transient sources.

The sensitivity of conventional Michelson interferometers to gravitational wave strain can be increased via several techniques. Methods used in the LIGO and Virgo detectors include km-scale arms, Fabry-Perot cavities in those arms to enhance phase sensitivity to the passage of a gravitational wave, and "recycling" of laser power via the insertion of a mirror between the laser and beam splitter, in order to increase effective light power and thereby decrease shot noise at high frequencies. The response of a "power-recycled Michelson interferometer" to gravitational waves is inherently broad-band (relative to the narrow bandwidth of a resonant bar detector [1]), allowing searches for broadband transients and reconstruction of waveforms.

There exists now a global interferometer network, consisting of not only the LIGO and Virgo interferometers (the most sensitive at present), which are the main focus of this report, but also the GEO 600 interferometer near Hannover, Germany [4] and the TAMA 300 interferomter in Tokyo [5]. In general, the larger the number of interferometers that can detect a source, the greater the confidence of that detection will be, and the better the precision with which the radiation parameters can be determined, *e.g.*, source direction and wave polarization.

2.2. The LIGO and Virgo Interferometers, Data Runs, and the Future

The LIGO detector consisted until summer 2009 of three power-recycled Michelson interferometers: two in Hanford, Washington (H1 with 4-km arms, H2 with 2-km arms) and one in Livingston, Louisiana (L1 with 4-km arms). The LIGO interferometers, as configured during the fifth science run (S5) from November 2005 through September 2007, are described in detail elsewhere [6]. Commissioning the interferometers to meet the 4-km design requirement of an integrated root-mean-square strain sensitivity below 10^{-21} over a 100-Hz band demanded many years of effort, but yielded detectors that surpassed this requirement by more than a factor of two. A more ambitious target strain sensitivity curve was also surpassed over most of the detector bandwidth, as shown in Figure 1.

During the LIGO commissioning phase, a large number of engineering and science data runs were taken, both for carrying out astrophysical searches and for evaluating detector performance and stability during sustained running. The nearly 2-year S5 run at design sensitivity has led to many publications that continue to appear. A selected sampling of these results will be presented below. The progress in LIGO commissioning can be seen in the ever improving sensitivities shown in Figure 1 from one science run to the next.

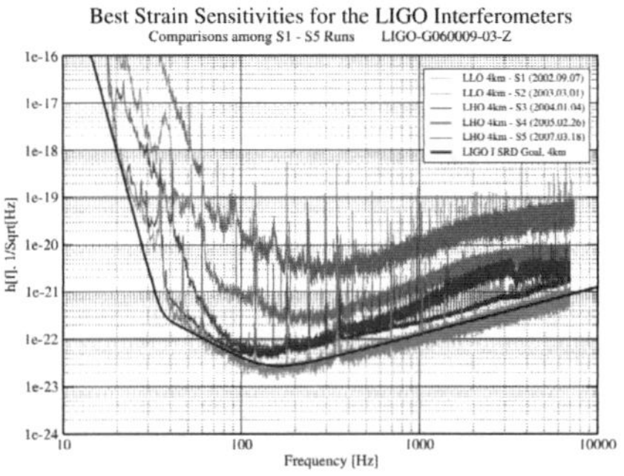

Figure 1. History of LIGO strain noise spectra.

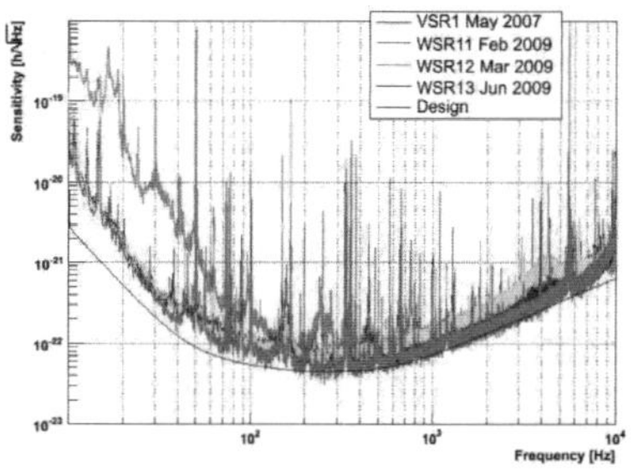

Figure 2. History of Virgo strain noise spectra.

The Virgo detector consists of a single power-recycled Michelson interferometer with 3-km arms in Cascina, near Pisa. The interferometer is described in detail elsewhere [7]. As with LIGO, many years of commissioning were required to bring the interferometer to its present sensitivity, shown in Figure 2, together with the sensitivity achieved in 2007 during the first Virgo science run VSR1 which was coincident and coordinated with the last several months of the LIGO S5 run. Note that the Virgo sensitivity is now better than the LIGO sensitivity at low frequencies, offering potentially better range for low-frequency young pulsars and binary black hole mergers.

New science runs (LIGO S6 and Virgo VSR2) are to begin in July 2009 and run until approximately the end of 2010, but with scheduled interruptions for commissioning. Virgo VSR2 sensitivity should be improved over the full band compared to that of VSR1 (see Figure 2), while LIGO S6 sensitivity should be improved by up to a factor of two over S5 sensitivity at frequencies above several hundred Hz. The shutdown for Advanced LIGO and Virgo upgrades will likely occur in late 2010 or early 2011, leading eventually to dramatic improvements in expected gravitational wave event rates for those 2nd-generation detectors [8].

3. Results from Searches for Gravitational Waves

A sampling of results to date from searches in the LIGO S5 data will be presented here. Analyses of LIGO and Virgo data are carried out by working groups focusing on four distinct source types: 1) binary inspirals, 2) unmodeled bursts, 3) continuous waves, and 4) stochastic background. Results from these four efforts will be discussed in turn below.

3.1. Binary Inspiral Searches

Searches for binary inspiraling neutron star systems enjoy the benefit of well understood expected waveforms, which allows matched Wiener filtering to be applied, giving optimal SNR. Binary black hole binaries, especially for rapidly spinning black holes, are not as well understood. Nonetheless a phenomenologically defined set of templates offers near-optimal SNR [9]. Searches have been carried out using matched filter template banks spanning a broad volume in binary system parameter space, ranging from primordial black holes with less than 1 $M_\odot$ to systems containing black holes with masses greater than 30 $M_\odot$. No evidence of excess events has been seen in an analysis of the 1st year S5 data, leading to upper limits on event rates for various binary systems [9]. Figure 3 shows some resulting rate limits (events per year per 10^{10} "blue"stars) *vs.* total system mass for BH-BH systems and *vs.* black hole mass for NS-BH systems.

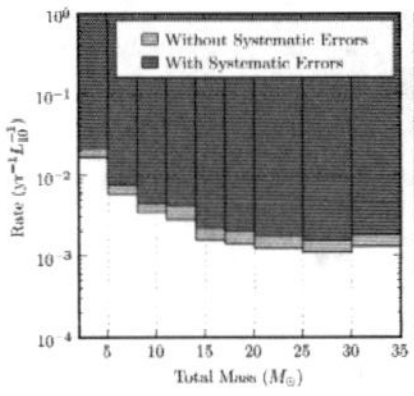

Figure 3. Upper limits on binary inspiral rates.

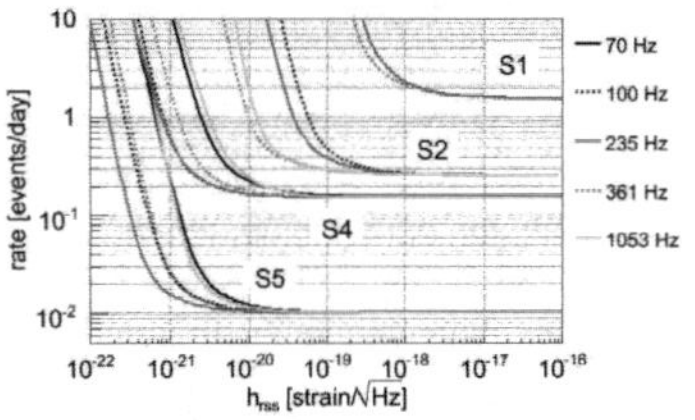

Figure 4. Upper limits on burst rates.

3.2. Unmodeled Burst Searches

Searches for unmodeled bursts have been carried out in the S5 data for both triggered sources, including gamma ray bursts detected by satellites, and untriggered sources (all-sky). An especially interesting triggered source was GRB 070201, a short hard gamma ray burst with a reconstructed position consistent with M31 (Andromeda). The absence of any plausible signal in LIGO data at the time of the burst and the nearness of M31 make it very unlikely that the GRB was a binary merger in M31 [10]. More likely it was an SGR giant flare in M31.

Searches for untriggered double- and triple-coincident burst events in the LIGO interferometers in the first year of S5 data were also unsuccessful in detecting a signal. Efficiencies have been determined from simulations using a variety of *ad hoc* phenomenological waveforms with the waveform strengths parametrized in terms of the "root sum square of h", $h_{\rm rss}$, defined according to

$$(h_{\rm rss})^2 \quad = \quad \int [|h_+(t)|^2 + |h_\times(t)|^2]\, dt, \tag{2}$$

where $h_+(t)$ and $h_\times(t)$ are the "+" and "×" quadrupolar polarizations of the detectable strain waveforms [1]. Figure 4 shows rate upper limits (events/day) for sine-Gaussian bursts *vs.* $h_{\rm rss}$ for the first year of S5 data, along with corresponding limits for earlier science runs [11].

3.3. Continuous Wave Searches

Searches have been carried out in S5 data for continuous gravitational waves from known pulsars using matched-filter templates based on precise radio or X-ray timing. These targeted searches with integration times longer than a year yield strain amplitude sensitivities below 10^{-25} across a broad band. Figure 5 shows S5 upper limits on strain for 116 known pulsars [12], along with the corresponding limits from S4 for a subset of 78 pulsars. An especially interesting result is that for the Crab pulsar, where the upper limit on energy loss rate due to gravitational radiation is about 2% of the Crab's total known energy loss (inferred from its slowing of spin).

Searches have also been carried out for unknown spinning neutron stars in any direction on the sky. To search every direction and frequency with long integration times is computationally intractable because of the need to make direction-dependent Doppler modulations due to the Earth's motion. So sensitivity must be sacrificed via incoherent methods, such as summing many strain powers measured over short integration times. All-sky upper limits on strain for circularly and linearly polarized continuous waves are shown in Figure 6 for the 50-1100 Hz band, based on the first eight months of S5 data [13] using thousands of summed powers with coherence times of 30 minutes. Comparable limits were found over a broader band (50-1500 Hz) using an early-S5 pilot run of the Einstein@Home distributed computing project looking at coincident outliers among 28 coherent observation times of 25-hour durations [14].

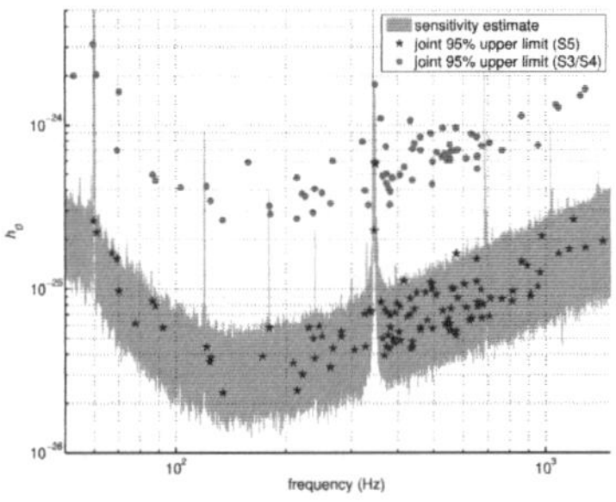

Figure 5. Limits on strain amplitude from known pulsars.

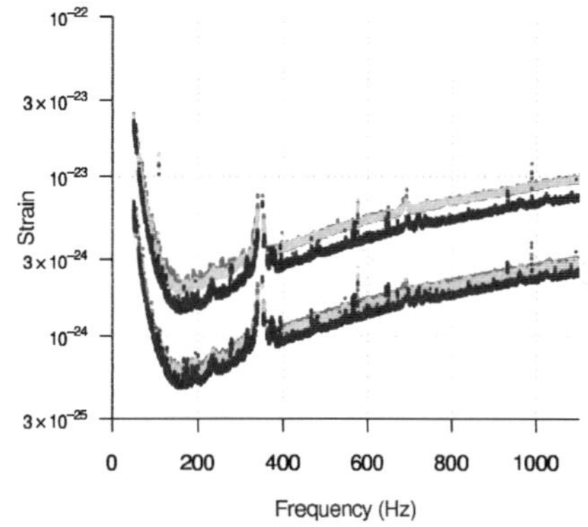

Figure 6. All sky limits on strain amplitude from unknown pulsars.

3.4. Stochastic Background Searches

A primordial isotropic gravitational wave background is predicted by most cosmological theories, although the predicted strengths of the background vary enormously. It is customary to parametrize the background strength *vs.* frequency f by its energy density normalized to the critical energy density $\rho_{\rm crit}$ of the universe:

$$\Omega_{\rm gw}(f) \quad = \quad \frac{1}{\rho_{\rm crit}} \frac{d\rho_{\rm gw}(f)}{d\,ln(f)} \tag{3}$$

Searches for this radiation are carried out via measured cross-correlations in pairs of interferometers, where large spatial separation reduces sensitivity to high frequencies, but mitigates disturbances due to common environmental noise sources. Analysis of the early-S5 LIGO data from the Hanford H1 and Livingston L1 interferometer correlations leads to a preliminary 90% upper limit on a constant $\Omega_{\rm gw} < 9.0 \times 10^{-6}$ in the band 42-177 Hz, improving upon previous indirect limits derived from analysis of big-bang nucleosythesis [15].

4. Searches with Pulsar Timing Arrays

An entirely different effort is underway in the radio astronomy community to detect stochastic gravitational waves by way of precise pulsar timing. Very-low-frequency waves (~nHz) in the vicinity of the Earth could lead to a quadrupolar pattern in the timing residuals from a large number of pulsars observed at different directions on the sky. Several collaborations have formed in recents years to carry out the precise observations required: 1) The Parkes Pulsar

Timing Array (Australia) [16], 2) the European Pulsar Timing Array (U.K., France, Netherlands, Italy) [17], and 3) Nano-Grav (U.S.A.) [18]. These searches aim to improve upon the existing limit of $\Omega_{\rm gw}(\rm nHz) < 10^{-8}$ [16] by two orders of magnitude and perhaps discover radiation from binary supermassive black holes in the cores of distant galaxies.

5. Summary

Unfortunately, gravitational waves have not yet been detected directly. But limits obtained have become astrophysically interesting, and not all S5 / VSR1 searches have been completed. Moreover, searches in the more sensitive S6 / VSR2 data are just beginning. Advanced LIGO and Virgo promise major sensitivity improvements with orders of magnitude increase in expected event rates, offering a near guarantee of discovery [8]. There is a distinct possibility, however, that the first direct detection will come not from interferometers operating in the audio band, but from precise pulsar timing measurements in the nHz band.

Acknowledgments

The authors gratefully acknowledge the support of the United States National Science Foundation for the construction and operation of the LIGO Laboratory and the Science and Technology Facilities Council of the United Kingdom, the Max-Planck-Society, and the State of Niedersachsen/Germany for support of the construction and operation of the GEO600 detector. The authors also gratefully acknowledge the support of the research by these agencies and by the Australian Research Council, the Council of Scientific and Industrial Research of India, the Istituto Nazionale di Fisica Nucleare of Italy, the Spanish Ministerio de Educación y Ciencia, the Conselleria d'Economia, Hisenda i Innovació of the Govern de les Illes Balears, the Royal Society, the Scottish Funding Council, the Scottish Universities Physics Alliance, The National Aeronautics and Space Administration, the Carnegie Trust, the Leverhulme Trust, the David and Lucile Packard Foundation, the Research Corporation, and the Alfred P. Sloan Foundation.

References

[1] Saulson P 1994 *Fundamentals of Interferometric Gravitational Wave Detectors* (Singapore: World Scientific)
[2] Weisberg J M and Taylor J H 2005 in *Binary Radio Pulsars* ed Rasio F A and Stairs I H, *Astron. Soc. Pacific Conf. Ser.* vol 328 (San Francisco: Astron. Soc. Pacific)
[3] Shaughnessy R O, Kalogera V and Belczynski K 2009 Binary compact object coalescence rates: The role of elliptical galaxies *Preprint* arXiv:0908.3635
[4] Willke B 2002 *Class. Quant. Grav.* **19** 11
[5] Takahashi R 2004 *Class. Quant. Grav.* **21** S403
[6] Abbott B P *et al.* 2009 *Rep. Prog. Phys.* **72** 076901
[7] Acernese F *et al.* 2005 *Class. Quant. Grav.* **22** S869
[8] Losurdo G in these proceedings
[9] Abbott B P *et al.* 2009 *Phys. Rev. D* **79** 122001
[10] Abbott B P *et al.* 2008 *Ap. J.* **681** 1419
[11] Abbott B P *et al.* 2009 Search for gravitational-wave bursts in the first year of the fifth LIGO science run *Preprint* arXiv:0905.0020
[12] Abbott B P *et al.* 2009 Searches for gravitational waves from known pulsars with S5 LIGO data *Preprint* arXiv:0909:3583
[13] Abbott B P *et al.* 2009 *Phys. Rev. Lett.* **102** 111102
[14] Abbott B P *et al.* 2009 *Phys Rev. D* **80**— 042003
[15] The preliminary result presented at TAUP has been recently superseded by a final S5 result $\Omega_{\rm gw} < 6.9\times10^{-6}$ in Abbott B P *et al.* 2009 *Nature* **460** 990
[16] Manchester R N 2008 *AIP Conf. Proc.* **983** 584
[17] Janssen G H, Stappers B W, Kramer M, Purver M, Jessner A and Cognard I 2008 *AIP Conf. Proc.* **983** 633 in *40 years of pulsars, millisecond pulsars, magnetars, and more*
[18] Jenet F *et al.* 2009 The north american nanohertz observatory for gravitational waves *Preprint* arXiv:0909.1058

Journal of Physics: Conference Series **203** (2010) 012003 doi:10.1088/1742-6596/203/1/012003

Results from the DAMA/LIBRA experiment

R Bernabei [1,2], P Belli [2], F Cappella [3,4], R Cerulli [5], C J Dai [6], A d'Angelo [3,4], H L He [6], A Incicchitti [4], H H Kuang [6], X H Ma [6], F Montecchia [2,7], F Nozzoli [1,2], D Prosperi[3,4], X D Sheng[6] and Z P Ye[6,8]

[1] Dip. di Fisica, Università di Roma "Tor Vergata", I-00133 Rome, Italy
[2] INFN, sez. Roma "Tor Vergata", I-00133 Rome, Italy
[3] Dip. di Fisica, Università di Roma "La Sapienza", I-00185 Rome, Italy
[4] INFN, sez. Roma, I-00185 Rome, Italy
[5] Laboratori Nazionali del Gran Sasso, I.N.F.N., Assergi, Italy
[6] IHEP, Chinese Academy, P.O. Box 918/3, Beijing 100039, China
[7] Laboratorio Sperimentale Policentrico di Ingegneria Medica, Università degli Studi di Roma "Tor Vergata"
[8] University of Jing Gangshan, Jiangxi, China

E-mail: `pierluigi.belli@roma2.infn.it`

Abstract. The DAMA project is an observatory for rare processes and it is operative deep underground at the Gran Sasso National Laboratory of the I.N.F.N. In particular, the DAMA/LIBRA (Large sodium Iodide Bulk for RAre processes) set-up consists of highly radiopure NaI(Tl) detectors for a total sensitive exposed mass of $\sim$ 250 kg. The results, obtained by this set-up by exploiting the model independent annual modulation signature of Dark Matter (DM) particles, have confirmed and improved those obtained by the former DAMA/NaI experiment. A model independent evidence for the presence of Dark Matter particles in the galactic halo is cumulatively obtained at 8.2 σ C.L.. No systematics or side reactions able to account for the measured modulation amplitude and to contemporaneously satisfy all the many specific requirements of the signature have been found or suggested by anyone over more than a decade. Future perspectives are shortly addressed.

1. Introduction

With the present technology the only reliable signature able to point out – in a model independent way – the presence of Dark Matter (DM) particles in the galactic halo is the DM annual modulation signature. This signature – originally suggested in the middle of '80 in ref. [1] – exploits the effect of the Earth revolution around the Sun on the number of events induced by the Dark Matter particles in a suitable low-background set-up placed deep underground. In fact, as a consequence of its annual revolution, the Earth should be crossed by a larger flux of Dark Matter particles around $\sim$ 2 June (when its rotational velocity is summed to the one of the solar system with respect to the Galaxy) and by a smaller one around $\sim$ 2 December (when the two velocities are subtracted). In fact, the expected differential rate as a function of the energy (see also ref. [2, 3, 4, 5, 6, 7, 8, 9, 10] for some discussions) also depends on the DM particle velocity distribution and on the Earth's velocity in the galactic frame. This offers an efficient model independent signature, able to test a large interval of cross sections and of halo densities. Thus, the contribution of the signal to the counting rate in the k–th energy interval

Journal of Physics: Conference Series **203** (2010) 012003 doi:10.1088/1742-6596/203/1/012003

can be written as $S_k = S_{0,k} + S_{m,k} \cos\omega(t - t_0)$, where: i) $S_{0,k}$ is the constant part of the signal; ii) $S_{m,k}$ is the modulation amplitude; iii) $\omega = \frac{2\pi}{T}$ with period T; iv) t_0 is the phase.

The DM annual modulation signature is very distinctive since the DM signal must simultaneously satisfy all the following requirements: the rate must contain a component modulated according to a cosine function (1) with one year period (2) and a phase that peaks roughly around $\simeq 2^{nd}$ June (3); this modulation must only be found in a well-defined low energy range, where DM particle induced events can be present (4); it must apply only to those events in which just one detector of many actually "fires" (*single-hit events*), since the DM particle multi-interaction probability is negligible (5); the modulation amplitude in the region of maximal sensitivity must be 7% for usually adopted halo distributions (6), but it can be larger in case of some possible scenarios such as e.g. those in refs. [11, 12]. Only systematic effects or side reactions able to fulfil these requirements and to account for the whole observed modulation amplitude could mimic this signature; thus, no other effect investigated so far in the field of rare processes offers a so stringent and unambiguous signature. It is worth noting that the DM annual modulation is not – as often naively said – a "seasonal" variation and it is not a "winter-summer" effect. In fact, the DM annual modulation is not related to the relative Sun position, but it is related to the Earth velocity in the galactic frame. Moreover, the phase of the DM annual modulation (roughly 2nd June) is well different than those of physical quantities (such as temperature of atmosphere, pressure, other meteorological parameters, cosmic rays flux, ...) instead correlated with seasons.

In the following, we will summarize just some of the main aspects of the first results obtained by DAMA/LIBRA exploiting over four annual cycles (exposure: 0.53 ton$\times$yr) the model independent DM annual modulation signature. These data and those previously collected by DAMA/NaI over 7 annual cycles (0.29 ton$\times$yr) correspond to 11 annual cycles for a total exposure of 0.82 ton$\times$yr. All details can be found in ref. [13, 14] and references therein.

2. The DAMA project

The DAMA project is mainly based on the development and use of low background scintillators [2, 3, 4, 5, 6, 7, 8, 9, 13, 14, 15, 16, 17, 18, 19]; the main aim is the direct detection of DM particles in the galactic halo by investigating the DM model independent annual modulation signature. Profiting from the low background features of these set-ups, other rare processes are also investigated obtaining often competitive results.

The main experimental set-ups are: i) DAMA/NaI ($\simeq$ 100 kg of highly radiopure NaI(Tl)) which completed its data taking on July 2002 [2, 3, 4, 5, 6, 7, 8, 9, 15, 17]; ii) DAMA/LXe ($\simeq$ 6.5 kg liquid Kr-free Xenon enriched either in ^{129}Xe or in ^{136}Xe) [20, 21] (which is again in operation since december 2007 after an upgrading and the waiting for the restoring of the underground laboratory water plant, necessary for the compressor refrigeration); iii) DAMA/R&D, devoted to tests on prototypes and to small scale experiments [22]; iv) the new second generation DAMA/LIBRA set-up ($\simeq$ 250 kg highly radiopure NaI(Tl)) in operation since March 2003 [13, 14, 19] and firstly upgraded in fall 2008. Moreover, in the framework of devoted R&D for radiopure detectors and photomultipliers, sample measurements are carried out by means of the low background DAMA/Ge detector (installed deep underground since more than $\simeq$ 15 years); this detector is also used for small scale experiments [23].

The DAMA/NaI set up and its performances are described in ref.[2, 3, 16, 24], while DAMA/LIBRA set-up and its performances in ref. [14]. Here we just summarized that: i) the detectors' responses range from 5.5 to 7.5 photoelectrons/keV; ii) the hardware threshold of each photomultiplier (PMT) is at single photoelectron (each detector is equipped with two low background PMTs working in coincidence); iii) energy calibration with X-rays/γ sources are regularly carried out down to few keV; iv) the software energy threshold of the experiment is 2 keV [14]. The DAMA/NaI experiment collected an exposure of 0.29 ton$\times$yr over 7 annual cycles [2, 3, 4, 5, 6, 7, 8, 9], while DAMA/LIBRA [13, 14, 19] has released so far an exposure of 0.53 ton×yr collected over 4 annual cycles [13]. Thus, the total exposure of the two experiments is 0.82 ton×yr, which is orders of magnitude larger than the exposures typically collected in the field.

3. The model independent DM annual modulation results

Several analyses on the model-independent investigation of the DM annual modulation signature have been performed (see ref. [13] and references therein); here just few arguments are reminded.

Fig. 1 shows the time behaviour of the experimental residual rates of the *single-hit* events collected by DAMA/NaI and by DAMA/LIBRA in the (2–6) keV energy interval. The superimposed curve represents the cosinusoidal function behavior $A \cos \omega(t - t_0)$ with a period $T = \frac{2\pi}{\omega} = 1$ yr and with a phase $t_0 = 152.5$ day (June 2nd), while the modulation amplitude, A, has been obtained by best fit over the DAMA/NaI and DAMA/LIBRA data. When the period and the phase parameters are released in the fit, values well compatible with those expected for a DM particle induced effect are obtained [13]. In particular, the cumulative analysis of the *single-hit* residual rate favours the presence of a modulated cosine-like behaviour with the proper features at 8.2 σ C.L.[13].

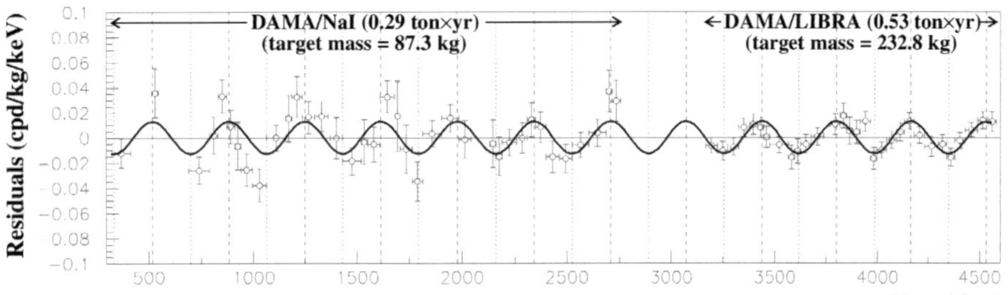

Figure 1. Experimental model-independent residual rate of the *single-hit* scintillation events, measured by DAMA/NaI and DAMA/LIBRA in the (2 – 6) keV energy interval as a function of the time. The zero of the time scale is January 1st of the first year of data taking of DAMA/NaI. The experimental points present the errors as vertical bars and the associated time bin width as horizontal bars. The superimposed curve is the cosinusoidal function behavior $A \cos \omega(t - t_0)$ with a period $T = \frac{2\pi}{\omega} = 1$ yr, with a phase $t_0 = 152.5$ day (June 2nd) and with modulation amplitude, A, equal to the central value obtained by best fit over the whole data, that is (0.0129 ± 0.0016) cpd/kg/keV. The dashed vertical lines correspond to the expected maximum of the signal (June 2nd), while the dotted vertical lines correspond to the expected minimum. The total exposure is 0.82 ton × yr [13].

The same data of Fig.1 have also been investigated by a Fourier analysis, obtaining a clear peak corresponding to a period of 1 year [13]; the same analysis in other energy regions shows instead only aliasing peaks. Similar result is obtained when comparing the *single-hit* residuals in the (2–6) keV with those in other energy interval; in fact, a clear modulation is present in the lowest energy interval, while it is absent just above [13]. In particular, in order to verify absence of annual modulation in other energy regions and, thus, to also verify the absence of any significant background modulation, the energy distribution measured during the data taking periods in energy regions not of interest for DM detection has also been investigated. In fact, the background in the lowest energy region is essentially due to “Compton” electrons, X-rays and/or Auger electrons, muon induced events, etc., which are strictly correlated with the events in the higher energy part of the spectrum. Thus, if a modulation detected in the lowest energy region would be due to a modulation of the background (rather than to a signal), an equal or larger modulation in the higher energy regions should be present. The data analyses have allowed to exclude the presence of a background modulation in the whole energy spectrum at a level much lower than the effect found in the lowest energy region for the *single-hit* events [13].

A further relevant investigation has been done by applying the same hardware and software procedures, used to acquire and to analyse the *single-hit* events, to the *multiple-hits* ones. In fact, since the probability that a DM particle interacts in more than one detector is negligible, a DM signal can be present just in the *single-hit* residual rate. Thus, this allows the study of the background behaviour in the same energy interval of the observed positive effect. In particular, while a clear modulation with proper features is present in the (2–6) keV *single-hit* events, the modulation amplitude of the *multiple-hits* ones in the same energy interval is well compatible with zero [13] (see Fig. 2). Similar results were previously obtained also for the DAMA/NaI

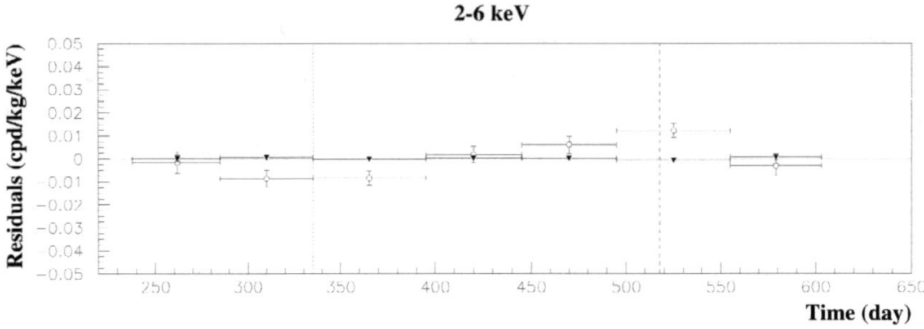

Figure 2. Experimental model-independent residual rates of the (2 – 6) keV *single-hit* events (open circles) (class of events to which DM events belong) and of the (2 – 6) keV *multiple-hits* events (filled triangles) (class of events to which DM events do not belong), measured in the four DAMA/LIBRA annual cycles (as collected in a single annual cycle; the initial time of the scale is taken on August 7th). The same identical hardware and the same identical software procedures have been applied in both cases. The experimental points present the errors as vertical bars and the associated time bin width as horizontal bars. For details see ref. [13].

case [3]. Thus, again evidence of annual modulation with the proper features required by the DM annual modulation signature, is present in the *single-hit* residuals (events class to which the DM particle induced events belong), while it is absent in the *multiple-hits* residual rate (event class to which only background events belong). Since the same identical hardware and the same identical software procedures have been used to analyse the two classes of events, the obtained result offers an additional strong support for the presence of DM particles in the galactic halo further excluding any side effect either from hardware or from software procedures or from background.

The annual modulation present at low energy has also been shown by depicting the differential modulation amplitudes, $S_{m,k}$, as a function of the energy, as obtained by maximum likelihood method over the data, considering T =1 yr and t_0 = 152.5 day; positive signal is present in the (2–6) keV energy interval, while $S_{m,k}$ values compatible with zero are present just above. It has also been verified that the measured modulation amplitudes are statistically well distributed in all the crystals, in all the annual cycles and in the energy bins; these and other discussions can be found in ref. [13].

The data have also been analysed [13] releasing in the maximum likelihood procedure the assumption of a phase t_0 = 152.5 day, that is considering: $S_k = S_{0,k} + S_{m,k}\cos\omega(t-t_0) + Z_{m,k}\sin\omega(t-t_0) = S_{0,k} + Y_{m,k}\cos\omega(t-t^*)$. Obviously, DM induced signal requires: i) $Z_{m,k} \sim 0$ (because of the orthogonality between the cosine and the sine functions); ii) $S_{m,k} \simeq Y_{m,k}$; iii) $t^* \simeq t_0 = 152.5$ day. The obtained results (see Fig. 3) further confirm those already achieved by other kinds of analyses. In particular, a modulation amplitude is present in the *single-hit* events in the lower energy intervals and the period and the phase agree with those expected for DM induced signals. For more discussions see ref. [13].

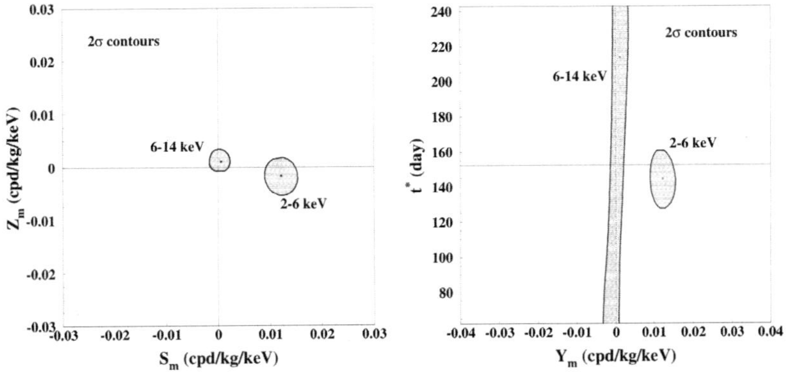

Figure 3. 2σ contours in the plane (S_m, Z_m) (*left*) and in the plane (Y_m, t^*) (*right*) for the (2–6) keV and (6–14) keV energy intervals. The contours have been obtained by the maximum likelihood method, considering the seven annual cycles of DAMA/NaI and the four annual cycles of DAMA/LIBRA all together. A modulation amplitude is present in the lower energy intervals and the period and the phase agree with those expected for DM induced signals. For details see ref. [13].

As previously done for DAMA/NaI [2, 3, 24], careful investigations on absence of any significant systematics or side reaction effect in DAMA/LIBRA have been quantitatively carried out and reported in details in ref. [13]. In order to continuously monitor the running conditions, several pieces of information are acquired with the production data and quantitatively analyzed. No systematics or side reactions able to account for the measured modulation amplitude and to simultaneously satisfy all the requirements of the signature have been found or suggested by anyone over more than a decade. For all details and for discussions see ref. [13] and references therein.

Just as an example we remind here the case of muons, whose flux has been reported by the MACRO experiment to have a 2% modulation with phase around mid–July (that is different than expected for the DM annual modulation signature and measured by the DAMA experiments). In particular, it has been shown [13] that not only this effect would give rise in the DAMA set-ups to a quantitatively negligible contribution, but some of the six requirements necessary to mimic the annual modulation signature – namely e.g. the conditions of presence of modulation just in the *single-hit* event rate and of the phase value – would also fail. Moreover, even the pessimistic assumption of whatever hypothetical (even exotic) cosmogenic product – whose decay or de-excitation or whatever else (with mean-life τ) might produce: i) only events at low energy; ii) only *single-hit* events; iii) no sizeable effect in the *multiple-hits* counting rate – cannot give rise to any side process able to mimic the investigated DM signature. In fact, not only this latter hypothetical process would be quantitatively negligible, but in addition e.g. its phase – as it can be easily derived – would be even (much) larger than July 15th, and therefore again well different from the one measured by the DAMA experiments and expected from the DM annual modulation signature ($\simeq$ June 2nd). Thus, also any possible effect from muons can be safely excluded.

Summarizing, DAMA/LIBRA has confirmed the presence of an annual modulation satisfying all the requirements of the DM annual modulation signature, as previously pointed out by DAMA/NaI; in particular, the evidence for the presence of DM particles in the galactic halo is cumulatively supported at 8.2 σ C.L..

It is worth noting that no other experiment exists, whose result can be directly compared in a model-independent way with those by DAMA/NaI and DAMA/LIBRA. In particular, let us also point out that results obtained with different target materials and/or different approaches cannot intrinsically be directly compared among them even when considering the same kind of candidate and of coupling, although apparently all the presentations generally refer to cross sections on nucleon. Therefore, claims for contradictions made by experiments insensitive to the DM annual modulation signature, using different target materials and approaches, having well different sensitivities to various DM candidate and interactions, etc. have by the fact no impact even in the single arbitrary scenario they usually consider without accounting for experimental and theoretical uncertainties, using often crude approximation in the calculation, etc. Moreover, (see for example [2, 25, 26]), some critical points exist in those activities, claiming for exclusion, on important experimental aspects (energy threshold, energy scale, multiple selection procedures, stabilities, etc.). A relevant argument is also the methodological robustness [27]. Thus, the long-standing claims for controversy are arbitrary and have no serious scientific basis. Finally, the boost effect of DAMA activities and results in the field of direct Dark Matter investigation can be easily derived.

It is also worth noting that, whenever an experiment using the same identical target material and methodological approach would be available in future, as usual in whatever field of Physics a serious comparison would require – in every case – e.g. a deep investigation of the radiopurity of all the part of the different set-ups, of their specific performances in all the aspects, of the detailed procedures used by each one, of the used exposures, of the stability parameters, etc.

Finally, as regards the indirect detection searches, let us note that also no direct model-independent comparison can be performed between the results obtained in direct and indirect activities, since it does not exist a biunivocal correspondence between the observables in the two kinds of experiments. Anyhow, if possible excesses in the positron to electron flux ratio and in the γ rays flux with respect to an assumed simulation of the hypothesized contribution, which has to be expected from standard sources, might be interpreted in terms of Dark Matter (but huge and still unjustified boost factor and new interaction types are required), this would be not in conflict with the effect observed by DAMA experiments, as also discussed in literature at some extent.

4. Already performed and planned upgradings

During September 2008 a first upgrade of the DAMA/LIBRA set-up has been realized and the shield has been opened in HP Nitrogen atmosphere. This has allowed the increase of the exposed mass since one detector has been recovered by replacing a broken PMT [13]. Moreover, a new optimization of some PMTs and HVs has been done. Finally, a total replacement of the used transient digitizers with new ones, having better performances, has been realized and a new DAQ with optical fibers has been installed and put in operation. The data taking has been restarted on October 2008. Moreover, in order to further increase the experimental sensitivity, it has been pointed out the relevance to lower the software energy threshold of the experiment and to improve some other aspects; thus, the replacement of all the PMTs with new ones with higher quantum efficiency has been planned.

A larger exposure collected by DAMA/LIBRA (or possibly by DAMA/1ton) and the lowering of the 2 keV energy threshold in the data analysis will improve the sensitivity of the experiment and the corollary information on the nature of the DM candidate particle(s) and on the various related astrophysical, nuclear and particle Physics scenarios. In addition, it is worth noting that ultra low background NaI(Tl) scintillators can also offer the possibility to achieve significant results on several other rare processes as already done e.g. by the former DAMA/NaI apparatus [17] and by DAMA/LIBRA so far [19].

Finally, we also mention that a third generation R&D effort towards a possible NaI(Tl) ton set-up, DAMA proposed in 1996, has been funded by I.N.F.N. and is in progress. Let us point out that large exposed target masses is a key point for the investigation of the DM particles in the galactic halo with higher sensitivities, but it is not sufficient. In fact, a competitive DM experiment studying the annual modulation signature must also have a high duty cycle, long exposure time and very high stability for the long time of data taking, etc.. Many of these requirements are not satisfied by proposed and planned experiments in this field (see e.g. for some cases [25, 26]).

References

[1] K.A. Drukier et al., Phys. Rev. D 33 (1986) 3495; K. Freese et al., Phys. Rev. D 37 (1988) 3388.
[2] R. Bernabei el al., La Rivista del Nuovo Cimento 26 n.1 (2003) 1-73.
[3] R. Bernabei et al., Int. J. Mod. Phys. D 13 (2004) 2127.
[4] R. Bernabei et al., Eur. Phys. J. C 47 (2006) 263.
[5] R. Bernabei et al., Int. J. Mod. Phys. A 21 (2006) 1445.
[6] R. Bernabei et al., Int. J. Mod. Phys. A 22 (2007) 3155.
[7] R. Bernabei et al., Eur. Phys. J. C 53 (2008) 205.
[8] R. Bernabei et al., Phys. Rev. D 77 (2008) 023506.
[9] R. Bernabei et al., Mod. Phys. Lett. A 23 (2008) 2125.
[10] P. Belli et al., Phys. Rev. D 61 (2000) 023512.
[11] D. Smith and N. Weiner, Phys. Rev. D 64 (2001) 043502; D. Tucker-Smith and N. Weiner, Phys. Rev. D 72 (2005) 063509.
[12] K.Freese et al. Phys. Rev. D 71 (2005), 043516; Phys. Rev. Lett. 92 (2004) 11301.
[13] R. Bernabei et al., Eur. Phys. J. C 56 (2008) 333.
[14] R. Bernabei et al., Nucl. Instr. & Meth. A 592 (2008) 297.
[15] R. Bernabei et al., Phys. Lett. B 389 (1996) 757; R. Bernabei et al., Phys. Lett. B 424 (1998) 195; R. Bernabei et al., Phys. Lett. B 450 (1999) 448; R. Bernabei et al., Il Nuovo Cim. A 112 (1999) 545; P. Belli et al., Phys. Rev. D 61 (2000) 023512; R. Bernabei et al., Phys. Lett. B 480 (2000) 23; R. Bernabei et al., Eur. Phys. J. C 18 (2000) 283. R. Bernabei et al., Phys. Lett. B 509 (2001) 197; R. Bernabei et al., Eur. Phys. J. C 23 (2002) 61; P. Belli et al., Phys. Rev. D 66 (2002) 043503.
[16] R. Bernabei et al., Il Nuovo Cim. A 112 (1999) 545.
[17] R. Bernabei et al., Phys. Lett. B408 (1997) 439; P. Belli et al., Phys. Lett. B460 (1999) 236; R. Bernabei et al., Phys. Rev. Lett. 83 (1999) 4918; P. Belli et al., Phys. Rev. C60 (1999) 065501; R. Bernabei et al., Il Nuovo Cimento A112 (1999) 1541; R. Bernabei et al., Phys. Lett. B 515 (2001) 6; F. Cappella et al., Eur. Phys. J.-direct C14 (2002) 1; R. Bernabei et al., Eur. Phys. J. A 23 (2005) 7; R. Bernabei et al., Eur. Phys. J. A 24 (2005) 51.
[18] R. Bernabei et al., Astrop. Phys. 4 (1995) 45; R. Bernabei, in the volume "The identification of Dark Matter", World Sc. Pub. (1997) 574.
[19] R. Bernabei et al., Eur. Phys. J. C 62 (2009) 327.
[20] P. Belli et al., Astropart. Phys. 5 (1996) 217; P. Belli et al., Nuovo Cim. C 19 (1996) 537; P. Belli et al., Phys. Lett. B 387 (1996) 222; Phys. Lett. B 389 (1996) 783 (err.); P. Belli et al., Phys. Lett. B 465 (1999) 315; P. Belli et al., Phys. Rev. D 61 (2000) 117301; R. Bernabei et al., New J. of Phys. 2 (2000) 15.1; R. Bernabei et al., Phys. Lett. B 493 (2000) 12; R. Bernabei et al., Eur. Phys. J. direct C 11 (2001) 1; R. Bernabei et al., Nucl. Instr. & Meth A 482 (2002) 728; R. Bernabei et al., Phys. Lett. B 527 (2002) 182; R. Bernabei et al., Phys. Lett. B 546 (2002) 23; R. Bernabei et al., in the volume *Beyond the Desert 2003*, Springer, Berlin (2003) 365; R. Bernabei et al., Eur. Phys. J. A 27, s01 (2006) 35.
[21] R. Bernabei et al., Phys. Lett. B 436 (1998) 379.
[22] R. Bernabei et al., Astropart. Phys. 7 (1997) 73; R. Bernabei et al., Nuovo Cim. A 110 (1997) 189; P. Belli et al., Astropart. Phys. 10 (1999) 115; P. Belli et al., Nucl. Phys. B 563 (1999) 97; R. Bernabei et al., Nucl. Phys. A 705 (2002) 29; P. Belli et al., Nucl. Instr. & Meth A 498 (2003) 352; R. Cerulli et al., Nucl. Instr. & Meth A 525 (2004) 535; R. Bernabei et al., Nucl. Instr. & Meth A 555 (2005) 270; R. Bernabei et al., Ukr. J. Phys. 51 (2006) 1037; P. Belli et al., Nucl. Phys. A 789 (2007) 15; P. Belli et al., Phys. Rev. C 76 (2007) 064603; P. Belli et al., Phys. Lett. B 658 (2008) 193; P. Belli et al., Nucl. Phys. A 789 (2007) 15; P. Belli et al., Nucl. Phys. A 826 (2009) 256.
[23] P. Belli et al., Nucl. Instr. & Meth. A 572 (2007) 734; Nucl. Phys. A 806 (2008) 388; Nucl. Phys. A 824 (2009) 101.
[24] R. Bernabei et al., Eur. Phys. J. C 18 (2000) 283.
[25] R. Bernabei et al., ISBN 978-88-95688-12-1, pages 1-53 (2009) Exorma ed. (arXiv:0806.0011[astro-ph]).
[26] A. Benoit et al., Phys. Lett. B 637 (2006) 156.
[27] R. Hudson, Found. Phys. 39 (2009) 174-193.

Results from the CDMS II Experiment

Jodi Cooley for the CDMS II Collaboration
Department of Physics, Southern Methodist University, Dallas, Texas 75275, USA
E-mail: cooley@physics.smu.edu

Abstract.
I report recent results and the status of the Cryogenic Dark Matter Search (CDMS II) experiment at the Soudan Underground Laboratory in Minnesota, USA. A blind analysis of data taken by 30 detectors between October 2006 and July 2007 found zero events consistent with WIMPs elastically scattering in our Ge detectors. This resulted in an upper limit on the spin-independent, WIMP-nucleon cross section of 6.6 x 10^{-44} cm^2 (4.6 x 10^{-44} cm^2 when combined with our previous results) at the 90% C.L. for a WIMP of mass 60 GeV/c^2. In March 2009 data taking with CDMS II stopped in order to install the first of 5 SuperTowers of detectors for the SuperCDMS Soudan project. Analysis of data taken between August 2007 and March 2009 is ongoing.

1. Introduction

The Cryogenic Dark Matter Search (CDMS II) experiment was designed to detect the passage of Weakly Interacting Massive Particles (WIMPs) by making simultaneous measurements of ionization and athermal phonons. The experiment was operated at the Soudan Underground Laboratory in Soudan, Minnesota between June 2006 and March 2009.

The experiment contained a total of 30 z-sensitive ionization and phonon mediated (ZIP) detectors, 19 of which were germanium crystals (~250 g each) and 11 of which were silicon crystals (~100 g each) stacked vertically into five towers. The detectors were 7.6 cm diameter, 1 cm thick disks which were photolithographically patterned on their top face with aluminum fins and tungsten transition-edge sensors (TESs) to collect athermal phonon signals. The bottom face contained two concentric electrodes, an inner and outer electrode covering ~85% and ~15% of the face respectively. The detectors are operated at superconducting temperatures with a 3 V/cm voltage applied across. Further details of the detectors and their operation can be found in Refs. [1] [2] [3].

When a particle interacts in our detectors it produces ionization and phonons. WIMPs and neutrons will produce nuclear recoils while most backgrounds (electrons and photons) will produce electron recoils. The ratio of ionization to phonon recoil energy, called "ionization yield", is strongly dependent on event type, providing us the ability to discriminate between nuclear and electron recoils by a factor $> 10^6$.

Electron recoils occurring within the first ~10μm of the surface of our detectors suffer from a suppressed ionization signal which is sufficient to misclassify these events as nuclear recoils. Surface events produce phonon pulses with faster timing properties than events occurring in the bulk of the crystal. Hence, parameterization of the phonon pulse timing properties was used to further discriminate between signal and background events. Discrimination based on both the ionization yield and the timing parameters of the phonon pulse of events are demonstrated in Fig. 1.

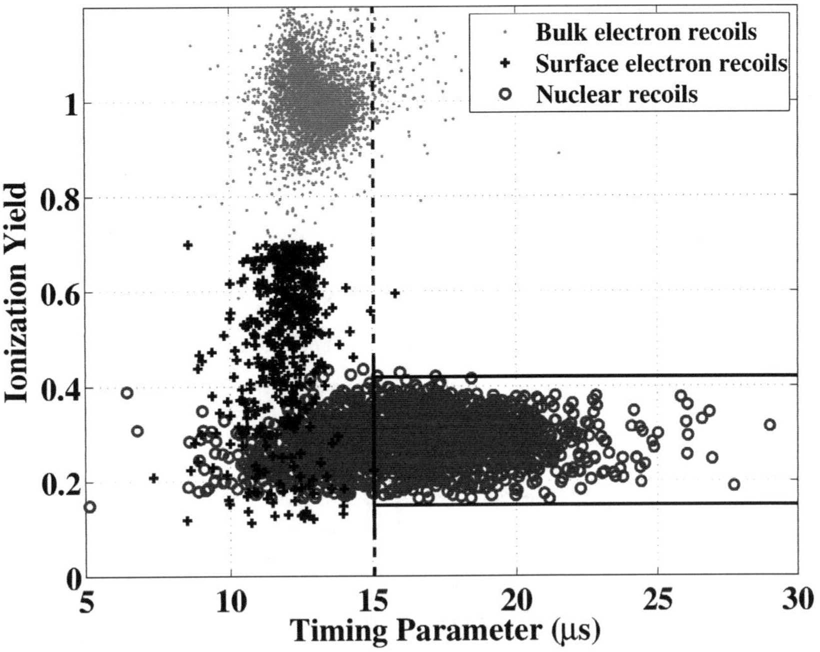

Figure 1. Ionization yield versus timing parameter for calibration data in one of our Ge detectors. Bulk electron recoils (red dots) and low yield surface events (black +) from a ^{133}Ba source, and neutron induced nuclear recoil events (blue ∘) from a ^{252}Cf source show strong dependence on ionization yield and timing parameter. The vertical dashed line indicates minimum timing parameter for this analysis. The boxed region shows the approximate signal region. Taken from Ref. [4]

In order to reduce the overall external γ and neutron backgrounds the detectors are shielded using layers of copper, polyetheyene, lead and an active muon veto which are described elsewhere [3]. To reduce the cosmogenic neutron background, the experiment is located at a depth of 2040 meters water equivalent (m.w.e.) in the Soudan Underground Laboratory, Minnesota, USA.

2. Analysis and Results

Here I summarize the analysis and results from data taken with all five towers operated during two periods between October 2006 and July 2007 (runs R123-R124). A full accounting of the analysis and results can be found in Ref. [4]. The data taken in this period resulted in a raw exposure of 397.8 kg days in our Ge detectors. To prevent bias, we performed a blind analysis where event selection and efficiencies were calculated from calibration data and WIMP-search data outside signal region. Events inside the signal region met the following criteria: they were inside the detector fiducial volume, in anti-coincidence with our active muon veto, scattered in only one detector, where within the 2σ nuclear recoil yield band and passed the phonon pulse timing cut.

The expected background due to surface interactions in this analysis was $0.6^{+0.5}_{-0.3}(stat.)^{+0.3}_{-0.2}(syst.)$ event. The neutron background was estimated using Monte Carlo simulations of cosmic ray muons and subsequent neutron production and transport using FLUKA, MCNPX and GEANT4. The results were normalized to the observed veto-coincident, multiple scatter event rate. This lead to an estimate of <0.1 event from neutrons in this analysis.

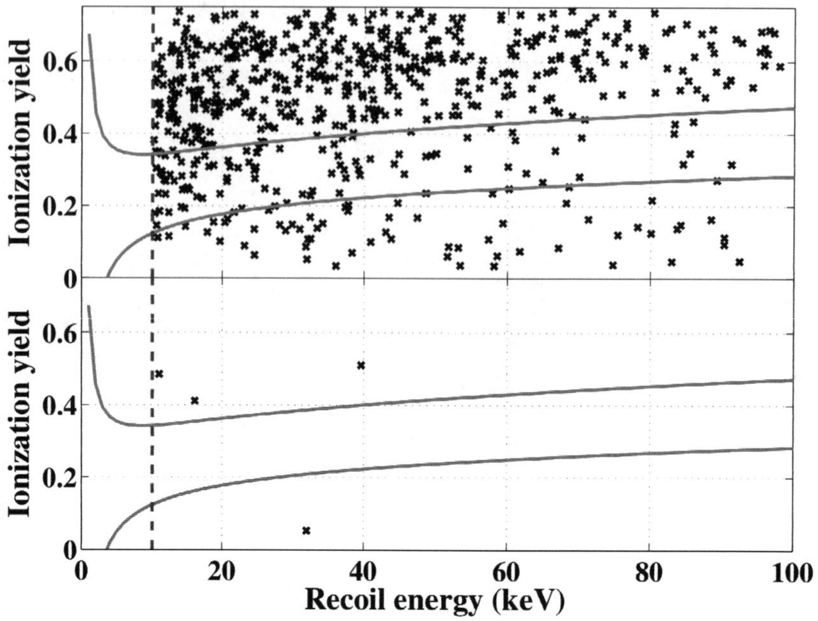

Figure 2. . Top: Ionization yield versus recoil energy for all detectors included in this analysis before timing cuts were applied. The signal region between 10 and 100 keV (between horizontal red solid lines) was defined using neutron calibration data taken from a ^{252}Cf source. Bottom: Same as top after timing cuts were applied. No events were found in the signal region. Taken from Ref. [4]

After applying all selection criteria, no events were observed in the WIMP-search region as shown in Fig. 2. Under the assumptions of WIMP-nucleon spin-independent couplings and standard assumptions about our galactic halo, this resulted in a minimum upper limit of 6.6 x 10^{-44} cm^2 at the 90% confidence level for a WIMP of mass 60 GeV. Taking into account analyses of previous data, a combined limit of 4.6 x 10^{-44} cm^{-2} for a WIMP of mass 60 GeV. Figure 3 compares the result from this work, other recent experiments and representative parameter space from SuperSymmetric models.

3. Future

In March 2009, the CDMS II experiment stopped taking data. The analysis of data taken between August 2007 and March 2009 (R125-R128) is on going. The raw exposure of data taken during this time period is ~1.7 times larger than the analysis reported here. Improvements for this analysis include a faster processing package written in C++, refined timing based position correction algorithms and improved optimization of discrimination criteria based on timing of the phonon pulse. Histograms of the phonon pulse timing parameter of calibration data indicate consistency between this data and the previous runs as shown in Fig. 4. Results are expected to be announced in December 2009.

In order to increase the sensitivity of the the CDMS program, an increase in total exposure is needed while keeping the backgrounds under control. The SuperCDMS Soudan project will

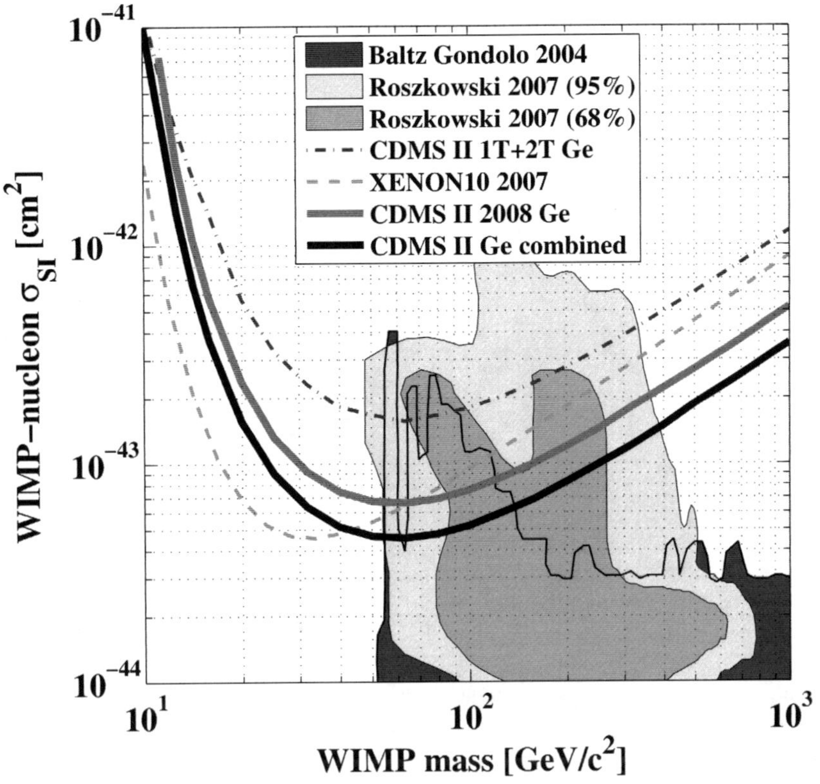

Figure 3. WIMP-nucleon spin-Independent cross section versus WIMP mass upper limits at the 90% C.L. from this work and other recent experiments. The shaded region represents parameter ranges expected from supersymmetric models. Taken from Ref. [4]

deploy 15 kg of Ge detectors, five towers each consisting of five vertically stacked 1-inch thick detectors, in the existing Soudan setup. For this project, the new 1-inch thick detectors with redesigned phonon sensors were developed, fabricated [5] and tested . The new detector thickness provides an increased mass of 2.54 over the CDMS II detectors and the redesigned phonon sensors optimize the phonon collection area [6]. The first of the five SuperTowers has been deployed and is taking data in the Soudan Underground Laboratory.

4. References

[1] Irwin K D, Nam S W, Cabrera B, Chugg B and Young B A 1995 *Rev. Sci. Instrum.* **66** 5322
[2] Saab T *et al.* 2002 *AIP Conf. Proc.* **605** 497–500
[3] Akerib D S *et al.* (CDMS) 2005 *Phys. Rev.* **D72** 052009
[4] Ahmed Z *et al.* (CDMS) 2009 *Phys. Rev. Lett.* **102** 011301
[5] Brink P L *et al.* (SuperCDMS) *To be published in LTD 13, AIP Conf. Proc.*
[6] Ahmed Z *et al.* (SuperCDMS) *To be published in LTD 13, AIP Conf. Proc.*

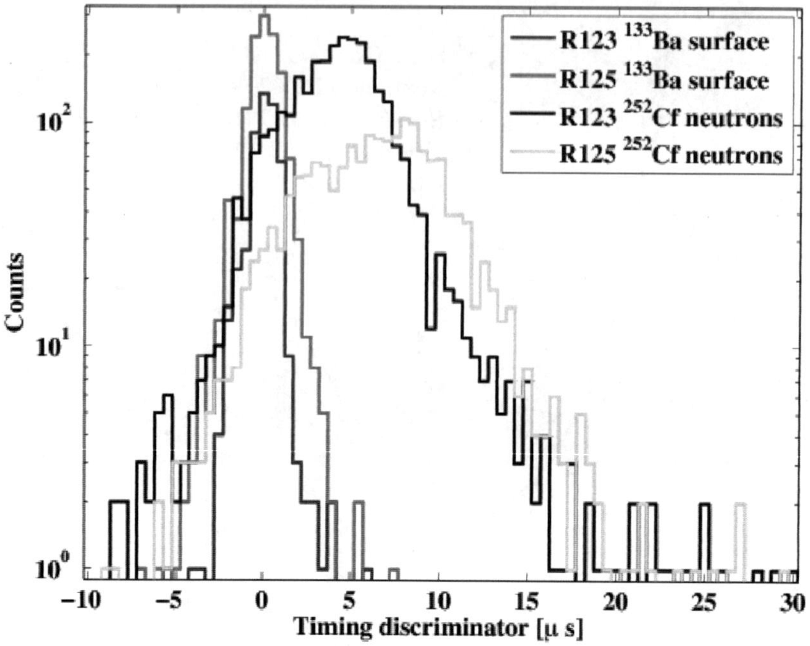

Figure 4. Histogram of timing parameter for calibration data taken in previous analysis and current analysis. The timing discriminator behaves similarly for surface events (solid blue current, solid red ongoing) and neutrons (solid black current, solid green ongoing) between the two runs.

Topics in Astroparticle and Underground Physics (TAUP 2009) IOP Publishing
Journal of Physics: Conference Series **203** (2010) 012005 doi:10.1088/1742-6596/203/1/012005

The XENON100 Dark Matter Experiment at LNGS: Status and Sensitivity

Elena Aprile (on behalf of the XENON Collaboration)
Columbia University,Physics Department, New York,NY, USA
E-mail: age@astro.columbia.edu

Abstract. The XENON100 experiment aims at the direct detection of cold dark matter particles via their collisions with xenon nuclei. The detector is a two-phase xenon time projection chamber, with an active target mass of 65 kg, shielded by 105 kg of liquid xenon as scintillator veto. Additional passive shielding surrounds the target to reduce the gamma and neutron induced background. The experiment is located underground in the Gran Sasso National Laboratory (LNGS) and is in the final stage of commissioning. I will summarize the status and sensitivity reach of the experiment, as well as the current plan for an upgrade.

1. Introduction

There is an increasing number of astrophysical and astronomical observations pointing to the existence of a non-luminous, non-baryonic and cold (i.e. non-relativistic) matter component in the universe, called Cold Dark Matter (CDM) [1, 2, 3]. Weakly Interactive Massive Particles (WIMPs) predicted by Supersymmetric theories (SUSY), models with extra dimensions and little Higgs models [4, 5, 6, 7] are favorite CDM candidates. Three different approaches are being pursued for WIMP detection: direct production of WIMPs in the highest energy particle collisions expected at the the Large Hadron Collider; indirect detection of WIMP annihilation signals from the center of the Sun or from the center of the galaxy, and direct detection of nuclear recoils produced by WIMPs scattering off normal matter.

The XENON project is one of several direct detection experiments worldwide, using a noble liquid as target and detector medium. After the successful results of the first 10 kg scale liquid xenon (LXe) detector XENON10 [8, 9], the collaboration has designed and built a second-generation detector, with the goal of significantly improving the sensitivity by lowering the background and increasing the target mass. Like XENON10, the new detector is a two-phase time projection chamber (TPC) with about a factor of 10 more target mass and about a factor of 100 less radioactive background. The sensitivity reach of the XENON100 experiment is a factor of 50 better than that of XENON10.

2. Detector Concept

The XENON100 detector is a double-phase (liquid-gas) time projection chamber (TPC). A particle interacting with the target generates scintillation light and ionization electrons (Figure 1). The primary light (S1) is detected immediately by the two photomultiplier (PMT) arrays above and below the target. An electric field ($\sim$ 1 kV/cm) across the TPC drifts the free ionization electrons upwards, where they are extracted into the gas phase by a strong extraction

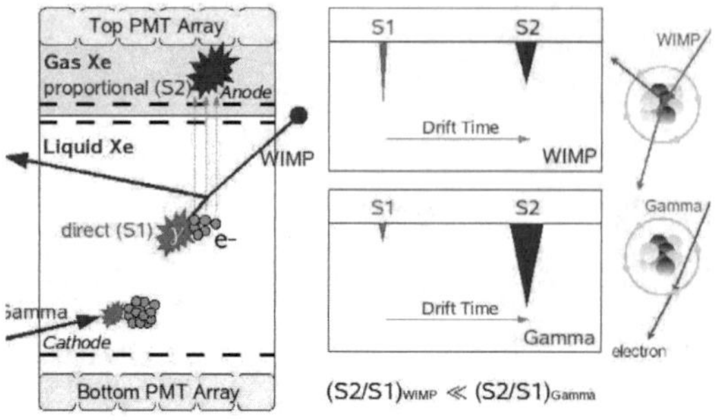

Figure 1. Principle of a two-phase liquid xenon TPC. A particle generates primary scintillation light (S1) and ionization electrons. These are drifted upwards by an electric field and detected via secondary scintillation light in the gas phase (S2). The S2 hit pattern (xy) and the drift time (z) give the three-dimensional information on the position of events. Additionally, the ratio S2/S1 allows event discrimination between nuclear recoils (WIMPs, neutrons) and electron recoils (γ, β).

field. In the gas phase, the electrons generate very localized secondary scintillation light (S2), that can be used to determine the xy-position of the interaction point. Since the z-position is known from the drift time, the event positions can be fully reconstructed and used to fiducialize the target volume in order to drastically reduce the radioactive background from external sources.

The high ionization density of nuclear recoils in liquid xenon leads to larger S1 and smaller S2 signals compared to electron recoils. The simultaneous measurement of charge and light provides a powerful discrimination between signal (nuclear recoils) and background events (electron recoils) via the ratio S2/S1. A discrimination of 99.5-99.9% was achieved in XENON10.

3. Experimental Setups

In XENON100, the LXe target volume of 65 kg (30 cm drift height and 30 cm diameter), is surrounded by 105 kg of LXe as active scintillator, for additional background rejection. PTFE is used as an effective UV light reflector in order to optically isolate the target from the surrounding LXe veto. A set of 40 field shaping rings and their resistive divider chain are mounted on the panels and create a homogenous drift field.

The light readout is based on 1" × 1" Hamamatsu R8520-06-Al low-radioactive PMTs with quantum efficiencies up to ~ 35%. 98 PMTs in the top array are arranged to improve position reconstruction, 80 PMTs on the bottom to optimize the light collection, and 64 PMTs in the LXe veto to give another factor of 4 in background reduction.

An Iwatani PC150 Pulse Tube Refrigerator (PTR) is used to liquefy xenon (170 W cooling power) and to keep the detector at the operating temperature of -100°C. The refrigerator provides excellent stability of operation, with temperature variations below 0.1°C and pressure changes of less than 1%. In order to minimize the radioactive background in the detector, the cryogenic system is located outside the passive shield. This shield consists of 5 cm copper, 30 cm polyethylene and 20 cm lead. The cryostat and detector production was completed in early 2008 and the detector was installed underground at LNGS in Spring 2008. A picture of the XENON100 detector in front of its shield and the schematic view of the inner TPC are shown in Figure 2 and Figure 3, respectively.

3.1. XENON Purification System

Successful operation of the xenon TPC requires long VUV light attenuation length and effective charge drift (long electron lifetime in liquid xenon). We are using a purification system based on continuous xenon gas circulation with purification through a high temperature metal getter (SAES) to reduce the impurities (water, oxygen etc.) in commercially available xenon to a level below 1 part per billion (ppb) O_2 equivalent.

Figure 2. XENON100 after cleaning the cryostat and the shield cavity from dust, grease, stains, etc. right before closing the shield. The circular copper pipe around the detector is used for the insertion of calibration sources from outside the shield. The Pb brick visible in the front is necessary to block gamma rays from the AmBe source during a neutron calibration.

Figure 3. The XENON100 TPC. Left: Photo showing the PTFE panels as well as the veto PMTs on the top and bottom. Right: Drawing of the structure to show the location of the inner top and bottom PMT arrays.

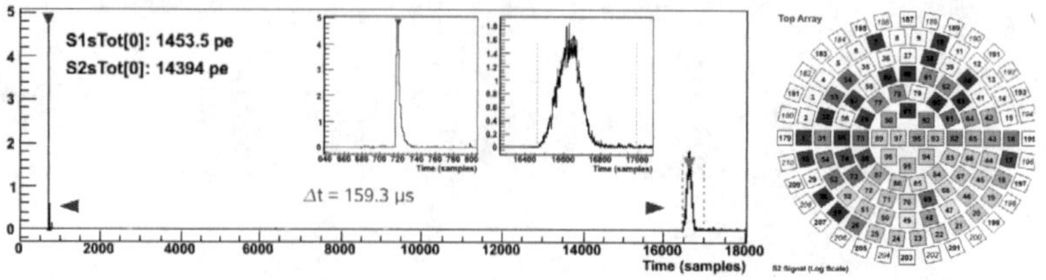

Figure 4. Example of a XENON100 event with almost maximum drift time. The interaction happened close to the cathode. The insets show the enlarged S1 and the S2 pulses. The narrow pulse at 720 is the prompt S1 scintillation light, the broader peak at 16600 is the charge signal (S2). The xy position of the event can be reconstructed using the hit pattern of the S2 signal on the top PMT array.

Xenon can be purified from most radioactive impurities, except ^{85}Kr. Krypton is present in commercial xenon gas at the few parts per million (ppm) level. The beta decay of ^{85}Kr (E_{max} = 687 keV, $T_{1/2}$ = 10.6 yr) is a significant source of background for any xenon based dark matter search experiment. In order to reduce the ^{85}Kr level to below 50 parts per trillion (ppt) as required by the XENON100 sensitivity goal, we have installed and tested underground a cryogenic distillation column made by Taiyo-Nippon Sanso. The xenon has been purified with this system and data indicates a satisfactory reduction of ^{85}Kr.

3.2. Readout Electronics and Data Acquisition
The XENON100 data acquisition system (DAQ) is used to generate a trigger signal for the TPC, digitize the waveforms of the 242 PMTs, and store this data. The PMT signals are first amplified by a factor of 10 (Phillips 776 amplifiers) and then digitized by CAEN V1724 Flash ADCs with a sampling rate of 100 MHz. A typical XENON100 waveform from a ^{137}Cs calibration is shown in figure 4. The DAQ system is fully operational and used continuously to characterize the detector. Detector calibration is performed with several gamma sources, and the PMTs single photoelectron response is measured regularly with blue LEDs.

Apart from general performance tests, the first phase after installation was devoted to characterize the detector with S1 signals. The measured S1-background agrees with the MC simulation, suggesting that the envisaged background reduction by two order of magnitudes was achieved. After some unexpected delays due to a leak in the cryostat seal, which lead to a limited LXe purity, we started to operate in dual-phase mode early 2009 and are now characterizing the fully operational detector with gamma sources.

4. Background
4.1. Materials Screening
To reduce the radioactivity from the detector and the shielding materials in order to improve the sensitivity of the experiment, radioactivity screening was performed with a high purity germanium spectrometer at LNGS. The radioactivity of all the components has been measured and is used as input for detailed Monte-Carlo simulations of the γ and neutron background of the experiment. The radioactive contaminations in a selection of materials screened is shown in Table 1.

Table 1. Results from some of the materials screened for radioactivity with high purity germanium spectrometers.

Sample	Unit	^{232}Th	^{238}U	^{40}K	^{60}Co
Innermost Pb (shield)	[mBq/kg]	<0.55	<0.66	<1.46	<0.11
Outermost Pb (shield)	[mBq/kg]	<0.43	<0.92	<14 ± 36	<0.12
Polyethylene (shield)	[mBq/kg]	<0.094	<0.23±0.05	<0.7 ±0.4	N.A
Stainless Steel	[mBq/kg]	<1.9	<1	8.5±0.9	10.5 ±4.2
PTFE	[mBq/kg]	<0.16	<0.31	<2.25	<0.11
22 QE PMTs	[mBq/PMT]	0.18±0.06	<0.24	11 ±2	0.6 ±0.1

4.2. Nuclear Recoil Background

Spontaneous fission and (α, n) reactions in the detector, shielding materials, and surrounding concrete/rock are the major contributions to the nuclear recoil background events (neutron scattering). Using neutron yields and energy spectra calculated with a modified SOURCES4A code [10], we estimate the total neutron rate from PMT materials as 3.7 neutrons/year. For this neutron rate the expected number of single nuclear recoils in liquid xenon is 1.4×10^{-7} events/kg/day/keV (dru = differential rate unit) at 10 keVr.

The muon flux at the 3700 mwe (meter water equivalent) Gran Sasso depth is 12 muons/m^2/day. Muons will produce neutrons in the lead and polyethylene of the shield due to electromagnetic and hadronic showers and through direct spallation. High energy muons interacting in the rock can produce highly penetrating neutrons with energies up to a few GeV. The total expected neutron background in XENON100 is 1.3×10^{-5} dru and thus well below the expected WIMP rate of 2×10^{-4} dru for a cross section of 9×10^{-45} cm ^{2}.

4.3. Electron Recoil Background

Gammas from the decay chains of the radioactive contaminants in the detector materials (mostly ^{238}U, ^{232}Th, ^{40}K, ^{60}Co) dominate the electron recoil background. The background rate was estimated using measured detector material activities listed in Table 1 as input parameters for GEANT4 [11] Monte Carlo simulations. The activity of the PMTs is the major contributor to the current XENON100 electron recoil background. The total electron recoil background in XENON100 is 6×10^{-3} dru. After 99% electron recoil rejection we obtain a conservative background rate of 6×10^{-5} dru.

To summarize, the expected rate of single scattering electron recoil signals is $(24.05 \pm 0.17) \times 10^{-3}$ events/kg/day/keV and $(9.07 \pm 0.14) \times 10^{-3}$ events/kg/day/keV for 50 kg and 30 kg fiducial mass, respectively. The total neutron induced single nuclear recoil rates are 1.62 n/year and 0.60 n/year for 50 and 30 kg fiducial mass, respectively.

5. XENON100 Sensitivity

A WIMP search with the current XENON100 will be first performed in a 50 kg fiducial volume with 40 live-days of exposure, to reach a sensitivity for a spin-independent WIMP cross section of $\sigma \sim 6 \times 10^{-45}$ cm^2 for WIMPs with a mass of 100 GeV/c^2. An improvement by a factor of three ($\sigma \sim 2 \times 10^{-45}$ cm^2) can be achieved after 200 days of data taking and a 30 kg fiducial volume, to remain background free. The expected sensitivity of XENON100, compared with the current limits, is shown in Figure 5.

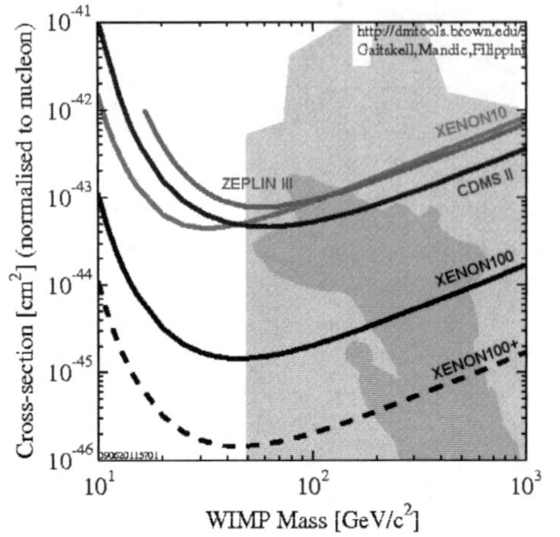

Figure 5. Exclusion plots for spin independent WIMP-nucleon interactions. The current limits are the curves around 10^{-43} cm^2, given by XENON10 (red, [8]), CDMS II (blue, [13]), and ZEPLIN III (green, [12]). The projected sensitivity for XENON100 is one order of magnitude lower (black) and even better when upgraded to XENON100+. The shaded regions are theoretical expectations of some models.

6. XENON100+

The XENON100 gamma background is dominated by the PMTs and the PMT bases, followed by the polyethylene of the shield and the stainless steel of the cryostat. In order to achieve reduced background for the next step topwards an even more enhanced sensitivity, the bottom PMT array will be replaced by 19 QUPIDs [14], novel photosensors with an extremely low intrinsic radioactivity, developed by UCLA and Hamamatsu for this experiment. Additionally, a copper cryostat and an improved shield with an active muon veto will be implemented. With such upgrade, a spin-independent WIMP-nucleon limit of 2×10^{-46} cm^2 could be achieved by 2012.

References

[1] W. Freeman and M. Turner, Rev. Mod. Phys.,**75**, 1433 (2003).
[2] D. Clowe et al., ApH, **648**, L109 (2006).
[3] M. J. Jee et al., http://xxx.lanl.gov/abs/0705.2171
[4] A. Bottino et al., *Phys. Rev.* **D 69**, 037302 (2004).
[5] J. Ellis et al., *Phys. Rev.* **D 71**, 095007 (2005).
[6] H. C. Cheng et al., *Phys. Rev. Lett.* **89** 211301 (2002).
[7] A.Birkedal-Hansen and J. G. Wacker, *Phys. Rev.* **G 69** 065022 (2004).
[8] J. Angle et al., (XENON10 collaboration), *Phys. Rev. Lett.*,**100**, 021303 (2008).
[9] J. Angle et al., (XENON10 collaboration), *Phys. Rev. Lett.*,**101**, 091301 (2008).
[10] R. Lemrani et al., *Nucl. Instr, Meth.* **A 560**, 454 (2006).
[11] S. Agostinelli et al., (GEANT4 collaboration), *Nucl. Instr. Meth.* **A 506** 250 (2003).
[12] V. N. Lebedenko et al., (Zeplin III collaboration), arXiv:0901.4348 (2009).
[13] Z. Ahmed et al. (CDMS II collaboration), *Phys. Rev. Lett.* **102**, 011301 (2009).
[14] K. Arisaka et al., arXiv:0808.3968 (2008).

The WArP Experiment

R. Acciarri[c], M. Antonello[b], B. Baibussinov[f], P. Benetti[a], F. Calaprice[e], E. Calligarich[a], M. Cambiaghi[a], N. Canci[c], C. Cao[e], F. Carbonara[d], F. Cavanna[c], S. Centro[f], M.B. Ceolin[f], A. Chavarria[e], A. Cocco[d], F. Di Pompeo[bc], G. Fiorillo[d], C. Galbiati[e], L. Grandi[bc], B. Loer[e], G. Mangano[d], A. Menegolli[a], G. Meng[f], C. Montanari[a*], O. Palamara[b], L. Pandola[b], F. Pietropaolo[f], G.L. Raselli[a], M. Roncadelli[a], M. Rossella[a], C. Rubbia[b], R. Saldanha[e], E. Segreto[b], A.M. Szelc[b], S. Ventura[f], C. Vignoli[a].

[a] Università degli Studi di Pavia e INFN, Pavia, Italy

[b] INFN - Laboratori Nazionali del Gran Sasso, Assergi, Italy

[c] Università degli Studi dell'Aquila e INFN, L'Aquila, Italy

[d] Università degli Studi di Napoli e INFN, Napoli, Italy

[e] Princeton University Department of Physics, Princeton, New Jersey, USA

[f] Università degli Studi di Padova e INFN, Padova, Italy

[*] Corresponding author, e-mail: claudio.montanari@pv.infn.it

Abstract. Cryogenic noble liquid detectors are presently considered one of the best options for WIMP Dark Matter searches, especially when extensions to multi ton scale sensitive masses are foreseen. The WArP experiment is the first one that exploits the unique characteristics of liquid Argon to make a highly sensitive search for WIMP Dark Matter candidates. In 2008, a double phase detector has been assembled in the Gran Sasso National Laboratory with 140 kg sensitive mass and a discovery potential in the range of 5 x 10^{-45} cm^2 in the spin-independent WIMP-nucleon cross-section. In addition to standard neutrons and gamma-rays passive shields, WArP implements an 8 ton liquid Argon active shield with 4π coverage. The detector was commissioned and put into operation during the first half of 2009 for a first technical run. Detector design, construction and assembly are described, together with the very first results of this technical run.

1. Introduction

Increasing efforts are being made to unveil the nature of Dark Matter, especially since its existence was definitively certified by astrophysical and cosmological observations [1]. Several possibilities for Dark Matter constituency are seriously considered although a central role is presently played by direct WIMP searches due to the potential link with the SUSY hypothesis.

In the last ten years several technologies have grown up from the laboratory scale to full size experiments running or to be run in several underground locations [2]. Despite the increasing efforts made by the present experiments, the positive signal claimed by the DAMA Collaboration in terms of seasonal variations [3] still stands without an independent confirmation or clear disproof. In this scenario there are already many proposals for upgrades of the presently adopted technologies to sensitive masses of the order of few tons with increasing complexity in terms of shielding and background control in order to fully explore the predicted regions of the minimal SUSY hypothesis. Noble liquid detectors play a central role in this future scenario due to their potentiality for the instrumentation of very large sensitive masses. WArP is the very first experiment that exploits the characteristics of Liquid Argon (LAr) for direct Dark Matter searches.

The WArP technology was established in several years of R&D studies. They were started at CERN at the beginning of the nineties with pioneering works with liquid Xenon in the framework of the general R&D activities of the Icarus experiment [4]. When the concept of double phase chamber as a tool for direct Dark Matter searches was put forward [5], around the mid of the nineties, we opted for argon as our primary choice for a number of reasons [6].

As a liquid scintillator, argon has a very good light yield: about 20 eV are required, for a relativistic electron, to produce a scintillation photon, just slightly lower than the one of liquid xenon [7]. Scintillation light is emitted at 128 nm wavelength, in the Vacuum Ultraviolet Region, and specific methods have to be implemented to wave shift and efficiently collect it. Contrary to liquid xenon the fast (singlet) and the slow (triplet) components of the scintillation light are very widely separated: the fast component is emitted with a decay time of about 7 ns while the slow component has a decay time of about 1400 ns [8]. This separation allows for very efficient identification of nuclear recoils against gamma and beta backgrounds by means of pulse shape discrimination with methods similar to the ones commonly in use with specific organic scintillators. Liquid argon has a very high electron mobility, it is available in large quantities at low cost and can be easily purified from electronegative impurities allowing for very long drift distances, of the order of several meters, for free electrons as amply shown by the results obtained by the Icarus Collaboration [9]. On the other side, argon is contaminated by ^{39}Ar, a β active isotope with 269 years half-life and end point at 565 keV; ^{39}Ar is produced by interaction of cosmic rays in the upper atmosphere and is giving a β activity in commercially available liquid argon of about 1 Bq / kg [10]. Argon nuclei have no spin, so that spin-independent only interactions are allowed. The relatively low atomic mass of argon results in a lower integrated cross-section for elastic scattering off nuclei, which is proportional to A^2 due to the coherence effect, when compared to other materials like xenon and germanium. On the other hand the recoil spectrum is more energetic because of the nuclear form factor, so that at a recoil energy threshold of about 30 keV, the integrated rates in argon, xenon and germanium are almost the same but with recoil energies in argon that extend well above the ones from the other targets.

After the pioneering works done in CERN, the R&D activities were also carried on in the laboratories of the Collaboration groups: in Pavia, Padova, Naples, LNGS and later in Princeton. Various technical issues have been investigated in details: electrons extraction from liquid to gas, secondary light production and its proportionality to the extracted charge, light wave-shifting, methods for light collection and their optimization, photomultipliers (PMTs) behavior at cryogenic temperatures, materials behavior and their radioactivity, effects of argon pollutants, procurement of low activity argon either by isotopic separation of commercial argon or by extraction of "aged" argon from underground sources [11]. A key role has been played by a 2.3 liters prototype built at the beginning of years 2000. This device has been widely used (and it is continuously used) mainly to qualify the technical solutions to be implemented in the 100 liters detector. In 2004 it was taken to LNGS to extend the R&D activity. It has been run underground in several conditions. With full passive shields it was able to provide a very convincing limit on spin-independent WIMP-nucleon cross-section [12], even though the internal backgrounds were intentionally not kept at optimal levels, thus proving the goodness of the reached technological advances.

Design and construction of the 100 liters started in 2004 soon after the approval of the experiment proposal by INFN. In 2006 also NSF decided to approve the WArP proposal and provided the funds necessary to complete the detector instrumentation.

2. Design and construction of the 100 liters detector

The basic scheme of the 100 liters detector is shown in figure 1. Starting from the center, the detector consists in:

a) the inner detector: a double phase chamber with 100 liters of liquid volume surmounted by a gas volume where the readout PMTs are placed;
b) the active shield completely surrounding the inner detector, with a minimum thickness of 60 cm, for a total liquid argon volume of about 5600 liters, and readout by a set of PMTs placed on the external surface;
c) an internal passive shield made of polyethylene with 10 cm thickness, to filter neutrons coming from the main cryostat walls;
d) the main cryostat, made of stainless steel, with double walls, super-insulated under vacuum, with an internal volume of 30 m^3 of which 23 m^3 are to be filled with LAr;
e) an external passive shield made of 10 cm thick lead, encapsulated in stainless steel boxes, plus a 70 cm thick layer of polyethylene;
f) an external, anti-seismic, sustaining structure made of carbon steel;
g) circuits for the cryostat cooling and for the purification and recirculation of the liquid argon.

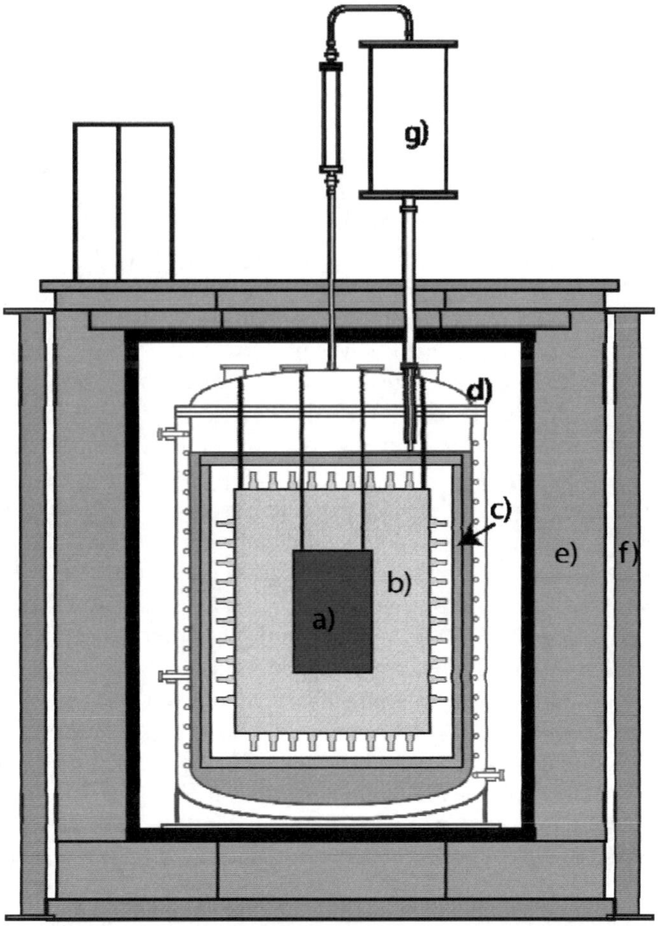

Figure 1

All the internal surfaces of the inner detector and the active shield are covered with a reflector / wave-shifting layer made of a highly reflective plastic foil onto which a highly efficient wavelength shifter (TetraPhenylButadiene – TPB) is deposited. For optimal uniformity and efficiency, the wavelength shifter deposition is made by thermal evaporation under vacuum, with a specially designed evaporation system. The inner detector and the active shield are optically separated, but the argon volume is the same so that no independent filling and purification systems are required.

The inner detector can be subdivided into three regions: the drift region, in liquid phase; the extraction and multiplication region, in gas phase; the light detection region, where the PMTs are placed, in the gas phase. The drift region is delimited by a cathode (an Oxygen Free copper disk with about 50 cm diameter) to be supplied to -75 kV, a stainless steel grid with inner diameter of about 50 cm, placed 60 cm above the cathode, to be put at -15 kV, and a set of circular stripes (copper stripes printed on a Kapton^{TM} substrate) placed every 2 cm between the cathode and the grid, acting as field shaping electrodes (race-tracks) and ensuring in the drift region a uniform field that drives ionization electrons through the grid to the liquid surface. Race-tracks are arranged in a truncated conical shape to prevent ionization electrons from being trapped between the race tracks themselves. Supply voltage is distributed from the cathode to each race-track and to the grid, by means of a resistive divider chain. Above the first grid a second and a third grids are placed, in the gas phase. The field between the first and the second grid is arranged in such a way to provide complete extraction from the liquid of the ionization electrons and to accelerate them in the gas phase in order to produce a suitable secondary, proportional, light signal. Electrons are then decelerated above the second grid and collected on the third one. Voltage is supplied to the second and third grid by means of dedicated high voltage channels. A set of 31, 3" PMTs plus 6, 2" PMTs is placed 3 cm above the third grid. All the PMTs, including the ones of the active shield, are specifically designed to work at LAr temperature, they have glass windows and peak sensitivity (in the range of 20%) at 430 nm wavelength, well matched with the peak emission of the wavelength shifter. The photocathode coverage of the inner detector is about 10%. All PMTs have their divider chains mounted directly on them; divider chains are made of circuits printed on a Kapton^{TM} substrate with discrete components (resistors and capacitors) selected for operation at LAr temperature. All PMTs are negatively biased; the anode signals are readout at ground. PMTs of the inner detector are coated with TPB in a polystyrene matrix to ensure good localization of the secondary signal in the horizontal plane. The upper part of the inner detector is contained into a vacuum super-insulated dewar placed upside down. This dewar is acting as a diving bell with a set of 12 small holes placed in correspondence of the desired liquid level (5 mm above the first grid). A set of heating resistances is placed just below the liquid surface; they continuously boil off the liquid and the produced gas in excess is evacuated through the holes, thus ensuring that the liquid level is precisely positioned. A copper sheet completely surrounds the inner detector. This sheet is grounded, ensuring that the drift fields are confined in the central volume.

The active shield volume is approximately of the shape of a rotational ellipsoid with a vertical axis of 220 cm and an horizontal axis of 180 cm for a total volume (excluding the volume occupied by the inner detector) of 5600 liters (corresponding to 7800 kg of LAr). The active shield mechanical structure is made of Oxygen Free copper pipes and plates and sustains the photomultipliers and the wave-shifting / reflector layers. Both the active shield and the inner detector are suspended to the main cryostat upper flange with adjustable mechanics that allow positioning and alignments to 0.5 mm precision. A total of 300 PMts (264 of 3" diameter plus 36 of 2" diameter) are placed on the external surface of the active shield for a total photocathode coverage of about 7% of the active shield surface. Two different types of electronics are used to readout the PMTs. The one for the active shield was manufactured by CAEN on a custom design made by the Princeton group and is essentially based on signal digitization and multiplexing after charge integration. The one for the inner detector is based on fast digitization (1 GHz) of the non-integrated signals by means of high performance digitizers manufactured by Aquiris.

Construction of the 100 liters device started in 2005. All construction materials have been subject to strict selection for low radioactive contamination. Samples of all the employed materials have been measured in the LNGS Low Activity Facilities prior to their acceptance for the final construction. After the selection, the main source of unvetoed background for the WArP experiment, resulting from materials measurements and from Montecarlo data, is coming from the photomultipliers and is accounted for less than 1 event / year of exposure.

The detector mechanical components were first delivered in Pavia where they have been completely assembled in order to control and qualify the assembly procedures and the mechanical tolerances. The detector was then dismounted and the components were sent to a specialized industry for cleaning before the final delivery to LNGS. PMTs were subject to very severe testing procedure: each PMT have been tested by the manufacturer in liquid nitrogen for stability and at room temperature for compliance to the specifications (quantum efficiency, dark counts, gain). The phototubes were then sent to Napoli where they underwent an additional set of tests (operational stability and intrinsic noise) in liquid nitrogen. The PMTs were then sent to LNGS for pre-assembly operations installation.

Final detector assembly started in July 2007, when the underground installation area, in LNGS Hall B, became available. The external shield basement, the main cryostat and the anti-seismic supporting structure were first put in place. The assembly clean and dehumidified room was then built and put in operation. Assembly of the internal passive shield followed. In parallel several pre-assembly operations took place in the external laboratories above ground, they consisted in: pre-assembly of the PMTs with Kapton™ bases, cutting and pre-assembly of the reflector foils, thermal evaporation of the wavelength shifter on the reflector foils and quality control of evaporated samples. All the pre-assembled components were taken in the clean room underground for the final installation. The final detector assembly started in June 2008 from the inner detector and was completed in December of the same year. All the PMTs were tested in dark conditions before that the detector was finally moved in the main cryostat around the mid of December. Vacuum pumping started soon after. The detector plant was completed in January 2009 with the installation of the external cooling and purification systems. In parallel, the external shield lead walls were installed together with a first part of the lateral polyethylene external shield.

Figure 2

3. Detector Commissioning

The first phase of the Detector Commissioning was started on Dec. 22nd, 2008 with the vacuum pumping of the cryostat internal volume, combined with a number of cycles of inert gas flushing (N_2 at room temperature) for the reduction of the residual H_2O content of the internal materials. This phase lasted about five months (from Dec. 23rd, 2008 to May 5th, 2009).

Initially, leak search, localization and repairs required about one month and successively vacuum pumping lasted up until the design residual pressure was reached as measured by the pressure probes located in various points of the cryostat. During this period external cabling of the PMT's and of the slow-control sensors (LAr levels, Temperature, Pressure) has been completed. A full test of the PMT signals and HV mapping has also been then performed and validated.

In this phase of the commissioning the composition of the residual gas has been continuously monitored by mass spectrometry. Air components (from residual leaks) and water (from material outgassing) have been identified and monitored during their progressive decrease (e.g. H_2O outgassing rate was, in the last period of the vacuum pumping, $\approx 6 \times 10^{-5}$ mbar lt/sec). The vacuum pumping period was assumed as completed when the residual pressure reached the design value in the range $2 - 6 \times 10^{-6}$ mbar (end of March, 2009).

The second phase of the commissioning procedure (detector Cooling and LAr Filling) was started on May 5th, the first day of resuming into normal operation at LNGS underground – after the suspension of the activities due to the earthquake (during the month of April, vacuum pumping indeed continued without interruptions).

The *Cooling phase* lasted less than 3 days – via low-quality LAr forced recirculation in the piping system inserted in the vacuum jacket of the cryostat, in thermal contact with the internal wall. This produced a smooth cooling down of the materials inside the cryostat to less than 200 K (in the innermost element of the detector), thus stopping the water outgassing (as seen by the drop of the H_2O partial pressure). The ultimate residual pressure just below 1×10^{-6} mbar was then reached.

The following step with the "LAr filling" lasted four more days. Good-quality LAr (equivalent to 5.5 grade) at a speed of about 300 lt/hr was conveyed inside the cryostat through a dedicated vacuum insulated filling line from a movable LAr storage. The line included a large two-stages filter (Oxygen reactant for O_2 removal and molecular sieves for H_2O trapping).

During the first day of the filling phase all the inner components (internal Polyetilene shield and detector) were cooled down to LAr temperature and successively the LAr level inside the cryostat increased to the detector operation design level (full immersion of the active veto detector).

On May 13th, the WArP-100 detector Technical Commissioning was successfully completed.

A technical run of the WArP 100lt detector started with the activation of the various components of the detector and of the cryogenic system (May-July, 2009):

- GAr recirculation (liquefaction and purification) activated, allowing direct measurement of the actual total heat load of the system (estimated at a level of ≤ 1 kW, in agreement with design value).
- GAr/LAr precise level configuration for the two-phase 100lt inner detector successfully established (gas Argon pocket generated and maintained by resistive heaters inside the "bell-shaped" immersed cryostat).
- PMT activation with voltage bias at nominal value with slow HV ramp and individual test.
- DAQ & Trigger for 100lt detector, after initial debugging phase, in run with progressive tuning and optimization of the performance.
- Off-line event reconstruction successfully tested and used for first data analysis.
- HV system for the Inner Detector (connection from external power supply to inner detector cathode): while ramping up the HV a major discharge occurred inside the cryostat along the connection line. This unforeseen event prevented to establish the EF in the LAr volume of the 100 lt detector, necessary for operation in two-phase configuration.

The technical run was then continued in the summer to complete the analysis of the detector response.

Journal of Physics: Conference Series **203** (2010) 012006 doi:10.1088/1742-6596/203/1/012006

References

[1] D. N. Spergel et al., (WMAP Collaboration), Astrophys. J. Suppl. **170** (2007) 377:
M. Tegmark et al., (SDSS Collaboration), Phys. Rev. **D74** (2006) 123507;
D. N. Schramm and M. S. Turner, Rev. Mod. Phys. **70** (1998) 303;
R. H. Cyburt, B. D. Fields, and K. A. Olive, Phys. Lett. **B567** (2003) 227.

[2] For a review see for example: N.J. Spooner, "*Direct Dark Matter Searches*", J. Phys. Soc. Jap. **76** (2007) 111016, also as arXiv:0705.3345v1[astro-ph].

[3] R. Bernabei et al., "*First results from DAMA/LIBRA and the combined results with DAMA/NaI*" Eur. Phys. J. **C 56** (2008) 333-355.

[4] P. Benetti, et al., NIM **A327** (1993) 203-206;
P. Benetti, et al., NIM **A329** (1993) 361-364;
F. Arneodo, et al., NIM **A449** (2000) 147-157.

[5] D. Cline, et al., Astropart. Phys., **12** (2000) 373-377.

[6] C. Rubbia, et al., Letter of Intent, "*A programme to search for WIMP particles in liquid argon at the LNGS*", University of Pavia, July 1999.

[7] See for example T.Doke, "*Fundamental Properties of Liquid Argon, Krypton and Xenon as Radiation Detector Media*", Portug. Phys., **vol.12** (1981) 9-48.

[8] See for example: A. Hitachi et al. Phys. Rev. B, **27** (1983) 5279 and references therein

[9] See for example: S. Amerio, et al., NIM **A527** (2004) 329-410.

[10] H. H. Loosli, Earth and Planetary Science Letters **63** (1983) 51;
P. Benetti, et al., NIM **A574** (2007) 83-88.

[11] For a list of the papers of the WArP Collaboration see the site: http://warp.lngs.infn.it

[12] P. Benetti, et al., Astropart. Phys., **28** (2008) 495-507.

Journal of Physics: Conference Series **203** (2010) 012007 doi:10.1088/1742-6596/203/1/012007

Particle Dark Matter Interpretations of Direct Searches

Alessandro Bottino

Istituto Nazionale di Fisica Nucleare, Sezione di Torino
Dipartimento di Fisica Teorica, Università di Torino
via P. Giuria 1, I–10125 Torino, Italy

E-mail: bottino@to.infn.it

Abstract. The recent results of direct search experiments for dark matter particles (DMPs) have been discussed by many authors in terms of numerous, quite disparate, theoretical models. Here, after some general considerations related to the possible distributions of DMPs in the galactic halo, the discussion is focussed on a few models: inelastic dark matter, mirror dark matter and relic neutralinos.

1. Introduction

The interpretation of the experimental results derived from direct searches for particle dark matter depends critically on two main aspects: a) one concerns the distribution of the DMPs in the galactic halo, b) the other refers to the nature of the DMP and its related interaction mechanism with the target material employed in the experimental apparatus. Indeed, a detection rate is a convolution of the phase-space distribution function (DF) of the DMP in the halo with the cross-section of the DMP with the particle constituent of the target material, $\sigma^{DMP}_{targetparticle}$. Thus, the physical information which is derived on properties of a putative DMP from experimental results of direct searches depends on the form of the DF assumed in the analysis. On both aspects (a) and (b) there have recently been considerable developments.

In this brief overview I first underline some of the major novelties in the field of the phase-space distribution functions, then I discuss some of the most interesting particle models which can provide a viable interpretation of the present data from the direct detection experiments.

2. Phase-space distribution functions

In most of the experimental and theoretical analyzes it is customary to employ the isothermal sphere model to describe the galactic phase-space distribution of DMPs, yet this model has more the merit of being structurally simple than the one of being realistic. Actually the DF could present deviations from isotropy either in spatial or in velocity distributions; some specific examples of these categories, and their consequences in the interpretation of direct detection data, are discussed for instance in Ref. [1].

A feature which would modify considerably the standard picture based on thermalized distribution functions is the possible presence of streams in the halo [2]. Also an effect of clumpiness could sizably affect the detection rates.

These aspects have been addressed by recent analyzes carried out with new high-resolution simulations. This is the case of the investigation in Ref. [3], which reports the results obtained with high-resolution simulations of the galaxy halo taken from the Aquarius Project. These authors conclude that the DM local density distribution turns out to be very smooth, implying that bound clumps are very unlikely; the local velocity is also rather smooth with no indication of important effects due to streams. However, this velocity distribution shows some specific features, *i. e.* it is anisotropic with some deviations from a multivariate Gaussian distribution: mainly some depletion around the maximum and some enhancement over the high-velocity tail (Fig. 1 from Ref. [3] displays these properties).

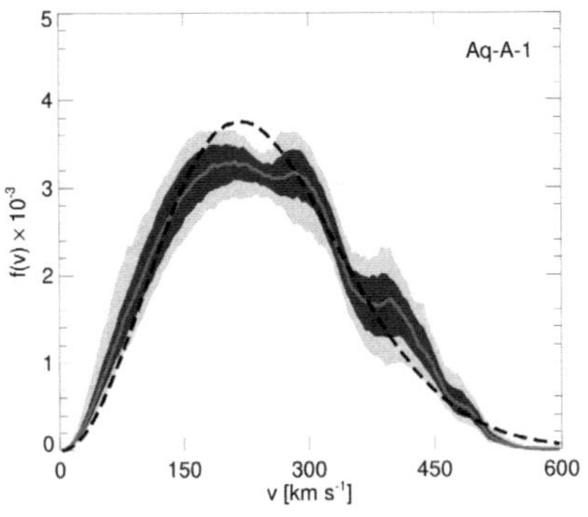

Figure 1. Velocity modulus distribution in one of the halo simulations of Ref. [3]. The dashed line denotes a fit obtained with a multivariate Gaussian model. For other details see Ref. [3] from where this picture is taken.

Of particular interest are also the new high-resolution simulations which, together with dark matter, include baryonic matter too. In Ref. [4] it is shown that in these evaluations stellar/gas disk drags the merging satellites toward the disk plane where they are torn apart by tidal forces. This *dark disk* would be co-rotating with the galactic stellar disk, implying a modification in the velocity distribution usually employed for DMPs (see also Ref. [5]).

Particular attention should be devoted to these last developments in numerical cosmological simulations, which however, for the time being, appear to still suffer from limited statistics and model-dependence.

3. Particle candidates and interaction mechanisms

The most remarkable result in the experimental search for DMPs is represented by the measurement of an annual-modulation effect at 8.2 sigmas, obtained by the DAMA/NaI and DAMA/LIBRA experiments (with a total exposure of 0.82 ton yr) [6]. This effect is just one of the typical signatures expected for DMPs [7].

All other experiments for direct detection, which, on the other hand, do not have the prerequisites for a measurement of the yearly effect, report upper bounds on the DM detection rate [8].

The annual-modulation effect can be generated by a wide number of possible DM candidates and interaction mechanisms between the DM particle and the target material [6]. The most natural generic candidate is a WIMP, with the neutralino as its most attractive physical realization. Many other possibilities have been examined by various authors; they consists for instance either of candidates which have dominant interactions on atomic electrons, or of inelastic transitions within the DM candidate itself or of a conversion of the impinging particle energy into electromagnetic radiation. These possibilities have been considered in a large number of recent papers. Limits of space allow us only to illustrate the case of the most popular candidate, the neutralino, and a couple of more exotic models: inelastic dark matter and mirror dark matter [9].

3.1. Inelastic dark matter

This model arises from the idea that the DM particle χ might have a companion exited state χ^* with a mass splitting $\delta = m_{\chi^*} - m_\chi$ with the property that elastic scattering of χ off a nucleus is suppressed as compared to the inelastic scattering which would excite χ into χ^* [10].

A DM particle with these properties would interact with a target nucleus only if its velocity is larger than $(2m_N E_R)^{-1/2}(m_N E_R/\mu + \delta)$, where E_R is the deposited energy, m_N is the nuclear mass and μ is the reduced mass of the DM particle/nucleus system. From these prerequisites some characteristic features follow; in particular, heavier targets are favored over lighter ones and the signal is sizably hindered at low recoil energies as compared to the one expected in the case of elastic scattering of a DMP with the target nucleus.

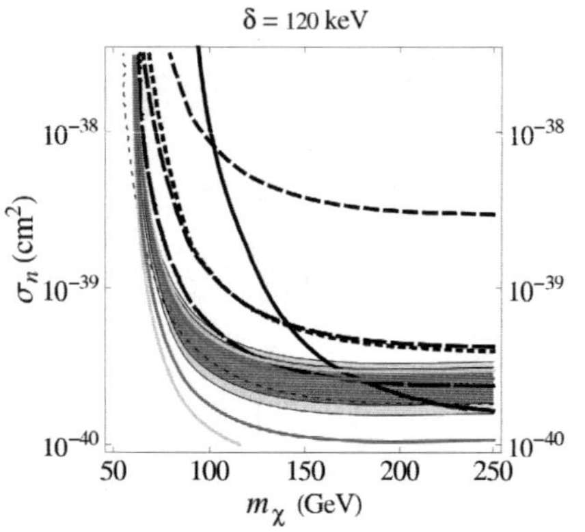

Figure 2. Cross section for a scattering of an inelastic DMP off a nucleon in the model of Ref. [10] as a function of the DMP mass for a representative value of the δ parameter. The top five lines and the faint short-dashed line on the left denote upper bounds, the other lines enclose the regions at 99 % CL, 90 % CL, for the DAMA/LIBRA results (the two inner curves refer to an analysis including DAMA/LIBRA and DAMA/NaI, the two outer ones contain only the DAMA/LIBRA data). For other details see Ref. [10] from where this picture is taken.

These properties turn out to be of relevance when $\delta \sim v^2 m_\chi \sim 100$ keV (v being the DMP velocity in units of c). Indeed, in Ref. [10] it is shown that for masses of the DMP of order 100 GeV, no contradiction exist among the DAMA data and results of other direct experiments. This property is for instance displayed in Fig. 2 taken from [10].

Various possible physical realizations of the inelastic DM particle exist. One of these is represented by sneutrinos in models which contain a lepton-number violation lagrangian: in this case sneutrinos would mix with anti-sneutrinos generating a mass splitting in the mass eigenstates [11]. Another realization could be a fourth generation Dirac neutrino constituted by 2 Majorana states originally degenerate under an U(1) symmetry; the breaking of this U(1) would cause a splitting between the two Majorana fermions [12]. Another viable model could be a fourth generation Dirac neutrino constituted by 2 Majorana states originally degenerate under an U(1) symmetry; a breaking of this U(1) would produce a splitting between the two Majorana fermions, giving rise to a possible inelastic DMP [12]. Finally, it is worth mentioning a model proposed in Ref. [13] and consisting of a Majorana particle with a transition electric/magnetic moment.

3.2. Mirror dark matter

One can entertain the possibility that in nature there is an exact unbroken parity symmetry under the transformation: $\vec{x} \to -\vec{x}, t \to t$, with the usual sector of ordinary particles paralleled by a mirror sector constituted by mirror particles in a one-to-one correspondence with ordinary particles. Thus, for example, the electron would have its mirror partner in a mirror electron with the same mass. Couplings of the particles in the mirror sector would be the same as the corresponding couplings in the ordinary sector.

Within this model, which possesses a remarkable pedigree [14], more recently was also developed the idea that stable mirror particles can play the role of dark matter [15, 16, 17], though the relevant cosmology is rather involved. Ordinary particles and mirror particles

Journal of Physics: Conference Series **203** (2010) 012007 doi:10.1088/1742-6596/203/1/012007

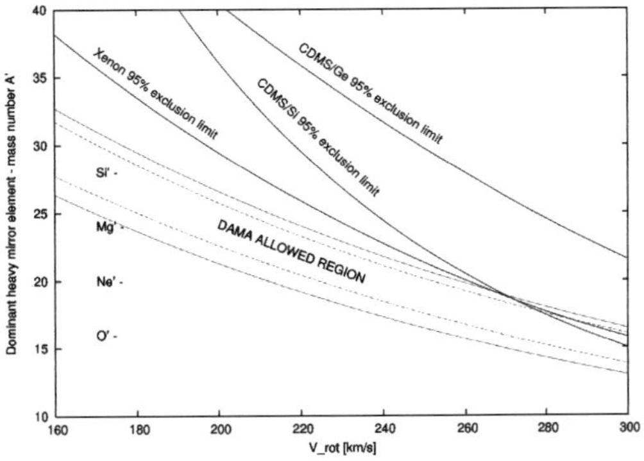

Figure 3. The 3σ (dashed-line) and the 5σ (solid-line) DAMA/LIBRA region is represented in the A'- v_{rot} (rotational velocity) parameter space together with upper bounds from other experiments. For other details see Ref. [16] from where this picture is taken.

interact with each other via gravity. Yet, other interactions are possible, e.g. via a photon-mirror photon interaction Lagrangian $L = (\epsilon/2)F^{\mu\nu}F'_{\mu\nu}$, $F^{\mu\nu}$ and $F'_{\mu\nu}$ being the field tensors for the electromagnetic and mirror electromagnetic fields, respectively. This term would entail that a charged mirror particle couples to an ordinary photon with an effective charge equal to the electric charge times ϵ. Signals in experiments of DMP direct detection would be caused by halo mirror nuclei which scatter off an ordinary nucleus in the detector by a standard Rutherford scattering.

In Ref. [16] the expected detection rates for various target nuclei have been evaluated by assuming that the halo dark matter is dominated by a mirror nucleus of mass number A', and the results of different experiments of direct DMP detection have been analyzed in this model. The comparison between the various results is reported in Fig. 3 from Ref. [16]: this figure shows that the region related to the DAMA/LIBRA annual modulation results has only a marginal overlap with the upper bounds of other experiments.

3.3. Neutralinos

In Ref. [18] it is proved that the DAMA/LIBRA annual modulation data agree very well with the theoretical expectations within an effective Minimal Supersymmetric extension of the Standard Model (effMSSM) at the electroweak scale without gaugino masses unification at a grand unification scale [19]. This is displayed in Figs. 4-5, where the quantity $\xi\sigma_{\rm scalar}^{\rm (nucleon)}$ is displayed as a function of the DMP mass (ξ is the DMP fractional local density and $\sigma_{\rm scalar}^{\rm (nucleon)}$ is the DMP-nucleon cross section). The region covered by a (red) slant hatching denotes the DAMA annual modulation region, under the hypothesis that the effect is due to a DMP with a coherent interaction with nuclei, the channeling effect [20] being disregarded in Fig.4 and included in Fig.5. This region represents the domain where the likelihood-function values differ more than 6.5 σ from the null hypothesis (absence of modulation). It has been derived by the DAMA Collaboration by varying the WIMP galactic distribution function over the set considered in Ref.[1] and by taking into account other uncertainties of different origins. The scatter plot represents supersymmetric configurations calculated with the model summarized in [18], at a fixed representative set of values for the couplings of the Higgs bosons or of the squarks with nucleons. The (red) crosses denote configurations with a neutralino relic abundance which matches the WMAP cold dark matter amount ($0.098 \leq \Omega_\chi h^2 \leq 0.122$), while the (blue) dots refer to configurations where the neutralino is subdominant ($\Omega_\chi h^2 < 0.098$). The (blue)

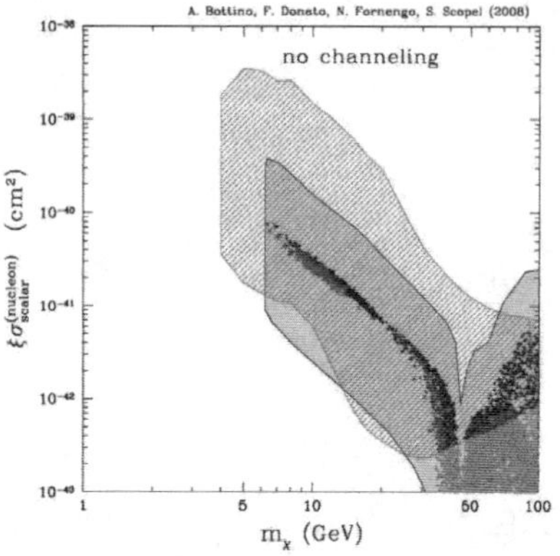

Figure 4. $\xi\sigma^{(\mathrm{nucleon})}_{\mathrm{scalar}}$ as a function of the DMP mass (see text for explanations). No channeling effect is included here in the analysis of the experimental data.

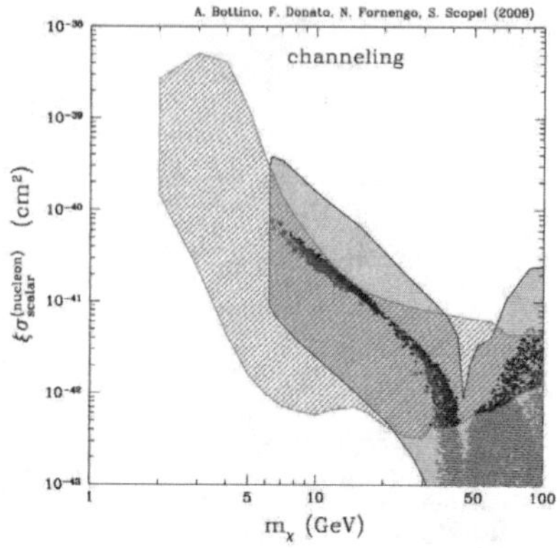

Figure 5. $\xi\sigma^{(\mathrm{nucleon})}_{\mathrm{scalar}}$ as a function of the DMP mass (see text for explanations). The channeling effect is included here in the analysis of the experimental data.

uniformly–shaded region represents the extension of the scatter plot upwards and downwards, when the uncertainties affecting the aforementioned couplings are included.

Notice that the scatter plot in Figs. 4-5 starts at $\simeq$ 6 GeV, which is the lower bound to the neutralino mass determined by the cosmological upper bound on cold DM [19]. The lower bound on m_χ can be smaller in Next-to-MSSM models [21].

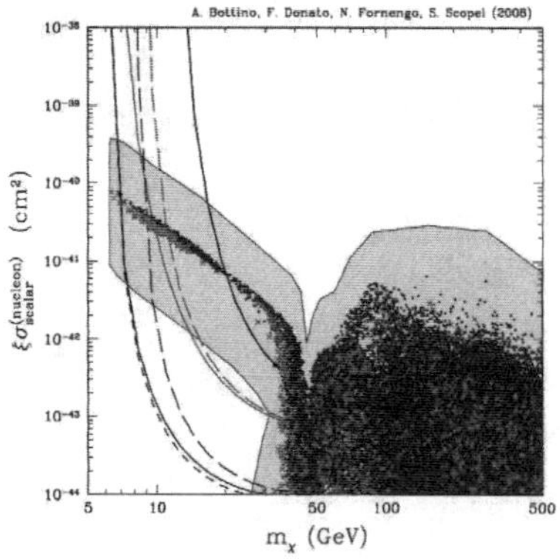

Figure 6. $\xi\sigma^{(\mathrm{nucleon})}_{\mathrm{scalar}}$ as a function of the DMP mass. The theoretical features are the same as in Figs. 4-5. The other lines represent the upper bounds derived from the data of Ref. [22] by varying the form of the DF (for full explanation of notations see [18]).

The comparison between the theoretical predictions and the DAMA annual modulation region in case of specific forms of DFs is discussed in [18]. In this paper it is also shown that, altough in general light DMP are severely constrained by cosmic antiprotons, a large set of the light neutralino population discussed above is compatible with the cosmic antiproton bounds especially for values of local dark matter density and local rotational velocity in the low side of their physical ranges and for values of the diffusion parameters not too close to the values of their maximal set.

Obviously, also in the case of direct experiments which only provide upper bounds the dependence of the halo DF has to be taken into account. In Fig. 6 we show for instance what is the effect of changing the DF in the case of the CDMS experiment [22].

As for the comparison among the results of various direct searches, it is worth stressing that, apart from DAMA/NAI and DAMA/LIBRA experiments, other experiments [8] are not sensitive to the annual modulation effect which is the peculiar signature of the DM signal. In the derivation of upper bounds, event discrimination procedures, not based on the peculiar signature of the effect, are applied; this makes the comparison of these upper limits with DAMA results somewhat uncertain. However, these upper bounds, even when taken at their face values, are not in conflict with the annual-modulation data and with the neutralino interpretation for masses in the range $m_\chi \simeq 7 - 10$ GeV [18]. Comparisons of DAMA/LIBRA data with results of other direct experiments has been largely discussed in recent literature [23].

We finally wish to mention that the whole population of light neutralinos (6 GeV $\leq m_\chi \leq$ 50 GeV) might produce sizable signals at neutrino telescopes [24] and in searches in space for antideuterons [25], and most notably can be searched for at Large Hadron Collider [26].

References

[1] Belli P, Cerulli R, Fornengo N and S. Scopel, 2002 *Phys. Rev.* D **66** 043503.
[2] Freese K, Gondolo P, Newberg H J and Lewis M, 2004 *Phys. Rev. Lett.* **92** 111301; Bernabei R et al., 2006 *Eur. Phys. J* C **47** 263; Duffy L D and Sikivie P, 2008 *Phys. Rev.* D **78** 063508.
[3] Vogelsberger M *et al*, preprint arXiv:0812.0362 [astro-ph].
[4] Read J I, Mayer L, Brooks A M, Governato F and Lake G, preprint arXiv:0902.0009 [astro-ph].
[5] Ling F-S, Nezri E, Athanassoula and Teysser R, preprint arXiv:0909.2028 [astro-ph].
[6] Bernabei R *et al*, 2008 *Eur. Phys. J. C* **56** 333; Belli P, invited talk presented at TAUP 2009.
[7] Drukier A K, Freese K, and Spergel D N, 1986 *Phys. Rev. D* **33** 3495; Freese F, Frieman J A, and Gould A, 1988 *Phys. Rev. D* **37** 3388.
[8] For a thorough review of recent activity in this field see the invited talks given by P. Belli, J. Cooley-Sekula, E. Aprile, C. Montanari and the contributions in the workshop sessions on "Dark matter searches" at TAUP 2009.
[9] Other theoretical models for DMPs not specifically discussed in the present overview have been presented at TAUP 2009 in contributions by C. Arina, F. Bazzocchi, D.G. Cerdeno, B. Dutta and L. Visinelli.
[10] Chang S, Kribs G D, Tucker-Smith and Weiner N., 2009 *Phys.Rev.D* **79** 043513.
[11] Hall L J, Moroi T and Murayama H, 1998 *Phys. Lett. B* **424** 305; Tucker-Smith D and Weiner N, 2001 *Phys. Rev. D* **64** 043502; Arina C and Fornengo N, 2007 *JHEP* 0711:029.
[12] Tucker-Smith D and Weiner N, 2005 *Phys. Rev. D* **72** 063509.
[13] Masso E, Mohanty S and Rao S, 2009 *Phys. Rev. D* **80** 036009.
[14] Lee T D and Yang C N, 1956 *Phys. Rev.* **104** 256; Kobzarev I, Okun L and Pomeranchuk I, 1966 *Sov. J. Nucl. Phys.* **3** 837.
[15] Blinnikov S I and M. Yu. Khoplov M Yu, 1982 *Sov. J. Nucl. Phys.* **36** 472.
[16] R. Foot, 2008 *Phys. Rev. D* **78** 043529.
[17] Berezhiani Z, Pilo L and Rossi N, arXiv:0902.0146 [astro-ph.CO].
[18] Bottino A, Donato F, Fornengo N and Scopel S, 2008 *Phys. Rev. D* **78** 083520.
[19] Bottino A, Donato F, Fornengo N and Scopel S, 2003 *Phys. Rev. D* **68** 043506.
[20] Bernabei R *et al.* (DAMA Collaboration), 2008 *Eur. Phys. J. C* **53** 205.
[21] Flores R, Olive K A and Thomas D, 1991 *Phys. Lett. B* **263** 425; Abel S A, Sarkar S and Whittingham I B, 1993 *Nucl. Phys. B* **392** 83; Cerdeno D G, Hugonie C, Lopez-Fogliani D E, Munoz C and Teixera A M, 2004 *JHEP* 0412:048; Gunion J, Hooper D, and McElrath B, 2006 *Phys. Rev. D* **76** 015011.
[22] Ahmed Z et al., 2009 *Phys. Rev. Lett.* **120** 011301.
[23] Petriello F J and Zurek K M, 2008 *JHEP* 0809:047; Feng J L, Kumar J and Strigari L E, 2008 *Phys.Lett.B* **670** 37; Fairbairn M and Schwetz T, 2009 *JCAP* 0901:037; Savage C, Gelmini G, Gondolo P and K. Freese K, arXiv:0808.3607 [astro-ph].
[24] Niro V, contribution at TAUP 2009; Niro V, Bottino A, Fornengo and Scopel S, arXiv:0909.2348 [hep-ph].
[25] Donato F, Fornengo N and Maurin D, 2008 *Phys. Rev. D* **78** 043506.
[26] Bottino A, Fornengo N, Polesello G and Scopel S, 2008 *Phys. Rev. D* **77** 115026; Bottino A, Choi S, Fornengo N and Scopel S (work in progress).

Topics in Astroparticle and Underground Physics (TAUP 2009) IOP Publishing
Journal of Physics: Conference Series **203** (2010) 012008 doi:10.1088/1742-6596/203/1/012008

Dark-Matter QCD-Axion Searches

Leslie J Rosenberg

University of Washington, Department of Physics, Seattle, Washington USA

ljrosenberg@phys.washington.edu

Abstract. The axion is a hypothetical elementary particle appearing in a simple and elegant extension to the Standard Model of particle physics that cancels otherwise huge CP-violating effects in QCD; this extension has a broken U(1) axial symmetry, where the resulting Goldstone Boson is the axion. A light axion of mass $10^{-(6-3)}$ eV (the so-called "invisible axion") would couple extraordinarily weakly to normal matter and radiation and would therefore be extremely difficult to detect in the laboratory. However, such an axion would be a compelling dark-matter candidate and is therefore a target of a number of searches. Compared to other dark-matter candidates, the plausible range of axion dark-matter couplings and masses is narrowly constrained. This restricted search space allows for "definitive" searches, where non-observation would seriously impugn the dark-matter QCD-axion hypothesis. Axion searches employ a wide range of technologies and techniques, from astrophysical observations to laboratory electromagnetic signal detection. For some experiments, sensitivities are have reached likely dark-matter axion couplings and masses. This is a brief and selective overview of axion searches. With only very limited space, I briefly describe just two of the many experiments that are searching for dark-matter axions.

1. What makes up the dark matter?

As is well known, the vast majority of all matter in the Universe is comprised of something other than dust, planets, gas…the ordinary matter we see around us. From observations of velocities in rich galactic clusters, plus those clusters' gravitational lensing of background "candles", as well as precision measurements of cosmological parameters, we now have a robust estimate of the dark matter fraction of the Universe of around one-quarter of its total energy density. Almost all the remaining mass-energy density of the Universe is the mysterious "dark energy", a uniformly-distributed energy driving outward acceleration of the universe. Here on Earth, the nearby dark matter amounts to around ½ of proton mass per cc; this is adduced from observations of our Milky Way dynamics, plus micro-lensing rates of MACHOs observed towards the galactic center.

From constraints imposed by structure formation in the Universe, plus the remnant light-isotope abundance left from the Big Bang, our good guess is that the dark matter is some particle relic left over from the Big Bang. A number of particle candidates have been proposed, two of which have properties that fit all the observations. These two are WIMPS and axions. WIMPs are neutral particles with mass in the 100 GeV range and having about weak-interaction couplings. Such particles have the properties of dark matter and would have frozen out in the early universe with about the right density to be dark matter. Much of the interest in WIMPs is driven by interest in the theory of Super

Symmetry, which provides a WIMP candidate, the neutralino. Equally well-motivated, but less well-known, is the axion. (See reference [1] for an overview of the axion.)

2. What is the axion?

Quantum Chromo Dynamics (QCD) was firmly established in the early 1970's as the underlying quantum theory of the nuclear interactions. It was remarkably successful, predicting, e.g., the structure of hadronic jets and the "running" of the strong coupling constant. However, by the mid-1970's it was realized this same theory predicts huge CP-violating interactions from "instanton" (multiple degenerate vacuua) effects. One such effect would be a large permanent electric dipole moment in a spinning non-degenerate object bound by the strong interaction, e.g., the neutron.

In one of the most pleasing series of measurements in modern physics, the neutron is observed to have a vanishingly small upper bound to its permanent electric dipole moment. This strongly suggests QCD, despite its successes, is not the whole story of the strong interactions. In 1977, Roberto Peccei and Helen Quinn proposed a hidden U(1) axial symmetry of the quarks that had the effect of cancelling the unphysical strong-CP effects. Steven Weinberg and Frank Wilczek shortly thereafter realized such a symmetry, unseen in nature and therefore manifestly broken, implies the existence of a new pseudoscalar Goldstone-boson, the axion [2].

The properties of a light (in the μeV mass range) axion make for an ideal dark-matter candidate. In the Big Bang, axions appear as a zero-temperature Bose condensate, remaining at near-zero temperature and never in thermal equilibrium with the rest of the Universe. The axion, having the same quantum numbers as the π^0, can similarly decay into two photons, but its lifetime is vastly longer than the age of the Universe. Most importantly, sufficiently light axions would have density to be the dark matter. Much lighter (<<μeV) and axions would have severely overclosed the Universe. Much heaver (>meV) and axions produced in supernova sn1987a would have efficiently transported energy out of the explosion, thereby observably shortening the neutrino arrival pulse length recorded on Earth. These bounds result in an allowed axion mass window of $10^{-(6-3)}$ eV. Such light axions, while possibly supplying the dominant form of matter in the Universe, have interactions so feeble as to render them nearly "invisible" to normal matter and radiation.

3. Overview: Present limits on dark-matter axions

The axion has a wide variety of couplings and decay modes; think of the π^0 and its complexity; add to that a new U(1) axial symmetry with its new charges, and the model-space of couplings is large. However, there's a great simplification for the axion decay $a \rightarrow \gamma\gamma$. This two-photon decay rate, characterized by an effective coupling constant $g_{a\gamma\gamma}$, contains the ratio of color and electromagnetic anomalies of the new U(1) symmetry. Hence, this rate doesn't depend explicitly on the value of the axion's U(1) couplings since they cancel in the ratio. Even those measurements that depend on other decays and interactions other than with two photons are often cast into terms of constant $g_{a\gamma\gamma}$ to allow differing measurements and limits to be compared.

Figure 1 shows selected limits on axion couplings and masses from a variety of techniques. The horizontal axis is the putative axion mass. The vertical axis is the effective coupling of the axion into two photons. "KSVZ" and "DSVZ" refer to two benchmark classes of axion models commonly targeted by searches; in a sense they represent the extremes of allowing axions to couple with optional "full" or zero QCD strength to leptons. Dark-matter axions have properties that put them somewhere between the "KSVZ" and "DFSZ" couplings and in the mass range 1-10 μeV (or possibly 1-100 μeV). Not shown in figure 1 is the very restrictive upper bound to the coupling inferred by sn1987a, which is a horizontal line around a coupling strength of 10^{-13}/GeV. Notice the wide variety of measurements. At the upper left, terrestrial measurements look for effects of interactions of photons with virtual axions in the propagation of light. Just below in the figure, axions can affect the properties of the Sun in several ways, including its seismic signature and energy output. As well, those same solar axions could scatter off a germanium crystal at the appropriate Bragg angle and convert into x-rays. To the right, axions in halos of astrophysical objects could spontaneously decay into pairs of optical photons, subsequently detected in telescopes. None of these methods are sensitive to dark-matter QCD axions of the expected couplings and masses. In the dark-matter band, next in sensitivity are astrophysical bounds. Here, axion emission from astrophysical objects would observably affect the evolution of those objects. Such objects include stars along the red giant horizontal branch and white dwarfs. More recently, the CERN Axion Solar Telescope, aiming to detect axions emitted from the Sun, achieved sensitivities superior to the usual astrophysical bounds. One important bound is that from axion emission in supernovae. As I mentioned, such emission would have shortened the neutrino burst duration detected on Earth seen from sn1987a. This gives by far the astrophysical coupling-constant constraint of around 10^{-13}/GeV. But even there, the sn1987a bounds lack adequate sensitivity to dark matter axions. However, a technology of converting nearby axions remaining from the big bang into microwave photons ("microwave cavity" on the figure) is indeed sensitive to plausible dark-matter axion masses and couplings.

Figure 1 shows only a sampling of the limits and technologies. Much more information on axions and limits is contained in the summary[1] and references therein. But since the topic here is dark-matter QCD axions, a simplification results since most technologies are by far too insensitive. While those searches may be sensitive to unusual axion variants, they don't speak to the well-motivated QCD dark-matter axion hypothesis. I therefore restrict further discussion to the astrophysical bounds and the RF cavity technique and I'll pick one experiment from each as an example. While from figure 1 it appears the astrophysical bounds don't achieve adequate sensitivity, this may not be so. In particular, axion emission in those objects may not depend directly on the axion-to-two-photon coupling, so the emission rate is highly uncertain. For instance, axions emitted via nuclear bremsstrahlung, unlike the two-photon coupling, does depend on the unknown new U(1) axial charge of the quarks. It may well be that those astrophysical bounds are more sensitive than heretofore assumed. One may then wonder whether the astrophysical bounds ("HB Stars" on the figure) are already well excluded by the considerably more restrictive sn1987a bound. This may be, but plasma effects in supernovae explosions are important and difficult to calculate. Hence, the supernovae axion emission may be suppressed, which considerably softens the supernovae bound, which in turn increases the discovery potential of the other astrophysical bounds.

4. Example search: The CERN Axion Solar Telescope (CAST)

Axions would be emitted by the Sun with their kinetic energies in a broad distribution centered at 3 keV. The idea of detecting these axions via their conversion into x-rays goes back several decades, to an paper by Pierre Sikivie. The first serious implementation of this was an effort at Brookhaven National Laboratory. A larger and more sensitive instrument was built, and recently recommissioned, at the University of Tokyo. Figure 2 shows a sensitive instrument based on this method, the CERN Axion Solar Telescope (CAST)[2]. It features state-of-the art x-ray optics and detectors at the ends of a long, high-field LHC prototype dipole magnet. The entire assembly rides on a telescope mount that keeps the magnet axis aligned with the Sun for several hours at dawn and dusk. Axions would be signaled by an excess of counts in the x-ray detectors. As the axion-to-photon conversion rate is quadratic in field-strength and conversion length (until the onset of decoherence), and the x-ray detection efficiency is high, CAST represents a significant advance in the sensitivity of this technology.

The first phase of this experiment operated the magnet bore in a vacuum, results of which are shown in figure 2. No axion signal was observed. This configuration was sensitive to axions in the mass range up to a few 10's of meV, beyond which the axion mass was poorly matched to the photon dispersion relation with result the axion-to-photon conversion rate plummets. This result was notable for being at or more sensitive to the astrophysical bounds, and perhaps even sensitive to plausible dark-matter axion couplings at the more massive end of the search range.

At higher axion masses, in order to match the axion and photon dispersion relations, the magnet bore is backfilled with ^{4}He or ^{3}He gas at controlled pressure. At higher pressures, the peak of the conversion efficiency shifts to higher axion masses. This next experiment phase is in operation and the goal is to have sensitivity to axions with masses up to around an eV with couplings near that of the astrophysical bounds.

5. Example search: The Axion Dark-Matter eXperiment (ADMX)

In the same paper where Pierre Sikivie described the solar-axion search, he as well conceived of an RF-cavity search for Milky Way halo axions. The technique is to thread a high-Q microwave cavity with a large static magnetic field. Nearby halo axions scatter of the field and thereby convert into microwave photons. The resulting photon energy is that of the total energy of the axion. The microwave photons are detected in what is in essence an ultra low noise double-heterodyne radio receiver. The resonant frequency of the cavity is tunable across a search bandwidth. At each cavity tuning setting, the cavity power is averaged until the putative signal-to-noise ratio exceeds a threshold for realistic axions, and the power spectrum examined for excess power. The cavity is then re-tuned and search repeated until the cavity tuning range is exhausted. A schematic of ADMX realization of this[3] is shown in figure 3. The sensitivity of the technique is very good since the small axion-to-two-photon coupling appears only once, at the axion-to-photon conversion step in the cavity. Other techniques, not relying in already present axions, must as well produce those axions, which necessitates another factor of the very small, e.g., axion-to-two-photon coupling. The main challenge of the RF-cavity technique is that the expected microwave signal is very small, around 10^{-22} watts or less. Detecting such small electromagnetic RF power levels requires liquid helium temperatures to reduce the cavity blackbody photon backgrounds and electronic noise.

Low-noise microwave amplification is the key technology of this search. Early versions of ADMX used cryogenic field-effect-transistor (FET) amplifiers of modified radio telescope design, having noise temperatures in the neighborhood of 2K. The latest phase of this experiment replaces the FET amplifiers with dc SQUID amplifiers. These devices, when cooled with a dilution refrigerator, have noise temperatures near the quantum limit, approximately 50 mK at signal frequencies near 1 GHz. The averaging time it takes to achieve a certain signal-to-noise ratio depends on the square of the noise power. Hence, the reduction in noise from 2 K to 50 mK in replacing transistor amplifiers by SQUIDs yields a potential speed-up of over 1000. In practice, some of this gain will be used to improve sensitivity rather than simply speed up the search. Figure 1 shows the early ADMX results with transistor amplifiers ("Microwave Cavity" limits to the far left). Notice this is the only technique with sensitivity to realistic dark matter axion masses and couplings. Recently, ADMX announced limits from the experiment retrofitted with SQUID amplifiers and will shortly retrofit their instrument with a dilution refrigerator to reach near quantum-limited low-noise operation.

6. Summary and outlook

This overview only barely touched on dark-matter axion detection. I would be remiss without mentioning the axion model uncertainties. Although the calculation of the axion-to-two-photon coupling is claimed to be robust, there could be surprises. Also, precise mass limits at the extremes of the search range are not well known. Surprises could lurk there, as well. It's therefore prudent to search for axions with "unusual" couplings and masses. Finally, this very short discussion necessarily left out many important searches and developments, for which I apologize. Perhaps the main result is that sensitivity to dark-matter QCD axions has at last been achieved with the RF cavity technique, and we may know soon whether the dark matter is made of axions

Acknowledgments
I thank the generous support of the U.S. Department of Energy, Office of High Energy Physics, for sponsoring this research under contract number DE-FG02-96ER40956.

References

[1] Particle Data Group C.Amsler et al., Phys.Lett. B667 (2008) 459, "Axions and other very light bosons".
[2] K.Zioutas et al., Phys.Rev.Lett. 94 (2005) 121301.
[3] S.Asztalos et al., Phys.Rev. D64 (2003) 092003.
[4] S. Asztalos et al., Ann.Rev.Nucl.Part.Sci. 56 (2006) 293.

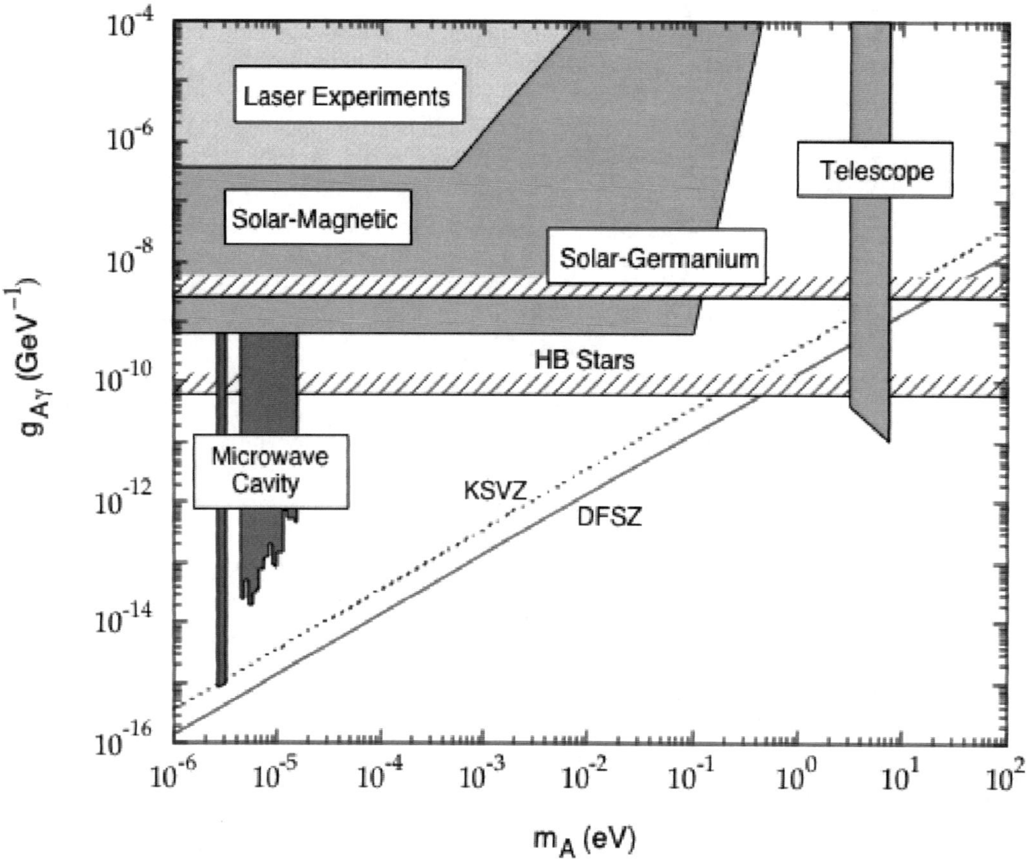

Figure 1. Selected limits on axion masses and couplings from a variety of techniques. The horizontal axis is the putative axion mass. The vertical axis is the effective coupling of the axion to two photons. "KSVZ" and "DSVZ" refer to two classes of axion models commonly targeted by searches. Dark-matter axions lie between the "KSVZ" and "DFSZ" models in the mass range 1-100 μeV. Not shown is the very restrictive upper bound to the coupling from sn1987a, which is a

horizontal line around a coupling of 10^{-13}/GeV. (Figure from reference [4].)

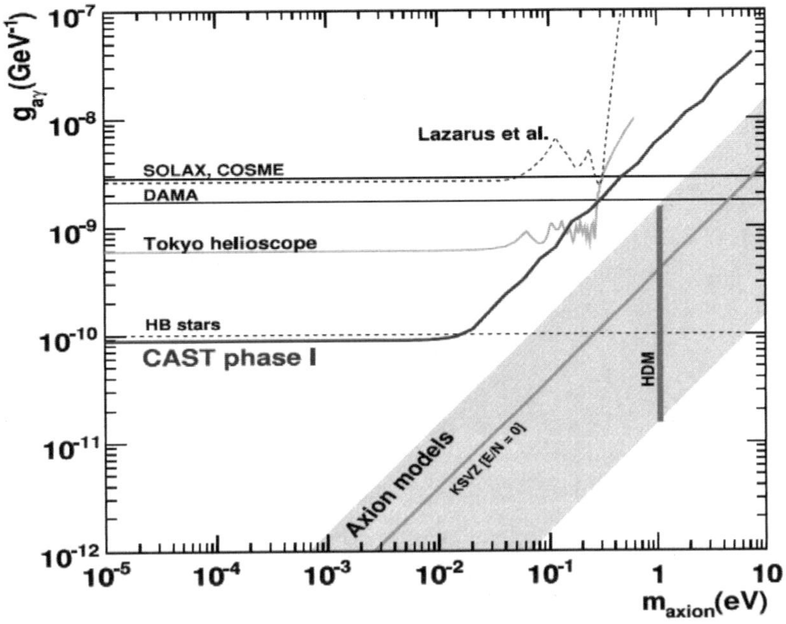

Figure 2. Limits from CAST Phase I operations ("CAST phase I"). The horizontal and vertical axes are those of figure 1. Note the CAST sensitivity in this phase is comparable to the astrophysical bounds ("HB stars"). The limit "Tokyo helioscope" are earlier results from a Japanese solar-axion search; this project has recently returned to data-taking. "SOLAX, COSME" and "DAMA" are limits from solar-axion searches using axion-to-photon conversion in a crystal. "Lazarus et al." are limits from axion effects on propagation of laser light. (Figure from ref. [2])

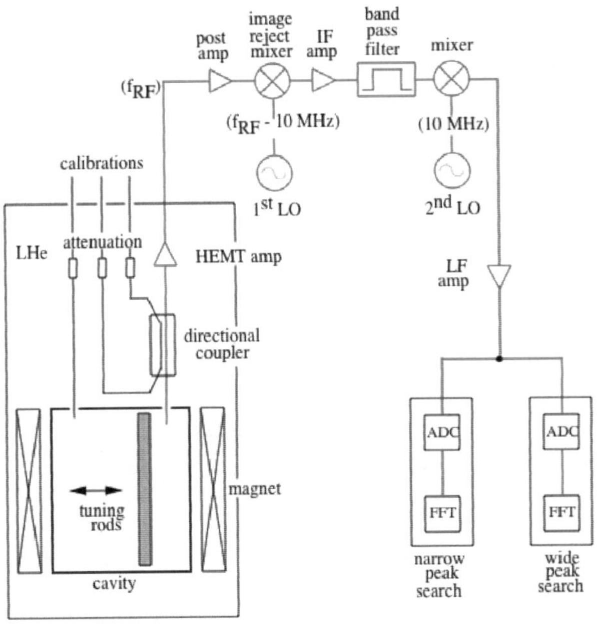

Figure 3. Schematic of the ADMX axion detector. The RF cavity, 0.5 m diameter by 1.0 m long, is in the bore of an 8.5 tesla solenoid magnet. Microwave power is amplified by a low-noise cryogenic amplifier and mixed-down to near audio. The result is digitized and processed with FFT electronics and the power spectrum is searched for axion signals. Two frequency resolutions, wide and narrow, are optimized for thermalized or non-thermalized (respectively) axions in our Milky Way halo.

Topics in Astroparticle and Underground Physics (TAUP 2009) IOP Publishing
Journal of Physics: Conference Series **203** (2010) 012009 doi:10.1088/1742-6596/203/1/012009

Progress in the understanding of neutrino properties

J. W. F. Valle

AHEP Group, Instituto de Física Corpuscular, C.S.I.C. – Universitat de València
Edificio de Institutos de Paterna, Apartado 22085, E–46071 València, Spain

E-mail: valle@ific.uv.es

Abstract. I briefly summarize neutrino oscillation results and discuss their robustness. I mention recent attempts to understand the pattern of neutrino mixing within various seesaw mechanisms, with or without supersymmetry and/or flavor symmetries. I also mention the possibility of intrinsic supersymmetric neutrino masses in the context of broken R parity models, showing how this leads to clear tests at the LHC.

1. Neutrino Oscillations

Neutrino flavors inter-convert during propagation both *in vacuo* and in matter. Uncontroversial evidence for this follows from solar and atmospheric, as well as reactor and accelerator neutrino studies [1]. The basic concept needed to describe oscillations is the lepton mixing matrix, the leptonic analogue of the quark mixing matrix. In its simplest 3×3 unitary form it is given as [2]

$$K = \omega_{23}\omega_{13}\omega_{12} \tag{1}$$

where each factor is effectively 2×2 with an angle and a corresponding CP phase. Since current experiments are insensitive to CP violation, we neglect phases, so that oscillations depend only on the three mixing angles $\theta_{12}, \theta_{23}, \theta_{13}$ and on the two squared-mass splittings $\Delta m^2_{21} \equiv m^2_2 - m^2_1$ and $\Delta m^2_{31} \equiv m^2_3 - m^2_1$ associated to solar and atmospheric transitions [3]. To a good approximation, one can also set $\Delta m^2_{21} = 0$ in the analysis of atmospheric and accelerator data, and Δm^2_{31} to infinity in the analysis of solar and reactor data. The resulting neutrino oscillation parameters are summarized in Figs. 1 and 2, taken from [4].

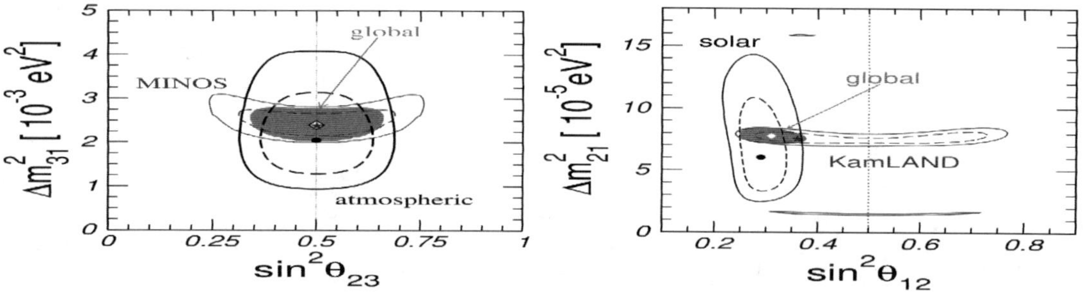

Figure 1. Neutrino oscillation parameters from a global analysis of the world's current data [4].

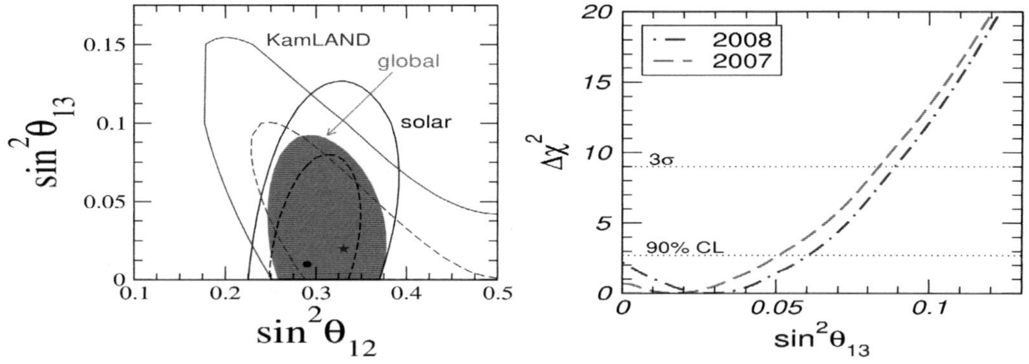

Figure 2. Constraints on $\sin^2\theta_{13}$ from different parts of the global data given in Ref. [4].

The left and right panels in Fig. 1 give the "atmospheric" and "solar" oscillation parameters, θ_{23} & Δm^2_{31}, and θ_{12} & Δm^2_{21}, respectively, minimizing with respect to the undisplayed parameters in each case, and including always all relevant data. The dot, star and diamond indicate the best fit points of atmospheric MINOS and global data, as well as solar, KamLAND and global data, respectively. Note the complementarity between data from artificial and natural neutrino sources: reactors and accelerators give the best determination of squared-mass-splittings, while mixings are mainly determined by solar and atmospheric data. Current data slightly prefer a nonzero value of the remaining angle θ_{13}, as seen in Fig. 2. Since the hint is not yet significant, we prefer to interpret this as a weaker bound on θ_{13} [1]:

$$\sin^2\theta_{13} \le \begin{cases} 0.060\ (0.089) & \text{(solar+KamLAND)} \\ 0.027\ (0.058) & \text{(CHOOZ + atm + K2K+MINOS)} \\ 0.035\ (0.056) & \text{(global data)} \end{cases} \qquad (2)$$

If confirmed by future data, a nonzero θ_{13} would encourage the search for CP violation in upcoming neutrino oscillation experiments [5, 6]. The expected CP asymmetries are small, as they are suppressed both by θ_{13} and by the small ratio α of solar/atmospheric squared mass splittings, currently determined as $\alpha = 0.032\,, \quad 0.027 \le \alpha \le 0.038 \quad (3\sigma)\,.$

2. Are oscillations robust?

Many effects may distort the "celestial" neutrino fluxes reaching the detectors, and hence affect the determination of oscillation parameters. In this connection several regular [7–9] and random [10, 11] convective zone solar magnetic field models have been considered. These fields would induce neutrino spin-flavor precession [12–14] and hence modulate the originally produced standard solar model neutrino fluxes. Similarly, the presence of radiative zone random magnetic fields could also induce density fluctuations deep inside the Sun [15] and substantially modify the energy-dependence of the solar neutrino survival probability [16, 17], with a potentially important impact on the determination of oscillation parameters.

Reactor neutrino data from KamLAND have played a crucial role in probing the robustness of solar oscillations against astrophysical uncertainties, such as radiative [18, 19] or convective zone [7–9] magnetic fields. The stringent limits on solar anti-neutrinos from KamLAND lead to strong limits on neutrino magnetic transition moments, especially in the case of turbulent fields [10, 11]. The result is that oscillations constitute the only viable explanation of the data,

[1] Note: the bounds in Eq. (2) are given for 1 dof, while the regions in Fig. 2 (left) are 90% CL for 2 dof

Journal of Physics: Conference Series **203** (2010) 012009 doi:10.1088/1742-6596/203/1/012009

and of the oscillation solutions allowed by solar data [20], only the large mixing angle solution is consistent with the spectrum measurements at KamLAND [21].

In Standard Model language neutrino mass is described by an effective dimension-five operator, shown in the left panel in Fig. 3.

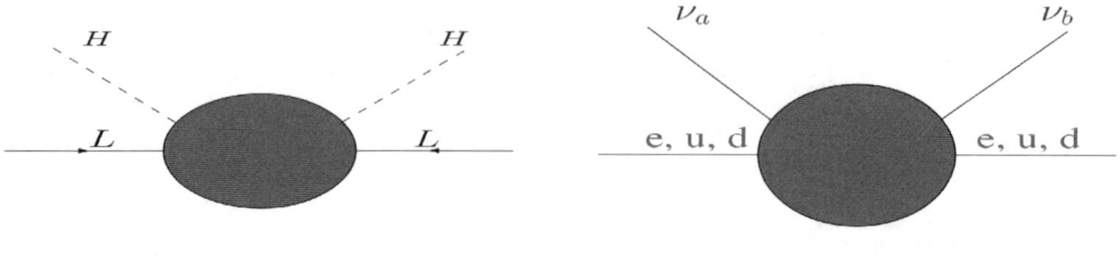

Figure 3. Neutrino mass and corresponding NSI operators.

Most neutrino mass generation schemes also induce sub-weak strength ($\sim \varepsilon G_F$) dimension-6 operators depicted in the right panel in Fig. 3. They can be of two types: flavour-changing (FC) and non-universal (NU). For example, the presence of NSI is expected low-scale seesaw models, such as the inverse [22–24] and the linear [25] seesaw models, where the non-unitary piece [2] of the lepton mixing matrix can be sizeable, hence the induced non-standard interactions. Relatively sizable NSI strengths may also be induced in models with radiatively induced neutrino masses [26–29]. The strength of the NSI operators will play an important role in elucidating the origin of neutrino mass, as it will help discriminate between high and low-scale schemes.

As noted in Ref. [30], current determination of solar neutrino parameters is not yet robust against the existence of non-standard neutrino interactions. In fact, in the presence of such interactions, there is a new "dark side" solution [30], as shown in the left panel in Fig. 4, taken from Ref. [31]. This solution is almost degenerate with the usual one, and survives the inclusion of reactor data. As seen in the right panel in Fig. 4, the lines are compared with the experimental rates for the pp neutrino flux (from the combination of all solar experiments), the 0.862 MeV ^{7}Be line (from Borexino) and two estimated values of the ^{8}B neutrino flux from Borexino and SNO (third phase). Distinguishing between these solutions poses a challenge for low energy solar neutrino experiments.

As already noted, the slight tension between current solar and KamLAND data is alleviated allowing for a non-zero value of the mixing angle θ_{13}. Likewise, non-standard flavor-changing interactions may also alleviate this tension and hence constitute a source of confusion in the determination of θ_{13}. Indeed, a full three-flavor analysis of solar and KamLAND data in the

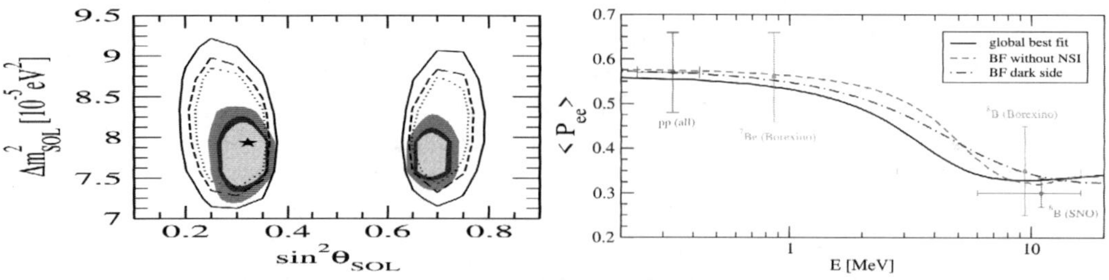

Figure 4. Left: normal and "dark-side" solutions in the presence of non-standard neutrino interactions (NSI). Right: solar neutrino survival probabilities averaged over the ^{8}B neutrino production region for three reference points, see Ref. [31] for details. Vertical bars correspond to experimental errors, while horizontal ones indicate the energy range relevant for each experiment.

presence of flavor-changing NSI [32] reveals a degeneracy between θ_{13} and the vectorial coupling ϵ_{13} characterizing the non-standard transitions between ν_e and ν_τ in the forward scattering process with d-type quarks.

Non-standard neutrino interactions may also be probed at high energies. One can show that already even a small residual NSI strengths may have dramatic consequences for the sensitivity to θ_{13} at a neutrino factory [33, 34]. Improving the sensitivities on NSI constitutes a necessary step and opens a window of opportunity for neutrino physics in the precision age. The presence of NSI also leads to novel effects in supernova neutrino propagation [35], for example, the possibility of internal resonant neutrino conversions even in the absence of neutrino masses [36–38].

In contrast to the "solar" sector, thanks to the large statistics of atmospheric data over a wide energy range, the determination of Δm^2_{31} and $\sin^2\theta_{\rm ATM}$ is hardly affected by the presence of NSI, at least within the 2–neutrino approximation [39]. Future neutrino factories will substantially improve this bound [40].

3. How do neutrinos get mass?

Neutrino oscillations provide the first sign of physics beyond the Standard Model (SM). Pinning-down the ultimate origin of neutrino mass remains a challenge. Apart from any detailed mass generation mechanism, since they carry no electric charge they should have Majorana-type masses [2]. This is indeed what happens in specific neutrino mass generation schemes, markedly different from those of charged fermions. The latter come in two chiral species and get mass linearly in the electroweak symmetry breaking vacuum expectation value (vev) $\langle\Phi\rangle$ of the Higgs scalar doublet. In contrast, as illustrated in the left panel in Fig. 3, neutrino masses arise as an effective lepton number violating dimension-five [2] operator $\mathcal{O} \equiv \lambda L\Phi L\Phi$ (where L denotes a lepton doublet) [43]. It follows that a simple way to account for the smallness of neutrino masses, is that the coefficient characterizing the strength of $\mathcal{O}$ is suppressed either by a high-scale M_X in the denominator or, alternatively, it may involve a low-mass-scale in the numerator. The search of NSI will help discriminate which of the two pathways is chosen by nature.

Gravity is often argued to break global symmetries [44, 45], and would induce Weinberg's operator $\mathcal{O}$, with M_X identified to the Planck scale M_P. The resulting masses are too small, implying the need for physics beyond the SM, typically at a high scale below M_P [46]. Alternatively $\mathcal{O}$ may be suppressed by small scales, Yukawa couplings and/or loop-factors [47]. This way one may have tree level, radiative, and hybrid mechanisms, all of which may have high- or low-scale realizations. Depending on whether it is gauged or not, spontaneous lepton-number violation implies either an extra neutral gauge boson or a Nambu-Goldstone boson coupled to neutrinos. One may construct models of both types. However the most basic and general seesaw description is in terms of the $SU(3) \otimes SU(2) \otimes U(1)$ gauge structure [2]. In such a framework the relevant scale can either be large or small, depending on model details, with a fair chance that the origin of neutrino mass may be probed at accelerators like the LHC.

3.1. Minimal seesaw

Here the operator $\mathcal{O}$ is induced by the exchange of heavy states with masses close to the "unification" scale, as indicated in Fig. 5, for a review see Ref. [47]. In type-I or type-III seesaw the exchanged states are fermions singlets or triplets, while Type-II seesaw is induced by triplet scalar exchange. The "complete seesaw" has both fermion and scalar intermediate states and its phenomenology has been thoroughly studied in [2] including the general seesaw diagonalization method [48]. The latter is obtained using the hierarchy of vevs $v_3 \ll v_2 \ll v_1$ needed in order to account naturally for the small neutrino masses and which follows by minimizing the scalar potential.

[2] Lepton number violation may also show up at higher dimension [41, 42].

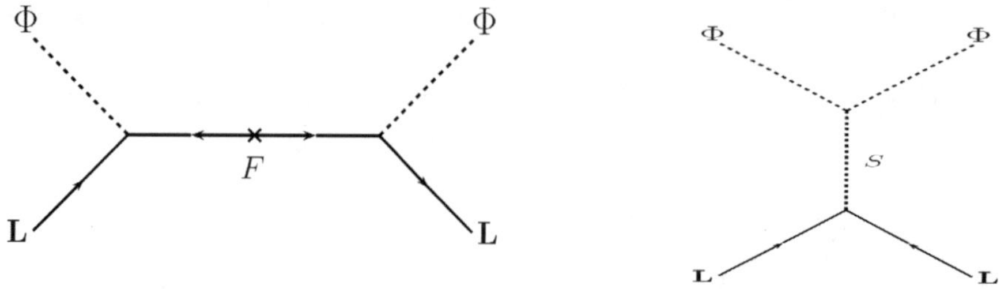

Figure 5. Type-I and III (left) and Type-II (right) realizations of the seesaw mechanism.

3.2. "Non-minimal" seesaw

The seesaw is not a model but a general paradigm. One may implement it with different gauge groups and multiplet contents, with gauged or ungauged B-L, broken explicitly or spontaneously, at a high or at a low energy scale, with or without supersymmetry. An attractive class of extended seesaw schemes employs extra gauge singlet neutrinos in addition to the ν^c which are present in the 16 of $SO(10)$ group [22]. One may implement such schemes with low-scale breaking of B-L [25,49], leaving a TeV-scale Z' [50]. However, whatever the symmetry is, it must ultimately break to $SU(3) \otimes SU(2) \otimes U(1)$, hence the most general seesaw description must be formulated at this level [2]. Such low-energy description is crucial in accurately describing low-scale variants of the seesaw mechanism, where the basic lepton-number-violating parameter may be naturally small and calculable due to supersymmetric renormalization group evolution [23].

3.3. Radiative schemes

The operator $\mathcal{O}$ may arise as loop effects [26, 27], with no need for a large scale. In this case its coefficient λ is suppressed by small loop-factors, Yukawa couplings and possibly by a small scale parameter characterizing the breaking of lepton number, leading to naturally small neutrino masses. Like low-scale seesaw models discussed above, radiative schemes typically have new TeV states opening the door to phenomenology at the LHC. [51].

3.4. R parity violation

The origin of neutrino masses may be intrinsically supersymmetric in models where the R parity symmetry breaks [52–54]. This may happen spontaneously, driven by a nonzero vev of an $SU(3) \otimes SU(2) \otimes U(1)$ singlet sneutrino [55–57], leading to effective bilinear R-parity violation [58, 59]. This provides the minimal way to break R parity in the Minimal Supersymmetric Standard Model [59]. The induced neutrino mass spectrum is hybrid, with one scale (typically the atmospheric) induced by neutralino-exchange at the tree level, and the other scale (solar) induced by *calculable* one-loop corrections [60].

Unprotected by any symmetry, the lightest supersymmetric particle (LSP) decays, typically inside detectors at the Tevatron or the LHC [60, 61]. The LSP decay-length is a measure of the neutrino mass and can be experimentally resolvable, leading to a displaced vertex [62, 63]. More strikingly, its decay properties correlate with the neutrino mixing angles. Indeed, the LSP decay pattern is predicted by the low-energy measurement of the atmospheric angle θ_{23} [64–66]. Such a prediction will be clearly tested at the LHC. Similar correlations hold in schemes based on other supersymmetry breaking mechanisms and, correspondingly, featuring other states as LSP [67].

Journal of Physics: Conference Series **203** (2010) 012009 doi:10.1088/1742-6596/203/1/012009

4. Understanding lepton mixing with flavor symmetries

As seen above current neutrino oscillation data indicate solar and atmospheric mixing angles which are unexpectedly large when compared with quark mixing angles. This places a challenge to the understanding of the flavor problem in unified schemes where quarks and leptons are related. It has been noted that the neutrino mixing angles are approximately tri-bi-maximal [68]. There have been many schemes suggested in the literature in order to reproduce this pattern using various discrete flavor symmetry groups containing mu-tau symmetry, e. g. [69–78]. One expects the flavor symmetry to be valid at high energy scales. Deviations from the tri-bi-maximal ansatz [79] may be calculable by renormalization group evolution [80–82].

A specially simple ansatz is that, as a result of a given flavor symmetry such as A4 [69, 70], neutrino masses unify at high energies M_X [83], the same way as gauge couplings do in the presence of supersymmetry [84]. Such quasi-degenerate neutrino scheme predicts $\theta_{23} = \pi/4$ and $\theta_{13} = 0$, leaving the solar angle θ_{12} unpredicted, though Cabibbo-unsuppressed. If CP is violated θ_{13} becomes arbitrary and the Dirac phase is maximal [72]. One can show that lepton and slepton mixings are related and that at least one slepton lies below 200 GeV, within reach of the LHC. The lower bound on the absolute Majorana neutrino mass scale $m_0 \gtrsim 0.3$ eV ensures that the model will be probed by future cosmological tests and $\beta\beta_{0\nu}$ searches. Expected rates for LFV processes BR($\mu \to e\gamma$) $\gtrsim 10^{-15}$ and BR($\tau \to \mu\gamma$) $> 10^{-9}$ typically lie within reach of upcoming experiments. Note that flavor symmetries, such as our A4, may also be implemented in low-scale seesaw schemes, both type-I [85] and type-III [24], leading to different neutrino mass spectra.

5. Lepton flavor violating (LFV) effects

The unequivocal evidence that neutrinos oscillate suggests that, at some level, flavor violation should also show up as transitions involving the charged leptons, since these are their electroweak doublet partners. There are two basic mechanisms: (i) neutral heavy lepton exchange [86–88] and (ii) supersymmetry exchange [89–91]. Due to the sizeable admixture of right-handed

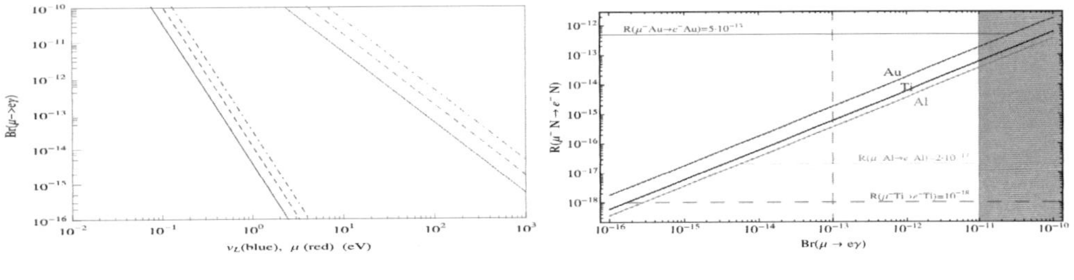

Figure 6. Left: $Br(\mu \to e\gamma)$ versus the LNV scale in the inverse seesaw in top (red color), and linear seesaw in bottom (blue color). The heavy lepton mass M is fixed as $M = 100\,GeV$ (continuous line), $M = 200\,GeV$ (dashed line) and $M = 1000\,GeV$ (dot-dashed line), from [85]. Right: typical correlation between mu-e conversion in nuclei and $Br(\mu \to e\gamma)$, from [92].

neutrinos in their leptonic charged current, low-scale seesaw schemes induce potentially large LFV rates [86, 87] and CP violating processes as well [93, 94]. This can happen in the absence of supersymmetry and in the limit of massless neutrinos, hence their magnitude is unrestricted by the smallness of neutrino masses. In Fig. 6 we display $Br(\mu \to e\gamma)$ versus the small lepton number violating (LNV) parameters μ and v_L for two different low-scale seesaw models, the inverse and the linear seesaw, respectively. Clearly the LFV rates are sizeable in both cases. Similarly one can see that in low-scale seesaw models the nuclear $\mu^- - e^-$ conversion rates lie within planned sensitivities of future experiments such as PRISM [95]. Note that in type-III versions of such low-scale seesaw schemes [24], the TeV RH neutrinos would not only induce LFV processes but also be copiously produced at the LHC.

In contrast, barring fine-tunings, high-scale seesaw models require supersymmetry in order to have sizeable LFV rates. Here lepton flavour violation is expected to show up in the most direct way in the production of supersymmetric particles at the LHC, as seen in Fig. 7.

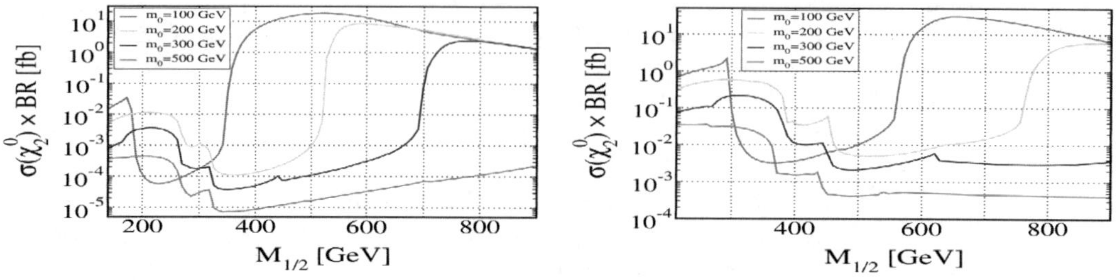

Figure 7. LFV rate for μ-τ lepton pair production from χ_2^0 decays versus $M_{1/2}$ for the indicated m_0 values, assuming minimal supergravity parameters: $\mu > 0$, $\tan\beta = 10$ and $A_0 = 0$ GeV, for type-I (left) and for type-II seesaw (right). Here $\lambda_1 = 0.02$ and $\lambda_2 = 0.5$ are Type-II seesaw parameters, and we imposed the constraint $\text{Br}(\mu \to e + \gamma) \leq 1.2 \cdot 10^{-11}$, from Ref. [96].

Both supersymmetric and RH neutrino contributions to lepton flavour violation exist in supersymmetric seesaw schemes, and their interplay depends on the seesaw scale [97].

6. Lepton number violation and $\beta\beta_{0\nu}$

The Dirac or Majorana nature of neutrinos is manifest only through the observation of LNV processes, such as $\beta\beta_{0\nu}$ [2] [3], whose current status and perspectives was reviewed by Schoenert [98]. A nonzero rate for $\beta\beta_{0\nu}$ implies that, within a gauge theory, at least one neutrino gets a Majorana mass, this argument is known as the "black-box" theorem [99], illustrated in Fig. 8, and recently discussed in [100].

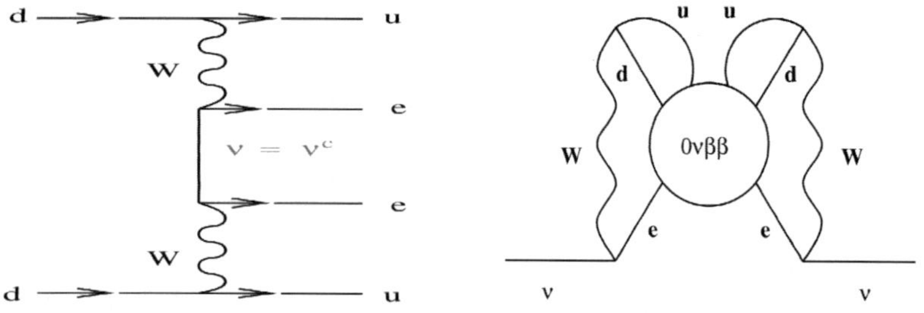

Figure 8. Neutrino mass mechanism for $\beta\beta_{0\nu}$ (left), and black box theorem (right) [99].

Given the neutrino oscillation data it follows that light Majorana neutrino exchange will induce $\beta\beta_{0\nu}$ and the corresponding rate is a measure of the absolute scale of neutrino mass, complementary to the one probed in beta decay studies [101], and cosmological observations [102]. It is also sensitive to the Majorana CP violation [103] which drops out of oscillations. Using current neutrino oscillation parameters and state-of-the-art nuclear matrix elements [104] one can determine the average mass parameter $\langle m_\nu \rangle$ characterizing the neutrino exchange contribution to $\beta\beta_{0\nu}$.

[3] Electromagnetic interactions of neutrinos can also probe their Majorana nature.

Inverted hierarchy implies a generic lower bound for the $\beta\beta_{0\nu}$ amplitude, while for normal hierarchy neutrino spectra the three neutrinos can interfere destructively, so that no generic lower bound exists. Specific flavor models may provide a lower bound for $\beta\beta_{0\nu}$ even with normal hierarchy, as discussed in [85] [105, 106]. Quasi-degenerate neutrinos [69, 70] give the largest possible $\beta\beta_{0\nu}$ signal.

Acknowledgments:
Work supported by Spanish grants FPA2008-00319/FPA and PROMETEO/2009/091 and by European Union network UNILHC (PITN-GA-2009-237920).

[1] Ranucci G, Dierckxsens M, Kajita K, Dracos M and Klein J 2009, Talks at this Conference
[2] Schechter J and Valle J W F 1980 *Phys. Rev.* **D22** 2227
[3] Maltoni M, Schwetz T, Tortola M A and Valle J W F 2004 *New J. Phys.* **6** 122
[4] Schwetz T, Tortola M and Valle J W F 2008 *New J. Phys.* **10** 113011 (*Preprint* 0808.2016)
[5] Bandyopadhyay A *et al.* (ISS Physics Working Group) 2009 *Rep. Prog. Phys.* **72** 106201
[6] Nunokawa H, Parke S J and Valle J W F 2008 *Prog. Part. Nucl. Phys.* **60** 338–402
[7] Miranda O G *et al.* 2001 *Nucl. Phys.* **B595** 360–380
[8] Guzzo M *et al.* 2002 *Nucl. Phys.* **B629** 479–490 (*Preprint* hep-ph/0112310 v3 KamLAND-updated version)
[9] Barranco J *et al.* 2002 *Phys. Rev.* **D66** 093009 (*Preprint* hep-ph/0207326 v3 KamLAND-updated version)
[10] Miranda O G *et al.* 2004 *Phys. Rev. Lett.* **93** 051304 (*Preprint* hep-ph/0311014)
[11] Miranda O G, Rashba T I, Rez A I and Valle J W F 2004 *Phys. Rev.* **D70** 113002 (*Preprint* hep-ph/0406066)
[12] Schechter J and Valle J W F 1981 *Phys. Rev.* **D24** 1883 err. D25, 283 (1982)
[13] Akhmedov E K 1988 *Phys. Lett.* **B213** 64–68
[14] Lim C S and Marciano W J 1988 *Phys. Rev.* **D37** 1368
[15] Burgess C P *et al.* 2004 *Mon. Not. Roy. Astron. Soc.* **348** 609 (*Preprint* astro-ph/0304462)
[16] Loreti F N and Balantekin A B 1994 *Phys. Rev.* **D50** 4762–4770 (*Preprint* nucl-th/9406003)
[17] Nunokawa H *et al.* 1996 *Nucl. Phys.* **B472** 495–517 (*Preprint* hep-ph/9602307)
[18] Burgess C *et al.* 2003 *Astrophys. J.* **588** L65 (*Preprint* hep-ph/0209094)
[19] Burgess C P *et al.* 2004 *JCAP* **0401** 007
[20] Gonzalez-Garcia M C *et al.* 2001 *Phys. Rev.* **D63** 033005 (*Preprint* hep-ph/0009350)
[21] Pakvasa S and Valle J W F 2003 Proc. of the Indian National Acad. of Sci. on Neutrinos, Vol. 70A, No.1, p.189 - 222 (2004), Eds. D. Indumathi, M. Murthy and G. Rajasekaran (*Preprint* hep-ph/0301061)
[22] Mohapatra R N and Valle J W F 1986 *Phys. Rev.* **D34** 1642
[23] Bazzocchi F, Cerdeno D G, Munoz C and Valle J W F 2009 (*Preprint* 0907.1262)
[24] Ibanez D, Morisi S and Valle J W F 2009 *Phys. Rev.* **D80** 053015 (*Preprint* 0907.3109)
[25] Malinsky M, Romao J C and Valle J W F 2005 *Phys. Rev. Lett.* **95** 161801
[26] Zee A 1980 *Phys. Lett.* **B93** 389
[27] Babu K S 1988 *Phys. Lett.* **B203** 132
[28] Aristizabal Sierra D and Restrepo D 2006 *JHEP* **08** 036 (*Preprint* hep-ph/0604012)
[29] Aristizabal Sierra D and Hirsch M 2006 *JHEP* **12** 052 (*Preprint* hep-ph/0609307)
[30] Miranda O G, Tortola M A and Valle J W F 2006 *JHEP* **10** 008
[31] Escrihuela F J, Miranda O G, Tortola M A and Valle J W F 2009 (*Preprint* 0907.2630)
[32] Palazzo A and Valle J W F 2009 *Phys. Rev.* **D80** 091301(R) (*Preprint* 0909.1535)
[33] Huber P, Schwetz T and Valle J W F 2002 *Phys. Rev. Lett.* **88** 101804 (*Preprint* hep-ph/0111224)
[34] Huber P, Schwetz T and Valle J W F 2002 *Phys. Rev.* **D66** 013006 (*Preprint* hep-ph/0202048)
[35] Dighe A, 2009 Talk at this Conference
[36] Valle J W F 1987 *Phys. Lett.* **B199** 432
[37] Nunokawa H *et al.* 1996 *Phys. Rev.* **D54** 4356–4363 (*Preprint* hep-ph/9605301)
[38] Esteban-Pretel A, Tomas R and Valle J W F 2007 *Phys. Rev.* **D76** 053001
[39] Fornengo N *et al.* 2002 *Phys. Rev.* **D65** 013010 (*Preprint* hep-ph/0108043)
[40] Huber P and Valle J W F 2001 *Phys. Lett.* **B523** 151–160 (*Preprint* hep-ph/0108193)
[41] Gogoladze I, Okada N and Shafi Q 2009 *Phys. Lett.* **B672** 235–239 (*Preprint* 0809.0703)
[42] Bonnet F, Hernandez D, Ota T and Winter W 2009 (*Preprint* 0907.3143)
[43] Weinberg S 1980 *Phys. Rev.* **D22** 1694
[44] Coleman S R 1988 *Nucl. Phys.* **B310** 643
[45] Kallosh R, Linde A D, Linde D A and Susskind L 1995 *Phys. Rev.* **D52** 912–935 (*Preprint* hep-th/9502069)
[46] de Gouvea A and Valle J W F 2001 *Phys. Lett.* **B501** 115–127 (*Preprint* hep-ph/0010299)
[47] Valle J W F 2006 *J. Phys. Conf. Ser.* **53** 473–505 review lectures at Corfu (*Preprint* hep-ph/0608101)
[48] Schechter J and Valle J W F 1982 *Phys. Rev.* **D25** 774

[49] Hirsch M *et al.* 2007 *Phys. Rev.* **D75** 011701 (*Preprint* hep-ph/0608006)
[50] Valle J W F 1987 *Phys. Lett.* **B196** 157
[51] Aristizabal Sierra D, Hirsch M and Kovalenko S G 2008 *Phys. Rev.* **D77** 055011
[52] Hall L J and Suzuki M 1984 *Nucl. Phys.* **B231** 419
[53] Ross G G and Valle J W F 1985 *Phys. Lett.* **B151** 375
[54] Ellis J R and et al 1985 *Phys. Lett.* **B150** 142
[55] Masiero A and Valle J W F 1990 *Phys. Lett.* **B251** 273–278
[56] Romao J C, Santos C A and Valle J W F 1992 *Phys. Lett.* **B288** 311–320
[57] Romao J C, Ioannisian A and Valle J W F 1997 *Phys. Rev.* **D55** 427–430 (*Preprint* hep-ph/9607401)
[58] Diaz M A, Romao J C and Valle J W F 1998 *Nucl. Phys.* **B524** 23–40 (*Preprint* hep-ph/9706315)
[59] Hirsch M and Valle J W F 2004 *New J. Phys.* **6** 76 (*Preprint* hep-ph/0405015)
[60] Hirsch M *et al.* 2000 *Phys. Rev.* **D62** 113008 err-ibid. **D65**:119901,2002 (*Preprint* hep-ph/0004115)
[61] Diaz M A *et al.* 2003 *Phys. Rev.* **D68** 013009 (*Preprint* hep-ph/0302021)
[62] de Campos F *et al.* 2005 *Phys. Rev.* **D71** 075001 (*Preprint* hep-ph/0501153)
[63] de Campos F *et al.* 2008 *JHEP* **05** 048
[64] Porod W *et al.* 2001 *Phys. Rev.* **D63** 115004
[65] Romao J C *et al.* 2000 *Phys. Rev.* **D61** 071703 (*Preprint* hep-ph/9907499)
[66] Mukhopadhyaya B, Roy S and Vissani F 1998 *Phys. Lett.* **B443** 191–195
[67] Hirsch M and Porod W 2003 *Phys. Rev.* **D68** 115007 (*Preprint* hep-ph/0307364)
[68] Harrison P F, Perkins D H and Scott W G 2002 *Phys. Lett.* **B530** 167
[69] Babu K S, Ma E and Valle J W F 2003 *Phys. Lett.* **B552** 207–213 (*Preprint* hep-ph/0206292)
[70] Hirsch M *et al.* 2004 *Phys. Rev.* **D69** 093006 (*Preprint* hep-ph/0312265)
[71] Harrison P F and Scott W G 2002 *Phys. Lett.* **B547** 219–228 (*Preprint* hep-ph/0210197)
[72] Grimus W and Lavoura L 2004 *Phys. Lett.* **B579** 113–122 (*Preprint* hep-ph/0305309)
[73] Altarelli G and Feruglio F 2005 *Nucl. Phys.* **B720** 64–88 (*Preprint* hep-ph/0504165)
[74] Mondragon A, Mondragon M and Peinado E 2007 *Phys. Rev.* **D76** 076003 (*Preprint* 0706.0354)
[75] Bazzocchi F, Merlo L and Morisi S 2009 *Phys. Rev.* **D80** 053003 (*Preprint* 0902.2849)
[76] Altarelli G, Feruglio F and Merlo L 2009 *JHEP* **05** 020 (*Preprint* 0903.1940)
[77] Grimus W, Lavoura L and Radovcic B 2009 *Phys. Lett.* **B674** 117–121 (*Preprint* 0902.2325)
[78] Joshipura A S, Kodrani B P and Patel K M 2009 *Phys. Rev.* **D79** 115017 (*Preprint* 0903.2161)
[79] King S F 2009 *Phys. Lett.* **B675** 347 (*Preprint* 0903.3199)
[80] Antusch S, Kersten J, Lindner M and Ratz M 2003 *Nucl. Phys.* **B674** 401–433 (*Preprint* hep-ph/0305273)
[81] Plentinger F and Rodejohann W 2005 *Phys. Lett.* **B625** 264–276 (*Preprint* hep-ph/0507143)
[82] Hirsch M *et al.* 2007 *Phys. Rev.* **D75** 053006 (*Preprint* hep-ph/0606082)
[83] Chankowski P *et al.* 2001 *Phys. Rev. Lett.* **86** 3488 (*Preprint* hep-ph/0011150)
[84] Raby S, Ratz M and Schmidt-Hoberg K 2009 (*Preprint* 0911.4249)
[85] Hirsch M, Morisi S and Valle J W F 2009 *Phys. Lett.* **B679** 454 (*Preprint* 0905.3056)
[86] Bernabeu J *et al.* 1987 *Phys. Lett.* **B187** 303
[87] Gonzalez-Garcia M C and Valle J W F 1992 *Mod. Phys. Lett.* **A7** 477–488
[88] Ilakovac A and Pilaftsis A 1995 *Nucl. Phys.* **B437** 491 (*Preprint* hep-ph/9403398)
[89] Borzumati F and Masiero A 1986 *Phys. Rev. Lett.* **57** 961
[90] Casas J A and Ibarra A 2001 *Nucl. Phys.* **B618** 171–204 (*Preprint* hep-ph/0103065)
[91] Antusch S, Arganda E, Herrero M J and Teixeira A M 2006 *JHEP* **11** 090 (*Preprint* hep-ph/0607263)
[92] Deppisch F, Kosmas T S and Valle J W F 2006 *Nucl. Phys.* **B752** 80–92 (*Preprint* hep-ph/0512360)
[93] Branco G C, Rebelo M N and Valle J W F 1989 *Phys. Lett.* **B225** 385
[94] Rius N and Valle J W F 1990 *Phys. Lett.* **B246** 249–255
[95] Kuno Y 2000 *AIP Conf. Proc.* **542** 220–225
[96] Esteves J N *et al.* 2009 *JHEP* **05** 003
[97] Deppisch F and Valle J W F 2005 *Phys. Rev.* **D72** 036001 (*Preprint* hep-ph/0406040)
[98] Avignone Frank T I, Elliott S R and Engel J 2008 *Rev. Mod. Phys.* **80** 481–516 (*Preprint* 0708.1033)
[99] Schechter J and Valle J W F 1982 *Phys. Rev.* **D25** 2951
[100] Hirsch M, Kovalenko S and Schmidt I 2006 *Phys. Lett.* **B642** 106 (*Preprint* hep-ph/0608207)
[101] Drexlin G (KATRIN collaboration) 2005 *Nucl. Phys. Proc. Suppl.* **145** 263–267
[102] Lesgourgues J and Pastor S 2006 *Phys. Rep.* **429** 307–379 (*Preprint* astro-ph/0603494)
[103] Schechter J and Valle J W F 1981 *Phys. Rev.* **D23** 1666
[104] Faessler A, Fogli G, Lisi E, Rodin V, Rotunno and Simkovic 2009 *Phys. Rev.* **D79** 053001
[105] Hirsch M *et al.* 2005 *Phys. Rev.* **D72** 091301
[106] Hirsch M, Morisi S and Valle J W F 2008 *Phys. Rev.* **D78** 093007, and 2009 *Phys. Rev.* **D79** 016001

Frontiers of Low Energy Neutrino Physics and Astrophysics

G. Ranucci
Istituto Nazionale di Fisica Nucleare, via Celoria 16, 20133 Milano, Italy

E-mail: gioacchino.ranucci@mi.infn.it

Abstract. Low energy neutrino physics and astrophysics has represented one of the most active field of particle physics research over the past decade, accumulating important and sometimes unexpected achievements. After reviewing some of the most recent impressive successes, the future perspectives of this exciting area of neutrino research will be discussed.

1. Introduction

Under the broad characterization of the title of this talk three very active areas of neutrino physics are grouped: solar neutrinos, geo-neutrinos and supernova neutrino detection. In the following I will try to characterize the current status and achievements of each of them, besides highlighting the relevant perspectives in the imminent and long term future.

Solar neutrinos research is a mature field of investigation which over the past four decades has accumulated fundamental successes in the particle physics arena. Originally conceived as a powerful tool to deeply investigate the core of our star, solar neutrinos underwent a very successful detour to particle physics, heavily concurring to the successful demonstration of the neutrino oscillation phenomenon, according to the MSW mechanism [1][2]. The solar experiments were also paramount to determine with high accuracy the Δm_{12}^2 and $\sin^2\theta_{12}$ parameters of the PMNS mixing matrix governing the oscillation in the solar sector [3], with the recent fascinating promise, stemming from the joint analysis of the reactor and solar data, of an early indication of a non zero subleading θ_{13} angle [4]. Hence this extremely rich and successful chapter of particle physics produced a large part of the solid foundations upon which the next era of precision measurements of the neutrino oscillation parameters will be built. Furthermore, once completed the mission of unveiling the oscillating nature of neutrinos, solar experiments are now back to the original concept of testing the functioning mechanism of the Sun; in this context important insights are awaited from the running and future experiments, of special relevance to address the current issue regarding the surface metallic content of the Sun [5].

Geo-neutrino science, instead, is a much more recent experimental field, though rich of profound potential implications. The suggestion of the possible detection of antineutrino from the intrinsic radioactivity of the Earth dates back to the early times of the first neutrino detection from Cowan and Reines and is due to Gamow. Only 50 years later the reactor antineutrino experiment KamLAND [6] provided the first, and for the moment the only, indication of the observation of antineutrinos from Earth. This pioneering measurement, proving the technical feasibility of geoneutrino detection, paves the way for a series of planned future experiments with the ambitious goal to use the accurate measurements of the antineutrino flux to probe the characteristics of the interior of our planet.

The observation of the neutrino burst from the Supernova 1987A, performed by the Kamiokande [7], IMB [8] and Baksan [9] detectors, represented in the last century a fundamental breakthrough: for the first time the last moments of a dying star were observed through the copious neutrino flux which accompanies such a cosmic thunderstorm. After that, a great deal of studies showed how the accurate determination of the supernova neutrinos can represent an extraordinarily powerful tool to investigate the evolution of the supernova mechanism; this led to the establishment of an ambitious program for the efficient capture of neutrinos from a galactic star explosion, which not only sees a growing number of experiments potentially able to perform such a task, but also their strategic alliance in a worldwide network, SNEWS, which will greatly enhance the probability of a successful detection of the eagerly awaited "galactic firework", when it will occur.

2. Solar neutrinos

It is now well known, after several decades of theoretical and experimental investigations, that neutrinos are abundantly produced in the core of the Sun. They originate from the nuclear reactions that power our star, producing the energy required to sustain it over the billions of years of its life. Two different chain reactions occur at the temperatures characteristic of the core of the Sun, the so called pp chain and CNO cycle, respectively. Actually for the sun the vast majority of the energy (>98%) is coming from the pp chain, while the CNO contribution is estimated to less than 1.6 %.

The effort to produce a model able to reproduce fairly accurately the solar physical characteristics, as well as the spectra and fluxes of the several produced neutrino components, was led for more than forty years by John Bahcall [10]; this effort culminated in the synthesis of the so called Standard Solar Model (SSM), which represents a true triumph of the physics of XXth century, leading to extraordinary agreements between predictions and observables. Such a beautiful concordance, however, has been somehow recently spoiled as a consequence of the controversy risen regarding the surface metallic content of the Sun, stemming from a more accurate 3D modeling of the Sun photosphere. Therefore, there are now two versions of the SSM, according to the adoption of the old (high) or revised (low) metallicity of the surface [11].

From the experimental side, solar neutrino experiments also represent a successful 40 years long saga, commenced with the pioneering radiochemical experiments, i.e Homestake, Gallex/GNO and Sage, continued with the Cerenkov detectors Kamiokande/Super-Kamiokande in Japan and SNO in Canada, and with a last player which entered very recently the scene, Borexino at the Gran Sasso Laboratory, which introduced in this field the liquid scintillation detection approach.

It is well known that for more than 30 years the persisting discrepancy between the experimental results and the theoretical predictions of the Solar Model formed the basis of the so called Solar Neutrino Problem, which in the end culminated with a crystal clear proof of the occurrence of the neutrino oscillation phenomenon, via the MSW effect. In particular, the joint analysis of the results from the solar experiments and from the KamLAND antineutrino reactor experiment pin points with high accuracy the values of the oscillations parameter within the LMA (large mixing angle) region of the MSW solution [3].

So, given this epochal achievement, which surely makes the solar neutrino study one of the more productive particle physics area over the past decade, one may wonder if there are future interesting perspectives for this field, and if yes, which they are. In the following I will illustrate the answer to this question by browsing through the main existing and planned solar experiments, underlining the future plans of each of them.

2.1 Borexino

I start this review with the most recent project, Borexino at the Gran Sasso Laboratory [12], a scintillator detector which employs as active detection medium 300 tons of pseudocumene-based scintillator. The intrinsic high luminosity of the liquid scintillation technology is the key toward the goal of Borexino, the real time observation of sub-MeV solar neutrinos through νe elastic scattering, being the ^{7}Be component the main target. However, the lack of directionality of the method makes it impossible to distinguish neutrino scattered electrons from electrons due to natural radioactivity, thus leading to the other crucial requirement of the Borexino technology, e.g. an extremely low radioactive contamination of the detection medium, at fantastic unprecedented levels.

The active scintillating volume is observed by 2212 PMTs located on a 13.7 m diameter sphere and is shielded from the external radiation by more than 2500 tons of water and by 1000 of hydrocarbon equal to the main compound of the scintillator (pseudocumene), to ensure zero buoyancy on the Inner Vessel containing the scintillator itself. Of paramount importance for the success of the experiment are also the many purification and handling systems, which were designed and installed to ensure the proper manipulation of the fluids at the exceptional purity level demanded by Borexino.

When data taking started in May 2007, it appeared immediately that the daunting task of the ultralow radioactivity was successfully achieved, representing *per se* a major technological breakthrough, opening a new era in the field of ultrapure detectors for rare events search.

The exceptional purity obtained implies that, once selected by software analysis the design fiducial volume of 100 tons and upon removal of the muon and muon-induced signals, the recorded experimental spectrum is so clean to show spectacularly the striking feature of the ^{7}Be scattering edge, i.e. the unambiguous signature of the occurrence of solar neutrino detection.

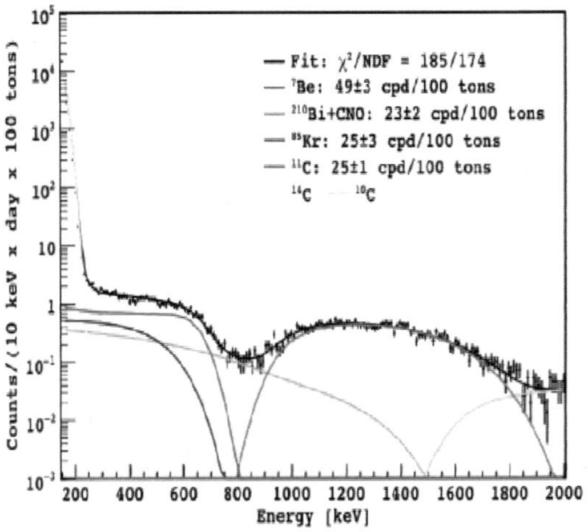

Figure 1. Fit of the experimental Borexino spectrum

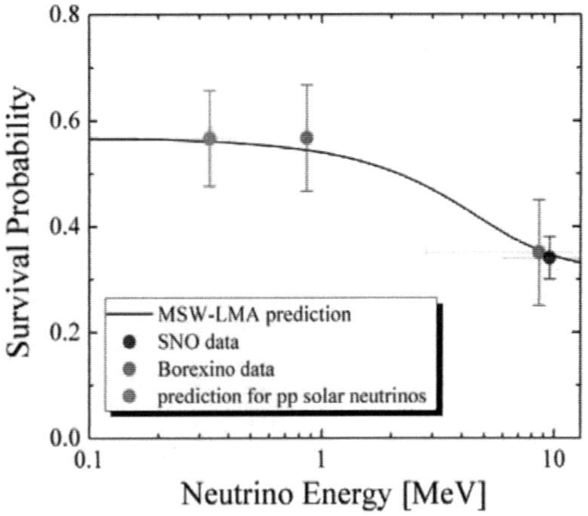

Figure 2. Low energy validation of the MSW-LMA solution provided by Borexino

For the quantitative extraction of the ^{7}Be flux, the spectrum is fitted to a global signal-plus-background model. The fit output for the 192 days data sample released up to now by the Collaboration [13] is reported in Fig. 1 (the results in the legenda are conventionally expressed in counts/day/100 tons of scintillator). Taking into accounts the systematic uncertainties, the ^{7}Be evaluation is $49 \pm 3_{stat} \pm 4_{sys}$ counts/day/100 tons, hence a 10% precision measurement, which translates into a ^{7}Be flux of $(5.08 \pm 0.25) \cdot 10^9\ cm^{-2}s^{-1}$, very well in agreement with the prediction of the BPS08(GS98) Standard Solar Model [11]. For comparison, the detected count rate in case of absence of oscillations would have been 74 ± 4 counts/day/100 tons. The resulting electrons survival probability at the ^{7}Be energy is P_{ee}=0.56± 0.10.

In Fig. 2 the MSW predicted P_{ee} is shown, together with three experimental points, i.e the ^{8}B from the previous Cerenkov data, the ^{7}Be Borexino point and the pp datum as drawn by the comparison of Borexino with the Gallium experiments: altogether from this figure we can conclude that Borexino on one hand spectacularly confirms the MSW-LMA solar neutrino oscillation scenario, and on the other provides the first direct measurement of the survival probability in the low energy vacuum MSW regime.

Other measurements provided by the experiment in its first two years of operation, and discussed in [14], are the ^{8}B spectrum with a threshold as low as 2.8 MeV, the absence of day-night asymmetry (as predicted by the MSW-LMA solution) in the ^{7}Be region, and a limit on the neutrino magnetic moment at the level of $5.4 \times 10^{-11}\ \mu_B$.

Many further solar measurements are prospectively possible in the next years of running of the detector. ^{7}Be can be pin pointed to an accuracy of 5% (with respect to the 10% uncertainty of the measurement reported here), providing the final, high accuracy, low energy validation of the MSW-LMA solution; moreover the statistics of the ^{8}B neutrino study will be increased, and the possibility to investigate the extremely challenging pep and CNO fluxes will be pursued. This last task will require to cope with the background represented by the ^{11}C cosmogenic signals, exploiting the triple fold coincidence strategy already devised by the Collaboration in [15].

Finally, the extremely low ^{14}C level, coupled to the good achieved energy resolution, opens also a possible exploration window between 200 and 240 keV in which the observation of the fundamental pp flux can be attempted.

2.2 Solar Neutrino Observatory

The SNO (Solar Neutrino Observatory) experiment [16], located in Sudbury, Canada, is now completed, with the heavy water being returned to the owner until the very last drop. Based on multiple CC, NC and elastic scattering detection of solar neutrinos, the key advance provided by SNO in the field was the model independent proof that neutrinos from the Sun undergo flavour conversion. The experiment evolved through three phases, characterized by different detection procedures of the neutrons signalling the occurrence of the neutral current reactions: a pure heavy water phase, a salt phase and the final ^{3}He counters stage.

The estimated NC, CC and ES fluxes over the three phase are statistically in agreement, with the exception of the ES measure of the ^{3}He stage, lower than the previous results, but consistent with being a downward statistical fluctuation. Also, in the latest ^{3}He measurement procedure the ratio NC/CC resulted equal to 0.301 ±0.033, slightly lower than the previous phases. While including these updated data in the global solar + KamLAND analysis, the SNO collaboration finds $\Delta m^2 = 7.59^{+0.19}_{-0.21} \times 10^{-5} eV^2$, $\phi_{8B} = 4.91 \times 10^6 cm^2 s^{-1} (\pm \approx 7\%)$ and $\theta_{12} = 34.4^{+1.3}_{-1.2}$ deg [3]. It has to be noted, in particular, that the error on the θ_{12} angle has been reduced of almost a factor two by the ^{3}He measurement, as effect of the different systematic and minimal correlation of the NC and CC measures in this phase.

Despite the fact that the SNO measurements are over, still very important outputs are expected from the further study of the accumulated data. In particular the Collaboration is now focused to the so called LETA (low energy threshold) analysis, whose purpose is to push the analysis threshold down to 3.5 MeV (electron kinetic energy). To do so it is required to model accurately down to that energy the several background components, they are 17, that can be identified within the data and which accompany the three CC, NC and ES signal components. In Fig.3 how such a modelling can be accomplished is shown on a basis of a Monte Carlo estimate. More insights in this exciting possibility are expected from the talk of SNO [17].

It should be added that the LETA analysis will be extremely beneficial on one hand in further tightening the error both on θ_{12} and on ϕ_{8B} , and on the other, in the context of a full 3 ν analysis, in shedding more light on the recently emerged hint [4] for a non zero θ_{13}.

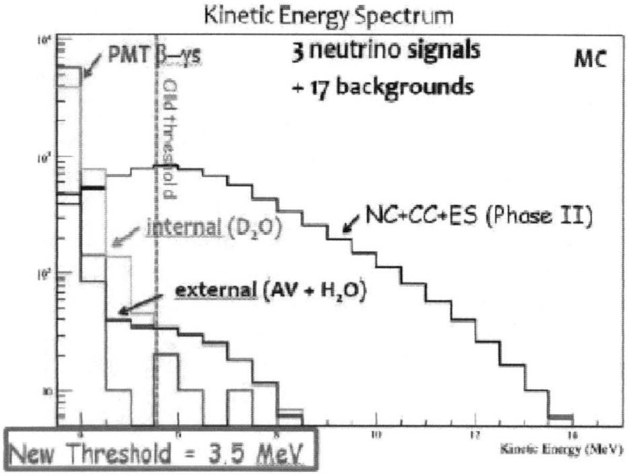

Figure 3. Perspectives for threshold reduction in the SNO analysis

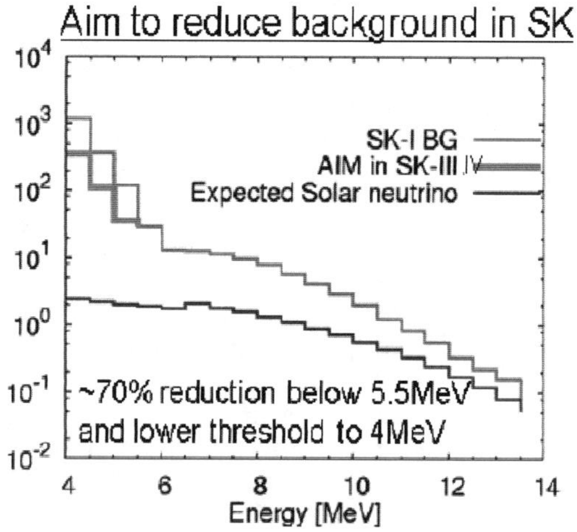

Figure 4. Perspectives for threshold reduction in the future SK data taking

2.3 Super-Kamiokande

Super-Kamiokande [18] is still currently taking data. The long history of this detector started in 1996 and evolved through four phases: the first phase lasted until the PMT incident of November 2001 and produced the most accurate measure up to now of the ^{8}B flux via the ES detection reaction. The phase II with reduced number of PMTs, from the end of 2002 to the end 2005, confirmed with larger error the phase I measurement. After the refurbishment of the detector back to the original number of PMTs, the third phase lasted from the middle of 2006 up to the middle of 2008. After that, an upgrade of the electronics brought the detector into its fourth phase, ready for the T2K beam experiment. It is important to highlight the evolution of the energy threshold (total electron energy) in all the phases: 5 MeV in phase I, 7 MeV in phase II, 4.5 MeV in phase III and 4 MeV targeted for phase IV.

The reduction of the threshold for ^{8}B acquisition and analysis is crucially linked to the radiopurity of the detector, being the radon in the Fiducial Volume the main limitation in this respect. Analysis of the data of the phase III is very promising in term of finally achieving the 4.5 MeV threshold, while the Collaboration is working further on the background to reach the ultimate goal of 4 MeV in the phase IV. Fig. 4 exemplifies the background suppression already achieved in the FV.

The low energy threshold results that Super-Kamiokande will eventually produce will be, similarly to what illustrated for SNO, of paramount importance to further probe the LMA-MSW solution, in particular if the expected upturn of the low energy portion of the ^{8}B spectrum will be identified. It is worth remind that such an up-turn is the expected imprint on the spectrum of the transition of the electron survival probability from the high energy matter-dominated region to the low energy vacuum-driven regime. Said in other words, the low threshold determination of the ^{8}B spectrum is a way to probe directly the transition portion of the survival probability curve.

In the framework of this effort, the Collaboration plan to reach a 2 sigma level of discovery (or exclusion) of the spectrum upturn after two years of data taking in the phase IV with reduced background [19].

2.4 Gallium experiments

In this review of the current status of the solar experiments it cannot be forgotten the important role still played by one of the pioneering Gallium experiments, Sage, which is continuing taking data at the Baksan Laboratory. Based on 50 tons of metallic Gallium, after 18 year of data taking the cumulative result is of $65.4^{+4.0}_{-4.1}$ SNU (Solar Neutrino Units), mainly due to the contribution of pp neutrinos. Extractions continue regularly every 4 weeks. The Sage results can be combined with those of the Gallex and GNO experiments at Gran Sasso (now over) 67.5 ± 5.1 SNU , giving as grand result of the Gallium experiments 66.1 ± 3.1 SNU, to be compared with the SSM prediction of $126.^{+9}_{-7}$ SNU [20][21].

The experiments have been calibrated respectively with ^{51}Cr (Gallex) and ^{37}Ar (Sage) sources. Interestingly, in both cases the ratio of the measured over predicted rates is substantially less than unity, e.g. 0.88 ± 0.08 for Gallex and 0.86 ± 0.08 for Sage. The combined value of 0.87 ± 0.05 deviates from 1 more than two sigmas. Although not statistically conclusive, the combination of these experiments suggests that the predicted rates is overestimated. The most likely hypothesis is that the cross sections for neutrino capture to the lowest two states in ^{71}Ge, both of which can be reached using either ^{51}Cr or ^{37}Ar sources, have been overestimated.

A new source experiment in SAGE will shed light on such an inconsistency. For this it is necessary to arrange a ^{51}Cr neutrino source with activity of 2 MCi or higher, an optimized two-zone spherical target in the detector, and to carry out the measurement with an accuracy of 4 – 4.5%. These goals are part of future experimental program in SAGE [22].

2.5 Future experiments

A series of new solar experiments are being planned or prepared, with some of them likely ready to produce very interesting new results in the imminent future, e.g. KamLAND and SNO+.

KamLAND, after the epochal results achieved with the reactor antineutrino experiment [23], is aggressively moving towards the solar phase of its program. To do so the Collaboration requires to suppress the otherwise overwhelming low energy background, which would definitively prevent the identification of solar neutrinos. The achievement of this target is based upon the on-line purification of the liquid scintillator via a dedicated purification system: two purification cycles have been already accomplished, bringing the background to a level close to that needed specifically for the detection of ^{7}Be neutrinos. The collaboration is actively debating whether to continue with the purification or start the measure without further actions on the scintillator, resorting to the analysis for the final discrimination between the solar signal and the background. Anyhow, independently from this decision, KamLAND will be very likely the second experiment after Borexino to produce sub-MeV real time detection of solar neutrinos.

The first target of the detector are the ^{7}Be neutrinos, followed by the pep and CNO neutrinos, for which a special deadtimeless electronics has been installed, able to detect with high efficiency neutrons and thus providing the basis for the triple fold coincidence mentioned above, in the framework of Borexino, for ^{11}C suppression.

SNO+ [24] is the new experiment that will replace SNO. It is a liquid scintillator detector, as well, that can go on line soon thanks to the massive re-use of the SNO hardware. In particular the adoption of a new liquid scintillator, linear alkyl benzene, featuring the great advantage of being acrylic-compatible, will allow the re-use of the SNO acrylic vessel. Only the support system will have to be built from scratch in order to cope with the different buoyancy condition with respect to the heavy water situation. The other major piece of equipment to realize is the purification system. The project has been funded for these major items on June 2009, with the perspective to start data taking in 2011. The solar phase will be targeted mainly to pep and CNO detection, profiting of the depth of SNOLAB which suppress enormously the cosmogenic ^{11}C background. It should be, however, highlighted that the main goal of SNO+ will be the double beta decay with Neodymium dissolved in the scintillator, and currently it is not yet decided whether the solar phase will come before or after the Neodymium phase.

While KamLAND and SNO+ represent the imminent and near future perspectives for new sub-MeV solar neutrino results, it should be mentioned that there are other experiments in the R&D phase with great potentialities for future interesting outputs in this field. In this framework a special mention is due to LENS [25], which is the modern version of the old brilliant idea for an Indium experiment. The experimental tool used for the detection of solar neutrinos is the tagged capture of ν_e's on ^{115}In via charged current : $\nu_e + ^{115}In \rightarrow ^{115}Sn^* + e^- \rightarrow ^{115}Sn + 2\gamma$. The low Q of the reaction (114 keV) allows in principle the full solar neutrino spectroscopy, including the pp neutrinos, see Fig. 5. Time and space coincidence, the former intrinsic to the detection reaction and the latter realized via a granular detector design, will be the key to suppress the radioactive backgrounds as well as the inherent background from the beta decay of ^{115}In. R&D results very promising in term of stability of the Indium loaded scintillator and background rejection have been recently presented [26].

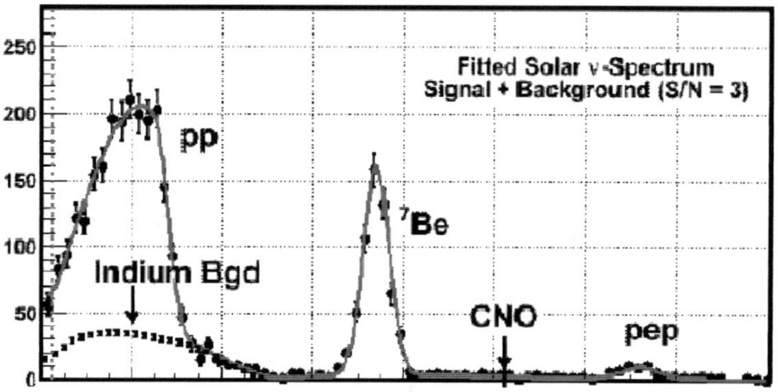

Figure 5 – Solar spectrum as would be detected by LENS

Other three projects that deserve to be mentioned are CLEAN, XMASS and MOON. They are multidisciplinary projects which have within their potential physics reaches double beta decay, dark matter and solar neutrino investigations.

The basic idea of CLEAN [27] is to take advantage of the fact that both neon or argon scintillate when exposed to radiation. When an energetic charged particle (such as the recoil from a neutrino-electron scattering event) passes through the liquid noble gas, it will produce scintillation light in the extreme ultraviolet range, that can be detected using a combination of a wavelength shifting fluor and a photomultiplier tube. Liquid neon has a number of advantages for detecting neutrinos or other weakly interacting particles. Because neon has no long-lived isotopes, it has no inherent radioactivity to create backgrounds in a sensitive detector. This separates neon from heavier noble gases which do contain radioactive isotopes and from organic scintillator materials which inevitably contains beta radiation from ^{14}C. Neon also has very low binding energies to a variety of surfaces, allowing it to be effectively purified of radioactive contaminants using cryogenic traps. Another advantage is its relative density compared to liquid helium; neon can be used efficiently as a self-shielding medium, while also allowing a smaller detector volume. These characteristics render neon a well suited choice as a neutrino detection medium.

The current effort of the Collaboration is towards the installation within 2009 of a 360 kg detector for dark matter investigation at SNOLAB. In perspective, a 50 tons device targeted especially to the pp solar neutrino detection could be installed at the planned DUSEL underground facility in US.

The original concept of XMASS was proposed in 2000 [28] as a multi-purpose astroparticle and neutrino experiment. A liquid Xenon detector with a 10 ton effective mass can look for dark matter and double beta decay and can also measure solar pp-neutrinos through neutrino electron scattering, as much as CLEAN. As a phase-I experiment, a small scale detector dedicated to look for dark matter, where the mass of the inner volume surrounded by PMTs is 857kg and the fiducial mass is 100~200kg, is being prepared and installed at the Kamioka mine in Japan.

The double beta decay and solar neutrino project MOON [29] is based on ^{100}Mo. The merits of using ^{100}Mo for solar ν's study are the large capture rate and the high selectivity for the low energy ν's. The low threshold energy of 0.168 MeV and the large response for the solar-ν absorption allow observation of low energy sources such as pp and ^{7}Be. The pp and ^{7}Be ν's are captured only into the ground state of ^{100}Tc. The solar ν signal can be selected by requiring delayed coincidence with the successive β decay of ^{100}Tc, and thus natural and cosmogenic backgrounds are reduced substantially.

Research and development programs of solid and liquid scintillators for the MOON detector are in progress. A possible option for the solid scintillator is a supermodule of hybrid plate and fiber scintillators. One module consists of a plate scintillator and two sets of X-Y fiber scintillator planes, between which a thin ^{100}Mo film of about 20 - 50 mg cm^{-2} is interleaved. The fiber scintillators coupled with multi-anode photomultiplier tubes (PMT's) enable one to get the necessary position resolution, the scintillator plate (X-Y plane) with multiple PMT's at both X and Y sides provides an adequate energy resolution to satisfy the physics goals.

3. Geoneutrinos

Geo-neutrinos, the antineutrinos from the progenies of U, Th and ^{40}Kdecays in the Earth, bring to the surface information from the whole planet, concerning its content of radioactive elements. Their detection can shed light on the sources of the terrestrial heat flow, on the present composition, and on the origins of the Earth.

Although geo-neutrinos were conceived very long ago, only recently they have been considered seriously as a new probe of our planet, as a consequence of two fundamental advances that occurred in the last few years: the development of extremely low background neutrino detectors and the progress on understanding neutrino propagation.

Geoneutrinos potentially can unlock many interesting unanswered questions regarding our planet: what is the radiogenic contribution to terrestrial heat production? How much is the content of U and Th in the crust and in the mantle, respectively? What is hidden in the Earth's core, e.g a geo-reactor or ^{40}K, etc? And finally, is the standard geochemical model (denoted as BSE) consistent with geo-neutrino data?

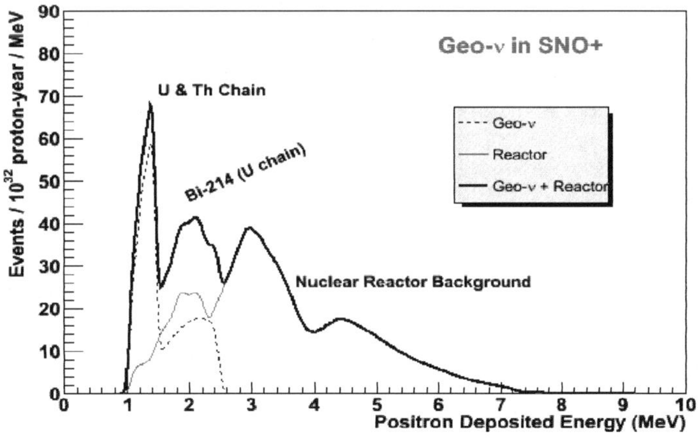

Figure 6. Example of a typical expected geoneutrino spectrum in a scintillation detector, together with the reactor background

In particular this last question would ideally require multiple measurements carried out in several locations of the Earth's surface, to be compared with the theoretical predictions of geoneutrino flux (different groups performed such calculations, which are each other in agreement at the 10% level [30][31][32]).

A dangerous potential background for geoeneutrinos are anti-neutrinos from reactor; in Fig. 6 the global spectral expectation for a standard scintillator detector is shown (the example is drawn for SNO+).

The first and for the moment the only published geoneutrino measure comes from the KamLAND scintillator experiment [6]. The detection occurs via the classical inverse beta reaction $\overline{\nu_e} p \rightarrow e^+ n$, signaled by the delayed 2.2 MeV gamma stemming from the subsequent neutron capture on protons. In terms of Terrestrial Neutrino Unit, a geoneutrino unit similar to the solar SNU unit, the flux measured by KamLAND is equal to $38.9^{+14.4}_{-14.2}$ TNU, corresponding to 75 ± 27 geoneutrino events both from the U and Th chain. This results compares well with the model prediction of 36.9 TNU, though the uncertainty is large, at the level of 37%.

The main virtue of the pioneering result of KamLAND is the demonstration that geoneutrinos can be detected, opening the door to the a completely new branch of geosciences. In the future other experiments are expected to confirm the neutrino detection from KamLAND. Borexino is supposed to be the first, being already running and thus accumulating geo-neutrino events, with the Collaboration intending to release the data when a solid three sigma evidence will be achieved.

But many other planned large scintillator detectors (e.g. SNO+, LENA, HANOHANO, EARTH…) have geo-neutrino measurements among their primary goals (a recent review is presented in [33], se also [34]), thus the coming years will likely see the rapid growth of this innovative way to study our planet.

4. Supernova neutrinos

On February 23, 1987 the Kamiokande [7], IMB [8] and Baksan [9] detectors recorded for the first (and up to now the only) time a pulse of neutrinos from a supernova, the famous SN1987A. Despite the small number of recorded signals, 11, 8 and 5 respectively, plenty of information were squeezed from them, providing substantial confirmations and new insights on the supernova mechanism. On the basis of this success, it can be anticipated that further deep insights and details of the supernova birth and evolution will be gained when the next burst will be observed, especially if the supernova will occur at the typical galactic distance from the Earth of 10 kpc, compared to the 50 kpc distance of the SN19877A, which indeed was located in the Large Magellanic Cloud.

A number of multipurpose detectors, some already running and some still in the construction phase, will detect the next supernova neutrinos. Among the current detectors, Superkamiokande is surely the largest with its 32000 tons of water. For a 10 kpc distant supernova it will observe about 7300 $\overline{\nu_e} p \to e^+ n$ events, ~300 $\nu + e \to \nu + e$ scattering events, ~360 ^{16}O neutral current gamma events and ~100 ^{16}O charged current events. The scattering events will identify the supernova direction with an accuracy of about 5°. The large statistics of inverse beta events, instead, will allow the precise energy spectrum measurement, enabling the possibility to discriminate among various proposed models.

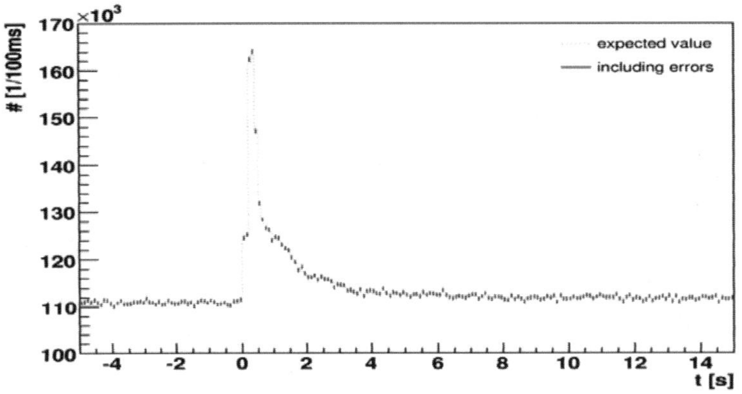

Figure 7. Simulation of a neutrino burst at the IceCube detector for a 10 kpc distant supernova

At Gran Sasso since 1992 the LVD detector is in operation: it is a segmented liquid scintillator type apparatus with 840 counters and a total target mass of 1000 tons. It can detect about 300 $\overline{\nu_e} p \to e^+ n$ events for the reference 10 kpc supernova . The energy threshold of the counter is 4 MeV and it can tag neutrons with an efficiency of 50 ±10%. The online trigger efficiency is more than 90% for a distance less than 40 kpc.

KamLAND, Borexino and the future SNO+ are single volume liquid scintillation detectors with 1000, 300 and 1000 tons of detection medium, respectively. They feature several interactions channels; by normalizing the expected number of events for a 10 kpc supernova to a target mass of 1000 tons, the predictions are ~300

$\overline{\nu}_e p \rightarrow e^+ n$ events, several tens of CC and NC events on ^{12}C and about 300 $\nu p \rightarrow \nu p$ NC events. This last interaction, being not affected by neutrino oscillations, has the very interesting capability to measure the initial spectrum originated by the supernova explosion.

The South Pole IceCube detector has a giga-ton target volume, with the photosensors arranged in a three dimensional structure. The entire inventory of strings (75) will be deployed by 2011. In this detector the observation of a supernova burst will not occur via individual neutrino observation, but through the coherent increase of the PMT dark noise. As a consequence, the detector can measure the time variation of the energy release fairly accurately, see Fig. 7.

Finally, it has to be highlighted that Super-Kamiokande, LVD, IceCube and Borexino are part of the SNEWS consortium (Supernova Early Warning System), which, upon detection of a neutrino burst, sends prompt alerts to astronomers to trigger the optical observation of the potential identified source. More detailed information on this topic can be found in [35].

5. Conclusion

Low energy neutrino physics and astrophysics is a field that, despite the many successes already accumulated, is promising for the future many more exciting results.

Solar neutrino investigation is reaching the stage in which the precise spectroscopy of the whole low energy spectrum is a real possibility, up to a level that makes it feasible the check of the SSM, hence paving the path for resolving the discrepancy between high Z and low Z models

An entire new area of investigation is represented by the geoneutrinos study, which exploits the more powerful detectors aimed at the solar neutrino detection to open a completely new window on the mysteries of the interior of the Earth

Supernova neutrino experiments are already a powerful large family with new members expected to join in the near future, and hence if Nature will arrange for us a "galactic firework", surely we will not miss the fun!

6. Acknowledgements

The author acknowledges M.Nakahata, V. Gavrin, R, Raghavan, D. McKinsey, M. Chen, A, Sukuzi, T. Kirsten, Y. Suzuki, E. Ejiri, G. Fiorentini and W. Fulgione for supplying material for this review.

7. References

[1] L. Wolfenstein, Phys. Rev. D 17, 2369 (1978)
[2] S.P. Mykheyev and A.Yu. Smirnov, Sov. J. Nucl. Phys. 42, 913 (1985)
[3] B. Aharmim et al , Phys. Rev. Lett 101, 111301 (2008)
[4] G. Fogli et al., Phys. Rev. Lett 101, 141801 (2008)
[5] M. Asplund et al., Annu. Rev. Astron. Astrophys. 47, 481 (2009)
[6] T. Araki et al., Nature 436, 499 (2005)
[7] K. Hirata et al., Phys. Rev. Lett 58, 1490 (1987)
[8] R. M. Bionta et al., Phys Rev. Lett 58, 1494 (1987)
[9] E. N. Alexeyev et al., Phys. Lett. B205, 209 (1988)
[10] http://www.sns.ias.edu/~jnb/ John Bahcall home page
[11] C. Peña-Garay and A.M. Serenelli arXiv:0811.2424
[12] G. Alimonti et al., Nucl. Instr. and Meth A 600, 568 (2009)
[13] C. Arpesella et al., Phys. Rev. Lett. 101, 091302 (2008)
[14] L. Oberauer, these proceedings
[15] H. Back et al., Phys. Rev. C 74, 045805 (2006)
[16] J. Boger et al., Nucl. Instr. and Meth. A449, 172 (2000)
[17] J. Klein, these proceedings
[18] J. Hosaka et al., Phys. Rev. D 73, 112001 (2006)
[19] M. Smy, these proceedings
[20] J.N. Abdurashitov et al., Physi. Rev. C 80, 015807 (2009)
[21] T. Kirsten, Journal of Physics: Conference Series 120, 052013 (2008)
[22] Gorbachev, these proceedings
[23] S. Abe et al, Phys. Rev. Lett. 100, 221803 (2008)
[24] M. ChenNucl. Phys. B (Proc. Suppl.) 154, 65 (2005)

[25] R. S. Raghavan, Journal of Physics: Conference Series 120, 052014 (2008).
[26] R. S. Raghavan, talk at the Physun workshop at LNGS, http://www.mi.infn.it/PHYSUN/Raghavan.pdf
[27] D. N. McKinsey and K.J. Coakley, Astroparticle Physics 22, 335 (2005)
[28] Y. Suzuki, Proc. of Science 001(idm2008)
[29] H. Ejiri, Journal of Physics: Conference Series 173, 012009 (2009)
[30] S. Enomoto, Ph.D. thesis, Tohoku University (2005)
[31] G. Fogli et al, Phys. Lett. B 623, 80 (2005)
[32] F. Mantovani et al., Phys. Rev. D 69, 013001 (2004)
[33] G. Fiorentini et al., Physics Reports 453, 117 (2007)
[34] S. Dye, these proceedings
[35] K. Schoelberg, these proceedings

Status and perspectives of short baseline studies

Mark Dierckxsens
Department of Physics, University of Chicago, Chicago, IL 60637, U.S.A.
Now at: Université Libre de Bruxelles, Science Faculty CP230, B-1050 Brussels, Belgium
E-mail: `mdier@hep.uchicago.edu`

Abstract.
The study of flavor changing neutrinos is a very active field of research. I will discuss the status of ongoing and near term experiments investigating neutrino properties at short distances from the source. In the next few years, the Double Chooz, RENO and Daya Bay reactor neutrino experiments will start looking for signatures of a non-zero value of the mixing angle θ_{13} with much improved sensitivities. The MiniBooNE experiment is investigating the LSND anomaly by looking at both the $\nu_\mu \rightarrow \nu_e$ and $\bar{\nu}_\mu \rightarrow \bar{\nu}_e$ appearance channels. Recent results on cross section measurements will be discussed briefly.

1. Introduction
The relation between the flavor and mass eigenstates of the neutrinos can be described by the PMNS matrix [1, 2]. In case of 3-neutrino mixing, the parameters relevant to oscillations describing this matrix are three mixing angles, θ_{12}, θ_{23} and θ_{13}, and a CP-violating phase $\delta_{\rm CP}$. The frequencies of oscillations between the different states are governed by the energy, distance and the two possible differences between the masses squared, Δm^2_{21} and Δm^2_{31}.

Since the confirmation of neutrino oscillations a decade ago [3], great progress has been made in the measurements of these parameters. Two of the mixing angles, θ_{12} and θ_{23}, and the size of the two mass splittings, Δm^2_{21} and Δm^2_{31}, are well measured (for a recent combination of global data, see e.g. [4]). The most stringent limit on the third angle comes from the CHOOZ reactor neutrino experiment [5]: $\sin^2 2\theta_{13} < 0.16$ (90% C.L.) at $\Delta m^2_{31} = 2.5\ 10^{-3}$eV. However, small contributions from various experiments add up to prefer a non-zero value of $\theta_{13} \simeq 8^o$ at the 2σ level [4, 6]. Several experiments currently under construction will probe well into this region. Those observing neutrinos at nuclear reactors will be covered in the next section.

However, not all measurements fit clearly into the above picture. The LSND experiment has observed an excess of 87.9 ± 23.2 (3.8σ) events above background in the $\bar{\nu}_\mu \rightarrow \bar{\nu}_e$ appearance search [7]. The L/E value requires a mass splitting different than the solar and atmospheric Δm^2 to explain this effect with neutrino oscillations. Three mass differences need mixing with a unobserved fourth type of neutrino with different mass, suggesting the existence of sterile neutrinos. The MiniBooNE experiment was designed to investigate this excess in detail, and will be covered in Section 3.

2. Reactor neutrino experiments

2.1. Neutrino oscillations at nuclear reactors

Nuclear reactors are a pure and isotropic source of anti-electron neutrinos produced at a rate of about $2 \cdot 10^{20}/\mathrm{GW_{th}}/s$. The particles are produced through the decay of the fission products in the decay chain of predominantly ^{235}U, ^{239}Pu, ^{241}Pu and ^{238}U. The β-spectra from the first three isotopes are measured to a precision of 1.8% [8, 9]. These spectra are converted into a predicted neutrino spectrum resulting in a final error of 2.5 – 4% depending on the energy. Plans are underway to decrease this error significantly by measuring the ^{238}U β-spectrum and performing improved conversion calculations including more branches in the decay chain.

The survival probability for anti-electron neutrinos is given by

$$P(\bar{\nu}_e \to \bar{\nu}_e) \simeq 1 - \sin^2 2\theta_{13} \sin^2(\frac{1.27 L \Delta m^2_{31}}{E}) - \cos^4\theta_{13} \sin^2 2\theta_{12} \sin^2(\frac{1.27 L \Delta m^2_{21}}{E}), \quad (1)$$

with L measured in m, E in MeV and Δm^2 in eV. In case of a non-zero θ_{13}, the first oscillation node is around 2 km for typical energies in nuclear reactor experiments. An observation of neutrino disappearance at this distance will constitute a clean measurement of $\sin^2 2\theta_{13}$: the last term containing the solar parameters can be neglected and the probability only depends on Δm^2_{31}, a well measured quantity. The probability is not affected by matter effects due to the short distances and is insensitive to the CP-violating phase. Several experiments have searched unsuccessfully for the disappearance of rector neutrinos at various distances due to a non-zero θ_{13}. The best limit was set by the Chooz experiment to be $\sin^2 2\theta_{13} < 0.16$ (90% C.L.) [5].

Three experiments are currently under construction with a much improved sensitivity: Double Chooz, RENO and Daya Bay. The statistical uncertainty will be lowered by building larger detectors, running for a longer time and by being located near more powerful reactors. The reduction in systematics uncertainty is mainly obtained by placing one or more identical detectors close to the reactor cores. A ratio measurement will eliminate the cross section, the neutrino flux and some detector uncertainties. There is also a gain from an improved detector design with lower thresholds, increased efficiencies and implementation of a very detailed calibration program. Further enhancements are obtained by lowering the background contamination with better veto systems, shielding, higher overburden and the use of materials with improved radio-purity.

2.2. Detection technique

The anti-electron neutrinos are measured through the inverse β-decay process $\bar{\nu}_e + p \to e^+ + n$, with a threshold energy of 1.8 MeV. Folding the neutrino flux from the reactor with the cross section results in an observable spectrum peaking just below 4 MeV.

The three experiments employ similar detector designs with three regions containing different types of liquids, separated by optically transparent acrylic vessels. The central region is filled with liquid scintillator doped with a small amount of Gadolinium, typical 0.1% in weight. The produced positron annihilates instantly and deposits a total amount of visible energy closely related to the neutrino energy: $E_{\mathrm{vis}} \simeq E_\nu - 0.8$ MeV. The neutron slows down through thermalization before being captured by a nucleus. This will most often happen on Gadolimium due to its high cross section for neutron capture. The typical capture time is $30\mu s$ and releases a total amount of energy around 8 MeV in the form of several gammas. The target region is surrounded by the gamma catcher: a volume with undoped scintillator to capture any gamma escaping the target. The outside region contains the photo-multiplier tubes (PMT) and is filled with non-scintillating oil. This design suppresses a large fraction of the radioactivity coming from the PMTs and surrounding material. The detectors are also surrounded by active veto detectors and passive shielding to further suppress natural radioactivity and backgrounds related to cosmic muon interactions.

	Power $[GW_{th}]$	L_{Near} [m]	L_{Far} [m]	M_{target} [ton]	σ_{stat} [%]	σ_{syst} [%]	$\sin^2 2\theta_{13}$ > (90% C.L.)
Double Chooz	8.6	400	1050	8.3	0.5	0.6	0.03
RENO	17.3	290	1380	16	0.3	0.5	0.02
Daya Bay	17.4	360 (500)	1990 (1620)	80	0.2	0.4	0.01

Table 1. Total reactor power, distance L to near and far detectors and far detector target mass M_{target} of the Double Chooz, RENO and Daya Bay experiments. The distance for Daya Bay is to the DB plant, the ones between brackets to the LA and LA II plants. The expected statistical and systemic uncertainties and sensitivity to $\sin^2 2\theta_{13}$ at 90% confidence level are given on the right.

The typical signature of a prompt signal plus a delayed energy deposit consistent with neutron capture on Gadolinium is used to select neutrino candidates. The background events can be categorized into two classes. The accidental backgrounds consist of a positron-like signal from natural radioactivity present in the PMTs, the detector materials and the surrounding rock, followed close in time by an independent neutron-like signal. These can be actual neutrons generated by cosmic muons interacting in the material surrounding the detector, or high energy gammas mimicking a neutron capture on Gadolinium. The correlated backgrounds are produced by one single event with multiple energy deposits. A fast neutron produced by cosmic muons slows down by recoiling on the protons in the scintillator producing electron-like signals until it gets captured by the Gadolinium. Other sources are long lived isotopes produced in high energetic showering cosmic muon interactions undergoing a β-n decay. The isotopes have typical livetimes of $\mathcal{O}$(100ms) and are impossible to eliminate by vetoing every cosmic muon. The largest contribution comes from ^{9}Li, but other isotopes like ^{8}He and ^{11}Li also contribute.

2.3. Future experiments

The three experiments currently under construction are described below and summarized in Table 1. The expected sensitivities mentioned are always upper limits on $\sin^2 2\theta_{13}$ at 90% confidence level in case no signal is observed.

Double Chooz. The experiment [10] is located near the Chooz-B Power Plant in Northern France, which consists of two reactor cores operating at a total power of 8.6 GW_{th}. The near detector will be located at a distance of about 400 m from the cores. The far detector is being constructed in the same hall as the CHOOZ experiment at a distance of 1050 m. The target regions of both detectors will be 8.3 ton.

The far detector is expected to be operational by spring 2010 for the first phase with far detector only running. After one year, the expected sensitivity to $\sin^2 2\theta_{13}$ will be 0.06. The second phase of the experiment with both detectors operational will start in 2011, with a sensitivity to $\sin^2 2\theta_{13}$ of 0.03 after three years of data taking. The expected sensitivity over time can be seen in the left plot of Figure 1. More details about and the status of the experiment are covered in [11].

RENO. The experiment [12] is being built near the YeongGwang power plant in the south-west of South Korea. The 6 operational cores lie on one axis and produce a total power of 17.3 GW_{th}. The target regions of both detectors will be 20 ton. The near detector hall is located at 290 m from the closest point to the reactor cores axis and the far detector hall is at a distance of 1380 m.

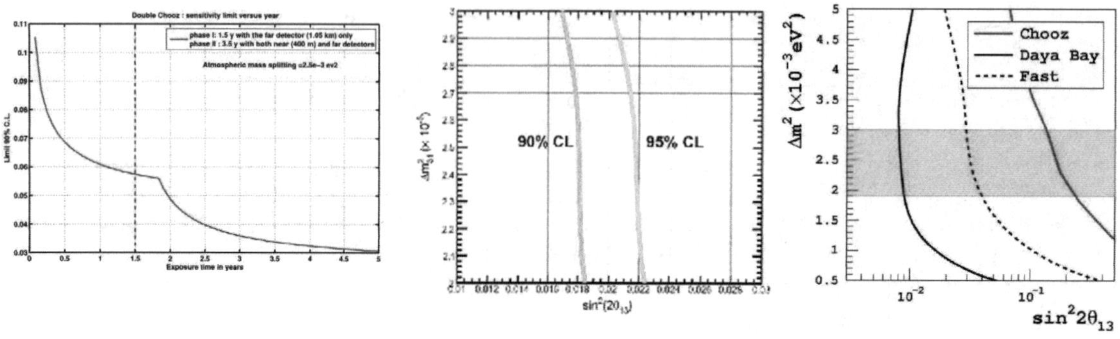

Figure 1. The expected sensitivity to $\sin^2 2\theta_{13}$ at 90% C.L.: as a function of running time for Double Chooz (left) where the near detector becomes operational after 1.5 years of far detector only running; as function of Δm^2_{31} for RENO (middle) and Daya Baye (right) after three years of data taking with all detectors and reactors fully operational.

Both detectors will be ready for data taking during the early months of 2010. The expected sensitivity to $\sin^2 2\theta_{13}$ after 3 years will be 0.02, and can be seen as function of Δm^2_{31} in the middle plot of Figure 1.

Daya Bay. The experiment [13] will use neutrinos produced by the Daya Bay (DB) and Ling Ao (LA) Power plants located in just north-east of Hong Kong. In 2011, the LA II power plant will be operational resulting in a total power of 17.4 $\mathrm{GW_{th}}$ produced by six cores. A total of 4 identical modules measuring 20 ton each will be placed at the far hall, 1990 m and 1620 m away from the DB and LA reactors, respectively. Two modules will be placed at a distance of about 360m from the DB plant, and another two at about 500m from the LA plants.

The far hall will be ready for data taking in Summer 2011. After 3 years with all detector modules operational and 6 reactor cores at full power, the sensitivity to $\sin^2 2\theta_{13}$ is expected to reach 0.01, which can be seen from the right plot in Figure 1. More details about and the status of the experiment are covered in [14].

3. The MiniBooNE experiment

3.1. Experimental setup

The MiniBooNE experiment [15] is designed to address ν_e appearance in a ν_μ beam with the same L/E value as LSND but under different experimental conditions. A 8.9 GeV proton beam from the Booster at FNAL hits a beryllium target. The produced hadrons are focused in the forward direction by a magnetic horn. They are allowed to decay in a tunnel followed by dirt to absorb all particles but neutrinos. This results in a ν_μ beam peaked around 700 MeV. The polarity of the horn can be reversed to select a $\bar{\nu}_\mu$ beam. An 800 ton mineral oil Čerenkov detector is placed at about 500 m downstream the neutrino beam. A short overview of the results are given in the next section. More details are covered in [16].

3.2. Results

$\nu_\mu \rightarrow \nu_e$ appearance. The full data sample of $6.5 \cdot 10^{20}$ protons-on-target (PoT) taken in neutrino mode is analyzed, selecting charged current quasi-elastic ν_e interactions. No excess of events was observed in the signal region above 475 MeV corresponding to the LSND results [17]. This rules out the 2-neutrino oscillation hypothesis as source of the LSND excess (assuming no CP & CPT violation). The resulting exclusion limit can be seen in the left plot of Figure 2.

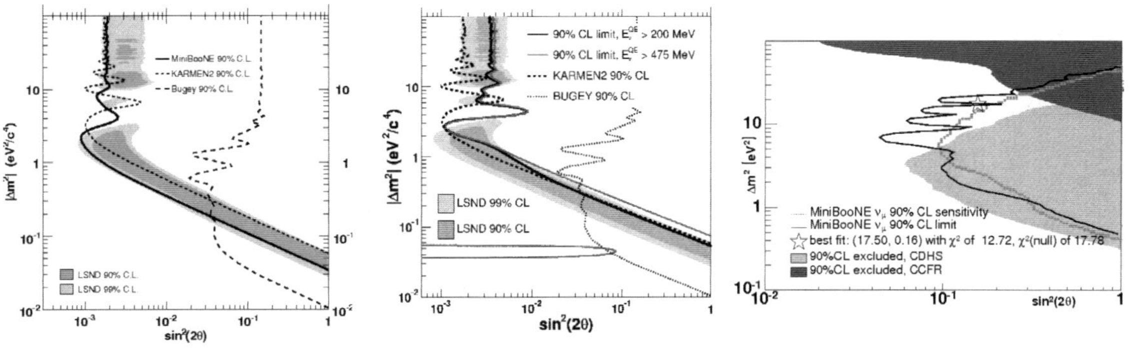

Figure 2. The observed MiniBooNE exclusion limits on $\nu_\mu \to \nu_e$ (left) and $\bar{\nu}_\mu \to \bar{\nu}_e$ (middle) appearance and on ν_μ disappearance (right).

However, an excess of 129 ± 43 events was observed above background at energies between 200 and 475 MeV [18]. The size of the signal is similar to the LSND result but the shape of the energy spectrum is not consistent with 2 neutrino oscillations. The effect is also observed looking at far off-axis neutrinos coming from the NuMI beam [19]. An improvement in the latter analysis is expected, including a larger data sample and reduced systematic uncertainties.

$\bar{\nu}_\mu \to \bar{\nu}_e$ appearance. The anti-neutrino mode is a direct test of the LSND result. Results from a limited data sample of $3.4 \cdot 10^{20}$ PoT show no significant excess over background, neither in the LSND region, nor at lower energies [20]. The obtained exclusion limits are show in the middle plot of Figure 2. In total $5.1 \cdot 10^{20}$ PoT are recorded in anti-neutrino mode and $5 \cdot 10^{20}$ PoT more are expected. Such a threefold increase in the data sample should shine more light on these inconclusive results.

ν_μ and $\bar{\nu}_\mu$ dispearance. Looking at this channel set limits in previously unexplored parameter space [21], which can be seen in the right plot of Figure 2 for the ν_μ channel. The results will be updated including the data from the SciBooNE experiment, a fine-grained tracking detector located about 100m from the target and which took data from June 2007 until August 2008.

3.3. Beyond MiniBooNE

The MicroBooNE collaboration [22] proposes to build a 70 ton Liquid Argon TPC near MiniBooNE as an advanced R&D project and to investigate the observed low energy excess. If all funding is assured, the project can start to take data as early as 2011. Currently, a 170L LAr TPC, called ArgoNeut, is taking data in the NuMI beam in front of the MINOS detector [23].

OscSNS [24] is a proposal to place a MiniBooNE-like detector at 60m from the target of the Spalation Neutron Source at ORNL, a 1 GeV, pulsed (60 Hz), 1.4 MW proton beam produces π^+s which decay at rest. The experiment is expected to have a 15 times better sensitivity than the LSND experiment.

4. Cross section measurements

The intense neutrino beams also allow for much improved cross section measurements. These are in particular important for future neutrino oscillation experiments operating in the few GeV range. The currently ongoing experiments are MiniBooNE and SciBooNE [25] in the Booster neutrino beam (0.4 - 2 GeV) and MINOS [26] in the NuMI beam (1 - 20 GeV). One of the many updated and new measurements reported on at the recent NuInt09 workshop [27] is the quasi-elastic (CCQE) cross section. All ongoing experiments prefer a value of M_A around 1.35 GeV[28, 29, 30], where M_A is the axial form factor in the relativistic Fermi Gas Model. These measurements on carbon and iron are higher than previous results on D_2, and on carbon at higher energies by NOMAD [31]. The MINERvA experiment [32] will be able to study these discrepancies in much more detail when it starts operating early 2010. This dedicated neutrino scattering experiment, located in front of the MINOS near detector, has the ability to measure cross sections on different target materials in the energy range of 1 - 20 GeV. Also the T2K [33] and NOvA [34] near detectors will provide valuable cross section measurements around 0.8 GeV and 2 GeV, respectively.

References

[1] Z. Maki, M. Nakagawa and S. Sakata, Prog. Theor. Phys. **28** (1962) 870.
[2] B. Pontecorvo, Sov. Phys. JETP **26** (1968) 984 [Zh. Eksp. Teor. Fiz. **53** (1967) 1717].
[3] Y. Fukuda *et al.* [Super-Kamiokande Collaboration], Phys. Rev. Lett. **81** (1998) 1562.
[4] G. L. Fogli, E. Lisi, A. Marrone, A. Palazzo and A. M. Rotunno, Nucl. Phys. Proc. Suppl. **188** (2009) 27.
[5] M. Apollonio *et al.* [CHOOZ Collaboration], Eur. Phys. J. C **27** (2003) 331.
[6] A. Palazzo, these proceedings.
[7] A. Aguilar *et al.* [LSND Collaboration], Phys. Rev. D **64** (2001) 112007.
[8] K. Schreckenbach, G. Colvin, W. Gelletly and F. Von Feilitzsch, Phys. Lett. B **160** (1985) 325.
[9] A. A. Hahn *et al.* Phys. Lett. B **218** (1989) 365.
[10] F. Ardellier *et al.* [Double Chooz Collaboration], *Preprint* arXiv:hep-ex/0606025.
[11] J. Maricic, these proceedings.
[12] S. B. Kim [RENO Collaboration], AIP Conf. Proc. **981** (2008) 205 [J. Phys. Conf. Ser. **120** (2008) 052025].
[13] X. Guo *et al.* [Daya-Bay Collaboration], *Preprint* arXiv:hep-ex/0701029.
[14] K. M. Heeger, these proceedings.
[15] A. A. Aguilar-Arevalo *et al.* [MiniBooNE Collaboration], Nucl. Instrum. Meth. A **599** (2009) 28.
[16] M. H. Shaevitz, these proceedings.
[17] A. A. Aguilar-Arevalo *et al.* [The MiniBooNE Collaboration], Phys. Rev. Lett. **98** (2007) 231801.
[18] A. A. Aguilar-Arevalo *et al.* [MiniBooNE Collaboration], Phys. Rev. Lett. **102** (2009) 101802.
[19] P. Adamson *et al.* [MiniBooNE and Minos Collaboration], Phys. Rev. Lett. **102** (2009) 211801.
[20] A. A. Aguilar-Arevalo *et al.* [MiniBooNE Collaboration], Phys. Rev. Lett. **103** (2009) 111801.
[21] A. A. Aguilar-Arevalo *et al.* [MiniBooNE Collaboration], Phys. Rev. Lett. **103** (2009) 061802.
[22] http://www-microboone.fnal.gov/
[23] J. Spitz, these proceedings.
[24] http://physics.calumet.purdue.edu/~oscsns/
[25] A. A. Aguilar-Arevalo *et al.* [SciBooNE Collaboration], *Preprint* arXiv:hep-ex/0601022.
[26] D. G. Michael *et al.* [MINOS Collaboration], Nucl. Instrum. Meth. A **596** (2008) 190.
[27] 6th International Workshop on Neutrino-Nucleus Interactions in the Few-GeV Region (NuInt09), http://nuint09.ifae.es/Welcome.html
[28] J. L. Alcaraz-Aunion and J. Walding, *Preprint* arXiv:0909.5647, to appear in the proceedings of NuInt09.
[29] T. Katori [MiniBooNE Collaboration], *Preprint* arXiv:0909.1996, to appear in the proceedings of NuInt09.
[30] M. Dorman, to appear in the proceedings of NuInt09.
[31] V. Lyubushkin *et al.* [NOMAD Collaboration], Eur. Phys. J. C **63** (2009) 355.
[32] D. Drakoulakos *et al.* [Minerva Collaboration], *Preprint* arXiv:hep-ex/0405002.
[33] Y. Itow *et al.* [The T2K Collaboration], *Preprint* arXiv:hep-ex/0106019.
[34] D. S. Ayres *et al.* [NOvA Collaboration], *Preprint* arXiv:hep-ex/0503053.

Topics in Astroparticle and Underground Physics (TAUP 2009) IOP Publishing
Journal of Physics: Conference Series **203** (2010) 012012 doi:10.1088/1742-6596/203/1/012012

Status and Prospect of Atmospheric Neutrinos and Long Baseline Studies

Takaaki Kajita

ICRR and IPMU, University of Tokyo, Kashiwa-no-ha 5-1-5, Kashiwa Chiba 277-8582, Japan

E-mail: kajita@icrr.u-tokyo.ac.jp

Abstract. Experimental studies of neutrino oscillations by atmospheric and long baseline experiments are discussed. Present data from these experiments are explained well with 2 flavor $\nu_\mu \to \nu_\tau$ oscillations. Status and prospect for the non-zero θ_{13} searches are discussed. Once a finite θ_{13} is discovered, future experiments will study the CP violation and the mass hierarchy. Possible future experiments on these topics are discussed.

1. Introduction

Small but finite neutrino masses are believed to be related to physics at a very high energy scale [1], [2], [3]. The observed large neutrino mixing angles might be a hint for understanding physics at the very high energies [4]. Furthermore, the physics of neutrino masses might be related to the origin of the baryon asymmetry of the Universe [5].

One of the most sensitive methods to observe small neutrino masses is to study neutrino flavor oscillations [6], [7]. If neutrinos have finite masses, each flavor eigenstate (for example, ν_μ) can be expressed by a combination of mass eigenstates (ν_1, ν_2 and ν_3). By measuring the neutrino oscillations, it is possible to get information on the mass-squared difference of the neutrino mass eigenstates and the mixing angle.

It is already more than 10 years since neutrino oscillations were discovered [8]. In this paper, we discuss the present status and the future prospect of neutrino oscillation studies with atmospheric and long baseline experiments. First, we discuss 2 flavor $\nu_\mu \to \nu_\tau$ oscillations. Later we extend our discussion to 3 flavor oscillations.

2. Status of atmospheric neutrino experiments

Atmospheric neutrino experiments have played a fundamental role in the discovery of neutrino oscillations [9], [10], [11], [8], [12], [13]. As of this writing, the statistics of the atmospheric neutrino events are dominated by the data from Super-Kamiokande (Super-K). Therefore, we discuss the atmospheric neutrino results from Super-K. [1]

Super-K have several stages. In the summer 2009, data from Super-K-III are used for the final physics analysis for the first time. The total exposure of the detector for fully-contained atmospheric neutrino analysis is 173 kton·yr. In the Super-K data, zenith-angle and energy

[1] After the conference, the author learned that the new results from SNO on upward going muons [14] have been published. The neutrino induced down-going muon flux has been measured, due to the great depth of the detector. They constrain the flux of high energy atmospheric neutrinos.

dependent deficit of ν_μ charged current (CC) events has been observed. See Fig. 1 (left), which shows some sub-samples of the zenith angle distributions from Super-K. By fitting the zenith-angle distributions from various sub-samples [15], the oscillation parameters are accurately measured.

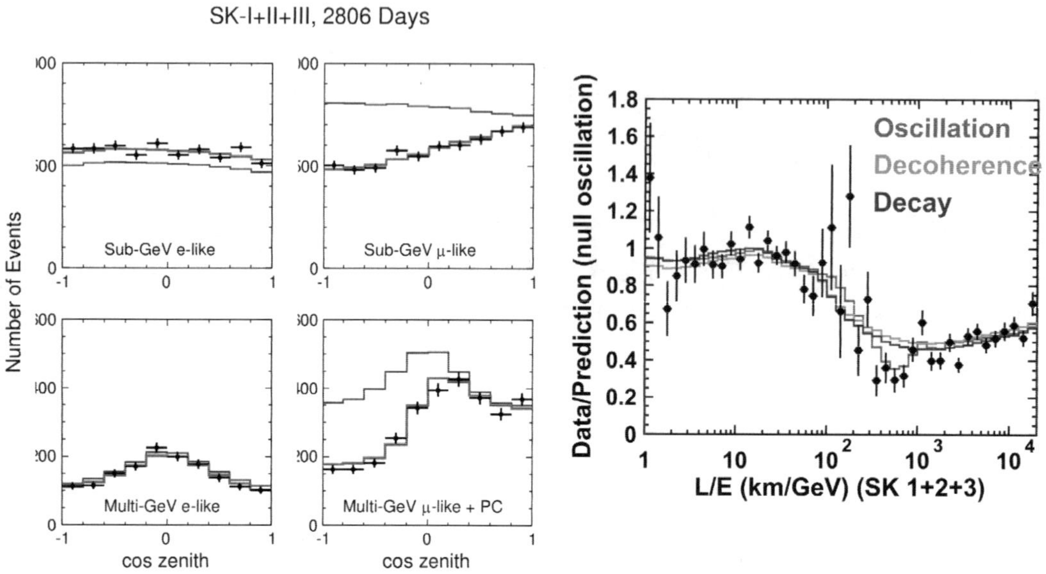

Figure 1. Left: Zenith angle distributions observed in Super-K-I+II+III (173 kton·year exposure, preliminary). Top panels show fully-contained (FC)) sub-GeV e-like (left) and μ-like (right) events. The bottom panels show FC multi-GeV e-like (left) and FC multi-GeV μ-like plus partially-contained events (right). Sub- and multi-GeV are defined to be less and more than 1.33 GeV visible energy, respectively. The blue histograms show the prediction without oscillations. The red histograms show the best fit $\nu_\mu \to \nu_\tau$ expectation. Right: L/E distributions observed in Super-K-I+II+III. The red, blue and green histograms show the best-fit predictions based on neutrino oscillation, decay and decoherence models, respectively.

Although, the Super-K zenith-angle data are explained well by oscillations, the sinusoidal oscillation pattern is not seen in the left panel of Fig. 1 due to the smearing. Therefore, Super-K carried out a dedicated L/E analysis using only high L/E resolution events [16]. Figure 1 (right) shows the L/E distributions from Super-K-I+II+III. The L/E distribution shows a dip around $L/E = 500$ km/GeV that corresponds to the first oscillation minimum. The distribution is used to compare the oscillation and other hypotheses. The neutrino decay [17] and decoherence [18] models are excluded at about 5 standard deviations. Due to the observation of the dip, the allowed oscillation parameter, especially Δm^2_{23}, region is also constrained strongly.

3. Long baseline experiments

Long baseline experiments have a fixed, well-known neutrino flight length. Therefore, in principle, long baseline neutrino oscillation experiments are much suited to estimate the Δm^2_{23} value accurately than the atmospheric neutrino experiments.

K2K was the first long baseline neutrino oscillation experiment. Neutrinos were produced using a 12 GeV proton beam at KEK. These neutrinos were detected in Super-K, which was located 250 km distant from the neutrino production point. K2K observed an energy dependent deficit of muon neutrino events. The data were consistent with neutrino oscillations, excluding no oscillation at 4.3 σ [19].

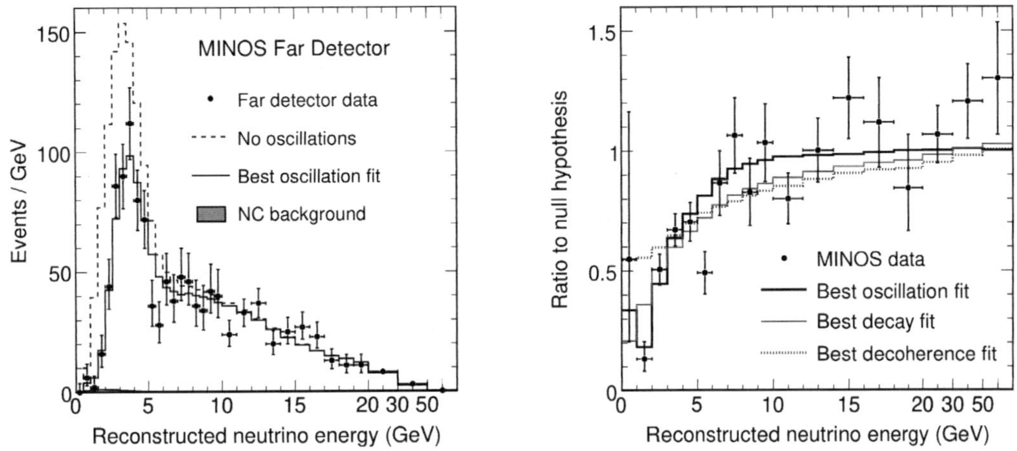

Figure 2. Left: Reconstructed neutrino energy spectrum in MINOS using selected CC ν_μ events. The dashed and solid histograms show the predicted spectrum for no oscillation and $\nu_\mu \to \nu_\tau$ oscillations, respectively. Right: Data/(no oscillation prediction) as a function of the reconstructed neutrino energy. (From Ref. [20])

The next long baseline neutrino oscillation experiment is MINOS. Neutrinos are produced by the 120 GeV Main Injector at Fermilab. The MINOS far detector is a 5.4 kton magnetized tracking detector, which is located 735 km away from the neutrino production point. In this experiment, the mean neutrino energy is tunable by adjusting the target position relative to the horn. Based on the presently known Δm^2_{23} value, MINOS uses the low-energy beam in order to get the best sensitivity.

The MINOS beam experiment started in 2005. The MINOS results have been reported using data until May 2007 [20]. The total number of protons on target used in the analyses so far is approximately 3.4×10^{20}.

Figure 2 shows the reconstructed neutrino energy spectrum for the selected CC ν_μ candidates. A very clear deficit of events is observed with the dip position at around 1.5 to 2.0 GeV. The observed dip position makes it possible to measure Δm^2_{23} accurately. The observed energy spectrum also indicates that the neutrino disappearance probability obeys sinusoidal function as predicted by neutrino oscillations.

Figure 3 shows the estimated allowed regions of 2-flavor $\nu_\mu \to \nu_\tau$ oscillation parameters from the atmospheric and long baseline neutrino oscillation experiments. We find that the allowed regions overlap very well, suggesting further that the standard neutrino oscillation scenario is valid. From this figure, we find that Δm^2_{23} is most accurately measured by the MINOS experiment, while the mixing parameter is still measured most accurately by the Super-K atmospheric neutrino experiment.

Finally, it should be mentioned that MINOS is a magnetized detector, and is able to measure neutrino and anti-neutrino events separately. Using this feature, MINOS has estimated the allowed region of the oscillation parameters based on the anti-neutrinos in the beam. The obtained result is essentially consistent with that based on the neutrino events [21].

4. Search for non-zero θ_{13}

We discuss the 3 flavor oscillation effects. For simplicity, we neglect the effect of the oscillations related to Δm^2_{12}. Under this approximation, there are only three oscillation parameters; θ_{13}, θ_{23}

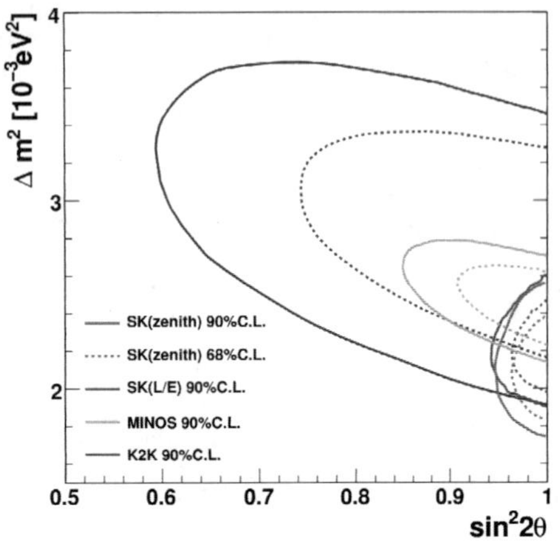

Figure 3. Estimated allowed regions of $\nu_\mu \to \nu_\tau$ oscillation parameters from the atmospheric and long baseline neutrino oscillation experiments at 68%C.L. (dotted) and 90%C.L. (solid).

and Δm^2_{13} $(= \Delta m^2_{23})$. In this case, for example, the $\nu_e \to \nu_e$ and $\nu_\mu \to \nu_e$ oscillation probabilities in vacuum can be written as:

$$P(\nu_e \to \nu_e) = 1 - \sin^2 2\theta_{13} \sin^2\left(\frac{1.27\Delta m^2_{13} L}{E_\nu}\right), \tag{1}$$

$$P(\nu_\mu \to \nu_e) = \sin^2\theta_{23} \sin^2 2\theta_{13} \sin^2\left(\frac{1.27\Delta m^2_{13} L}{E_\nu}\right). \tag{2}$$

The CHOOZ experiment searched for $\overline{\nu}_e$ disappearance in the reactor $\overline{\nu}_e$ flux based on Eq.1. The observed flux was consistent with the no-oscillation expectation. The upper limit on $\sin^2 2\theta_{13}$ was ~0.15 [22].

Recently, there are considerable interest in the experimental results on θ_{13}. There is a slight tension between the solar and reactor measurements of θ_{12}. If a finite value of θ_{13} is allowed, the tension disappears [23], [24]. Also, the MINOS results on the ν_e appearance search show slight excess above the expected background [21]. These results suggest a relatively large θ_{13} angle. However, due to the limited sensitivities, it is not possible to draw any clear conclusion with the present data. Much more sensitive experiments might be required to observe non-zero θ_{13} unambiguously. If a finite value of θ_{13} is observed, we understand the overall structure of the neutrino mixing matrix (except for the CP phase). Thus, θ_{13} should be measured with a high priority. Furthermore, the measurement of θ_{13} is very important for future neutrino oscillation experiments, which will be discussed in the next section.

Atmospheric neutrino experiments are also sensitive to θ_{13}. The new results from Super-K do not show any evidence for non-zero θ_{13} [25], and the 90%C.L. allowed regions for the normal and inverted mass hierarchies are still weaker than the CHOOZ limit. In any case, future atmospheric neutrinos with much larger exposure are expected to be sensitive to θ_{13} and other sub-dominant oscillation effects [26], [27]. These studies reviled that atmospheric neutrino experiments are very important if the value of θ_{13} is relatively large (as suggested by some recent data).

The accelerator long baseline experiments search for ν_e appearance as the evidence for non-zero θ_{13}. See Eq.2 for the approximate appearance probability. The T2K experiment [28] and the NOvA experiment [29] have high sensitivities to the ν_e appearance.

T2K uses a high intensity, sub-GeV neutrino beam produced by the 30 GeV PS in the J-PARC accelerator complex. The designed beam power is 0.75 MW. The far detector is Super-

Kamiokande. The baseline length is 295 km. The neutrino energy is tuned to the maximum oscillation energy using the off-axis beam technique. The proton beam was successfully accelerated to 30 GeV and extracted to the neutrino beam line hitting the neutrino production target. The T2K experiment will start the physics run either at the end of 2009 or in early 2010.

NOvA is the long baseline experiment in the United States. This experiment plans to use the existing NuMI beam line at Fermilab. A beam power will be upgraded to about 1 MW. The far detector is a 15 kton fully-active, finely-segmented liquid scintillator detector. The baseline length is 810 km. In order to get the high flux at the oscillation maximum, this experiment also uses the off-axis technique. The peak neutrino energy is about 2 GeV. The NOvA experiment is under construction and is expected to start taking data before the mid 2010's.

These experiments have similar sensitivities in $\sin^2 2\theta_{13}$. Assuming 5 years of data taking with the planed beam intensities, these experiments can find clear evidence for a non-zero θ_{13}, if the true value of $\sin^2 2\theta_{13}$ is larger than ~ 0.02. If there is no evidence for a non-zero θ_{13}, they can set the upper limit of $\sim$0.01 on $\sin^2 2\theta_{13}$.

5. Oscillation Physics Beyond θ_{13}

We assume that the next generation oscillation experiments will observe evidence for a non-zero θ_{13}, suggesting $\sin^2 2\theta_{13}$ to be larger than $\sim$0.01. In this case, we know the approximate values of the three mixing angles. However, we believe that this should not be the end of the neutrino oscillation studies. In fact, there are still important and unknown neutrino oscillation parameters. One is the CP phase. According to the leptogenesis scenario [5], the seed of the baryon asymmetry in the Universe is generated by the CP violating decay of the heavy Majorana particles of the see-saw mechanism. Therefore, it is generally believed that the measurement of the CP violation in the neutrino sector is a very important step forward for the understanding of the baryon asymmetry in the Universe. In addition to the CP violation, we should note that we do not know whether the ν_3 mass eigenstate is the heaviest or the lightest. This is also an important question to be addressed experimentally.

If θ_{13} is non-zero, future large-scale long-baseline experiments can address these questions. Various possibilities for these experiments have been studied. One possibility for such experiments could be the Phase-2 of the T2K project [28]. The CP violation phase can be measured by observing the difference in the neutrino oscillation probabilities between $\nu_\mu \rightarrow \nu_e$ and $\overline{\nu}_\mu \rightarrow \overline{\nu}_e$. To observe this effect, a huge detector (0.54 Mton fiducial mass, Hyper-Kamiokande), and a beam power upgrade of J-PARC will be required. Due to the relatively short baseline length of 295 km and therefore due to a relatively small matter effect, the mass hierarchy may not be determined in this setup. The T2K experiment uses an off-axis beam. It is noticed that, with the present T2K beam-line configuration, the beam is simultaneously available in Kamioka (which is 295 km away from the target) and Korea (which is more than 1000 km away from the target). Studies have been carried out assuming the two detectors in Kamioka and Korea with 0.27 Mton fiducial mass for both detectors [30], [31].

The other study has been carried out in the United States, which assumed a 1300 km baseline between Fermilab and DUSEL at Homestake [32]. The beam power could be as high as 2 MW. The detectors used in this study are a 300 kton water Cherenkov detector and a 100 kton liquid Argon detector.

From these studies, it is understood that the CP phase and mass hierarchy can be determined by these future experiments if $\sin^2 2\theta_{13}$ is larger than $\sim$0.01 and if the baseline length is 1000 km or longer.

6. Summary

Studies of atmospheric neutrinos led to the discovery of neutrino oscillations in 1998. Since then, many atmospheric and long-baseline studies have contributed to the understanding of oscillations. Dominant $\nu_\mu \to \nu_\tau$ is already studied accurately. However, it is known that there are many interesting physics in the sub-dominant oscillations. They include θ_{13}, the CP violated and the mass hierarchy. Therefore there are intense activities to study these effects. The immediate goal is the search for non-zero θ_{13}. Experiments to search for non-zero θ_{13} will start within a year or so. The subsequent big goals are the measurement of the CP violation and the determination of the mass hierarchy. The world community is working hard for the best strategy for these measurements.

Acknowledgments

We would like to thank the organizers of this conference for the kind invitation. Also, we would like to thank many people who provided the information for the talk; A.Habig, M.Messier, N.Mondal, R.Wendel, C.Ishihara, F.Dufour, K.Okumura, A.Ichikawa, J.Maricic. This work is partly supported by the Grant-Aid in Scientific Research of JSPS.

References

[1] Minkowski P 1977 *Phys. Lett.* **B67** 421
[2] Yanagida T In Proceedings of the Workshop on Unified Theories and Baryon Number in the Universe, eds. O. Sawada and A. Sugamoto (KEK report 79-18, 1979)
[3] Gell-Mann M, Ramond P and Slansky R In Supergravity, eds. P. van Nieuwenhuizen and D.Z. Freedman (North Holland, Amsterdam, 1979)
[4] Valle J W F In these Proceedings.
[5] Fukugita M and Yanagida T 1986 *Phys. Lett.* **B174** 45
[6] Maki Z, Nakagawa M and Sakata S 1962 *Prog. Theor. Phys.* **28** 870
[7] Pontecorvo B 1968 *Sov. Phys. JETP* **26** 984–988
[8] Fukuda Y *et al.* (Super-Kamiokande) 1998 *Phys. Rev. Lett.* **81** 1562–1567 (*Preprint* hep-ex/9807003)
[9] Hirata K S *et al.* (KAMIOKANDE-II) 1988 *Phys. Lett.* **B205** 416
[10] Becker-Szendy R *et al.* 1992 *Phys. Rev.* **D46** 3720–3724
[11] Fukuda Y *et al.* (Kamiokande) 1994 *Phys. Lett.* **B335** 237–245
[12] Sanchez M C *et al.* (Soudan 2) 2003 *Phys. Rev.* **D68** 113004 (*Preprint* hep-ex/0307069)
[13] Ambrosio M *et al.* (MACRO) 2004 *Eur. Phys. J.* **C36** 323–339
[14] Aharmim B *et al.* (SNO) 2009 *Phys. Rev.* **D80** 012001 (*Preprint* 0902.2776)
[15] Ashie Y *et al.* (Super-Kamiokande) 2005 *Phys. Rev.* **D71** 112005 (*Preprint* hep-ex/0501064)
[16] Ashie Y *et al.* (Super-Kamiokande) 2004 *Phys. Rev. Lett.* **93** 101801 (*Preprint* hep-ex/0404034)
[17] Barger V D *et al.* 1999 *Phys. Lett.* **B462** 109–114 (*Preprint* hep-ph/9907421)
[18] Lisi E, Marrone A and Montanino D 2000 *Phys. Rev. Lett.* **85** 1166–1169 (*Preprint* hep-ph/0002053)
[19] Ahn M H *et al.* (K2K) 2006 *Phys. Rev.* **D74** 072003 (*Preprint* hep-ex/0606032)
[20] Adamson P *et al.* (MINOS) 2008 *Phys. Rev. Lett.* **101** 131802 (*Preprint* 0806.2237)
[21] Habig A f t M c In these Proceedings.
[22] Apollonio M *et al.* (CHOOZ) 2003 *Eur. Phys. J.* **C27** 331–374 (*Preprint* hep-ex/0301017)
[23] Fogli G L, Lisi E, Marrone A, Palazzo A and Rotunno A M 2008 *Phys. Rev. Lett.* **101** 141801 (*Preprint* 0806.2649)
[24] Schwetz T, Tortola M and Valle J W F 2008 *New J. Phys.* **10** 113011 (*Preprint* 0808.2016)
[25] Wendell R f t S K c In these Proceedings.
[26] Kajita T, Ishitsuka M, Okumura K, Obayashi Y and Shiozawa M Prepared for 5th Workshop on Neutrino Oscillations and their Origin (NOONE2004), Tokyo, Japan, 11-15 Feb 2004
[27] Sajjad Athar M and others (INO collaboration) 2006
[28] Itow Y *et al.* (The T2K) 2001 Hep-ex/0106019 (*Preprint* hep-ex/0106019)
[29] Ayres D S *et al.* (NOvA) 2004 Hep-ex/0503053 (*Preprint* hep-ex/0503053)
[30] Ishitsuka M, Kajita T, Minakata H and Nunokawa H 2005 *Phys. Rev.* **D72** 033003 (*Preprint* hep-ph/0504026)
[31] Dufour F 2008 Talk presented at the 4th International Workshop on Nuclear and Particle Physics at J-PARC (NP08), Mito, Japan, (Paper in preparation)
[32] Barger V *et al.* 2007 ArXiv:0705.4396 [hep-ph] (*Preprint* 0705.4396)

Neutrinos from CERN To Gran Sasso: The CNGS Project

Marcos Dracos

IPHC, Université de Strasbourg, CNRS/IN2P3, F-67037 Strasbourg, France

E-mail: marcos.dracos@ires.in2p3.fr

Abstract. The CNGS project (CERN Neutrinos to Gran Sasso) has been approved to make an unambiguous and direct observation of $\nu_\mu - \nu_\tau$ oscillations. An intense and relatively pure ν_μ beam is generated at CERN and directed towards LNGS (Laboratori Nazionali del Gran Sasso) in Italy. Two experiments, OPERA already started in 2007 and ICARUS T600, will try to detect tau neutrinos using large and complex detectors. These "appearance" experiments will use the τ decay topology as signature of the presence of ν_τ in the initial neutrino beam. An overview of the CNGS beam facility is given. Results from the first observed OPERA events and performances are presented. Few elements are given concerning the ICARUS T600 starting.

1. Introduction

Neutrino oscillations have proved that neutrinos have a non zero mass. Up to now, to prove that neutrinos oscillate, experiments have observed the disappearance of neutrinos from their production up to their detection. The CNGS project have been proposed to observe for the first time the appearance of ν_τ in a ν_μ beam and prove without ambiguity that the neutrino disappearance of one flavour eigenstate is due to the oscillation to another flavour eigenstate. This direct observation of neutrino oscillations will remove theoretical speculations about other possible mechanisms to explain the neutrino disappearance already observed by several experiments.

2. CNGS facility

For the CNGS facility [1, 2, 3], the neutrino beam is produced using a high intensity 400 GeV proton beam extracted from the CERN SPS accelerator. The protons collide with a graphite target producing mostly pions and kaons which decay through $\pi^{+(-)} \rightarrow \mu^{+(-)} + \nu_\mu(\bar{\nu_\mu})$ and $K^{+(-)} \rightarrow \mu^{+(-)} + \nu_\mu(\bar{\nu_\mu})$. The positively charged π/K are guided using two focusing lenses (horn, reflector) in the direction towards Gran Sasso (Fig. 1). These mesons decay mainly into muon-neutrinos and muons in the 992 m long decay vacuum tube. The remaining particles are stopped by a hadron stop located at the end of the decay tunnel. Two muon detector stations are used to monitor the intensity and profile of the neutrino beam produced.

The graphite target has been designed to afford 7×10^{13} 400 GeV protons per 6 s cycle. Two fast extractions of the beam are performed per cycle spaced by 50 ms and lasting 10.5 μs each. This pattern can be repeated several times in a super–cycle. Typically, 3 cycles are operated during a super–cycle of about 40 s.

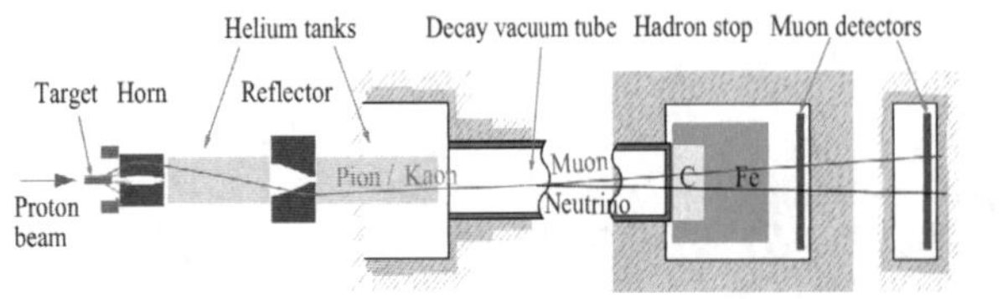

Figure 1. Schematic view of the main CNGS elements (proton beam, target, horn, reflector, decay tunnel, hadron stop and monitoring muon detectors).

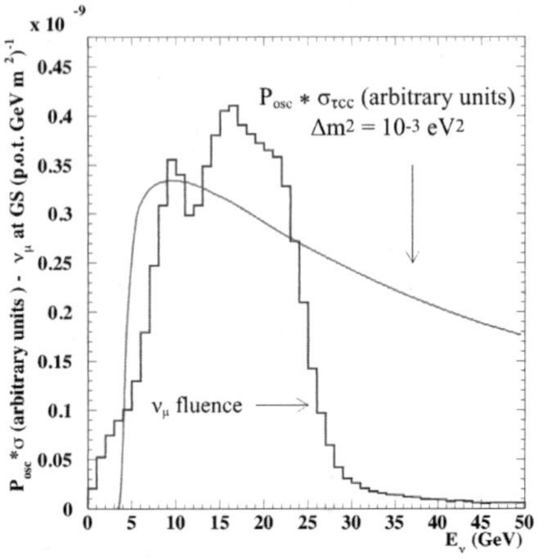

Figure 2. CNGS neutrino energy spectrum and oscillation probability multiplied by the ν_τ cross section.

The beam is strongly focused on the target ($\sigma = 0.5$ mm) and it is inclined by 3.4^o downwards in the direction of the LNGS. The target station contains a magazine equipped with 5 target units. Only one is used to produce the neutrino beam, the other four provide in-situ spares with the possibility to replace remotely the irradiated target in case of failure. Each target unit is made of 13 graphite rods (10 cm long, 5 mm diameter for the 2 first and 4 mm for the remaining 11).

Downstream of the target station, a hadron collection system has been installed to direct the extracted mesons from the target towards LNGS. This magnetic focusing system consists of 2 magnetic horns (the second horn is called reflector), pulsed during each proton beam extraction with a current of 150 kA for the fist horn and 180 kA for the second one.

Fig. 2 shows the obtained neutrino energy spectrum. On the same figure, the oscillation probability multiplied by the charged current ν_τ cross section is also depicted. The two distributions have their maximum more or less at the same position showing the good beam optimization for ν_τ appearance experiments. The mean neutrino energy is of the order of 17 GeV while the distance between CERN and LNGS is 732 km leading to an oscillation parameter L/E of the order of 43. The produced ν_μ beam contains a fraction of ν_e's and $\bar{\nu_\mu}$'s producing a charged current event contamination at the level of the experiment of the order of 0.8% and 2.0% respectively.

CNGS commissioning has been done in 2006 where 8.5×10^{17} protons have been sent on the target (p.o.t.) before experiencing a water leak problem on the second horn and few other minor problems. Repairing has been performed during the 2006 winter shutdown. In 2007, after a 6 weeks run (7.9×10^{17} p.o.t.), a failure of ventilation system installed in the target area has been observed due to radiation effects in the control electronics of this system. Due to this problem, the beam has been stopped and extra radiation shielding has been installed in order to protect more efficiently the related electronics. This run has allowed OPERA experiment to observe the first neutrino interaction in lead/emulsion bricks in Gran Sasso.

In 2008, a full run has been performed with 1.78×10^{19} p.o.t. to be compared to the optimal value of 4.5×10^{19} p.o.t. The low integrated efficiency of the neutrino facility is mainly due to frequent failures of the relatively old CERN accelerator complex. During this run, about 2×10^{13} p.o.t. were routinely obtained per fast extraction close to the nominal one of 2.4×10^{13} p.o.t.

3. OPERA experiment

The aim of OPERA experiment [4] is to observe ν_τ appearance in the CNGS ν_μ using nuclear emulsions providing the necessary granularity for this delicate detection. The detector, located in Hall C of the LNGS underground laboratory, is divided into two identical parts called super–modules (Fig. 3). Each super–module is composed by a target and a spectrometer.

The target has 31 lead/emulsion brick walls, giving a total mass of 0.64 kton, interleaved with 31 electronic Target Tracker walls. The Target Tracker walls are made of horizontal and vertical scintillating strips and their main role is to find the right bricks to extract where the neutrino interactions occur. They also provide calorimetric information about the overall deposited energy in the OPERA detector.

Each super–module's spectrometer consists of a dipolar magnet in which 22 RPC layers are installed. Inside and outside of the two magnet arms, 4 layers of drift tubes (HPT) are installed to measure precisely the charged tracks bending produced by the spectrometers magnetic field. One layer of crossed strip RPC's, called XPC's, is placed just after the last Target Tracker wall of each super–module. A VETO detector composed of two layers of glass RPC's is installed in front of the OPERA detector to reject muons (and cosmic tracks) produced mainly in the rock in front of the experiment by CNGS neutrino interactions. A detailed description of the detector can be found in [5].

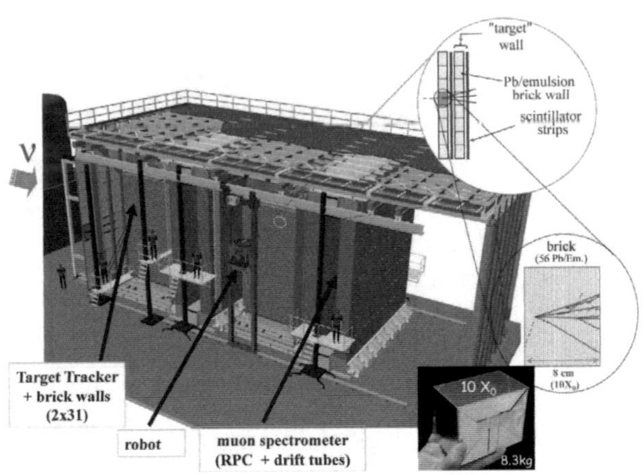

Figure 3. Schematic view of the OPERA detector.

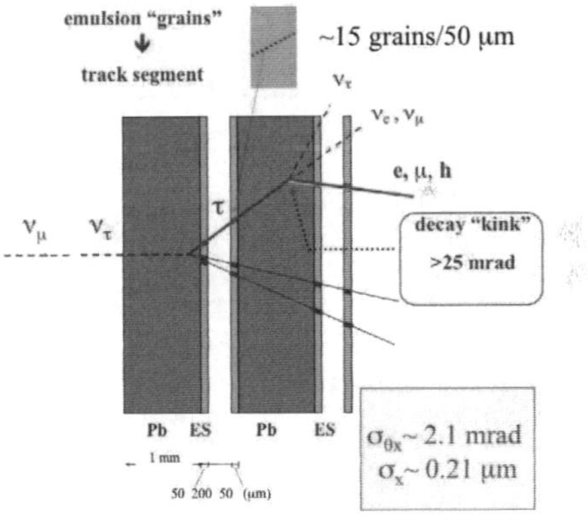

Figure 4. OPERA detection technique using layers of lead (target) and nuclear emulsions (detector).

To reconstruct the ν_τ interactions and especially the produced τ decay kink which is the signature of this kind of events (Fig. 4), a micro–metric accuracy is needed (τ path of the order of 1 mm). The solution chosen for this purpose, able to cover large surface and volume providing the needed accuracy in a cost effective way, is the utilization of nuclear emulsions.

DONUT experiment [6] has already demonstrated that the concept of stacking emulsion films with interleaved metallic plates called Emulsion Cloud Chamber (ECC), is able to detect ν_τ interactions by observing the resulting τ decays. The OPERA ECC consists of 57 emulsion sheets 300 microns thick each and 56 lead sheets of 1 mm thickness. The surface covered by each brick is 128×102 mm^2 with a total thickness of 79 mm and a weight of about 8 kg. The nuclear emulsion sheets used in OPERA are made of a transparent plastic base film for mechanical support (200 μm thick) on which are deposited two 44 μm thick layers of photosensitive gel. The intrinsic emulsion coordinate resolution is 0.2 μm.

A similar topology to τ decays is produced by the decay of charmed mesons, having comparable lifetime. This background is mainly reduced by rejecting events containing "wrong" sign muons (muons coming directly from ν_μ interactions or τ^- decays are negatively charged while those coming from charmed meson decays have a positive charge). Another way to remove this kind of events is to reject all events where a muon is found to belong to the primary vertex.

3.1. Brick processing

After the electronic detectors have indicated the candidate brick to contain a neutrino interaction, this brick is extracted from the detector by an automatic system. This system runs permanently during and after the data taking periods. The extracted bricks are not replaced leading to a total decrease of the target mass of about 15% after 5 years of data taking.

On each brick, a changeable sheet doublet (CSD) of emulsions, placed in a box attached to its downstream face, is scanned to confirm the electronic detectors prediction before opening the brick itself. If at least one track is found in the CSD, then the brick is disassembled and sent for further processing, otherwise it is equipped with a new CSD and returned back to the detector. Processing includes the deposition of fiducial marks on the emulsions by X–ray exposure, 12 hours exposure to cosmic muons (outside the underground laboratory) in order to improve the film to film alignment, development of the emulsions and scanning using optical high–magnification microscope systems.

The development of automated emulsion scanning systems during the last years allows now a scanning of about 20 cm^2/hour of emulsion surface in fully automated mode. 25 scanning tables are devoted to this scanning in Europe while 5 other high speed tables scanning with a speed higher than 50 cm^2/hour are used in Japan.

After a scan–back of tracks found in the CSD has been performed and a candidate primary vertex has been found, a volume scan around this position is performed in order to find all attached tracks and secondary vertices. After this stage, a particle identification, electromagnetic shower reconstruction and measurement of particle momentum using the Coulomb multiple scattering can be done. Combining the scanning and electronic detectors information, a decision about the nature of the observed interaction is then made.

Fig. 5 presents a charged current event detected during the short run of 2007. The left part shows a top view of the event recorded by the electronic detectors while the right part presents the region around the neutrino interaction point as reconstructed in the associated lead/emulsion brick. On this part of the figure, a decay kink is very well seen (red track) very probably produced by the decay of a charged charm meson (this particle left a visible short track in the emulsion layers before decaying). A muon attached to the primary vertex has been found proving well that this interaction has been produced by a ν_μ and thus excluding a ν_τ interaction and τ decay. A clear electromagnetic shower is also seen on this figure showing the ability of detecting this kind of topologies using the emulsion technique.

3.2. OPERA running

During the short 2007 CNGS run, OPERA collected 38 neutrino interactions while during the 2008 run 1700 interactions have been observed [7]. Up to July 2009, about 1000 bricks have been scanned from the 2008 run where no ν_τ candidates have been found, while the expected number of candidates for the whole collected events is 0.6. Already, 10 "charm" candidate events have been found showing well that the OPERA detector is able to detect this kind of topologies. Some delay observed concerning the brick scanning is mainly due to the earthquake which occurred last April in the L'Aquila region causing the LNGS closure for more than a month.

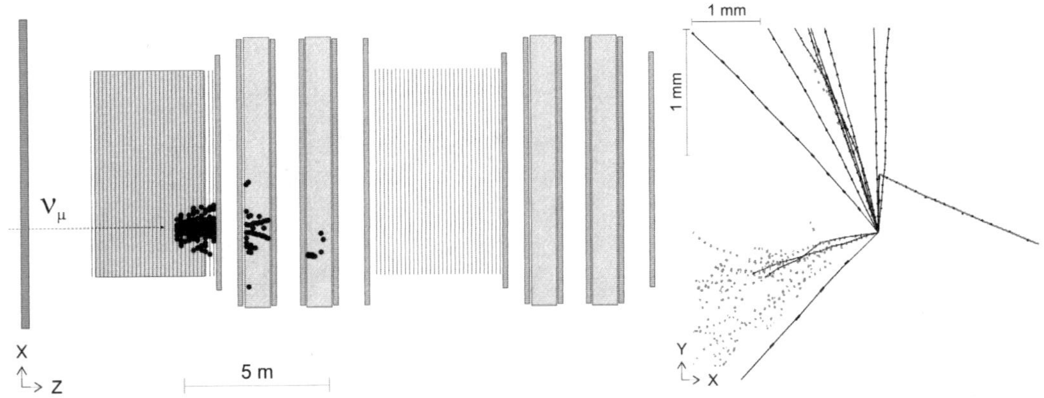

Figure 5. A candidate charm event as seen by the electronic detectors (left) and the nuclear emulsions (right). The red track is produced by a short lifetime particle supposed to be a charged charm meson.

For the 2009 run already started beginning of June, 3.2×10^{19} p.o.t. are expected to be delivered to CNGS corresponding to about 3000 events observed by OPERA. This value is by 30% lower than the nominal one and will oblige the collaboration to run more than 5 years (duration initially expected) in order to accumulate the necessary statistics corresponding to 22.5×10^{19} p.o.t. that would allow to extract statistically significant results. For 22.5×10^{19} p.o.t. and $\Delta m^2_{13} = 2.5 \times 10^{-3}$ eV^2, OPERA expects to observe 10.4 ν_τ interactions (2.9 $\tau \rightarrow \mu$, 3.5 $\tau \rightarrow e$ and 4 $\tau \rightarrow$ *hadrons*) with less than one background event.

4. ICARUS T600

ICARUS T600 [8] is a large cryostat divided in two identical, adjacent half–modules of internal dimensions $3.6 \times 3.9 \times 19.9$ m^3, each containing 300 t of liquid argon (LAr). It was initially designed to be the first part of a larger detector.

Each half–module is composed by two Time Projection Chambers (TPC), the field shaping system, monitors, probes and PMTs. They are externally surrounded by a set of thermal insulation layers (Fig. 6 and 7). Each TPC is formed by three parallel planes of wires, spaced by 3 mm, oriented at $0, \pm 60°$ angles. The parallel wires pitch is 3 mm, and their role is to detect the ionization electrons produced inside the liquid argon volume. The cathode plane is parallel and equidistant to the wire planes in each TPC. A high voltage system produces a uniform electric field, perpendicular to the wire planes, forcing the drift of the ionization electrons (the maximum drift path is 1.5 m).

The installation of ICARUS T600 is now near to completion in the LNGS underground laboratory in Hall B. The detector, the cryogenic plant and the data acquisition are almost ready. The underground laboratory infrastructures, required for the detector filling, have been completed and the T600 start–up procedures, logistics and authorizations have been almost defined. It is expected that the filling of the detector with 600 tons of liquid Argon should be done before the end of the current CNGS run allowing the experiment to detect the first CNGS neutrinos.

With its fiducial mass of 480 t liquid argon, ICARUS T600 is expected to observe about 1200 ν_μ and 7 ν_e charged current interactions per year for 4.5×10^{19} p.o.t. For 5 years data taking, 7 ± 2.6 $\tau \rightarrow e$ events are expected to be observed out of the background in a statistical way.

This detector operation will be a significant achievement opening the way to the utilization

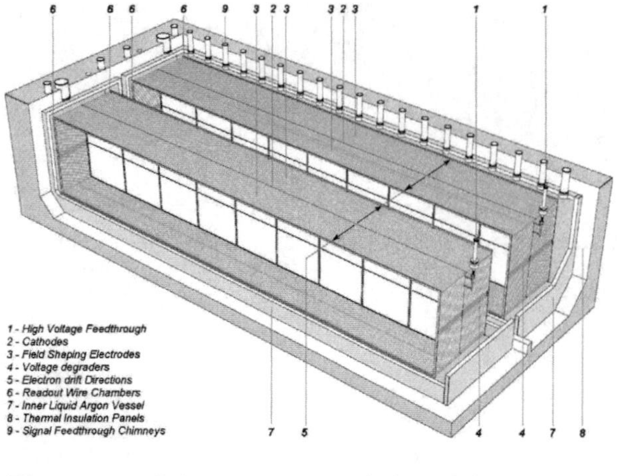

Figure 6. Schematic view of the ICARUS T600 detector.

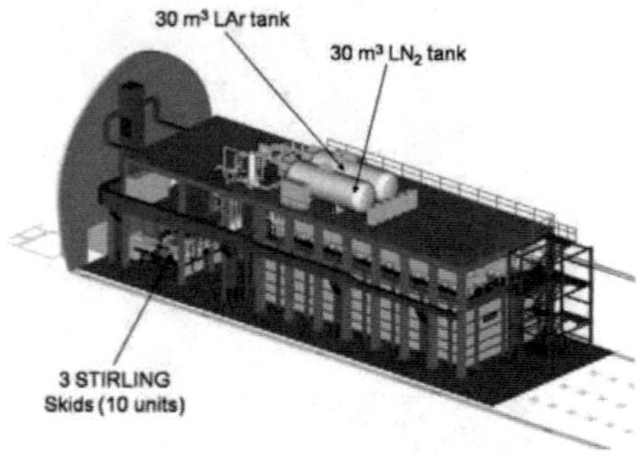

Figure 7. T600 as installed in LNGS Hall B.

of liquid argon for neutrino detection by larger detectors.

5. Conclusions

CNGS neutrino facility operates now in stable conditions providing in 2008 1.8×10^{19} p.o.t. while 3.2×10^{19} p.o.t. are expected in 2009. It has to be recalled that the nominal number of protons on the target is 4.5×10^{19}/year, significantly higher than what was delivered up to now.

OPERA experiment runs now in optimal conditions and has recorded 1700 neutrino interactions during the 2008 run. The collaboration has already scanned 1000 lead/emulsion bricks allowing to observe 10 "charm" candidate events having a similar topology to the ν_τ interactions. While no ν_τ events have been yet detected, the charm events observation proves well that the chosen detection technique is able to detect this kind of event topologies. For 2009, 3000 neutrino interactions are expected to be observed.

ICARUS T600 detector is now almost ready to be filled with liquid argon that will allow to observe soon the first CNGS neutrinos.

References

[1] General Description of the CNGS Project, 2000, http://projcngs.web.cern.ch/proj-cngs/.

[2] K.Elsener (Ed.) et al., The CERN Neutrino Beam To GRAN SASSO, Conceptual Technical Design, CERN-98-02, INFN/AE-98/05.

[3] R. Bailey et al, The CERN Neutrino To GRAN SASSO, Addendum to report CERN-98-02, INFN/AE-98/05, CERNSL/99-034 (DI), INFN/AE-99/05.

[4] OPERA proposal, *"An appearance experiment to search for $\nu_\mu \leftrightarrow \nu_\tau$ oscillations in the CNGS beam"*, CERN/SPSC 2000–028, SPSC/P318, LNGS P25/2000, July 10, 2000.

[5] OPERA collaboration, R. Acquafredda et al., The OPERA experiment in the CERN to Gran Sasso neutrino beam, 2009 JINST 4 P04018.

[6] DONUT collaboration, K. Kodama et al., Observation of tau neutrino interactions, Phys. Lett. B 504 (2001) 218 [hep-ex/0012035].

[7] OPERA collaboration, N. Agafonova et al., The detection of neutrino interactions in the emulsion/lead target of the OPERA experiment, 2009 JINST 4 P06020.

[8] ICARUS Collaboration, A second–generation proton decay experiment and neutrino observatory at the Gran Sasso Laboratory, LNGS-94/99-I (1994) and references therein.
S. Amerio et al., Nucl. Instr. Meth. A527 (2004), 329.

Neutrinoless Double Beta Decay

Stefan Schönert

Max Planck Institut für Kernphysik, Heidelberg Saupfercheckweg 1, 69115 Heidelberg, Germany

E-mail: stefan.schoenert@mpi-hd.mpg.de

Abstract. The search for neutrinoless double beta decay is the only practical way to test whether neutrinos are Majorana or Dirac particles. The next generation of experiments aim to probe the effective Majorana neutrino mass down to few 10 meV, as predicted by oscillation experiments in case of the inverse mass hierarchy. According to recent nuclear matrix calculations, the predicted decay rates per mass of double beta isotope are varying within a factor of few when comparing them within the same theoretical model framework. The sensitivity of the upcoming experiments depend therefore primarily on the available mass of double beta isotopes and the experimental conditions. In particular, the achievable background suppression and the detection efficiency will be decisive for their success.

1. Introduction

The search for neutrinoless double beta decay is a central research topic in fundamental physics. Excellent review articles are available at [1] and in the references therein. Detailed reports on the ongoing experimental and theoretical activities have been given at this conference and are part of this conference proceedings. This paper will first give an introduction to neutrinless double beta decay which has been adapted from our recent Aspera roadmap paper [2], then focus on the question whether current nuclear matrix element calculations favor one double beta isotope over an other one, subsequently give a brief overview on the current experimental projects and their time lines, and conclude with a short summary on the experimental road map. Different to the oral presentation, I will not discuss here details of the experimental projects, as the interested reader can find this in much more detail in the dedicated articles in this proceedings.

2. The physics case

The discovery of neutrino oscillations in solar and atmospheric neutrino experiments have established neutrinos as particles with non-zero rest mass. Long-baseline experiments with neutrinos from nuclear power reactors and accelerators have confirmed these findings and, in conjunction with solar and atmospheric neutrino experiments, provide precision data and constraints on the mass differences and mixing angles. Neutrino masses and mixing are commonly considered as the first manifestation of physics beyond the Standard Model of electroweak interaction.

The central questions today in neutrino physics concern the nature of the neutrino particle (Dirac of Majorana), the absolute mass scale, the mixing angle θ_{13} and the CP violation in the leptonic sector. Establishing whether neutrinos are Dirac fermions and therefore different from their antiparticle, or Majorana fermions and therefore identical to their antiparticles, is of paramount importance for understanding the underlying symmetries of particle interactions and the origin of ν masses.

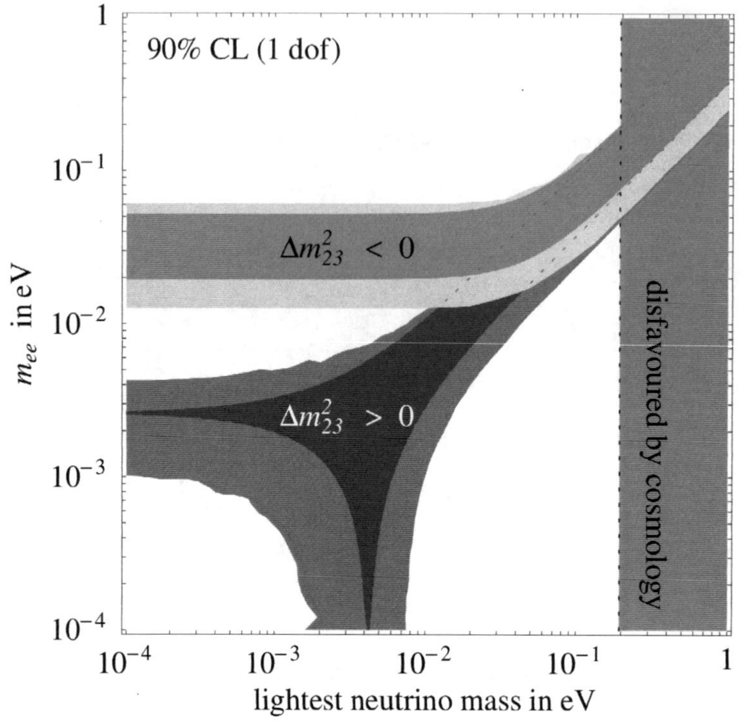

Figure 1. Allowed effective neutrino mass vs. mass of the lightest mass state predicted by neutrino oscillation experiments [3].

The only practical way to test whether neutrinos are Majorana particles is to search for the neutrinoless double beta decay ($0\nu\beta\beta$). The neutrinoless double beta decay of a nucleus consists of the simultaneous transition of two neutrons into two protons with the emission of two electrons: (A,Z) $\rightarrow$ (A,Z+2) + e^- + e^-. This process is forbidden in the Standard Model because it violates lepton number conservation ($\Delta L = 2$). The experimental signature of this decay is a peak in the distribution of the electron sum energy at the transition energy. On the contrary, a continuous spectrum extending from 0 to the transition energy is observed for the allowed 2nd order process of the electroweak interaction: (A,Z) $\rightarrow$ (A,Z+2) + e_1^- + e_2^- + $\bar{\nu}_1$ + $\bar{\nu}_2$. The latter decay mode has been measured for several nuclei.

There are several possible mechanisms leading to the $0\nu\beta\beta$ process: exchange of a light neutrino, right-handed weak currents, exchange of super-symmetric particles, and other non-standard interactions. Independent of the leading term, the observation of $0\nu\beta\beta$ decay would unambiguously establish the Majorana nature of neutrinos. Once the $0\nu\beta\beta$ decay has been observed experimentally, the nature of the leading term could be studied by measuring the energy and angular distribution of the single electrons, by the branching ratios of $0\nu\beta\beta$ decays to excited levels, and by comparison of the decay rates of different nuclei.

In the case of light neutrino exchange, the expression of the half-life depends on the effective neutrino mass: $[T^{0\nu}_{1/2}(0^+ \rightarrow 0^+)]^{-1} = G^{0\nu}\,|M^{0\nu}|^2\,\langle m_\nu\rangle^2$, where $G^{0\nu}$ is a calculable phase space factor and $M^{0\nu}$ is the nuclear matrix element (NME) of the process. $\langle m_\nu\rangle =$

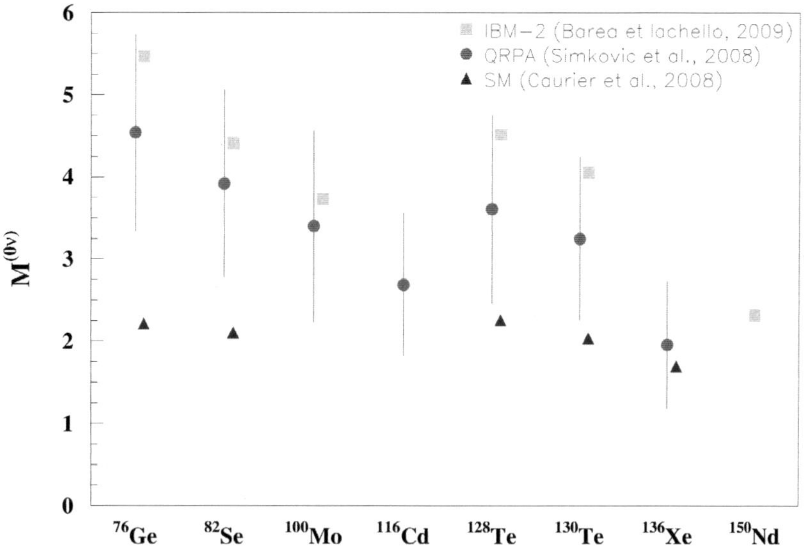

Figure 2. Compilation of three recent nuclear matrix element ($M^{(0\nu)}$) calculations for neutrinoless double beta decay [4, 6, 7].

$||U_{e1}|^2 m_1 + |U_{e2}|^2 m_2 e^{i\phi_1}||U_{e3}|^2 m_3 e^{i\phi_2}$ is the effective Majorana neutrino mass, m_1, m_2, m_3 are the masses related to the three neutrino mass eigenstates, U_{e1}, U_{e2}, U_{e3} are the elements of the first row of the neutrino mixing matrix and ϕ_1, ϕ_2 are the Majorana CP phases (± 1 if CP is conserved).

Fig. 1 summarizes the sensitivity required on $\langle m_\nu \rangle$ to determine the neutrino mass scheme. The present experiments with about 10 kg of double beta isotopes are sensitive to the quasi degenerate mass scheme ($m_1 \cong m_2 \cong m_3$). The next generation of experiments, with approximately 100 kg of isotopes, will explore part of the mass range predicted by the inverted hierarchy ($m_1 > m_2 \gg m_3$). These experiments will as well serve as bench tests for the following one ton scale experiments which are required to explore completely the inverted mass hierarchy. The meV mass range, predicted by the normal hierarchy ($m_1 \ll m_2 < m_3$), is beyond the reach of the current technologies.

3. Comparison of decay rates of double beta decay isotopes

Central to the comparison of isotopes are the values of the nuclear matrix elements. Recent calculation using quasi-particle random phase approximations (Q)RPA [4, 5] and interacting Boson model (IBM-2) [6] show reasonable agreements, while shell model (SM) [7] calculations show discrepancies of about a factor of two, apart for ^{136}Xe. Recently, the authors of [8] pointed out that the discrepancy in ^{76}Ge is reduced when the wave functions are constrained to reproduce the experimental occupancies of the two nuclei involved in the transition. Fig. 5 displays the values of nuclear matrix elements taken from [4, 6, 7]. It should be noted that different nuclear radii have been used by the authors to evaluate the NME [9] which requires a (not included) rescaling for comparison. To illustrate how the different factors influence the overall decay rate per kg of isotopes for a given neutrino mass, the factors normalized to the respective ones from

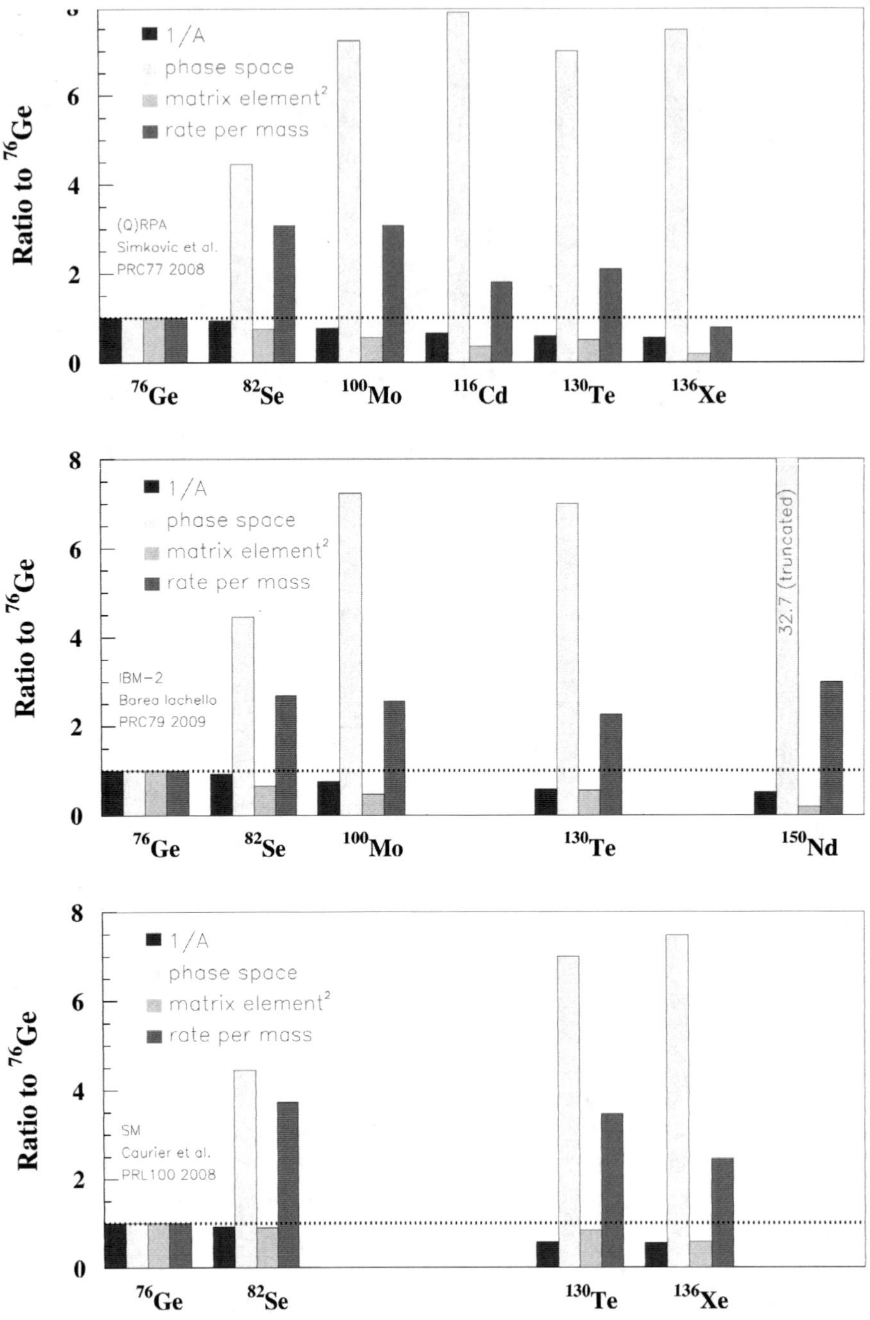

Figure 3. Comparison of the specific decay rates of double beta decay isotopes for a given effective neutrino mass are shown for three recent nuclear matrix element calculations: QRPA (top), IBM-2 (middle), SM (bottom). The contributing quantities ($1/A$, A: atomic number, phase space factor, (squared) nuclear matrix element) and the decay rates per mass are normalized to those of ^{76}Ge.

Journal of Physics: Conference Series **203** (2010) 012014 doi:10.1088/1742-6596/203/1/012014

^{76}Ge are displayed in Fig. 3. A re-scaling because of different nuclear radii is not required in this normalized representation.

It is noteworthy, that the decay rates per unit mass of isotopes vary within less than a factor of four when comparing them within the same theoretical framework. The absolute decay rates of ^{76}Ge for a 50 meV effective neutrino mass and per ton×year of exposure are 9.1 ([4]), 13.2 ([6]), and 2.2 ([7]) which have a stronger spread than the scatter between the different nuclei within one theoretical model. As mentioned above, recent calculations for ^{76}Ge NME taking into account the accupancies of the involved nuclei increases the SM and reduces the QRPA values, thus reducing the tension amongst the two theoretical approaches [8].

The current status of nuclear matrix element calculations do not favor a particular isotope over an other because the decay rates per mass of isotopes scatter only within a factor of four within a theoretical framework. Experimental consideration as detection efficiency and background suppression will determine the final sensitivity of the different experimental approaches.

Name	Nucleus	Mass*	Method	Location	Time line
		Running & recently completed experiments			
CUORICINO	Te-130	11 kg	bolometric	LNGS	2003-2008
NEMO-3	Mo-100/Se-82	6.9/0.9 kg	tracko-calo	LSM	until 2010
		Experiments with construction funding			
CUORE	Te-130	200 kg	bolometric	LNGS	2012
EXO-200	Xe-136	160 kg	liquid TPC	WIPP	2009 (comiss.)
GERDA I/II	Ge-76	35 kg	ionization	LNGS	2009 (comiss.)
SNO+	Nd-150	56 kg	scintillation	SNOlab	2011
		Substantial R&D funding / prototyping			
CANDLES	Ca-48	0.35 kg	scintillation	Kamioka	2009
Majorana	Ge-76	26 kg	ionization	SUSL	2012
NEXT	Xe-136	80 kg	gas TPC	Canfranc	2013
SuperNEMO	Se-82 or Nd-150	100 kg	tracko-calo	LSM	2012 (first mod.)
		R&D and/or conceptual design			
CARVEL	Ca-48		scintillation	Solotvina	
COBRA	Cd-116, Te-130		ionization	LNGS	
DCBA	Nd-150		drift chamber	Kamioka	
EXO gas	Xe-136		gas TPC	SNOlab	
MOON	Mo-100		tracking	Oto	
		Other decay modes			
TGV	Cd-106		ionization	LSM	operational

Figure 4. Summary of experimental projects, type and mass of isotopes (not including detection and analysis efficiencies), experimental method, financial status, location and planned year of start-up.

4. Experimental projects

An overview of the experimental projects are given in Fig. 4. The masses of isotopes listed do neither include detection nor analysis efficiencies. They can range from e.g. 18% for the NEMO-3 up to 90% for GERDA or Cuore experiments. Essential to achieve the intended sensitivities of the different experiments is to operate the experiments free of background within the envisioned exposure. How to achieve this goal is central to the different experiments and the interested reader is referred to the respective articles in this conference proceeding. The first of the next generation experiments to become operational are GERDA at LNGS, Italy and EXO at WIPP, USA. Both plan to start their detector commissioning phase in 2009. Not included in this table is the recent project to load ^{136}Xe into a liquid scintillator in the KamLAND detector.

5. The future strategy

The future development of the field will strongly depend on the results of the upcoming GERDA [11] and EXO [12] experiments. If neutrinoless double beta decay is observed at the 1 eV scale, as claimed by part of the Heidelberg Moscow experiment [10], the decay could be studied with high precision with many different isotopes and different techniques. The effective mass could be measured with accuracy and the leading term governing the decay mode identified. In case that the claim is refuted, at least two experiments with about one ton of isotopes and zero background in the region of interest for several year×ton of exposure are required to cover the full mass range down to 10 meV predicted by oscillation experiments for the inverse mass hierarchy. Yet, there is no experimental concept how to explore the mass range down to 1 meV which is expected, if the normal mass hierarchy was realized by nature.

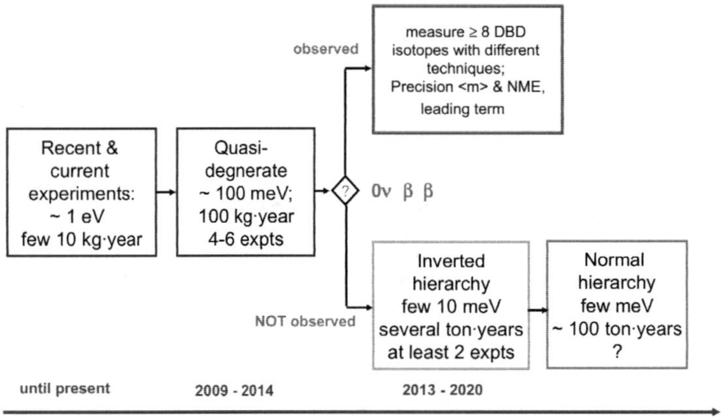

Figure 5. Time line and branching point for future double beta decay searches.

References

[1] F. T. Avignone, III, S. R. Elliott, J. Engel, *Rev. Mod. Phys.* **80** 481-516 (2008)
[2] F.Piquemal, A. Giuliani, S. Schönert in http://www.aspera-eu.org/images/stories/files/Roadmap.pdf; also NuPECC working group report double beta decay 2009 (unpublished)
[3] F. Feruglio, A. Strumia, F. Vissani *Nucl.Phys.* B **637** 345-377 (2002)
[4] F. Simkovic, A. Faessler, V. A. Rodin, P. Vogel, and J. Engel, *Phys. Rev. C* **77**, 45503 (2008)
[5] M. Kortelainen and J. Suhonen, *Phys. Rev. C* **75**, 51303 (2007)
[6] J. Barea, F. Iachello, *Phys.Rev.* **C** 79:044301 (2009)
[7] E. Caurier, J. Menendez, F. Nowacki, and A. Poves, *Phys. Rev. Lett.* **100**, 52503 (2008)
[8] J. Menendez, A. Poves, E. Caurier, F. Nowacki, *Phys.Rev.* **C**80, 048501 (2009)
[9] P. Grabmayr and A. Smolnkivo, GERDA internal report GSTR-09-004, (to be submitted for publication).
[10] H.V. Klapdor-Kleingrothaus, A. Dietz, I.V. Krivosheina, O. Chkvorets, *Nucl. Instrum. Meth.* **A** 522, 371-406 (2004)
[11] I. Abt et al. (GERDA collaboration), hep-ex/0404039; I. Abt et al. (GERDA collaboration), Proposal to the LNGS (2004) http://www.mpi-hd.mpg.de/gerda/; S. Schönert et al. (GERDA collaboration) *Nucl. Phys. Proc. Suppl.* 145:242-245 (2005) Nucl.Phys.Proc.Suppl.138:224-226,2005
[12] D. Akimov et al. (EXO collaboration), *Nucl. Phys. Proc. Suppl.* 138:224-226 (2005)

Supernova neutrino oscillations: what do we understand?

Amol Dighe

Tata Institute of Fundamental Research, Homi Bhabha Road, Colaba, Mumbai 400005, INDIA

E-mail: amol@theory.tifr.res.in

Abstract.
We summarize our current understanding of the neutrino flavor conversions inside a core collapse supernova, clarifying the important role played by the "collective effects" in determining flavor conversion probabilities. The potentially observable ν_e and $\bar{\nu}_e$ spectra may help us identify the neutrino mixing scenario, distinguish between primary flux models, and learn more about the supernova explosion.

1. Introduction

Neutrino energy spectra exiting a core collapse supernova (SN) are determined by the primary neutrino fluxes, the neutrino mixing scenario as well as the densities encountered by the neutrinos along their path. (For recent overviews, see [1, 2].) In this talk, we explain how the collective effects due to ν-ν interactions and MSW matter effects result in distinctive features in the ν_e and $\bar{\nu}_e$ energy spectra at the detectors. We emphasize some of the recent results involving the collective effects.

2. Primary neutrino fluxes

Neutrinos from a SN are emitted in roughly three phases. The first $\sim$ 10 ms of the SN neutrino signal is the neutronization burst, composed entirely of ν_e. Neutrinos and antineutrinos of all species are emitted during the next $\sim$ 10 s, initially in the accretion phase and later in the Kelvin-Helmholtz cooling phase [3].

The SN core acts essentially like a neutrino black body source with flavor dependent fluxes. Since the primary fluxes of $\nu_\mu, \bar{\nu}_\mu, \nu_\tau, \bar{\nu}_\tau$ are almost identical, it is convenient to work in terms of the three "flavors": ν_e , $\nu_x \equiv \cos\theta_{23}\,\nu_\mu - \sin\theta_{23}\,\nu_\tau$, $\nu_y \equiv \sin\theta_{23}\,\nu_\mu + \cos\theta_{23}\,\nu_\tau$, which are also the mass eigenstates at large matter densities. Clearly, the primary fluxes of $\nu_x, \nu_y, \bar{\nu}_x, \bar{\nu}_y$ are identical and we denote them as $F^0_{\nu_x}$.

The primary fluxes $F^0_{\nu_e}, F^0_{\bar{\nu}_e}, F^0_{\nu_x}$ are parameterized by the total number fluxes Φ_0, average energies E_0, and the "pinching parameters" that characterize their spectral shapes [4]. The values of the parameters are highly model dependent, as can be seen from Table 1. In the Table, M1 fluxes represent the predictions of the Livermore model [5]. The predictions of the Garching model [6], on the other hand, are close to M1 during the accretion phase and close to M2 during the cooling phase. As will be seen later, the neutrino flavor conversion probabilities depend on certain features of these primary spectra.

Table 1. Representative predictions of neutrino flux models. E_0 values are given in MeV.

Flux Model	$E_0(\nu_e)$	$E_0(\bar{\nu}_e)$	$E_0(\nu_x)$	$\frac{\Phi_0(\nu_e)}{\Phi_0(\nu_x)}$	$\frac{\Phi_0(\bar{\nu}_e)}{\Phi_0(\bar{\nu}_x)}$
M1	12	15	24	2.0	1.6
M2	12	15	18	0.85	0.75

3. Neutrino flavor conversions

3.1. Collective effects at large neutrino densities

The extremely high neutrino and antineutrino densities near the neutrinospheres make the ν-ν interactions in this region significant [7, 8] and give rise to "collective effects". Analytic studies of these effects reveal distinctive flavor conversion phenomena that are qualitatively different from the usual vacuum or MSW oscillations. These include "synchronized oscillations" [9] where ν and $\bar{\nu}$ of all energies oscillate with the same frequency, "bipolar/pendular oscillations" [10, 11] that correspond to pairwise conversions $\nu_e\bar{\nu}_e \leftrightarrow \nu_y\bar{\nu}_y$, and "spectral split" [12, 13, 14, 15, 16] where ν_e and ν_y ($\bar{\nu}_e$ and $\bar{\nu}_y$) spectra interchange completely within certain energy ranges. The dynamics in three generations can be factorized into a superposition of multiple two-flavor phenomena [17]. New three flavor effects also emerge: for example in early accretion phase, large μ-τ matter potential may cause interference between MSW and collective effects [18].

The dependence of the flavor evolution on the direction of propagation of the neutrino may give rise to direction-dependent evolution, or "multi-angle effects" [7, 8, 19, 20] that may lead to decoherence [21, 22, 23]. However, for a realistic asymmetry between neutrino and antineutrino fluxes, such multi-angle effects are likely to be small [24, 25] and the "single-angle" approximation can be used. Even for non-spherical geometries, one can study the single-angle evolution along stream lines of neutrino flux, as long as coherence is maintained [26].

The propagation of the neutrinos can be rather cleanly separated into regions where various collective effects dominate individually and hence the neutrinos experience these effects sequentially [17, 24]. When the neutrinos emerge from the region where collective effects dominate, the net effect is a swap between the spectra of ν_e and ν_y ($\bar{\nu}_e$ and $\bar{\nu}_y$) in some energy ranges, as illustrated in Fig. 1. The neutrino (antineutrino) spectra at this point can be divided into two energy regimes: the "swapped" regime where the ν_e and ν_y ($\bar{\nu}_e$ and $\bar{\nu}_y$) spectra have completely interchanged, and the "unswapped" regime where the spectra have remained unchanged. At the boundary between two such regimes, the survival probability changes sharply: this is the "spectral split". This boundary need not be sharp: the spectral split is typically smeared out over a small energy range. In general, there can be zero, one, or multiple spectral splits in ν as well as $\bar{\nu}$ channels [16, 27], and their positions can be approximately predicted depending on the mass ordering and primary spectra [16]. Although the exact evolution of the neutrino state leading to the splits can be calculated numerically (for animated simulations, see `http://www.mppmu.mpg.de/supernova/multisplits`), the current analytic understanding of the dynamics is a bit unsatisfactory. A clear direct mapping between the primary fluxes and split positions is also still lacking.

When the fluxes obey an approximate hierarchy, i.e. $F^0_{\nu_e} > F^0_{\bar{\nu}_e} > F^0_{\nu_x}$ except at high energies, the flux-split mapping is particularly straightforward. All energies in ν as well as $\bar{\nu}$ are unswapped in normal hierarchy, while in inverted hierarchy, only the energy range $E < E_c$ for neutrinos below a certain critical energy E_c remains unswapped [12, 14]. The fluxes in the accretion phase (e.g. M1) typically fall under this catagory. The results in [2, 17, 24, 28], in particular, have been obtained with this flux scenario.

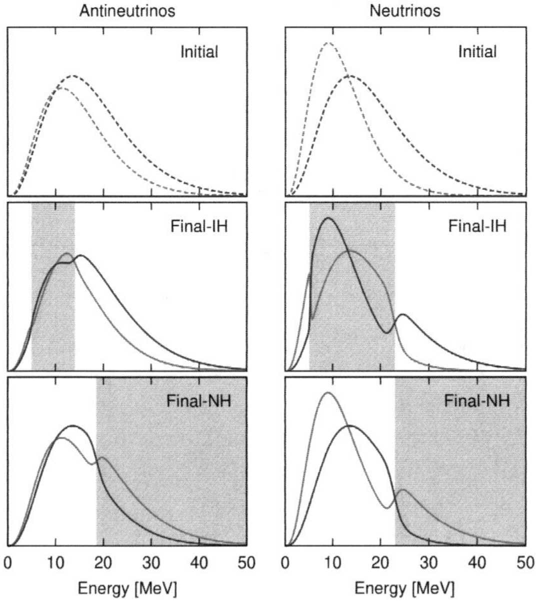

Figure 1. The fluxes of antineutrinos and neutrinos (light grey/ red: e flavor, dark grey/ blue: y flavor) before and after the action of collective effects, in the normal hierarchy (NH) and inverted hierarchy (IH). The shaded regions correspond to the swapped energy regime. Here M2 fluxes from Table 1 have been used. The Figure has been adapted from [16].

3.2. MSW resonances inside the SN and propagation through vacuum

In iron core supernovae, the collective effects have already become insignificant when neutrinos enter the MSW resonance regions. SN neutrinos must pass through two resonance layers: the H-resonance layer at $\rho_{\rm H} \sim 10^3$ g/cc characterized by $(\Delta m^2_{\rm atm}, \theta_{13})$, and the L-resonance layer at $\rho_{\rm L} \sim 10$ g/cc characterized by $(\Delta m^2_{\odot}, \theta_{12})$. The outcoming incoherent mixture of vacuum mass eigenstates travels to the Earth without any further conversions, and is observed at a detector as a combination of primary fluxes of the three neutrino flavors:

$$F_{\nu_e} = pF^0_{\nu_e} + (1-p)F^0_{\nu_x} \;, \qquad F_{\bar{\nu}_e} = \bar{p}F^0_{\bar{\nu}_e} + (1-\bar{p})F^0_{\nu_x} \;,$$

where p and $\bar{p}$ are the survival probabilities of ν_e and $\bar{\nu}_e$ respectively.

The neutrino survival probabilities are governed by the adiabaticities of the resonances traversed, which are directly connected to the neutrino mixing scheme [29]. Table 2 shows the survival probabilities in swapped and unswapped energy regions in various mixing scenarios. For intermediate values of θ_{13}, i.e. $10^{-5} \lesssim \sin^2\theta_{13} \lesssim 10^{-3}$, the survival probabilities depend on energy as well as details of the SN density profile. Note that though p and $\bar{p}$ entries in scenarios C and D are identical, since the swapped regime is determined by the mass hierarchy, even these scenarios lead to different net flavor conversions. The spectral swaps can occur for values of θ_{13} as low as 10^{-10} [13, 28]. Collective effects are thus sensitive to extremely small (but nonzero) θ_{13}, and can in principle allow identification of hierarchy.

3.3. Propagation through the shock wave near MSW resonances

The passage of the shock wave through the H-resonance ($\rho \sim 10^3$ g/cc) a few seconds after the core bounce may break adiabaticity, thereby modifying the flavor evolution of neutrinos that are emitted during this time interval [30, 31, 32, 33, 34, 35]. Such a situation is possible, even in principle, only in certain mixing scenarios: for neutrinos (antineutrinos), shock effects can be present only in the mixing scenario A (B). As a result, the mere identification of shock wave effects is enough to identify these scenarios, irrespective of the collective effects.

Table 2. Survival probabilities p and $\bar{p}$ in the swapped and unswapped energy regimes, in various mixing scenarios. Terms of $\mathcal{O}(\theta_{13}^2)$ have been neglected.

	Hierarchy	$\sin^2\theta_{13}$	p unswapped	p swapped	$\bar{p}$ unswapped	$\bar{p}$ swapped
A	Normal	$\gtrsim 10^{-3}$	0	$\sin^2\theta_{12}$	$\cos^2\theta_{12}$	0
B	Inverted	$\gtrsim 10^{-3}$	$\sin^2\theta_{12}$	0	0	$\cos^2\theta_{12}$
C	Normal	$\lesssim 10^{-5}$	$\sin^2\theta_{12}$	0	$\cos^2\theta_{12}$	0
D	Inverted	$\lesssim 10^{-5}$	$\sin^2\theta_{12}$	0	$\cos^2\theta_{12}$	0

The shock wave effects can be diluted by stochastic density fluctuations [36] or turbulence [37]. However, recent hydrodynamic simulations [38, 39] suggest that some of the shock wave effects survive in spite of these smearing factors.

3.4. Oscillations inside the Earth matter
If neutrinos travel through the Earth before reaching the detector, they undergo flavor oscillations and the survival probabilities change [40, 41, 42]. This change however occurs only in the energy ranges where $p \neq 0$ (for neutrinos) or $\bar{p} \neq 0$ (for antineutrinos). The presence or absence of Earth effects in the neutrino or antineutrino channels would therefore help in identifying some flux-mixing combination, using Table 2.

4. Observable signals at neutrino detectors
If a SN explodes in our galaxy at $\sim$ 10 kpc from the Earth, we expect $\mathcal{O}(10^4)$ events of $\bar{\nu}_e$ at Super-Kamiokande (SK). With future larger water Cherenkov detectors, large scintillation detectors with better energy resolutions, and liquid Argon detectors sensitive to ν_e, it should be possible to reconstruct the ν_e and $\bar{\nu}_e$ spectra to a good accuracy. SN neutrinos would also enable pointing at the SN in advance of the optical signal [43], with [44] or even without [45, 46, 47] any information on neutrino mixing, and in reconstructing the SN bounce time accurately [48].

4.1. Suppression of ν_e in the neutronization burst
Since the primary signal during the neutronization burst is pure ν_e, the entire energy range is unswapped. Since the model predictions for the energy and luminosity of the burst are fairly robust [49], the observation of the burst signal gives direct information about the survival probability of ν_e. This probability is $\mathcal{O}(\theta_{13}^2)$ in scenario A and $\sin^2\theta_{12}$ in all the other scenarios [29]. Thus, the strong suppression of ν_e burst would be a smoking gun signal for scenario A.

In an O-Ne-Mg supernova, the MSW resonances may lie deep inside the collective regions during the neutronization burst, when the neutrino luminosity is even higher. In such a situation, neutrinos of all energies undergo the MSW resonances together, with the same adiabaticity [50]. As long as this adiabaticity is nontrivial, one gets the "MSW-prepared spectral splits", two for normal hierarchy and one for inverted hierarchy [51, 52, 53]. The positions of the splits can be predicted from the primary spectra [53]. The splits imply ν_e suppression that is stepwise in energy. Such a signature may even be used to identify the O-Ne-Mg supernova, in addition to identifying the hierarchy.

4.2. Shock wave effects
Observables like the number of events, average energy, or the width of the spectrum may display dips or peaks for short time intervals, while the shock wave is passing through the H resonance.

The positions of the dips or peaks in the number of events at different neutrino energies would also allow one to trace the shock propagation while the shock is in the mantle, around densities of $\rho \sim 10^3$ g/cc [34]. This information, by itself or in combination with the corresponding gravitational wave signal, will yield valuable information about the SN explosion, in addition to identifying the scenario A or B.

If light sterile neutrinos exist, they may leave their imprints in the shock wave [54, 55]. For an O-Ne-Mg supernova, passage of the shock wave through the sharp density profile at the resonance leads to distinctive effects [56].

4.3. Spectral split in ν_e

A split in ν_e or $\bar{\nu}_e$ spectrum would manifest itself as a sharp jump at the split energy [13]. It is therefore a smoking gun signal for collective effects and the corresponding flux-mixing combination. However, the sharp split in p is smeared to some extent by the multi-angle effects [24] and its actual observation may be a challenging task.

4.4. Earth matter effects

Earth matter effects can be identified by the comparison of signals at two detectors, only one of which is shadowed by the earth. This could be achieved through the $\bar{\nu}_e$ spectra at two large water Cherenkov detectors [28] or through the time dependent ratio of luminosities at IceCube and Hyper-Kamiokande [57]. The Earth effects can be identified even at a single detector as long as it is capable of determining the neutrino energy. For example, Fourier transform of the "inverse energy" spectrum of $\bar{\nu}_e$ [58] gives peaks (multiple ones if the neutrinos traverse the Earth core) whose positions are independent of the primary neutrino spectra. These would reveal the presence of Earth matter effects [59]. For the typical accretion phase scenario (M1) and $\sin^2\theta_{13} \lesssim 10^{-5}$, the detection of Earth effects in $\bar{\nu}_e$ can identify normal hierarchy [28].

The presence of spectral splits implies that the Earth effects may be present in only certain energy regimes. The identification of these energy regimes can play a major role in the identification of the corresponding flux-mixing combination.

5. Concluding remarks

The flavor conversion probabilities of neutrinos inside a SN are sensitive to the neutrino mixing scenarios, in particular to the mass hierarchy even at extemely small θ_{13}, thanks to the collective effects and MSW matter effects. In addition, since the collective effects can change the neutrino flavor composition deep inside the core, they may affect the dynamics of the SN explosion.

Smoking gun signals of neutrino mixing scenarios as well as information on primary fluxes and shock wave propagation can be independently obtained through observations like the suppression of neutronization burst, time variation of the signal during shock wave propagation, and Earth matter effects. More reliable predictions of primary fluxes, better understanding of the collective effects, and improved limits on θ_{13} from terrestrial experiments will make our deductions more robust.

Acknowledgments

I would like to thank Basudeb Dasgupta, Alessandro Mirizzi, Georg Raffelt, and Alexei Smirnov for fruitful collaborations, the organizers of TAUP 2009 for their hospitality, and the Max Planck - India Partnergroup for financial support.

References

[1] Duan H and Kneller J P 2009 *J. Phys.* G **36** 113201 (*Preprint* arXiv:0904.0974 [astro-ph.HE])
[2] Dighe A 2008 *Preprint* arXiv:0809.2977v3 [hep-ph]
[3] Raffelt G G 1996 *Stars as Laboratories for Fundamental Physics* (Chicago: The University of Chicago Press)

[4] Keil M T, Raffelt G G and Janka H T 2003 *Astrophys. J.* **590** 971 (*Preprint* astro-ph/0208035)
[5] Totani T, Sato K, Dalhed H E and Wilson J R 1998 *Astrophys. J.* **496** 216 (*Preprint* astro-ph/9710203)
[6] Raffelt G G, Keil M T, Buras R, Janka H T and Rampp M 2003 *Preprint* astro-ph/0303226
[7] Duan H, Fuller G M, Carlson J and Qian Y Z 2006 *Phys. Rev.* D **74** 105014 (*Preprint* astro-ph/0606616)
[8] Duan H, Fuller G M, Carlson J and Qian Y Z 2006 *Phys. Rev. Lett.* **97** 241101 (*Preprint* astro-ph/0608050)
[9] Pastor S, Raffelt G G and Semikoz D V 2002 *Phys. Rev.* D **65** 053011 (*Preprint* hep-ph/0109035)
[10] Hannestad S, Raffelt G G, Sigl G and Wong Y Y Y 2006 *Phys. Rev.* D **74** 105010 (*Preprint* astro-ph/0608695)
Hannestad S, Raffelt G G, Sigl G and Wong Y Y Y 2007 *Phys. Rev.* D **76** 029901 (*Erratum*)
[11] Duan H, Fuller G M, Carlson J and Qian Y Z 2007 *Phys. Rev.* D **75** 125005 (*Preprint* astro-ph/0703776)
[12] Raffelt G G and Smirnov A Yu 2007 *Phys. Rev.* D **76** 081301 (*Preprint* arXiv:0705.1830 [hep-ph])
Raffelt G G and Smirnov A Yu 2008 *Phys. Rev.* D **77** 029903 (*Erratum*)
[13] Duan H *et al* 2007 *Phys. Rev. Lett.* **99** 241802 (*Preprint* arXiv:0707.0290 [astro-ph])
[14] Raffelt G G and Smirnov A Yu 2007 *Phys. Rev.* D **76** 125008 (*Preprint* arXiv:0709.4641 [hep-ph])
[15] Fogli G L *et al* 2008 *Phys. Rev.* D **78** 097301 (*Preprint* arXiv:0808.0807 [hep-ph])
[16] Dasgupta B *et al* 2009 *Phys. Rev. Lett.* **103** 051105 (*Preprint* arXiv:0904.3542 [hep-ph])
[17] Dasgupta B and Dighe A 2008 *Phys. Rev.* D **77** 113002 (*Preprint* arXiv:0712.3798 [hep-ph])
[18] Esteban-Pretel A *et al* 2008 *Phys. Rev.* D **77** 065024 (*Preprint* arXiv:0712.1137 [astro-ph])
[19] Duan H, Fuller G M and Carlson J 2008 *Comput. Sci. Dis.* **1** 015007 (*Preprint* arXiv:0803.3650 [astro-ph])
[20] Sawyer R F 2009 *Phys. Rev.* D **79** 105003 (*Preprint* arXiv:0803.4319 [astro-ph])
[21] Pantaleone J 1998 *Phys. Rev.* D **58** 073002
[22] Raffelt G G and Sigl G 2007 *Phys. Rev.* D **75** 083002 (*Preprint* hep-ph/0701182)
[23] Esteban-Pretel A *et al* 2008 *Phys. Rev.* D **78** 085012 (*Preprint* arXiv:0807.0659 [astro-ph])
[24] Fogli G L *et al* 2007 *J. Cosmol. Astropart. Phys.* JCAP12(2007)010 (*Preprint* arxiv:0707.1998 [hep-ph])
[25] Esteban-Pretel A *et al* 2007 *Phys. Rev.* D **76** 125018 (*Preprint* arXiv:0706.2498 [astro-ph])
[26] Dasgupta B *et al* 2008 *Phys. Rev.* D **78** 033014 (*Preprint* arXiv:0805.3300 [hep-ph])
[27] Fogli G *et al* 2009 *J. Cosmol. Astropart. Phys.* JCAP10(2009)002 (*Preprint* arXiv:0907.5115 [hep-ph])
[28] Dasgupta B, Dighe A and Mirizzi A 2008 *Phys. Rev. Lett.* **101** 171081 (*Preprint* arXiv:0802.1481 [hep-ph])
[29] Dighe A S and Smirnov A Yu 2000 *Phys. Rev.* D **62** 033007 (*Preprint* hep-ph/9907423)
[30] Schirato R C and Fuller G M 2002 *Preprint* astro-ph/0205390
[31] Takahashi K *et al* 2003 *Astropart. Phys.* **20** 189 (*Preprint* astro-ph/0212195)
[32] Lunardini C *et al* 2003 *J. Cosmol. Astropart. Phys.* JCAP06(2003)009 (*Preprint* hep-ph/0302033)
[33] Fogli G L *et al* 2003 *Phys. Rev.* D **68** 033005 (*Preprint* hep-ph/0304056)
[34] Tomàs R *et al* 2004 *J. Cosmol. Astropart. Phys.* JCAP09(2004)015 (*Preprint* astro-ph/0407132)
[35] Dasgupta B and Dighe A 2007 *Phys. Rev.* D **75** 093002 (*Preprint* hep-ph/0510219)
[36] Fogli G L *et al* 2006 *J. Cosmol. Astropart. Phys.* JCAP06(2006)012 (*Preprint* hep-ph/0603033)
[37] Friedland A and Gruzinov A 2006 *Preprint* astro-ph/0607244
[38] Kneller J P *et al* 2008 *Phys. Rev.* D **77** 045023 (*Preprint* arXiv:0705.3835 [astro-ph])
[39] Gava J *et al* 2009 *Phys. Rev. Lett.* **103** 071101 (*Preprint* arXiv:0902.0317 [hep-ph])
[40] Dighe A S 2001 *Preprint* hep-ph/0106325
[41] Lunardini C and Smirnov A Yu 2001 *Nucl. Phys.* B **616** 307 (*Preprint* hep-ph/0106149)
[42] Takahashi K and Sato K 2002 *Phys. Rev.* D **66** 033006 (*Preprint* hep-ph/0110105)
[43] Scholberg K 1999 *Preprint* astro-ph/9911359
[44] Scholberg K, Burgmeier A and Wendell R 2009 *Preprint* arXiv:0910.3174 [astro-ph.IM]
[45] Beacom J F and Vogel P 1999 *Phys. Rev.* D **60** 033007 (*Preprint* astro-ph/9811350)
[46] Ando S and Sato K 2002 *Prog. Theor. Phys.* **107** 957 (*Preprint* hep-ph/0110187)
[47] Tomàs R *et al* 2003 *Phys. Rev.* D **68** 093013 (*Preprint* hep-ph/0307050)
[48] Halzen F and Raffelt G G 2009 *Preprint* arXiv:0908.2317 [astro-ph.HE]
[49] Kachelrieß M *et al* 2005 *Phys. Rev.* D **71** 063003 (*Preprint* astro-ph/0412082)
[50] Wong Y Y Y 2003 *AIP Conf. Proc.* **655** 240 (*Preprint* arXiv:hep-ph/0211045)
[51] Duan H *et al* 2008 *Phys. Rev. Lett.* **100** 021101 (*Preprint* arXiv:0710.1271 [astro-ph])
[52] Duan H, Fuller G M and Qian Y Z 2008 *Phys. Rev.* D **77** 085016 (*Preprint* arXiv:0801.1363 [hep-ph])
[53] Dasgupta B *et al* 2008 *Phys. Rev.* D **77** 113007 (*Preprint* arXiv:0801.1660 [hep-ph])
[54] Choubey S, Harries N P and Ross G G 2006 *Phys. Rev.* D **74** 053010 (*Preprint* hep-ph/0605255)
[55] Choubey S, Harries N P and Ross G G 2007 *Phys. Rev.* D **76** 073013 (*Preprint* hep-ph/0703092)
[56] Lunardini C, Muller B and Janka H T 2008 *Phys. Rev.* D **78** 023016 (*Preprint* arXiv:0712.3000 [astro-ph])
[57] Dighe A S *et al* 2003 *J. Cosmol. Astropart. Phys.* JCAP06(2003)005 (*Preprint* hep-ph/0303210)
[58] Dighe A S *et al* 2003 *J. Cosmol. Astropart. Phys.* JCAP06(2003)006 (*Preprint* hep-ph/0304150)
[59] Dighe A S *et al* 2004 *J. Cosmol. Astropart. Phys.* JCAP01(2004)004 (*Preprint* hep-ph/0311172)

Journal of Physics: Conference Series **203** (2010) 012016 doi:10.1088/1742-6596/203/1/012016

Observations of ultra high energy cosmic rays

G. Matthiae

University and Sezione INFN of Roma Tor Vergata

matthiae@roma2.infn.it

Abstract. The Auger Observatory was designed to study the highest-energy cosmic rays by measuring the properties of the showers produced in the atmosphere. The most important results are the observation of the suppression of the high-energy events due to the GZK effect and the discovery of the angular correlation with nearby extragalactic objects.

1. The Auger Observatory

The Auger Observatory, located in the province of Mendoza (Argentina) at the latitude of about 35^0 S and altitude of 1400 m above see level, is a hybrid system, a combination of a large surface array and a fluorescence detector [1].

The surface detector (SD) is a large array of more than 1600 water Cherenkov detectors spaced at a distance of 1.5 km and covering a total area of 3000 km^2. Each detector is a plastic tank filled with purified water [2]. The tanks activated by a cosmic ray shower record the particle density and the time of arrival. This information is used to derive the direction of the axis of the shower and the point of impact at ground. An example of an event of high energy as observed by the SD is shown in Fig. 1. The signal S of each water Cherenkov detector is expressed in units of Vertical Equivalent Muons (VEM) which represents the signal produced by a muon traversing the tank vertically.

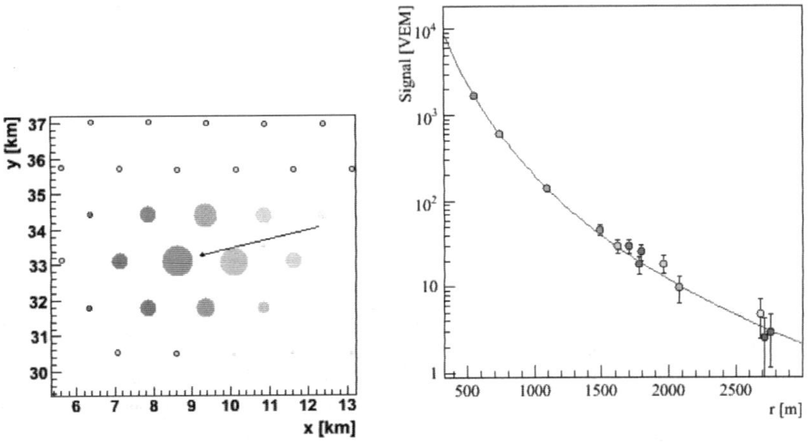

Figure 1. Example of an event of high energy as observed by the SD. As shown in the left panel the shower has activated 13 water Cherenkov detectors distributed over an area of about 20 km^2 . The area of the circles is proportional to the logarithm of the observed signals which are plotted in the right panel as a function of the distance from the reconstructed shower axis. The signals expressed in units of VEM are shown together with the results of the LDF fit.

The dependence of S on the distance r to the axis of the shower is described with a simple analytical expression known as Lateral Distribution Function (LDF) which is fitted to the data. The fit provides a value of the signal which would be observed at the distance of 1000 m from the shower axis. This interpolated quantity, S(1000), is a good energy estimator in the sense that it is well correlated with the energy of the primary.

The fluorescence detector (FD) of the Auger Observatory consists of 24 telescopes located in four stations which are built on small elevations on the perimeter of the site. The telescopes measure the longitudinal development of the showers by observing the fluorescence light produced by the interaction of the charged particles of the showers with the nitrogen molecules of the atmosphere. The field of view of each telescope is 30^0 x 30^0. The FD may operate only in clear moonless nights and therefore with an uptime of only about 13%. The method relies on the knowledge of the fluorescence yield and its dependence on pressure and temperature [3].
The absolute calibration of the FD telescopes is done using calibrated light sources. The attenuation of the fluorescence light due to Rayleigh and aerosol scattering along the path from the shower to the telescopes is continuously monitored.
The energy deposited by the shower particles as a function of the depth is obtained from the observed light profile. Examples of longitudinal profiles of showers are shown in Fig.2 together with fits using the Gaisser-Hillas analytical expression.

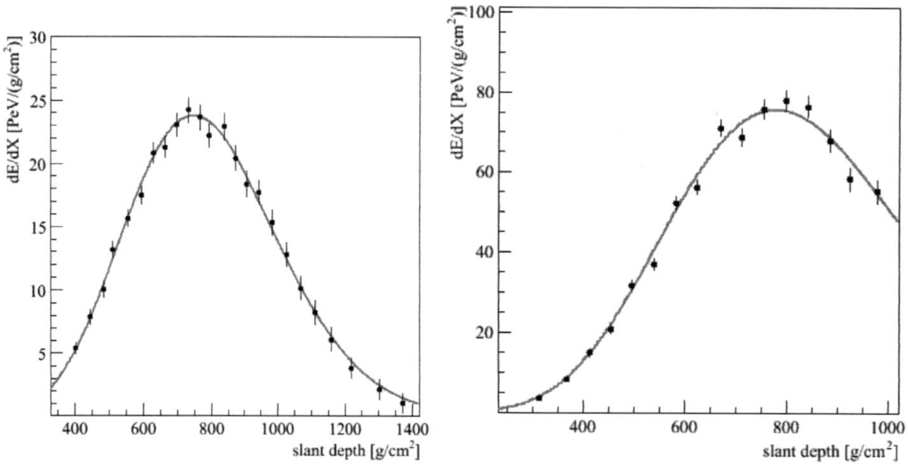

Figure 2. Examples of measured longitudinal profiles of high-energy showers. The energy deposited by the particles of the shower is plotted as a function of the atmospheric slant depth together with fits using the Gaisser-Hillas form.
Left panel: energy ≈ 1.5×10^{19} eV, zenith angle ≈ 55^0. Right panel: energy ≈ 4.5×10^{19} eV, zenith angle ≈ 36^0

2. The Energy Calibration.

The method used by the Auger collaboration to measure the energy spectrum exploits the hybrid nature of the Observatory using the data itself rather than simulations [4]. For each event, the energy estimator S(1000) is obtained from the LDF fit. The estimator S(1000) depends on the zenith angle because the effective atmospheric thickness seen by the showers before reaching ground changes with the zenith angle. The value of S(1000) corresponding to the median zenith angle of 38^0 is used as a reference and the zenith angle dependence of the energy estimator is determined assuming that the arrival directions are distributed isotropically. The estimator S_{38} is corrected for the missing energy (neutrinos and muons) using the mean value between proton and iron as derived from simulations of the showers. At the energy of 10^{19} eV the correction is 10% (about 8% for protons and 12% for iron nuclei). A sample of good quality hybrid events was selected to establish the correlation between the fluorescence detector energy E_{FD} and the energy estimator S_{38} as shown in Fig. 3.

Journal of Physics: Conference Series **203** (2010) 012016 doi:10.1088/1742-6596/203/1/012016

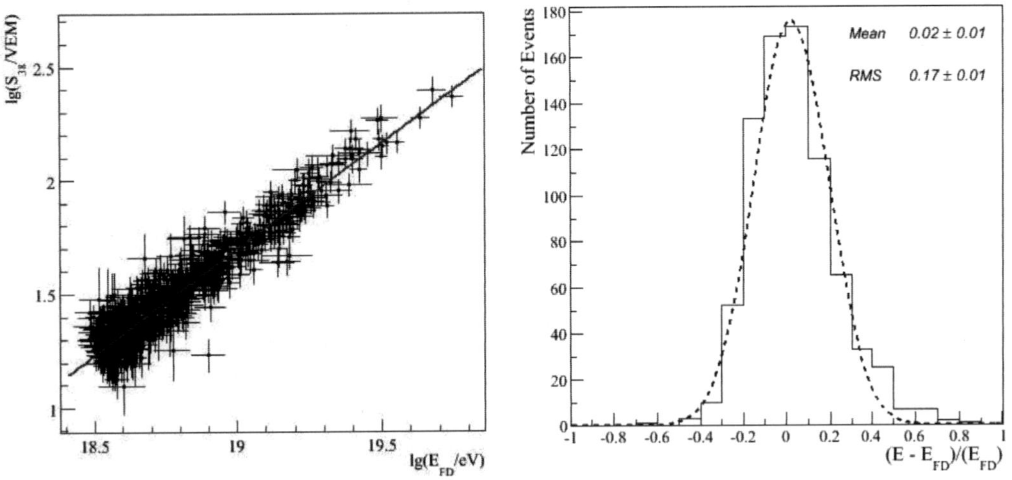

Figure 3. The calibration of the energy estimator S_{38} using the calorimetric energy from the FD is shown in the left panel together with the result of the fit. The right panel shows the energy resolution (17%).

The absolute energy calibration of S_{38} as obtained from the fit represented by the line in the left panel of Fig.3 is then used for the full set of events with higher statistics which is measured by the SD only.

The Auger method for the absolute calibration of the showers energy is at present affected by a systematic error of about ± 20%, mainly due to the uncertainty on the reconstruction of the shower profile, on the calibration of the FD telescopes and on the fluorescence yield.

3. The Energy Spectrum.

The data presented here refer to showers with zenith angle below 60^0 because the analysis of more inclined showers requires a more complex and sophisticated treatment. Two different methods were used to measure the energy spectrum

- SD events with energy calibration as explained in Sec 2. (E > $3\text{x}10^{18}$ eV).
- Hybrid events with at least one surface detector activated (E > 10^{18} eV).

Hybrid events enable lower energies to be reached. The two sets of data have the same energy calibration and therefore can be combined together. The resulting spectrum [5] is shown in Fig.4.

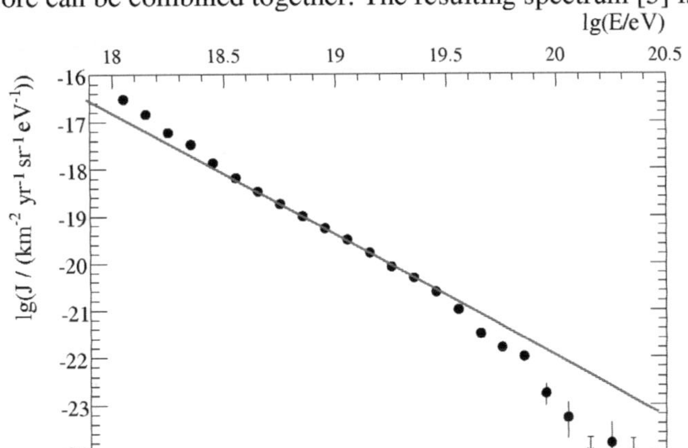

Figure 4. The combined energy spectrum from the Auger Observatory. The total number of events is about 37,000. The red line drawn through the data in the central region is only meant to give a qualitative indication of the two features (the ankle and the GZK suppression).

A simple way of describing the energy dependence of the spectrum in the three regions separated by the two breaking points is provided by a fit with three power law $E^{-\gamma}$ which is done leaving as free parameters also the two values of the energy where the spectral index γ changes (see Fig. 5).

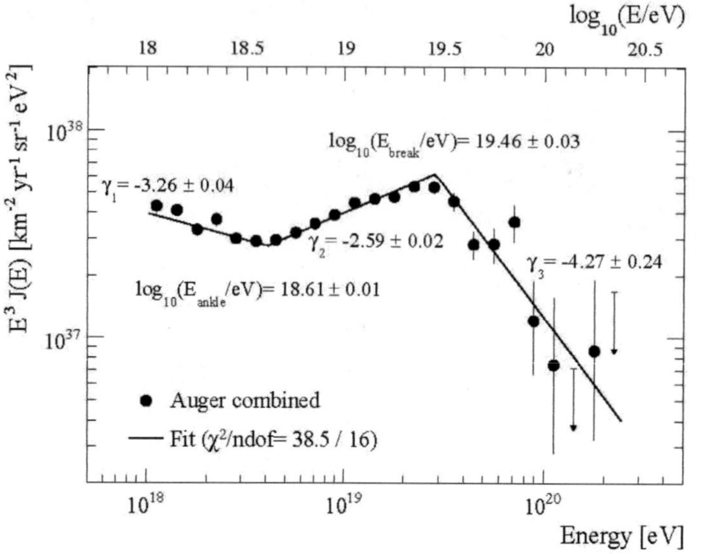

Figure 5. The Auger data are presented as Flux x E^3 together with the fit described in the text. The results of the fit are also shown. The ankle and the GZK suppression are clearly seen.

The Auger spectrum is shown again in Fig.6 together with the HiRes data. The disagreement between the two experiments seems to be due to different energy scale. A relative shift of the energy scale by an amount close to the quoted systematic error would bring the two sets of data in agreement.

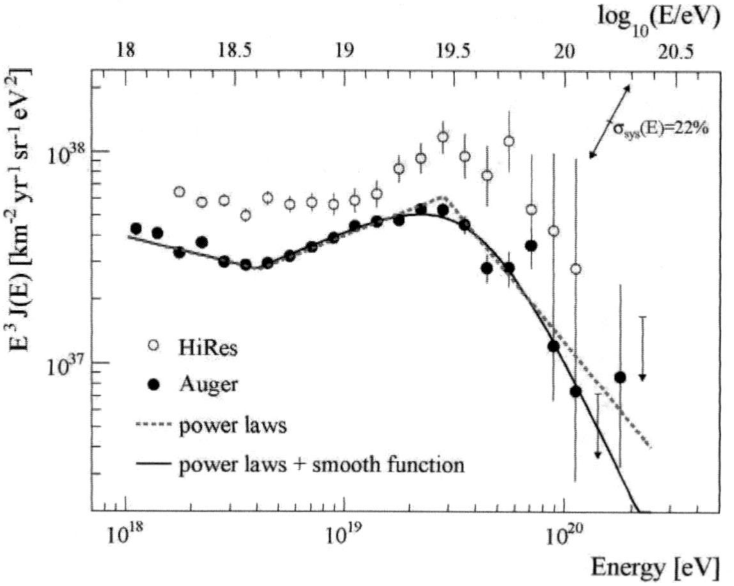

Figure 6. The Auger data are presented as Flux x E^3 together with the three power law fit (red dots) and a fit with smoothing in the GZK region (black line). The systematic error on the Auger energy scale is indicated. The HiRes data are also shown.

4. Correlation with nearby galaxies.

The observation of the GZK cutoff implies that the events close to the high-energy limit of the spectrum are of extragalactic origin. The Auger collaboration has been trying to identify the actual sources of these very high energy particles. A first attempt comparing the observed directions of cosmic rays with energy above 55 EeV with the position on the sky of the AGN galaxies listed in the Véron Cetty & Véron (VCV) catalogue at distances less than 75 Mpc was published in 2008 [6]. Table 1 updates that early study. The correlation increases significantly when the data around the galactic plane, where the catalogue is incomplete, are discarded. The probability that a "a priori" isotropic distribution would, by chance, simulate the observed distribution is at the level of a per cent.

	Number of events **E > 55 EeV**	**Correlated with AGN** **ψ = 3.1 degree**	**Expected for isotropy**
January 2004 - March 2009	**58**	**26**	**12.2**
Excluding band on the galactic plane (|b| > 12 degree)	**45**	**25**	**11.3**

Table 1. Results of the study of correlation of the high-energy Auger events with respect to the AGN galaxies listed in the VCV catalogue.

The distribution of the angular separation between the 58 events with E > 55 EeV and the closest AGN within 75 Mpc is shown in Fig. 7.

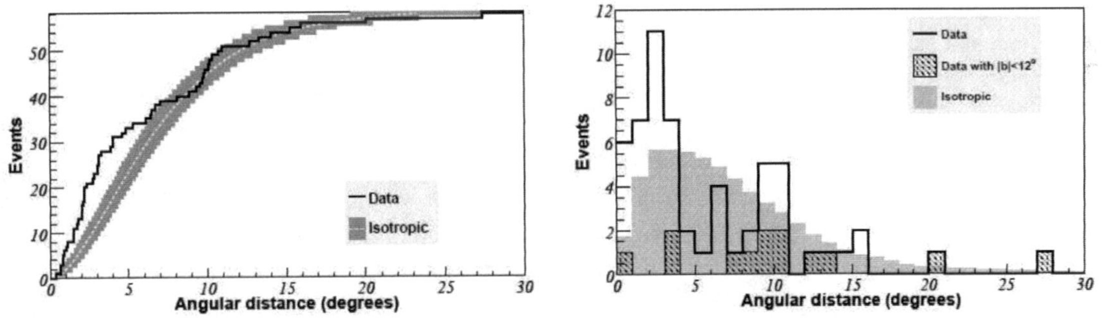

Figure 7. The distribution of the 58 events with E > 55 EeV and the closest AGN within 75 Mpc.
Left panel. Cumulative distribution. The blue band represents the expectation for isotropy at the 68% confidence level.
Right panel. Ordinary histogram.

These results provide a strong hint for the identification of the sources of very high energy cosmic rays as special galaxies of the AGN type.

The values of the parameters which were determined empirically as providing the best correlation with catalogues of galaxies are consistent with simple expectations. In fact the events with energy above 55 EeV are at the limit of the energy spectrum in the region of the GZK suppression which operates on extragalactic particles. In Fig. 8 the radius of the GZK sphere for protons is shown as a function of their observed energy on Earth for different fractions of the total number of events. The radius of the GZK sphere shrinks as the energy increases. For E > 55 EeV one expects that more than 40% of the events would come from distances less than 75 Mpc. This is consistent with the correlation results.

On the other hand, as shown in Fig. 9, the expected deflection in the galactic magnetic field of protons with energy of 60 EeV is around 3^0 and therefore a correlation angle of 3.1^0 appears plausible. Therefore the overall picture, GZK suppression and correlation with AGN galaxies appears consistent.

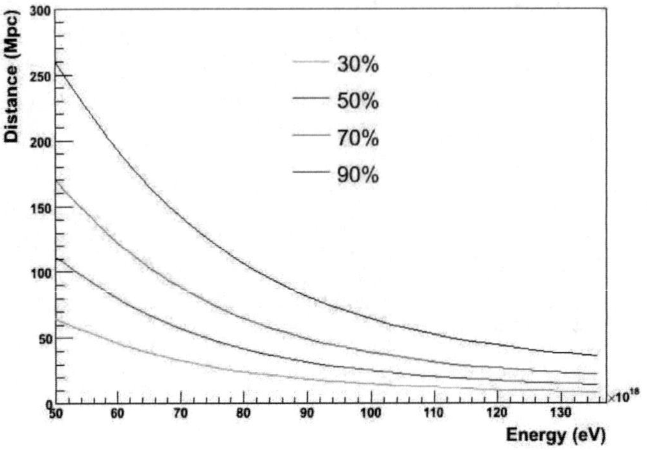

Figure 8. Calculation of the radius of the GZK sphere as a function of the observed energy of protons on Earth. The lines indicate the fraction of events which is actually observed for a given energy and a given distance of the source.

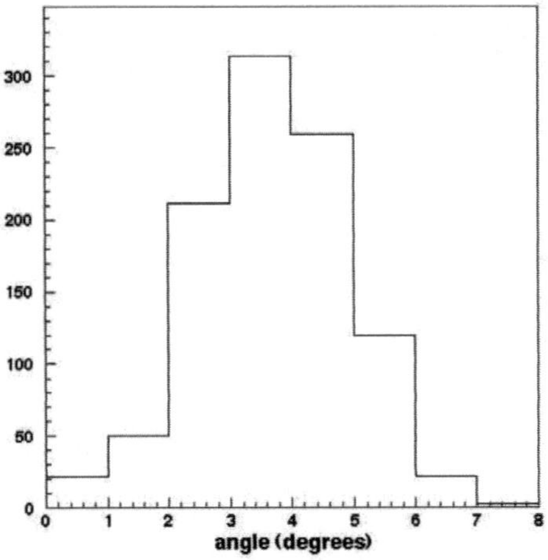

Figure 9. The calculated deflection of protons of 60 EeV in the galactic magnetic field [6].

References

[1] J. Abraham *et al.*, *Nucl. Instr. and Methods* **A523** (2004) 50
[2] I. Allekotte *et al., Nucl. Instr. and Methods* **A586** (2008) 409
[3] F. Arqueros *et al.*, *New Journal of Physics* **11** (2009) 065011 and references therein
[4] C. Di Giulio, *arXiv:*0906.2189v2 [astro-ph], Proc. 31st ICRC, Łódź, Poland (2009)
[5] F. Schüssler F 2009 *presentation at the* 31st ICRC, Łódź, Poland (2009)
see also *arXiv:*0906.2189v2 [astro-ph]
[6] J. Abraham *et al.,Astropart. Phys.* **29** (2008)188
J. Aublin and J.D.Hague *arXiv:*0906.2347v2 [astro-ph], Proc. 31st ICRC, Łódź, Poland (2009)

On the origin of high energy cosmic rays

Pasquale Blasi

INAF/Osservatorio Astrofisico di Arcetri, Largo E. Fermi, 5 50125 Firenze, Italy

E-mail: blasi@arcetri.astro.it

Abstract. I summarize some recent observational and theoretical developments in the investigation of the origin of cosmic rays. The implications of the supernova remnant paradigm for the origin of Galactic cosmic rays as well as for the transition to extragalactic cosmic rays will be emphasized. I will also discuss the complex observational situation in the ultra high energy region.

1. Introduction

In the last few years several new ideas and new observational facts have contributed to enrich our view of the problem of the origin of cosmic rays. X-ray observations of several supernova remnants (SNRs) have shown the presence of spatially very narrow filaments of X-ray emission which is thought to be produced by synchrotron emission of high energy electrons behind the forward shock, where particles (electrons and presumably nuclei) are accelerated. The observed thickness of these filaments leads to infer that strong magnetic field amplification, to values of $\sim 100\mu G$, is at work at these shocks. This is important for many reasons: on one hand it suggests that effectively accelerated particles may be responsible for this magnetic field amplification; on the other hand, particles are expected to be effectively accelerated (even up to the knee region) if magnetic field amplification is at work. In fact the discovery of such magnetic fields has been considered by many researchers in the field as the missing piece that may finally allow us to establish a direct connection between SNRs and cosmic rays, at least up to the knee, the so-called *SNR paradigm*. Amplification of seed magnetic fields in the shock proximity as induced by accelerated particles has been subject of much investigation for several decades and the interest in this phenomenon has recently been revived because of some new theoretical insights (e.g. [1]) and the observational situation illustrated above. The theoretical picture that arises from these new developments, together with the data on chemical composition in the knee region, suggest that the knee may be due to the superposition of spectra of different nuclei with maximum energies proportional to the charge of the nucleus. This picture has however as a natural consequence that the spectrum of CRs accelerated in SNRs should end at around $\sim 10^{17}$ eV. In the absence of additional sources in the Galaxy, this would also represent the end of the spectrum of Galactic CRs. This situation is at odds with the traditional picture in which the transition from Galactic to extragalactic CRs occurs at the *ankle*, at $\sim (0.5-1)\times 10^{19}$ eV. Two new ideas have been put forward in the last few years on the description of the transition: in the dip scenario [2, 3, 4] the transition is completed at $\sim 10^{18}$ eV and the feature known as ankle is in fact due to the combined action of adiabatic energy losses (induced by the expansion of the Universe) and proton pair production. In the mixed composition scenario [5, 6] the transition is rather smooth and is due to the superposition of the contributions of several components with different nuclear mass. The two scenarios lead to very different and potentially testable predictions for the chemical composition change as a function of energy.

Many new observational elements have been found in the ultra high energy regime. Both HiRes [7] and the Pierre Auger Observatory (PAO) [8] have confirmed the detection of the GZK feature in the CR spectrum. Moreover, the Pierre Auger Collaboration has claimed the first detection of anisotropies in the arrival directions of UHECRs in the form of correlations with AGNs in a specific catalog. Both the PAO and HiRes have also measured the mean penetration depth of the showers as a function of energy and the strength of the fluctuations on this quantity, which allow to infer the chemical composition of UHECRs.

In the following I will discuss some of these developments and their implications for the field. In §2 I will summarize the implications of recent X-ray and gamma ray observations for the origin of CRs. In §3 I will illustrate some viable approaches to the description of the transition between Galactic and extragalactic CRs. In §4 I will discuss the origin of ultra high energy cosmic rays (UHECRs) in the light of the new measurements carried out by HiRes and the PAO. I will conclude in §5.

2. X-rays, γ-rays and Cosmic Rays

The observation of narrow bright rims in X-rays with Chandra in several SNRs has refueled the interest in SNRs as potential sources of CRs up to the knee, since the simplest explanation of these filaments is that they reflect the synchrotron loss length of high energy electrons behind the shock. The magnetic field required for this interpretation to work is $\sim 100 - 1000\mu G$, and if this field is in the form of a flat enough power spectrum it may explain how SNRs can accelerate CR protons up to $\sim 10^{15}$ eV through diffusive particle acceleration at the shock front. From the theoretical point of view, it is expected that if particle acceleration is efficient, then the accelerated particles may generate their own scattering centers through resonant streaming instability [9, 10]. In other words, the presence of intense magnetic fields, in excess of the ones measured in the interstellar medium, would be an indirect proof of efficient acceleration of protons in SNRs. More recently in Ref. [1] it was proposed that non resonant modes could be amplified more effectively, though on spatial scales which are not immediately relevant for acceleration of the same particles that produced the waves in the first place. The discussion on which modes are relevant for particle scattering is very active and ongoing. In fact the debate is even wider in that the magnetic field could be amplified to the levels necessary to explain X-ray observations also through other mechanisms, which do not directly involve CRs (see for instance [11]). In order for particle acceleration to be efficient up to $\sim 10^{15}$ eV in these alternative scenarios the shock is required to be quasi-perpendicular [12]. On the other hand, if the magnetic field is amplified by the streaming of the energetically dominant hadronic component, the observation of the X-ray rims would also suggest that efficient CR acceleration is taking place in SNRs. In this case gamma ray emission is also expected if enough target for pp collisions is present in the surrounding medium. The importance of the detection of this emission has justified intensive observational campaigns with Cherenkov gamma ray telescopes as well as the ongoing search with the Fermi satellite.

Discriminating between these different possibilities requires a detailed investigation of the multi-frequency spectra of the individual SNRs in which these phenomena are observed. The most up-to-date calculations are based on the so-called non-linear diffusive shock acceleration (NLDSA) which allows one to describe the acceleration process taking into account the dynamical reaction of accelerated particles on the shock and the magnetic field amplification as due to streaming instability. These calculations seem to confirm that efficient CR acceleration may lead to strong magnetic field amplification which in turn can explain the X-ray rims and in principle explain the acceleration of protons up to the knee region. At least for some SNRs gamma ray emission has been measured by H.E.S.S. and more recently by Fermi [13], and for some of these cases observations are best explained in terms of production and decay of neutral pions. If these findings will be confirmed by ongoing observations they might represent the first evidence for (hadronic) CR acceleration in SNRs. In Figs. 1 and 2 (from [14]) we show, as an example of the power of this type of observations, the observed multi-frequency spectrum for the SNR RX J1713.7-3946 compared with the predicted one for a scenario involving efficient acceleration (Fig. 1) and one in which acceleration is inefficient (Fig. 2). In both cases the X-ray emission is due to synchrotron emission of relativistic electrons, but the magnetic field in which the radiation is emitted is very different in the two cases: when acceleration is efficient the magnetic field is boosted to values of $\sim 100-200\mu G$ downstream of the shock, so that the X-ray emission is spatially confined to narrow rims by severe synchrotron losses. In the inefficient case, the magnetic field behind the shock is only $\sim 10-20\mu G$ and the X-ray emission is broad. In the latter case an alternative explanation for the spatial distribution of the X-ray emission is required. The gamma ray radiation observed by H.E.S.S. [15, 16] is well described by production and decay of neutral pions in the case in which the acceleration is efficient. In this scenario, the highest energy of protons is high enough that the corresponding gamma ray emission extends to $\sim$ 100 TeV, as observed. In the inefficient scenario, the gamma ray emission is expected to be the result of inverse Compton scattering (ICS) of high energy electrons, but observations require a large optical photon background in the SNR (about 20 times the Galactic average) and in any case the last few H.E.S.S. data points are badly fit. In this latter case, protons are also accelerated but they carry a small fraction of the kinetic energy of the SNR (about 2%). The efficient scenario also has some problems: if protons and electrons share the same temperature behind the shock, then a large flux of thermal X-ray emission should be expected, which is not observed. In fact even if electrons are heated to a temperature much smaller than that of the protons, Coulomb scattering might be sufficient to heat electrons enough to start exciting emission lines. The non-detection of these lines (in fact no thermal emission is observed from this SNR) seems to score against a scenario with efficient CR acceleration, although the evidence is rather marginal at the present time due to the uncertainties in the values of environmental parameters. The ratio of electrons to protons spectrum at injection in the two (efficient and inefficient) scenarios illustrated above is respectively $\sim 10^{-4}$ and $\sim 10^{-2}$. One should be careful in not mistaking this ratio with that observed at Earth which is the result of the superposition of the injection in a given SNR over its temporal evolution and over many SNRs in the Galaxy, and is affected by propagation.

The detection of gamma rays both in the GeV and in the TeV energy range is crucial to test the SNR paradigm for the origin of CRs: first, the spectrum of gamma rays from π^0 decays is very different from that due to ICS; second, if SNRs are efficient CR accelerators then the spectrum of accelerated particles (and hence of the gamma rays from π^0 decays) is predicted by NLDSA to have a concave shape. This concavity (solid line on the right in Fig. 1) should be observable if the gamma ray spectrum from GeV to TeV is observed from an individual SNR. Both these observational issues will hopefully be accessible in the near future.

It is worth recalling that in addition to all these possible indirect evidences of hadronic acceleration in SNRs there is a unique way to prove or disprove that hadrons are accelerated efficiently in these astrophysical objects, namely detection of neutrinos (e.g. [17]) by upcoming km^3 neutrino telescopes.

3. Transition from Galactic to extragalactic CRs

The spectrum of protons as measured by ground detectors shows a sharp drop at energies $\sim 3\times 10^{15}$ eV (see for instance [18]). It is plausible (but not demonstrated) that this drop may correspond to a cutoff in the spectrum of accelerated protons. The situation illustrated above suggests that nuclei with charge Z may be accelerated to energies of order $3Z\times 10^{15}$

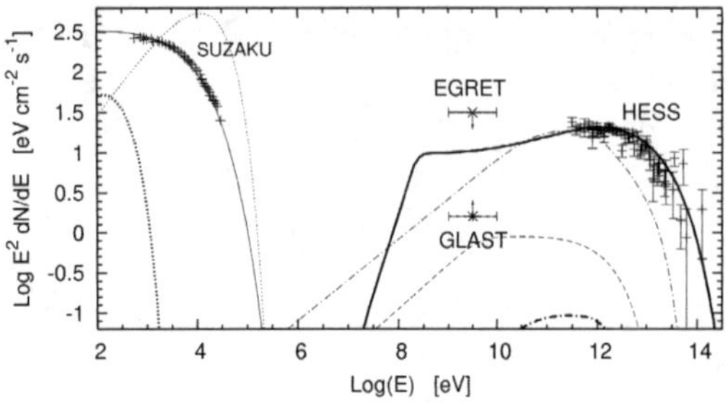

Figure 1. Multi-frequency spectrum for an efficient acceleration scenario. The following components are plotted: synchrotron emission (thin solid line), thermal emission (dotted line), ICS on the CMB (dashed line) and on Opt+IR background (dot-dashed line). The contribution from pion decay is shown as a thick solid line. HESS data taken from 2003 to 2005 are plotted together with Suzaku data in the X-ray band. Also EGRET upper limit and GLAST sensitivity are shown.

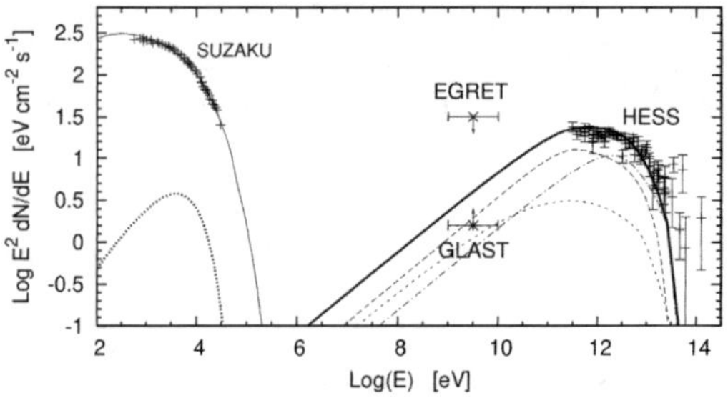

Figure 2. Multi-frequency spectrum for an inefficient acceleration scenario. The following components are plotted: synchrotron (thin solid line) and thermal (thick dotted line) electron emission, ICS component on the CMB (dashed line), optical (dotted line) and IR (dotdashed line) photons, and the sum of the three (thick solid line).

eV and SNRs appear to be the most plausible sources of these CRs. In this case we may infer that Galactic CRs should somehow *end* at energies of order $\sim 3Z_{Fe} \times 10^{15} eV \sim 10^{17}$ eV. This picture is clearly at odds with the traditional *ankle* scenario, where the transition from Galactic to extragalactic CRs occurs around 10^{19} eV. This picture, in which the knee seems to correspond to a transition to a gradually heavier chemical composition of CRs appears to be also confirmed by KASCADE data (see [18] for a recent review).

There are currently two models of the transition region that satisfy current observational constraints: the dip model [2, 3, 4] provides an excellent fit to the observed CR spectrum, simply invoking adiabatic and pair production losses of protons propagating on cosmological distances. The alternative explanation of the transition is provided by the mixed composition model [5, 6], in which the observed CR spectrum is explained as a superposition of the spectra of different elements propagated over cosmological distances. The two models lead to very different predictions on the chemical composition in the transition region. In the dip model the chemical composition changes rather sharply from a heavy dominated composition to a light composition (mainly protons) at 10^{18} eV, where the transition is already completed. In the mixed composition model the transition is much smoother and only at energies above 10^{19} eV the composition is dominated by protons [26]. This difference should be used as the crucial observational tool to distinguish the two models of the transition from Galactic to extragalactic CRs. It is worth stressing that in both models the chemical composition at the highest energies is dominated by protons, therefore in both cases the all-particle spectrum is characterized by the presence of the GZK feature, as due to photopion production on the cosmic microwave background photons (CMB). In the mixed composition model there is also a small contamination of lighter nuclei in the GZK region.

4. Ultra high energy cosmic rays

The propagation of protons from a homogeneous distribution of sources distributed on cosmological distances leaves an imprint on their spectrum in the form of two features: the dip and the GZK feature, the former induced by the competition between adiabatic and pair production losses and the latter induced by photopion production on the CMB. The energy at which these features appear is well defined because of the known kinematics of the processes involved and the well known spectrum of the CMB. The most recent data by HiRes [7] and the Pierre Auger Observatory (PAO) [8] confirm the detection of the GZK feature. The dip has been seen by all experiments although at somewhat different measured energies: it was shown by [3] that by normalizing the energies by the position of the dip, the spectra of CRs by different experiments are also in agreement with each other. Although not a proof of the detection of the dip, this certainly suggests that the dip may be there and hence that extragalactic CRs are mainly made of protons. On the other hand the observed spectrum is well fit also by assuming a mixed composition, which makes the situation rather ambiguous, but also hints to the fact that the measurement of the chemical composition is, at this point, crucial.

A breakthrough [19] in the field happened in 2007 with the announcement by the Pierre Auger Collaboration of the first detection of anisotropy at energies 55 EeV in the form of a correlation with AGNs in a specific catalog. The correlation was found to be best by assuming a *shift* angle between arrival direction and position of the source in the catalog of ~ 3 degrees, possibly to be interpreted with the deflection of protons in the magnetic field of the Galaxy. It is important to realize that this result cannot be considered as a proof that UHECRs are accelerated in AGNs, since it would hold even if AGNs were tracking the distribution of the actual sources of UHECRs.

This finding did strongly hint to a proton dominated chemical composition in the relevant energy range, since higher charge particles would suffer too large deflections in the Galactic magnetic field to keep any correlation with the sources. This result was also consistent with the claim of detection of the GZK feature by the PAO [8] and by HiRes [7].

In this sense, special care should be used in referring to this result as a detection of anisotropies or detection of correlations with sources or classes of sources. An anisotropic signal could be caused by few nearby sources or many sources belonging for instances to the local supercluster, in principle even in the presence of angular deflections comparable with the angular size of the supercluster in the sky (say ~ 20 degrees). Correlations with specific sources in a catalog and within a given angle from the source would lead to much more severe constraints and more important implications.

The measurement of $< X_{max}(E) >$, the penetration depth, and of its rms fluctuations, carried out with the PAO [20] (see also [21] for a review) showed a mixed composition at all energies but with a puzzling trend at the highest energies, where both $< X_{max}(E) >$ and its fluctuations appear to be compatible with a transition towards higher mass composition. This bit of information is however rather difficult to fit in a general meaningful scenario: on one hand the detection of the GZK feature and the detection of anisotropies are compatible with each other, but in this case the information on the chemical composition should be neglected. On the other hand, a heavy chemical composition is hard to reconcile with the presence of anisotropies since an iron-like nucleus would suffer too large deflections in the Galactic magnetic field. Moreover in the case of a heavy composition at the highest energies the observed flux suppression is not the GZK feature, as due to photopion production of protons, but it should be a cutoff probably associated with the photodisintegration of iron nuclei. On the other hand it is intriguing that even in the case of a pure iron composition at the source, the spectrum at Earth is in general dominated by protons, as a result of photodisintegration. In other words, even in this extreme case the observed composition would at worst be mixed and a GZK-like feature would be observed. As pointed out in Ref. [22], a way out of this conclusion would be that the observed cutoff is not the GZK feature but rather the cutoff in the source spectrum and that at lower energies the composition is mixed because of the superposition of the cutoffs of lighter elements modified by their photodisintegration. Also in this case the correlations observed by the PAO are hard to interpret since most particles at the relevant energies would be iron nuclei.

It is worth pointing out that neither the anisotropy nor the chemical composition measured by the PAO are confirmed by the HiRes collaboration. HiRes detected a trend in $< X_{max}(E) >$ which hints to an almost pure proton composition at $E > 10^{18}$ eV [23], in agreement with the prediction of the dip model. No evidence for anisotropies has been found in the HiRes data [24], though it is not clear if this may be due to limited statistics.

5. Summary and Discussion

A major breakthrough in the field of CR research has come from X-ray astronomy with the observation of narrow rims of synchrotron non-thermal X-ray emission from several young SNRs. This rims, most easily interpreted as due to the severe synchrotron losses of electrons close to their maximum energy, hint to the presence of intense magnetic fields at SNR shocks, much in excess of interstellar fields. Strong magnetic field amplification is in fact expected if efficient particle acceleration occurs at the shock front, so that the detection of the rims has been considered by many as indirect evidence of efficient CR acceleration in SNRs.

Whether this field is sufficient to explain CR acceleration up to the knee is still matter of much debate, mainly because of the difficulties in describing the diffusive properties of accelerated particles only based upon the strength of the observed magnetic field. Moreover PeV energies are expected to be reached only in SNRs at the very beginning of the Sedov-Taylor phase, and for a relatively short time compared with the overall lifetime of a remnant. It follows that the probability of actually *seeing* a pevatron in the acting of behaving as such is rather low.

From the observational point of view a sharp drop in the proton spectrum at $E \sim 3 \times 10^{15}$ eV seems to be observed (see for instance [18]). For heavier nuclei the observational uncertainties are such to make the situation less clear. Theoretical inference would suggest that the knee may be plausibly due to the superposition of the spectra of gradually heavier nuclei which are accelerated to maximum energies which scale with the charge of the nucleus. A natural conclusion of this line of thought would be to expect that the spectrum of Galactic CRs would end at $\sim 10^{17}$ eV, at odds with the traditional interpretation of the ankle. The dip model [2, 3] and the mixed composition model [5, 6] appear to fit current observations but they make very different predictions on the chemical composition in the transition region [25, 26]. The issue of understanding this transition is crucial for the whole problem of the origin of UHECRs.

The main problem in this respect is in the complexity of the observational situation at the present time: both the HiRes and the PAO collaborations claimed the detection of the GZK feature at the energy that would correspond to photopion production of protons on the CMB photons. This is consistent with the HiRes claim of a proton dominated chemical composition at $E > 10^{18}$ eV but it seems at odds with the PAO measurement of the shower penetration depth, which suggests a mixed composition and a transition to heavy composition in the highest energy bins. The PAO collaboration has also claimed the detection of a statistical correlation of the arrival directions of the highest energy events with the positions in the sky of AGNs in the local universe. This correlation requires small deflections in the Galactic magnetic field and hence a proton dominated chemical composition at the highest energies, apparently inconsistent with the observed mean value and rms value of the penetration depth.

The understanding of the origin of UHECRs will have to await for clarity in this complex observational situation.

References

[1] Bell A R 2004 *MNRAS* 353 550
[2] Berezinsky V, Gazizov A Z, Grigorieva S I 2004 *Phys. Lett. B* 612 147
[3] Berezinsky V, Gazizov A Z, Grigorieva S I, hep-ph/0204357
[4] Berezinsky, V, Aloisio R, Blasi P, Gazizov A, Grigorieva S, Hnatyk B 2007 *Astropart Phys* 27 76
[5] Allard D, Parizot E, Khan E, Goriely S, Olinto A V 2005 *A&A* 443 L29

[6] Allard D, Parizot E, Olinto A V 2007 *Astropart Phys* 27 61
[7] Abbasi R U et al (HiRes Coll.) 2008 *Phys Rev Lett* 100 1101
[8] Abraham J et al 2008 *Phys Rev Lett* 101 061101
[9] Zweibel E G 2003 *ApJ* 587 625
[10] Achterberg A 1983 *A&A* 119 274
[11] Giacalone J, Jokipii J R 2007 *ApJ Lett* 663 41
[12] Jokipii J R 1987 *ApJ* 313 842
[13] Funk S, Review Talk at the 2009 Fermi Symposium, http://fermi.gsfc.nasa.gov/science/symposium/2009/abs/sfunk.html
[14] Morlino G, Amato E, Blasi P 2009 *MNRAS* 392 240
[15] Aharonian F et al (H.E.S.S. Coll.) 2006 *A&A* 449 223
[16] Aharonian F et al (H.E.S.S. Coll.) 2007 *A&A* 464 235
[17] Morlino G, Blasi P, Amato E 2009 *Astropart Phys* 31 376
[18] Bertaina M, Battistoni G, Muraro S, Navarra G, Stamerra A 2008 *J Phys C Suppl* 120 2023
[19] Abraham J et al 2007 (PAO Coll.) *Science* 318 188
[20] Abraham J et al (PAO Coll.) 2009, in the Proceedings of the 31st International Cosmic Ray Conference, Lodz, Poland, July 2009 (arXiv:0906.2319)
[21] Cronin J 2009, in the Proceedings of the Conference Blois 2009 (arXiv:0911.4714)
[22] Aloisio R, Berezinsky V 2009 *Ultra High Energy Cosmic Rays: The disappointing model* (arXiv:0907.5194)
[23] Abbasi R U et al (HiRes Coll.) 2009 *Evidence for proton-dominated cosmic ray composition above 1.6 EeV* arXiv:0910.4184
[24] Abbasi R U et al (HiRes Coll.) 2008 *Astropart Phys* 30 175
[25] Berezinsky V, Aloisio R, Blasi P, Ostapchenko S 2008 *Phys Rev D* 77 5007
[26] Allard D, Olinto A V, Parizot E 2007 *A&A* 473 59

Journal of Physics: Conference Series **203** (2010) 012018 doi:10.1088/1742-6596/203/1/012018

Ultrahigh Energy Neutrinos: Theoretical Aspects

M. Kachelrieß

Institutt for fysikk, NTNU, Trondheim, Norway

Abstract. Ultrahigh energy (UHE) neutrinos can be produced by cosmic rays (CR) interactions with gas or photons in their sources, with cosmic microwave background photons during propagation, and by top-down sources. We review the status of these predictions.

1. Introduction

The boundary between HE and UHE neutrino astronomy can be connected to a change in observational techniques, with underwater and ice detectors as IceCube and Baikal/ANTARES mainly suited for the HE neutrino range up to $E_\nu \sim 10^{17}\,\mathrm{eV}$ and new techniques like radio detection for the UHE part [1]. In this review, we prefer to define the border between HE and UHE as the transition from galactic to extragalactic astrophysical neutrino sources, while in the case of neutrinos from decays or annihilations of exotic particles we use $E_\nu \sim 100\,\mathrm{TeV}$ as lower limit. The latter choice effectively excludes WIMP annihilations or decays as neutrino sources that were discussed by P. Salati [2].

Any process involving hadronization leads mainly to the production of pions. Isospin symmetry fixes then the ratio of charged and neutral pions. For instance, close to the threshold, the two possible channels of photo-meson production, $p + \gamma \to p + \pi^0$ and $p + \gamma \to n + \pi^+$, have a ratio that is close to 2:1. While the π^0 decay produces two photons, the final state of the second channel contains after the decays two electrons and four neutrinos. Clearly, the production of neutrinos is thus intimately tied to the one of photons and electrons, and both depend in turn on the flux of primary cosmic rays. A similar argument holds for neutrinos produced in the decay or annihilation of heavy particles into quarks, $\bar{X}X \to \bar{q}q$: Here the ratio of neutrino:photon:nucleon fluxes at production is roughly $50:13:1$.

While UHE photons and electrons start electromagnetic cascades, either on photons in the source or from the extragalactic background light, neutrinos reach us suffering only from redshift losses. Therefore, the measurement of the diffuse extragalactic γ-ray flux can be used to derive a strict upper limit on possible high energy neutrino fluxes, as suggested first in Ref. [3]. The connection between the observed cosmic rays and neutrinos is more model-dependent. A popular set of assumptions is the following one: Protons are accelerated in a source region, confined by magnetic fields. They interact on background photons in the source, neutrons from $p + \gamma \to n + \pi^+$ escape and provide the observed cosmic rays, while the π^+ decays produce neutrinos. This argument does not exclude that (a subset of) sources exists with sufficiently large interaction depth τ such that also neutrons interact frequently inside the source. The cosmic ray flux from these sources would be strongly suppressed and therefore this type of sources has been coined "hidden sources." On the other hand, the escape probability of CRs in diffuse shock acceleration is already without escaping neutrons such that a power-law spectrum with $\sim 1/E^2$ results. Thus the interaction depth for photo-pion production in the source could be much less than one, with correspondingly reduced neutrino fluxes.

Before we consider in the next sections different sources of UHE neutrinos, we comment on the effect of neutrino oscillations. After the discovery of large ν_μ–ν_τ mixing, it was often taken for granted that the small oscillation length (of order of 10 kpc for $E_\nu \sim 10^{18}$ eV) together with $\nu_e : \nu_\mu : \nu_\tau = 1{:}2{:}0$ at generation leads to flavour equipartition at observation, $\nu_e{:}\nu_\mu{:}\nu_\tau = 1 : 1 : 1$. There exist however several examples of neutrino sources where at least in some energy range significant deviations from this canonical flavor ratio from pion decay can be expected [4, 5]. Specific examples are the cases where the neutrino spectrum is dominated by neutron [5] or kaon decays [6] or influenced by muon damping [7]. For a detailed discussion of the various possibilities see Ref. [8].

2. Astrophysical neutrinos

The idea that neutrinos and photons are produced as secondaries in CR interactions close to the core of active galactic nuclei (AGN) has a long history [9]. In particular the diffuse neutrino fluxes from all AGN have been studied in great detail [10–12]. Moreover, the expected neutrino fluxes from Centaurus A (Cen A) were discussed recently in view of the results [13] from the Pierre Auger Observatory (PAO) in Refs. [14, 15].

Gamma-ray bursts (GRB) are another promising neutrino source [18, 19]. It is doubtful if GRBs give the dominant contribution to the extragalactic UHECR flux, because the emissivity of all GRBs is probably too small and also the maximal CR energy achievable in a GRB is unclear. Nevertheless, the possibility to correlate the photon and the putative neutrino signal, thereby essentially reducing the background to zero, promotes GRBs to promising targets for neutrino astronomy.

2.1. Individual sources

The extremely large mean free path of neutrinos together with the relatively poor angular resolution of neutrino telescopes ($\sim 1^\circ$) and the small expected event numbers makes the identification of extragalactic sources challenging using only neutrinos. Performing neutrino astronomy beyond the establishment of a diffuse neutrino background requires therefore most likely additional input, either timing or angular information from HE photon or CR experiments.

Flaring AGN The AGN responsible for the UHECR flux produce also neutrinos, if the interaction depth in the source is non-negligible. Possible targets are the UV photons emitted by the accretion disk and Compton up-scattered X-rays, a possibly compact IR background in the core, the synchrotron field in the jet, or clumps and the torus as baryonic targets. Among the most prominent AGN sub-classes proposed as neutrino sources are TeV photon emitters. These sources often show large and violent variations in the electromagnetic spectrum. Such variability lead in turn to the possibility for correlation studies of neutrino and photon fluxes, thereby reducing the background from atmospheric neutrinos. An example for such studies is the search for neutrinos associated with orphan TeV flares, i.e. the enhanced high energy γ-ray emission without any contemporaneous X-ray flaring activity. At the time of writing, positive evidence for such correlations is still missing.

Centaurus A The Auger collaboration announced 2007 evidence [13] for a correlation of the arrival directions of UHECRs with AGN. In particular, Ref. [13] found two events within the search bin of 3.1° around the nearest active galaxy, Cen A, that is located close to the supergalactic plane. This result has prompted several studies, attempting to use Cen A as a test case for successful "multi-messenger astronomy." Even in this case—where the UHECR

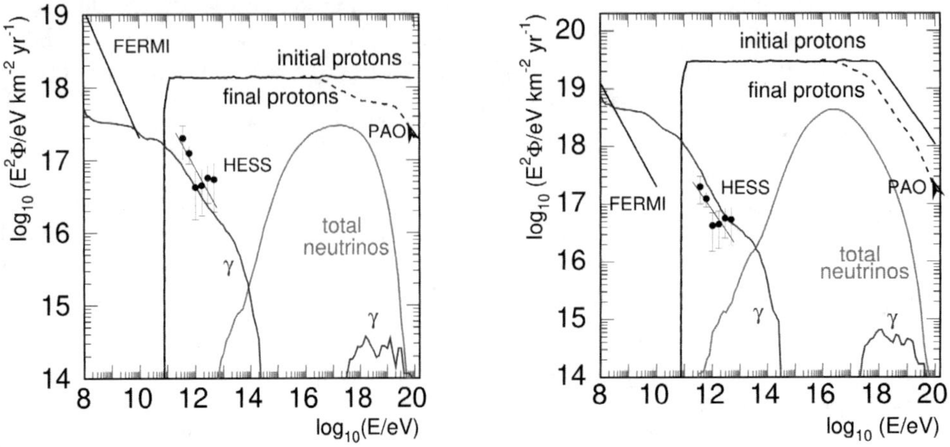

Figure 1. CR (black), neutrino (red) and photon (blue) fluxes from Cen A normalized to the PAO results assuming acceleration close to the core; left for an initial CR spectrum $1/E^2$, right for a broken power-law.

source flux could be fixed by the Auger data and much more detailed astronomical data about the source exists than for more distant AGN—the predictions for the secondary neutrino fluxes differ by up to three order of magnitude. We review here the results for the "core model" of Ref. [15] that has the merit of having predicted TeV gamma-ray fluxes from Cen A that are in agreement with subsequent H.E.S.S. measurements.

Figure 1 displays the particle fluxes from Cen A from [15] as function of the energy. The case of a CR injection spectrum $dN/dE \propto E^{-\alpha}$ with $\alpha = 2$ is shown by black solid lines on the left, while a broken power-law $\alpha = 2.7$ above 10^{17} eV is shown on the right. In this model, proton interact with IR and UV photons in the AGN core, with an interaction depth of a few above the threshold energy $\sim 10^{16}$ eV. The resulting secondary photon (blue solid) and neutrino fluxes (red solid) arriving on Earth are shown too, and compared with the recently published results from Fermi [17] and H.E.S.S. [16]. Remarkably, the photon flux in the Fermi and the H.E.S.S. energy range has approximately the same power-law exponent ($\alpha \sim 2.6$ and 2.7), but the latter requires a larger normalization constant. Such a behavior is expected, if a new component, e.g. of hadronic origin, sets in above 100 GeV, while at lower energies photons of electromagnetic origin dominate the spectrum. The shape of H.E.S.S.'s γ-ray spectrum agrees very well with the slope predicted in [15] for the core model. The Fermi measurements restrict additionally the source model, excluding the broken-power law case that leads to an excessive photon fluxes in the GeV range. The parameter space that leads to a (marginally) observable neutrino flux in a 1 km^3 neutrino telescope, is excluded by the γ-ray data too, leading to the pessimistic prediction of $\mathcal{O}(10^{-2})$ neutrino event numbers per year and km^3.

Gamma-ray bursts GRBs were proposed as main sources of UHECRs above few $\times 10^{19}$ eV [18, 19]. Synchrotron photons can provide a suitable target for $p\gamma$ reactions in GRBs and neutrino production. Stacking of optically identified GRBs for neutrino searches in IceCube provides the best limit so far, reaching within a factor of a few above the GRB neutrino flux prediction given in [19]. A single, bright GRB can be identified with a fully operational 1 km^3-detector like IceCube, producing 1 – 2 events at a much lower background level.

2.2. *Diffuse flux*
Having defined the properties of individual neutrino source and chosen its luminosity and source density evolution with cosmic time, the diffuse neutrino flux from these sources can be calculated in a straightforward way. Practically always, the resulting diffuse flux is easier to detect than the flux from individual sources, even if they are nearby as e.g. Cen A [20]. Most of the early, optimistic estimates [21] that normalized the neutrino emissivity by the observed X-ray flux are meanwhile excluded.

3. Cosmogenic neutrinos
Soon after the prediction of the GZK effect [22], Berezinsky and Zatsepin [23] realized that photo-pion interactions of UHECRs do not only lead to a cutoff in the CR spectrum but also to UHE neutrinos as secondaries. They showed that the UHE neutrino fluxes can be much higher than the observed UHECR flux in case of a strong cosmological evolution of the CR sources. Moreover, before the advent of gauge theories it was even plausible to speculate UHE neutrinos interact strongly and could thus mimic UHE protons[1].

For a known nuclear composition of UHECRs, the main ingredients for the calculation of the cosmogenic neutrino flux are the spectral shape of the extragalactic flux and the cosmological evolution of the CR sources. Assuming that the CR injection spectrum at the source is a power-law $dN/dE \propto E^{-\alpha}$, the choice of the exponent α is equivalent with the choice of the transition energy between the galactic and the extragalactic component of the UHECR flux.

There are three main approaches to explain the observed spectral shape of UHECRs: In the first one, the ankle is identified with the transition from a steep galactic, usually iron-dominated component to extragalactic protons with injection spectrum between $1/E^2$ and $1/E^{2.2}$ as suggested by shock acceleration. In the second approach, the dip is a feature of e^+e pair production and one is able to fit the UHECR spectrum down to $E = 10^{18}$ eV using only extragalactic protons and an injection spectrum between $1/E^{2.6}$ and $1/E^{2.7}$. The third one uses a mixed composition and aims also at explaining the spectral shape down to a few$\times 10^{18}$ eV. The first possibility requires the existence of Galactic sources accelerating up to few$\times 10^{19}$ eV and a relatively strong Galactic magnetic field to isotropize CRs. The second explains with a minimal set of parameters the UHECR spectrum over a large range, but is challenged by the recent Auger composition results [26]. Note that the steep *average* $\alpha = 2.6$ can be explained e.g. by a distribution of maximal source energies $\propto 1/E_{\max}^{1.6}$ [27].

Since in the case of nuclei primaries the main source for cosmogenic neutrinos are neutrinos from neutron decay with average energy $E_\nu \sim (m_e/m_n)E_0$, the cosmogenic neutrino flux peaks in this case between 10^{16} and 10^{17} eV and not at 10^{20} eV as for proton primaries. Thus the observation of cosmogenic neutrinos would be an important tool not only to learn the transition energy between the galactic and extraglactic CR flux [28], but also to study the nuclear composition of the extragalactic CR component.

4. Neutrinos from top-down sources
Topological defects such as (superconducting) cosmic strings, monopoles, and hybrid defects can be effectively produced in non-thermal phase transitions during the preheating stage [29]. Therefore the presence of TDs is not in conflict with an inflationary period of the early Universe. They can naturally produce particles with high enough energies but have problems to produce large enough fluxes of UHE primaries, because of the typically large distance between TDs. Then the flux of UHE particles is either exponentially suppressed or strongly anisotropic if a TD is nearby by chance.

[1] For modern versions of this speculation see Sec. 5.

The suggestion [30, 31] of superheavy metastable relic particles as UHECR sources was originally motivated by the AGASA excess, although it was soon realized that stable or metastable particles in this mass range are generically good DM candidates.

What are the particle candidates for SHDM? A well-suited candidate for superheavy dark matter is the lightest neutralino within the scenario of superheavy supersymmetry [32]. This is a unique case where the SHDM has weak interactions and respects perturbative unitarity despite the large mass hierarchy $m_X \gg m_Z$. Since the annihilation cross section of a (point) particle is bounded by unitarity, $\sigma_{\rm ann} \propto 1/m_X^2$, the resulting flux of UHE neutrinos for stable SHDM is too small to be observable without an additional enhancement of the annihilation rate. The flux of UHE neutrinos can be observable either if the SHDM particle is meta-stable with suitable life-time or the DM is much more clumpier than predicted by the simplest inflationary scenario.

5. UHE Neutrino interactions as test for new physics

The calculation of the UHE neutrino-nucleon cross section requires an extrapolation of parton distribution functions to yet unexplored values of Bjorken x and center-of-mass energies. The error estimate of these predictions at $E_\nu \sim 10^{20}\,$eV is however only of order 30%, making it possible to search for new processes that enhance cross section from its Standard Model (SM) value $\sigma_{N\nu} \sim 10^{-31}{\rm cm}^2$. Among the models suggested leading to large $\sigma_{N\nu}$ are string theories with large extra dimension, where an exchange of Kaluza-Klein gravitons and the production of black holes can increase the neutrino-nucleon cross section [33]. Even within the SM, it has been suggested that non-perturbative electroweak instanton contributions strongly increase the neutrino-nucleon interactions [34].

While the flux of horizontal air showers due to neutrino interactions in the atmosphere increases with $\sigma_{N\nu}$, the flux of up-going charged leptons produced by neutrino interactions below the surface is inversely proportional to the cross section [35]. Taken together, up-going and horizontal rates events offer the potential to measure UHE neutrinos both for SM-like and for deviating cross sections. Although the observed number of neutrino-induced events determines only the product of the neutrino scattering cross section and the incoming cosmic neutrino flux, the two quantities can be disentangled using the angular dependence of the event rate.

6. Summary and Outlook

In the near future one may expect an improved measurement of the diffuse extragalactic γ-ray flux by FERMI. Compared to EGRET, the subtraction of previously unresolved point sources will lead to a reduction of the diffuse γ-ray flux and thereby to a stronger cascade limit on the UHE neutrino flux. This suggests, together with the non-observation of astrophysical HE and UHE neutrinos hitherto, that experimentalists should aim at even larger detector volumes in the future.

References

[1] For an overview of experimental neutrino astronomy see this volume.
[2] For an overview of indirect dark matter searches see P. Salati, this volume.
[3] V. Berezinsky and A. Smirnov, Astrophys. Space Sci. **32**, 461 (1975).
[4] P. Bhattacharjee and N. Gupta, hep-ph/0501191.
[5] P. D. Serpico and M. Kachelriess, Phys. Rev. Lett. **94**, 211102 (2005) [arXiv:hep-ph/0502088].
[6] S. Ando and J. F. Beacom, Phys. Rev. Lett. 95, 061103 (2005) [astro-ph/0502521]. [12] K. Asano and S. Nagataki, Astrophys. J. 640, L9 (2006) [astro-ph/0603107].
[7] T. Kashti and E. Waxman, Phys. Rev. Lett. 95, 181101 (2005) [astro-ph/0507599
[8] M. Kachelrieß, S. Ostapchenko and R. Tomas, Phys. Rev. D **77**, 023007 (2008) [arXiv:0708.3047 [astro-ph]].
[9] V. Berezinsky, Proc. of Neutrino 1977, Elbros; D. Eichler, Astrophys. J. **232**, 106 (1979).

[10] F. W. Stecker, C. Done, M. H. Salamon and P. Sommers, Phys. Rev. Lett. **66**, 2697 (1991) [Erratum-ibid. **69**, 2738 (1992)], see also F. W. Stecker, Phys. Rev. D **72**, 107301 (2005) [astro-ph/0510537]; A. P. Szabo and R. J. Protheroe, Astropart. Phys. **2**, 375 (1994) [astro-ph/9405020].

[11] K. Mannheim, Astropart. Phys. **3**, 295 (1995). A. M. Atoyan and C. D. Dermer, Astrophys. J. **586**, 79 (2003) [astro-ph/0209231]; A. M. Atoyan and C. D. Dermer, New Astron. Rev. **48**, 381 (2004) [astro-ph/0402646].

[12] D. V. Semikoz and G. Sigl, JCAP **0404**, 003 (2004) [hep-ph/0309328].

[13] J. Abraham *et al.* [PAO Collaboration], Science **318**, 939 (2007), arXiv:0711.2256 [astro-ph]; arXiv:0712.2843 [astro-ph].

[14] A. Cuoco and S. Hannestad, Phys. Rev. D **78**, 023007 (2008) [arXiv:0712.1830 [astro-ph]].

[15] M. Kachelrieß, S. Ostapchenko and R. Tomas, New J. Phys. **11**, 065017 (2009) [arXiv:0805.2608 [astro-ph]].

[16] F. Aharonian *et al.* [HESS Collaboration] Astrophys. J. Lett. **695**, L40.

[17] T. Cheung *et al.* [Fermi-LAT Collaboration], PASA, to appear 2009.

[18] M. Vietri, Astrophys. J. **453**, 883 (1995) [arXiv:astro-ph/9506081].

[19] E. Waxman and J. N. Bahcall, Phys. Rev. Lett. **78**, 2292 (1997) [arXiv:astro-ph/9701231].

[20] H. B. J. Koers and P. Tinyakov, Phys. Rev. D **78**, 083009 (2008) [arXiv:0802.2403 [astro-ph]].

[21] L. Nellen, K. Mannheim and P. L. Biermann, Phys. Rev. D **47**, 5270 (1993); F. W. Stecker and M. H. Salamon, Space Sci. Rev. **75**, 341 (1996); J. Alvarez-Muniz and P. Meszaros, Phys. Rev. D **70**, 123001 (2004).

[22] K. Greisen, Phys. Rev. Lett. **16**, 748 (1966), G. T. Zatsepin and V. A. Kuzmin, Pisma Zh. Experim. Theor. Phys. **4**, 114 (1966).

[23] V. S. Berezinsky and G. T. Zatsepin, Phys. Lett **B 28**, 423 (1969); V. S. Berezinsky and G. T. Zatsepin, Soviet Journal of Nuclear Physics **11**, 111 (1970).

[24] V. Berezinsky, A. Z. Gazizov and S. I. Grigorieva, Phys. Rev. D **74**, 043005 (2006) [arXiv:hep-ph/0204357]. astro-ph/0210095.

[25] D. Allard, E. Parizot, E. Khan, S. Goriely and A. V. Olinto, Astron. Astrophys. **443**, L29 (2005) [arXiv:astro-ph/0505566].

[26] For a discussion see P. Blasi, this volume.

[27] M. Kachelrieß and D. V. Semikoz, Phys. Lett. B **634**, 143 (2006) [astro-ph/0510188].

[28] D. Seckel and T. Stanev, Phys. Rev. Lett. **95**, 141101 (2005) [arXiv:astro-ph/0502244]. M. Ahlers *et al.*, Phys. Rev. D **72**, 023001 (2005) [arXiv:astro-ph/0503229].

[29] For a recent review of the status of top-down models see M. Kachelrieß, arXiv:0810.3017 [astro-ph].

[30] V. Berezinsky, M. Kachelrieß and A. Vilenkin, Phys. Rev. Lett. **79**, 4302 (1997) [astro-ph/9708217].

[31] V. A. Kuzmin and V. A. Rubakov, Phys. Atom. Nucl. **61**, 1028 (1998) [astro-ph/9709187].

[32] V. Berezinsky, M. Kachelrieß and M. A. Solberg, Phys. Rev. D **78**, 123535 (2008) [arXiv:0810.3012 [hep-ph]].

[33] S. Dimopoulos and G. Landsberg, Phys. Rev. Lett. **87**, 161602 (2001) [hep-ph/0106295]; S. B. Giddings and S. Thomas, Phys. Rev. D **65**, 056010 (2002) [hep-ph/0106219].

[34] A. Ringwald, JHEP **0310**, 008 (2003) [arXiv:hep-ph/0307034]; M. Ahlers, A. Ringwald and H. Tu, Astropart. Phys. **24**, 438 (2006) [arXiv:astro-ph/0506698].

[35] A. Kusenko and T. J. Weiler, Phys. Rev. Lett. **88**, 161101 (2002) [arXiv:hep-ph/0106071].

Gamma ray observation from ground

M. Mariotti

Dipartimento di Fisica dell'Università di Padova, via F. Marzolo 8, I-35131 Padova, ITALY

E-mail: mariotti@pd.infn.it

Abstract. Nowadays the Gamma ray observation of the sky is opening an important window to understand the nature of high-energy phenomena in the Universe and provides important tools to test fundamental physics that is constrained in the production and propagation of high-energy photons. The observation of gamma ray from ground is playing an important role: the success of the running ground based experiments shows that the large collection area and the high sensitivity to the high energy domains makes the gamma ray observation from ground complementary to the satellite one, and in many cases fundamental for a large class of scientific investigation. In this paper the main ground based gamma ray running experiment are reviews and a glance of the perspective in this field is given.

1. Introduction

Thanks to MAGIC [1] and VERITAS [2] (in the northern hemisphere), H.E.S.S. [3] and CANGAROO [4] (in the southern hemisphere) a new wavelength domain is opened for astronomy, the domain of very high energy gamma rays with energies between about 100 GeV to about 100 Tera electronvolts TeV, energies which are a millions times higher than the energy of visible light. The high-energy gamma rays probe a "non-thermal" Universe [5] [6], where mechanisms other than thermal emission by hot bodies allow the concentration of large amounts of energy into a single quantum of radiation. High-energy gamma rays can be produced in a top-down fashion by decays of heavy particles such as the hypothetical dark matter particles or cosmic strings, both relics that might be left over from the Big Bang. In a bottom-up fashion, gamma rays can be generated when high-energy nuclei - accelerated for example in the gigantic shock waves created in stellar explosions - collide with ambient gas particles. The flux and energy spectrum of the gamma rays reflects the flux and spectrum of the high-energy nuclei. They can therefore be used to trace these cosmic rays in distant regions of our own Galaxy or even in other galaxies. The first images of the Milky Way in very high-energy gamma rays [7] were obtained in recent years, and reveal a chain of gamma-ray emitters lining the Galactic equator (Fig. 1.1) demonstrating that sources of high-energy radiation are ubiquitous in our Galaxy. Sources of this radiation include supernova shock waves, where atomic nuclei are presumably accelerated and generate the observed gamma rays.

Another important class of objects in this context are "nebulae" surrounding pulsars, where giant rotating magnetic fields give rise to a steady flow of high-energy particles. Some of the objects discovered are binary systems, where a black hole or neutron star orbits a massive star. Along the elliptical orbit, the conditions for particle acceleration vary and hence the radiation intensity is modulated with the orbital period. These systems are particularly interesting in that they allow probing how particle acceleration processes respond to the varying ambient conditions. One of several

surprises was the discovery of "dark sources", objects which emit high-energy radiation, but have no obvious counterpart in other wavelength regimes. In other words, there are objects in the Galaxy, which are so far only visible and detectable in high-energy gamma rays.
Beyond our Galaxy, well over a dozen extragalactic sources of high-energy radiation have been discovered, located in active galaxies, where a supermassive black hole at the centre is fed by a steady stream of gas and is releasing enormous amounts of energy. Gamma rays are believed to be emitted from the vicinity of these black holes, allowing the study of the processes occurring in this violent and as yet poorly understood environment.

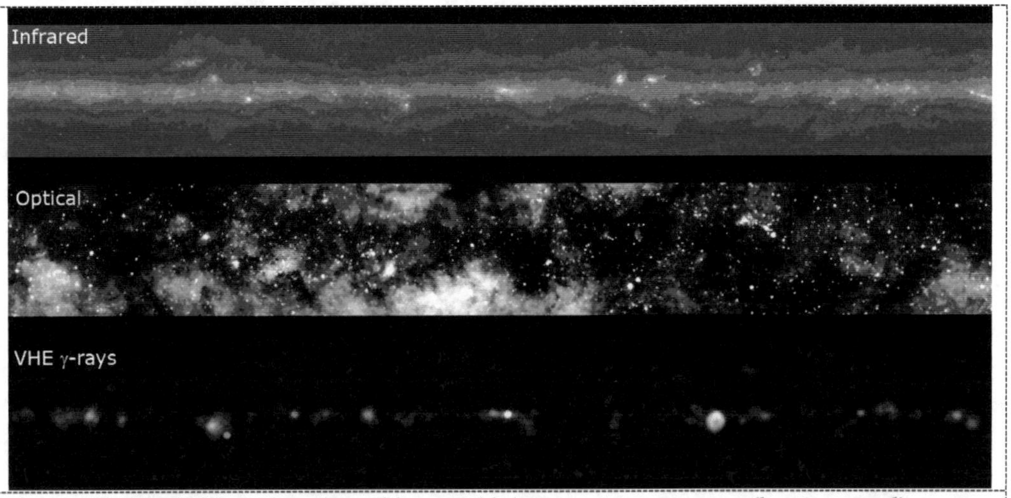

Figure 1.1 The Milky Way viewed in very high energy gamma rays (bottom panel), compared to its appearance in visible and infrared light

High-energy gamma rays trace populations of high-energy particles in distant regions of our own Galaxy or in external galaxies.

2. Ground based gamma ray experiments
Ground based detectors reveal Gamma rays by their interaction with the atmosphere that becomes an essential part of the detector itself. Because of this interaction, just high-energy photons (few tenth of Gev) can be detected. However thanks to the large effective area of the ground based experiments the ground installation are unique for the high sensitivity to the high-energy component of the gamma ray spectrum.

2.1. EAS detectors
The EAS detectors, such as MILAGRO and ARGO, are made by a large array of detectors sensitive to charged secondary particles generated by the atmospheric showers. They have high duty cycle and a large FoV, but a low sensitivity. Since the maximum of a photon-initiated shower at 1 TeV typically occurs at 8 km a.s.l., the energy threshold of such detectors is rather large. Direct sampling of the charged particles in the shower can be achieved:
• either by using an array of sparse scintillator-based detectors, as for example in the Tibet AS instrument (located at 4100 m a.s.l. to reduce the threshold). For an energy of 100 TeV there are about 50 000 electrons at mountain-top altitudes, so sampling is possible;
• or by effective covering of the ground to ensure efficient collection and hence lower energy threshold.

– The ARGO-YBJ [9] detector (see Fig. 2.1) at the Tibet site follows this approach. It is made of an array of resistive plate counters. Its energy threshold lies in the 0.5 TeV-1 TeV range. The first results show that ARGO can detect the Crab Nebula with a significance of about 5 standard deviations (σ) in 50 days of observation.
– MILAGRO [10] (see Fig. 2.2) is a water-Cherenkov based instrument near Los Alamos (about 2600 m altitude). It is made of photomultipliers in water. It detects the Cherenkov light produced by the secondary particles of the shower when they pass through the water.

Figure 2.1 View of the ARGO surface detector

The energy threshold of EAS detectors is at best in the 0.5 TeV-1 TeV range, and it also depends on where the first interaction of the atmospheric shower occurred, so such detectors are built to detect UHE photons as well as the most energetic VHE gammas. At such energies fluxes are small, so they need to have large surfaces, of order of 10^4 m^2 .

Figure 2.2 View of the MILAGRO surface detector

EAS detectors are possibly provided with a muon detector devoted to hadron rejection; otherwise the discrimination from the background can be done based on the reconstructed shower shape. The direction of the detected primary particles is computed by taking into account their arrival times, and the angular precision is about 1 degree. Energy resolution is also poor. The calibration can be performed by studying the shadow in the reconstructed directions caused by the presence of the moon. Thanks to an effective method of background rejection (based on a deep water Cherenkov light emission as a veto) MILAGRO made the discoveries of 3 extended sources, 4 less signficant hot-spots, and evidence for diffuse emission along the Galactic plane[13] [14].

2.2. Imaging Atmospheric Cherekov Tepescopes (IACT) detectors

The recent breakthroughs in very high-energy gamma-ray astronomy where achieved with Cherenkov telescopes. When a gamma ray enters the atmosphere, it interacts with atmospheric nuclei and generates a cascade of secondary particles, with electrons and positrons. Moving through the atmosphere at speeds higher than the speed of light in the gas, these particles emit a beamed bluish light, the Cherenkov light. On the ground, this Cherenkov light illuminates an area of about 250 m diameter, where it can be captured with optical elements and be used to image the trajectory of the particle cascade, which vaguely resembles the trajectory of a shooting star. Reconstructing this trajectory in space and tracing it back on to the sky, the celestial origin of the gamma ray can be determined.

Figure 2.3 View From top to bottom, the HESS array of four telescopes located in Namibia; the MAGIC array of two 17 m diameter telescopes located in the Canary island La Palma; the recently completed four IACT array VERITAS at Mt Hopkins, Arizona.

Accumulating many gamma rays, an image of the gamma-ray sky such as shown in Fig. 1.1 is created. Since Cherenkov light is very faint, large optical reflectors with areas in the 100 m^2 range and beyond are required to collect enough light, and the instruments can only be operated at night at dark sites. The Whipple Telescope in the United States pioneered this Imaging Atmospheric Cherenkov

Technique; after more than 20 years of development, the first source of very high-energy gamma rays, the Crab Nebula, was discovered in 1989 [6]. The Crab Nebula is among the strongest sources of very high-energy gamma rays, and is often used as a "standard candle". Modern instruments, using multiple telescopes to track the cascade from different perspectives and employing fined-grained photon detectors for improved imaging, can detect sources down to 1% of the flux of the Crab Nebula. Finely-pixellated imaging was first employed in the French CAT telescope, and the use of "stereoscopic" telescope systems to provide images of the cascade from different viewing points was pioneered by the European HEGRA instrument.
Currently, the world largest ground-based IACTs are HESS, MAGIC and VERITAS (Figure 2.3). HESS is an array of 4 clone telescopes, each of 12 m diameter, located in the Gamsberg Mountain in Namibia and is operating since 2003 with a very high scientific impact. A fifth telescope, dubbed HESS- II, of 28 m diameter, is under construction at the center of the array, and its completion is foreseen for 2010. MAGIC has operated since 2004 with a single dish of 17 m diameter and parabolic profile in the Canary island La Palma in Spain [5]. Despite the use of a single reflector does not guarantee the sensitivity of an array, its world-largest dish allowed to reach the lowest energy threshold of the IACT technique, performing for the first time observation below 100 GeV with this technique. Recently MAGIC has detected the first ever observed GR pulses at 25 GeV from the Crab pulsar. Recently this year, a second clone MAGIC telescope, dubbed MAGIC II, has been inaugurated. The use of the stereoscopic system will allow MAGIC phase-II to reach the sensitivity of larger arrays. A more recent experiment was started in the Arizona desert in USA, following the successful experience of the Whipple experiment. VERITAS has soon reached the expected performance, with sensitivity comparable (higher actually because of aging) to HESS and is starting to collect important scientific results.

3. The Future

The most advanced project for the EAS detectors is HAWC (High Altitude Water Cherenkov experiment). The aim of this project is to use high energy gamma rays probe the most extreme astrophysical environments including those that produce the highest energy cosmic-ray particles. The Milagro observatory has demonstrated that a detector with a wide field of view (2sr) and nearly 100% duty cycle can discover new sources of TeV gamma rays at energies between 10 and 100 TeV, and map the diffuse emission from the plane of our Galaxy. The HAWC (High Altitude Water Cherenkov) observatory builds on the experience and technology of Milagro to make a second-generation high-sensitivity detector. This unique detector will be capable of continuously surveying the TeV sky for steady and transient sources from 100 GeV to 100 TeV.

Concerning the Cherenkov telescopes, despite the promising achievements from the current generation of IACTs, there are a number of limitations that the future generation will overcome:

a) IACTs of current generation are sensitive in a limited energy range, from 100 GeV to 50 TeV. At the lower end, IACTs are limited by the background for low energy. At the high end, the limit is posed by insufficient statistics;

b) Due to the lack of a calibrated cosmic GR source, IACT spectral reconstruction is limited by systematic bias and statistical uncertainties on the energy reconstruction;

c) They have a limited aperture, with typical field of view (FOV) of the order of 3–5 deg diameter;

d) They have a limited angular resolution, which currently states around few arc-min;

e) They have limited collection area;

f) They are rather poorly automatized.

On the other hand, from a physical point of view, there are strong arguments to improve in the following aspects:

1. Decrease the energy threshold to few tens of GeV
2. Acquire sensitivity beyond 50 TeV

3. Increase sensitivity in the bulk range 100 GeV, 50 TeV
4. Improve energy and angular resolution

The accomplishment of the above goals is intimately related to the overcome of the technological questions discussed at the beginning of this section. For this reason, a new generation of IACT is under design now with expected performance well above the current generation. We believe that with CTA we are heading toward a new era of *"precision GR astronomy"*.

4. Conclusions

The field of VHE γ-ray astrophysics has gone through a dramatic evolution since 2004, thanks to the high sensitivity of the new generation of ground based gamma ray experiment IACTs.

The HESS, MAGIC and VERITAS experiments represents a major step in that it has revealed, beyond the large number of sources, diverse classes of γ-ray emitting galactic objects and acceleration sites: young shell-type SNRs [11], SNRs interacting with molecular clouds, middle-aged offset PWNe, very young composite PWNe and γ-ray binaries. Given the large number of still unidentified sources, other potential classes of sources could emerge, including the promising case of massive star clusters. The increasing number of blazars sources in the extragalactic domain allows now for population studies, and one non-blazar source. Also, while the early attempts to constrain the intergalactic radiation field suffered from the very limited number of sources and a reduced range in red-shift, the growing number of objects, and especially the detection of 3C 279 [12] obtained at a low energy threshold by MAGIC, have definitely opened the path towards the cosmological application of γ -ray astrophysics. There is no doubt that VHE γ -ray astronomy is now a genuine branch of astronomy with multiple connections to cosmology and fundamental physics.

References

[1] http://wwwmagic.mppmu.mpg.de/
[2] http://veritas.sao.arizona.edu/
[3] http://www.mpi-hd.mpg.de/hfm/HESS/
[4] http://icrhp9.icrr.u-tokyo.ac.jp/
[5] Baade, W. & Zwicky, F. ,1934, "Cosmic Rays from Super-novae", Proc. Nat. Acad. Sci. 20(5), 259
[6] Koyama, K. et al., 1995, Nature 378, 255.
[7] Aharonian, F. A., et al., 2004, Nature 432, 75.
[8] Weekes, T.C., et al. 1989, ApJ, 342, 379
[9] http://argo.na.infn.it/
[10] http://www.lanl.gov/milagro/
[11] Aharonian, F. A., et al., 2008, A&A477, 353
[12] J. Albert et all, Science 320 (2008) 1752
[13] Abdo, A. A., et al., 2007, ApJ, 664, L91
[14] Huentemeyer, P. et al., 30th ICRC, Merida, Mexico, 2007

Journal of Physics: Conference Series **203** (2010) 012020 doi:10.1088/1742-6596/203/1/012020

Space-based observations of gamma-rays

Philippe Bruel

Laboratoire Leprince-Ringuet, École polytechnique, CNRS/IN2P3, Palaiseau, France

E-mail: Philippe.Bruel@llr.in2p3.fr

Abstract. The γ-ray space telescopes AGILE and Fermi have been successfully launched on April 2007 and June 2008. Thanks to better capabilities than their predecessor EGRET, and in particular to their considerably improved angular resolution, they are able to provide a much more detailed view of the γ-ray sky above ~20 MeV. They have already reported on a wide variety of astrophysical phenomena and sources such as diffuse galactic emission, pulsars and other galactic sources, active galactic nuclei, gamma-ray bursts, and cosmic-ray electrons.

1. Introduction

After the EGRET experiment [1] on CGRO ended in June 2000, there was no space based experiment able to look at the γ-ray sky above 100 MeV, until the AGILE and the Fermi satellites were launched, in April 2007 and June 2008 respectively. Since their designs and performance are significantly advanced compared to their predecessor, these two experiments reach an unprecedented sensitivity and allow a better and deeper understanding of the γ-ray sky.

2. From EGRET to AGILE and Fermi

The EGRET mission was very successful, with 271 sources reported in the Third EGRET Catalog, far more than the 25 sources detected by its predecessor COS-B. Among these sources, 6 were pulsars, 90 were blazars and 170 were unidentified. These numbers show the limitations of EGRET: 6 pulsars are not enough to perform a real population study, and it is nearly impossible to have a complete understanding of the sky when almost two thirds of the sources are unidentified. The relatively small number of detected sources and the lack of identification was due to the limited sensitivity of the detector, arising from a rather poor angular resolution (5.8° at 100 MeV and 0.5° at 10 GeV) and a relatively small field of view. The successors of EGRET had to overcome these limitations in order to reach a higher sensitivity.

In order to detect γ-rays and measure their directions and energies, a space-based detector needs three sub-detectors: a tracker, in which γ-rays pair-convert, to measure their directions, a calorimeter to measure their energies and an anti-coincidence detector (ACD) to reject charged particles. EGRET used a spark chamber to track e^+e^- pairs. By using the silicon strip technique with tungsten convertors, AGILE and Fermi are able to achieve a finer precision in the measurement of the direction of the incoming γ-ray. This technique also leads to a better aspect ratio: the height of the tracker is reduced and thus the field of view is greatly enhanced. The ACD of EGRET was a monolithic dome, which caused problems above ~10 GeV: electromagnetic showers in the calorimeter produced backsplash which made a signal in the ACD, thus tagging high energy γ-rays incorrectly as charged particles. To overcome this vetoing problem, the ACDs of AGILE and Fermi are segmented. Another limitation of

EGRET was its large deadtime, which was a serious drawback for measurement of gamma-ray bursts (GRBs). Thanks to fast electronics, AGILE and Fermi have a very small deadtime.

3. AGILE and Fermi

3.1. AGILE
AGILE [2] is an Italian Space Agency (ASI) mission launched on April 23 2007 and capable of observing cosmic sources simultaneously in X-ray (18–60 keV) and γ-ray (30 MeV–30 GeV) bands with the coded-mask hard X-ray imager (SuperAGILE) and the Gamma-Ray Imaging Detector (GRID), respectively. The GRID instrument is the core of the AGILE mission. It is a pair conversion telescope composed of a tungsten-silicon tracker (12 (x,y) planes) and a CsI(Tl) calorimeter (30 CsI(Tl) scintillator bars arranged in two orthogonal layers, 1.5 X_0). AGILE is a small and light instrument: its weight is 100 kg and its sensitive surface is 38x38cm^2.

3.2. Fermi
The Fermi Gamma-ray Space Telescope (formerly known as Gamma-ray Large Area Space Telescope, GLAST) was successfully launched on June 11 2008. It consists of two instruments: The Large Area Telescope (LAT) [3] and the Gamma-ray Burst Monitor (GBM) [4]. The GBM covers the energy range from 8 keV to 40 MeV and is largely devoted to the detection of GRBs. The LAT is the main instrument of Fermi and was made by an international collaboration (USA, Italy, Japan, France, Sweden) of scientists and engineers. It detects γ-rays from ~20 MeV to above 300 GeV. It is a pair-conversion telescope with 16 identical towers (tracker+calorimeter), covered by an ACD made of 89 plastic scintillator tiles and 8 flexible scintillator ribbons. The tracker consists of 18 tungsten-silicon (x,y) planes. The hodoscopic calorimeter is an array of 1536 CsI(Tl) crystals, arranged in 8 alternating orthogonal layers, which allows the measurement of the position and the shape of the showers. On axis, the thickness of the LAT is 1.5 X_0 (tracker)+8.8 X_0 (calorimeter).

3.3. Event reconstruction and selection for the LAT
The reconstruction of the direction of the incoming γ-ray is based on the tracker information. Knowing the direction, the energy of the γ-ray can be reconstructed. Below ~1 GeV, a large fraction of the energy is deposited in the tracker, so the energy reconstruction relies on the tracker and the calorimeter information and their correlation. Above ~1 GeV, a large fraction of the energy escapes the calorimeter, so the energy is reconstructed from the longitudinal profile of the shower, which is measured thanks to the longitudinal segmentation of the calorimeter. The direction information is used to extrapolate the trajectory of the particle back to the ACD in order to determine whether the particle is charged. This information, in addition to the topology of the shower in the tracker and in the calorimeter, is used for the selection of the γ-ray candidates. The reconstruction and selection are based on classification trees and have been optimized using a detailed GEANT4 simulation. The performance of the instrument has been carefully checked during beam tests and is monitored on orbit.

3.4. Orbit characteristics and observation mode
AGILE and Fermi both have an almost 1.5 hour orbit period, but do not have the same inclination. With a 2.5° inclination, AGILE does not pass through the South-Atlantic anomaly (SAA). In contrast, with a 25.6° inclination, Fermi loses 14% of its time in the SAA. AGILE operates exclusively in pointing mode, while Fermi's main observation mode optimized to survey the full sky: the vertical axis of the instrument is pointed to 35° above and below the orbital plane on alternate orbits. In this mode, Fermi can see the whole sky in 2 orbits and the exposure is almost uniform over the full sky. In special cases, particularly for follow-up observations of GRBs, the observatory can be inertially pointed.

Journal of Physics: Conference Series **203** (2010) 012020 doi:10.1088/1742-6596/203/1/012020

3.5. Performance comparison

The improvement of sensitivity from EGRET to AGILE and Fermi is mainly due to the improvement of the angular resolution and the increase of the field of view. The angular resolution at 100 MeV improves from 5.8° for EGRET to 3.5° for AGILE and Fermi. At 10 GeV, the improvement is much more important: from 0.5° to 0.1°. The product of the effective area times the field of view improves from 750 cm^2 sr to 1500 cm^2 sr for AGILE and 20000 cm^2 sr for Fermi. Fig. 1 shows the same portion of the sky as seen by EGRET (left), AGILE (center) and Fermi (right). One can see that after 9 months AGILE, with a much smaller sensitive surface, does almost as well as EGRET in 9 years. Fermi, with a larger sensitive surface and observing efficiency, reveals the γ-ray sky above 100 MeV with an unprecedented precision.

4. AGILE and Fermi bright source lists

With one year of data and a conservative analysis, AGILE [5] detected 40 sources above 4σ significance: 21 pulsars, 13 AGNs, and 8 remained unassociated. Using 3 months of data, Fermi [6] detected 206 sources above 10σ: 30 pulsars, 119 AGNs, and 37 remained unassociated. Compared to the results of EGRET after 9 years of operation (6 pulsars, ~90 AGNs and 170 unidentified), the differences are striking. Thanks to their better sensitivity, AGILE and Fermi are able to detect far more sources with much less non-identification (from ~60% for EGRET to 20% for AGILE and Fermi). The fraction of sources being pulsars jumped from 2% for EGRET to 15% for Fermi, enabling population studies of this important class of γ-ray sources.

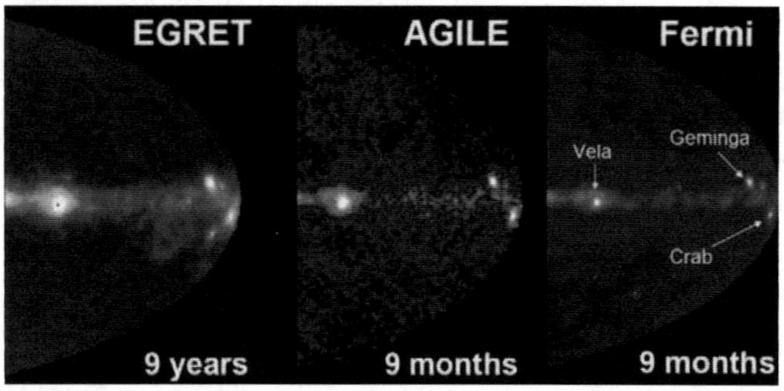

Figure 1. The same part of the sky seen by EGRET, AGILE and Fermi.

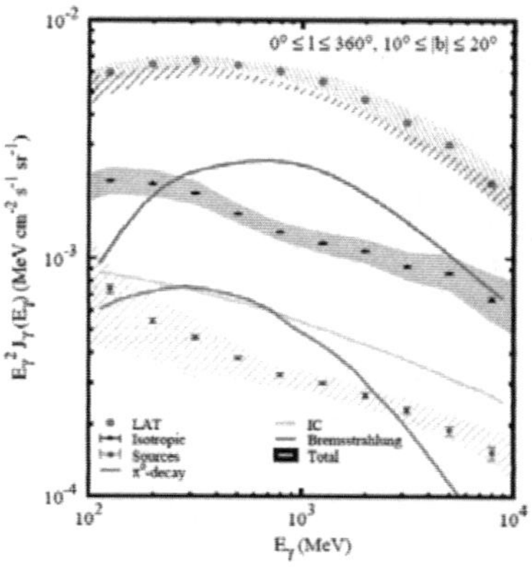

Figure 2. Comparison of Fermi DGE measurement with model prediction [7].

5. The diffuse Galactic emission

The diffuse Galactic emission (DGE) dominates the γ-ray sky. It is produced by interactions of cosmic rays (CRs), mainly protons and electrons, with the interstellar gas (via π^0 production and Bremsstrahlung) and radiation field (via inverse Compton scattering). It is a direct probe of CR fluxes in distant locations, and may contain signatures of physics beyond the Standard Model, such as dark matter annihilation or decay. Measurements by EGRET indicated an excess of γ-ray emission above 1 GeV relative to diffuse Galactic γ-ray emission models, which were constructed to be consistent with directly measured CR spectra. The excess emission was observed in all directions on the sky and has remained unexplained for a decade. Fermi reported on the DGE at intermediate Galactic latitudes [7], where most of the emission comes from interactions of CRs with the local gas. Fig. 2 shows that the Fermi spectrum is well described by the sum of three components : an a priori model for DGE emission, a point source contribution and an isotropic diffuse emission (extragalactic diffuse emission and residual background). Fermi does not detect the EGRET GeV excess.

6. Galactic sources

6.1. Pulsars

Among the wide variety of γ-ray sources, pulsars were the first objects identified in our Galaxy. EGRET detected only 6 pulsars with high confidence, despite the few thousand observed in radio. Five of them were normal radio pulsars, discovered using the ephemerides measured by radio astronomers, and the last one, Geminga, was a radio-quiet pulsar. AGILE has detected 21 pulsars [5] and Fermi has detected 46 pulsars [8].

Pulsars are fast-rotating magnetic neutron stars. They come with different flavors : radio loud and radio quiet pulsars, young and old recycled pulsars (millisecond pulsars). In addition to the 6 pulsars seen by EGRET, Fermi has detected 16 young radio pulsars, 16 new pulsars discovered in a blind search and 8 millisecond pulsars.

The light curves measured by AGILE and Fermi for all these different kinds of pulsars are very similar. As a consequence they seem to host the same γ-ray emission mechanism. An important question not answered by EGRET is whether the γ-ray emission arises near the surface of the neutron star, close to the classical radio emission (polar cap model), or at a significant fraction of the light cylinder distance (outer gap, slot gaps models). The peaks of the γ-ray light curve are appreciably offset from the radio peak, suggesting that the γ-rays arise at high altitude. This is confirmed by the spectral shape of the pulsars: all spectra are well modeled with a simple power law with an exponential cut-off, as shown in Fig. 3. Low altitude emission would imply an hyper-exponential cut-off because of pair attenuation with the strong magnetic field.

With much more γ-ray statistics for individual sources, Fermi is able to perform a phase-resolved spectral analysis. In [9] Fermi reported that in the light curve of Vela there is a third peak, evolving very clearly with energy as shown in Fig. 4.

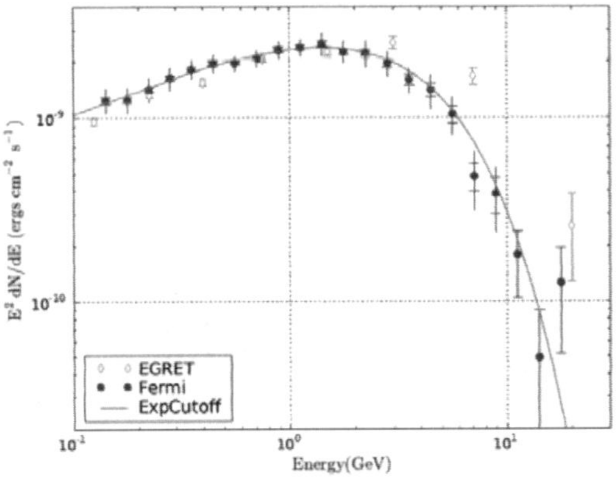

Figure 3. Vela spectrum measured by Fermi fitted with a power law with an exponential cut-off [9].

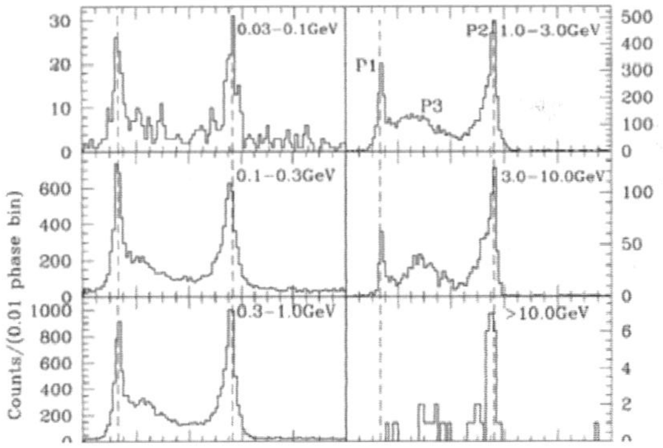

Figure 4. Counts as function of pulsar phase for different energy bands for Vela, showing that the third peak P3 evolves with energy [9].

6.2. Other Galactic results

Related to these pulsars discoveries is the first detection at γ-ray energies of a globular cluster, 47 Tucanae, by Fermi [10]. Globular clusters contain from tens to several hundreds of millisecond pulsars (MSPs). Since Fermi had discovered a population of MSPs, it seemed possible to finally detect a globular cluster. So far 23 MSPs have been detected in 47-Tuc by radio and/or X-ray observations and the total population is estimated at 30-60. The spectrum measured by Fermi is well fitted by a power law with exponential cut-off, as are all pulsars seen by Fermi. The spectral index and cut-off are very similar to the mean spectral index and cut-off of the 8 MSPs detected by Fermi. Fermi reported an estimation of the total number of MSPs in 47 Tuc which is compatible with previous estimates.

Another result related to galaxy emission is the detection of the Large Magellanic Cloud by Fermi [11]. The LMC is an excellent target for studying the link between CR acceleration and γ-ray emission since it is nearby, has a large angular extent, and is seen at a small inclination angle that avoids source confusion along the lines of sight. In addition, the LMC is relatively active, housing many supernova remnants, bubbles and superbubbles and massive star forming regions that are all potential sites of CR acceleration. The close correlation found by Fermi between CR density and massive star tracers supports the idea that CRs are accelerated in massive star forming regions as a result of the large amounts of kinetic energy that are input by the stellar winds and supernova explosions of massive stars into the interstellar medium.

7. Active Galactic Nuclei

About 1% of galactic centers are active and emit radio, optical, UV, Xray and γ-ray radiation. Rapidly varying fluxes and large luminosities of AGN are best explained if the γ-rays are emitted from collimated jets of charged particles moving at relativistic speeds. Most of the AGN seen at γ-ray energies are blazars in which the jet points directly to us. Their spectra are usually well described by two broad humps. This is explained in purely leptonic models by an accelerated population of electrons in a magnetic field: the low energy peak is due to the synchrotron emission and the high energy peak is due to Inverse Compton scattering. There are two kinds of blazars: Flat Spectrum Radio Quasars (FSRQs), having strong emission lines and an intense radiation field due to the accretion disk and clouds, BLLacs (BLs), which have nearly featureless spectra and have a low radiation field. The synchrotron peaks of FSRQs lie at IR energies and their IC peak lies below 1 GeV. The IC emission is large because the intense radiation field provides a large density of IC targets. But this large density prevents the γ-rays from reaching very high energy, which explains why the IC peak does not reach higher energies. For BLs, the lower radiation field allows the IC peak to reach hundreds of GeV but the IC emission is less intense. BLs are subclassified into low/intermediate/high-frequency peaked BLs (LBLs/IBLs/HBLs), depending on the position of the synchrotron peak. With their broad energy ranges, AGILE and Fermi are very well suited to study AGN. AGILE [12] has detected 13 AGNs and Fermi [13] has detected 119 AGNs in 3 months.

Thanks to its good energy resolution, Fermi is able to perform a precise spectral measurement of AGN. The spectral index of an AGN as measured by Fermi is correlated with the position of the IC peak: Fermi measures the rising (falling) part of the IC peak of FSRQs (BLs), so the spectral index is greater (lower) than 2. Fermi has found that the spectral index changes continuously from FSRQs to HBLs. Since FSRQs show cosmic evolution but the BLs do not, it seems to indicate that the young FSRQs with standard accretion disk evolve to old BLs with radiatively inefficient accretion disk. Fermi also discovered that many AGN exhibit a spectral break around a few GeV. This is observed for essentially all FSRQs and some LBLs, but not for HBLs.

AGN are strongly variable. Thanks to their sensitivity, AGILE and Fermi are able to perform much better variability measurements than EGRET, as shown in Fig. 5 and 6, which is essential in order to understand the emission mechanism in AGN.

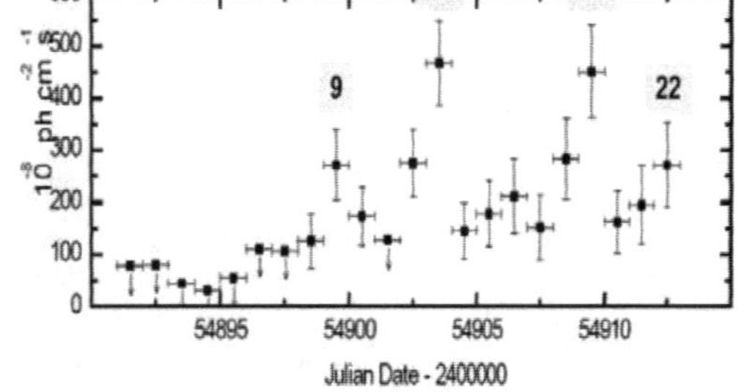

Figure 5. AGILE γ-ray light curve above 100 MeV of PKS 1510-089 during March 2009 [14].

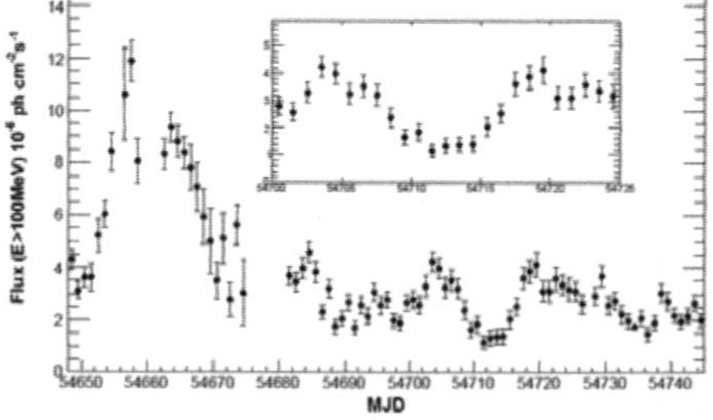

Figure 6. Fermi γ-ray light curve above 100 MeV of 3C454.3 [15].

Journal of Physics: Conference Series **203** (2010) 012020 doi:10.1088/1742-6596/203/1/012020

8. Gamma-Ray Bursts

GRBs are bright and distant flashes of γ-rays, taking place at a rate of about one per day, briefly shining as the most luminous objects in the Universe. EGRET detected 4 GRBs in its spark chamber. As of May 2009, AGILE has localized 21 GRBs with SuperAGILE (X-rays), has detected 103 GRBs using only its calorimeter (350 keV-100 MeV) and 3 GRBs with the full detector [16]. The Fermi-GBM (<40 MeV) has detected 192 GRBs and the Fermi-LAT (>20 MeV) has detected 8 GRBs. The latter are the ones with the highest measured γ-ray energies. For instance, the highest energy event in GRB 080916C [17] (z=4.35) had 13.2 GeV and arrived 16.5s after the start of the burst. In some quantum gravity models, high-energy photons travel more slowly through the Universe and therefore arrive later than low-energy photons, so this GRB was used to set a lower limit on the quantum gravity mass, $M_{QG} > 1.3 \times 10^{18}$ GeV/c^2. Using both the GBM and the LAT, Fermi covers a large energy range: from 8 keV to 300 GeV. So far, Fermi has not detected any departure from the Band function. That is to say, as of May 2009, no new high energy component has been seen in GRBs.

9. Cosmic-ray electrons

Prior to Fermi, available data on high energy electrons were obtained mainly in balloon-borne experiments (except AMS-01) and had limited statistics. Since the only important difference between e^+e^- and γ-rays in the Fermi-LAT instrument is a signal detected by the ACD, the Fermi team carefully investigated the capability of the LAT to measure cosmic-ray electrons. As a result, a high statistics spectrum of electrons for the energy range 20 GeV - 1 TeV was obtained [18]. This spectrum and the recent PAMELA positron fraction result suggests that there is one or several nearby sources of e^+e^- (believed to be pulsars or SNRs) responsible for the excess with respect to the conventional model (based on contribution from quasi-uniformly distributed distant sources).

10. Conclusions

Thanks to their performance, AGILE and Fermi have opened a new era of space-based observation of γ-rays. Compared to EGRET, they have a much better angular resolution, which allows a better source identification, and they have a better sensitivity, which allows them to detect more sources and to study the brightest ones with much more precision. The detection and identification of more sources allows population studies that were not conceivable a decade ago. AGILE and Fermi will certainly continue to improve our understanding of the high energy sky in the forthcoming years.

References

[1] Thompson, D. J. et al. 1993, ApJS, 86, 629
[2] Tavani, M. et al. 2009, A&A, 502, 3
[3] Atwood, W. B. et al. 2009, ApJ, 697, 1071
[4] Meegan, C. et al. 2009, ApJ, in press (arXiv:0908.0450)
[5] Pittori, C. et al. 2009, A&A, in press (arXiv:0902.2959)
[6] Abdo, A. A. et al. 2009, ApJS, 183, 46
[7] Abdo, A. A. et al. 2009, PRL, submitted (or arXiv:0907.0294)
[8] Abdo, A. A. et al. 2009, ApJ, submitted (arXiv:0910.1608)
[9] Abdo, A. A. et al. 2009, ApJ, 696, 1084
[10] Abdo, A. A. et al. 2009, Science, 325, 845
[11] Abdo, A. A. et al., in preparation (or arXiv:0907.0293)
[12] D'Ammando, F. et al. 2009, in Proc. of 44th Rencontres de Moriond (arXiv:0906.1451)
[13] Abdo, A. A. et al. 2009, ApJ, 700, 597
[14] D'Ammando, F. et al. 2009, in Proc. of "Accretion and Ejection in AGN" (arXiv:0910.0966)
[15] Abdo, A. A. et al. 2009, ApJ, 699, 817
[16] Marisaldi, M. et al 2009, in Proc. of ICRC 2009 (arXiv:0906.1446)
[17] Abdo, A. A. et al. 2009, Science, 323, 1688
[18] Abdo, A. A. et al. 2009, Phys. Rev. Lett., 102, 181101

Detection of Antimatter in our Galaxy

Piergiorgio Picozza and Roberta Sparvoli

University of Rome Tor Vergata and INFN Roma Tor Vergata, Rome, Italy

E-mail: piergiorgio.picozza@roma2.infn.it, roberta.sparvoli@roma2.infn.it

Abstract. Searching heavy antimatter and dark matter signals has considerably progressed in the recent years with the PAMELA space mission and the long duration balloon-borne BESS Polar experiments. The AMS-02 instrument is almost ready to be installed on board of the ISS at the end of July 2009. Interesting results have been obtained from PAMELA experiment on the antiproton to proton and positron to all electron ratios. The positron fraction shows an increasing at energies above 10 GeV, mostly interpreted as due to contributions from exotic sources, as dark matter or pulsars.

1. Introduction

Secondary Cosmic Rays in the terrestrial atmosphere have been for forty years the only tool to study physical processes at energies exceeding several orders of magnitude those available from natural radioactivity. Particle Physics born and developed studying cosmic rays, and positrons, muons, pions and strange particles were discovered detecting directly cosmic radiations or products of their interaction with matter targets. It took four decades of technological efforts to reproduce cosmic rays energies with accelerators and to compete in intensity. Three decades later technical developments of space particle detectors allowed the study of cosmic rays with energies largely exceeding those supplied by accelerators, opening new possibilities for the study of one of the most important issue facing astrophysics and cosmology today: what was the role of matter and antimatter in the early Universe? Is there antimatter on a cosmological scale? Is the present Universe baryon-symmetric? Although theoretical arguments based on the cosmic diffuse gamma-ray spectrum constrain the distance to any hypothetical domains of antimatter to be comparable to the horizon scale, the apparent baryon asymmetry in the Universe can be probed by searching for antinuclei in the cosmic radiation.

A wide program of direct antimatter search was triggered from the results obtained in 1979 by the balloon borne experiments of R. Golden [1] and E. Bogomolov [2] that identified the first antiprotons in cosmic rays. The detected antiproton spectrum largely exceeded the expected antiproton flux produced in interaction of cosmic rays with the interstellar matter. These data were interpreted in terms of primordial antimatter coming from antimatter domains in a baryonic symmetric Universe, of evaporation by Hawking effect of primordial mini black holes, or of exotic particles annihilation. Several experiments, focused on antimatter detection, followed the first ones, mainly performed by the WiZard, BESS and HEAT collaborations on board stratospheric balloons, and by AMS-01 on board the Space Shuttle. The core of these instruments was a magnetic spectrometer associated with some detectors for the hadronic and electromagnetic component identification. The experimental limits obtained by these missions for the $\overline{He}/He$ ratio are presented in fig. 1: the current lowest limit is obtained combining all BESS flights

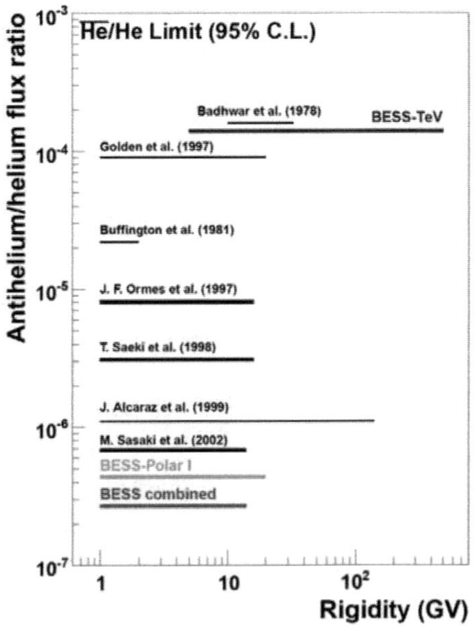

Figure 1. *The ratio of anti-helium to helium in the cosmic radiation shown as a function of rigidity. No observation of anti-helium has been made to date and so upper limits are shown.*

reaching a value of the order of 3×10^{-7}. As for the antiproton/proton ratio, the first pioneer results were not confirmed later.

At the same time experimental cosmology was progressing and another issue became very intriguing and fascinating: what kind of matter the Universe is made of? Actually, the WMAP data strongly suggest that the energy budget of the Universe is shared among baryonic matter (4%), dark (unknown) matter (23%) and dark energy (73%). The favourite candidates for the non-baryonic matter component are electrically-neutral weakly interacting massive particles (WIMP's) with a mass in the range between 10's of GeV to TeV. The most studied is the lightest neutralino χ, in R-parity conserving supersymmetric models. Another interesting WIMP candidate is the lightest Kaluza-Klein particle in the Universal Extra Dimension framework. WIMPS should pervade the Milky Way halo and be concentrated at the galactic centre. As they mutually annihilate, they should produce high energy photons and matter and antimatter particles, mixed with a huge background produced by interaction of cosmic rays with the ISM. In figs. 2 and 3 the experimental available results for the antiproton to proton and positron to electron ratios, obtained by the balloon borne and AMS-01 experiments, are reported together with different calculations accounting for a pure secondary component [3, 4, 5, 6]. No clear conclusion can be obtained, showing that experiments at higher energies, better knowledge of the background, higher statistics and continuous monitoring of solar modulation are mandatory to disentangle possible exotic components from the standard production. The space missions PAMELA and AMS-02 and the long duration flights of the balloon borne BESS experiments were conceived to realize this venture.

2. The PAMELA and AMS-02 Space Missions

The primary scientific objectives of PAMELA (fig. 4) and AMS-02 (fig. 5) are the search for heavy antimatter and nonbaryonic particles outside the Standard Model. In particular they can search for anti-helium (primordial antimatter), heavy anti-nuclei (anti-stars), new matter in the

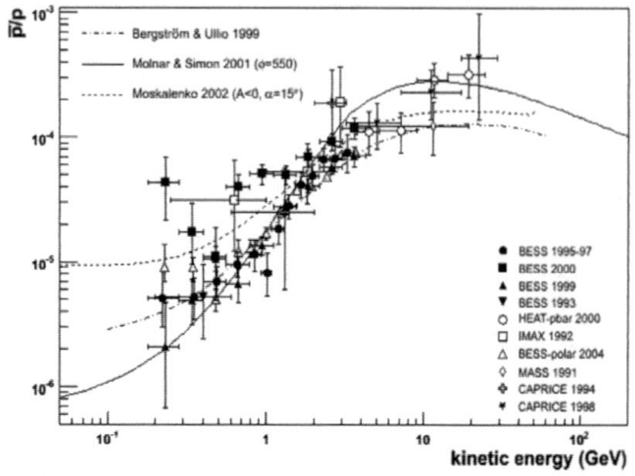

Figure 2. *Recent experimental results for the antiproton to proton ratio along with theoretical calculations for pure* $\overline{\mathrm{p}}$ *secondary production [3, 4, 5, 8].*

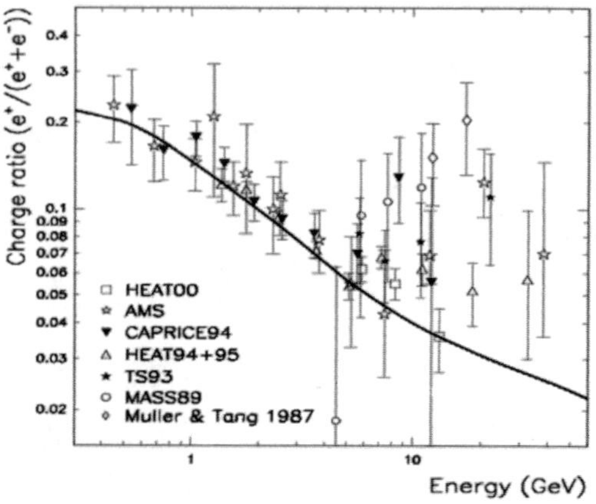

Figure 3. *The positron fraction as a function of energy measured by several experiments. The solid line is a calculation of the secondary positron fraction [6, 9].*

Universe (strangelets?). They can perform also precise measurements of the antiparticle energy spectrum to unravel possible contributions from exotic sources and do accurate studies of nuclei and their isotopes to test cosmic-ray propagation models. Concomitant goals include the study of solar physics and solar modulation, the investigation of the interaction of cosmic-rays with the Earth's magnetosphere and the search for high energy electrons to discover local sources.

2.1. The PAMELA mission

PAMELA - a Payload for Matter-Antimatter Exploration and Light Nuclei Astrophysics [7] - is an instrument realized by an Italian, Russian, German and Sweden collaboration and launched into space on board of the Russian satellite DK1 by a Soyuz-U rocket from the Baikonur cosmodrome in Kazakhstan on June 15th 2006. The satellite orbit is elliptical and semi-polar, with an altitude varying between 350 km and 610 km, at an inclination of 70°. Since July 2006 PAMELA is daily transmitting 16 Gigabytes of data to the Moscow ground segment. The instrument, shown in fig. 4, is constituted by a permanent magnet with a microstrip silicon tracker, a plastic scintillator time-of-flight system, a silicon-tungsten electromagnetic imaging calorimeter, a neutron detector, an anticoincidence system, a shower tail catcher scintillator. The combination of these devices allows antiparticles to be reliably identified from a large background of other charged and their fluxes to be measured in a wide range of energy.

PAMELA has recently published results concerning the antiproton to proton flux ratio [8] and the positron to all electrons fraction [9]. The $\overline{p}/p$ ratio, measured in the energy interval between 1 GeV and 100 GeV, is shown in fig. 6 and compared with recent results from other experiments and with theoretical calculations assuming pure secondary production of antiprotons during the propagation of cosmic rays in the galaxy [8]. The ratio increases smoothly from about 4×10^{-5} at a kinetic energy of 1 GeV and levels off at about 1×10^{-4} for energies above 10 GeV. The results on the $\overline{p}/p$ ratio do not show any particular feature to be referred to some relevant exotic source contribution. However, they allow tight constraints to be placed on many dark matter models, on parameters relevant for secondary production calculations and on contribution of hypothetical local cosmic-ray components.

The measured positron to all electron fraction in the range 1-100 GeV is shown in fig. 7,

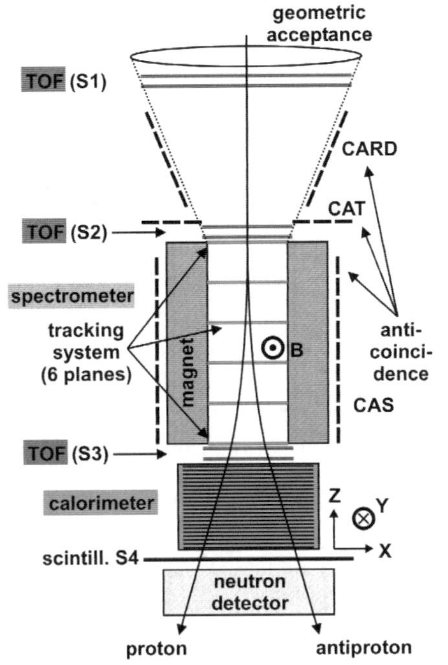

Figure 4. *Schematic overview of the PAMELA apparatus.*

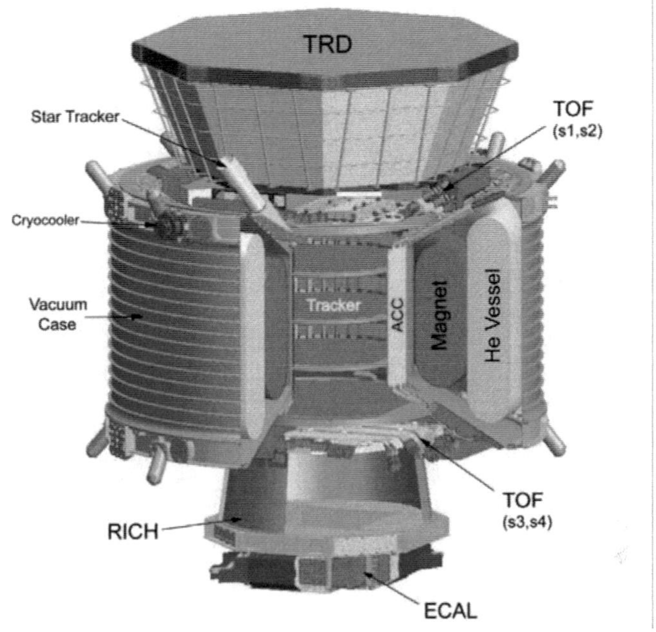

Figure 5. *Schematic overview of the AMS-02 apparatus.*

compared with other recent measurements and a calculation for pure secondary production of positrons [9]. Below 5 GeV PAMELA data are lower than most of the other data and are in agreement with the results from the AESOP balloon-borne experiment, which flew in June 2006. At higher energies the PAMELA positron fraction is compatible with other measurements, but the significantly higher statistics clearly shows that the positron fraction increases with energy, in contrast with the theoretical decrease expectation. This excess of positrons has led to many speculations about its origin, as annihilation or decaying of dark matter, cosmic strings, young pulsar positron contribution, some inhomogeneity in SNR distribution in the Galaxy, standard astrophysics background.

A rise in the positron fraction at high energy has been postulated for the annihilation of dark matter particles in the galactic halo. The most problematic theoretical challenge posed by the PAMELA results is the asymmetry between leptonic (positron fraction) and hadronic (antiproton-proton ratio), difficult to explain in the framework where the neutralino is the dominant dark matter component [10]. In this case suitable explanation requires a very high mass (M=10 TeV) neutralino which, however, is not in a favoured part of the supersymmetric phase space. Better explanations are obtained in terms of leptonic annihilation channel for a wide range of the WIMP mass [10]. This, however, would disfavour the neutralino, since, as a Majorana particle, it cannot annihilate directly into light fermions with large rates, due to helicity suppression. More preferred would be the Kaluza-Klein lightest stable particle that, as a boson, can directly annihilate into e^+e^- pairs with no helicity suppression factor. It is worth noting, however, that all explanations in terms of dark matter annihilation request a boost factor for the annihilation standard rate ranging between 10^2 to 10^4. Calculations [11, 12] for both hypotheses are shown in figs. 8 and 9.

Another explanation asks for a contribution from nearby and young pulsars. Primary

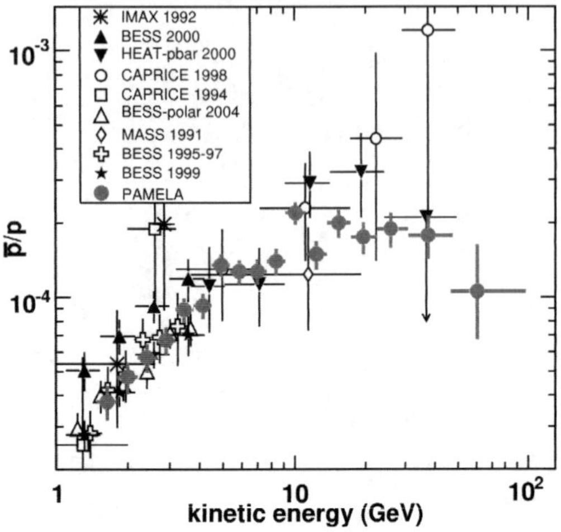

Figure 6. *PAMELA antiproton-to-proton flux ratio compared with previous measurements [8].*

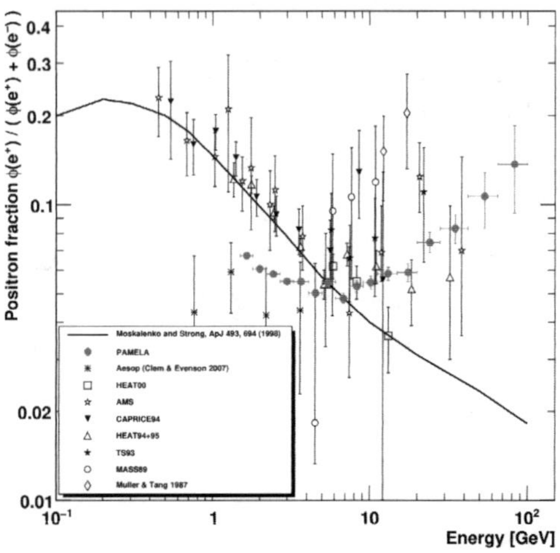

Figure 7. *The positron fraction measured by PAMELA compared with other recent experimental data [9].*

electrons are accelerated in the magnetosphere of pulsars in the polar cup and in the outer gap along the magnetic field lines, emitting gamma rays by synchrotron radiation, gammas that in presence of the pulsar's gigantic magnetic field can evolve in positron and electron pairs. These, after an additional acceleration, are trapped in the cloud surrounding the pulsar, and about 10000 years later escape into the interstellar medium, giving a further contribution to the electron and positron components [13, 14].

On the other hand, standard secondary production to explain this anomalous increasing of the ratio would demand high modifications in the current knowledge of the electrons, protons and helium spectra, although some papers report an explanation of this increasing in terms of few nearby SNR or of secondary production taking place in the same region where cosmic rays are being accelerated [15, 16]. PAMELA is performing accurate measurements of the absolute fluxes of electrons, positrons, protons and light nuclei, that will tightly constrain the secondary production models. The difference in the lower energy part of the spectrum between PAMELA data and almost all previous measurements is interpreted as a consequence of time and charge dependent solar modulation effects. PAMELA data have been collected in solar phase A^-, where the magnetic dipole projection on the solar rotational axis and the rotational axis itself are anti-parallel. In the A^+ phase, the situation is inverted. The charge sign-dependent effect of the solar modulation is different between the two phases, due to a systematic deviation from reflection symmetry of the interplanetary magnetic field [17].

We remind that also Fermi satellite [18] and ATIC balloon borne [19] experiments reported very interesting results on the total electron flux, that have been interpreted in terms of dark matter or pulsar contributions, but their instruments do not allow to distinguish between matter and antimatter components.

2.2. AMS-02 on the ISS

The AMS-02 instrument [20] has been realized by a worldwide collaboration and will be installed outside the International Space Station at the end of July 2010. The detector (fig. 5) has a size of $\sim 3\ \mathrm{m}^3$ and the weight of 7 tons. At present, it is in final assembling and testing phase at CERN.

The main detector of AMS-02 is a magnetic spectrometer composed of a magnet constituted

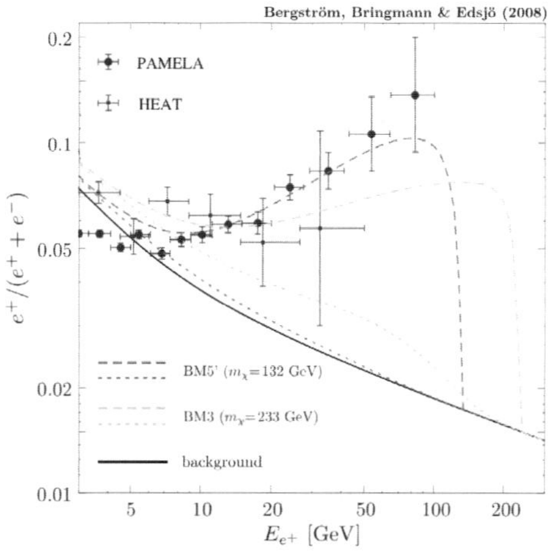

Figure 8. *The solid line is the expected flux ratio $e^+/(e^+ + e^-)$ as calculated following standard secondary production. The data points are the combined HEAT and PAMELA data. The expected flux ratio is shown without (dotted lines) and after taking into account radiative corrections (dashed lines) [11].*

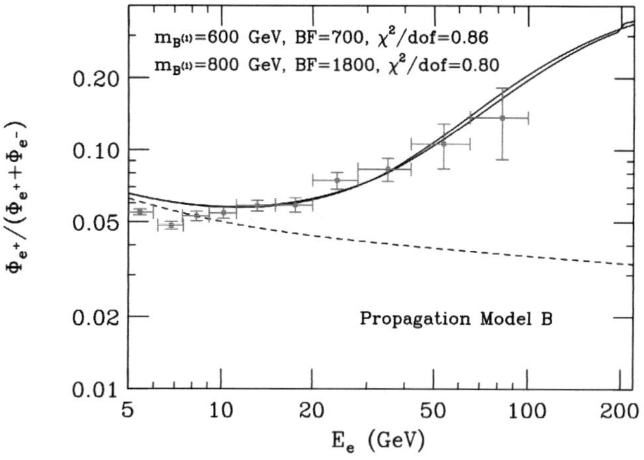

Figure 9. *The positron fraction as a function of energy including contributions from Kaluza-Klein dark matter annihilations, compared to the measurements of the PAMELA experiment. Results are shown for dark matter masses of 600 GeV and 800 GeV, and for one propagation model. The dashed line denotes the positron fraction with no contribution from dark matter (secondary positron production only) [12].*

of 12 superconducting coils with a magnetic field of $\sim$ 0.9 T and a bending power of $\sim$ 0.8 T$\times$ m^2, and of a tracking system made of 8 layers, each of 0.8 m^2, for a particle rigidity measurement up to few TeV. Complementary detectors are a Transition Radiation Detector, composed of 328 modules made of fleece radiators and straw tube arranged in 20 layers assembled in an octagonal shape structure, a trigger and Time- of-Flight system, consisting of 4 scintillator planes for a total of 34 crossed scintillator paddles, a Ring Imaging Detector, constituted by 2 different radiators - respectively an Aerogel and a Sodium fluoride - a conical reflector and a matrix of 680 photomultipliers. To improve the discrimination between the hadronic and the leptonic components, an electromagnetic calorimeter of 17 radiation lengths is added at the bottom of the apparatus; it is made of 9 superlayers of lead scintillating fiber sandwiches disposed, alternatively, along the X and Y directions. The calorimeter measures gamma-rays, electrons and positrons and separate leptons from hadrons with a rejection power of $10^3 - 10^4$ in the energy range from 1 GeV up to 1 TeV.

The combination of all these detectors allows high precision and high statistics in the search for light and heavy antimatter in space, and in the measurement of cosmic rays spectra and chemical composition up to 1 TeV . In particular AMS-02 will improve the PAMELA positron and electron data, extending the measurements up to 800 GeV with high statistics. The detector is also a powerful gamma detector. A unique feature of AMS-02 is the combined searches in different channels - antiprotons, positrons, antideuterons, gammas - that will considerably increase sensitivity to SUSY DM signals detection.

3. Polar Ballon Flights

Long duration polar flights represent a new interesting opportunity for cosmic rays research, at a low cost with respect to space missions.

On December 2004 the BESS (Ballon Borne Experiment with a Superconducting Spectrometer) Polar I experiment, an American-Japanese collaboration, made its first successful polar flight in Antarctica, 8.5 days long, detecting 1520 antiprotons in the energy range 0.1-4.2 GeV. The instrument was composed by a superconducting solenoid magnet, with 1 T magnetic field, combined with internal JET and drift chambers giving an MDR of $\sim$ 200 GeV, a Time-Of-Flight hodoscope with 3 sets of scintillators placed, respectively, one on the top and two on the bottom of the solenoid vessel, and an aerogel Cherenkov counter for e/μ rejection. The scientific objectives were the search for antiparticle and antimatter in the low energy region and the precise measurements of various primaries cosmic rays. The main results in the $\overline{He}/He$ and $\overline{p}/p$ ratios are shown in fig. 1 and 2, respectively.

A real quality improvement on long duration flights has been done on 23rd of December 2008, when the BESS Polar II experiment was launched from William Field, McMurdo, in Antarctica. The floating time was of 29.5 days with 24.5 days of data taking. Promising results on $\overline{p}/p$ ratio and on the $\overline{D}$ and $\overline{He}$ search are expected from the analysis of the collected data.

Other long duration flights are planned for the next years. PEBS (Positron Electron Balloon Experiment) has been designed to measure positrons and electrons in the energy range from 0.5 GeV to 200 GeV and will be launched from Arctica Swalbard base. GASP from Antarctica and DbarSUSY from Arctica experiments will be dedicated to anti-deuteron search at very low energy.

4. Summary

The search of antimatter and dark matter signals in space has largely developed in the last years by the ATIC and BESS balloon-borne and PAMELA and Fermi satellite experiments. Results from the PAMELA instrument show very interesting features in the positron to all electron fraction, compared to expectation from the standard secondary production. Such features are mainly interpreted in terms of positron primary sources, as dark matter annihilation or nearby pulsars contributions. Explanations in terms of nearby SNR's or non standard processes in the secondary positrons production are also reported. The antiproton to proton spectrum appears in agreement with the standard secondary production. AMS-02 is now ready to start, and will complete - through different observation channels - the exploration at higher energy and with better statistics.

5. References

[1] R. L. Golden et al., Phys. Rev. Lett., 43 (1979), 1196.
[2] E. Bogomolov et al., XX Int. Cosmic Ray Conf., 2 (1987), 72.
[3] A. Molnar and M. Simon, XXVII Int. Cosmic Ray Conf., (2001), 1877.
[4] L. Bergstrom, J. Edsjo and P. Ullio, ApJ, 526 (1999), 215.
[5] I. V. Moskalenko et al., ApJ, 565 (2002), 280.
[6] I. V. Moskalenko and A. W. Strong, ApJ, 493 (1998) 694.
[7] P. Picozza et al., Astrop. Phys., 27 (2007), 296.
[8] O. Adriani et. al., PRL, 102 (2009), 051101.
[9] O. Adriani et. al., Nature, 458 (2009), 697.
[10] M. Cirelli et al., Nucl. Phys. B, 813 (2009), 1.
[11] L. Bergstrom, T. Bringmann and J. Edsjo, Phys. Rev. D, 78 (2008), 103520.
[12] D. Hooper and K. M. Zurek, arXiv:0902.0593 (2009).
[13] H. Yuksel, M. D. Kistler and T. Stanev, arXiv:0810.2784 (2008).
[14] D. Hooper, P. Blasi and P. D. Serpico, JCAP, 0901 (2009), 025.
[15] N. J. Shaviv, E. Nakar and T. Piran, arXiv:0902.0376 (2009).
[16] P. Blasi, arXiv:0903.2794 (2009).
[17] J. Clem et al., ApJ, 464 (1996), 507.
[18] A. A. Abdo et al., PRL, 102 (2009), 181101.
[19] J. Chang et al., Nature, 456 (2008), 362.
[20] B. Beischer et al., New Journal of Physics, 11 (2009), 105021.

Topics in Astroparticle and Underground Physics (TAUP 2009) IOP Publishing
Journal of Physics: Conference Series **203** (2010) 012022 doi:10.1088/1742-6596/203/1/012022

Dark matter signals in space

Pierre Salati

LAPTH, Université de Savoie, CNRS, B.P.110, F-74941 Annecy-le-Vieux Cedex, France

E-mail: salati@lapp.in2p3.fr

Abstract. The confirmation by the PAMELA collaboration of a positron excess above 10 GeV has triggered a lot of excitement in the field of particle astrophysics. This excess could be the first long waited hint of the presence of massive and weakly interacting species in the halo of the Milky Way. If so, the nature of the astronomical dark matter is about to be unveiled after more than seventy years of unsuccessful searches. This review summarizes the state of the art, a year of bubbling activity after the PAMELA announcement. The dark matter candidates which can potentially lead to a positron excess have quite special properties. They are severely constrained by radio and gamma observations unless they are tightly packed inside unprobable or bizarre dark matter clumps. These species could also be unstable with abnormally long lifetimes. Although the positron excess could be generated by annihilating and/or decaying dark matter particles, William of Ockham would warn us that a more natural explanation is to be found in pulsars for instance, and that *entia non sunt multiplicanda praeter necessitatem*.

1. Evidence for primary cosmic ray positrons

The ATIC [1], HESS [2, 3] and Fermi [4] collaborations have recently reported measurements of the total cosmic ray (CR) electron and positron flux at high energy. The Fermi-LAT instrument in particular has collected high precision data between 20 GeV and 1 TeV which can be fitted by the simple power-law spectrum [5]

$$\Phi_{e^\pm} = \left\{175.40 \pm 6.09 \ \mathrm{GeV}^{-1}\,\mathrm{m}^{-2}\,\mathrm{s}^{-1}\,\mathrm{sr}^{-1}\right\} \, \epsilon^{-(3.045\pm0.008)} \;\; , \tag{1}$$

where ϵ denotes the electron or positron energy expressed in units of GeV. Primary CR electrons are believed to originate from the interstellar medium – like most of the primary CR nuclei – and are accelerated by supernovae driven shock waves which inject them with a spectral index $\alpha \sim 2.2 \pm 0.1$ inside the Milky Way disc. They subsequently diffuse on the inhomogeneities of the galactic magnetic field with an effective space diffusion coefficient K proportional to ϵ^{δ}. In the GeV-TeV regime, CR electrons lose energy through synchrotron radiation as they spiral in the magnetic fields and by inverse Compton scattering (ICS) on the CMB and on galactic stellar light. A detailed investigation [6] leads to a CR electron flux at the earth of the form

$$\Phi_{e^-} \propto \epsilon^{-\alpha - 0.5 - \delta/2} \;\; , \tag{2}$$

with a spectral index of order 3.05 ± 0.15, in reasonable agreement with the Fermi-LAT measurement. The value of $\delta = 0.7$ used here may seem too large to some colleagues [7]. It suffices then to increase α in order to match the Fermi-LAT spectrum. The latter also contains a small admixture of secondary electrons and positrons which are produced by CR primary protons and helium nuclei impinging on the interstellar gas. These spallation reactions generate many charged pions which eventually decay into positrons and in a lesser extent into electrons. The source spectral index is set by the CR proton and helium nuclei fluxes. With a value $\alpha = 2.7$, we expect a spectral index of 3.55 ± 0.05 for that secondary component which is nonetheless swamped in the total flux.

The PAMELA collaboration has measured [8] the positron fraction $e^+/e^- + e^+$ on a large energy range. From the above mentioned arguments, we would naively expect that fraction to decrease with energy as $\epsilon^{-1/2}$. A more refined analysis [6] confirms that a CR electron spectral index as hard as 3.045 is associated to a decrease of the positron fraction above 10 GeV. The PAMELA measurements establish on the contrary that the positron fraction increases at high energy. This very important observation is in contradiction with a pure secondary origin for the positrons. Primary sources which directly inject these particles in the interstellar medium are necessary.

2. DM species with quite special properties

The positron excess, clearly established by the PAMELA collaboration [8] a year ago, has triggered since then an intense activity in our field. The excess may be related to the astronomical dark matter (DM), an essential component of the universe known to be non baryonic and the nature of which is still unresolved. DM could be made of weakly interacting massive particles (WIMPs) yet to be discovered at accelerators. According to what is now a well accepted working hypothesis, the DM species pervade the galactic halo where they mutually annihilate into a host of lighter particles, including positrons and antiprotons. WIMP annihilations are a source of primary positrons [9] with a production rate given by

$$q_{e^+} = \frac{1}{2} \, \langle \sigma_{\rm an} v \rangle \left\{ \frac{\rho_\chi}{m_\chi} \right\}^2 f(\epsilon) \; . \tag{3}$$

If the WIMPs are thermally produced during the big bang, their relic abundance matches the WMAP value [10] of $\Omega_\chi h^2 = 0.1131 \pm 0.0034$ provided that their annihilation cross section at the time of decoupling is equal to $\langle \sigma_{\rm an} v \rangle \sim 3 \times 10^{-26}$ cm^3 s^{-1}. The energy distribution $f(\epsilon)$ of the positrons produced in a single annihilation is set by the WIMP type. High energy positrons (and electrons) cannot diffuse on long distances and those detected at the earth must have been produced locally, hence a value of $\rho_\odot = 0.3$ GeV cm^{-3} for the WIMP density ρ_χ. Baring in mind these bench mark values, one may estimate q_{e^+} and eventually the DM positron flux Φ_{e^+}. Alas, the values derived for a large choice of DM candidates and for various DM galactic distributions are way too small to account for the observed excess. For a WIMP mass m_χ of 1 TeV, the DM positron production rate q_{e^+} needs to be enhanced by a factor of $\sim 10^3$ in order to match the PAMELA measurements, as shown in the left panel of figure 1.

To save the WIMP explanation of the PAMELA positron anomaly, a first remedy is the possibility that $\langle \sigma_{\rm an} v \rangle$ is much larger than what is currently assumed. Three directions at least have been explored so far. To commence, WIMP decoupling in the early universe takes place conventionally during a period of radiation domination. If the universe at that time is dominated by another component, like a scalar field rolling down its potential [11, 12], the expansion rate is significantly increased and the same WIMP relic abundance Ω_χ requires now a much larger annihilation cross section. Another possibility is that WIMPs do not decouple from the primordial plasma. In this class of less natural scenarios, a non-thermal production takes place through the decay of heavier species like gravitinos or moduli fields [13]. The WIMP relic abundance Ω_χ is no longer determined by $\langle \sigma_{\rm an} v \rangle$. It depends on the production rate and lifetime of these unstable heavier states. A final solution relies on the so-called Sommerfeld effect [14]. If the DM species are massive enough and interact through the exchange of a light particle ϕ, they are attracted towards each other at sufficiently low velocities. Their wave function is focalized at the interaction point and the annihilation cross section can be significantly enhanced. Scenarios based on that effect have been recently proposed [15–17], reviving the idea of secluded dark matter [18]. WIMPs are still thermally produced in the early universe. Their annihilation cross section at that time is equal to the conventional value of 3×10^{-26} cm^3 s^{-1}. As the universe expands, the average WIMP velocity β decreases. The Sommerfeld effect sets in as soon as the WIMP kinetic energy is dominated by the interaction energy, i.e. when β no longer exceeds the effective WIMP coupling constant α'. The cross section increases and may be today large enough to account for the PAMELA positron excess. The enhancement saturates when the WIMP de Broglie wavelength $\sim 1/m_\chi \beta$ exceeds the range $\sim 1/m_\phi$ of the interaction. This occurs for $\beta \leq m_\phi / m_\chi$. The annihilation cross section is resonantly increased whenever $\alpha' m_\phi \sim \alpha'^2 m_\chi / n^2$ where n is an integer [19].

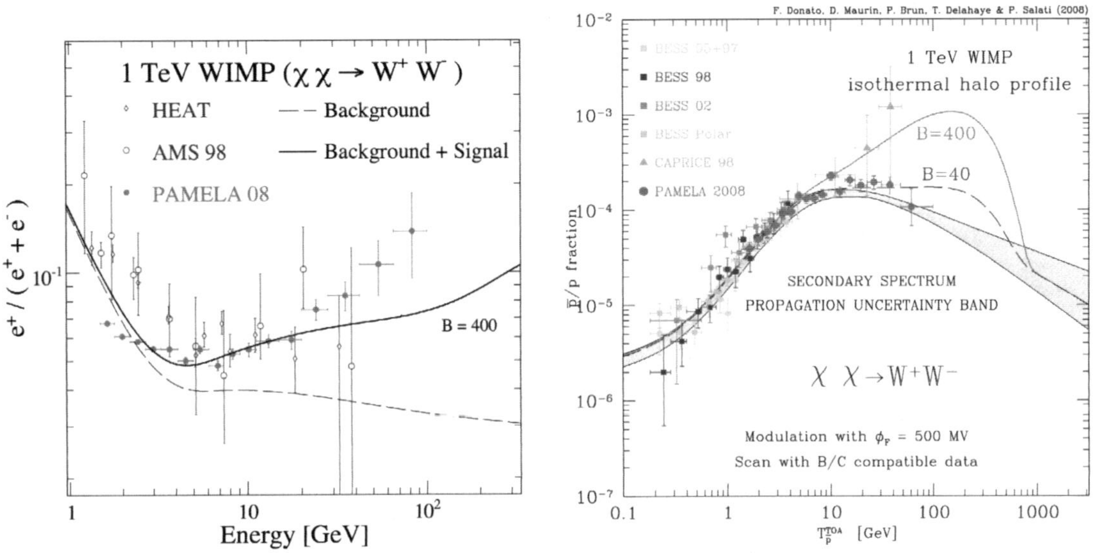

Figure 1. The pedagogical case of a 1 TeV LSP annihilating into a W^+W^- pair is presented. In the left panel, the positron signal which this DM species yields has been increased by a factor of 400, hence the solid curve and a marginal agreement with the PAMELA data. Positron fraction data are from HEAT [20], AMS-01 [21, 22] and PAMELA [8]. If the so–called Sommerfeld effect [14] is invoked to explain such a large enhancement of the annihilation cross section, the same boost applies to antiprotons and leads to an unacceptable distortion of their spectrum as indicated by the red solid line of the right panel. Figures and caption are borrowed from [23]

Irrespective of the mechanism enhancing $\langle \sigma_{\rm an} v \rangle$, WIMPs should not overproduce antiprotons [23, 24]. The annihilation channels through which these antimatter species are produced must be significantly suppressed. The antiproton to proton ratio measured by the PAMELA collaboration [25] does not show actually any excess and is consistent with a pure secondary origin as featured in the right panel of figure 1. Therefore, on top of an abnormally large annihilation cross section, WIMPs preferentially annihilate into charged leptons, a feature which is unusual in supersymmetry but more typical of Kaluza-Klein theories. From a phenomenological point of view, WIMP annihilations must proceed either directly to lepton pairs or, in some models, through the production of the above mentioned light mediator species ϕ which subsequently decay into leptons via

$$\chi + \chi \rightarrow \phi + \phi \rightarrow l^+ l^- l^+ l^- \ . \tag{4}$$

The possibility to explain the PAMELA positron excess by leptophilic DM particles with enhanced annihilation cross section has been extensively studied over the past year. Depending on its magnitude, the injection of high energy positrons and electrons is associated to clear astrophysical signatures [26]. As they spiral along the galactic magnetic field lines, these particles produce a radio emission through synchrotron radiation. They also inverse Compton scatter photons from the CMB and stellar light. Finally, the charged leptons released by the WIMP annihilations can radiate high energy gamma rays. After an extensive scan [27–29] of the various possibilities, the region of the $(m_\chi, \langle\sigma_{\rm an} v\rangle)$ plane compatible with the PAMELA and Fermi-LAT measurements is found to be excluded for most of the leptophilic DM candidates. The most stringent limit arises from the synchrotron radio emission produced at the galactic center, a region where the DM distribution may be cuspy and where magnetic fields should be strong. That limit weakens considerably though as soon as a cored isothermal DM distribution is preferred to the usual Einasto or NFW profile. In that case, leptophilic WIMPs are not excluded and could be a viable explanation of the positron anomaly. A more reliable tool is provided by the ICS of the positrons and electrons produced by WIMP annihilation at high galactic latitude, a region which is less subject to uncertainties than the center of the Milky Way. A careful investigation [30], performed only in the case where the WIMPs annihilate directly into lepton pairs, excludes values of m_χ larger than 1 TeV. Another model independent analysis implies the injection in the intergalactic medium at redshift ~ 1000 of the secondary particles produced by WIMP annihilation. The recombination process is affected and an imprint is left on the CMB, which may be searched for in the high precision WMAP data [31]. Most of the PAMELA compatible leptophilic DM candidates still survive that test [32] but will be probed by the Planck mission.

3. The effect of clumpiness on DM annihilation

Another way to increase the positron production rate (3) relies on the existence of clumps inside the smooth DM galactic distribution. Because $\langle\rho_\chi^2\rangle$ is always larger than $\langle\rho_\chi\rangle^2$, inhomogeneities tend to enhance q_{e^+}, a quantity proportional to the square of the DM density. How large is that enhancement depends on the position of the substructures with respect to the DM smooth component of the Milky Way. The inner structure of clumps as well as their mass distribution are also key factors. The lower mass cut-off and the concentration-to-mass relation of the DM substructures come finally into play. A statistical analysis is mandatory since we live inside one particular realization of the clump distribution to be taken out of an infinite set of similar realizations. Using the tools specifically forged for computing the odds of that galactic lottery [33], a comprehensive analysis [34] indicates that the boost factor to be applied to q_{e^+} in the case of a ΛCDM universe cannot exceed at most a factor of 20 at high energy. The correct value is presumably much smaller, i.e. of order unity [35].

Although a clumpy DM halo does not lead on average to an increase of the positron flux at the earth, the statistical variance of the boost factor is large at high energy, hence the existence of configurations where q_{e^+} could be significantly enhanced. The possibility that a DM clump lies for instance in the vicinity of the solar system has been suggested [36, 37] as an explanation for the PAMELA positron anomaly. Such a local substructure would actually outshine the rest of the galactic DM distribution. Alas, the probability that such a situation occurs in the DM halo of the Milky Way is vanishingly small as demonstrated in a recent analysis [38] based on the results of the cosmological N-body simulation Via Lactea II [39]. In the favorable case of a 100 GeV DM species annihilating into $e^\pm$ pairs, the PAMELA positron excess is best fitted by a subhalo with annihilation volume

$$\xi = \int_{\rm clump} \left\{\rho_c(\mathbf{x})/\rho_\odot\right\}^2 d^3\mathbf{x} \sim 114 \ {\rm kpc}^3 \ , \tag{5}$$

located at 1.22 kpc from the earth. That configuration has a probability of only 0.37 percent to occur. Other arrangements are even less probable.

According to scenario B of [40], when a DM halo with mass 10^7 $M_\odot$ forms at redshift ~ 20, a fraction of ~ 3 percent of the gas which it contains cools and collapses as a pressure supported disc. A baryonic mass of 10^5 $M_\odot$ looses its angular momentum and is transferred at the center of that disc where it forms an intermediate mass black hole (IMBH). During the process, DM is dragged inwards and is adiabatically compressed onto the central object around which a large density enhancement dubbed minispike develops. Although minispikes have an outer radius of only 3 pc on average, their inner DM density reaches the annihilation limit close to the IMBH, hence typical annihilation volumes ξ as large as 3.3×10^6 kpc^3 for a 1 TeV WIMP with thermal annihilation cross section. A hundred minispikes are expected today in the halo of the Milky Way. The probability that one of these objects lies close enough to the earth to generate a positron excess matching the PAMELA measurements reaches 84 percent [41] for the above mentioned DM species and a pure $e^\pm$ annihilation channel.

4. Decaying dark matter

Finally, the possibility of unstable DM candidates with very large lifetimes has also been explored [42, 43]. Irrespective of their production mechanism in the early universe, unstable species could in principle generate the PAMELA signal. The production rate (3) may be written now as

$$q_{e^+} \; = \; \tau_{\rm dec}^{-1} \left\{ \frac{\rho_\chi}{m_\chi} \right\} \, f(\epsilon) \;\; , \tag{6}$$

where the quantity $1/2 \times \langle \sigma_{\rm an} v \rangle \times \{\rho_\chi / m_\chi\}$ has been replaced by the decay rate $\tau_{\rm dec}^{-1}$. For a 1 TeV WIMP, a lifetime of order 2×10^{26} sec is required to reproduce the positron excess. The synchrotron radio emission from the galactic center is no longer a problem even in the case of a NFW or Einasto profile since the positron production rate from decaying DM is proportional to ρ_χ and not to ρ_χ^2. The lifetime needs to be fine tuned though. Higher dimension operators must be invoked to explain the large value of $\tau_{\rm dec}$ required by PAMELA. If a four-fermion point-like coupling is responsible for the WIMP instability, we anticipate [44] a decay rate $\propto m_\chi^5 / M^4$ where M denotes the typical scale of the underlying high energy theory. If factors of 2 and π are neglected, a GUT mass $M \sim 2.3 \times 10^{16}$ GeV could lead to the required lifetime. Notice that the DM species must also be leptophilic otherwise antiprotons are also overproduced.

5. Ockham's razor

Although the PAMELA excess can be explained by annihilating or decaying WIMPs, William of Ockham would refresh our enthusiasm by warning us that more natural explanations are provided by existing astrophysical objects. Pulsars for instance are highly magnetized neutron stars where strong electric fields can extract charged particles from the surface. Stripped electrons generate electromagnetic showers inside the magnetosphere and produce positrons which are eventually released in the interstellar medium [45]. An injection spectrum $\alpha = 1.5$ is necessary to match the PAMELA observations [46].

The possibility that CR spallations could take place during the acceleration process has also been proposed [47, 48]. As particles are accelerated by a supernova driven shock wave, they also interact with the material flowing through the shock and generate secondary species like positrons. The latter are then accelerated and replenish the high energy tail of the positron spectrum. According to that scenario, we should also expect an increase with energy of the B/C ratio as well as an antiproton excess above ~ 100 GeV [49].

Acknowledgments
Pierre Salati would like to thank the organizers for the inspiring and exciting atmosphere of this very nice meeting.

References
[1] Chang J *et al.* 2008 *Nature* **456** 362–365
[2] Aharonian F *et al.* (H.E.S.S.) 2008 *Phys. Rev. Lett.* **101** 261104 (*Preprint* 0811.3894)
[3] Aharonian H E S S C F 2009 (*Preprint* 0905.0105)
[4] Abdo A A *et al.* (The Fermi LAT) 2009 *Phys. Rev. Lett.* **102** 181101 (*Preprint* 0905.0025)
[5] Grasso D *et al.* (FERMI-LAT) 2009 *Astropart. Phys.* **32** 140–151 (*Preprint* 0905.0636)
[6] Delahaye T *et al.* 2009 *Astron. Astrophys.* **501** 821–833 (*Preprint* 0809.5268)
[7] Strong A W, Moskalenko I V and Reimer O 2004 *Astrophys. J.* **613** 962–976 (*Preprint* astro-ph/0406254)
[8] Adriani O *et al.* (PAMELA) 2009 *Nature* **458** 607–609 (*Preprint* 0810.4995)
[9] Delahaye T, Lineros R, Donato F, Fornengo N and Salati P 2008 *Phys. Rev.* **D77** 063527 (*Preprint* 0712.2312)
[10] Hinshaw G *et al.* (WMAP) 2009 *Astrophys. J. Suppl.* **180** 225–245 (*Preprint* 0803.0732)
[11] Salati P 2003 *Phys. Lett.* **B571** 121–131 (*Preprint* astro-ph/0207396)
[12] Profumo S and Ullio P 2003 *JCAP* **0311** 006 (*Preprint* hep-ph/0309220)
[13] Kane G, Lu R and Watson S 2009 (*Preprint* 0906.4765)
[14] Hisano J, Matsumoto S and Nojiri M M 2004 *Phys. Rev. Lett.* **92** 031303 (*Preprint* hep-ph/0307216)
[15] Cirelli M, Franceschini R and Strumia A 2008 *Nucl. Phys.* **B800** 204–220 (*Preprint* 0802.3378)
[16] Arkani-Hamed N, Finkbeiner D P, Slatyer T R and Weiner N 2009 *Phys. Rev.* **D79** 015014 (*Preprint* 0810.0713)
[17] Pospelov M and Ritz A 2009 *Phys. Lett.* **B671** 391–397 (*Preprint* 0810.1502)
[18] Pospelov M, Ritz A and Voloshin M B 2008 *Phys. Lett.* **B662** 53–61 (*Preprint* 0711.4866)
[19] Lattanzi M and Silk J I 2009 *Phys. Rev.* **D79** 083523 (*Preprint* 0812.0360)
[20] Barwick S W *et al.* (HEAT) 1997 *ApJ* **482** L191+ (*Preprint* astro-ph/9703192)
[21] Aguilar M *et al.* (AMS-01) 2007 *Phys. Lett.* **B646** 145–154 (*Preprint* astro-ph/0703154)
[22] Alcaraz J *et al.* (AMS) 2000 *Phys. Lett.* **B484** 10–22
[23] Donato F, Maurin D, Brun P, Delahaye T and Salati P 2009 *Phys. Rev. Lett.* **102** 071301 (*Preprint* 0810.5292)
[24] Cirelli M, Kadastik M, Raidal M and Strumia A 2009 *Nucl. Phys.* **B813** 1–21 (*Preprint* 0809.2409)
[25] Adriani O *et al.* 2009 *Phys. Rev. Lett.* **102** 051101 (*Preprint* 0810.4994)
[26] Borriello E, Cuoco A and Miele G 2009 *Astrophys. J.* **699** L59–L63 (*Preprint* 0903.1852)
[27] Bertone G, Cirelli M, Strumia A and Taoso M 2009 *JCAP* **0903** 009 (*Preprint* 0811.3744)
[28] Bergstrom L, Bertone G, Bringmann T, Edsjo J and Taoso M 2009 *Phys. Rev.* **D79** 081303 (*Preprint* 0812.3895)
[29] Meade P, Papucci M, Strumia A and Volansky T 2009 (*Preprint* 0905.0480)
[30] Cirelli M and Panci P 2009 *Nucl. Phys.* **B821** 399–416 (*Preprint* 0904.3830)
[31] Galli S, Iocco F, Bertone G and Melchiorri A 2009 *Phys. Rev.* **D80** 023505 (*Preprint* 0905.0003)
[32] Slatyer T R, Padmanabhan N and Finkbeiner D P 2009 *Phys. Rev.* **D80** 043526 (*Preprint* 0906.1197)
[33] Lavalle J, Pochon J, Salati P and Taillet R 2007 *A&A* **462** 827–840
[34] Lavalle J, Yuan Q, Maurin D and Bi X 2008 *A&A* **479** 427–452 (*Preprint* 0709.3634)
[35] Diemand J 2009 private communication
[36] Cumberbatch D T and Silk J 2007 *Mon. Not. Roy. Astron. Soc.* **374** 455–465 (*Preprint* astro-ph/0602320)
[37] Hooper D, Stebbins A and Zurek K M 2009 *Phys. Rev.* **D79** 103513 (*Preprint* 0812.3202)
[38] Brun P, Delahaye T, Diemand J, Profumo S and Salati P 2009 *Phys. Rev.* **D80** 035023 (*Preprint* 0904.0812)
[39] Diemand J *et al.* 2008 *Nature* **454** 735–738 (*Preprint* 0805.1244)
[40] Bertone G, Zentner A R and Silk J 2005 *Phys. Rev.* **D72** 103517 (*Preprint* astro-ph/0509565)
[41] Bringmann T, Lavalle J and Salati P 2009 *Phys. Rev. Lett.* **103** 161301 (*Preprint* 0902.3665)
[42] Ibarra A and Tran D 2009 *JCAP* **0902** 021 (*Preprint* 0811.1555)
[43] Ibarra A, Tran D and Weniger C 2009 (*Preprint* 0906.1571)
[44] Chen C R, Nojiri M M, Takahashi F and Yanagida T T 2009 *Prog. Theor. Phys.* **122** 553–559 (*Preprint* 0811.3357)
[45] Profumo S 2008 (*Preprint* 0812.4457)
[46] Hooper D, Blasi P and Serpico P D 2009 *JCAP* **0901** 025 (*Preprint* 0810.1527)
[47] Blasi P 2009 *Phys. Rev. Lett.* **103** 051104 (*Preprint* 0903.2794)
[48] Ahlers M, Mertsch P and Sarkar S 2009 (*Preprint* 0909.4060)
[49] Blasi P and Serpico P D 2009 *Phys. Rev. Lett.* **103** 081103 (*Preprint* 0904.0871)

Journal of Physics: Conference Series **203** (2010) 012023 doi:10.1088/1742-6596/203/1/012023

Underground Laboratories: cosmic silence, loud science

Eugenio Coccia
Department of Physics, University of Rome "Tor Vergata"
and INFN Gran Sasso National Laboratory

E-mail: coccia@lngs.infn.it

Abstract. Underground laboratories provide the low radioactive background environment necessary to host key experiments in the field of particle and astroparticle physics, nuclear astrophysics and other disciplines that can profit of their characteristics and of their infrastructures. The cosmic silence condition existing in these laboratories allows the search for extremely rare phenomena and the exploration of the highest energy scales that cannot be reached with accelerators. I briefly describe all the facilities that are presently in operation around the world.

1. Introduction

In the Bertolt Brecht's "Leben des Galilei" (Galilei's life), the scientist declares his thirst for knowledge: "*Sometimes I think I would imprison myself deep underground, where no ray of light arrives, if in exchange I could discover the light nature*!". Scientists working in underground laboratories certainly do like this exclamation of the man whose observations, 400 years ago, made original and profound contributions toward understanding the physical laws of the universe. These scientists are observing today the cosmos with new telescopes, very different from Galilei's one, but with the same aim and enthusiasm for discovering our place in the Universe.

We would like to be "imprisoned" deep underground to respond to vital scientific questions: What is the universe made of? What is the nature of dark matter? What happened to the antimatter? What are neutrinos telling us about the fundamental symmetries of nature, and about the stars and the Earth?

Very high energy phenomena, such as proton decay and neutrinoless ββ decay, happen spontaneously, but at extremely low rates. The study of neutrino properties from natural and artificial sources and the detection of dark matter candidates require capability of detecting extremely weak effects. Underground laboratories provide the necessary low radioactive background environment to investigate these processes. So these laboratories appear complementary to those with particle accelerators in the basic research of the elementary constituents of matter, of their interactions and symmetries. Underground experiments have been successful in discovering the first clear evidence for physics beyond the standard model (SM). The discoveries that neutrinos are massive and flavor lepton numbers are not conserved have been made in underground laboratories using neutrino natural sources as the Sun and the cosmic rays.

In general one can consider two classes of motivations to push beyond the SM. There are "particle physics" reasons: the SM does not truly unify the elementary interactions, it leaves the problem of fermion masses and mixings unsolved and it exhibits the gauge hierarchy problem in the scalar sector. The second class of reasons finds its origin in "astroparticle and neutrino physics" issues: the problems of the solar and atmospheric neutrinos deficits, Majorana or Dirac neutrinos, baryogenesis, dark matter. These astroparticle and neutrino physics issues can be faced by contemporary experimental physics and constitute a formidable motivation for the underground laboratories activity.

I report here concise information on the underground laboratories and facilities in the world.

2. Underground laboratories

In underground physics the struggle to advance the high energy frontier and to go beyond the standard model is the struggle for background control and reduction. Environment is the principal source of background. The environmental backgrounds of the laboratory depend on the depth and on the nature of the surrounding rocks and, as a consequence, may differ in the different facilities.

The high-energy cosmic rays muons flux decreases almost exponentially with increasing depth, this being the main reason to go underground. The neutron flux at low energies is mainly due to fission and α from U and Th in the rocks. As such, it depends on local geology and becomes depth-independent already at shallow depth. Muon spallation processes are negligible at low (MeV) energies but produce a depth-dependent flux of high-energy (GeV) neutrons. The gamma flux, including radon and its progeny, depends again on local geology and is practically depth independent.

Other sources of backgrounds, the ultimate contribution in some cases, are the detector materials, supports, shielding, electrical connections, etc. Cosmic rays may produce traces of radioactive nuclides (a process called cosmogenesis) both during the construction phase of the detectors and of its materials on the surface - often a period of several months is needed before the data taking can start - and during the operational phase underground.

Frontier experiments do not require to the laboratory only a low background environment, but also technological support, easy and safe access and general as well as specific support structures.

The reported information on the underground labs is based (and updates) the paper by Sandro Bettini: "*The world underground science facilities. A compendium*" arXiv:0712.1051. The facilities are reported in geographical order, from W to E, starting from Europe.

2.1 Europe

BUL. Boulby Palmer Laboratory (UK). http://www.hep.shef.ac.uk/research/dm/boulby/boulby.php

The site was developed starting in 1988 by Neil Spooner and collaborators from RAL. An active potash mine at 1000 m depth under a flat surface hosts it. The access is through a shaft. The salt environment limits the cavities width to about 5 m. A clean area of approximately 1500 m2 is available to experiments. The neutron flux with $E > 0.5$ MeV is 1.7×10^{-2} m^{-2} s^{-1}, the muon flux is 4.5×10^{-2} m^{-2} s^{-1}. A building on the surface (200 m^2) hosts laboratories for computing, electronics and chemistry, offices, conference room, changing rooms, mess rooms, mechanical workshop, storage and construction rooms. The scientific program is focused on dark matter search: ZEPLIN II is completed, ZEPLIN III is to be completed by 2010 and DRIFT II is in R&D phase. There are low radioactivity measurements and geophysics research. About 30 scientists work at the laboratory. Excellent potential for expansion.

LSC. Laboratorio Subterráneo de Canfranc (Spain). http://www.lsc-canfranc.es/

The first underground facility under the Pyrenees, close to a dismissed railway tunnel, was created in the 1980s by Angel Morales and the Nuclear and High-Energy Physics Department of the Saragossa University. Taking profit of the excavation of a parallel road tunnel, the new laboratory was later built. The underground structures have been completed in 2005. Soon after, several design and construction defects have emerged and the necessary reparation works are now being made. LSC is managed by a Consortium between the Spanish Ministry for Education and Science, the Government of Aragon and the University of Saragossa. The surface building has been funded. It will contain headquarters, administration, library, meeting room, offices, laboratories, storages and a mechanical workshop,

Journal of Physics: Conference Series **203** (2010) 012023 doi:10.1088/1742-6596/203/1/012023

safety structures and management, for a total of approximately 1500 m2. A dozen of employees are being hired.
The access is horizontal, via one of the tunnels. Entrance must be communicated to the freeway tunnel control. The available spaces underground are: Hall A measuring 40x15x12(h) m^3, Hall B of 15x10x8(h) m^3, Clean room of 45 m^2 and Services for 215 m^2. The old lab area is 100 m^2. The maximum rock coverage is 850 m. The muon flux is between $2x10^{-3}$ and $4x10^{-3}$ m^{-2} s^{-1} depending of the location, the n flux is $2x10^{-2}$ m^{-2} s^{-1}. The Rn activity in the air is 50-80 Bq/m^3 with a ventilation of 11 000 m^3/h, i.e. one lab volume in 40'.
The scientific program is now defined with the advice to the director of a newly appointed International Scientific Committee.

LSM. Laboratoire Subterrain de Modane (France). http://www-lsm.in2p3.fr/
The Laboratory is operated jointly by the CNRS/IN2P3 and CEA/DSM. The excavation of the Laboratory started in 1979 and was completed by 1982, to host a 900 t iron tracking calorimeter to search for proton decay. This "Frejus" experiment finished in 1988.
The access is horizontal through the Frejus roadway tunnel. Intervention of the tunnel control is needed to stop the traffic at the entrance or exit of a vehicle from the lab. The Main Hall is 30x10x11(h) m^3, the Gamma Hall has an area of 70 m^2, two smaller halls have 18 m^2 and 21 m^2 areas, for a total of 400 m^2. The surface building includes offices (100 m^2), a warehouse and workshop (150 m^2) and a flat. The personnel are 8 technicians and engineers and one post doc.
The rock overburden is 1700 m. The muon flux is 4.7 x 10^{-5} m^{-2} s^{-1}. The n flux is 5.6 x 10^{-2} m^{-2} s^{-1} . A low Radon activity in the air, 15 Bq/m^3, is obtained by in taking fresh air at the rate of 1.5 lab volumes/hour. An "antiradon factory" produces 150 m^3/hr of air with 10 mBq/m^3. The laboratory is almost full with NEMO 3 ($\beta\beta$ decay), EDELWEISS (dark matter), which should run at least up to 2010, and a low radioactivity counting facility. About 100 scientists work at the lab.
An extension of the lab of 60000 m^3 is planned (Ulisse project), profiting of the unique opportunity given by the construction of a new tunnel approved by the French and Italian Governments to increase the safety conditions of the traffic. Two large halls are foreseen: Hall A of 24x100 m^2 and Hall B of 18x50 m^2. An extremely low background environment will be obtained in Hall B by surrounding its central volume with a water shield and by artificially producing a very low Rn content atmosphere (0.1 mBq/m^3).

LNGS. Laboratori Nazionali del Gran Sasso (Italy). http://www.lngs.infn.it/
LNGS is a national laboratory of the INFN. It is the largest in the world, serving the largest and most international scientific community. In 1979 the President of the INFN Antonino Zichichi proposed to the Parliament to build a large underground laboratory close to the Gran Sasso freeway tunnel then under construction (an opportunity that reduced substantially the cost). In 1982 the Parliament approved the construction, which was completed by 1987.
Access is horizontal, through the freeway. The underground laboratory consists of three main halls (called A, B and C), about 100x20x18(h) m^3 plus ancillary tunnels, providing space for services and small scale experiments. Two 90 m long tunnels were built for two Michelson interferometers for geology studies. The total area is 17 300 m^2, the total volume 180 000 m^3.
Services hosted on the surface campus include offices, mechanical workshop, storage facilities, chemical lab, electronic workshop, assembly hall, computer and networking, library, canteen, sleeping rooms, conference rooms, headquarters, administration. Special care has been given to the development of structures, instrumentation, procedures and training activities in matter of safety, of the users and of the citizens, and environmental impact. A number of outreach activities and visits to the lab are systematically organized by a dedicated Service. A Training Service has been created to exploit the technological and scientific transfer of the lab expertise to young people of the Abruzzo region. Personnel (physicists, engineers, technicians, administration) include a permanent staff of 76 and about 20 non-permanent positions.

The rock overburden is 1400 m. Muon flux is $3x10^{-4}$ m^{-2} s^{-1}. *n* flux is $3.78x10^{-2}$ m^{-2} s^{-1}. Radon in the air is 50-120 Bq/m^3 with a ventilation system providing one lab volume of fresh air in 3.5 hr. Major civil engineering works have been performed in 2004-2007 to upgrade the safety conditions of the interacting structures of the free-way, the water collection systems and the laboratory.
LNGS is operated as an international laboratory. An International Scientific Committee, appointed by INFN, advises the Director. The rich experimental program includes the CERN to Gran Sasso neutrino beam experiments OPERA and ICARUS; dark matter search, with LIBRA, CRESST2, XENON, WARP; neutrinoless ββ decay with COBRA, CUORICINO, GERDA; Solar neutrinos (and geoneutrinos) with BOREXINO; Supernova neutrinos with LVD; Nuclear astrophysics with LUNA2. A special facility is dedicated to low radioactivity measurements. The laboratory also supports several experiments on geology, biology and environmental issues. Almost all the experiments are second generation ones and are approved for several years of data taking. The scientific user community consists of 752 scientist from 26 countries.

Coordination in Europe
The directors of the four Western Europe underground laboratories (Gran Sasso, Modane, Canfranc and Boulby) initiated actions to co-ordinate activities and to optimise the use of the available resources, taking into account the different characteristics of the infrastructures. In the 6th "framework programme" the ILIAS project (Integrated Large Infrastructures for Astroparticle Science, putting together the underground and the gravitational wave communities) was submitted successfully to European Commission (EC) by ApPEC, the European inter-agency committee for astroparticle physics.
ILIAS was launched on April 2004 as a 5 years project, and got a total EC financial contribution of € 7,5 Million. ILIAS contained several elements: the co-ordination of a number of activities of the laboratories, such as environmental background measurement and control, safety procedures, outreach activities, and the funding for access of new users to the labs.
The continuation of this program (a new proposal called ILIAS-Next is being submitted to EC) and the continuation of the coordination among the four underground labs (a new agreement called EuLAbs is being signed by the agencies Presidents) has been favored by ASPERA. Created by ApPEC and funded by the EC through the 7th Framework Programme, ASPERA is a network of national government agencies responsible for coordinating and funding national research efforts in Astroparticle Physics in Europe. See http://www.aspera-eu.org/ .
In the same Framework Programme, EC also funded a Design Study project (LAGUNA) focusing on the design of an European infrastructure able to host underground new large-volume instruments with a fiducial mass from several tens of ktons up to 1 Mton for low-energy neutrino astronomy and proton decay search.

CUPP. Centre for Underground Physics in Pyhäsalmi (Finland). http://cupp.oulu.fi/
The Centre is hosted in a working mine. Several cavities, dismissed by the mine, are available at different depths down to 980 m, for a total area of more than 1000 m^2. Presently the mine works between 1000 m and 1400 m depth. Access is both via shaft and via an inclined tunnel. The EMMA experiment is being installed. Small lab and office space is available in a surface building. A guesthouse is also available. The personnel consist of bout 3 people on site and 3 in Oulu University.

SUL. Solotvina Underground Laboratory (Ukraine). http://lpd.kinr.kiev.ua/LPD_SUL.htm
The Laboratory was constructed in 1984 under the leadership of Yuri Georgievich Zdesenko by the Lepton Physics Department of the Institute for Nuclear Research (Ukrainian National Academy of Sciences) in a salt mine. The laboratory space is divided in a Main Hall: 25x18x8(h) m^3 and four chambers 6x6x3(h) m^3. The total is area about 1000 m^2. On surface, three living rooms are available. Staffs consist of 14 technicians and engineers. Access is vertical by the mine cage, taking into account the time-table of the mine.

The lab is 430 m deep in salt (≈1 km w.e.). The muon flux is $1.7x10^{-2}$ m^{-2} s^{-1}. The *n* flux is $2.7x10^{-2}$ m^{-2} s^{-1}. Radon concentration in air is 33 Bq m^{-3}. Eleven researchers and PhD students of the LPD work at SUL mainly on ββ decay, preparing a new 116Cd experiment using 1-2 kg 116CdWO4 higher quality crystal scintillators and developing R&D projects on scintillators and for SuperNEMO.

Baksan Neutrino Observatory (Russia). http://www.inr.ac.ru/INR/
The Laboratory is operated by the INR of the Russian Academy of Sciences. It is managed as an observatory, with very long duration experiments. It is the oldest facility in the world built specifically for scientific research. M. Markov, the Head of Nuclear Physics Division of the Academy of Sciences of the USSR, obtained in 1966 a special Decree of the Soviet Government and construction of the Baksan Neutrino Observatory started under Mount Andyrchi in the Caucasus. A new village, called 'Neutrino', was built as a part of the original project in a previously empty space with personnel providing all necessary services (heating station, water supply system, first medical help, transportation, safety, etc.). The staff directly related to science is 50-60. The scientific activity started under the leadership of Alexander Chudakov and George Zatsepin. The access is horizontal via two dedicated tunnels, with train transportation.
A large hall, 24x24x16 m^3 in volume, 300 m deep, hosts the Baksan Underground Scintillation Telescope. BUST is ready to observe neutrinos from galactic supernovae since 1978. Another hall, 60x10x12 m^3 at a vertical depth of 2100 m, hosts SAGE, the Gallium Germanium Neutrino Telescope. In this lab the muon flux is $3x10^{-5}$ $m^{-2}s^{-1}$. The neutron flux (*E*>1 MeV) is $1.4x10^{-3}$ m^{-2} s^{-1}. The Rn activity is 40 Bq/m3 with a fresh air input of 60 000 m3/h.
The construction of a larger and deeper hall, about 40 000 m^3 in volume, was started in 1990 and stopped in 1992, when the Soviet Union collapsed. The fate of this unfinished construction is under discussion. Low Background Chambers with volume from 100 to 300 m^3 are used for R&D of dark matter and neutrinoless ββ decay search as well as for gravitational wave search and for some geophysics measurements. The number of users is 30-35.

For the sake of completeness, I report information on two mines in East Europe, which may turn of interest in the near future as underground facilities.

The Polkowice-Sieroszowice mine in Poland. Situated near Wrocław, south-west of Poland, is easily accessible from the Wroclaw airport and from the A4 motor-way, 950 km from CERN. The Sieroszowice mine (178 km^2 of underground excavation area), belongs to the KGHM holding of copper mines and metallurgic plants. Existing big chambers in salt: volume: 85 x 15 x 20 m^3, at a depth of about 950 m from the surface (2200 m.w.e.). Very low humidity, temperature ~35^0, and very low natural radioactivity of the salt rock. Work together with the mine management staff is going on to start the initial laboratory in 2010.

The Unirea salt mine, in Romania. It is situated in Unirea salt mine, in Slanic Prahova town, in Prahova County, in sub-Carpathians hills, about 100 km N from Bucharest. In this environment, a Low Background Radiation Laboratory has been situated at a depth of 208 m beneath the surface at water equivalent thickness of ~600 m. This mine consists of a hive-like structure composed of more galleries 32 or 36 meters wide, 52 to 57 m height and hundreds of meter long. Also it must be pointed out the remarkable stability of the microenvironment characterized by a constant temperature all over the years of 12^0 C and relative humidity of 60 to 65 %.

2.2 Asia

Y2L. YangYang Laboratory (Korea). http://dmrc.snu.ac.kr/
The lab is operated by the Dark Matter Research Centre (DMRC) of Seoul National University. The presently available area is 100 m^2. Expansion to 800 m^2 is planned but not yet funded. 100 m^2 space for offices, computing and detector test facility is available on surface. Access is horizontal by car.

The lab uses the space in the tunnel of the host YangYang Pumped Storage Power Plant. Researchers are requested to obey the safety and security regulation of the Plant.
Rock overburden is 700 m with a muon flux of $2.7\text{x}10^{-3}$ m^{-2} s^{-1}. Neutron flux is $8\text{x}10^{-3}$ m^{-2} s^{-1} for 1.5 MeV < *En* < 6.0 MeV. Radon activity is 40-80 Bq/m^3. The underground space is mostly occupied by the Korea Invisible Mass Search (KIMS) experiment, currently data taking for WIMP search with 100 kg CsI(Tl) crystal detectors. Other activities include R&D for ββ decay and background measurements with HPGe counter. Scientific users are about 30.

Oto Cosmo Observatory (Japan). http://wwwkm.phys.sci.osaka-u.ac.jp/info/syoukai/oto-e.html.
The laboratory has been developed by H. Ejri of the Osaka University and collaborators. Its area consists of Lab. 2 (50 m^2) hosting ELEGANT V and MOON-1 on dark matter search with NaI and neutrinoless ββ decay of 100Mo and Lab. 1 (33 m2) hosting ELEGANT VI on neutrinoless ββ decay of 48Ca and dark matter search with CaF2. The rock coverage is 470 m, with a muon flux of $4\text{x}10^{-3}$ m^{-2} s^{-1}. The neutron flux is $4\text{x}10^{-2}$ m^{-2} s^{-1}. The Radon inside "radon free" containers is 10 Bq m^{-3}. The access is horizontal through non-used railway tunnel, which provides also the non-forced ventilation.

Kamioka Observatory (Japan). http://www-sk.icrr.u-tokyo.ac.jp/index_e.html
The Kamioka Observatory is operated by the Institute for Cosmic Ray Research, University of Tokyo. It was established in 1983 by M. Koshiba as Kamioka Underground Observatory. The original purpose of this observatory was the KamiokaNDE experiment, to which Super-Kamiokande followed, the largest existing underground experiment. The KamLAND experiment is operated by the Neutrino Centre, Tohoku University.
Recently, a process of enlargement has started, in order to accommodate more experiments.
Buildings for offices and computer facilities are available on surface. The staffs are 13 scientists, 2 technical support units, one for administration.
The coverage is 1000 m and the muon flux is $3\text{x}10^{-3}$ m^{-2} s^{-1}. The thermal neutron flux is 8x 10^{-2} m^{-2} s^{-1}, the non-thermal is 11x 10^{-2} m^{-2} s^{-1}. The ventilation is 3000 m3/h. The access is horizontal by car, with no interference with mining activity.
The underground structures are as follows: Hall SK (50 m diameter) hosting Super-Kamiokande, to be continued 10 years, at least. Clean room (10x5 m^2) with XMASS prototype. Hall 40 (L-shape, 40 m x 4 m arms) hosting the purification tower for XMASS and the NEWAGE experiment on Dark matter. Hall 100 (L-shape, 100 mx4 m arms) with CLIO, a prototype of Gravitational Antenna (to be terminated in 2013) and a Laser displacement detector. The new Hall A (15x21 m^2) hosting XMASS 800 kg (till 2012) with space available for another experiment. The new Hall B (6x11 m^2) hosting CANDLE on ββ decay, to be occupied till 2012. Small areas are available in the dismissed mine. The scientific users are more than 200 in number. Budget request for the underground large cryogenic gravitational antenna LCGT has been submitted.
This year the T2K experiment is starting. It is a second generation long baseline neutrino oscillation experiment. The J-PARC facility will produce an intense off-axis beam of muon neutrinos. The beam is directed towards the Super-Kamiokande detector, which is 295 km away. The main goal of T2K is to measure the oscillation of ν_μ to ν_e and to measure the value of θ_{13}.

INO. India based Neutrino Observatory (India). http://www.imsc.res.in/~ino/
One of the two experiments that first observed the atmospheric neutrinos in 1964 was located at 2700 m depth in the Kolar Gold Mine in India. The India based Neutrino Observatory is the project to create an underground laboratory in southern India. It will be located near the PUSHEP hydroelectric pumping station, under 1 300 m rock overburden.
The lab will be organized with international laboratory standards with a Scientific Advisory Committee, services to the users, environmental, safety, security and outreach activities. Two main underground cavities are foreseen, Lab1 with a volume of 26x 135x 25(h) m^3 and Lab 2 with 53.4x 12.5x 8.6(h) m^3 plus connection tunnels and services. Access will be horizontal through dedicated 2

km tunnel. On surface it is planned to have: a 1400 m^2 building for administration, offices, shops, etc., a 2750 m^2 building with lecture hall and guest house and a residential complex with 20 quarters. Personnel will be 50 to 100.
The main foreseen experiment is ICAL, a 50 kt magnetized Fe tracking calorimeter for atmospheric and very long base-line accelerator neutrinos. It will occupy only a fraction of Lab1.

2.3 North America

SNO-Lab (Canada). http://www.snolab.ca/ also http://www.sno.phy.queensu.ca/
The SNO experiment has completed its glorious life and its cavity, 200 m^2 area, is now being freed for further experimental activity. New structure have been constructed: a Main hall of volume 18x 15x (15 to 19.5 height) m^3, a service hall of about 180 m^2 and a number of narrow (6-7 m) volumes called "ladder labs". The volume for a further structure, called *cryopit*, has also been excavated. This hall is designed to cope with the safety issues surrounding large volumes of cryogenic fluids. The total area is 7 215 m^2, of which 3 055 m^2 available for the experiments, the total volume is 46 648 m^3 of which 29 555 m^3 available for the experiments. The access is vertical, through the shaft of the working mine, available daily. All the laboratory will be clean, class 1500. On the surface a 3 159 m^2 building will host clean room, laboratories, staging and assembly areas, office space (60 users), meeting rooms, control rooms, IT server room, emergency generator, high speed network link off site, high speed network link surface/underground, safety structures and management. Staff will be of 30 people full time. Rock coverage is 2000 m under flat surface. Muon flux $3x10^{-6}$ $m^{-2}s^{-1}$, thermal neutron flux is $4.7x10^{-2}$ m^{-2} s^{-1}, fast neutron flux is 4.6x 10^{-2} m^{-2} s^{-1}. Radon in the air is high, 120 Bq/m^3. The ventilation in the smaller lab spaces provides 10 air changes per hour, in the larger ones 5 air changes per hour.
The full PICASSO detector, searching for dark matter with the super-heated bubbles technique, is running. SNO+ is hosted in the former SNO cavity; it is based on liquid scintillator for low energy solar neutrinos, geoneutrinos and ββ decay, by dissolving 150Nd in the liquid. Dark Matter search includes DEAP/CLEAN with noble liquids, which is operating with a prototype, and the installation of superCDMS with bolometers. More LoIs are expected to be reviewed by the Experimental Advisory Committee.

SUL. Soudan Underground Laboratory (USA). http://www.soudan.umn.edu/
The underground structures include: the Soudan lab (20x7x10(h) m^3) that hosts: a) CDMS, expected to run until 2010-2011; b) a low-background counting facility that currently occupies 5x5x3 m^3 and will expand to 25x14x14(h) m^3, if funded. The MINOS lab, in which MINOS occupies 35x16x14(h) m^3 and is expected to run until 2010-2011. Possibility of a LAr medium sized prototype to be replace MINOS to bask in the NuMI beam.
The users are 265 in number. The access is vertical via a two-compartment slightly angled shaft. Diameters in excess of 1m and lengths in excess of 10m pose a problem. Access outside normal operating hours is possible. There is an access charge paid to the host institution, Soudan Underground Mine State Park. Normal laboratory safety requirements are in place. The laboratory coexists with an historic State Park, which offers mine tours during the summer months to the public, and winter tours to school groups. Some tours utilize a visitor's gallery available in the MINOS laboratory. There is no active mining activity.
The overburden is 700 m of rock. The muon flux is $2x10^{-3}$ m^{-2} s^{-1}. The neutron interaction rates are approximately 10 kg^{-1} d^{-1} (from U/Th, low energy) or 0.01 kg^{-1} d^{-1} (muon generated in the rock). The radon concentration is seasonal, varying from 300 Bq/m^3 in the winter to 700 Bq/m^3 in the summer.
The mine has natural ventilation, about 550 m^3/h for the level of the laboratories. Half of this is diverted to ventilate the MINOS and Soudan spaces. This results in a complete air change every 110'.
The major facility on the surface is a building of approximately 650 m^2 with offices, kitchen and sanitary facilities. The laboratory has a staff of 9, including secretarial and accounting assistance and

network and computer maintenance personnel. It is staffed 10 hours/day, 5 days per week, but the staff is on-call during the balance of the time and responds to requests for emergency access.

DUSEL. Deep Underground Science and Engineering Laboratory (USA).
http://www.lbl.gov/nsd/homestake/
Solar neutrino physics started in the Homestake mine in South Dakota with the experimental work of Ray Davis and the theoretical sun model developed by John Bahcall. After a long and complex process, in spring 2007, NSF selected amongst several proposals the Homestake mine in South Dakota as the site in which the Deep Underground Science (physics, biology, and geology) and Engineering Laboratory (DUSEL) should be designed. The project foresees a funding by NSF of about 250 M$ for the facility plus a contribution to the initial set of experiments, costing about 250 M$. In addition, to prepare the site, SD provides 46 M$ on its own and 70 M$ from a donor: T. D. Sanford. NSF is expected to fund with 15 M$ the design of the facility. For the time being, the facility is a SD State Laboratory, funded primarily from SD controlled money. Water constantly flows into the mine at a rate of 1.2 Mt/yr. Dewatering and site preparation is in progress, as well as the pre-costrution design. This phase should last until 2013, when the DUSEL construction should start to end in 2019.
On the surface several existing buildings will be rehabilitated for about 10 000 m^2 to host offices, support structures and laboratories. A major science education centre has been funded by Sanford. The initial staff is estimated 30-50, to increase to 100-150 when the State Lab will become DUSEL.
The Scientific Committee has been established with an USA composition in 2006. Laboratory spaces will be built separately for biology, geology and physics. Service and R&D structures (e.g. electroforming) will be available for physics at 100 m deep level. Two main campuses are foreseen for physics about 1450 m and 2200 m deep. Each will contain a number (4 and 3 respectively) of standard modules of 50x20x15 m^3 plus service areas. A staged construction of the upper campus using SD and private funding is foreseen. For the lower campus NSF and Congress approval will be necessary.

3. More science

Recently a new community, belonging to the astroparticle field, got interest in an underground environment: the gravitational wave researchers. The evolution of the current (first generation) gravitational wave detectors is well defined: the laser interferometers will evolve toward their second generation: the advanced (Virgo and LIGO) detectors. According to the current gravitational wave sources modelling, when these apparatuses will reach their nominal sensitivity, the detection of the gravitational waves seems assured in few months of data taking. But the sensitivity needed to test the Einstein's gravity in strong field condition or to realize a precision gravitational wave astronomy goes beyond the expected performances of the advanced detectors. The fundamental limitations at low frequency of the sensitivity of the 2nd generation detectors are given by the seismic noise, the related gravitational gradient noise (so-called Newtonian noise) and the thermal noise of the suspension last stage and of the test masses. To circumvent these limitations new infrastructures are necessary: an underground site for the detector, to limit the effect of the seismic noise, and cryogenic facilities to cool down the mirrors to directly reduce the thermal vibration of the test masses. ET (Einstein Telescope http://www.et-gw.eu/) is a Design Study project supported by the EC under the Framework Programme 7. It concerns the study and the conceptual design for a future European third generation gravitational wave detector to be realized underground. Specific tests of seismic and gravity gradient noise have started, or are planned, in Homestake and Gran Sasso.
There are other fundamental questions increasing the interest in underground science. Questions are raising from life sciences: How deeply in the Earth does life extend? What makes life successful at extreme depth and temperature? What can life underground teach us about how life evolved on earth and about life on other planets? And other questions are raising from Earth sciences: How rock mass strength depends on length and time scales? Can we understand slippage mechanisms in high stress environment, in conditions as close as possible to tectonic faults/earthquakes? Which are the mechanisms behind the constant Earth evolution? And there are more questions on rock mechanics at

large scales and on geo-hydrology. Some activities in these fields are going on in some labs, like Gran Sasso. More attention should be put on these activities in future. The program of DUSEL is taking seriously this issue.

4. Conclusions

Experiments in underground laboratories have opened a new window in neutrino physics, beyond the standard model. The laboratories present activities give an idea of how brilliant are the perspectives for underground physics in the next decade. Let me try to list the main challenges:

- Determine the nature of dark matter. Requires ton-scale detectors with reduced background and good stability to be sensitive to the annual modulation signatures.
- Determine neutrinos absolute masses and nature (Majorana vs. Dirac), searching for neutrinoless $\beta\beta$ decay. Ton to multi-ton scale detectors and drastic background reduction are needed.
- Solar Be and pp neutrino fluxes should be measured with flavour sensitivity and high (few percent) precision. Weak Interactions may be still richer than what we know today (remember lessons of J. Bahcall and of neutrino oscillations).
- Search for p-decay. Needs Megaton scale Cherenkov detector or tens of kiloton of liquid argon TPC.
- Measuring or limiting θ_{13}. Requires a very intense muon neutrino beam at low, but not too low, energy (of the order of 1 GeV) and a massive detector. May be at shallow depth. This step is necessary before seriously launching a project for a neutrino factory.
- Supernova neutrinos may give information on the sign of $m_3^2 - m_2^2$ and on the size of θ_{13} . Present detectors are mainly sensitive to electron antineutrino. Do we need a new one?

A new large-volume detector with the fiducial mass from several tens of ktons up to 1 Mton will provide unique scientific opportunities in low-energy neutrino astronomy and Grand Unification physics. Such a detector can drive the extension of a present laboratory or the realization of a new (shallow depth) laboratory and exploit dedicated neutrino beams.
Finally, I think we have to favor the increasing interest for underground facilities from other communities. There is more than physics, and it is certainly a benefit to widen the underground frontier and look to our labs from a global multidisciplinary perspective.

I would like to warmly thank Sandro Bettini and all the “underground” colleagues for providing much of the information reported here.

The Argon Dark Matter Experiment (ArDM)

Christian Regenfus
Physik-Institut der Universität Zürich, CH–8057 Zürich, Switzerland
on behalf of the ArDM collaboration
E-mail: `regenfus@cern.ch`

Abstract.
The ArDM experiment, a 1 ton liquid argon TPC/Calorimeter, is designed for the detection of dark matter particles which can scatter off the spinless argon nucleus, producing nuclear recoils. These events will be discerned by their light to charge ratio, as well as the time structure of the scintillation light. The experiment is presently under construction and commissioning on surface at CERN. Cryogenic operation and light detection performance was recently confirmed in a test run of the full 1 ton liquid argon target under purely calorimetric operation and with a prototype light readout system. This note describes the experimental concept, the main detector components and presents some first results.

1. Introduction

The best limit on spin-independent WIMP cross-sections is presently given by the CDMS experiment [1], which records phonon and ionisation signals of particle interactions in cryogenic semiconductor targets [1] with high sophistication. A row of noble liquid dark matter experiments are currently striving for a similar kind of experimental perfection, actually starting to challenge these best limits [2]. While noble liquids are equally blessed with high scintillation and ionisation yields, they feature a high potential to be scaled up in the ton or even multi-ton target range, which makes this technology so popular and rapidly progressing. Large target sizes require very high radiopurity of the target, not only to limit feedthrough of background in the nuclear recoil band, but also to comply with a maximal tolerable trigger rate. This is a main concern in the case for liquid argon (LAr) due to the long lived ^{39}Ar isotope. Recent progress in liquifying underground gases and extracting substantial quantities of ^{39}Ar depleted liquid argon gives however a promising outlook for this technology. Self-shielding [2], one of the strongest motivations to go to large masses, is very much improving for larger and larger target sizes. Further on a large detector is able to determine, on a statistical base, the background of single scattering neutrons in the data from the multiplicity distribution of neutron interactions. Figure 1 shows

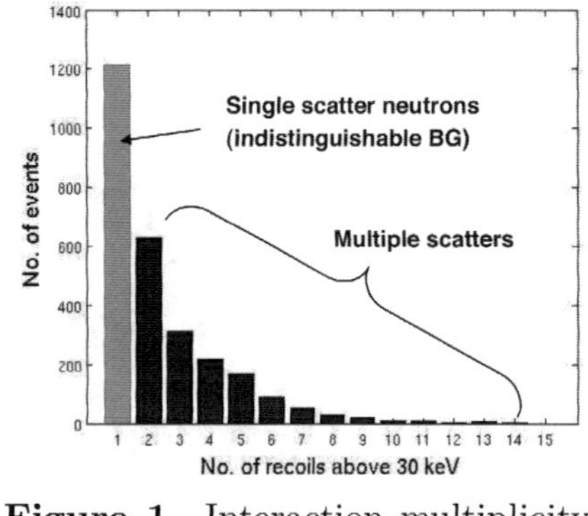

Figure 1. Interaction multiplicity of background neutrons in ArDM (MC).

[1] Total target mass is around 6 kg
[2] against external radiation, above all gammas, but not neutrons

Journal of Physics: Conference Series **203** (2010) 012024 doi:10.1088/1742-6596/203/1/012024

a frequency distribution of neutron interactions on the example of the ArDM geometry [3]. A hydrogen rich shield will nevertheless be used to protect the detector in its final working location from background neutrons (after a measurement of the spectrum).

In liquid argon, both, the scintillation light to charge ratio and the temporal structure of the light emission [4, 5, 6] can be used for electron- to nuclear-recoil discrimination. This is due to the ionisation density dependent ratio of the two argon excimer ground states ($^1\Sigma_u^+$ and $^3\Sigma_u^+$), which are responsible for the VUV luminescense of LAr. The large difference of their radiative lifetimes ($\approx 10^3$) allows for excellent recoil separation down to 20 keV$_{ee}$ (electron equivalent scale) [7]. This effect alone is used to classify recoils by some single phase experiments (e.g. DEAP/CLEAN, XMASS [8, 9]). The two phase configuration (as we employ it in ArDM) allows for an additional measurement of the ionisation charge (e.g. XENON, WARP, LUX [10, 11, 12] experiments). A drawback of the argon technology is the short wavelength of the scintillation light (128 nm) and the already mentioned presence of the ^{39}Ar β-emitter. However, because of form factors, argon is less sensitive to the threshold of the nuclear recoil energy, than is e.g. xenon. For the same reason the recoil energy spectra of argon and xenon are quite different. These liquids are therefore complementary in providing a crosscheck once a WIMP signal has been found.

2. Conceptual design

ArDM [13] was projected as a ton scale liquid argon spectrometer aiming at the detection of nuclear recoils above 30 keV$_r$ (i.e. ≈ 10 keV$_{ee}$). At the same time it serves as a prototype unit for future large LAr detectors, for more sensitive DM experiments or next generation neutrino observatories. Three dimensional imaging and event by event interaction type identification will be used to reach high background suppression. About 400 VUV photons and a couple of ionisation charges[3] are typically produced in LAr by a WIMP interaction at 30 keV. The background rejection will be achieved by the combination of cuts on the fiducial volume, the event topology (excl. multiple scatter), the light to charge ratio and the temporal structure of the light emission. The technical requirements comprise a large volume electric field, a large area position sensitive charge readout (3rd dimension from drift time), a fast and large area light readout and an efficient liquid argon purification system. The event trigger is generated from the light signal. Figure 2 shows a sketch of the two-phase operating mode of the detector. In a particle interaction excited and ionised argon atoms form the argon excimer states [5] which decay under the emission of 128 nm VUV radiation. This light can not be absorbed by neutral argon atoms and hence propagates to the side walls of the detector which are coated with the wave shifting material tetraphenyl-butadiene (TPB). VUV light is absorbed and with high efficiency re-emitted at around 430 nm, the region of high quantum efficiency of borosilicate windowed bialkali PMTs. By diffusive reflection on the side walls, the light is transported to the bottom of the apparatus to an array of 14 hemispherical 8" PMTs. The strong electric field is capable of preventing some free electrons in the densely ionised region around a nuclear recoil from recombining and sweeps them to the surface of the liquid, where they are extracted into the gaseous phase of the detector. The

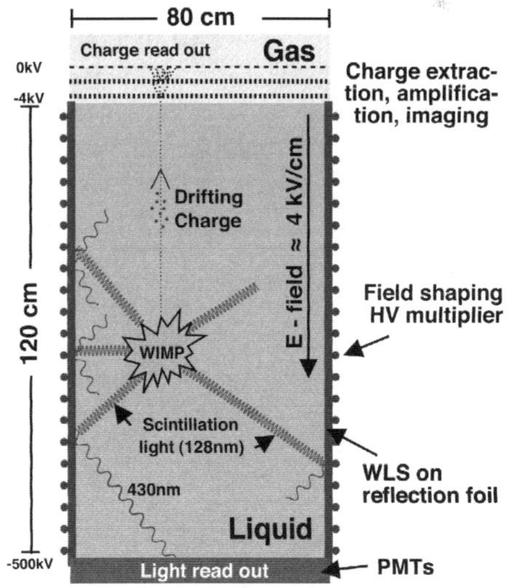

Figure 2. Conceptual design of ArDM.

[3] if the electrical field is above 1 kV/cm

charges are then multiplied by means of a rigid large gas electron multiplier (LEM) and finally recorded by a position sensitive read out plane [14].

3. The experimental components

Figure 3 left shows the mechanical arrangement of the cryogenic cooling and cleaning system of the setup together with the main stainless steel dewar (containing roughly 1800 kg of liquid argon). An inner cylindrical volume of 80 cm diameter and 120 cm height is delimited by round ring electrodes (field shapers) constituting the 850 kg active LAr target in a vertical TPC configuration (Fig. 3 middle). The field shaper rings are connected to a 210 stage HV diode-capacitor charge pump system (Cockroft-Walton circuit) which is fully immersed in the liquid

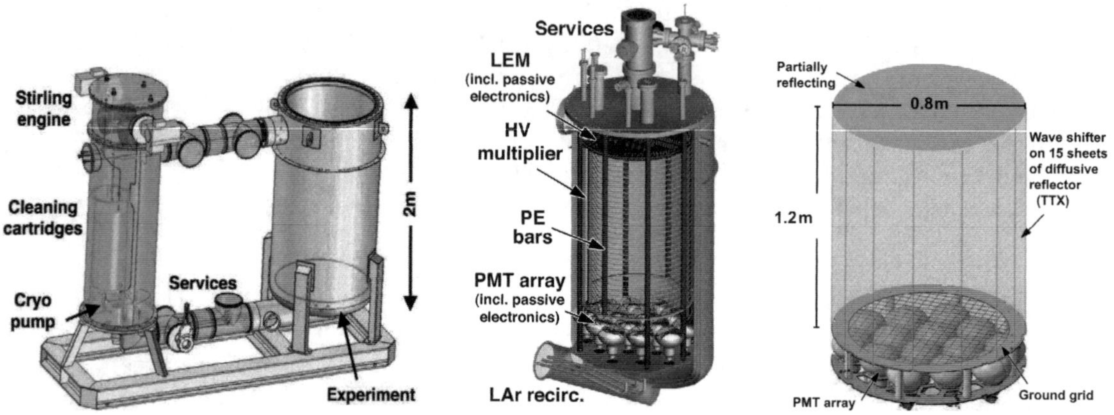

Figure 3. Left: cryo-system and main dewar; middle: detector components mounted to the top flange by polyethylene bars; right: arrangement of the wavelength-shifter/reflector foils in light diffusion cell arrangement inside the field shaper rings.

argon and designed to reach up to -500 kV ($\approx$4 kV/cm) at the bottom cathode. Another unique feature of the experiment is the use of a LEM, which is placed 5 mm above the liquid level in the gas. The positional readout is achieved by segmenting the upper LEM surface and the anode plate with 1.5 mm wide x and y-strips respectively. In total there are 1024 readout channels which are AC coupled to charge sensitive preamplifiers located externally on the top flange of the apparatus. Because the LEM is operated in very pure argon gas, which cannot quench charge avalanches, it has to be built with considerable attention to HV discharges. To read scintillation light homogeneously over the full volume we adopted a (wave shifting) diffusion cell design with the PMT array at the bottom in the liquid. This keeps the system simple [4] and scalable. The shifting of the VUV into the range of high quantum efficiency of the bialkali PMTs is done by a thin layer ($\approx$1 mg/cm^2) of tetraphenylbutadiene (TPB) evaporated onto 15, cylindrically arranged, 25 cm wide reflector sheets which are located in the vertical electric field. These sheets, made out of the PTFE fabric Tetratex^{TM} (TTX), are clamped to the upper- and lowermost field shaper rings. The PMTs (Hamamatsu R5912-02MOD-LRI) were made from particularly radiopure borosilicate glass and are sensitive to single photons. They are equipped with with a Pt-underlay under their photocathodes for operation at LAr temperatures. This ineluctably reduces their quantum efficiency by roughly one third to a value between 15 and 20% [15]. The PMT glass windows are also coated with TPB to convert directly impinging VUV photons. The average number of recorded photoelectrons (n_{pe}) is expected (laboratory measurements) to be in the order of 1 pe/keV$_{ee}$. The development of the light detection system and particularly

[4] large area VUV sensitive photosensors, e.g. MgF_2 windowed PMTs are commercially not available

the operation of gaseous argon test cells with α particle excitation were described in earlier work [16, 17, 18].

4. First test run of ArDM on the full 1 ton LAr target

A first run of the experiment was undertaken in May 2009 to test cryogenic functionality (incl. various safety installations) with a prototype light read out system (7 PMTs out of 14 were mounted). No internal electric field was yet present but background events and more over the response of the detector to external gamma sources could be explored for the first time. Cryogenics proofed to work well, showing stable operation at high argon purity over the full 2 weeks of operation despite no liquid recirculation and cleaning. LAr purity was monitored by the lifetime of the second scintillation component (see [17] for details on this technique), recording a stable high value of ≈ 1500 ns over the entire test period. Figure 4 shows the response of the detector to external gamma sources ^{137}Cs (190 kBq) and ^{22}Na (20 kBq), where the emission of the latter was controlled and triggered by an external crystal scintillator. The shoulders of full gamma absorption are marked with arrows. Dashed spectra stem from data triggered on large PMT signals at a threshold of roughly 150 keV$_{ee}$ (^{22}Na data was taken in coincidence with the crystal). The full, grey spectrum was recorded triggering on the external crystal only and shows an unbiased 511 keV response (event reconstruction down to 50 keV$_{ee}$ possible). We find an average yield for the prototype light read out of ≈ 0.5 pe/ keV$_{ee}$, well in agreement with the expected number of around 1 pe/ keV$_{ee}$ of a fully equipped detector. A first comparison of the spectra with MC confirms an average value of 18% for the light yield of the Pt-underlay PMTs.

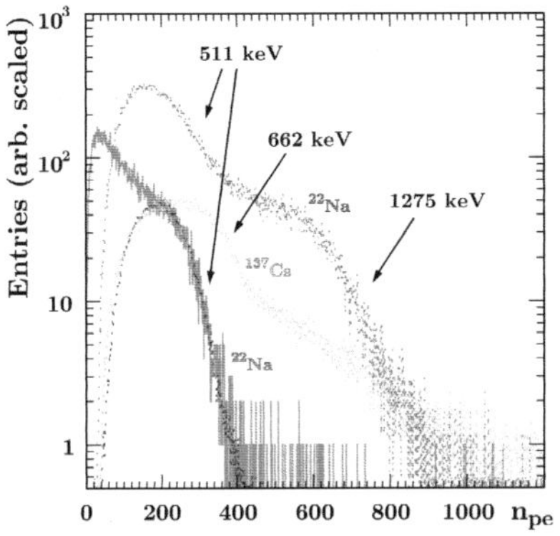

Figure 4. gamma spectra in ArDM.

While R&D work for sub detector parts is now finalising, main mechanical components are set together at CERN. Following a successful commissioning at surface we consider a deep underground operation.

Acknowledgments

This work is supported by grants from the Swiss National Science Foundation and ETH Zürich.

References

[1] Ahmed Z *et al.* (CDMS collaboration) 2009 *Phys. Rev. Lett.* **102** 011301
[2] Angle J *et al.* (XENON10 collaboration) 2008 *Phys. Rev. Lett.* **100** 021303
[3] Kaufmann L 2008 *Doctoral Diss. ETH Zürich* No. 17806
[4] Kubota S *et al.* 1978 *J. Phys. C* **11** 2645
[5] Suzuki M *et al.* 1982 *NIM* **192** 565
[6] Hitachi A *et al.* 1983 *Phys. Rev. B* **27** 9 5279
[7] Lippincott W *et al.* 2008 *Phys. Rev. C* **78** 035801
[8] DEAP/CLEAN: http://deapclean.org
[9] Abe K *et al.* 2008 *J. Phys. Conf. Ser.* **120** 042022
[10] XENON100: http://www.physik.uzh.ch/groups/groupbaudis/xenon/
[11] WARP: http://warp.lngs.infn.it/papers/proposal.pdf
[12] LUX: http://www.luxdarkmatter.org
[13] ArDM, Rubbia A 2006 *J. Phys. Conf. Ser.* **39** 129
[14] Badertscher A *et al.* 2009 *NIMA A* doi:10.1016/j.nima.2009.10.011
[15] Bueno A *et al.* 2008 *JINST* **3** P01006
[16] Regenfus C 2007 *Proc. of IDM2006, Rhodes Greece, World Scientific*, page 325
[17] Amsler C *et al.* 2008 *JINST* **3** P02001
[18] Boccone V *et al.* 2009 *JINST* **4** P06001

Journal of Physics: Conference Series **203** (2010) 012025 doi:10.1088/1742-6596/203/1/012025

Results from the first science run of ZEPLIN-III

M I Lopes for the ZEPLIN-III Collaboration

LIP-Coimbra, Department of Physics, Rua Larga, 3004-516 Coimbra, Portugal

E-mail: isabel@lipc.fis.uc.pt

Abstract. ZEPLIN-III is a two-phase (liquid/gas) xenon time projection chamber designed to search for dark matter in the form of weakly interacting massive particles (WIMPs). Its active volume contains 12 kg of liquid above an array of 31 2-inch photomultipliers. The detector uses both scintillation and ionization produced in the liquid in order to discriminate between electron recoil background interactions and the nuclear recoils expected from the elastic scattering of WIMPs off the xenon nuclei. The first science run of ZEPLIN-III acquired 847 kg.days of raw data. An analysis of those data has excluded a WIMP-nucleon elastic scattering spin-independent cross section above 8.1×10^{-8} pb at 60 GeV/c^2 WIMP mass and a WIMP-neutron spin-dependent cross section above 1.9×10^{-2} pb at 55 GeV/c^2 both with a 90% confidence two-sided limit.

1. Introduction

ZEPLIN-III is a galactic dark matter experiment deployed 1100 m underground at the Palmer Underground Laboratory at Boulby, UK. It is based on a two-phase (liquid/gas) xenon detector intended to detect weakly interacting massive particles (WIMP) via their elastic scattering from xenon nuclei. The recoiling nuclei will produce vacuum ultraviolet primary scintillation light (S1) and ionization charge in the liquid. Under an applied electric field, the electrons escaping recombination drift towards the top of the liquid being extracted into the gas layer where they produce proportional electroluminescence (S2). Event by event discrimination between nuclear recoils (expected from WIMP and neutron interactions) and electron recoils (from γ-ray background) is obtained from S2/S1 as this ratio is very different for these two classes of events.

ZEPLIN-III has 31 2-inch diameter photomultiplier tubes (PMTs) immersed in the liquid to detect the prompt S1 light signal and the delayed S2 pulse with good light collection efficiency. Its design and performance were described in detail elsewhere [1,2].

2. ZEPLIN-III first science run

For the first science run (FSR) the ZEPLIN-III detector was completely surrounded by a shield of 30 cm thick polypropylene and 20 cm thick lead that provide ~ 10^5 attenuation factors for both γ–rays and neutrons from the cavern walls.

WIMP-search data were collected over 83 days of underground continuous operation with 84% live time during which some 847 kg.days of raw data were acquired from the 12 kg target volume. This exposure is reduced to 453.6 kg.days after applying the cuts that define the fiducial volume: drift time window to limit the height of liquid and a radial limit coming from the S2 position reconstruction in the horizontal plane.

Energy calibrations of the S1 and S2 responses, as well as the flat-fielding of the PMT array, were obtained with ^{57}Co γ-rays. The primary and the secondary light yields were measured: S1 light yield was found to be 1.8 p.e./keVee at the operating field of 3.9 kV/cm and 5 p.e./keVee without electric field; the S2 light yield at the instrument operation conditions was measured to be 6.5×10^2 p.e./keVee. The data collected daily with the ^{57}Co source were also used to assess several operational parameters, such as the liquid purity, detector tilt, liquid level and pressure effects, over the entire run.

Calibration measurements were performed with ^{137}Cs and Am-Be (α,n) sources at the start and at the end of the science run. The Compton scattering events from the 662 keV γ-rays from the ^{137}Cs allow establishing the electron recoil response at low energies (down to 2 keVee, where 'keVee' unit refers to the equivalent mean S1 signal produced by 122 gamma-rays). The neutron elastic scattering events from the Am-Be provide calibration of the response to nuclear recoils in the energy range of interest to WIMP signals.

The low-energy Compton-scattered events from the ^{137}Cs were also used to investigate the linearity of response of each light readout channel and to determine the mean single photoelectron response of each photomultiplier [4].

Details on the calibration, data processing and selection are provided elsewhere [3].

3. Results

Clear separation between nuclear and electron recoils was achieved as shown in Fig.1 where electron-recoil data from ^{137}Cs and elastic nuclear recoil data from Am-Be are plotted.

The WIMP-search box boundary was defined as $2 < E < 16$ keVee and $(\mu_n-2\sigma) < \log_{10}(S2/S1) < \mu_n)$ where μ_n is the energy-dependent mean of the nuclear recoils (acceptance of 47.7%). This region was defined before unblinding and was kept for the subsequent analysis. The effective total exposure within this box after taking account all the efficiencies is 127.8 kg.days.

The single-scatter neutron background in the data set is estimated at 1.2±0.6 in the WIMP-search box with 90% originated in the PMTs. This estimate is obtained with a full GEANT4 simulation [1]. The electron-recoil background below hundreds keV was measured and was found to be in good agreement with the GEANT4 simulation [1], according to which the dominant contribution is 10.5 dru (1 dru= 1 evt/kg/day/keVee) from the photomultipliers.

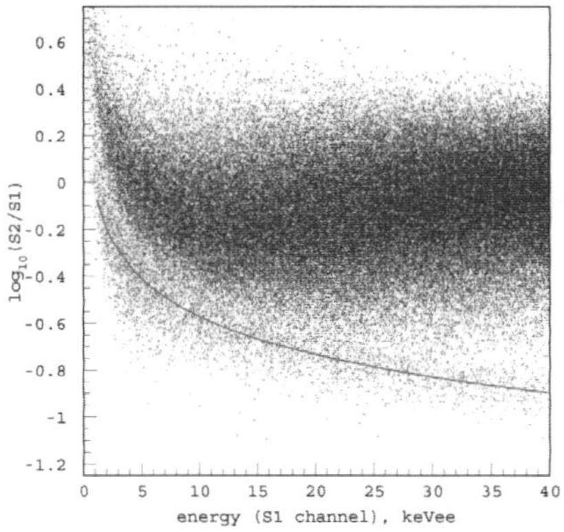

Figure 1. ^{137}Cs (blue) and Am-Be (red) calibration data.

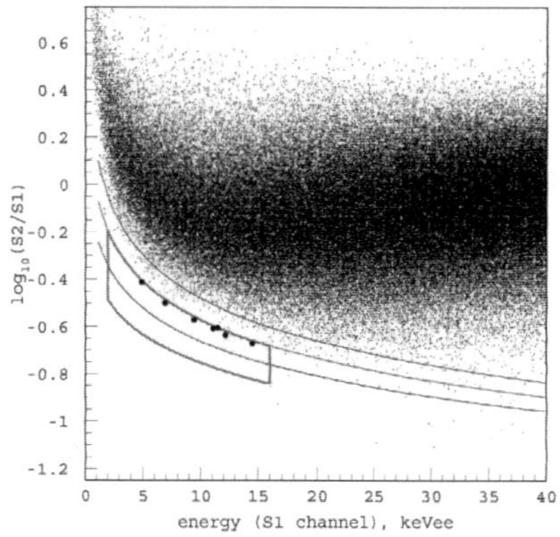

Figure 2. First science run data. The thick red box delimits the WIMP-search region.

The science data contain 7 events within the WIMP-search box, as shown in Fig. 2. To compute the WIMP-nucleon cross-section limits, the energy scale has to be converted from keVee into kVnr and the energy-dependent detector efficiency for nuclear recoils must be determined. The latter was found by comparing Am-Be data sets with very different trigger thresholds in both hardware and software [3]. The energy scale conversion is then obtained by comparing the experimental nuclear recoil differential spectrum with that obtained by simulation. In so doing, we find evidence for a significant non-linearity in the conversion between S1 keVee and S1 keVnr [3]. Recently, such non-linearity was also observed, though less pronounced, by other authors in beam measurements [5].

Following the procedure detailed in [3], it was found that the data is consistent with no WIMP signal and the 90% confidence upper limit was set as 3.05 events. This limit was then converted into an upper limit on the WIMP-nucleon spin-independent elastic scattering cross section as a function of the WIMP mass using the standard spherical isothermal galactic halo model [3]. The upper limit on WIMP-nucleon spin-independent cross-section from ZEPLIN-III is shown in Fig. 3 together with the results from the other presently leading experiments. The analysis of the data for obtaining the upper limit of the WIMP-neutron/proton spin-dependent cross-sections is presented elsewhere [6].

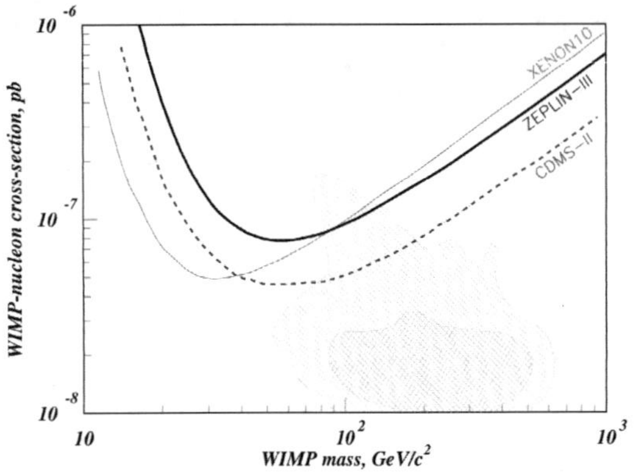

Figure 3. Limits on WIMP-nucleon spin-independent cross-sections from the first science run of ZEPLIN-III. The results published by XENON10 [7] and CDMS-II [8] are also shown. Note that the XENON10 curve is a one-sided limit while CDMS-II and our result are both 90% two-sided limits. For comparison, the XENON-10 two-sided limit is 1×10^{-7}pb at a WIMP mass of 30 GeV/c^2 [9]. Moreover the XENON10 result did not take into consideration the non-linearity between keVee and keVnr.

4. Second science run

ZEPLIN-III has been upgraded in view of a second science run. Since the photomultipliers were the dominant source of background in the first science run they were replaced by pin-by-pin compatible ones manufactured with materials of higher radiopurity. It is expected that the γ-ray and neutron backgrounds will be improved by a factor of ~30.

An active scintillator veto, surrounding the detector inside the lead castle, replaced the previously used passive neutron shielding. It consists of 52 plastic scintillator modules placed around and above the detector. Each module has a 3" photomultiplier at one end, a mirror at the other end and it is wrapped with PTFE along length followed by a black outer wrap. A 15 cm thick 1% gadolinium loaded polypropylene was placed between the scintillator and the target working as neutron moderator.

The second science run is expected to start by the end of the year.

5. Acknowldgements

We wish to pay tribute to our colleague and friend V.N. Lebedenko (1939-2008), who had a key role in designing and building the instrument. LIP-Coimbra acknowledges funding from Fundação para a Ciência e Tecnologia (FCT) through the project-grant CERN/FP/83501/2008.

References

[1] Araújo H M *et al* 2006 *Astroparticle Physics* **26** 140
[2] Akimov D Yu *et al* 2007 *Astroparticle Physics* **27** 46
[3] Lebedenko V N *et al* 2009 *Physical Review D* **80** 052010
[4] Neves F *et al* 2009 (accepted in *Astroparticle Physics*) arXiv:0905.2523
[5] Manzur A *et al* 2009 arXiv:0909.1063
[6] Lebedenko V N *et al* 2009 *Phys. Rev. Lett* **103** 151302
[7] Angle J *et al* 2008 *Phys. Rev. Lett* **100** 021303
[8] Ahmed Z *et al* 2009 *Phys. Rev. Lett* **102** 011301
[9] Aprile E and Doke T arXiv:0910.4956

The LUX Dark Matter Search

D. N. McKinsey*[,k], D. Akerib [b], S. Bedikian [k], A. Bernstein [c], A. Bolozdynya[d], A. Bradley [b], J. Chapman [a], K. Clark [b], T. Classen [f], A. Curioni [k], E. Dahl [a], S. Dazeley [c], M. Dragowsky [b], L. de Viveiros [a], E. Druszkiewicz [h], S. Fiorucci[a], R. Gaitskell [a], C. Hall [g], C. Hernandez Faham [a], L. Kastens [k], K. Kazkaz [c], R. Lander [f], D. Leonard [g], D. Malling [a], R. Mannino [e], D. Mei [j], J. Mock [f], J. A. Nikkel [k], P. Phelps [b], T. Shutt [b], W. Skulski [i], P. Sorensen [c], J. Spaans [j], T. Stiegler [e], R. Svoboda [f], M. Sweany [f], M. Tripathi [f], N. Walsh [f], R. Webb [e], J. White [e], F. Wolfs [h], M. Woods [f], C. Zhang [j]

[a]Brown University, Dept. of Physics, 182 Hope St., Providence RI 02912
[b]Case Western Reserve University, Dept. of Physics, 10900 Euclid Ave, Cleveland OH 44106
[c]Lawrence Livermore National Laboratory, 7000 East Ave., Livermore CA
[d]Moscow Engineering Physics Institute, 31 Kashirskoe shosse, Moscow 115409
[e]Texas A & M University, Dept. of Physics, College Station TX 77843
[f]University of California Davis, Dept. of Physics, One Shields Ave., Davis CA 95616
[g]University of Maryland, Dept. of Physics, College Park MD 20742
[h]University of Rochester, Dept. of Electrical and Computer Engineering, Rochester NY 14627
[i]University of Rochester, Dept. of Physics and Astron., Rochester NY 14627
[j]University of South Dakota, Dept. of Physics, 414E Clark St., Vermillion SD 57069
[k]Yale University, Dept. of Physics, 217 Prospect St., New Haven CT 06511

*Email: daniel.mckinsey@yale.edu

Abstract. The Large Underground Xenon (LUX) experiment is a liquid xenon time projection chamber designed for extremely low levels of radioactive background in its fiducial volume. The overall liquid xenon mass is 300 kg, with a 100 kg fiducial mass. LUX is currently under construction, and integration of the full detector will begin in Fall 2009 at the Sanford Underground Science and Engineering Laboratory in South Dakota. The LUX sensitivity to the WIMP-nucleon spin-independent scattering cross-section will be 7 x 10^{-46} cm^2 at 100 GeV after 300 days of low-background operation.

1. The LUX Experiment

The LUX dark matter experiment is a new 300-kg active mass two-phase Xe detector that will be used to extend current dark matter sensitivity two orders of magnitude beyond current best limits, to a WIMP-nucleon cross-section of approximately 7 x 10^{-46} cm^2 for a 100 GeV WIMP. A large detector is required not only to set such a sensitivity limit, but also to accumulate WIMP statistics in a reasonable time frame if a signal is detected. LUX is fully funded by the Department of Energy and National Science Foundation.

The LUX detector specifically exploits a number of features of liquid xenon (LXe) to drive backgrounds down. The active detector volume permits the full reconstruction of event energy deposition and position. An inner fiducial volume (100 kg LXe mass) can be clearly established with < 2 cm errors in position determination. Within the fiucial volume, the background rate is heavily suppressed due to the active shielding by the surrounding Xe. The electron recoil vs. nuclear recoil discrimination in two-phase Xe detectors has been shown to be 99.3–99.9%, as determined in LXe prototypes, and confirmed in XENON10 [1] and ZEPLIN-II [2] operation.

The basic design of LUX is very similar to that of XENON10 and ZEPLIN-II. Significant advances in the LUX design include a) use of an 8-m-diameter water tank to shield out gammas and neutrons from the cavern, instead of lead and polyethylene b) a cathode high voltage feedthrough located outside the shielding, connected with a Xe-filled umbilical to the detector and c) substantial use of titanium as a construction material. The cryostat housing the central detector is a simple pair of titanium vacuum insulated cans with the inner can housing LXe at a temperature of 165-190 K and a pressure of up to ~3 bar. Certain grades of titanium have recently been shown by the LUX collaboration to have especially low levels of uranium and thorium contamination. Liquid nitrogen is used to cool down the detector, liquefy Xe gas and maintain the LXe temperature at the desired value. Removal of impurities that would otherwise reduce the charge or light signal will be accomplished using a commercial getter.

An event in the LUX time projection chamber will be characterized by two signals, corresponding to detection of direct scintillation light (*S1*) and proportional light from ionization electrons (*S2*). Two arrays of photomultipliers, each with 61 PMTs, are used to detect both *S1* and *S2*. The pulses occur within the maximum drift time of 110 microseconds, for a saturated electron drift velocity of 2 mm/microsecond in LXe. Since electron diffusion in LXe is small, the proportional scintillation pulse is produced in a small spot with the same *x-y* coordinates as the interaction site, allowing 2D localization with an accuracy of 1 cm. With the more precise *z* information from the drift time measurement, the 3D event localization provides additional background discrimination via fiducial volume cuts.

The central, fully active Xe region is defined by 11 PTFE reflector panels and the cathode grid, which is located 55 cm below the Xe liquid surface. The maximum width is ~49 cm. The electric field to drift the ionization signal across the 55 cm gap will be 2 kV/cm. Field shaping rings, spaced by 2 cm, will be mounted on the outer PTFE vessel. A luminescence grid 0.5 cm above the surface, and a bottom grid 0.5 cm below the surface, are used to generate high fields (5 kV/cm in the liquid, 10 kV/cm in the gas) that extract the electrons and accelerate them, thus generating proportional scintillation light. Scintillation photons are produced almost uniformly along the gas gap, regardless of the electron extraction point location. Two arrays of 5 cm diameter R8778 (Hamamatsu) PMTs will detect both prompt and proportional scintillation light. One PMT array above the liquid surface is primarily used to image the *x-y* position of the proportional light pulse. Using likelihood pattern recognition techniques developed for XENON10 and ZEPLIN II we expect ~1 cm position accuracy at low energy. The second PMT array is in the liquid, below the cathode grid. Because of internal reflection at the liquid surface (LXe has a high refraction index of 1.69 at 170 K [3]), most prompt light is collected in this bottom array. Taking into account the quantum efficiency of the R8778, slightly taller detector geometry than in XENON10, more transparent grids, and use of PTFE reflectors between PMTs, we predict an S1 signal yield of ~6 photoelectrons/keVee (with drift field applied), about twice the signal yield achieved in XENON10.

LUX will be calibrated with a system that allows radioactive sources to be accurately deployed to specific positions. Planned sources include several monoenergetic gamma ray sources, including ^{57}Co, ^{137}Cs, and ^{133}Ba for light yield calibration, and ^{228}Th for accumulating statistics on the charge-to-light discrimination band. A well-collimated ^{137}Cs source will be used to produce a narrow beam of gamma rays for position resolution studies. An Am-Be fast neutron source will be used for nuclear recoil calibration. Gaseous sources, dissolved in the LXe, will also be used to calibrate the energy scale and *S2/S1* discrimination efficiency. These include ^{83m}Kr, recently demonstrated at Yale [4] and Zurich [5] and tritium, under development at the University of Maryland.

In the LUX WIMP search analysis we do not propose to subtract any backgrounds from the data to derive the residual number of WIMP events. Given the expected level of gamma ray background projected from simulations (4 x 10^{-4} events/keV/kg/day, dominated by low-angle Compton scattering of gamma rays from the PMTs), we expect to keep background leakage less than 1 event in 300 days of operation, which in the case of electron recoil events is achieved simply by keeping the total number of events less than 180 before discrimination. This should be contrasted with the data sample of XENON10 where ~1800 electron recoil events were recorded for a 58.6 live day exposure in the energy region of interest before applying S2/S1 rejection. Internal backgrounds are projected to be small, as most radioactive impurities are efficiently removed by the xenon purification system, and most impurities result in large energy depositions, well outside the energy range relevant for WIMP detection. Isotopes that decay emitting a "naked" beta can still cause unwanted background, and the LUX collaboration has done a full study of what beta decay impurities might be introduced into the detector by muon spallation or other means. Kr, which contains the naked beta decay isotope ^{85}Kr, is naturally present at ppm levels in commercial xenon gas, but will be removed to ppt levels from the LUX xenon using activated charcoal.

In late Fall 2009, LUX will be integrated at a newly renovated surface facility near the Homestake mine in South Dakota. Underground commissioning in the Davis cavern at the Homestake 4850 ft. level is planned to begin Summer 2010.

2. The LUX-ZEPLIN Program

The LUX and ZEPLIN collaborations are currently proposing to construct a 1.5 tonne LXe dark matter detector, known as LZS, to be installed at the 4850 level of the Sanford Laboratory in Lead, South Dakota. This detector will use the same technology as LUX, and the same infrastructure and water tank at the 4850 ft. level of SUSEL, allowing a rapid turn-around at the conclusion of LUX. With a fiducial mass of 750 kg, the experiment will probe WIMP-nucleon cross sections down to 3 x 10^{-47} cm^2 within 2 years of operation, about 1,000 times better than current results and covering a substantial range of theoretically-motivated candidates. This proposal is being submitted by the US groups within the LUX-ZEPLIN collaboration, and is being co-developed with our international partners for consideration by the European funding agencies. Established groups at Caltech, UC Berkeley and UC Santa Barbara will join current LUX collaborators.

Following LZS, we plan to propose and build a 10-20 tonne experiment, known as LZD. The LZD experiment would be incorporated in the initial suite of experiments at DUSEL, and be sensitive to dark matter at the level of 10^{-48} cm^2 for a 100 GeV WIMP. LZD represents the ultimate non-directional WIMP dark matter search, as below this cross-section neutrino-nucleus coherent scattering (an irreducible background) begins to dominate [6]. Development of the LZD design is newly funded by the National Science Foundation.

References

[1] J. Angle *et al.,* (XENON Collaboration, Phys. Rev. Lett. **100** (2008) 021303.
[2] G.J. Alner *et al.*, Astropart. Phys. **28**, 287 (2007).
[3] V. N. Solovov *et al.*, NIM A **516**, 462 (2004).
[4] L. Kastens *et al*, arXiv:0905.1766, accepted to Phys. Rev. C.
[5] A Manalaysay *et al.*, arXiv:0908.0616.
[6] L. Strigari, arXiv:0906.5361.

Journal of Physics: Conference Series **203** (2010) 012027 doi:10.1088/1742-6596/203/1/012027

Investigation of ambient and μ-induced neutron background in LSM

Klaus Eitel

Karlsruhe Institute of Technology, Institut für Kernphysik, Postfach 3640, 76021 Karlsruhe, Germany

E-mail: `klaus.eitel@kit.edu`

Abstract. A good particle candidate for Cold Dark Matter (CDM) is the supersymmetric neutralino or more generally a weakly interacting massive particle (WIMP). The expected interaction rate of WIMPs with the detector medium in the direct detection experiments is below 0.01 events/kg/day. This makes a good knowledge of the background conditions highly important, especially with ever increasing sensitivity of the detectors. One major component of the background is related to energetic neutrons induced by cosmic muons. However, the flux of low energy and thermal neutrons typically from (α, n) reactions in the rock surrounding underground experiments gives also valuable information about the overall environmental activity. Detailed studies carried out in the framework of the Edelweiss experiment are presented with first identification of muon-induced neutron candidates in Ge bolometers.

1. Introduction

The Edelweiss experiment is a direct Dark Matter search located in the Modane underground laboratory LSM in the French Alps which provides a shielding factor of $\sim$4800 m.w.e. The detection principle is based on measuring energy of the recoil nucleus originating from the WIMP elastic scattering. Bolometers of pure natural Ge are used both as the detectors and the target material. First results with a newly established readout technology have been presented recently [1]. Despite the excellent suppression of β and γ background [2], neutrons coming from the natural radioactivity or induced by the remaining cosmic muons can still mimic the nuclear recoil signal of WIMP events. Therefore, this type of background requires special careful investigation. Its knowledge also becomes highly important in view of large 1-tonne-scale experiments of the next generation like EURECA [3].

2. Ambient neutron flux measurements

To monitor the ambient neutron flux at LSM, a setup of 4 proportional counter tubes of 3 cm diameter and 1 m length filled with ^{3}He gas [4] to measure neutrons via the capture reaction n+^{3}He$\rightarrow$T+p (Q =0.764 MeV) was installed. In an initial measuring campaign, these tubes were surrounded by a massive block of polyethylene (PE) ($12\times20\,\text{cm}^2$) to moderate energetic neutrons. With an internal background of 0.2 counts per day in the region of interest, the measured average count rate of 160 counts/day is consistent with earlier measurements [5] and allowed to perform a long term monitoring of the overall neutron flux at LSM near the EDELWEISS setup. Preliminary results comparing the neutron flux measured through 2008 set an upper limit on an annual variation of less than 4 % (90 %C.L.).

To be sensitive to thermal neutrons only, the PE block was removed for a second measurement campaign in 2008, with one ^{3}He counter having a sensitivity of ≈ 250 counts/second for a thermal neutron flux of 1 neutron/cm^2/s. The thermal neutron flux was then measured in 7 different locations of LSM near the concrete wall of the lab as well as near the existing setups to account for shielding effects of the large experiments EDELWEISS and NEMO. Count rates of > 60 counts/day within the region of interest have been measured. Preliminary results show a variation of the neutron flux by a factor of up to 3 between different positions.

3. Muon-induced neutrons in EDELWEISS

Cosmic muons with a flux of $\approx 4.5\ \mu/\mathrm{m}^2/\mathrm{d}$ and consecutive μ-induced neutrons constitute an important background component for direct DM search. In order to understand this background in detail, detailed 3-dimensional GEANT4 simulations were developed [6]. These simulations, for instance, show that while a muon interaction leads to a production of both electromagnetic and hadronic showers, the induced bolometer events with a deposited energy of E≥50 keV are dominantly electron recoils while events with E≤50 keV are nuclear recoils induced by neutrons (Fig. 1 (left)). An analysis of bolometer data measured in coincidence with the muon veto identified for the first time μ-induced neutrons in the EDELWEISS Ge crystals [7], Fig. 1 (right). A rather small rate of ≈0.04 (neutron- and electron-type) coincidences/kg/d for E≤250 keV however limits a detailed further investigation of μ-induced neutrons due to limited statistics. To measure μ-induced neutrons in situ in LSM with sufficient statistics, a dedicated neutron

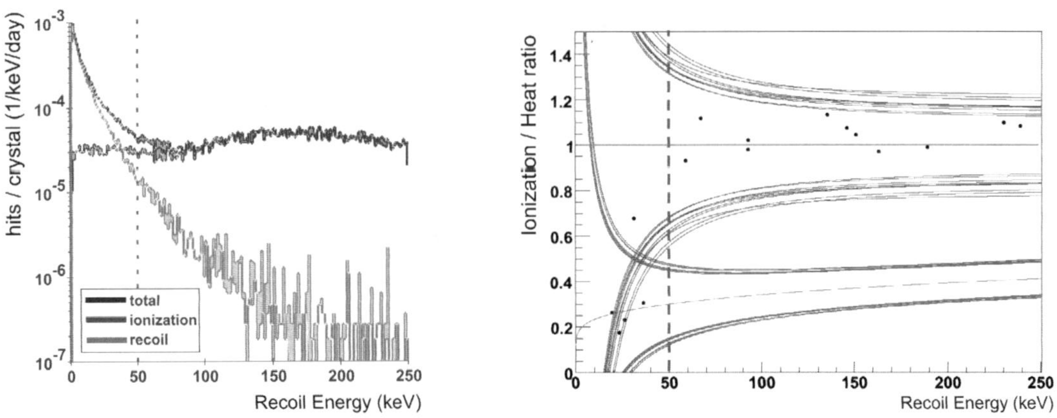

Figure 1. left: GEANT4 simulated energy deposit in bolometers for muon-induced events; right: measured bolometer events in coincidence with the muon veto system, data corresponding to 283 kg · day exposure.

counter has been designed, installed and is under operation since the end of 2008.

4. Measurement of muon-induced neutrons

The measuring principle of μ-induced neutrons is based on detecting the capture of thermal neutrons in delayed coincidence with the incoming muon or by detecting a multiple neutron event (secondary neutrons produced in a μ-induced particle shower). To efficiently observe neutrons, a liquid scintillator of 1 m^3 volume (50x100x200 cm^3) loaded with 0.2% Gadolinium (St. Gobain Bicron BC525) is used as a core of the detector. The neutron capture process on Gd results in several gammas with 8 MeV sum energy. The scintillator volume is viewed from each of the two module ends by 8 photomultiplier tubes (PMT) of 8-inch diameter (Fig. 1). These PMTs are optimized to register the light produced after the neutron capture. In addition, the system is equipped with 6 smaller PMTs (2-inch diameter) to register muons crossing the neutron detector (these muons create much more light and thus the 8-inch PMTs will saturate). The scintillator and PMTs are placed in one plexiglass container divided into three parts: the central one for the scintillator itself and two side ones filled with paraffin in which the PMTs are immersed. This plexiglass chamber is then placed in an aluminum vessel as secondary safety container. Finally, the system is surrounded by iron plates to reflect a fraction of neutrons back to the scintillator. In order to enhance the neutron production (up to a factor of 10 comparing to rock) a 10 cm thick layer of lead bricks is put underneath the detector. On top of the counter, a plastic scintillator module of 3.15m length (same type as the muon veto of Edelweiss) is installed. The complete system is positioned directly to the western wall of the Edelweiss muon veto. Based on the currently measured muon flux in the lab and first identified neutron cascades, the expected count rate of muon-induced neutrons is about one sequence per day. Data taking is expected to continue until end of 2010.

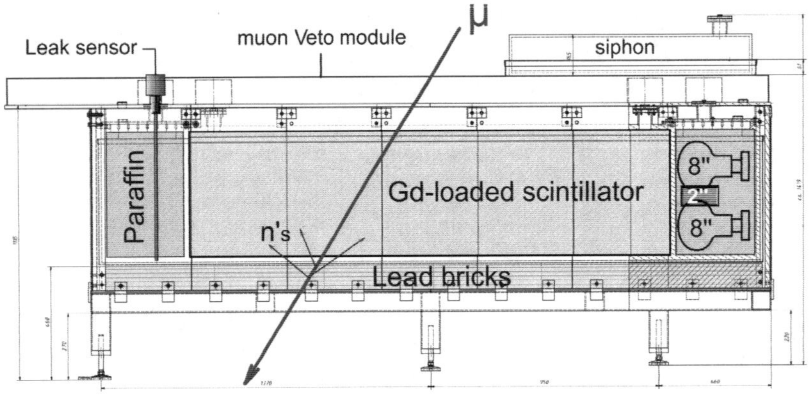

Figure 2. General scheme of the neutron counter (side view): 1-tonne of Gd-loaded liquid scintillator viewed by 16 PMTs of 8-inch and 6 PMTs of 2-inch diameter. A layer of lead bricks below the liquid scintillator volume acts as an effective source of secondary neutrons from muon interactions.

5. Acknowledgments

The studies of the muon-induced neutron background are in part funded by the German Research Foundation (DFG) through the Transregional Collaborative Research Center SFB-TR27; the dedicated neutron measurements (ambient and muon-induced) are partly supported by the EU contract RII3-CT-2004-506222.

References

[1] V. Sanglard for the Edelweiss Collaboration, *these proceedings*.
[2] A. Broniatowski et al., arXiv:0905.0753v1 [astro-ph.IM] and references there in; submitted to Phys. Lett. B.
[3] H. Kraus et al., Nucl. Phys. B (Proc. Suppl.) 173, 168 (2007).
[4] G.S. Vidyakin et al., Instruments and experimental techniques, 32(4), 809 (1989).
[5] V. Chazal et al., Astroparticle Physics 9, 163 (1998).
[6] M. Horn, Simulations of the muon-induced neutron background of the EDELWEISS-II experiment for Dark Matter search, Ph.D. thesis, Universität Karlsruhe (TH), 2007 and FZKA scientific report 7391 (http://bibliothek.fzk.de/zb/berichte/FZKA7391.pdf)
[7] A. Chantelauze for the Edelweiss Collaboration, SUSY2007 Karlsruhe, Germany, 2007; arXiv:0710.5849v1 [astro-ph]; Ph.D. thesis in preparation.
[8] V.Yu. Kozlov for the Edelweiss Collaboration, IDM08 Stockholm, Sweden, 2008, PoS (idm2008) 086 (http://pos.sissa.it); arXiv:0902.4858v1 [astro-ph.IM].

Topics in Astroparticle and Underground Physics (TAUP 2009) IOP Publishing
Journal of Physics: Conference Series **203** (2010) 012028 doi:10.1088/1742-6596/203/1/012028

Status and Prospects of a Deep Underground Laboratory in China

K.J. Kang[1], J.P. Cheng[1], Y. H. Chen[2], Y.J. Li[1], M. B. Shen[2], S. Y. Wu[2], Q. Yue[1]

[1]Department of Engineering Physics, Tsinghua University, Beijing, 100084, China
[2]Ertan Hydropower Development Company, Chengdu, 610051, China

E-mail. yueq@mail.tsinghua.edu.cn

Abstract. An excellent candidate location for a deep underground laboratory with more than 2500 m of rock overburden has been identified at Sichuan Province in China. It can be accessed through a road tunnel of length 17.5 km, and is supported by services and amenities near the entrance provided by the local Ertan Hydropower Plant. The particle physics community in China is actively pursuing the construction of an underground laboratory at this location, under the leadership of Tsinghua University. Memorandum has been signed with Ertan Hydropower Plant which permits access to and construction of the underground laboratory — China JinPing Deep Underground Laboratory (CJPL). The basic features of this underground site, as well as the status and schedules of the construction of the first laboratory cavern are presented. The immediate goal is to have the first experiment operational in 2010, deploying an Ultra-Low-Energy Germanium detector for WIMP dark matter searches, with emphasis on the mass range of 1-10 GeV. The conceptual design of the experiment, as well as the future plans and prospects of the laboratory, will be surveyed.

1. Introduction

The essence of dark matter direct detection is to survey possible parameter space for the most popular dark matter candidate WIMP (*Weakly interactive massive particles*) and has been a subject of intense. in particle physics, astrophysics and cosmology. The experimental approach is to study the recoil nuclei scattered off by an incident WIMP particle inside an active detector. The experiments require ultra-low background techniques to suppress background events, similar to those of double beta decay experiment, neutrino experiment, proton decay experiment, and so on.

Low background experiments require very good laboratory to reduce the background event rates which come mainly from the cosmic-ray muons, ambient radioactivity, as well as the internal backgrounds and the detector noise. The radioactive background from the ambient radioactive isotopes can be shielded with passive and/or active shielding system. The high energy muons, however, can penetrate the shielding system and interact with the detector, though this channel can be vetoed by an active veto system. Background due to cosmic-ray induced neutron and long-lived radioactive isotopes generated by the interactions of the incident muons with the materials in the vicinity of the detector are the most difficult to shield. So it will be important to find it is essential to perform dark matter experiments at underground laboratories where the muon flux is greatly reduced by the huge rock overburden.

There are many underground laboratories established or under construction in the world including the LNGS in Italy, Subdury in Canada, Modane in France, DUSEL in USA, and so on [1]. In this paper, we will give an introduction to a forthcoming deep underground laboratory in China which will have the deepest rock overburden and convenient conditions for work and lives. This deep underground laboratory is named China JinPing Deep Underground Laboratory (CJPL).

2. The condition of the CJPL

The Yalong River in Sichuan province, southwest China, changes it course due to the Jinping Mountain, creating a great river bend with length more then 200 km. Ertan Hydropower Company will build two hydropower plants at each side of the Jinping Mountain. Two parallel tunnels have been dug to pass through the Jinping Mountain for transportation and hydropower plant construction as shown in Figure1. A total of seven tunnels will be constructed including four headrace tunnels, one drainage tunnel, on top of the two transport tunnels where construction was completed in August 2008 [2]. The length of these two transport tunnels is 17.5km and the cross-section is about 6m*6m. The underground laboratory will be constructed in the central portion of one of two Jinping tunnels. The detailed position of CJPL is shown in Figure 2. The minimal rock overburden is more than 2500km.

The construction of CJPL will be done in stages as shown in Figure 3. In Phase-I, a hall with dimension of 6m×6m×40 will be established for dark matter search experiment by the end of 2009. This Hall A can be accessed by a right-angled tunnel which is 60 m in length and will be served by a second engineering laboratory (Hall B) before the end of 2009. At the same time, several tunnels have been constructed for the geological exploration to evaluate the construction of huge cavity for large-scale experiments in the future. Construction of the power supply, ventilation and drainage systems will be completed in the middle of 2010. The tunnel laboratory area is equipped with 3G wireless network and fiber access to broad-band internet. Apartments, restaurants, hotel and sport facilities are available at the tunnel entrance.

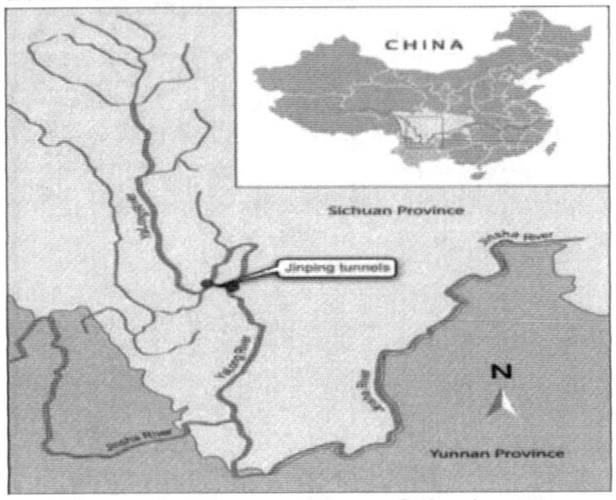

Figure 1 The position of Jinping Tunnel

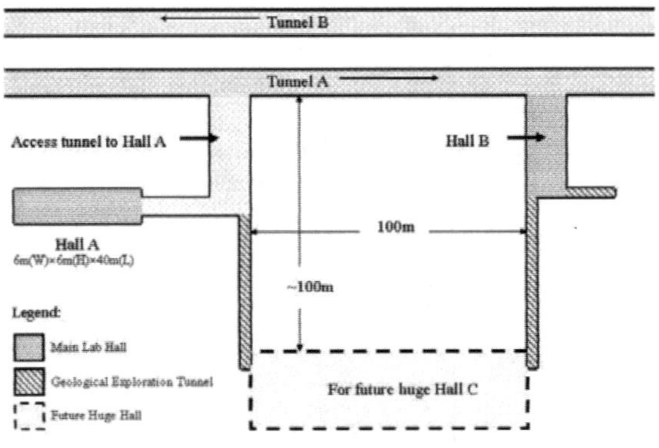

Figure 3 The layout of CJPL

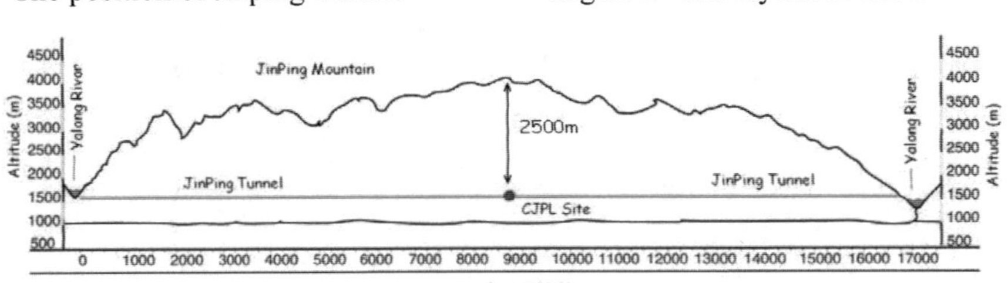

Figure 2 The cross-section plot of the Jinping Mountain along the transport tunnel. The site of CJPL is in the central portion of the JinPing tunnel and is directly accessible by cars.

3. The dark matter experiment at CJPL

By the end of 2009, the construction of Phase-I of CJPL infrastructure will be finished as Hall-A with dimension of 6m×6m×40m plus concrete coverage. The first experimental program will start to install its shielding system and main detector for dark matter search. More than 30 scientists from

Journal of Physics: Conference Series **203** (2010) 012028 doi:10.1088/1742-6596/203/1/012028

China have established a collaboration (China Dark matter EXperiment, CDEX) to conduct the experiment with the TEXONO [2] and KIMS collaborations. This experiment will use Ultra-Low-Energy High Purity Germanium (ULE-HPGe) detector as the target detector for WIMP search [4,5]. The complex shielding system covering the main ULE-HPGe detector includes, from inside out, 10 cm of copper, 10 cm of steel frame, 5 cm of plastic scintillator, 20 cm boron-loaded polythene, 20 cm of lead and 1 m of water as shown in Figure 4. Water and boron-loaded polythene are served as neutron decelerator and absorber. Lead and copper layers are served as gamma shielding and plastic scintillator is used as cosmic-ray veto detector. The steel frame is the supporting structure. The internal space with dimension of 1.5m(w)×2m(H)×2m(L) will be served as the main detector space, where the HPGe detector array will be located.

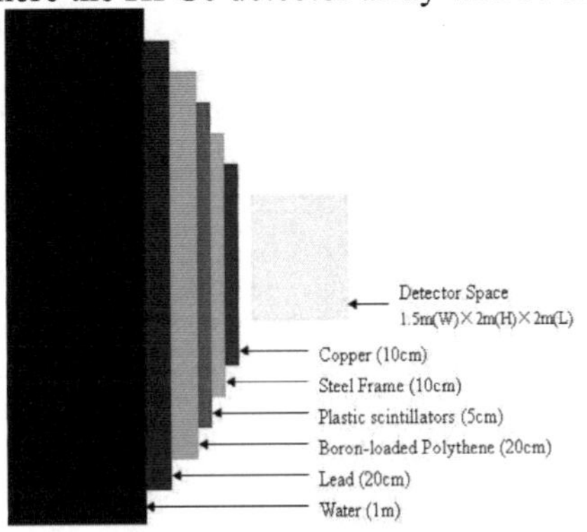

Figure 4 Schematic layout of the detector space system. and its shielding. The coverage is 4π, but only one face is shown

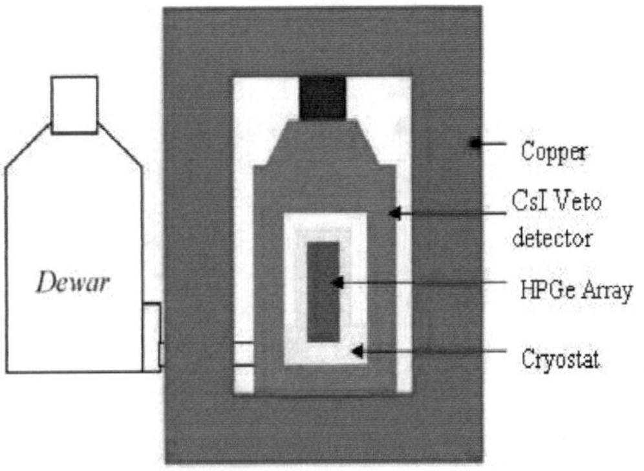

Figure 5 Schematic layout of the detector

The layout of the Ultra-Low-Energy High Purity Germanium (ULE-HPGe) detector and its active shielding system is shown in Figure 5. The HPGe detector array is surrounded by the cryostat. Outside the cryostat, the CsI(Tl) crystal detector serves as the active shielding for cosmic-ray muon and gamma background. The copper layer is used to shield the gamma background from the preamplifiers and the Dewar bottle which contains liquid nitrogen for Ge crystal cooling. The ULE-HPGe detector can achieve an energy threshold of less than 300eV and is sensitive to low-mass WIMP detection, especially for WIMP with mass less than 10GeV[3]. The Phase-I experiments will include WIMP searches using ULE-HPGe as well as Point-Contact Germanium detectors with mass will up to 500g.

4. Summary and Prospects

An underground laboratory named CJPL with minimal rock overburden of 2500m will be constructed in China and one of the experimental halls will be ready by the end of 2009. The first experiment for dark matter search with 500g ULE-HPGe will be run in the CJPL. Upgrades of the target mass of the ULE-HPGe detector to the scale of ~100 kg scale can be considered in the future, based on the results and experience of Phase-I.

References

[1] More information, see http://www.lngs.infn.it/, http://www.snolab.ca/, http://www-lsm.in2p3.fr/, http://www.lbl.gov/nsd/homestake/ and this preceedings.
[2] SCIENCE , 5 JUNE , 2009, VOL 324: 1246-1247
[3] S. T. Lin, H. B. Li, X. Li, et. al., PHYSICAL REVIEW D 79, 061101(R) (2009)
[4] YUE Qian, Cheng Jina-Ping, LI Yuan-Jing, et al., High Energy Physics and Nuclear Physics,V28(6), 2004:877—881
[5] LI Xin, YUE Qian, LI Yuan-Jing, et al., High Energy Physics and Nuclear Physics,V31(6), 2007:564—569

Topics in Astroparticle and Underground Physics (TAUP 2009) IOP Publishing
Journal of Physics: Conference Series **203** (2010) 012029 doi:10.1088/1742-6596/203/1/012029

Results from DMTPC 10-liter Detector

D. Dujmic, J. Battat, T. Caldwell, L. Fedus, P. Fisher, S. Henderson, R. Lanza, A. Lee, J. Lopez, A. Kaboth, G. Kohse, J. Monroe, R. Vanderspek, T. Sahin, G. Sciolla, I. Wolf, R. Yamamoto, H. Yegorian

Massachusetts Institute of Technology, Cambridge, MA 02421, USA

S. Ahlen, A. Inglis, K. Otis, H. Tomita

Boston University, Boston, MA 02215, USA

H. Wellenstein

Brandeis University, Waltham, MA 02454, USA
E-mail: ddujmic@mit.edu

Abstract. The known direction of motion of dark matter particles relative to the Earth may be a key for their unambiguous identification even in the presence of backgrounds. A direction-sensitive detector prototype using a low-density CF_4 gas with a 10 liter fiducial volume is operated for several weeks in a basement laboratory. We present initial results that confirm good detector performance and set preliminary limits on spin-dependent dark matter interactions.

1. Introduction

Observation of weakly interacting massive particles (WIMPs) may require a good background rejection and a correlation of the WIMP direction with the galactic motion through the dark matter halo [1]. Both requirements can be accomplished with detectors using low-pressure gas as the target material. However, the low density of gaseous detectors requires large detector volumes with fine granularity. We address this challenge by building a dark matter time-projection chamber (DMTPC) with optical readout that allows good sensitivity to WIMP-induced signals and is scalable to larger volumes.

2. DMTPC-10L Detector

A schematic of the DMTPC detector is shown in Figure 1. A stainless steel vessel is filled with 75 Torr of CF_4 gas and holds two back-to-back time-projection-chambers (TPC). Ionization electrons created by a recoiling nucleus are drifted in an electric field ($\sim$ 250 V/cm) toward the amplification region for detection. Each drift cage is made of stainless steel rings with the inner diameter of 25 cm and a total height of 25 cm. Attenuation of primary ionization electrons due to attachment to CF_4 gas molecules is measured to be consistent with zero.

The amplification region uses a solid copper for the anode and a stainless steel mesh with 256 μm pitch as the grounded electrode. A detailed description and performance evaluation of such amplification system is given in Ref. [2]. The charge gain in the detector at 75 Torr is around 10^5 with a spark rate less than 10 mHz for the nominal anode voltage (720 V). Scintillation photons created during the avalanche charge multiplication (see [3]) are collected by a Nikon photographic lens with f-stop ratio of 1.2 and focal distance of 55 mm, and are

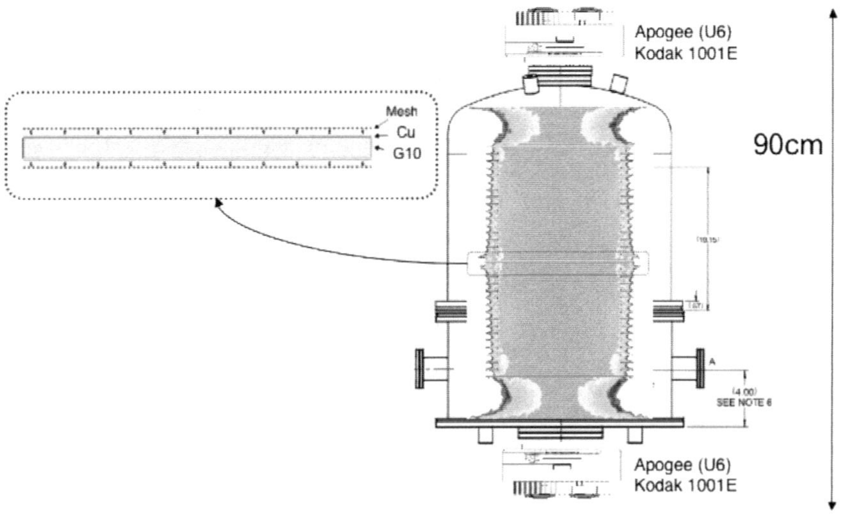

Figure 1. A vacuum vessel holding a double-sided TPC with 10-liter fiducial volume, and a calculation of field uniformity ($|E_{\perp}|/E_{tot}$), with each color shade corresponding to 1% of the change in the field ratio. The enlarged section of the amplification is shown in the left, consisting of a grounded mesh and a solid copper anode separated by 0.5 mm fluorocarbon wires. The scintillation light is recorded by CCD cameras, and charge is read out from the anodes.

recorded with Apogee U6 cameras using a Kodak KAF-1001E CCD chip. The total area imaged by each CCD is 16×16 cm^2 so the total active volume is approximately 10 liters.

The energy resolution of the CCD readout is measured using segments of tracks from the ^{241}Am source. The resolution is computed from the total light originating from the segment, and we find $\sigma_E/E = 14\%$ at 80 keV. The resolution can be further improved with charge readout from the anode. This is confirmed using 5.9 keV line from ^{55}Fe source to obtain the resolution of $\sigma_E/E = 10\%$.

The spatial resolution in the CCD readout plane is measured using the transverse spread of tracks from ^{241}Am source. The resolution is affected by the number of CCD pixels, n_{bin} merged into a readout bin ($98\mu\text{m} \cdot \text{n}_{\text{bin}}$), the spread of an electron avalanche in the amplification gap ($\sim 80\mu$m), and the mesh pitch ($256\mu\text{m}/\sqrt{12} \sim 74\mu$m). The measurement itself is affected by the imperfect collimation of the source (50 μm) and the straggling of alpha particles through CF_4 gas (40 μm). However, the dominant contribution to the spatial resolution uncertainty comes from the electron diffusion ($\sqrt{10^7 z_{drift}/E_{drift}}$ μm) where z_{drift} (cm) is the drift distance, and E_{drift} (V/cm) is the electric field.

3. Results from Surface Run

In order to evaulate the detector's background and the stability, we take data in a basement laboratory for several weeks. The total mass in the fiducial volume is 3.3 g and the total live time is 11 days. Gas is replaced every 24 h and we measure stability in the detector gain and the spatial resolution to be better than 1%. We take 5 second exposures and record every event.

Dominant sources of background are alpha tracks originating from the detector surface (10 mHz) and interactions with the CCD chip, so called 'worms' (10 mHz). The alpha background is entering the CCD viewfield from the sides and can be eliminated by requiring that a track not cross the edge of CCD field of view. The alpha rate is constant and originates from radioactivity inside the detector. We reduced this rate by a factor of five by replacing the

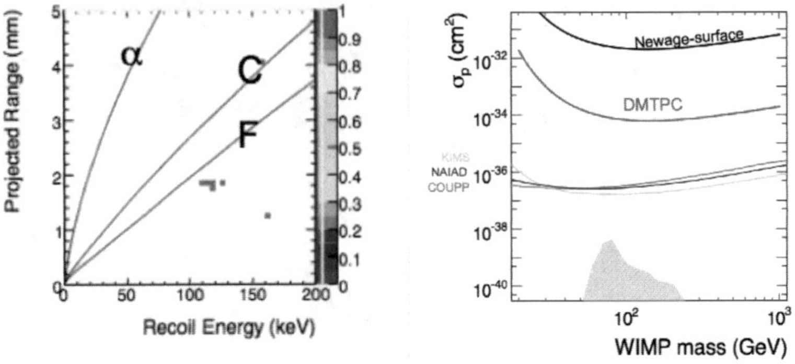

Figure 2. Left: energy and 2D range of remaining events. Curves represent relation between energy and 3D range for different species. Right: upper limit on spin-dependent interaction as a funcion of WIMP mass.

stainless-steel rings of the drift cage with rings made of off-the-shelf copper.

The worms consist of two sources: x-rays from materials inside the CCD and cosmic-induced radiation. Most of these events can be eliminated with cuts on the energy and shape of CCD pixels, but hardware-based remedies are under study, too.

We observe 7 events after the analysis cuts. The length and energy of these events are consistent with nuclear recoils induced by neutron scattering (Figure 2, left). Our preminary estimate for the number of expected recoils from cosmic-induced neutrons at the Earth's surface is twice as high, but more detailed calculations taking into account overburden of the basement lab are under way. We compute the upper limit on the spin-dependent cross section (Figure 2, right) using the Feldman-Cousins method and standard assumptions about the dark matter halo [4]..

4. Summary and Plans

A prototype detector with total fiducial volume of 10 liters and filled with low-pressure CF_4 gas has been built and operated for several weeks in a surface laboratory. We demonstrate excellent energy and spatial resolution in the detection of nuclear recoils. A preliminary analysis shows that the dominant background comes from alpha tracks inside the TPC materials and direct interactions with the CCD chip. These events can be eliminated with analysis cuts and the remaining background energy and range are consistent with interactions with cosmic-induced neutrons.

In order to improve the detector sensistivy and eliminate the neutron background we are making preparations for operation in an underground laboratory (WIPP). Completion of these tests will allow us to design a cubic-meter module that will be a building block of a large-scale detector.

5. Acknowledgments

We wish to thank the Office of Environment, Health & Safety and the Laboratory for Nuclear Science MIT for technical support. We acknowledge support by the Advanced Detector Research Program of the U.S. Department of Energy (contract number 6916448), the MIT Kavli Institute for Astrophysics and Space Research, and the Physics Department at MIT.

[1] D. N. Spergel, Phys. Rev. D **37**, 1353 (1988);
[2] D. Dujmic *et al.*, Astropart. Phys. **30** (2008) 58-64
[3] A. Kaboth *et al.*, [DMTPC Collaboration] Nucl. Instrum. Meth. A **592**, 63 (2008).
[4] J. D. Lewin and P. F. Smith, Astropart. Phys. **6**, 87 (1996).

Underground Low Flux Neutron Background Measurements in LSM Using a Large Volume (1m^3) Spherical Proportional Counter

I Savvidis[1], I Giomataris[2], E Bougamont[2], I Irastorza[4], S Aune[2], M Chapelier[2], Ph Charvin[2], P Colas[2], J Derre[2], E Ferrer[2], G Gerbier[2], M Gros[2], P Mangier[2], XF Navick[2], P Salin[5], J D Vergados[6] and M Zampalo[3]

1 : Aristotle University of Thessaloniki, Greece
2 : IRFU, Centre d'études de Saclay, 91191 Gif sur Yvette CEDEX, France
3. LSM, Laboratoire Souterrain de Modane, France
4: University of Saragoza, Spain
5 : LSBB, France
6: University of Ioannina, Greece

E-mail: savvidis@physics.auth.gr

Abstract: A large volume (1m^3) spherical proportional counter has been developed at CEA/Saclay, for low flux neutron measurements. The high voltage is applied to a small sphere 15mm in diameter, located in the center of the counter and the wall of the counter is grounded. Neutrons can be measured successfully, with high sensitivity, using ^{3}He gas in the detector. The proton and tritium energy deposition in the drift gaseous volume, from the reaction ^{3}He(n,p)^{3}H, can provide the neutron spectra from thermal neutrons up to several MeV.

The detector has been installed in the underground laboratory in Modane (LSM) to measure the neutron background. The sphere has been has been filled with gas mixture of Ar + 2% CH_4 +3gr He-3, at 275 mbar. The thermal neutron peak is well separated from the cosmic ray and gamma background, permitting of neutron flux calculation. Other potential applications requiring large volume of about 10 m in radius are described in detail in reference

1. Introduction

The investigation on the large volume spherical proportional counter resulted in the development of a new neutron detector, based on the ^{3}He(n,p)^{3}H reaction[1]. The low background of the detector and the possibility to separate the γ ray pulses from the protons and alpha particle pulses increases the sensitivity on the neutron detection. The detector can successfully measure very low neutron fluxes
(10^{-6}n/cm^2/s, for thermal neutrons), providing the neutron energy spectrum from thermal up to several MeV at ground and underground level.

The large spherical geometry drift (1m^3), the good energy resolution (<2% FWHM with alpha particles at 5.5 MeV) and the simple read out (one channel reading) are some of the advantages of the detector. Other potential applications of this device requiring large volume are described in detail in reference [2,3,4,5,6,7].

2. The detector

The detector consist of a copper sphere, 1.3 m in diameter and 6mm thick (figure 1). The spherical vessel is well pumped (up to 10^{-8} mbar) and then is filled with a gas mixture at a pressure from several hundreds of mbars up to 5 bars. Out gassing in the order of 10^{-9} mbars/s is necessary for the amplification stability, because the present of the O_2 in the drift volume changes the detector characteristics.

Figure 1. A photograph of the spherical vessel

A small stainless ball of 14 centimeter in diameter fixed in the center of the spherical vessel by a stainless steel rod, acts as an electrode with positive high voltage and as a proportional amplification counter. The detector was operated with positive bias applied to the anode (inner sphere) while the cathode (external sphere) remained at ground potential. A high voltage capacitor was decoupling the high voltage cable to protect the sensitive preamplifier.

3. The energy resolution

The energy resolution of the detector has been tested using ^{222}Rn gas and detecting the alpha particles from ^{222}Rn and ^{222}Rn daughters. Since the ^{222}Rn gas cover homogenously all the drift volume of the detector, we have alpha emission in every direction and in all the positions of the detector. The gas mixture consist of Ar (98%) and CH4(2%) at pressures, from 150 mbars up to 1 bar. The high voltage vary from 1.5 kV up to 5 kV depending on the gas pressure.

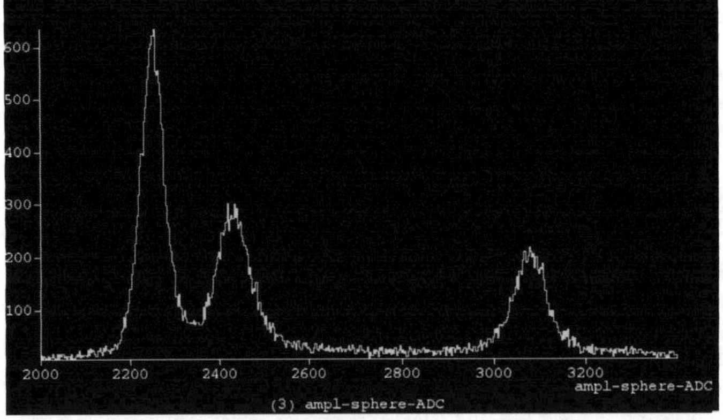

Figure 2. The peaks observed from a ^{222}Rn radioactive source. From left to right we observe the ^{222}Rn peak at 5.5 MeV, the ^{218}Po and ^{214}Po at 6.0 MeV and 7.7 MeV respectively.

In the figure 2 is shown the peaks observed from a ^{222}Rn radioactive source. From left to right we observe the ^{222}Rn peak at 5.5 MeV, the ^{218}Po and ^{214}Po at 6.0 MeV and 7.7 MeV respectively. The energy resolution was 2% FWHM at 200 mbar gas pressure and 2.8 kV High Voltage.

5. The thermal neutron flux in the LSM

In the present work we have been used the ^{3}He gas as converter for thermal and fast neutron detection up to several MeV. Neutrons interact with ^{3}He as follows, $n+^3He \rightarrow p+^3H+765$ keV. The signal is the sum of the p and ^{3}H energy deposition in the drift volume and depends on the neutron energy. In the case of thermal neutrons we measure one peak 765 keV and for the fast neutrons the energy peak is En + 765 keV, were En is the fast neutron energy.

The detector has been installed in the LSM, to measure the underground neutron flux. The gas mixture was Ar +2% CH4 at p=280 mbar with 3 gr of He-3

Thermal neutron capture rate was 0.0048 evts/s = 417 evts/d and the thermal neutron flux was $\Phi_{th} = 1.9\ 10^{-6}$ n/cm^2/s. In the figure 3 is shown the thermal neutron peak after rise time cut.

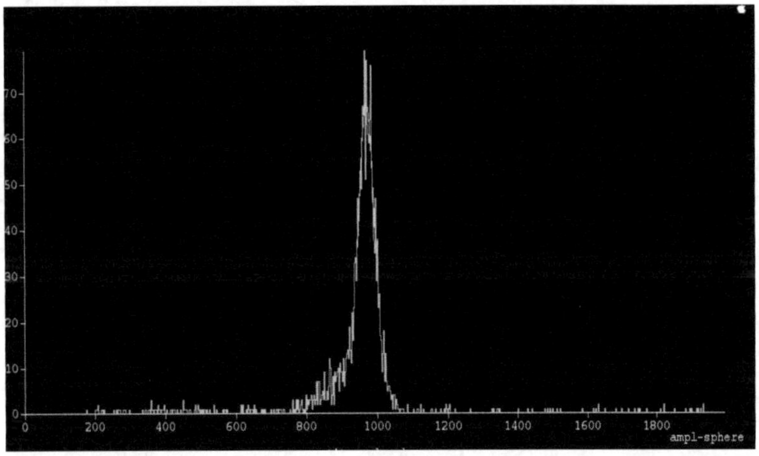

Figure 3. The thermal neutron peak after rise time cut.

6. Conclusions

The spherical proportional counter is a high sensitivity neutron detector. The sensor is stable for long time measurements and the decrease in gain is small (≈ 0.2% per day). The detector can measure low fluxes of the underground thermal neutrons and the seasonal variation of the flux.

REFERENCES

[1] I. Giomataris et al., JINST 3:P09007,2008.
[2] I. Giomataris and J.D. Vergados, Nucl.Instrum.Meth.A530:330-358,2004
[3] I. Giomataris *et al.* Nucl.Phys.Proc.Suppl.150:208-213,2006.
[4] S. Aune *et al.*, AIP Conf.Proc.785:110-118,2005.
[5] I. Giomataris and J.D. Vergados, Phys.Lett.B634:23-29,2006.
[6] J. D. Vergados et al., Phys.Rev.D79:113001,2009.
[7] A. Dedes et al., arXiv:0907.0758

Progress report on DRIFT-II

Mark Pipe for the DRIFT collaboration
Department of Physics and Astronomy, University of Sheffield, Sheffield S3 7RH, UK
E-mail: m.pipe@sheffield.ac.uk

Abstract. We present recent progress from the DRIFT (Directional Recoil Identification From Tracks) collaboration. We summarise recent work demonstrating 1) directional sensitivity to a neutron source in a 1 m^3 DRIFT-II detector in all 3 dimensions at energies of 50 keV 2) first measurements of the head-tail effect in ~keV/amu nuclear recoils using a DRIFT-II detector. We also summarise results of measurements of W-value, mobility and gas gain in CS_2-CF_4 gas mixtures from a single proportional counter experiment that suggest the potential of the DRIFT-II detector as a spin-dependant dark matter detector.

1. Introduction

Weakly Interacting Massive Particles (WIMPs) remain a particularly well-motivated candidate for cold dark matter. As discussed in [1], WIMP dark matter may form an isothermal sphere encompassing our galaxy, resolving the problem of flat galactic rotation curves. An observer on Earth would perceive a WIMP wind of several hundred kms^{-1}, reflecting the motion of the Earth's surface with respect to the galactic rest frame. DRIFT is pioneering operation of a directionally sensitive WIMP detector in a low-background, deep underground site, which is potentially capable of using the WIMP wind as a powerful discriminant between true astrophysical signals and laboratory based noise.

The DRIFT-II detector (described in detail in [2]) is a 1 m^3 low pressure gaseous negative ion time projection chamber (NI-TPC). WIMP-nucleon interactions in the drift region (0.8m^3 of CS_2 at 40 Torr - 134 g target mass) would create tracks of ionisation a few mm in length, and with an orientation biased in the direction of the incoming WIMPs [3]. Electrons liberated by the primary nuclear recoil bind to electronegative CS_2 molecules resulting in a track of negative CS_2 ions. These heavy ions drift with far less diffusion than electrons drifted alone, making it feasible to scale-up the drift region to meter dimensions whist retaining the ability to reconstruct the direction and charge distribution in the original track. There are currently two, essentially identical, DRIFT-II modules in operation. DRIFT-IId is located at a depth of 2805 m.w.e. at the Boulby Underground Laboratory, UK. DRIFT-IIc is located in a surface laboratory at Occidental College, Los Angeles, CA, USA.

2. Directional studies on DRIFT-IIc

A 202 μCi ^{252}Cf source was used to produce nuclear recoils similar to those expected from WIMP interactions. The source was placed on each of the principle dimensions of the detector at a distance of 3.31 m so that the angular spread of the neutron beams was collimated to $\sim$ 10% (excluding scattering). Range components of the nuclear recoils were calculated, with Δx and Δy calculated from charge deposition on the two perpendicular wire planes of the MWPC, and Δz coming from timing information. For each source position, it is found that the range component is largest for the axis on which the neutron source is placed [4]. This effect is greatest for the Δz component derived from timing, and least for the Δy component derived from signals on the MWPC grid wires, leading to an optimal orientation choice where the anode wires run parallel to lines of fixed latitude on the Earth's surface and the normal to the MWPC readout plane points to the zenith.

3. Head-tail effects in nuclear recoils

It has been shown that being able to determine the full vector of a WIMP-induced nuclear recoil, compared to the scaler direction, can reduce the number of events required for positive detection by an order of magnitude [5]. The differential ionisation of low energy (~keV/amu) nuclear recoils expected from a WIMP-nucleon interaction is unclear [6]. Determining whether these low energy recoils produce a head-tail effect, and whether this effect can be detected in a 1m^3 DRIFT-II detector are critical.

An experimental setup similar to that used for the directional measurements was used [7]. Analysis defined a head-tail asymmetry parameter derived from the ratio of charge deposited in the first and second half of ionisation tracks from nuclear recoils. Using this asymmetry parameter it is shown that the ionisation density was higher close to the primary collision than at the far end of the recoil track. Figure 1 shows that this head-tail effect increases with larger energy recoils and remains significant down to energies of 50 keV (~1.5 keV/amu).

4. Gas measurements with a proportional counter

A proportional counter experiment was set up such that UV photons from a xenon flashlamp were used to liberate single photoelectrons on the inner cathode wall of the proportional counter. This was used for measurements of W-value, mobility and gas gain for a variety of CS_2-CF_4 gas mixtures [8].

Measurements of W-value (mean energy required to create an ion pair) were made by using an ^{55}Fe source to produce X-rays of known energy (5.9 keV) in the proportional counter volume. Measuring the mean pulse size from X-ray events and combining this with the mean pulse size of single electron events provides a measure of the mean number of ion pairs produced in a 5.9 keV event, and hence the mean energy required to create a single ion pair. W-values were measured as (CS_2-CF_4 Torr): 24.9±0.8 (40-0), 25.2±0.6 (30-10), 29.2±1.0 (20-20) and 33.0±1.0 eV (10-30). The implication of this result for the operation of a $1\,\text{m}^3$ DRIFT-II detector with CS_2-CF_4 gas mixtures is that as CF_4 concentration increases relative to CS_2 the detector will become less sensitive to low energy events.

Mobility, μ, is defined by $v = (\mu E)/p$, where v is drift velocity, E is electric field and p is gas pressure. The drift velocity is deduced experimentally by measuring the average difference in time between excitation of a photoelectron on the cathode and arrival of charge at the anode. The average mobility was measured as (CS_2-CF_4 Torr): 0.54 ± 0.02 (40-0), 0.60 ± 0.02 (30-10), 0.69 ± 0.02 (20-20) and $0.81 \pm 0.03\,\text{cm}^2\,\text{atm}\,\text{V}^{-1}\text{s}^{-1}$ (10-30). At concentrations of CF_4 above 75% drift times fluctuate substantially, indicating fluctuating electron capture path lengths. This would result in large diffusion of ionisation before the negative ion track is formed, leading to loss of directional information. However, the measurements show that for concentrations up to 10-30 Torr CS_2-CF_4 negative ion drift is preserved, and the average mobility increases as CF_4 content is increased.

Gas gain (the charge multiplication that occurs in the avalanche region of the proportional counter) was calculated by measuring the ionisation produced by single electron events through an amplification chain of known gain. Figure 2 shows the gas gain curves for various CS_2-CF_4 mixtures as a function of applied voltage and shows that gas gain increases as CF_4 concentration is increased. The maximum applied voltage was reduced with increasing CF_4 concentration in order to prevent electrical breakdown in the proportional counter. These results show that increased gain, and hence improved signal-to-noise, is expected with the addition of CF_4 to the DRIFT-II detector, but that high-voltage stability will be reduced and lower MWPC operating voltages will be required for stable running.

5. Summary

The results from the directional and head-tail studies demonstrate key features of the DRIFT-II detectors. They are however, non-optimal. An upgrade of the detector electronics and DAq to a system with lower noise and less aggressive pulse shaping is in progress, expected to lower the energy threshold of the detector and increase sensitivity to the head-tail asymmetry. Improved parameterisation and track reconstruction techniques could also amplify these effects. The prospect of refining these techniques and combining these signals suggests DRIFT as a powerful tool for identification of WIMPs and further study of their properties.

The work with CS_2-CF_4 mixtures demonstrates the flexibility of the DRIFT-II detectors and offers an opportunity to run the detectors in a spin-dependent mode with the addition of CF_4. DRIFT-IId at the Boulby mine has been upgraded with a new gas mixing system that allows the continuous flow of CS_2-CF_4 gas mixtures (required to maintain gas purity in the 1.5m^3 stainless steel vacuum vessel). Stable running has been achieved with a 30-10 CS_2-CF_4 mixture. Data is currently being collected.

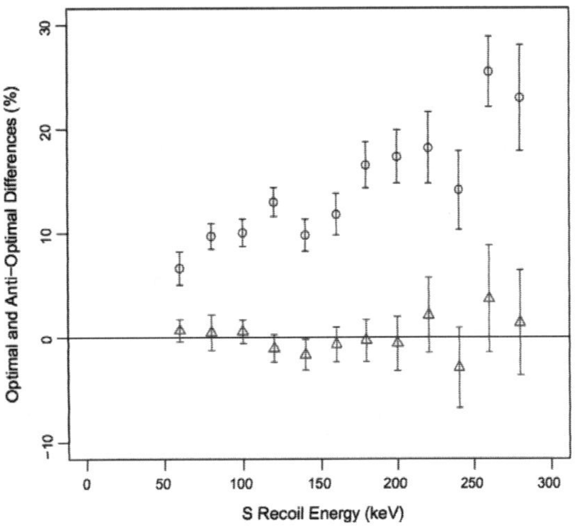

Figure 1. Pulse asymmetry parameter as a function of S recoil energy (blue circles optimal, brown triangles anti-optimal).

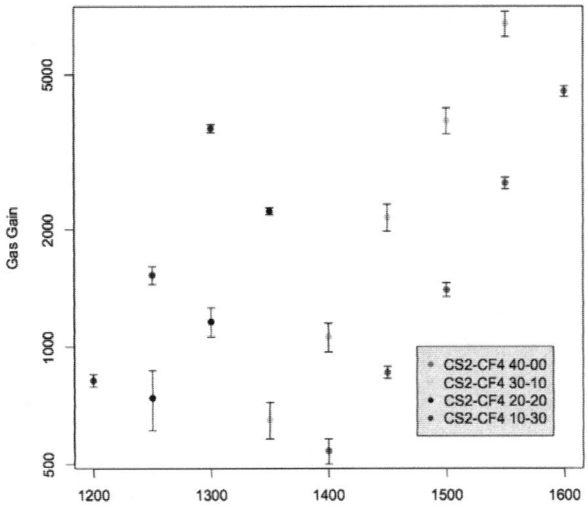

Figure 2. Gas gain as a function of applied voltage for CS_2-CF_4 gas mixtures.

References

[1] Vergados J D 2003 *Phys. Rev. D* **67**
[2] Alner G J *et al.* 2005 *Nucl. Instrum. Meth. A* **555** 173–183
[3] Snowden-Ifft D P, Martoff C J and Burwell J M 2000 *Phys. Rev. D* **61**
[4] Burgos S *et al.* 2009 *Nucl. Instrum. Meth. A* **600** 417–423
[5] Morgan B, Green A M and Spooner N J C 2005 *Phys. Rev. D* **71**
[6] Majewski P, Muna D, Snowden-Ifft D P and Spooner N J C 2009 ArXiv:0902.4430v1 [physics.ins-det]
[7] Burgos S *et al.* 2009 *Astropart. Phys.* **31** 261–266
[8] Pushkin K and Snowden-Ifft D 2009 *Nucl. Instrum. Meth. A* **606** 569–577

Topics in Astroparticle and Underground Physics (TAUP 2009) IOP Publishing
Journal of Physics: Conference Series **203** (2010) 012032 doi:10.1088/1742-6596/203/1/012032

Dark Matter Spin-Dependent Limits for WIMP Interactions on ^{19}F by PICASSO

Berta Beltran for the PICASSO collaboration
Department of Physics, University of Alberta, Edmonton, T6G 2G7, Canada
E-mail: bbeltran@phys.ualberta.ca

Abstract. The PICASSO experiment at SNOLAB uses super–heated C_4F_{10} droplets suspended in a gel as a target sensitive to WIMP–proton spin–dependent elastic scattering. The phase II setup has been improved substantially in sensitivity by using an array of 32 detectors with an active mass of ~65 g each and largely reduced background. First results are presented for a subset of two detectors with target masses of ^{19}F of 65 g and 69 g respectively and a total exposure of 13.75 ± 0.48 kgd. No dark matter signal was found and for WIMP masses around 24 GeV/c^2 new limits have been obtained on the spin–dependent cross section on ^{19}F of $\sigma_F = 13.9$ pb (90% C.L.) which can be converted into cross section limits on protons and neutrons of $\sigma_p = 0.15$ pb and $\sigma_n = 2.45$ pb respectively (90% C.L). The obtained limits on protons restrict recent interpretations of the DAMA/LIBRA annual modulations in terms of spin-dependent interactions.

1. Introduction

The PICASSO experiment (Project In Canada to Search for Super-Symmetric Objects) at SNOLAB searches for cold dark matter through the direct detection of neutralinos via their *spin-dependent* (SD) interactions with ^{19}F nuclei [1, 2]. The active detector material in PICASSO is liquid C_4F_{10}, which is in a metastable, superheated state at ambient temperature and pressure. In this condition a heat spike due to the energy deposited by an ionizing particle can cause the formation of a vapour bubble. This phase transition is explosive in nature and accompanied by an acoustic signal in the audible to ultrasonic frequency range. This can easily be registered by piezoelectric transducers.

By rising the temperature we can increase the superheat stored in the droplet. This results in less amount of energy needed to be deposit in order to trigger the phase transition [1]. Therefore each temperature corresponds to a well defined recoil energy threshold, and the spectrum of the particle induced energy depositions can be reconstructed by ramping the temperature. In such case, the response to different kinds of particle interactions depends on the respective specific energy losses, as shown in Fig. 1 (left).

Alpha particles are the main source of background events for PICASSO, but since the shapes of the WIMP and alpha curves differ substantially (see Fig. 1, left), the alpha background can be efficiently discriminated by measuring the temperature profile of the detector response.

2. Results

The new results presented in this work are based on a set of data collected with two PICASSO modules, detectors 71 and 72, which were deployed at SNOLAB on October 2006. Their ^{19}F target mass is 65.06 ± 3.2 g and 69.0 ± 3.5 g, respectively. The runs used for this analysis covered the period from June 2007 until July 2008, with three neutron source calibration periods interspersed, which allowed monitoring the stability over the period of data taking. The temperatures were varied in the range from 45°C to 18.5°C, which corresponds to a range of threshold energies from 2 keV to 200 keV. A total of 101.5 d (103.5 d) of data taking time has been used in this analysis for det. 71 (72), respectively, which results in a total exposure of 13.75 ± 0.48 kgd. Two independent analyses have been performed and yielded consistent results. For further details on the set up and the data analysis the reader is referred to [3].

The final count rates of det. 71 and 72 as a function of temperature can be seen in Fig. 1 (right). The two detectors exhibit temperature profiles identical to the response of the ^{226}Ra spiked detector shown in Fig. 1 (left) and the observed count rates are attributed to alpha particles.

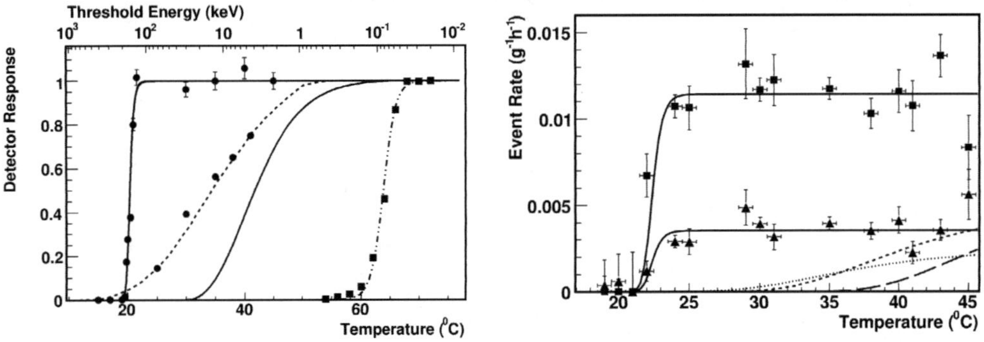

Figure 1: Left: Detector response to different types of particles as a function of temperature for detectors loaded with C_4F_{10} droplets of ~200μm in diameter. From left to right: alpha particles of 5.6 MeV in a detector spiked with ^{226}Ra (fit to data points represented by continuous line); nuclear recoils from fast neutrons of an AmBe source compared to simulations (dotted line); expected response for nuclear recoils following scattering of a 50 GeVc^{-2} WIMP (continuous line); response to 1.275 MeV gamma rays of a ^{22}Na source (dashed line). All responses are normalized to one at full detection efficiency. Temperatures are converted into threshold energies (upper x-axis). If not visible, experimental errors are smaller than the symbols. Right: Count rates of det. 71 and 72 as a function of temperature. The rates are normalized by the active mass of ^{19}F and indicate the different levels of α-background. Also shown are the expected response curves for WIMP induced nuclear recoils for M_W = 10 GeV c^{-2} (broken), 30 GeV c^{-2} (dashed) and 100 GeV c^{-2} (dotted); a cross section of σ_p = 1 pb was assumed for clarity.

In order to compare the data of detectors 71 and 72 to the expected ^{19}F recoil spectrum for interactions with WIMPs in our galactic halo, we follow the recommendations in [4]. Keeping the WIMP mass and the cross section as parameters, the expected WIMP induced count rate can be calculated as a function of temperature by applying the PICASSO threshold function [3] to the expected WIMP induced recoil spectrum. The resulting response curves are shown for several WIMP masses in Fig. 1 (right). They differ significantly in shape from the flat alpha background and therefore by fitting the two distributions to the data an upper bound can be set on the interaction cross section σ_F on ^{19}F for a given value of WIMP mass M_W.

The two free parameters of the fit are the scale factor of the alpha plateau and the cross section σ_F (M_W) for a given WIMP mass. The minima of the reduced chi-squares were found

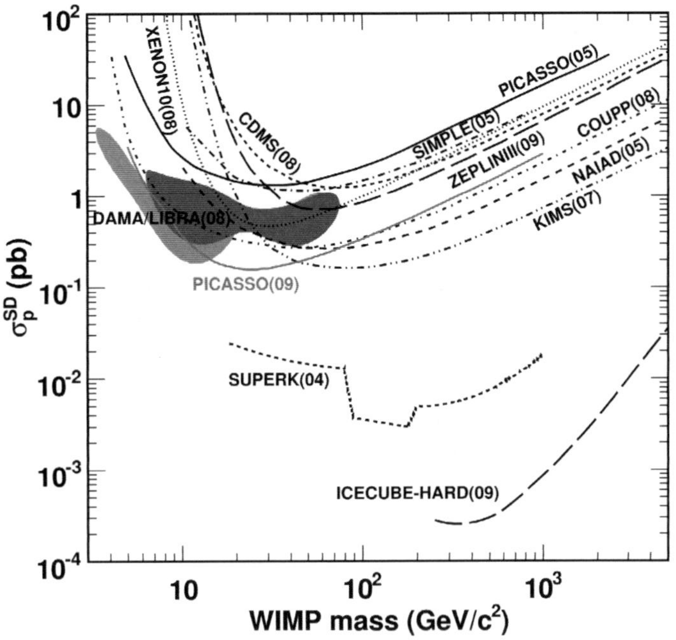

Figure 2: Upper limits on direct spin-dependent WIMP-protons from PICASSO (full lines) compared to other experiments.

to lie around $\chi^2 = 1.2$ and 1.1 for detectors 71 and 72, respectively over a mass range from $M_W = 7$ to 1000 GeVc^{-2}. By combining the fit results of both detectors in a weighted average we obtain the maximum sensitivity for a WIMP mass of $M_W = 24$ GeVc^{-2} and a cross section of $\sigma_F = -\ 0.446$ pb $\pm$ 10.83 pb $\pm$ 0.53 pb (1σ), compatible with no effect. This result can be converted into a limit on the cross section for WIMP induced reactions on ^{19}F of $\sigma_F = 13.9$ pb (90% C.L.). Following [4] we can translate the obtained fit results for σ_F into a cross section measurement on free protons of $\sigma_p = -\ 0.0051$ pb $\pm$ 0.124 pb $\pm$ 0.007 pb (1σ) which results in a limit of $\sigma_p = 0.15$ pb (90%C.L.) for a WIMP mass of 24 GeVc^{-2}. The resulting exclusion plot for the cross section on protons as a function of the WIMP mass is shown in Fig. 2, where we also compare the obtained limits with the most recent results of other experiments in the spin-dependent sector.

In the WIMP-proton spin-dependent sector the new limits rule out a substantial part of the allowed parameter space of the DAMA/LIBRA experiment if no ion channelling is assumed, but leave still some room if the ion channelling hypothesis is adopted [5].

References

[1] Barnabe-Heider M *et al.* (PICASSO Collaboration), *Nucl. Instrum. Meth. A* **555**, 184–204 (2005) (*Preprint* physics/0508098).
[2] Barnabe-Heider M *et al.* (PICASSO Collaboration), *Phys. Lett. B* **624**, 186–94 (2005) (*Preprint* hep-ex/0502028).
[3] Archambault S *et al.* (PICASSO Collaboration), Submitted to *Phys. Lett. B* (*Preprint* hep-ex/0907.0307).
[4] Lewin J D and Smith P, *Astrop. Phys.***6** 87–112 (1996)
[5] Savage C, Gelmini G, Gondolo P and Freese K, *JCAP* **0904**, 010 (2009) (*Preprint* astro-ph/0808.3607).

Topics in Astroparticle and Underground Physics (TAUP 2009) IOP Publishing
Journal of Physics: Conference Series **203** (2010) 012033 doi:10.1088/1742-6596/203/1/012033

Direct Dark Matter Search using CCDs

J. Estrada[1]
Fermilab Center for Particle Astrophysics
MS 127, POBox 500, Batavia, IL, 60510, USA

E-mail: estrada@fnal.gov

Abstract. There is currently vast evidence for Dark Matter (DM) from astronomical observations. However, in spite of tremendous efforts by large experimental groups, there is no confirmed direct detection of the dark matter in our galaxy. Recent experimental results and theoretical developments suggest the possibility of a DM particle with mass below 10 GeV, such a particle would escape most of the direct searches due to the large thresholds for the detection of nuclear recoils typically used. In this work we study the possibility of a new Dark Matter search with an unprecedented low threshold for the detection of nuclear recoils using high-resistivity CCD detectors (hr-CCD). Due their extremely low readout noise and the relatively large active mass, these detectors present a unique opportunity in this field.

The Weakly Interactive Massive Particle (WIMP) is the leading candidate dark matter particle. The search for dark matter in the halo of our galaxy by direct detection experiments has been a very active field in recent years, with the limits on WIMP-nucleon cross section improving significantly with time (for a review see Refs. [1] and [2]). However, typical searches of this kind have a poor sensitivity for low mass dark matter particles because of the high thresholds (~1 keV) established for the detection of the nuclear recoils. The experimental limits obtained for dark matter searches rise sharply as the particle mass approaches 10 GeV, as discussed below (see Fig. 5). One of the limitations for setting a lower threshold in the direct dark matter searches is the readout noise of the detectors used. There has been a continuing effort in reducing the readout noise for the detectors typically used in direct dark matter searches to overcome the high threshold issue [3].

The most popular models for dark matter particles predict masses above 50 GeV (for a review see Ref. [4]), a mass high enough that low energy thresholds would not be needed. However, there are models [5][6] where the dark matter particles have low mass and for which their detection will require a low energy threshold for nuclear recoils. A low mass dark matter candidate could also arise in the most simple extension of the Minimal Supersymmetric Standard Model as discussed in Ref. [7] and Ref. [8]. Models for which the typical velocity of the dark matter particles with respect to the earth is lower have been considered [9][10], and those cases also require a low threshold in a direct search experiment. In addition to the theoretical motivations for a low mass dark matter search, the recent results from the DAMA/Libra collaboration could also be interpreted as the evidence for a low mass dark matter particle. In particular, it has been pointed out that the DAMA/Libra annual modulation signal is consistent with a low mass dark matter particle if channeling is occurring in the detector. This channeling would also make all other null results reported for DM direct searches compatible with the DAMA/Libra observations [11].

We are developing a low threshold and low background experiment with CCD detectors for a dark matter search, such detectors will allow a search optimized for low mass particles. The CCD detectors

[1] To whom any correspondence should be addressed.

Journal of Physics: Conference Series **203** (2010) 012033 doi:10.1088/1742-6596/203/1/012033

the we propose to use are the high resistivity fully depleted CCDs developed for astronomy due to their higher efficiency in the near-IR [12]. Fermilab, is leading the construction of the DECam instrument for the Dark Energy Survey [13], this is currently the largest instrument being built using these type of detectors. In the process of building DECam we have acquired comprehensive expertise on the high resistivity CCDs, their readout electronics and their packaging.

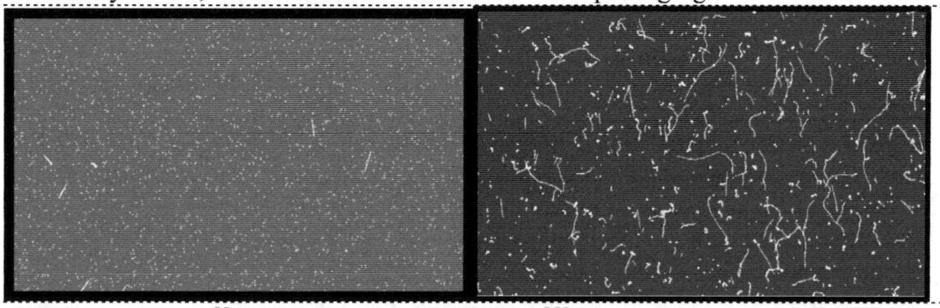

Figure 2. CCD exposure to ^{55}Fe X-Ray source (left), and ^{252}Cf neutron source (right). The top panel shows that the X-Ray produce point like hits, with size determined by the charge diffusion inside the detector. The bottom panel shows the diffusion limited hits from nuclear recoils produced by the neutrons, together with Compton scattered electrons producing tracks.

Recent advances in CCD technology [12] allow the fabrication of high resistivity (~10kΩ/cm2) detectors, up to ~300 μm thick which are fully depleted at relatively low bias voltages. These CCDs have a significantly higher quantum efficiency in the near infra-red wavelengths than conventional, thin CCDs. For this reason these CCDs are the optical detectors chosen by several groups building new mosaic cameras for astronomy, such as DECam [13] , SNAP [14] and HyperSuprime [15]. For a general description of CCDs see Ref [16]. We consider here the CCDs developed [12] by Lawrence Berkeley National Laboratory for DECam [13] [17] [18]. These devices are 63 x 33 mm2 and have square pixels 15 microns on each side. The CCDs weigh 1 gram. The detectors are p-channel devices thinned to 250 microns thickness. Application of bias voltage on the back side of the CCD establishes a depletion region. Interactions in the depletion region produce electron-hole pairs. Positive-charge holes traverse the silicon and are collected in the buried channels a few microns from the gate electrodes.

The noise performance of these detectors has been studied in detail as part of the characterization effort done by the DECam CCD team [17] [18]. The detectors have 8.4 million pixels arranged in 2048 columns and 4096 rows, and are read out by two amplifiers in parallel. Each amplifier sitting on opposite ends of a serial register towards which the charge is clocked from the center of the row. The detectors have an output stage with a electronic gain of ~2.5 μV/e-. The signal is digitized after correlated double sampling (CDS) of the output. The CDS removes the noise resulting from resetting the output stage after each pixel, and is also efficient rejecting the common mode noise (frequencies much lower than the pixel frequency). Each sample used for the CDS operation is the result of an integration during a time τ. The integration time can be selected freely and acts as a filter for high frequency noise. The noise observed for pixel readout times larger than 50 μs is $\sigma < 2$ e- (RMS). These results were obtained using a Monsoon [19] CCD controller.

As charge is produced on the back of the CCD, it has to travel to the potential minimum a few microns away from the gate electrodes, during this time it can diffuse to neighbor pixels. When considering these detectors for a dark matter search, lateral charge diffusion is an important parameter because it determines the size of the reconstructed nuclear recoil events. The nuclear recoils will produce a very localized charge cloud (<< 15 μm), and the signature in the CCD detectors will be a diffusion-limited charge deposition. Photons in the few keV energy range (X-rays) will also produce a diffusion limited hit, and for this reason we use X-ray hits in a back illuminated CCD to measure diffusion. Diffusion measurements were done for DECam CCDs using X-rays and optical methods [20], diffusion was also studied on similar CCDs by LBNL [21] [22]. The results indicate ~7 μm lateral charge diffusion.

Figure 2 shows an example of an X-ray image with diffusion limited hits (top) and an image obtained with a CCD exposure to neutrons (bottom) with diffusion limited hits, and other types of radiation. The size of the reconstructed hits on the CCD provides us with a tool for selection nuclear recoils in our images.

In most models, the interaction of a DM particle with a CCD detector will produce a nuclear recoil. For this reason is it very important to understand the ionization efficiency for nuclear recoils in these detectors. Some of the recoil energy is lost as acoustic vibrations or thermal excitations, the rest produces e-hole pairs. This so-called quenching factor must be accounted for when making measurements of nuclear recoil energy. The quenching factor has historically been well-described by the Lindhard theory [23]. The Lindhard theory has been applied in recent work [24] [25] and the quenching factor has been measured for silicon at energies above 4 keV [26]. However, at the low-energy region where low-mass DM candidates could be detected, the quenching becomes more dramatic and is increasingly energy-dependent. During the summer of 2009 we exposed a DECam CCD to a ^{252}Cf neutron source to produce a preliminary measurement quenching factor at low energies, the results are shown in Figure 3.

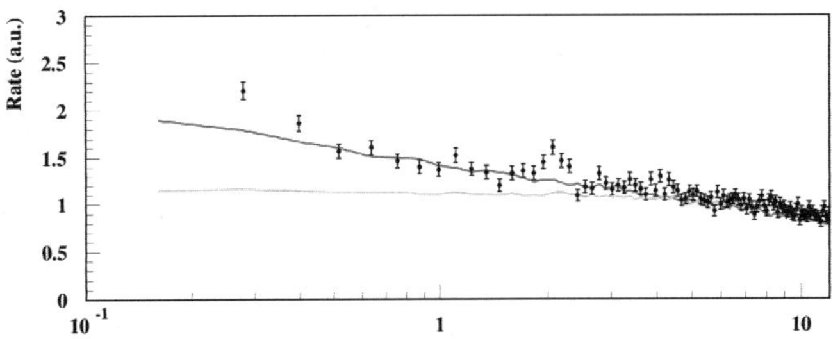

Figure 3. Fraction of the recoil energy that goes into ionization (quenching factor), measurements in Si, compared with the expectations from Lindhard theory (red solid curve), and the expectations for a constant quenching factor (green solid curve).

During the winter of 2009 we conducted a first test with DECam CCDs underground inside a lead shield. The idea for this test was to study the performance of the detectors in a low background environment. For this test we built an 8" lead shield using Fermilab surplus lead for the outer 6" and new lead, with low ^{210}Pb content, in the inner 2". We operated the detectors inside a vacuum vessel that was available at Fermilab from previous work, this vessel was made out of stainless steel and its shape made shielding very inefficient. The detectors were operated at -140 C, and kept at that temperature using a cryo-cooler available from Fermilab surplus. The electronics used for this experiment were engineering versions of the DECam electronics, including engineering grade DECam CCDs. We build a clean tent in the Minos near detector hall for the installation of this apparatus.

For this run, we achieved two orders of magnitude reduction in background compared with the tests of the CCDs done at the surface in Lab-A at Fermilab. The measured rate of radiation events was 10^4 cpd/kg/keV for nuclear recoils (diffusion limited hits) in the region 0.1-5 keV. This rate is still three orders of magnitude larger than what most low background experiments have achieved. This test was very useful to understand the background in our existing setup, and in this way establish a benchmark for the future development. We are also using the data obtained to understand the noise stability of the readout system, possible light leaks and the idea of rejecting non-DM events by looking at coincidences of hits in our 4 CCD array. For comparison with other experiments we present our results as a cross section limit as a function of dark matter particle mass. The limits, presented in Fig. 5, show that our 2009 run is about 2 orders of magnitude above the best existing limit for masses of 1 GeV. We also present in the figure the expectation for future DAMIC runs, assuming that we achieve 10 cpd/keV/kg and no improvements in the electronic readout noise (blue), and with a factor of 4 improvement in the electronic readout noise (yellow). These projections are calculated assuming 300 g-day exposure, which we could achieve with a 10 g detector (8 CCD array).

The tests performed underground during 2009 used a surplus vacuum vessel made out of stainless steel and used for earlier projects at Fermilab. The materials in this dewar were not selected considering their radio purity. In addition to this, the geometrical form of this vacuum vessel makes the shielding very inefficient, requiring a lot of shielding material and having to maintain inside the shield unnecessary components like the vacuum gauges and the cryogenic cooler head. We are now starting to build a low background dewar for this experiment and expect to start demonstrating the true potential of this technology. The recent results from DAMA/Libra have produced a generalized interest in the low mass DM searches. There results have been accompanied by the development of DM models which also point to low a mass dark matter particle. We are proposing here an experiment with CCDs to search for such low mass dark matter particle. Other groups are also conducting or planning dark matter searches for low mass particles [27][28]. The sensitivity of our experiments to a dark matter particle will depend on the results of the planned R&D tasks for lowering electronic readout noise and radiation background. We believe that this technique has great potential for low mass dark matter searches, and for other experiments requiring low threshold low background detectors.

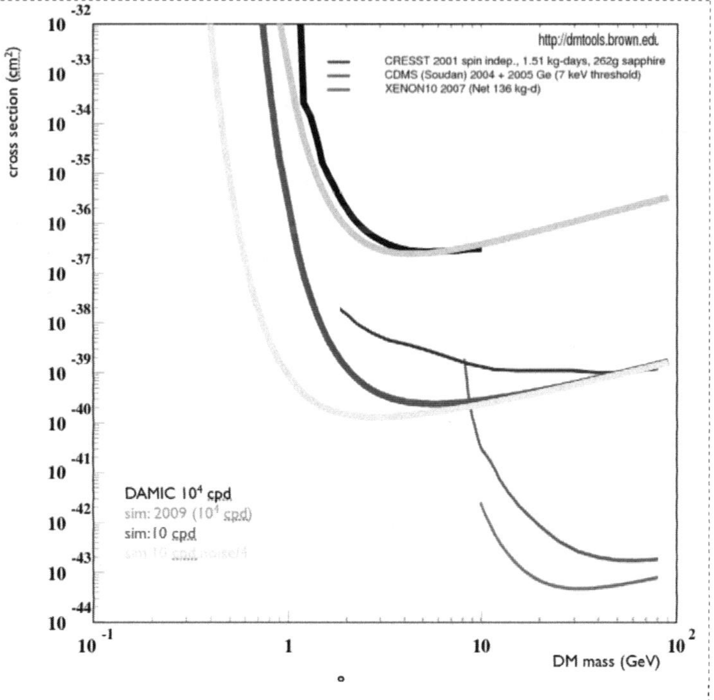

Figure 5. Cross section limit for DAMIC 2009 compared with existing results from other experiments. The blue curve correspond to the expectations for DAMIC with 10 cpd background. The yellow curve corresponds to expectations for DAMIC with 10 cpd and a factor of 4 reduction in electronic readout noise.

[1] R. J. Gaitskell, Annu. Rev. Nucl. Part. Sci., 54 315 (2004).
[2] G.Chardin, "Cryogenic Particle Detection", edited by Christian Enss, Springer (2005), arXiv:astro-ph/0411503
[3] P.S. Barbeau, J.I .Collar and O. Tench, Journal of Cosmology and Astroparticle Physics, 09 009 (2007).
[4] B. W. Lee and S. Weinberg, Phys. Rev. Lett. , 39 165 (1977).
[5] D. Hooper and K.M. Zurek, arXiv:0801.3686v1
[6] X. He, T. Li, X. Li and H. Tsai, Modern Physics Letters A, 22 2121 (2007).
[7] J.F. Gunion, D. Hooper, B. McElrath, Phys.Rev., D73 015011 (2006) .
[8] C. Bird, R. Kowalewski \& M. Pospelov, Modern Physics Letters A, 21 457 (2006).
[9] Gondolo, P. and Gelmini, G., Phys. Rev. , D71 123520 (2005).
[10] T. Damour and L.M. Krauss, ``Proceedings of the 3rd International Workshop on the Identification of Dark Matter", edited by N. J. C. Spooner & V. Kudryavtsev. World Scientific (2001), arXiv:astro-ph/9806165v3
[11] F. Petriello and K. M. Zurek, JHEP 0809, 047 (2008) [arXiv:0806.3989 [hep-ph]]; C. Savage, G. Gelmini, P. Gondolo and K. Freese, arXiv:0808.3607 [astro-ph].
C. Savage, K. Freese, P. Gondolo and D. Spolyar, arXiv:0901.2713 [astro-ph].
[12] S.E. Holland, D.E. Groom, N.P. Palaio, R. J. Stover, and M. Wei, IEEE Trans. Electron Dev., 50 225 (2003), LBNL-49992.
[13] Dark Energy Survey Collaboration, astro-ph/0510346; Flaugher, B., Ground-based and Airborne Instrumentation for Astronomy. Edited by McLean, Ian S.; Iye, Masanori. Proceedings of the SPIE, Volume 6269, (2006)
[14] ``Supernova / Acceleration Probe: A Satellite Experiment to Study the Nature of the Dark Energy", SNAP Collaboration, G.~Aldering et al., submitted to Publ. Astr. Soc. Pac., astro-ph/0405232; SNAP Collaboration, astro-ph/0507459.
[15] M. Satoshi et al., Ground-based and Airborne Instrumentation for Astronomy. Edited by McLean, Ian S.; Iye, Masanori. Proceedings of the SPIE, Volume 6269, (2006)
[16] J.R. Janesick, Scientific Charge Cupled Devices, SPIE press (2001).
[17] J. Estrada & R. Schmidt , Scientific Detectors for Astronomy 2005, Edited by J.E. Beletic, J.W. Beletic and P. Amico, Springer, (2006).
[18] J. Estrada et al. , Ground-based and Airborne Instrumentation for Astronomy. Edited by McLean, Ian S.; Iye, Masanori. Proceedings of the SPIE, Volume 6269, (2006).
[19] http://www.noao.edu/ets/new_monsoon
[20] H. Cease, H. T. Diehl, J. Estrada, B. Flaugher and V. Scarpine, Experimental Astronomy, Online First (2007).
[21] J.A. Fairfield , D. E. Groom, S. J. Bailey, C. J. Bebek, S. E. Holland, A. Karcher, W. F. Kolbe, W. Lorenzon, & N. A. Roe, Fairfield IEEE Trans. Nucl. Sci. 53 (6), 3877 (2006)
[22] A. Karcher, C.J. Bebek, W. F. Kolbe, D. Maurath, V. Prasad, M. Uslenghi, M. Wagner, IEEE Trans. Nucl. Sci. 51 (5), (2004) LBNL-55685
[23] Lindhard, J et al, 1963 *Mat. Fys. Medd. Dan. Vid. Selsk.* 33.10
[24] Chagani, H et al, 2008 *JINST* 3 P06003
[25] Lewin J & Smith, P, 1996 *P Astropart. Phys.* 6 87.
[26] Gerbier, G. et al., Phys.Rev.D. 42, 3211 (1990).
[27] S.T. Lin et al (Texono Collaboration), Phys.Rev.D79:061101,2009.
[28] Aalseth, C.E. et al, Review Letters, vol. 101, Issue 25, id. 251301

Journal of Physics: Conference Series **203** (2010) 012034 doi:10.1088/1742-6596/203/1/012034

New Limits on Low-Mass WIMP Dark Matter with sub-keV Germanium Detector

Henry T. Wong

Institute of Physics, Academia Sinica, Taipei 11529, Taiwan.

E-mail: htwong@phys.sinica.edu.tw

Abstract.
The current goals of the TEXONO research program are on the development of germanium detectors with sub-keV sensitivities to realize experiments on neutrino magnetic moments, neutrino-nucleus coherent scattering, as well as WIMP dark matter searches. A threshold of 220 eV was achieved with prototype detectors at the Kuo-Sheng Neutrino Laboratory. New limits were placed for the couplings of low-mass WIMPs with matter. Data are being taken with a 500 g Point Contact Germanium detector, where a threshold of ~350 eV was demonstrated. The dark matter program will evolve into a dedicated experiment at an underground laboratory under construction in Sichuan, China.

1. Introduction and Overview

A research program on low energy neutrino and dark matter physics is pursued at the Kuo-Sheng Neutrino Laboratory (KSNL) by the TEXONO Collaboration[1]. The laboratory is located at a distance of 28 m from a 2.9 GW reactor core and has an overburden of about 30 meter-water-equivalent. Results on neutrino magnetic moments[2] and neutrino-electron scattering cross-section have been obtained[3]. The present goals are to develop advanced detectors with kg-size target mass, 100 eV-range threshold and low-background specifications[4] for the searches of Weakly Interacting Massive Particles (WIMPs)[5] as well as the studies of neutrino-nucleus coherent scattering[6] and neutrino magnetic moments.

2. Results on Dark Matter Searches

A four-channel Ultra-Low-Energy Germanium (ULEGe) prototype detector with a total active mass of 20 g has collected low-background data at KSNL[5]. The trigger and analysis efficiencies are shown in Figure 1. An energy threshold of (220±10) eV was achieved at an efficiency of 50%. The background spectrum with 0.338 kg-day of exposure is displayed in Figure 2. Constraints on WIMP-nucleon spin-independent $[\sigma_{\chi N}^{SI}]$ and spin-dependent $[\sigma_{\chi N}^{SD}(n)]$ couplings as functions of WIMP-mass (m_χ) were derived, as depicted in Figures 3&4, respectively. Overlaid on the plots are results from experiments which define the current exclusion boundaries, the DAMA-allowed regions and that favored by SUSY models[5, 7]. The KSNL limits improve over previous results at $m_\chi \sim 3 - 6$ GeV. Sensitivities for full-scale experiments at 1 cpd background level are projected as dotted lines. The observable nuclear recoils at m_χ=5 GeV and $\sigma_{\chi N}^{SI}$=0.5×10^{-39} cm^2(allowed) and 1.5×10^{-39} cm^2(excluded) are superimposed with the measured spectrum in the inset of Figure 2 for illustrations.

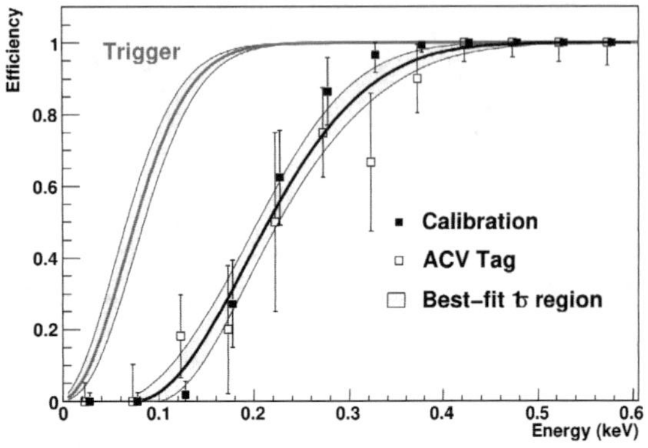

Figure 1. The trigger and analysis efficiencies of the 20 g ULEGe prototype detector, as derived by the test pulser and *in situ* background events, respectively.

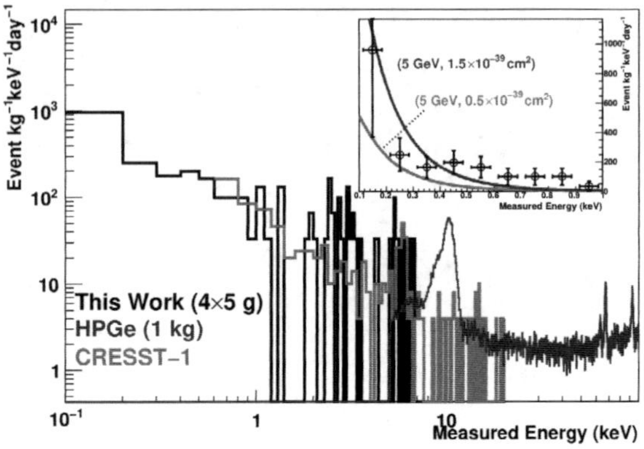

Figure 2. The measured spectrum of ULEGe with 0.338 kg-day of data, after various background suppression procedures. Background spectra of the CRESST-I experiment[7] and the HPGe[2] are overlaid for comparison.

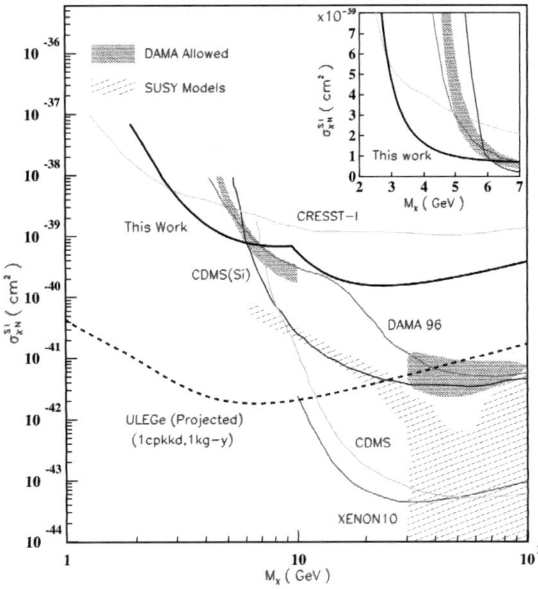

Figure 3. Exclusion plot of the spin-independent χN cross-section versus WIMP-mass.

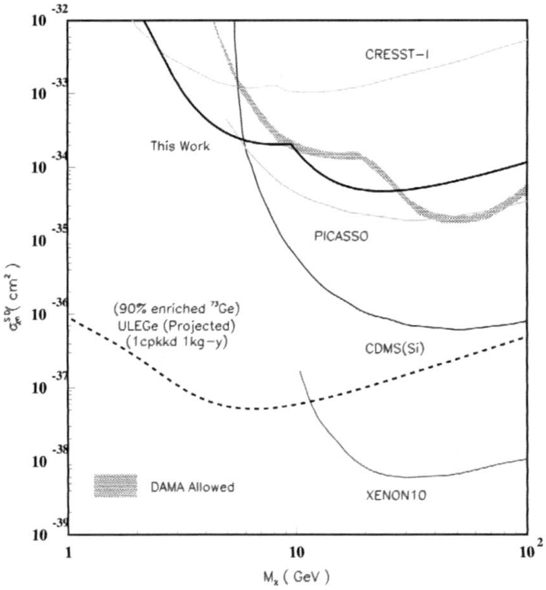

Figure 4. Exclusion plot of the spin-dependent χ-neutron cross-section versus WIMP-mass.

3. Performance of Point-Contact Germanium Detectors

The design of Point-Contact Germanium (PCGe) detectors was first proposed in the 1980's [8], offering the potential merits of sub-keV sensitivities with kg-scale target mass. There are intense recent interest triggered by successful realization and demonstration of the detector technique [9]. A PCGe of target mass 500 g was constructed and has been collecting data in KSNL since early 2009.

Similar procedures to those developed for the ULEGe were adopted to study the efficiency factors below the electronic noise edge. The results, analogous to those of Figure 1, are displayed

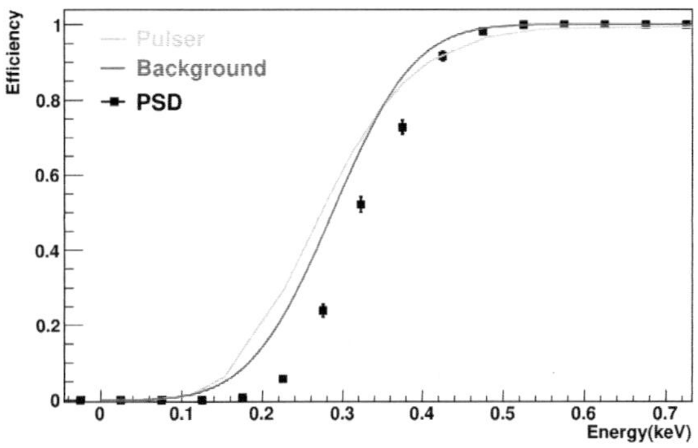

Figure 5. The trigger and analysis efficiencies of the 500 g PCGe detector, as derived from the test pulser and *in situ* events, respectively.

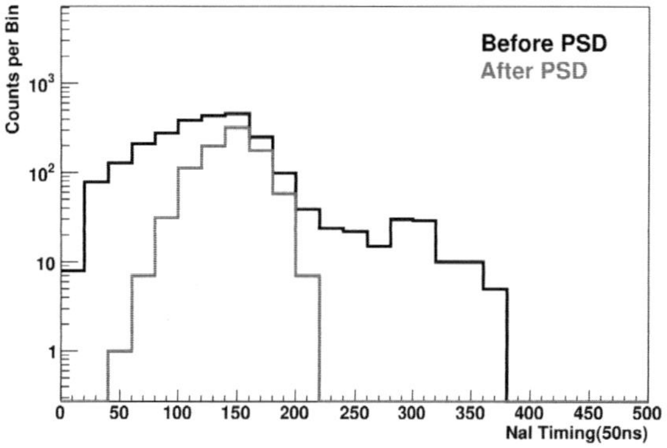

Figure 6. Events as a function of relative timing between ACV-NaI(Tl) and PCGe systems, before and after PSD selection.

in Figure 5. The trigger efficiencies were measured with two methods. The fractions of calibrated pulser events above the discriminator threshold provided the first measurement, while the studies on the amplitude distributions of *in situ* data contributed to the other. The relative timing between the PCGe and anti-Compton (ACV) NaI(Tl) detectors is shown in Figure 6, for "sub-noise edge" events at 200-400 eV before and after the pulse shape discrimination (PSD) selection processes. Events in coincidence with ACV at the "50−200 ns" window are due to multiple Compton scatterings, which are actual physical processes having similar pulse shapes as the neutrino and WIMP signals. It can be seen that only these events have substantial probabilities of surviving the cuts, and the fractions constitute to the PSD efficiencies. The threshold at ~50% combined efficiencies is ~350 eV. Intensive background and optimization studies with the PCGe at KSNL are underway.

4. Status and Plans

An excellent candidate site for a deep underground laboratory was recently identified in Sichuan, China where the China Jin-Ping Laboratory (CJPL) is being constructed [10]. The laboratory has more than 2500 m of rock overburden, is accessible by a road tunnel built for public traffic, and is supported by excellent infrastructures already available near the entrance. The first cavern of size 6 m(height)X6 m(width)X40 m(depth) is scheduled for completion in early 2010.

The low energy neutrino physics program will continue at KSNL, where a 900 g PCGe detector will be installed in 2010. Dedicated dark matter search with both 20 g ULEGe and 500 g PCGe detectors will be the first experimental program conducted at CJPL commencing 2010.

References

[1] Wong H T 2004 *Mod. Phys. Lett.* **A 19** 1207
[2] Li H B et al. 2003 *Phys. Rev. Lett.* **90** 131802 ; Wong H T et al. 2007 *Phys. Rev.* **D 75** 012001
[3] Li H B et al. 2001 *Nucl. Instrum. Methods* **A 459** 93 ; Deniz M and Wong H T 2008 *arXiv:0810.0809* ; Deniz M 2009 *these Proceedings*
[4] Wong H T 2008 *Mod. Phys. Lett.* **A 23** 1431
[5] Lin S T et al. 2009 *Phys. Rev.* **D 76** 061101(R), and references therein.
[6] Wong H T et al. 2006 *J. Phys. Conf. Ser.* **39** 266
[7] Drees M and Gerbier G 2008 *Phys. Lett.* **B 667** 241, and references therein.
[8] Luke P N et al. 1989 *IEEE Trans. Nucl. Sci.* **36** 926
[9] Barbeau P A, Collar J I and Tench O 2007 *JCAP* **09** 009
[10] Normile D 2009 *Science* **324** 1246 ; Yue Q 2009 *these Proceedings*

Topics in Astroparticle and Underground Physics (TAUP 2009) IOP Publishing
Journal of Physics: Conference Series **203** (2010) 012035 doi:10.1088/1742-6596/203/1/012035

Axion cold dark matter revisited

L Visinelli[†1], P Gondolo[‡]

Department of Physics, University of Utah, 115 S 1400 E rm.201, Salt Lake City, UT 84102, USA

E-mail: [†] visinelli@utah.edu

E-mail: [‡] paolo@physics.utah.edu

Abstract. We study for what specific values of the theoretical parameters the axion can form the totality of cold dark matter. We examine the allowed axion parameter region in the light of recent data collected by the WMAP5 mission plus baryon acoustic oscillations and supernovae [1], and assume an inflationary scenario and standard cosmology. We also upgrade the treatment of anharmonicities in the axion potential, which we find important in certain cases. If the Peccei-Quinn symmetry is restored after inflation, we recover the usual relation between axion mass and density, so that an axion mass $m_a = (85 \pm 3)$ μeV makes the axion 100% of the cold dark matter. If the Peccei-Quinn symmetry is broken during inflation, the axion can instead be 100% of the cold dark matter for $m_a < 15$ meV provided a specific value of the initial misalignment angle θ_i is chosen in correspondence to a given value of its mass m_a. Large values of the Peccei-Quinn symmetry breaking scale correspond to small, perhaps uncomfortably small, values of the initial misalignment angle θ_i.

1. Introduction

About 84% of the non-relativistic matter in the Universe is in the form of cold dark matter (CDM)[1], whose nature is still to be discovered. One of the most promising hypothetical candidates that could account for the CDM observed is the axion [2, 3].

The properties of the axion as the CDM particle have been studied in various papers [4, 5]. We examine the possibility that the axion accounts for 100% CDM in the light of the WMAP5 mission, baryon acoustic oscillations (BAO) and supernovae (SN) data. We also upgrade the treatment of anharmonicities in the axion potential, which we find important in a specific region of the parameter space. The axion parameter space is described by three quantities, the PQ energy scale f_a, the Hubble parameter at the end of inflation H_I and the axion initial misalignment angle θ_i.

2. Properties of the axion

The solution to the strong-CP problem by Peccei and Quinn [3] introduces a new $U(1)_{PQ}$ symmetry in the theory of strong interactions that spontaneously breaks at a temperature of order the PQ energy scale, $T \sim f_a$. The axion is the pseudo Nambu-Goldstone boson originating from the breaking of $U(1)_{PQ}$. We denote by $a(x)$ the axion field and $\theta(x) = a(x)/f_a$ the misalignment field. At a temperature of order the QCD transition $T \sim \Lambda_{QCD}$ an effective

[1] Speaker. Talk based on L Visinelli, P Gondolo, Phys. Rev. D **80**, 035024 (2009).

potential $V(\theta)$ arises through instanton effects [6] given by

$$V(\theta) = m_a^2(T)\, f_a^2\, (1 - \cos\theta)\,. \tag{1}$$

The temperature-dependent axion mass $m_a(T)$ is given by

$$m_a(T) = \begin{cases} m_a\, b\, \left(\frac{\Lambda}{T}\right)^4 & T > \Lambda_{QCD}, \\ m_a & T < \Lambda_{QCD}. \end{cases} \tag{2}$$

We set $\Lambda_{QCD} = 200\text{MeV}$, $b = 0.018$ [5]. Setting $f_{a,12} = f_a/10^{12}\text{GeV}$, the axion mass at zero temperature is

$$m_a = 6.2\mu\text{eV}\, f_{a,12}^{-1}/N. \tag{3}$$

The integer N represents the $U(1)_{PQ}$ color anomaly index; here $N = 1$.

3. Axion and cosmology

Setting $T_{GH} = H_I/2\pi$, axion cosmology is different if the PQ symmetry breaks after inflation, $f_a < T_{GH}$ (Scenario I), or during inflation, $f_a > T_{GH}$ (Scenario II). Over a Hubble volume, θ_i takes different values in Scenario I, while in Scenario II θ_i takes only one random value.

The subsequent evolution of the zero mode of the dynamical field $\theta(x)$ in a flat Friedmann-Robertson-Walker metric is given by

$$\ddot{\theta} + 3H(T)\,\dot{\theta} + \frac{1}{f_a^2}\frac{\partial V(\theta)}{\partial\theta} = 0. \tag{4}$$

Here a dot indicates a derivative with respect to time and $H(T)$ is the Hubble expansion rate. For $T \gg \Lambda_{QCD}$ the misalignment field is frozen at a constant value, $\theta \equiv \theta_i = \text{const}$, where θ_i is the initial misalignment angle. For $m_a(T) \approx 3H(T)$ the axion field begins to oscillate in the $V(\theta)$ potential. The axion energy density-to-critical density is computed as

$$\Omega_a^{mis}h^2 = \begin{cases} 0.236\langle\theta_i^2 f(\theta_i)\rangle\, f_{a,12}^{7/6}, & f_a < \hat{f}_a, \\ 0.0051\langle\theta_i^2 f(\theta_i)\rangle\, f_{a,12}^{3/2}, & f_a > \hat{f}_a, \end{cases} \tag{5}$$

with $\hat{f}_a = 9.91\times10^{16}\text{GeV}$. The function $f(\theta_i)$ accounts for anharmonicities in the axion potential, i.e., deviations from a pure harmonic potential when θ_i is no longer smaller than one. We look for an analytic function $f(\theta_i)$ that interpolates between previous numerical results [7]. We take

$$f(\theta_i) = \left[\ln\left(\frac{e}{1 - \theta_i^2/\pi^2}\right)\right]^{7/6}. \tag{6}$$

The function $f(\theta_i)$ in eq.(6) shows the correct asymptotic behaviors for $\theta_i \to \pi$ and $\theta_i \to 0$.

4. Results

Figure 1 shows the region of the axion parameter space in which the axion is 100% CDM, or $\Omega_a = \Omega_{CDM}$, where Ω_{CDM} is the CDM density in units of the critical density. The PQ scale f_a is bounded from below by measurements on white dwarfs cooling time, $f_a > 4\times10^8\text{GeV}$, while the Hubble rate at the end of inflation H_I is bounded from above by non-detection of tensor modes, $H_I < 6.3\times10^{14}\text{GeV}$. For $f_a < T_{GH}$ (Scenario I), the axion can be 100% CDM only if $f_a = (7.27\pm0.25)\times10^{10}$ GeV. The function $f(\theta_i)$ in eq.(6) affects the results by increasing the value of $\langle\theta_i^2 f(\theta_i)\rangle$. In Scenario II, $f_a > T_{GH}$, the parameter space on the top left is bounded by non-detection on axion isocurvature fluctuations that yield $f_a < 4.17 \times 10^{-5}\, H_I\, \theta_i$. The upper kink in the leftmost boundary is due to the change in the dependence of the axion mass with temperature, eq.(2), while the lower kink is due to the presence of anharmonicities, eq. (6). In the allowed region of Scenario II, for a given f_a the axion is 100% CDM only if a correspondent value $\theta_i = \theta_i(f_a)$ is selected. This value of θ_i does not depend on H_I. We show the function $\theta_i = \theta_i(f_a)$ in Figure 2, where we see that a value $\theta_i < \theta_i(f_a)$ for a given f_a makes the axion population sub-dominant compared to the CDM observed, while a value $\theta_i > \theta_i(f_a)$ is over abundant. We included the region excluded by ADMX1 axion direct search and future reaches by the Planck satellite and future cavity searches.

References

[1] E. Komatsu *et al.* [WMAP Collaboration], Astrophys. J. Suppl. **180**, 330 (2009).
[2] S. Weinberg, Phys. Rev. Lett. **40**, 223 (1978); F. Wilczek, Phys. Rev. Lett. **40**, 279 (1978).
[3] R. D. Peccei, H. R. Quinn, Phys. Rev. Lett. **38**, 1440 (1977).
[4] J. Preskill, M. Wise, F. Wilczek, Phys. Lett. B **120**, 127 (1983); L. Abbott, P. Sikivie, *ibid.*, 133; M. Dine, W. Fishler, *ibid.*, 137 (1983).
[5] M. Beltran *et al.*, Phys. Rev. D **75** 103507 (2007); M. P. Hertzberg *et al.*, Phys. Rev. D **78**, 083507 (2008).
[6] D. J. Gross, R. D. Pisarski, L. G. Yaffe, Rev. Mod. Phys. **53**, 43 (1981).
[7] M. S. Turner, Phys. Rev. D **33**, 889 (1986); D. H. Lyth, Phys. Rev. D **45**, 3394 (1992).

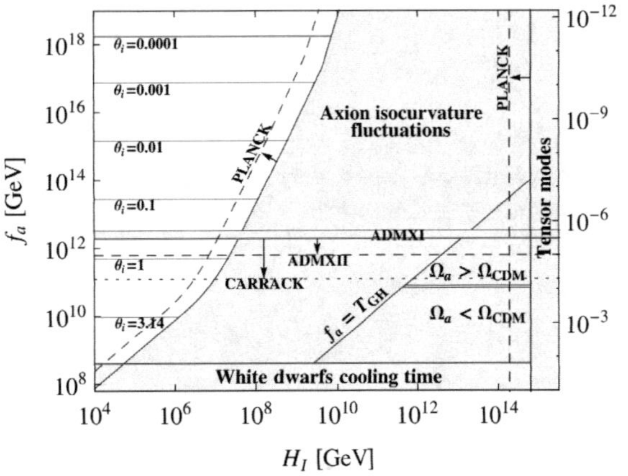

Figure 1. Region of axion parameter space where the axion is 100% of the cold dark matter. The axion mass scale on the right corresponds to eq. (3) with $N = 1$. When the PQ symmetry breaks after inflation ($f_a < H_I/2\pi$), the axion is 100%CDM if $f_a = (7.27 \pm 0.25) \times 10^{10}$ GeV, or $m_a = (85 \pm 3)\ \mu$eV, which is the narrow horizontal window shown on the right. If the axion is present during inflation ($f_a > H_I/2\pi$), axion isocurvature perturbations constrain the parameter space to the region on the top left, which is marked by the values of θ_i necessary to obtain 100% of the CDM density. Other bounds indicated in the figure come from observations of white dwarfs cooling times and the non-observation of tensor modes in the CMB spectrum. Dashed lines and arrows indicate the future reach of the PLANCK satellite and the ADMX and CARRACK cavity searches.

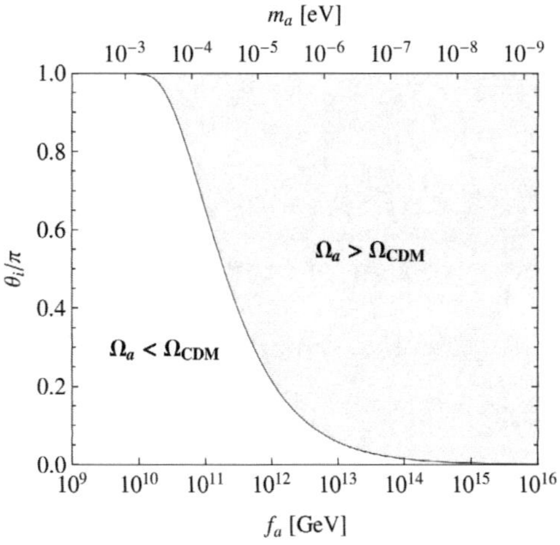

Figure 2. The misalignment angle θ_i necessary for the axion to be 100% of the cold dark matter in Scenario II ($f_a > H_I/2\pi$), as a function of f_a. Above the curve, $\Omega_a > \Omega_{CDM}$. For $f_a > 10^{17}$GeV, $\theta_i \approx 0.001(f_a/10^{17}\,\mathrm{GeV})^{-3/4}$.

Latest results from the CERN Axion Solar Telescope

Igor G Irastorza[1] ***for the CAST collaboration***
Laboratorio de Física Nuclear y Astropartículas, Universidad de Zaragoza, Zaragoza, Spain

Abstract. The CERN Axion Solar Telescope (CAST) experiment searches for axions from the Sun converted into few keV photons via the inverse Primakoff effect in the high magnetic field of a superconducting Large Hadron Collider (LHC) decommissioned test magnet. After results obtained with vacuum in the magnet pipes (phase I of the experiment) as well as with ^{4}He the collaboration is now immersed in the data taking with ^{3}He. The status of the experiment will be presented, including a preliminary indication of the axion parameter space explored up to now. CAST is currently sensitive to realistic QCD axion models at the sub-eV scale, and with axion-photon couplings in the range down to the $\sim 2 \times 10^{-10}$ GeV^{-1}, compatible with solar life limits.

1. Introduction

Axions are light pseudoscalar particles that arise in theories in which the Peccei-Quinn U(1) symmetry has been introduced to solve the strong CP problem [1]. They could have been produced in early stages of the Universe being attractive candidates to the cold Dark Matter (and in some particular scenarios to the hot Dark Matter) that could compose up to $\sim 1/3$ of the ingredients of the Universe.

The axion-photon coupling $g_{a\gamma}$, unlike other axion couplings[2], is automatically present in every QCD axion model, and allows for the conversion of axions into photons (and viceversa) in the presence of electromagnetic fields, being the base of almost every axion detection strategy.

A combination of astrophysical and nuclear physics constraints, restricts the allowed range for axion mass and couplings[3, 4, 5]. Pure cosmological arguments lead to a conservative, relatively model-independent allowed mass range of $10^{-6}\text{eV} \lesssim m_a \lesssim 1$ eV, the upper limit being recently set [6], by requiring thermal production of axions to be compatible with last CMB data. $g_{a\gamma}$ can be constrained independently by astrophysical observations, namely $g_{a\gamma} \lesssim 10^{-9}$ GeV^{-1} based on the solar standard model and helioseismological observations [7], or the so-called *globular cluster* limit of $g_{a\gamma} \lesssim 10^{-10}$ GeV^{-1} [8, 9]. This scheme is usually generalized to any axion-like particle of a given mass m_a and a photon coupling $g_{a\gamma}$[2].

Axions could be produced at early stages of the Universe by the so-called misalignment (or realignment) effect[2], plus in some cases also from decay of primordial topological defects like axion strings or walls. These relic axions could be a substantial part of the cold dark matter (CDM) if $m_a \sim 10^{-6} - 10^{-3}$ eV, and could be detected by microwave cavities as originally proposed in[10]. Such technique is followed by experiments like the Axion Dark Matter Experiment (ADMX)[11, 12], which is actively scanning the low axion mass range mentioned

[1] Attending speaker. Email: Igor.Irastorza@cern.ch
[2] One should not forget that some of the limits above are set using other properties of the QCD axions and therefore do not hold for axion-like particles.

above. Axions could also be copiously produced in the core of the stars by means of the Primakoff conversion of the blackbody photons in the fluctuating electric field of the plasma. Solar axions have energies of 1–10 keV, and could be detected by looking for x-rays of those energies coming from their conversion in a strong laboratory magnetic field. This *helioscope* concept[10, 13], was first experimentally applied in [14] and later on by the Tokyo helioscope [15], before CAST.

2. Status of CAST and latest results

The CAST experiment is making use of a decommissioned LHC test magnet that provides a magnetic field of 9 Tesla along its two parallel pipes of 2×14.5 cm^2 area and 10 m length, increasing the corresponding axion-photon conversion probability by a factor 100 with respect to the previous best implementation of the helioscope concept[16]. The magnet is able to track the Sun by about 3 hours per day, half in the morning and half in the evening. At its two ends x-ray detectors are placed, at the "sunrise" side, a Micromegas detector[17] and a CCD[18], and at the "sunset" side two additional Micromegas detectors, installed in 2007 replacing the former TPC[19]. All of the detector setups are conceived following low background techniques (shielding, radiopure materials). The CCD is coupled to a focusing X-ray device (X-ray telescope)[18] that enhances its signal-to-bakcground ratio by two orders of magnitude. Both the CCD and the X-ray telescope are prototypes developed for X-ray astronomy.

The experiment already released its phase I results from data taken in 2003 and 2004 with vacuum in the magnet bores[16, 20]. No signal above background was observed, implying an upper limit to the axion-photon coupling $g_{a\gamma} < 8.8 \times 10^{-11}$ GeV^{-1} at 95% CL for the low mass (coherence) region $m_a \lesssim 0.02$ eV (Fig. 1). Since 2006 the experiment runs its second phase, which makes use of a buffer gas inside the magnet bores to recover the coherence of the conversion for specific axion masses matching the effective photon mass defined by the buffer gas density. The pressure of the gas is changed in discrete small steps to scan the parameter space above $m_a \sim 0.02$ eV. The ^{4}He Run taken in 2006[21], allowed to scan axion masses up to 0.39 eV, for axion-photon couplings down to about 2.2×10^{-10} GeV^{-1}, entering into the QCD axion model band, as shown in Fig. 1. Due to gas condensation, in order to go to higher pressures, the experiment switched to ^{3}He as buffer gas in 2007. The experiment is currently inmersed in the ^{3}He Run since beginning of 2008. It should last until end of 2010 and should allow us to explore up to 1.2 eV in axion mass approximately, overlapping with the CMB upper limit on the axion mass disussed above.

At the moment of writing this paper, the experiment has explored a region of axion masses up to about $m_a \sim 0.70$ eV, as indicated in the plot 1. Everyday a new thin slice of untouched parameter space is being explored. Due to the sharp coherence effect, and to the fact that the parameter space to which we are sensitive now is populated by realistic QCD axion models and not excluded by previous experiments, a clear positive signal in CAST may appear at any moment.

The collaboration has performed by-product analysis of the data taken, to look for non-standard axion scenario to which CAST would also be sensitive. Using data from the TPC in the CAST phase I, a search for 14 keV axions coming from M1 transitions in the Sun has been recently released[22]. In addition, data taken with a calorimeter during the phase I, were use to search for high energy (MeV) lines from high energy axion conversion[23]. More recently a few days of data were taken with a visible detector coupled to one end of the CAST magnet, in search for axions with energy in the "visible" range, possibly produced in the surface magnetic field of the Sun. A preliminary result was presented recently [24], but more extended data taking is foreseen without interfering with the standard program of CAST by means of a permanent setup that will be installed in the next weeks in CAST.

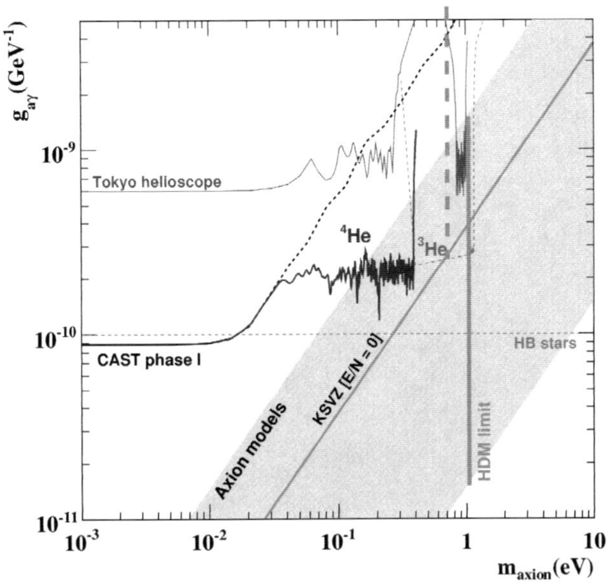

Figure 1. 95% CL exclusion line obtained from the published CAST data, both phase I (black line) and ^{4}He Run (red line), compared with other laboratory limits such as the Tokyo helioscope and the cosmological upper limit on the axion mentioned in the text. The shaded area indicates the region of theoretical preference for axion models. The thick vertical dashed red line indicates aproximately the current progress of CAST data taking with ^{3}He

3. Conclusions and prospects

The CAST experiment is going through the ^{3}He phase of data taking, and will fully complete its original program by the end of 2010. CAST will have explored a substantial region of axion(-like) parameter space beyond previous limits, including a part populated with realistic QCD models.

[1] R. D. Peccei and H. R. Quinn, Phys. Rev. Lett. **38**, 1440 (1977).
[2] M. S. Turner, Phys. Rept. **197**, 67 (1990).
[3] E. Kolb and M. Turner, *The Early Universe* (, 1990).
[4] A. Burrows, M. T. Ressell, and M. S. Turner, Phys. Rev. **D42**, 3297 (1990).
[5] J. Engel, D. Seckel, and A. C. Hayes, Phys. Rev. Lett. **65**, 960 (1990).
[6] S. Hannestad, A. Mirizzi, and G. Raffelt, JCAP **0507**, 002 (2005), hep-ph/0504059.
[7] H. Schlattl, A. Weiss, and G. Raffelt, Astropart. Phys. **10**, 353 (1999), hep-ph/9807476.
[8] G. G. Raffelt, Ann. Rev. Nucl. Part. Sci. **49**, 163 (1999), hep-ph/9903472.
[9] G. G. Raffelt, Phys. Rev. **D33**, 897 (1986).
[10] P. Sikivie, Phys. Rev. Lett. **51**, 1415 (1983).
[11] S. J. Asztalos *et al.*, Phys. Rev. **D64**, 092003 (2001).
[12] S. J. Asztalos *et al.*, Phys. Rev. **D69**, 011101 (2004), astro-ph/0310042.
[13] K. van Bibber, P. M. McIntyre, D. E. Morris, and G. G. Raffelt, Phys. Rev. **D39**, 2089 (1989).
[14] D. M. Lazarus *et al.*, Phys. Rev. Lett. **69**, 2333 (1992).
[15] S. Moriyama *et al.*, Phys. Lett. **B434**, 147 (1998), hep-ex/9805026.
[16] CAST, K. Zioutas *et al.*, Phys. Rev. Lett. **94**, 121301 (2005), hep-ex/0411033.
[17] P. Abbon *et al.*, New J. Phys. **9**, 170 (2007), physics/0702190.
[18] M. Kuster *et al.*, New J. Phys. **9**, 169 (2007), physics/0702188.
[19] D. Autiero *et al.*, New J. Phys. **9**, 171 (2007), physics/0702189.
[20] CAST, S. Andriamonje *et al.*, JCAP **0704**, 010 (2007), hep-ex/0702006.
[21] CAST, E. Arik *et al.*, JCAP **0902**, 008 (2009), 0810.4482.
[22] CAST, S. Andriamonje *et al.*, (2009), 0906.4488.
[23] CAST, . S. Andriamonje *et al.*, (2009), 0904.2103.
[24] CAST, G. Cantatore *et al.*, (2008), 0809.4581.

EDELWEISS-II Dark Matter Search : status and first results

Véronique SANGLARD for the EDELWEISS collaboration
IPNL, Université de Lyon, Université Lyon 1, CNRS/IN2P3, 4 rue Enrico Fermi, 69622 Villeurbanne, France
E-mail: sanglard@ipnl.in2p3.fr

Abstract. The EDELWEISS II experiment is devoted to the search for Weakly Interacting Massive Particles (WIMP) that would constitute the Dark Matter halo of our Galaxy. For this purpose, the experiment uses cryogenic germanium detectors, cooled down at 20 mK, in which the collision of a WIMP with an atomic nucleus produces characteristic signals in terms of ionization and elevation of temperature. We will present the preliminary results of the first operation of the detectors installed in the underground laboratory of the Frejus Tunnel (LSM), attesting to the very low radioactive background conditions achieved so far. New detectors, with a special electrode design for active rejection of surface events, have been shown to be suited for searches of WIMPs with scattering cross-sections on nucleon well below 10^{-8} pb. Preliminary results of WIMP search performed with a first set of these detectors are presented.

1. Introduction

Recent cosmological observations of the CMB show that most ($\sim$ 85%) of the matter in our Universe is dark and non baryonic. If non baryonic Dark Matter is made of particles, they must be stable, neutral and massive : WIMPs (Weakly Interactive Masssive Particles). In the MSSM (Minimal Supersymmetric Standard Model) framework, the WIMP could be the LSP (Lightest Supersymmetric Particle) called neutralino. Its mass is expected to lie between few tens and few hundreds of GeV/c^2, and a scattering cross section with a nucleon below 10^{-6} pb.
The EDELWEISS experiment is dedicated to the direct detection of WIMPs. The direct detection method (used also by other experiments like CDMS [1], XENON [2] and CRESST [3]) consists in the measurement of the energy released by nuclear recoils produced by the elastic collision of a WIMP from the galactic halo with an atomic nucleus. The main challenge is the expected extremely low event rate ($\leq$ 1 evt/kg/year) due to the very small interaction cross-section of WIMP with nucleons. The ability of the detector to unambiguously tag nuclear recoil interactions plays a major role in these searches. An other constraint is the relatively small deposited energy (a few keV to a few tens of keV).

2. Experimental set-up

The EDELWEISS experiment is located in the Modane Underground Laboratory (LSM) in the Fréjus tunnel connecting France and Italy under $\sim$ 1700 m of rock ($\sim$ 4800 mwe). In the laboratory, the muon flux is reduced down to 4 μ/m^2/d and the fast neutron flux has been measured to be $\sim 1.6\times 10^{-6}$ cm^2/s [4]. The EDELWEISS-II experiment installation (see

Journal of Physics: Conference Series **203** (2010) 012037 doi:10.1088/1742-6596/203/1/012037

Fig. 1) was completed end of 2005. Specific improvements have been made in order to reduce the possible background sources that have limited the sensitivity of the previous phase EDELWEISS-I [5, 6] To reduce the radioactive background in the cryostat all the materials were tested for

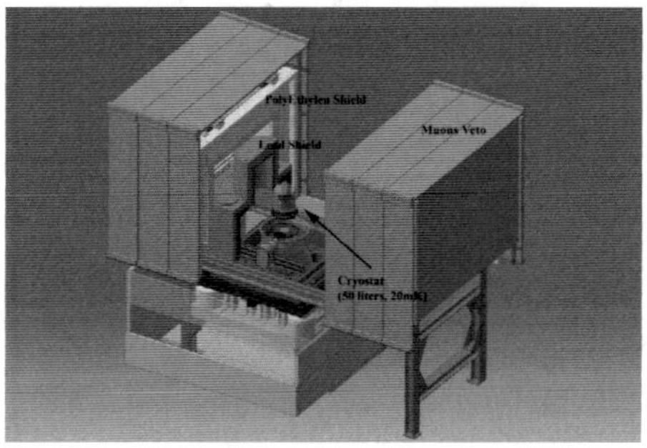

Figure 1. General scheme of the EDELWEISS-II experiment.

radiopurity in a HPGe dedicated detector with very low radon level. The experiment is installed in a class 10 000 clean room and the cryostat environment is submitted to a permanent flow of deradonized air. The gamma background is screened by a 20 cm thick lead shield. Concerning the low energy neutron background, due to the radioactive surrounding rock, it is attenuated by more than three orders of magnitude thanks to a 50 cm polyethylene shielding. In addition, a muon veto surrounding the experiment will tag muons interacting in the lead shielding [7]. The dilution cryostat is of inverted design, with the experimental detector volume on the top. The large volume, 50 ℓ, allows running 120 identical detectors in a compact arrangement, that will improve the possibility of detecting multiple interactions of neutrons and hence reject them. The detectors used in the experiment are cryogenic bolometers with simultaneous measurement of phonon and ionization signals, cooled at a temperature of $\sim$20 mK. They are made of a cylindrical high purity Ge crystal with Al electrodes to collect ionization signals and a NTD heat sensor glued onto one electrode to collect the phonon signal [8]. The simultaneous measurement of both heat and ionization signals provides an excellent event by event discrimination between nuclear recoils (induced by WIMP or neutron scattering) and electron recoils (induced by α, β or γ-radioactivity). The ratio of the ionization and heat signals depends on the recoiling particle, since a nucleus produces about one third of the ionization in a crystal than an electron does.

3. Rejection of surface event with ID detectors

The EDELWEISS experiment is currently running in its second phase. The first, EDELWEISS-I [6], has been limited by surface events generated by the β radioactivity of ^{210}Pb [5]. Because of diffusion, trapping and recombination, the charge induced by surface events is miscollected and can mimic nuclear recoils. Thus, they cannot be actively rejected with the standard EDELWEISS detectors. This may be seen in the data taken in 2008 with 11 detectors. An exposure of 93.5 kg.d was achieved with an analysis threshold set *a priori* to 30 keV. The α and β backgrounds were reduced in comparison to the EDELWEISS-I ones but, as it can be seen in the figure 2, it is still the limiting factor to improve sensitivity. To solve the problem of surface events, new detectors with active surface rejection were studied during last years inside the EDELWEISS facility. These new detectors, named InterDigits (ID) have a mass of the order of 400 g. The ionization electrodes are replaced by two sets of concentric, annular interleaved electrodes polarized differently [9, 10]. Surface events are tagged by the presence of charge on two electrodes on the same side of the detector. As shown in figures 3 and 4, this method is very efficient to reject surface events. Figure 3 shows the ionization yield as a function of recoil

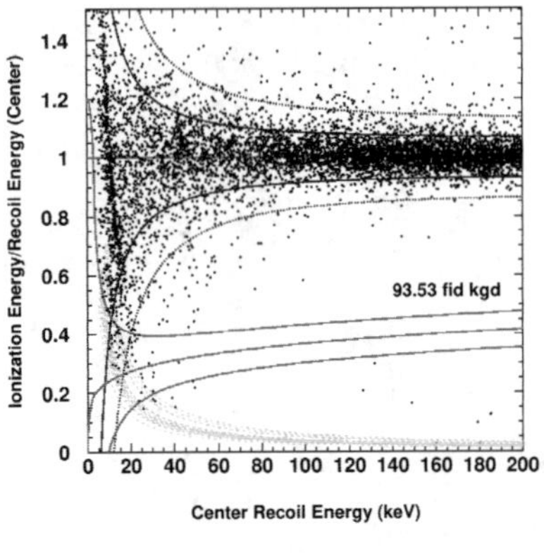

Figure 2. Ionization yield versus recoil energy for a 93.5 kg.d exposure using 11 standard EDELWEISS detectors. The leaking events down to the nuclear recoil band are attributed to β interactions.

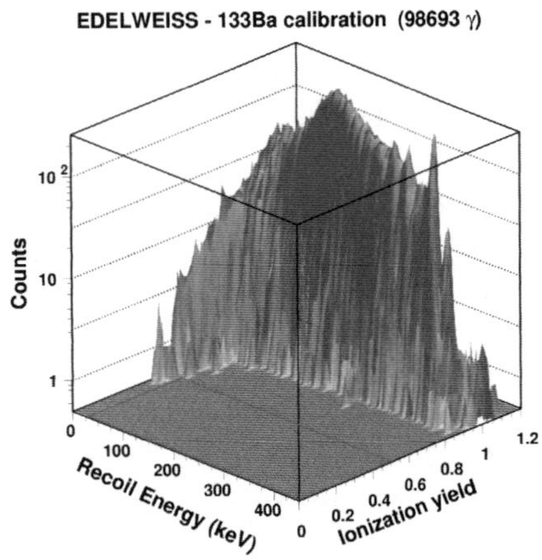

Figure 3. ID detectors calibration with a ^{133}Ba source.

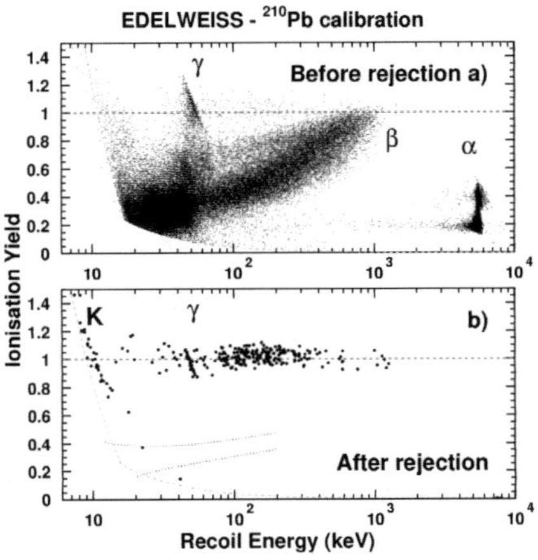

Figure 4. ID detectors calibration with a ^{210}Pb source.

energy for high-statistics calibration performed with a ^{133}Ba source. No events are present after the fiducial selection in the nuclear recoil band resulting in a rejection factor better than 1 in 10^5 for γ rays below 60 keV. Figure 4 shows the same kind of plot for a calibration with a ^{210}Pb source. Before the rejection of surface events, we see a lot of events in the nuclear recoil band but only one after giving a rejection factor of about 1 in 10^5. Such performances are described in details in [11] and open the way to reach sensitivities below 10^{-8} pb.

4. First results of the ID detectors

In 2008, a fiducial exposure of 18.3 kg.d was achieved using two 400 g ID detectors. Figure 5 shows the ionization yield as a function of recoil energy for background runs and no events have been recorded in the nuclear recoil band down to a threshold of 10 keV in recoil energy. These data were interpreted in terms of limits for spin-independent scattering cross-section for WIMPs as a function of their mass, as shown in figure 6. This limit is comparable to the one obtained

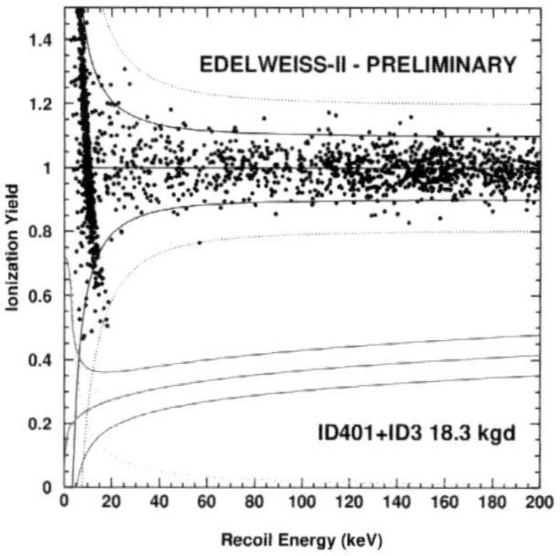

Figure 5. Ionization yield vs recoil energy recorded in ID detectors for a fiducial exposure of 18.3 kg.d.

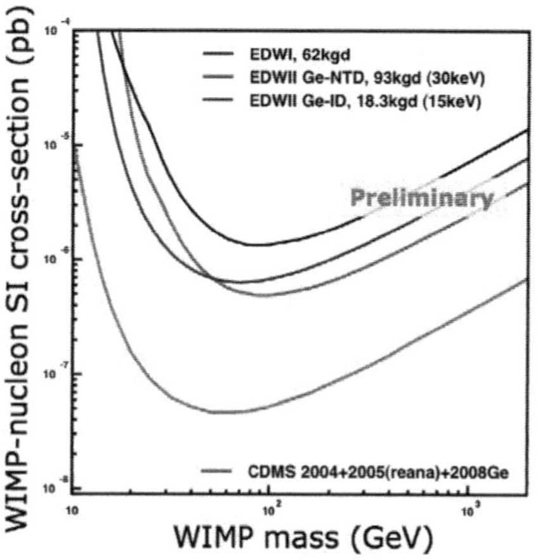

Figure 6. 90% C.L. limits for spin-independent scattering cross-section for WIMPs as a function of WIMP mass.

with the standard EDELWEISS detectors with an exposure of 93.5 kg.d despite the factor five in exposure. This demonstrates the importance of an active surface event rejection.
The EDELWEISS collaboration is presently operating ten 400 g ID detectors. A sensitivity of 4×10^{-8} pb is expected to be reached by 2010. Further improvements are expected using new detectors with an increased fiducial volume. On a longer timescale, ID detectors are well fitted for future larger scale experiments (100 kg to 1 ton) such as EURECA [12].

5. Conlusion

The EDELWEISS-II setup has been validated with both calibration and background runs. The results obtained with standard detectors show the importance of an active rejection of surface events. The new generation of ID detectors is also validated. With these new detectors showing an excellent β rejection, EDELWEISS-II is now aiming at sensitivities below 10^{-8} pb.

References

[1] Z. Ahmed et al., *Phys. Rev. Lett.*, 2009 **102**, 011301
[2] J. Angle et al., *Phys. Rev. Lett*, 2008 **100**, 021303
[3] G. Angloher et al., *Astropart. Phys.*, 2009 **31**, 270
[4] G. Chardin and G. Gerbier, in Proceedings of th 4th International Workshop on Identification of Dark Matter (IDM2002), eds N.J. Spooner and V. Kudryavstev
[5] S. Fiorucci et al., *Astropart. Phys.*, 2007 **28**, 143
[6] V. Sanglard et al., *Phys. Rev. D*, 2005 **71**, 122002
[7] K. Eitel, these proceedings
[8] X.F. Navick et al., *Nucl. Inst. Meth. A*, 2000 **444**, 361
[9] X. Defay et al., in the Proc. of the 12th Intl. Workshop on Low Temperature (LTD12), Paris 2007, *J. Low Temp. Phys.*, 2008 **151**, 896
[10] A. Broniatowski et al., in the Proc. of the 12th Intl. Workshop on Low Temperature (LTD12), Paris 2007, *J. Low Temp. Phys.*, 2008 **151**, 830
[11] A. Broniatowski et al., accepted in *Phys. Lett. B* [arXiv:0905.0753]
[12] H. Kraus et al., *Nucl. Phys. B (Proc. Suppl.)*, 2007 **173**, 168

Journal of Physics: Conference Series **203** (2010) 012038 doi:10.1088/1742-6596/203/1/012038

BGO scintillating bolometer: its application in dark matter experiments

N Coron[1], C Cuesta[2], E García[2], J Gironnet[1], J Leblanc[1], P de Marcillac[1], M Martínez[1], Y Ortigoza[2], A Ortiz de Solórzano[2], C Pobes[2], J Puimedón[2], T Redon[1], T Rolón[2], A Salinas[2], M L Sarsa[2], L Torres[1], J A Villar[2]

[1] Institut d'Astrophysique Spatiale (IAS), Bâtiment 121, Université Paris–Sud 11 and CNRS (UMR 8617), 91405 Orsay Cedex, France
[2] Laboratorio de Física Nuclear y Astropartículas, Universidad de Zaragoza, Pedro Cerbuna 12, 50009 Zaragoza, Spain

E-mail: `ortigoza@unizar.es`

Abstract. In the frame of the ROSEBUD (Rare Objects SEarch with Bolometers UndergrounD) collaboration, we have tested at surface level (Orsay) and underground (Canfranc) properties at low temperature of a BGO scintillating bolometer developed as a prototype for dark matter searches. The response of the detector to different particles, both in heat and light, using internal and external radioactive sources is reported. We have focused on its sensitivity as dark matter target and as γ–ray spectrometer to monitor external background. An algorithm implemented to analyze high energy events, which produce saturated pulses in low energy experiments (like dark matter searches), is also discussed.

ROSEBUD is a collaboration between the Institut d'Astrophysique Spatiales (IAS) and the University of Zaragoza started 10 years ago to develop bolometers for rare events searches [1] that in the last years has focused on scintillating bolometers [2]. A 46 g BGO scintillating bolometer has been tested both at surface in IAS (Orsay) and in an ultralow background environment underground at the Laboratorio Subterráneo de Canfranc (LSC). Details of the experimental set-up (detectors, dilution unit, shield, electronics and acquisition system, etc.) are described in [3, 4]. In this paper we present a summary of some of the most relevant results.

The great difference (a factor 12.3 ± 0.6) [4] in light response of the crystal to β/γ and nuclear recoil events (like those that would produce the hypothetical WIMPs) has been used to discriminate these particles down to an energy of 23.5 keV (90% C.L.) and to obtain at LSC a heat spectrum of events in the nuclear recoils band, that is, events compatible with being produced by nuclear recoils (see figure 1). The registered background level, shown in the figure 1, is very high for being due only to neutrons and it suggests other possible origins for these practically non–scintillating events. For a dark matter experiment lower particle discrimination threshold and lower background in the nuclear recoils band would be required.

In the heat spectrum, we have obtained full width at half maximum (FWHM) energy resolutions ranging from ~3 keV (at 15 keV) to 12 keV (at 1120.3 keV) (see [4] for details). These values, together with the high atomic number of Bi, make the BGO scintillating bolometer a good γ–ray spectrometer to monitor background in cryogenic dark matter experiments using low atomic mass targets. Since dark matter experiments search for signals at very low energies,

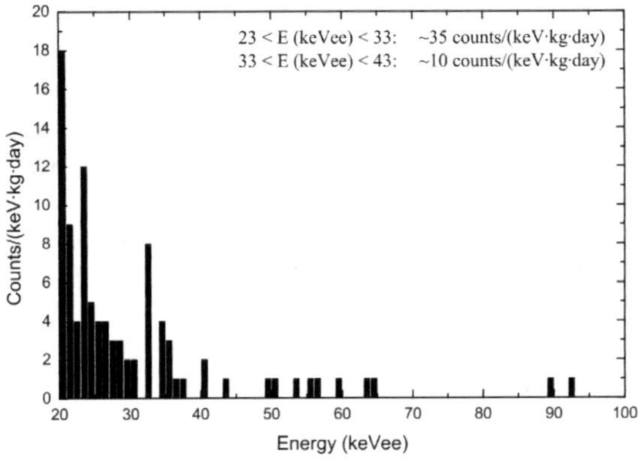

Figure 1. Heat background spectrum of events in the nuclear recoils band obtained in the LSC in 71.94 hours of live time. Background levels in two consecutive energy regions are given. Energies are electron equivalent ones.

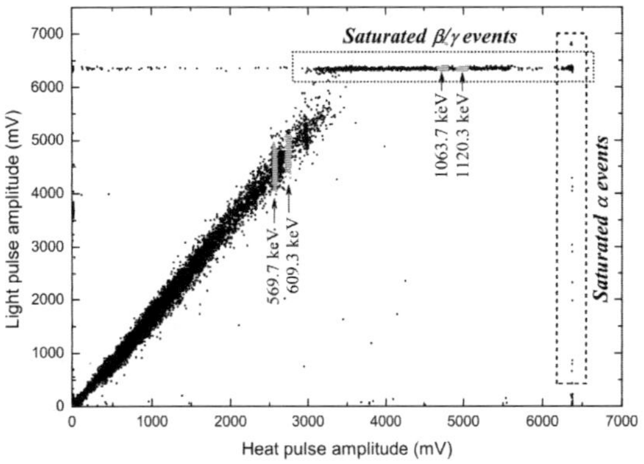

Figure 2. Light versus heat pulse amplitude plot in a low energy measurement. High energy β/γ and α events produce saturated pulses in light and/or heat.

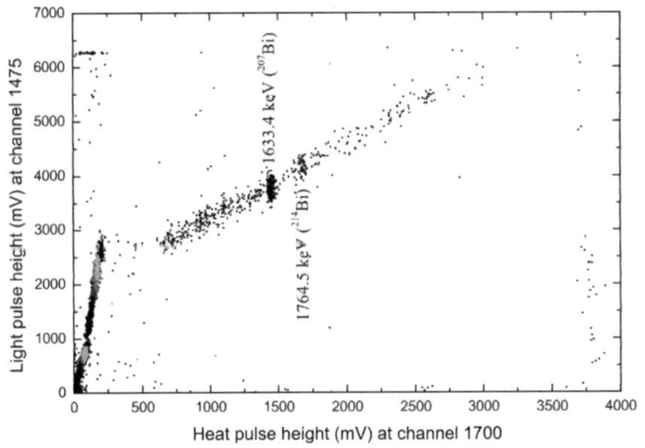

Figure 3. Light versus heat pulse height measured in a time position (channel) along the decay of the pulses selected to optimize the energy resolution of the spectrum of saturated β/γ pulses. It corresponds to the same measurement of figure 2. Note that the maximum of non–saturating pulses appears around channel 1323 and 1278 in the heat and light signal, respectively.

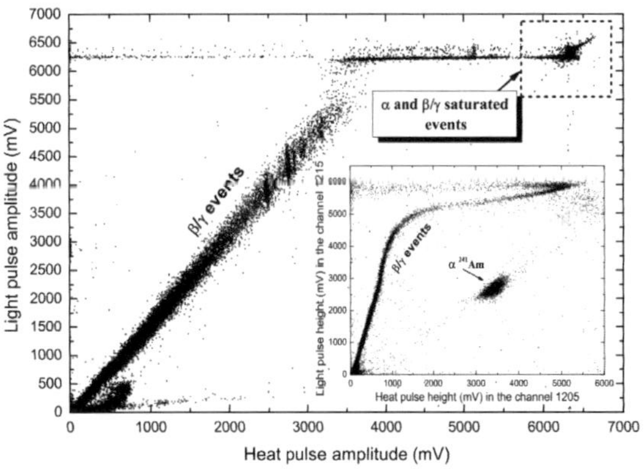

Figure 4. Light versus heat pulse amplitude of a low energy measurement. The inset shows the light versus heat pulse height measured in a time position (channel) along the pulse rise selected to optimize identification and discrimination of α saturated pulses.

the electronics and acquisition system are adjusted so that, in consequence, high energy β/γ and α events usually produce saturated pulses in heat and/or light (see figure 2). In these cases, the pulse amplitude cannot be used to estimate the energy deposited. Instead of that, we have implemented a simple method based on the selection of a time position along the rise and decay time which pulse height (used as energy estimator) optimizes energy resolution. That allows identification of high energy γ lines coming from saturated pulses like those at 1633.4 (^{207}Bi) and 1764.5 keV (^{214}Bi) that are shown in figure 3. Alpha events from an ^{241}Am source producing heat and light saturated pulses are also discriminated by these method (see figure 4).

The nuclear versus electron recoils thermal Relative Efficiency Factor (REF) has been estimated using data from the ^{241}Am source which has an activity per unit surface of 0.3 Bq/mm^2 and no coating at all. Recoiling ^{237}Np nuclei (of kinetic energy 92.4 ± 0.1 keV) exiting from the source are clearly identified in the light versus heat discrimination plot of figure 5. The heat nuclear recoils spectrum shows (see figure 6) a low energy tailing peak whose high energy side can be ascribed to full energy absorption of the recoiling ^{237}Np nuclei. This spectrum is fitted to a Gaussian (describing the high energy side of the peak), a lower exponential and a step function. Comparing the Gaussian mean with the BGO heat signal produced by a photon of 92.4 keV depositing all of its energy in the crystal, we derive a thermal REF for ^{237}Np versus electron recoils of $0.937 \pm 0.015_{stat}{}^{+0.302}_{-0.004 syst}$, compatible with 1.

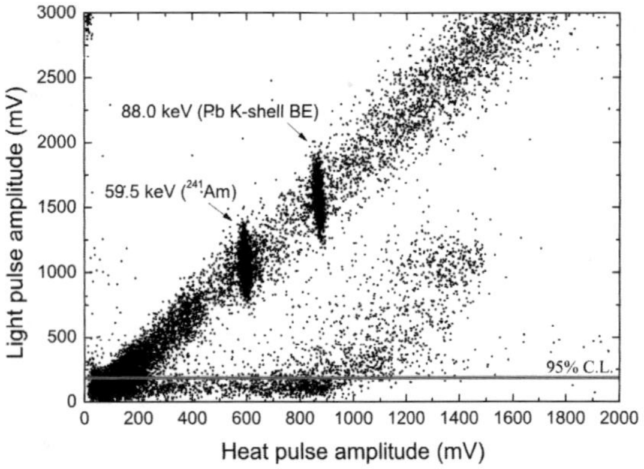

Figure 5. Light versus heat pulse amplitude of a measurement performed with a ^{241}Am source. Recoiling ^{237}Np nuclei are clearly visible (events with low scintillation and heat amplitude up to about 1000 mV). Lines of 59.5 keV and 88.0 keV are clearly identified in the β/γ band. Coincidences of recoiling ^{237}Np nuclei with 59.5 keV γ's are also identified (the intermediate region between the β/γ and the nuclear recoils bands).

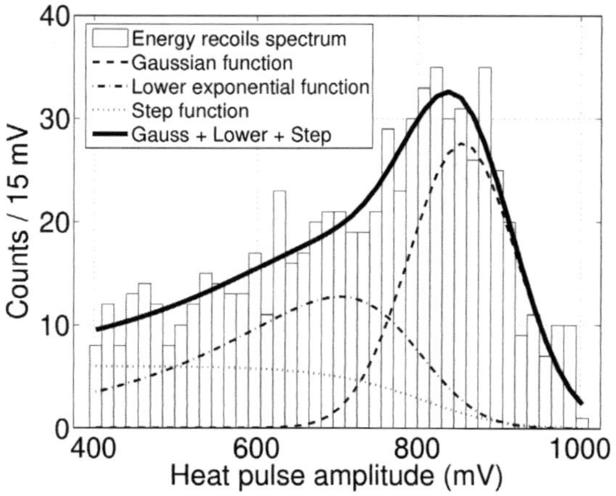

Figure 6. ^{237}Np heat nuclear recoil spectrum obtained by selected events below the 95% C.L. nuclear recoils line shown in the figure 5. See text for details about the fitting functions of the low energy tailing peak.

Acknowledgments

This work has been supported by the Spanish Ministerio de Ciencia e Innovación (FPA2008-03228), the French CNRS/INSU (MANOLIA and BOLERO projects) and the EU project ILIAS (RII3–CT–2004–506222). Y. Ortigoza and T. Rolón were also supported by a UZ/BSCH/Fundación Carolina Grant.

References

[1] Cebrián S et al. 1999 *Astroparticle Physics* **10** 361
[2] Cebrián S et al. 2004 *Astroparticle Physics* **21** 23
[3] Calleja A et al. 2008 *J. Low Temp. Phys.* **151** 848
[4] Coron N et al. 2009 *Optical Materials* **31** 1393

Journal of Physics: Conference Series **203** (2010) 012039 doi:10.1088/1742-6596/203/1/012039

The expected background spectrum in NaI dark matter detectors and the DAMA result

V. A. Kudryavtsev, M. Robinson, N. J. C. Spooner

Department of Physics and Astronomy, University of Sheffield, Sheffield S3 7RH, UK

E-mail: e-mail: v.kudryavtsev@sheffield.ac.uk

Abstract. Detailed Monte Carlo simulations of the expected radioactive background rates and spectra in NaI crystals are presented. The obtained spectra are then compared to those measured in the DAMA/LIBRA experiment. The simulations can be made consistent with the measured DAMA spectrum only by assuming higher than reported concentrations of some isotopes and even so leave very little room for the dark matter signal. We conclude that any interpretation of the annual modulation of the event rate observed by DAMA as a dark matter signal, should include full consideration of the background spectrum. This would significantly restrict the range of dark matter models capable of explaining the modulation effect.

1. Modelling of the background radiation in NaI

The DAMA group has reported a possible signal from dark matter particles based on observation of the annual modulation of the event rate at low energies [1, 2]. We have modelled the radioactive background in the DAMA/LIBRA experiment by considering three locations of radioactive sources: (i) external (PMTs and copper); ii) internal (NaI); and iii) surface sources with radioactive isotopes concentrated within the thin surface layer of the crystal. The background rate from the last source was found to be insignificant.

The setup of the DAMA/LIBRA experiment comprising 25 NaI(Tl) crystals, ~10 kg each, was modelled using the published geometry [1, 2]. For case (i) the source of background radiation was assumed to be in the 100 g envelopes of the photomultiplier tubes (PMTs) attached to the lightguides connected to the two flat surfaces of the crystals. The PMT envelopes were populated with radioactive isotopes of ^{238}U, ^{232}Th and ^{40}K. The isotopes were allowed to decay in secular equilibrium in our Monte Carlo code, based on the GEANT4 toolkit (version 9.2) [3]. We have also considered ^{60}Co in copper around the detectors and in a shield, as a possible source of background events. Figure 1 shows the spectra of energy depositions from electron recoils in the crystal. For normalisation we have used typical concentrations of radioactive isotopes in ultra-low background PMTs produced by ETL and used by DAMA: 30 ppb of U/Th, 60 ppm of K, and 10 mBq/kg of ^{60}Co in Cu. Only events in which a single crystal was hit, were included. The energy deposition spectra were convolved with Gaussian distributions describing the energy resolution of the DAMA/LIBRA detectors [2]. As can be seen from Figure 1 the energy deposition spectra from all decay chains at low energies from external sources are essentially flat, due to Compton scattering of high-energy gamma-rays in the crystal.

For case (ii) the sources of background radiation were distributed evenly in all the crystals. Figure 2 shows the energy deposition spectra from ^{238}U, ^{232}Th and ^{40}K decay chains, as

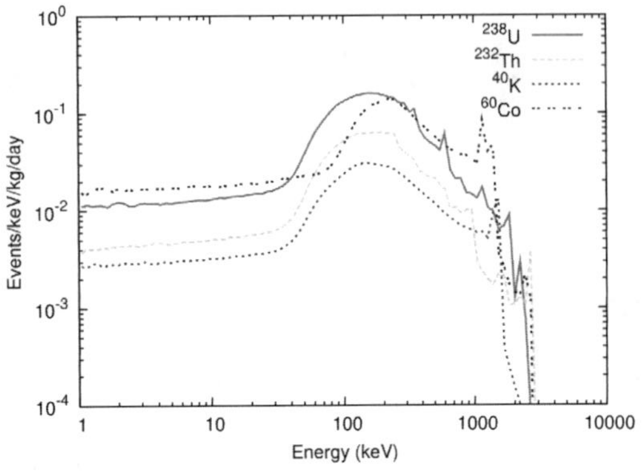

Figure 1. Spectra of energy depositions from electron recoils in the NaI crystals from external sources.

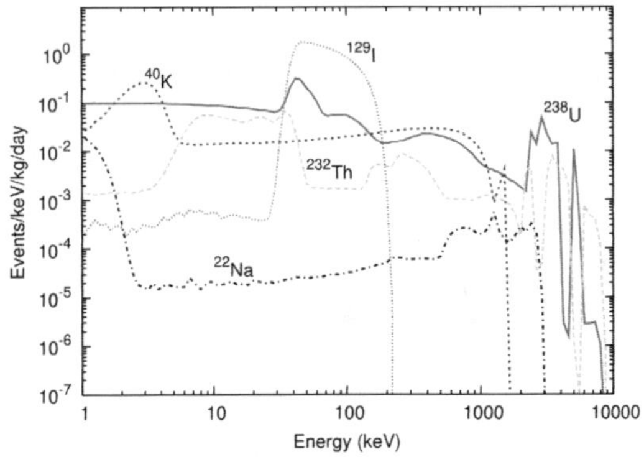

Figure 2. Spectra of energy depositions in the NaI crystals from intrinsic contamination.

well as from the decays of ^{129}I and ^{22}Na identified as possible sources of background in the DAMA/LIBRA experiment [1, 2]. The calculated spectra were normalised using the measured concentrations of radioactive isotopes or their decay rates as reported by DAMA [2]. Since the spectrum from ^{238}U chain does not have any features below 30 keV, the assumption of equilibrium should not affect the shape of the spectrum at low energies.

2. Implications for DAMA results

Figure 3 (solid curve) shows the sum spectrum of simulated events in NaI crystals due to internal and external sources of background as plotted in Figures 1 (except ^{60}Co) and 2. Assuming typical concentrations of U, Th and K in PMTs, this external source of background does not contribute substantially to the total spectrum. Figure 4 shows the measured spectra of background events at low energies. The simulated absolute rate below 10 keV (black solid curve in Figure 3) is significantly smaller than observed by DAMA (∼1 event/kg/day/keV, Figure 4). The broad peak at 40-100 keV is due to the decay of ^{129}I. The event rate in the peak is consistent with that reported by the DAMA Collaboration [2]. In an attempt to make the simulated spectrum compatible with the measurements we examined an effect of assuming higher concentration of some radioactive isotopes compared to the reported values.

The dashed curve in Figure 3 shows the spectrum calculated assuming higher concentrations of certain isotopes inside the crystals compared to the values used in Figure 2. An important feature of this spectrum is the presence of a peak from ^{210}Pb decay at ∼40-50 keV (β-decay accompanied by gamma-rays and X-rays) which dominates over ^{129}I decay. The dotted curve in Figure 3 shows the spectrum calculated using higher contaminations of potassium in the PMTs (12000 ppm or 1.2%) and in the crystals (20 ppb) to account for the flat part of the measured spectrum and absolute rate at low energies (including the peak at about 3 keV). An important feature of any spectrum dominated by external source, is the presence of a broad back-scatter peak at about 150 keV. However, the measured spectrum at 40-80 keV (for two individual crystals) can fully be explained by the sum of the ^{129}I and ^{210}Pb decays leaving too little room for the back-scattered gamma-rays.

Figure 4 shows simulated spectra below 10 keV with 'optimised' concentrations to match the measured spectrum at low energies. The simulated spectrum is compared with the measurements with and without signal subtraction. The minimum at 1.5-2 keV that should be present in the background spectrum due to the contribution of the WIMP signal to the observed event rate, is not seen in the simulated spectra independently of the assumptions made about radioactive

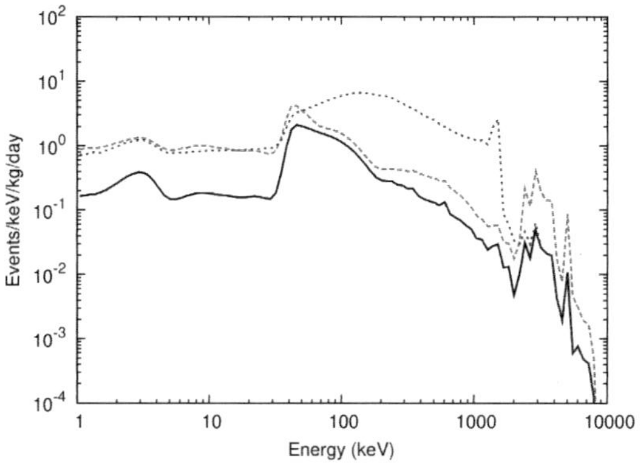

Figure 3. Combined spectra of energy depositions in the NaI crystals from radioactive background originated in the crystals and in the PMTs. Solid curve: PMTs – 30 ppb of ^{238}U, 30 ppb of ^{232}Th, 60 ppm of natural K in 100 g of the PMT envelope; NaI – 5 ppt of ^{238}U, 5 ppt of ^{232}Th, 10 ppb of natural K, 0.2 ppt of ^{129}I and 6.46×10^{-14} ppb of ^{22}Na. Dotted curve: PMTs – U/Th - the same as for solid curve, 1.2% of K; NaI – U/Th - the same as for solid curve, 20 ppt of K. Dashed curve: PMTs – the same as for solid curve; NaI – 20 ppb of natural potassium, 40 ppt of ^{238}U, 20 ppt of ^{232}Th, other isotopes - the same as for solid curve.

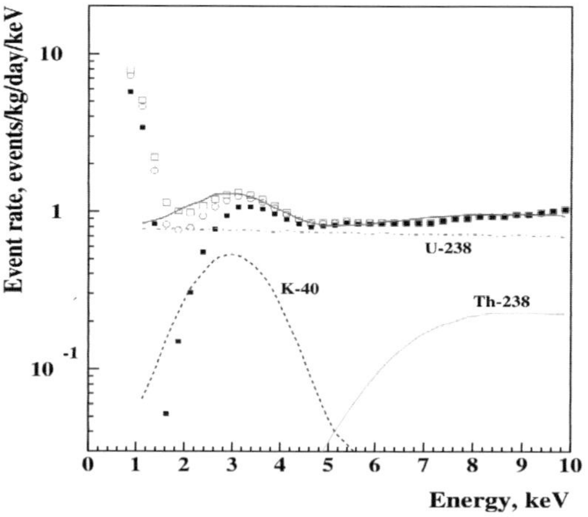

Figure 4. Simulated spectra of energy depositions in the NaI crystals from internal sources at low energies assuming 'optimised' concentrations of isotopes in the crystals: 40 ppt of ^{238}U, 20 ppt of ^{232}Th, 20 ppb of natural potassium, 0.2 ppt of ^{129}I. The measured spectrum without signal subtraction (open squares) and with signal subtraction (60 GeV mass WIMPs with $\sigma_{SI} = 7 \times 10^{-6}$ pb – filled squares, and $\sigma_{SI} = 2 \times 10^{-6}$ pb – open circles) are also shown.

contaminations in different materials in and around the detectors. Even for a model with small cross-section, the measured spectrum minus signal (open circles in Figure 4) does not match the predictions at 1.5-2.5 keV. A better match between the simulated background spectrum and the measurements minus signal would require a very small ratio of non-modulated to modulated parts of the signal spectrum (certainly less than a factor of 10 at 2-4 keV).

3. Conclusions

We have shown that any specific source of background radiation has a characteristic spectrum. We found that it is hard to match the measured spectrum of DAMA/LIBRA at low energies (even without signal) with the simulated background assuming measured concentrations of radioactive isotopes in NaI crystals. The assumption of higher concentrations of certain isotopes, although improving agreement with the low-energy part of the spectrum, results in some features that are not seen by DAMA at high energies. If a signal is present, then the background (measured spectrum minus signal) should have a minimum at about 1.5-2 keV that is not present in the simulated spectra. We do emphasise that any interpretation of the annual modulation of the event rate observed by DAMA, as dark matter signal, is incomplete without clear understanding of the background spectrum. Such integrated analysis would significantly restrict the dark matter models capable of explaining the modulation effect.

[1] R. Bernabei et al. *Eur. Phys. J. C* **56** (2008) 333; arXiv:0804.2741v1 [astro-ph].
[2] R. Bernabei et al. *Nucl. Instrum. & Meth. in Phys. Res. A*, **592** (2008) 297; arXiv:0804.2738v1 [astro-ph].
[3] S. Agostinelli et al. (GEANT4 Collaboration), *Nucl. Instrum. & Meth. in Phys. Res. A*, **506** (2003) 250.

Technical aspects and dark matter searches

R Bernabei [1,2], P Belli [2], F Cappella [3,4], R Cerulli [5], C J Dai [6], A d'Angelo [3,4], H L He [6], A Incicchitti [4], H H Kuang [6], X H Ma [6], F Montecchia [2,7], F Nozzoli [1,2], D Prosperi[3,4], X D Sheng[6] and Z P Ye[6,8]

[1] Dip. di Fisica, Università di Roma "Tor Vergata", I-00133 Rome, Italy
[2] INFN, sez. Roma "Tor Vergata", I-00133 Rome, Italy
[3] Dip. di Fisica, Università di Roma "La Sapienza", I-00185 Rome, Italy
[4] INFN, sez. Roma, I-00185 Rome, Italy
[5] Laboratori Nazionali del Gran Sasso, I.N.F.N., Assergi, Italy
[6] IHEP, Chinese Academy, P.O. Box 918/3, Beijing 100039, China
[7] Laboratorio Sperimentale Policentrico di Ingegneria Medica, Università degli Studi di Roma "Tor Vergata"
[8] University of Jing Gangshan, Jiangxi, China

E-mail: `rita.bernabei@roma2.infn.it`

Abstract. A variety of detectors has been proposed for dark matter direct detection, but most of them – by the fact – are still at R&D stage. In many cases, it is claimed that the lack of an adequate detectors' radio-purity might be compensated through heavy uses of MonteCarlo simulations, subtractions and handlings of the measured counting rates, in order to claim higher sensitivity (just for a particular scenario). The relevance of a correct evaluation of systematic effects in the use of MonteCarlo simulations at very low energy (which has always been safely discouraged in the field so far) and of multiple subtractions and handling procedures applied to the measured counting rate is shortly addressed here at some extent. Many other aspects would also deserve suitably deep investigations.

In this paper some arguments presented at the TAUP09 conference will be shortly summarized. More details, tables and figures can be found in the slides at the conference site[1].

Let us firstly comment the possibility of reliable evaluations of the background contributions at the keV energy region in the field of Dark Matter searches. As well known, it has been generally discouraged this procedure in the field of Dark Matter over more than twenty years. In fact, the estimation by a MonteCarlo simulation of the background component in the counting rate from the residual radioactivity requires a detailed knowledge of: i) the exact set-up geometry (detector or detectors' matrix, all materials, details of the assembling, of the shield layers, of the site, etc.); ii) the detector response function (e.g. energy resolution, α/β ratio, channeling, etc.); iii) the nature, the position and the concentration of all the existing radioactive contaminants; iv) etc.. Unfortunately, apart from the geometrical layout of the set-up that are generally well known by people inside the experimental group, all the other quantities necessary in the MonteCarlo simulation require dedicated measurements. Moreover, there are some quantities (such as concentration of residual contaminants, etc.) that can be poorly known and just upper/lower limits are available; in some cases these quantities can be even totally unknown. As an example, the experimental energy resolution as a function of the energy and the energy scale should be measured/verified down to the energy threshold, as done e.g. by the DAMA/LIBRA experiment where they are continuously measured by external/internal known sources from MeV down to the energy threshold [2]. On the contrary, in other experiments these quantities are instead extrapolated from calibrations at much higher energy (as done e.g. by liquid noble gas set-ups, where the energy threshold and the few keV energy scale are generally unproven, also because of position dependence, of non-uniform signal collection, etc.; see e.g. [3]). Regarding the presence of residual contaminants in the set-up, generally only limits on the contributions of the "standard" contaminants are given; these limits forbid any reliable estimation of the background (being unknown the exact values) and cannot be obviously exceeded (see also later). Moreover, possible presence of many non-standard contaminants should be also included. In addition, the MonteCarlo simulation also depends on the precise location of all the contaminants – that is generally unknown even for the "standard" ones – in complex set-ups. The situation is more complex for multi-detectors set-up and when the energy distribution refers to events where each detector has all the others in anticoincidence (*single-hit* events). Thus, it is trivial to conclude that a reliable precise simulation of the background counting rate – in particular at keV energy region – is not univocally determined and is a quite impossible task.

In addition, beyond the fore-mentioned arguments, we need to take into account that a MonteCarlo code cannot manage all the possible low energy atomic physical processes. This argument is still subject of improvements; in fact, as an example, non-negligible differences are also obtained by different versions of the same MonteCarlo code [4].

Some instructive examples are given by the trails in MonteCarlo simulations in ref. [5]. As shown there, these simulations noticeably differ from the measured energy distributions in the cases of XENON-10 and of ZEPLIN-III. In fact, they predict twice the measured rate for XENON-10 near 200 keV, and more than one order of magnitude the measured rate for ZEPLIN-III in the MeV range [5]. The same approach has been also pursued by the same authors [5] trying a MonteCarlo simulation of the background in the DAMA/LIBRA set-up. In particular, apart from errors in the details of the set-up geometry reconstruction and in the multiple-hit definition, many crude and arbitrary approximations in the nature and in the location of the residual contaminants have been arbitrarily assumed; in fact, e.g.: i) only standard contaminants, ii) only unbroken chains, iii) only uniform location of contaminants in the detectors, etc. have been taken into account. As a result of this rough, partial and arbitrary approach, the predicted rate has been estimated within a factor 10 lower than the measured one. Instead of refining the quality of the simulation or reasonably recognizing the impossibility of precise determination, the authors just pursued the exercise of arbitrarily increasing "by hand" the assumed values of the contaminants at levels much larger than the measured experimental limits [5]. This also implies an overestimate of the background in higher energy region with respect to the measured experimental rate. In conclusion, although the arbitrary and the erroneous adopted procedures, these authors do not succeed in reproducing either the low or the high energy spectra. Nevertheless, as a conclusion of this arbitrary exercise [5], this artificially-boosted simulated spectrum has been subtracted by the measured one, attempting to obtain a limit for the unmodulated Dark Matter signal component. This example shows how subtraction procedures using MonteCarlo simulations in the few keV energy region can give rise to erroneous conclusions; thus, any constraint on Dark Matter signal on this basis would be an artefact.

Furthermore, let us also note that the measured spectra e.g. of the existing/past NaI(Tl) detectors (such as e.g. ANAIS, Frejus, NAIAD, ELEGANT, etc...) do not support even the shape presented in ref. [5]. In addition, well different counting rates at keV energy region are present even for detectors of the same experimental group, as e.g. the case of NAIAD in 1996 and in 2003 [1]. In conclusion, it does not exist an unique recipe for a precise and reliable MonteCarlo simulation of whatever set-up, and for NaI(Tl) in particular.

Let us finally remind that a safer approach has been presented in this conference by DAMA collaboration; this shows that enough space is present in the measured counting rate of the DAMA/LIBRA keV energy spectrum for the unmodulated component of Dark Matter signal [1, 6].

In the second part of this contribution [1], problems related to the application of multiple subtraction procedures of the measured counting rate, as pursued by experiments trying to identify the presence of recoil nuclei in the measured energy spectrum, have been summarized. In fact, many of the existing Dark Matter candidates – also in the WIMP class – can give rise to signals that either have totally an electromagnetic nature (see e.g. [7, 8, 9]) or involve electromagnetic signals associated to nuclear recoils (see e.g. [10, 11]); obviously, approaches that are based on multiple subtraction procedure of the electromagnetic component of the counting rate are blind to similar scenarios. Moreover, well known side processes exist for recoils (such as recoils induced by neutrons, fission fragments, end-range alphas, surface electrons, etc.). This approach is generally pursued when the detectors suffer from a not-suitable radiopurity level in the sensitive target-material and in the surroundings.

Those activities generally apply a large number of cut procedures to the data; each one is affected by non-negligible systematic errors which are usually not suitable quantified. As an example, the XENON-10 experiment applies more than 10 different cuts to the data; the experiment collects $\sim 10^4$ events but only 10 are claimed to survive to the cuts and handling procedures [12]. Very high reduction factors following applied multiple cuts are dangerous because of the difficult precise estimate of all the involved systematics. For example, it has been shown in ref. [13] that ZEPLIN-I has claimed a sensitivity 3 orders of magnitude larger than the one properly obtained when accounting for systematics. Thus, the robustness of some results appeared in the "race for the best exclusion-plot" (valid just in a single set of assumptions for a certain kind of WIMP) should be considered "cum grano salis". For example, for some liquid noble gas set-ups, apart from the robustness of the applied cuts themselves, the very low energy scale and the energy threshold are determined by extrapolating from calibrations at much higher energy and applying some kind of corrections for relevant non-uniformity of the detector's response; with a light yield of about 2.2 photoelectrons/keV and not specific calibrations even $\approx$ 1.5-2 keV electron equivalent is claimed as energy threshold [12, 1]. Considering the energy threshold dependence of the exclusion plots, up to several orders of magnitude differences can be present between claimed and realistic evaluation of the experimental sensitivity.

Another crucial aspect is the proper accounting for the existing experimental and theoretical uncertainties in the calculation and in the comparison of experiments using different target materials and approaches.

References

[1] F. Nozzoli, presentation given at this conference (TAUP 2009) http://taup2009.lngs.infn.it/slides/jul3/nozzoli.pdf or http://people.roma2.infn.it/ dama/pdf/nozzoli_taup09.pdf.
[2] R. Bernabei et al., Nucl. Instr. & Meth. A 592 (2008) 297.
[3] R. Bernabei et al., ISBN 978-88-95688-12-1, pages 1-53 (2009) Exorma ed. (arXiv:0806.0011[astro-ph]).
[4] see e.g. R.M. Kippen, New Astr. Rev. 48 (2004) 221.
[5] V. Kundryavtsev, presentation given at this conference (TAUP 2009).
[6] R. Bernabei et al., Eur. Phys. J. C 56 (2008) 333.
[7] R. Bernabei et al., Int. J. Mod. Phys. A 21 (2006) 1445.
[8] R. Bernabei et al., Phys. Rev. D 77 (2008) 023506.
[9] R. Bernabei et al., Mod. Phys. Lett. A 23 (2008) 2125.
[10] R. Bernabei et al., Int. J. Mod. Phys. A 22 (2007) 3155.
[11] R. Bernabei et al., Eur. Phys. J. C 53 (2008) 205.
[12] J. Angle et al., Phys. Rev. Lett. 100 (2008) 021303.
[13] A. Benoit et al., Phys. Lett. B 637 (2006) 156.

Inert Doublet Model and DAMA: elastic and/or inelastic dark matter candidates

Chiara Arina

Service de Physique Théorique, Université Libre de Bruxelles, CP225, Bld du Triomphe, 1050 Brussels, Belgium

E-mail: `carina@ulb.ac.be`

Abstract. We investigate the compatibility of the DAMA results with the other direct detection experiments in the framework of a specific dark matter model, the Inert Doublet Model. In this model there are three different regimes for which the dark matter candidate is either light, heavy or in the middle mass range. We present our analysis for the light WIMPs cases with elastic interaction on nuclei. The heavy mass regime is instead an inelastic dark matter candidate, while in the middle mass range both interactions are considered. In the three regimes we exhibit regions of the parameter space in agreement with all the experiments, even though they appear to be significantly constrained.

1. Overview on the Inert Doublet Model

The Inert Doublet Model (IDM) is a simple extension of the Standard Model with dark matter (DM), consisting in two Higgs doublets and a Z_2 symmetry and it has been widely discussed in the literature. Here we focus our attention on the direct detection phenomenology in the light of the DAMA experiment.

The usual Higgs doublet is denoted by H_1, while the extra, or inert doublet, H_2, is the only field of the model that is odd under the Z_2 symmetry. This ensures the stability of the lightest member of H_2, which will be a DM candidate, and prevents from flavor changing neutral currents [1]. We will assume that Z_2 is not spontaneously broken and that H_2 does not develop a vacuum expectation value. We write $H_2 = (H^+ \quad (H_0 + iA_0)/\sqrt{2})^T$, similarly to the ordinary Higgs doublet, and $H_1 = (h^+ \quad (v_0 + h + iG_0)/\sqrt{2})^T$. The potential is written as

$$\begin{aligned} V(H_1, H_2) &= \mu_1^2|H_1|^2 + \mu_2^2|H_2|^2 + \lambda_1|H_1|^4 + \lambda_2|H_2|^4 \\ &+ \lambda_3|H_1|^2|H_2|^2 + \lambda_4|H_1^\dagger H_2|^2 + \frac{\lambda_5}{2}\left[(H_1^\dagger H_2)^2 + h.c.\right] . \end{aligned} \quad (1)$$

After electroweak symmetry breaking, $\langle H_1 \rangle = v = -\mu_1^2/\lambda_1 = 246$ GeV, the masses of the physical scalar fields are given by

$$\begin{aligned} M_h^2 &= 2\lambda_1 v^2 , \\ M_{H_0}^2(M_{A_0}^2, M_{H^+}^2) &= \mu_2^2 + \lambda_{H_0}(\lambda_{A_0}, \lambda_{H_c})v^2 , \end{aligned} \quad (2)$$

with $\lambda_{H_c} \equiv \lambda_3/2$ and $\lambda_{H_0,A_0} \equiv (\lambda_3 + \lambda_4 \pm \lambda_5)/2$. We will consider H_0 to be the DM candidate (*i.e.* $\lambda_5 < 0$) though the results would be exactly the same for A_0 changing the sign of λ_5. In the next section we discuss two particular limits of the mass relations, leading to an elastic scenario with a light DM candidate, and to an inelastic scenario in the multi-TeV DM mass range.

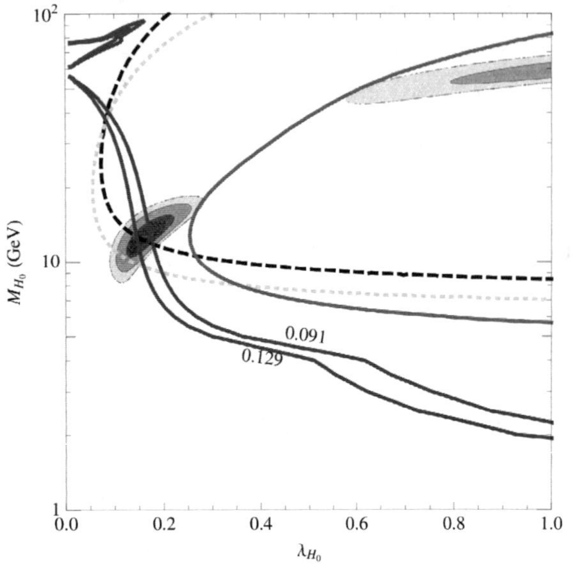

Figure 1. Elastic scenario: Allowed regions in the plane $|\lambda_{H_0}| - M_{H_0}$ consistent with the DAMA annual modulation signal at 90%, 99% and 99.9% CL. The solid (brown) curve is the total unmodulated rate for DAMA. The dashed (black) curve is the exclusions contour of CDMS-Ge, while the dotted (green) corresponds to XENON10, both at 99% CL. The velocity distribution is a standard Maxwellian distribution with escape velocity $v_{esc} = 650$ km/s. The solid (red) lines delimit the WMAP bounds at 3 σ CL.

2. Direct detection analysis – IDM and DAMA

The former DAMA/NaI [2] and present DAMA/LIBRA [3] experiments are designed to detect the dark matter recoil off nuclei through the model independent annual modulation signature due to the motion of the Earth around the Sun. The experimental results obtained by DAMA/LIBRA combined with the ones of DAMA/NaI show a modulated signal with a confidence level (CL) of 8.2 σ [3]. The analysis of the DAMA data is based on a goodness-of-fit method and considers the full 36 bins. In addition we take into account the constraint on the total event rate from the unmodulated signal. All the other collaborations for direct searches have reported no positive detection of DM recoils, setting upper bounds on the allowed dark matter scattering cross-section on nucleon. The exclusion limits are calculated with the standard Poisson statistics. A detailed discussion on the analysis procedure and on the astrophysical factors can be found in [4] and references therein.

2.1. Low mass regime ($M_{H_0} \ll m_W$) – elastic dark matter

If we consider $M_{A_0} = M_{H^+}$ the theory posses a global $SU(2)$ symmetry. Consequently it is possible to decouple the extra neutral component in such a way that $M_{H_0} \ll M_{A_0} \simeq M_{H^+}$, where M_{H_0} is now acting as singlet scalar dark matter [5]. The H_0 is a light DM candidate and its interaction on nucleon is through spin-independent (SI) elastic scattering exchanging an Higgs boson. In figure 1 we show the allowed regions compatible with the DAMA modulated signal, as a function of the model parameters $|\lambda_{H_0}|, M_{H_0}$. The strongest exclusion limits come from Xenon10 [6], CDMS-Ge [7] and the total unmodulated DAMA rate. Also shown is the prediction for the relic density; we underline the fact that no further adjustment of the parameters is needed to fix the relic abundance. The compatibility between DAMA and the other exclusion experiments, in this mass range, is ameliorated by the channelling on Iodine [8, 9], even though the model appears severely constrained. We remark however that the nuclear and astrophysical uncertainties can play a role in opening the available parameter space, see [4].

2.2. High mass regime ($M_{H_0} \gg m_W$) – inelastic dark matter

The second limit arises from $\lambda_5 \to 0$, in which case the neutral particles are degenerate [11] and the theory is invariant under a $U(1)_{\rm PQ}$ symmetry. The presence of an enhanced symmetry allows to consider a small mass splitting $\delta = M_{A_0} - M_{H_0} \ll M_{H_0}$ (corresponding to $\lambda_5 \simeq$

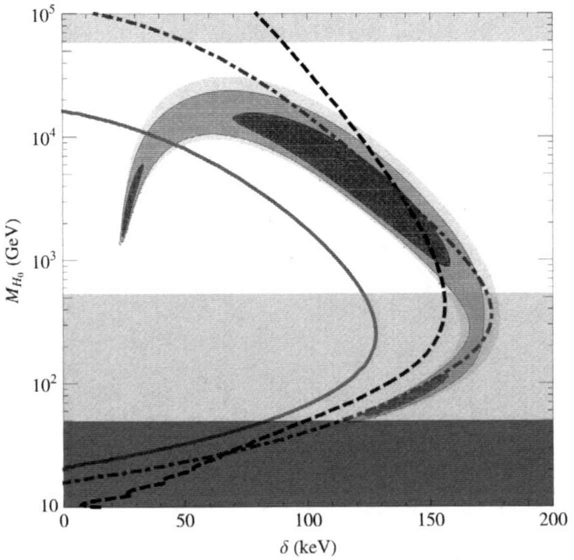

Figure 2. Inelastic scenario: Allowed regions in the plane $\delta - M_{H_0}$ consistent with the DAMA annual modulation signal at 90%, 99% and 99.9% CL. The solid (brown) curve is the DAMA total unmodulated rate. The dashed (black) curve is the exclusion contour of CDMS-Ge, while the dot-dashed (blue) corresponds to CRESST-II, both at 99% CL. The velocity distribution is as figure 1, with escape velocity v_{esc} = 600 km/s. The brown region (dark grey bottom) is excluded by LEPI measurements on the Z decay width, the grey region (top) denotes the unitarity limit [10, 11]. The red (light grey bottom) region represents subdominant dark matter halo components.

$\mathcal{O}(10^{-7} - 10^{-5})$) between the neutral components in a technically natural way. This small mass splitting leads to inelastic dark matter [12], with a SI cross-section on nucleon mediated by the Z boson. The predictions for the model parameters δ and M_{H_0} are shown in figure 2. The most constraining exclusion limits come from Cresst-II [13] and CDMS-Ge, while the DAMA total unmodulated rate is less restrictive respect to the elastic scenario. There exists a whole range of candidates, between $M_{H_0} \sim 535$ GeV and $M_{H_0} \sim 20$ TeV, which are compatible with DAMA and all other direct detection experiments and which have the relic density within the WMAP [14] bounds. The viable candidates in the mass range 50 – 100 GeV (light red/grey bottom) have a cosmic abundance below the WMAP value. Either they can be considered as subdominant components of the dark matter halo either one can infer an asymmetry mechanism in the early universe, provided by the smallness of the λ_5 coupling, to account for the suitable relic density amount. More details on the inelastic candidate and on the asymmetry are discussed in [4].

Acknowledgments
This work was supported by IISN and Belgian Science Policy (IAP VI/II).

References
[1] Deshpande N G and Ma E 1978 *Phys. Rev.* **D18** 2574
[2] Bernabei R *et al.* (DAMA) 2000 *Phys. Lett.* **B480** 23–31
[3] Bernabei R *et al.* (DAMA) 2008 *Eur. Phys. J.* **C56** 333–355 (*Preprint* arXiv:0804.2741)
[4] Arina C, Ling F S and Tytgat M H G 2009 (*Preprint* 0907.0430)
[5] Andreas S, Hambye T and Tytgat M H G 2008 *JCAP* **0810** 034 (*Preprint* arXiv:0808.0255)
[6] Angle J *et al.* (XENON) 2008 *Phys. Rev. Lett.* **100** 021303 (*Preprint* arXiv:0706.0039)
[7] Ahmed Z *et al.* (CDMS) 2009 *Phys. Rev. Lett.* **102** 011301 (*Preprint* arXiv:0802.3530)
[8] Bernabei R *et al.* 2008 *Eur. Phys. J.* **C53** 205–213 (*Preprint* arXiv:0710.0288)
[9] Bottino A, Donato F, Fornengo N and Scopel S 2008 *Phys. Rev.* **D77** 015002 (*Preprint* arXiv:0710.0553)
[10] Griest K and Kamionkowski M 1990 *Phys. Rev. Lett.* **64** 615
[11] Hambye T, Ling F S, Lopez Honorez L and Rocher J 2009 *JHEP* **07** 090 (*Preprint* 0903.4010)
[12] Tucker-Smith D and Weiner N 2001 *Phys. Rev.* **D64** 043502 (*Preprint* hep-ph/0101138)
[13] Angloher G *et al.* 2008 (*Preprint* arXiv:0809.1829)
[14] Komatsu E *et al.* (WMAP) 2009 *Astrophys. J. Suppl.* **180** 330–376 (*Preprint* 0803.0547)

Relevance of Ion-Channeling for Direct DM Detection

Graciela B. Gelmini

Department of Physics and Astronomy, UCLA, 475 Portola Plaza, Los Angeles, CA 90095, USA

E-mail: gelmini@physics.ucla.edu

Abstract. The channeling of the recoiling nucleus in crystalline detectors after a WIMP collision would produce a larger scintillation or ionization signal in direct detection experiments than otherwise expected. I present estimates of the importance of this effect for the total direct detection rate and the daily modulation of the signal using analytic models produced in the 1960's and 70's to describe the effects of channeling and blocking in crystals.

Channeling and blocking effects in crystals refer to the orientation dependence of positively charged ion penetration in crystals [2]. In the "channelling effect" ions incident upon a crystal along symmetry axis and planes suffer a series of small-angle scattering that maintain them in the open"channels" in between the rows or planes of lattice atoms and thus penetrate much further into the crystal than in other directions. Channeled incident ions do not get close to lattice sites, where they would experience large-angle scatterings. The "blocking effect" consists in a reduction of the flux of ions originating in lattice sites along symmetry axis and planes, creating what is called a "blocking dip" in the ions exiting from a thin enough crystal as function of the angle with respect to the symmetry axis or plane. These effects were first observed in the 1960's and are currently extensively used in crystallography, in the study of lattice disorder, ion implantation, surfaces, interfaces and epitaxial layers, short nuclear lifetimes etc.

The channeling effect in NaI (Tl) was first observed in 1973 by Altman *et al.* [3] who proved that channeled ions produce more scintillation light. The reason is that they loose most of their energy via electronic stopping rather than nuclear stopping. In 2007 Drobyshevski [4] and the DAMA collaboration [5] pointed out the relevance of this effect for direct detection experiments using NaI (Tl) crystals: when Na or I ions recoiling after a collision with a WIMP move along crystal axes and planes, their quenching factor is approximately $Q = 1$ instead of $Q_I = 0.09$ and $Q_{Na} = 0.3$. In 2008 Avignone, Creswick and Nussinov [6] raised the prospect of a daily modulation of the DM signal in direct detection due to channeling. In fact Earth's daily rotation makes the WIMP wind change direction with respect to the crystal, which produces a daily modulation in the measured recoil energy (equivalent to a modulation of the factor Q).

My collaborators, Nassim Bozorgnia and Paolo Gondolo, and I [1] are using classical analytic models of channeling developed in the 1960's and 70's, in particular Lindhard's model [7] to calculate the fraction of channeled ions as function of their energy and the expected daily modulation amplitudes due to channelling in NaI, Si and Ge. The results presented here are preliminary.

We use the continuum string and plane model, in which the screened Thomas-Fermi potential is averaged over a direction parallel to a row or a plane. This averaged potential U is considered to be uniformly smeared out along the row or plane of atoms, what is a good approximation if the propagating ion interacts with many lattice atoms in the row or plane by a correlated series of many consecutive glancing collisions with lattice atoms. Just one row, or one plane, is considered. We use the "Radon transform" of the WIMP velocity distribution (a truncated Maxwellian) to write the ion recoil momentum spectrum in terms of the average velocity of the WIMP's with respect to the laboratory, the velocity dispersion and escape speed [8]. We orient the crystal with respect to the galaxy and use the Hierarchical Equal Area isoLatitude Pixelization (HEALPix) method [9] to compute the integrals over the recoil direction. The HEALPix method uses an algorithm to deal with the pixelization of data on a sphere (used e.g. by cosmic microwave background experiments). It is useful to compute integrals over direction by dividing the surface of a sphere into many pixels, computing the integrand at each pixel (i.e. each direction, see Fig. 1) and finally summing up the value of all pixels over the sphere.

Our temperature dependent analytic results reproduce the channeled fraction of incoming Na ions as function of their energy given by the DAMA collaboration [5] (which does not include temperature effects), even with dechanneling effects included (shown in Fig. 2). We included only the dechanneling due to the first interaction with Tl doping atoms. A naive geometrical blocking brings our fractions to zero at very small energies in Fig. 2. In these calculations, however, the fact that the recoiling nuclei originate in lattice sites (and are not incident upon the crystal) thus blocking effects are important, was not properly taken into account.

Figure 1. HEALPix map of directions of 50 keV Na ions incident upon a NaI crystal showing in red those channeled. The axial and planar channels are clearly seen.

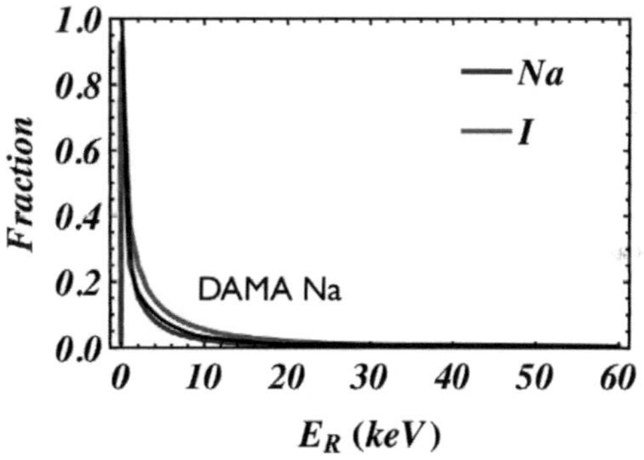

Figure 2. Preliminary results for the fraction of channeled Na and I ions incident upon a NaI crystal, including dechanneling, compared with the DAMA estimate (in black).

As argued originally by Lindhard [7], in a perfect lattice and in the absence of energy-loss processes the probability of a particle starting from a lattice site to be channeled would be zero. It is well known in statistical mechanics that the probability of particle paths related by time reversal is the same. In optics, if a source of radiation and a point of observation are interchanged, the intensity of the light measured at the new place of observation is the same as the old. In the same way, if the probability of an incoming channeled ion to get very close to a lattice site (where it would suffer a large angle scattering) is zero (channeled ions only undergo small angle correlated scatterings), the probability of the same ion moving in the time-reversal path, i.e. starting at a nuclear site and ending going out of the crystal in a path parallel to a channel, is zero too (this is Lindhard's *"Rule of Reversibility"*). However, any departure of the actual lattice from a perfect lattice, e.g. due to vibrations of the atoms in the lattice, violates the conditions of this argument and allow for some of the recoiling lattice nuclei to be channeled.

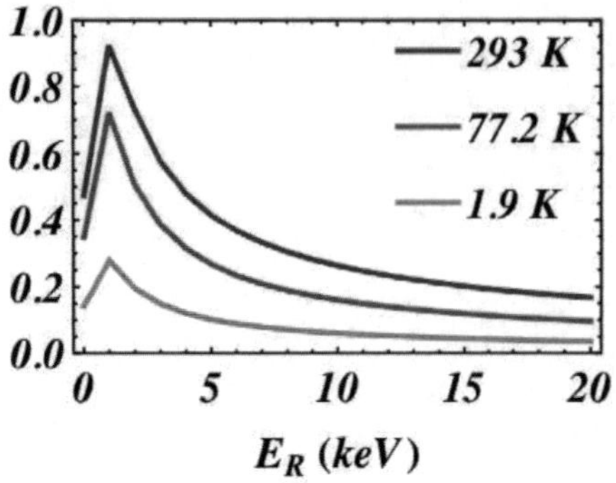

Figure 3. Preliminary results for the fraction of channeled recoiling Na nuclei in NaI (no including dechanneling). The effect is strongly temperature dependent.

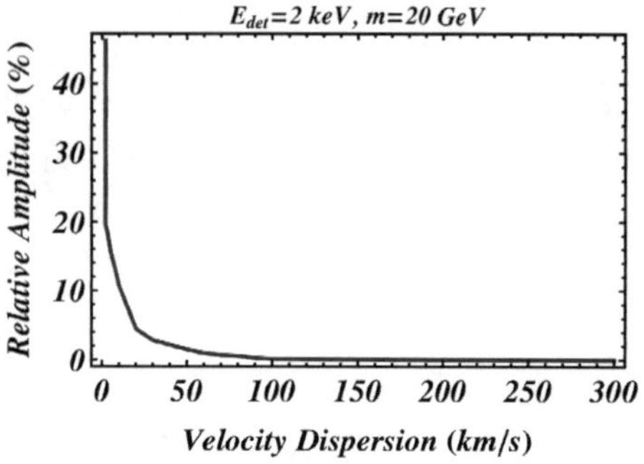

Figure 4. An example of the dependence of the daily modulation amplitude due to channeling as function of the WIMP velocity dispersion in the standard halo (changed to arbitrary unphysical values).

The channeling of particles emitted at lattice sites, such as protons scattered at large angles, due to lattice vibrations was measured and already understood in the 70's [10, 11]. Due to lattice vibrations the recoiling nucleus has a probability (which we take to be Gaussian) of being at a certain distance of the row of plane of atoms. The higher the temperature the higher the probability that the recoil starts outside a lattice site. We use the method of Komaki and Fujimoto [10] to compute the fraction of channeled recoils. They use the conservation of transverse energy of the recoiling ion with initial energy E_R, $E_\perp = E\sin^2\psi + U = E_R\sin^2\psi_R + U_{\text{initial}}$, valid in the continuum approximation. Here ψ is the angle between the tangent of the trajectory and the crystal axis and ψ_R is the recoil momentum after the collision. They distinguish two types of trajectories: those which can pass over the potential barrier U_{max} separating two parallel channels in the crystal (thus particles would not be trapped within one channel) and those which cannot pass over the barrier and are, thus, channeled. Using this approach, and taking U_{max} to be the value of the potential at the Thomas-Fermi screening distance, we preliminary find the fractions of channeled ions shown in Fig. 3, which clearly increase with increasing temperature.

Finally, we find that the amplitude of the daily modulation of the signal depends strongly on the velocity dispersion of WIMPs (see Fig. 4) and for the standard halo model is smaller than 0.1%. However it may be much larger for possible halo components with a smaller velocity dispersion, such as a thick disk.

Acknowledgments This work was supported in part by the US Department of Energy Grant DE-FG03-91ER40662, Task C. I thank S. Nussinov and F. Avignone for useful discussions.

References

[1] N. Bozorgnia, G. Gelmini and P. Gondolo, in preparation.
[2] D. S. Gemmell, Rev. Mod. Phys. **46**, 129 (1974).
[3] R. Altman, H. B. Dietrich, R. B. Murray, and T. J. Rock, Phys. Rev. B **7**, 1743 (1973)
[4] E. M. Drobyshevski, Mod. Phys. Lett. A **23** (2008) 3077 [arXiv:0706.3095 [physics.ins-det]].
[5] R. Bernabei *et al.*, Eur. Phys. J. C **53**, 205 (2008) [arXiv:0710.0288 [astro-ph]].
[6] F. T. Avignone, R. J. Creswick and S. Nussinov, arXiv:0807.3758 [hep-ph].
[7] J. Lindhard, Kongel. Dan. Vidensk. Selsk., Mat.-Fys. Medd.; **34** No. 14 (1965).
[8] P. Gondolo, Phys. Rev. D **66**, 103513 (2002).
[9] K. M. Górski *et al.*, ApJ **622** 756 (2005).
[10] K. Komaki and F. Fujimoto, Phys. Stat. Sol. (a) **2** 875 (1970).
[11] K.Komaki *el al.*, Phys. Stat. Sol. (a) **4** 495 (1971).

Elastic and Inelastic LSP-nucleus Scattering on Medium-heavy Nuclei

P Toivanen, M Kortelainen, J Suhonen and J Toivanen
Department of Physics, P.O. Box 35 (YFL), FI-40014 University of Jyväskylä, Finland
E-mail: jouni.suhonen@phys.jyu.fi

Abstract. Elastic and inelastic scattering rates of the lightest supersymmetric particle (LSP) off nuclei are derived for the stable iodine, xenon and cesium detectors. The parameters of the supersymmetric theory are decoupled from the nuclear structure. The involved nuclear wave functions are calculated in the nuclear shell model by using a realistic effective two-nucleon interaction in a realistic valence space. By using fitted nuclear gyromagnetic factors we successfully reproduce the relevant spectroscopic data in the discussed nuclei.

1. Introduction

The dark-matter component of the Universe exhausts some 30% of the energy balance and is mostly in the form of cold dark matter (CDM). The CDM is believed to consist of non-relativistic heavy particles called WIMPs (weakly-interacting massive particles). The WIMPs are believed to constitute the major component of dark matter in our own galactic halo. To access the nature of dark matter a large number of direct detection experiments are looking for recoil signals of the nucleus on which the WIMP has scattered elastically or inelastically. Our obtained results for the stable iodine, xenon and cesium nuclei are of interest e.g. for the DAMA [1, 2] (^{129}Xe and ^{127}I), KIMS [3] (^{127}I and ^{133}Cs), NAIAD [4] and COUPP [5] (^{127}I), ZEPLIN [6], XENON [7] and XMASS [8] (^{129}Xe and ^{131}Xe) experimental programs.

2. Nuclear-structure calculations

The large-scale shell-model calculations were performed in a valence space consisting of the single-particle orbitals $0g_{7/2}$, $1d_{5/2}$, $2s_{1/2}$, $1d_{3/2}$ and $0h_{11/2}$. This is the smallest realistic single-particle space to descripe the physics of the ^{127}I, ^{129}Xe, ^{131}Xe and ^{133}Cs dark-matter detector nuclei. For the nucleon-nucleon two-body interaction we used an effective interaction based on Bonn-CD G-matrix [9]. For the reason of the CPU time consumption the number of neutron configurations had to be truncated as described in [10, 11]. The resulting energy spectra are in reasonable correspondence with the experimental data [11].

Contrary to our earlier calculations [12] we have renormalized the magnetic dipole operator by introducing effective spin and orbital angular-momentum gyromagnetic factors (g factors in short). The four optimal effective g factors were determined by a fit to 11 known magnetic moments of the nuclei ^{127}I, ^{129}Xe, ^{131}Xe and ^{133}Cs. The use of the fitted effective g factors improves the calculated magnetic moments substantially both for the ground states and for the lowest excited states of the nuclei considered [11]. The results are superior to those of the earlier

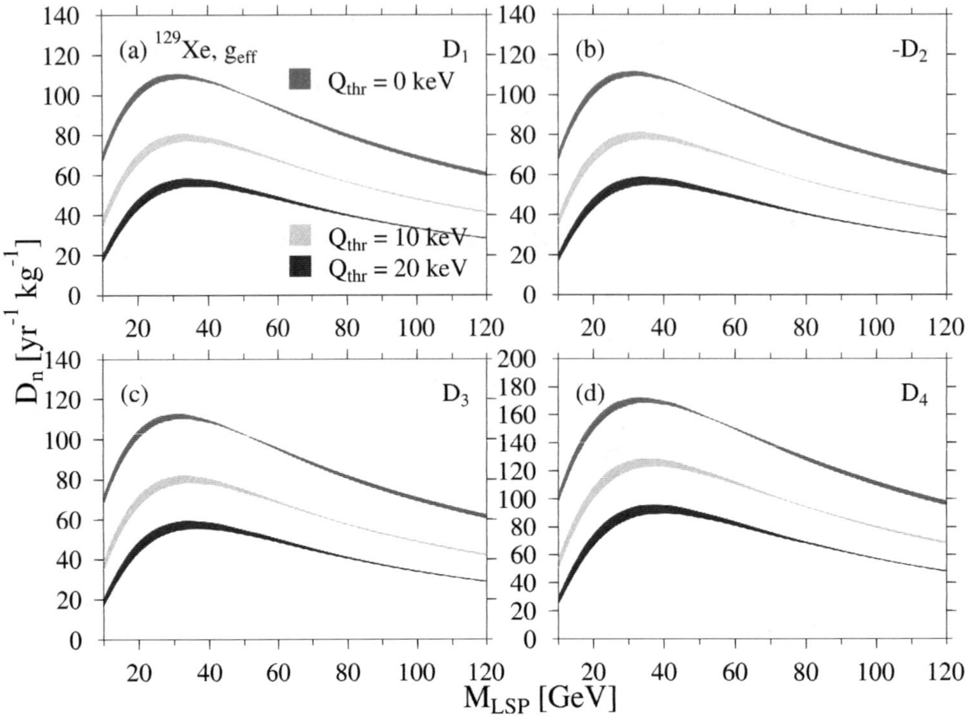

Figure 1. The D_n factors of (1), in units of $\mathrm{yr}^{-1}\mathrm{kg}^{-1}$, plotted as functions of the LSP mass for three different threshold energies in the case of elastic scattering of the LSP off a ^{129}Xe nucleus. Panel (a) shows D_1, panel (b) $-D_2$, panel (c) D_3, and panel (d) D_4. The thickness of the curves describes the annual modulation effect.

calculation [13]. Because of this success, we introduce effective proton and neutron spin operators with the renormalization factors $r_\mathrm{p} = 0.571$ for protons and $r_\mathrm{n} = 0.881$ for neutrons. Below we use these effective spin operators to evaluate the observables of the LSP-nucleus scattering.

3. Elastic and inelastic LSP-nucleus scattering rates

The event rate of the LSP scattering off an Earth-bound detector can be written as

$$\langle R\rangle = \left[\left(f_\mathrm{A}^0\right)^2 D_1 + 2f_\mathrm{A}^0 f_\mathrm{A}^1 D_2 + \left(f_\mathrm{A}^1\right)^2 D_3 + A^2\left(f_\mathrm{S}^0 - f_\mathrm{S}^1\frac{A-2Z}{A}\right)^2 D_4\right] m_\mathrm{det}[\mathrm{kg}]\ , \qquad (1)$$

where $m_\mathrm{det}[\mathrm{kg}]$ is the detector mass in units of kg, A is the mass number of the target nucleus and the f coefficients are specific to the chosen SUSY model [14]. The coefficients D_n are folded with the symmetric Maxwell–Boltzmann velocity distribution of the LSPs and are given in detail in [11], both for the elastic and inelastic channels. It should be noted that the spin-independent coefficient D_4, containing the nuclear form factor, vanishes for inelastic scattering. Thus only the spin-dependent coefficients $D_1 - D_3$ are active for both scattering channels.

As an example of the computed results in Fig 1 we plot the D_n in units of $\mathrm{yr}^{-1}\mathrm{kg}^{-1}$ and as functions of the LSP mass for the elastic scattering of the LSP off a ^{129}Xe nucleus. Here the detector threshold, Q_thr, is the lowest recoil energy detectable by the measurement apparatus. In Fig 2 the corresponding factors are plotted for the inelastic LSP-^{129}Xe scattering for $Q_\mathrm{thr} = 0$.

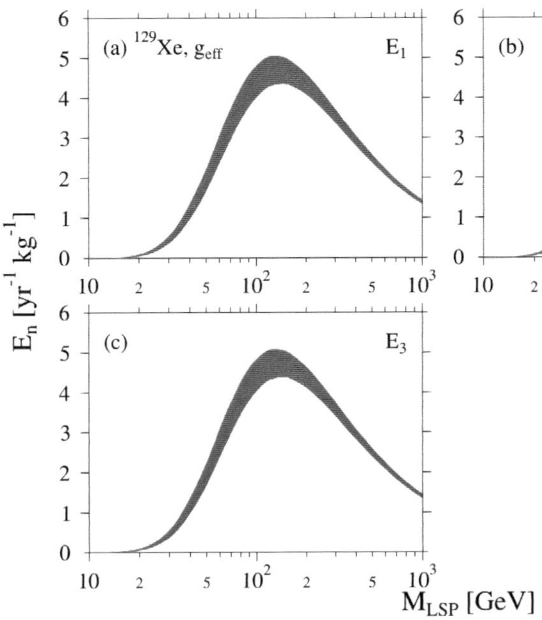

Figure 2. The D_n factors of (1), in units of $\text{yr}^{-1}\text{kg}^{-1}$, plotted as functions of the LSP mass for the inelastic scattering of the LSP off a ^{129}Xe nucleus. Panel (a) shows D_1, panel (b) $-D_2$, and panel (c) D_3. The thickness of the curves describes the annual modulation effect.

In [11] we have tabulated the D_n coefficients for various LSP masses and detector thresholds to enable easy use in the context of different SUSY parametrizations.

Acknowledgments

This work was supported by the Academy of Finland under the Finnish Center of Excellence Program 2006-2011 (Nuclear and Accelerator Based Program at JYFL).

References

[1] Bernabei R *et al.* 2000 *New J. Phys.* **2** 15.1
[2] Bernabei R *et al.* 2003 *Riv. Nuovo Cimento* **26N1** 1
[3] Lee H S *et al.* 2007 *Phys. Rev. Lett.* **99** 091301
[4] Spooner N J C *et al.* 2000 *Phys. Lett.* B **473** 330
[5] Behnke E *et al.* 2008 *Science* **319** 933
[6] Alner G J *et al.* 2005 *Astropart. Phys.* **23** 444
[7] Angle J *et al.* 2008 *Phys. Rev. Lett.* **100** 021303 .
[8] Abe Ko *et al.* 2008 *J. Phys.: Conf. Series* **120** 042022
[9] Hjorth-Jensen M, Kuo T T S and Osnes E 1995 *Phys. Rep.* **260** 125 ; M. Hjorth-Jensen, private communication.
[10] Toivanen P, Kortelainen M, Suhonen J and Toivanen J 2008 *Phys. Lett.* B **666** 1
[11] Toivanen P, Kortelainen M, Suhonen J and Toivanen J 2009 *Phys. Rev.* C **79** 044302
[12] Kortelainen M, Kosmas T S, Suhonen J and Toivanen J 2006 *Phys. Lett.* B **632** 226
[13] Ressell M T and Dean D J 1997 *Phys. Rev.* C **56** 535
[14] Holmlund E, Kortelainen M, Kosmas T S, Suhonen J and Toivanen J 2004 *Phys Lett.* B **584** 31

Journal of Physics: Conference Series **203** (2010) 012044 doi:10.1088/1742-6596/203/1/012044

ANAIS status report

J Amaré, S Borjabad, S Cebrián, C Cuesta, D Fortuño, E García, H Gómez, J Morales, Y Ortigoza, A Ortiz de Solórzano, C Pobes, J Puimedón, A Rodríguez, M L Sarsa and J A Villar

University of Zaragoza. C/Pedro Cerbuna 12, 50009. Zaragoza. Spain.

E-mail: `cpobes@unizar.es`

Abstract. We report the status of the ANAIS (Annual modulation with NAI Scintillators) experiment focusing on the developments towards the improvement of the background level of our detectors and the reduction of the energy threshold.

1. Introduction

ANAIS is the large scale conclusion of previous studies carried out with different prototypes by the University of Zaragoza group at the Canfranc Underground Laboratory (LSC). The complete experiment will use about 100 kg of NaI(Tl) crystals to study the expected annual modulation [1] in the galactic dark matter signal. A possible enlargement of target mass has been recently considered and funding asked for. DAMA/LIBRA [2], an experiment performed at the Gran Sasso National Laboratory, accumulating several annual cycles of data (also with NaI scintillators), obtained (and recently confirmed with larger statistics) a positive signal for the above mentioned annual modulation [3]. This result is difficult to reconcile with those obtained with other targets in different experimental approaches [4]-[9] but comparison is strongly model dependent. For that reason, ANAIS (that uses the same target and technique that DAMA/LIBRA) appears in the last roadmap of ApPEC/ASPERA as the experiment that would allow testing such result with an independent experimental set-up and in a model-independent way. For this goal, some of the experimental features of ANAIS should be improved and we are focusing efforts in the corresponding directions. Energy threshold below 2 keV should be achieved in order to improve sensitivity for the annually modulated WIMP signal. This goal is addressed by using new high quantum efficiency photomultiplier tubes and by developing new energy estimators at very low energies, as well as a better noise rejection. Background at low energy should be kept as low as possible. After characterization of the available NaI(Tl) crystals, the measured ^{40}K bulk crystal contamination appears as the most dangerous at the very low energies of interest for ANAIS and has forced ANAIS to look for more radiopure crystals, implying the development of new purification techniques.

2. ANAIS Status

We summarize the present status of the ANAIS experiment in the following topics:

- NaI(Tl) crystals radiopurity: In the case of NaI, radiopurity of the crystal is identified as one of the main issues to allow an important improvement in sensitivity. After characterization of existing ANAIS crystals (14 hexagonal 10 kg each NaI(Tl) detectors stored at LSC) background at low energy seems to be dominated by contaminations in the crystal itself, in particular ^{40}K. Activities in the range from 15 to 20 mBq/kg have been measured for these available detectors. Anticoincidence between modules has been found to be helpful to reject background although not enough to achieve the required sensitivity. New NaI(Tl) crystals with very low content in potassium should be used in ANAIS requiring the development of new purification techniques. In this direction ANAIS is starting to search for new possibilities and contacts with different enterprises and research groups at international level have been established. On the other hand, ^{210}Pb is also an important background source if the contamination is placed on the surface of the crystals or in the encapsulation materials and should be avoided by conveniently polishing the surface of the crystal and carefully selecting the encapsulation material. Efforts in this direction are also ongoing. Nonetheless, these set-ups with old crystals have allowed verification of the linearity of the response down to 3 keV (see figure 1) thanks to a special mylar window in the encapsulations developed at Zaragoza (see figure 2).

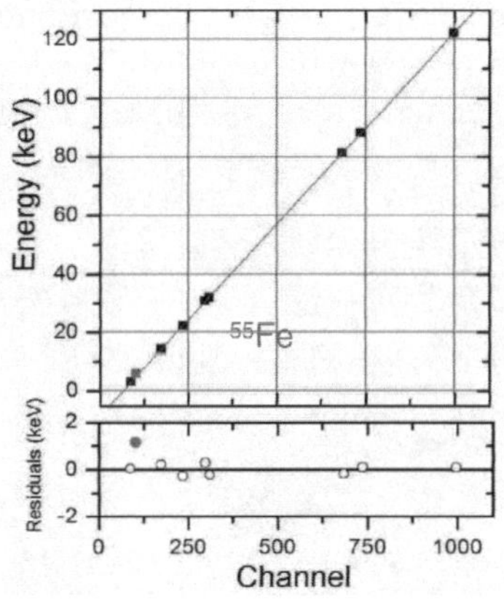

Figure 1. Linearity plot obtained with low energy sources. 5.9 keV from ^{55}Fe is seen at a lower energy indicating possible surface effects.

Figure 2. One of the ANAIS prototypes.

- High quantum efficiency photomultipliers: New photomultipliers with very high quantum efficiency (larger than 42%) are being tested in Zaragoza, and will be mounted in a prototype detector at Canfranc in the near future. The improvement in sensitivity with respect to the previously used in ANAIS could allow reducing the threshold of the experiment below 2 keV. On the other hand, radioactivity of these PMTs is being assessed at the low background test bench in Canfranc and seems to be too high in order to be used in the ANAIS final set-up (see table 1). More radiopure high efficient options are being searched for in collaboration with HAMAMATSU Photonics.

- Characterization of a new ultrapure NaI(Tl) crystal: A new ultrapure NaI(Tl) crystal (9.6 kg) has been very recently mounted (we will refer to it as Prototype IV) in order to better characterize ANAIS set-up background at low energy. An ETP (Electrolytic

Table 1. PMTs internal contamination measured at Canfranc for Electron Tubes ET 9302 and Hamamatsu high quantum effieciency and low background models.

Isotope	ET 9302B (mBq/PMT)	Ham High QE (Bq/PMT)[1]	Ham Low BKG (Bq/PMT)[2]
^{40}K	420±50	184.5±0.9	0.4
^{232}Th	24±4	0.42±0.04	0.3
^{238}U	220±12	0.51±0.03	0.2

[1] Values corrected with respect to online presentation.[2] Values reported by manufacturer

Tough Pitch) Copper encapsulation has been built for this prototype and screened at the low background radiopurity test bench at LSC. The crystal has been coupled to two low background PMTs (ET 9302B) and first tests are being carried out.

- A preliminary run: ANAIS-0. With ANAIS Prototype IV, after being completely characterized, we will start a long measurement in the best background conditions. Stability of the operation conditions should be conveniently monitored. New electronics and acquisition system may also be tested. Although 10 kg is not enough to confirm DAMA/LIBRA results, ANAIS-0 will be the final test bench for ANAIS.

3. Conclusions

ANAIS collaboration is working to improve the performances of the detectors, focusing on the reduction of the background level (towards 1 count/keV/kg/day) and of the energy threshold (below 2 keV). Dedicated set-ups at the old Canfranc facilities have allowed to measure the ^{40}K contamination of various NaI(Tl) crystals finding too high levels. We were able nonetheless to certify the linearity of the response down to 3 keV with an energy threshold at 2 keV. Cleaner crystals as well as new high QE and low background PMTs are being investigated at present. We are developing also new acquisition and monitorization systems which will be soon tested in an ANAIS-0 run with a single prototype detector. The collaboration aims to have the set-up for the whole experiment ready when the new Canfranc Underground Laboratory will be operative.

Acknowledgments
This project has been supported by the Spanish Agency of Science and Technology (CICYT) under the grants FPA2007-63777 and FPA2008-03228.

References

[1] Cebrián S *et al.* 2001 *Astroparticle Physics* **14** 339-350
[2] Bernabei R *et al.* 2008 *NIM A* **592, 3** 297-315
[3] Bernabei R *et al.* 2008 *EPJ C* **56, 3** 333-355
[4] Akerib D *et al.* [CDMS Collaboration] 2004 *Phys. Rev. Lett.* **93** 211301; Ahmed Z *et al.* 2009 *Phys. Rev. Lett.* **102** 011301
[5] Sanglard V *et al.* [EDELWEISS Collaboration] 2005, *Phys. Rev. D* **71** 122002
[6] G. Angloher *et al.* [CRESST Collaboration] 2005 *Astropart. Phys.* **23** 325-339
[7] Lebedenko V N *et al.* [ZEPLIN-III Collaboration] 2009 *Phys. Rev. Lett.* **103** 151302
[8] Benetti P *et al.* [WARP Collaboration] 2008 *Astropart. Phys.* **28** 495-507
[9] Angle J *et al.* [XENON10 Collaboration] 2008 *Phys. Rev. Lett.* **101** 091301

Topics in Astroparticle and Underground Physics (TAUP 2009) IOP Publishing
Journal of Physics: Conference Series **203** (2010) 012045 doi:10.1088/1742-6596/203/1/012045

Tidal effect in the radon-due neutron flux from the Earth's crust

V.V. Alekseenko[1], Yu .M. Gavriluk[1], V.V. Kuzminov[1] and Yu. V. Stenkin[2]

[1] Institute for Nuclear Research, Baksan Neutrino Observatory, Russian Academy of Sciences

[2] Institute for Nuclear Research, Moscow, Russian Academy of Sciences

vicalek@rambler.ru; yuri.stenkin@rambler.ru

Abstract. Baksan Neutrino Observatory's results on variations of thermal neutron flux below and above the ground surface measured with an unshielded scintillator detector are presented. Experimental evidences were obtained of the existence of seasonal wave in the long-term thermal neutron flux variations as well as correlation between this variations and lunar periods.

Introduction.

Well-known radioactive noble gas radon arises from decay chains of uranium and thorium (contained in terrestrial crust) and then diffuses into atmosphere. Simple but fruitful idea - to record a variation of radon concentration in strained terrestrial crust as an indicator of local seismic activity - is used during long time by geophysical community amongst a lot of other factors preceding or accompanying earthquakes. Information about concentration of radon in atmosphere is also a matter of interest in some applied scientific programs. Commonly used devices for measurement of radon concentration are based on recording of α,β-particles resulting from decay processes of radon nuclei in air. There are well known discomforts in using this method, for example humidity, atmospheric wind or ventilation of volume under control. But there is another possibility to monitor radon concentration, and directly in the crust. First it was assumed in ref. [1] that the terrestrial crust could be a possible source of neutrons, arising from (α, n)-reactions on the light nuclei of crust elements after α-decays of radon (^{222}Rn) and thoron (^{220}Rn), then thermalizing inside and diffusing out of crust to the atmosphere: radon-due neutron production mechanism. Hence, any variation of concentration of radon and thoron inside the crust will induce variation of neutron production, and that can be used for radon concentration monitoring with some specific neutron detector [2-5].

1. Detectors.

Instead of standard gas counters, we used solid state detecting compound ^{6}LiF+ZnS(Ag) to record thermal neutrons through the reaction ^{6}Li (n, α) ^{3}H + 4.78 MeV. Resulting α-particle and triton activate scintillator ZnS(Ag) viewed by a photomultiplier. Output signals from PMT are put to a digital oscilloscope. Such recording methods guarantees more reliable set of data in comparison with a traditional gas counter and integral discriminator counting rate. The method is free of above-mentioned demerits of the standard method and also has an advantage in comparison with neutron monitors that are insensitive to outer thermal neutron flux due to polyethylene moderator shield. Scintillator area was equal to 0.7 m^2. Detection efficiency is equal to ~20%. The experiment is carried out at Baksan Neutrino Observatory, North Caucasus, 1700 meters a. s. l., 43° N, 43° E both above ground level and underground at the depth of ~1000 m w. e. In the same room (without ventilation) a NaI detector of 8 cm x 8 cm was also installed to monitor gamma ray background above E_γ=0.6 MeV.

2. Experimental Results. Surface detector.

It is naturally to expect following the radon-due neutron production mechanism – the existence of variations of thermal neutron flux from terrestrial crust, similar to generally known periodical tidal wave in ocean, which occur twice during a Moon-day period. Hence, it was also naturally to apply a frequency analysis (FFT: Fast Fourier Transformation) to accumulated data to search for some well-known periodicities arising from well developed theory of Moon-Earth-Sun dynamics. Detailed results of this analysis can be found elsewhere [6].

As one can see from fig.1, with significance > 3 σ we can recognise semidiurnal periodicities: K2=11.95±0.01 h, P2=12.03±0.01 h, M2= 12.40±0.01 h, S2=12.00±0.01 h; where S2 is exactly a half of a solar day, while M2, P2, K2 are different modes of second harmonics of Moon day. All prominent peaks are in good agreement with theoretically predicted periodicities. On the other hand, more elaborate amplitude-phase analysis (out of the frame of this article) one must to do for, one must to subtract the trivial solar-time meteo-due waves from the observed S1 and S2 waves having in the mind to extract possible genuine tidal neutron S1 and S2 modes. And finally we would like to note that the absence of meteo partners for K2, P2, and M2 peaks gives a good chance to exclude a hypothesis in favour of meteo origin of these peaks.

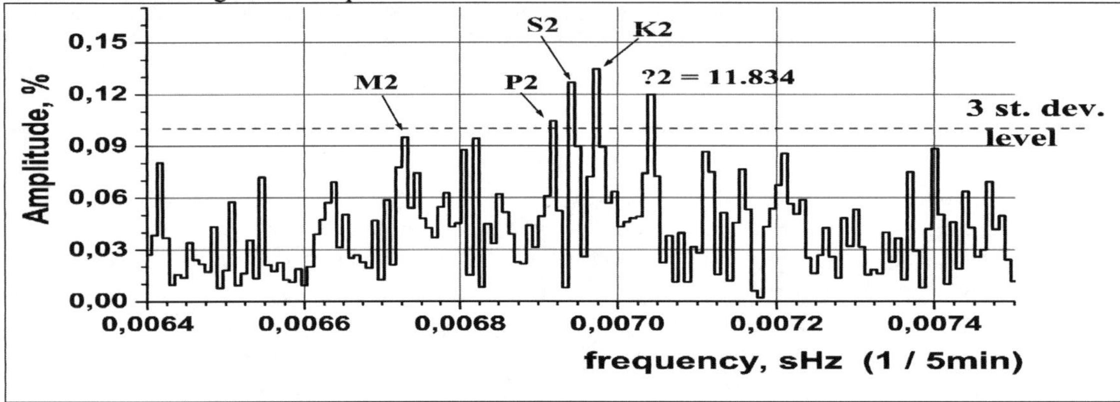

Fig.1. Moon's semidiurnal periodicities in thermal neutron flux.

Next step was a frequency analysis of day-to-day variations of counting rate. Now we have to remind here three most important basic Lunar month periods: *Anomalistic*, T_a = 27.554 day, $\omega_a=2\pi/T_a$, *Draconic*, T_d = 27.212 day, $\omega_d = 2\cdot\pi/T_d$ and *Synodic*, T_s = 29.531 day, $\omega_s = 2\cdot\pi/T_s$. Taking it into account we developed a simple model (ADS-model [3]) describing the experimental data. An essence of the model is an attempt of description of changing in time gravitational force in terms of amplitudes and phases of fixed basic periods of Moon rotation around Earth: Anomalistic, Draconic and Synodic. Hence we have tried to fit the observed day-after-day variations of counting rate (CR) by superposition of sinusoidal functional corresponding to basic periods taking into account also second harmonics: $CR(t) \sim (1+A(t,\omega_a)) \cdot (1+D(t,\omega_d)) \cdot (1+S(t,\omega_s)) - 1$. Amplitudes and phases in the model are free parameters leading to max ratio R/χ^2, where R is a coefficient of correlation between experimental curve and the model fit. $R\approx0.8$ has been obtained.

3. Results of underground measurements.

Underground measurements with neutron detector also shown various periodicities: seasonal, monthly, diurnal and semidiurnal. To show this we used superimposed epoch analysis. Fig.2 shows the seasonal variation of neutron flux underground along with variations of underground temperature and humidity. The amplitude of the neutron counting rate variation is close to 2% (in fact this is only a lower limit because the detector own background did not subtracted). Semidiurnal waves both in solar and in lunar time are shown in fig. 3. It is interesting that different delay exists between the solar or lunar culmination and the maximal counting rate.

The NaI detector data shown absolutely another behavior: it has clear diurnal solar wave with maximum at 6 h and counts around Th peak (2.6 MeV) has semidiurnal lunar wave similar to that in neutron detector, but with a different delay (phase) after the moon culmination.

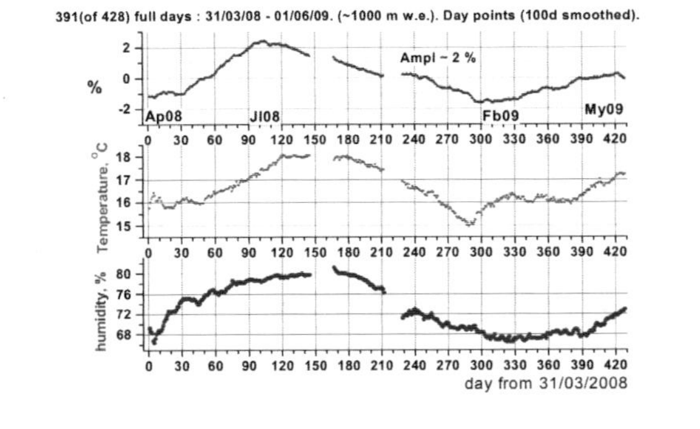

Fig.2. Seasonal variation of thermal neutron flux, temperature and relative humidity underground.

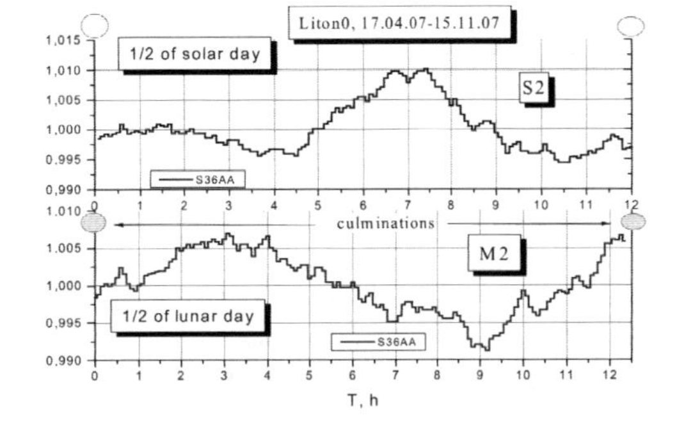

Fig.3. Semi-diurnal variations of thermal neutron flux deep underground.

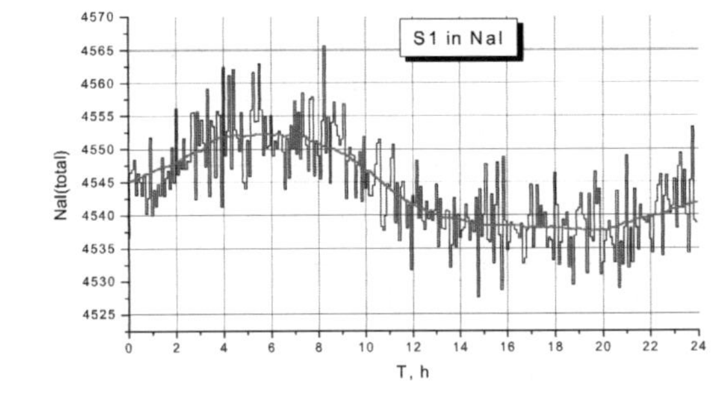

Fig.4. Solar diurnal wave in NaI detector.

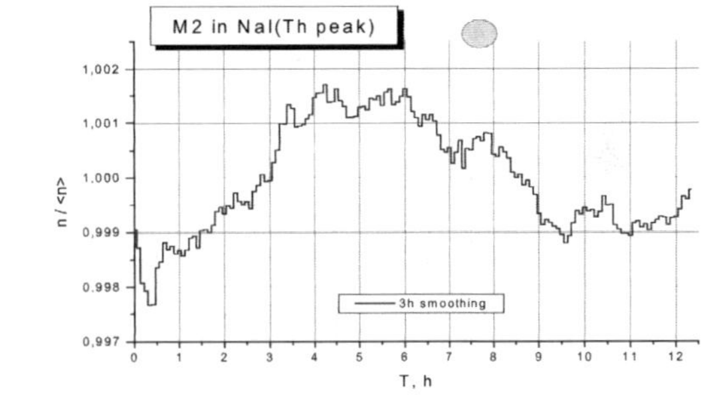

Fig.5. Moon's semidiurnal wave in gamma-rays.

5. Conclusion.

1. A novel experimental approach to record variations of radon concentration is presented. The method is based on the recording of thermal neutron flux from the Earth's crust.
2. The existence of seasonal, monthly and diurnal variations of neutron flux above and below the ground level has been established.
3. The method is based on nuclear physics and could be used in:
applied geophysical researches as a part component of the seismic station net;
radon-neutron monitoring of the environment and of global radon-neutron Earth's field;
low background experiments;
any applications where a control and monitoring of low intensity fluxes of neutrons is needed.

The work was supported in part by the Russian Fund for Basic Research grants No 08-02-01208, No 09-02-92426_CE-a and by the Russian Academy of Sciences Basic Research Program ``Neutrino Physics''.

References

[1] B.M. Kuzhevskij, O.Yu. Nechaev, et al.. Studies of the distribution of neutral and charged cosmic ray components in the atmosphere of the Earth. Preprint INP MSU 91-38/242, (1991)
[2] V.V. Alekeseenko et al. Bull. of the Russian Acad. of Sc.: Physics. Vol. 71, No. 7, (2007), p. 1080.
[3] V.V.Alekseenko et al. Proc. of 30th ICRC, Merida, (2007), rep. #932.
[4] Yu.V. Stenkin, D.D. Djappuev and J.F. Valdes-Galicia. Phys. of Atomic Nuclei. Vol. 70, No. 6, (2007), p. 1088.
[5] V.V. Alekseenko, Yu.M. Gavrilyuk et al. Phys. of the Solid Earth, v.45, No. 8, (2009), p. 709.

Dark matter search by means of segmented scintillator (PICO-LON)

K.Fushimi[1], Y.Kameda[1], K.Harada[1], S.Nakayama[1], H.Ejiri[2], T.Shima[2], K.Yasuda[3], S.Umehara[3], R.Hazama[4] and K.Imagawa[5]

1. Institute for Socio Arts and Sciences, The University of Tokushima, Tokushima 770-8502, Japan
2. Research Center for Nuclear Physics (RCNP), Osaka University, Ibaraki, Osaka 567-0047, Japan
3. Department of Physics, Osaka University, Toyonaka, Osaka 560-0047, Japan
4. Department of Engineering, Hiroshima University, Higashi Hiroshima, Hiroshima 739-8527 , Japan
5. Horiba Ltd. Kyoto city, Kyoto 601-8510, Japan

E-mail: `kfushimi@ias.tokushima-u.ac.jp`

Abstract. Thin and wide area inorganic crystal was applied to search for WIMPs dark matter. A multilayer NaI(Tl) detector has great advantages for WIMPs dark matter search. The performance of the thin (5mm thickness) and wide area (18cm×18cm) was developed. The performance tests were performed and the results were good for dark matter search.

1. PICO-LON detector system

Dark matter (DM) problem is a current important subject in astroparticle physics. WIMPs (Weakly Interacting Massive Particles) is one of the most promising candidates for DM. Many projects proposed the highly sensitive detectors with various target nucleus. DAMA has reported for several years annual modulation of the WIMPs signal [1, 2]. On the other hand, the allowed region of the cross section and the mass of WIMPs by DAMA has been excluded by other experiments using another target nuclei [3, 4, 5, 6].

We propose a highly sensitive detector system to extract the WIMPs signal. The project PICO-LON (Planar Inorganic Crystal Observation for LOw-background Neutr(al)ino) has great advantage. PICO-LON system distinguishes the type of interaction of WIMPs [7]. ^{127}I has a low energy excited state (57.6keV) which is excited via spin-coupled scattering. The 57.6keV gamma ray is emitted in coincidence with the nuclear recoil. The gamma ray signal and the nuclear recoil signal is separately observed by neighboring module as shown in Fig.1. In the case of elastic scattering of WIMPs, only one plate of NaI(Tl) makes the signal (S1). In the case of inelastic excitation, emitted gamma ray escapes from the original crystal and detected in the other crystal as S2 in Fig.1. The interaction type is clearly distinguished by segmentation of the detector system.

PICO-LON test module has been developed by Horiba Ltd. The dimension of the NaI(Tl) crystal is 18cm×18cm×0.5cm. The crystal was encapsulated with thin aluminum sheet and quartz light guides to avoid the humidity. Scintillation photons are guided to the edge of the

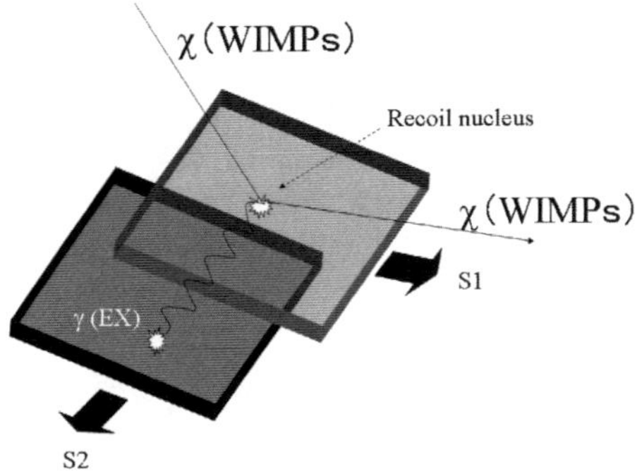

Figure 1. Multi NaI(Tl) plates system to distinguish the interaction type of WIMPs.

NaI(Tl) crystal by total reflection and a reflector with good reflective efficiency. The 12 modules of photomultiplier tubes (PMT) are contacted at the edge of quartz light guides by silicon oil. Each current pulses from the PMTs are individually integrated by analog-to-digital converter (ADC).

2. Performance of thin and large area NaI(Tl) scintillator

The most important performance for WIMPs detector is its low energy threshold and good energy resolution. Since the recoil energy due to WIMPs-nuclear interaction is less than several tens keV and the scintillation output response for nuclear recoil is factor 0.05 of the response of electron, the energy threshold must be less than 5keV electron equivalent energy.

In order to clearly distinguish the peak of 57.6keV gamma ray from the most serious background peak due to 46.5keV gamma ray from ^{210}Pb, the energy resolution around 60keV is needed to be less than 20% in FWHM.

The energy threshold and the energy resolution were measured by irradiating the gamma ray from ^{241}Am and ^{133}Ba sources. The scintillation light output spectrum are shown in Fig.2(A) and (B).

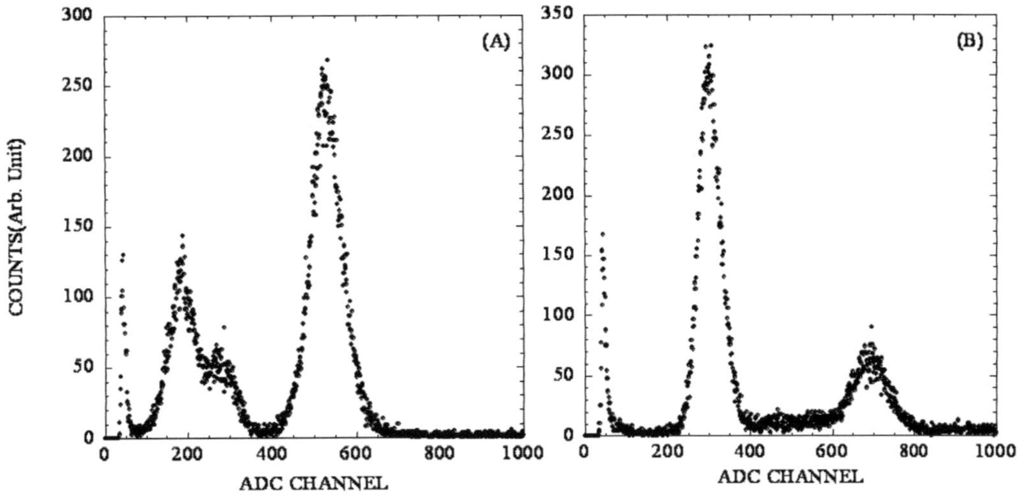

Figure 2. The energy spectra of thin NaI(Tl) plate with (A)^{241}Am. and (B)^{133}Ba.

Table 1. Energy resolution of prominent peaks of gamma rays and X ray.

Energy(keV)	$\Delta E/E(\%)$
31	17
59.5	14
81.0	10

The energy resolution was calculated by gauss fitting. In the Fig.2(A), 59.5keV gamma ray, 17keV X-ray and 31keV escape peak were clearly observed. The obtained energy resolutions of each peaks are listed in Table 1. It has been shown that the energy resolution is enough for observing the inelastic excitation.

The energy threshold was measured without any radioactive sources. The electric noise was reduced down to about 3keV, which was enough for observing the nuclear recoil.

3. Future prospect

The thin and wide area NaI(Tl) crystal was developed for studying nuclear rare processes. The PICO-LON detector system has the great advantages for dark matter search; selection power of the events, good energy resolution and low energy threshold. The purification of NaI crystal and selection of surrounding materials are now undergoing.

This work was supprted by Yamada Science Foundation and by Grant-in-Aid of Scientific Research of Japan Society for the Promotion of Science No.20540294.

References

[1] R.Bernabei et al., Nucl. Instr. and Meth. in Phys. Res.**A592** (2008) 297-315.
[2] P.Belli et al., Paper in this proceedings.
[3] Z.Ahmed et al., Phys. Rev. Lett. **102** (2009) 011301.
[4] J.Angle et al., Phys. Rev. Lett. **100** (2008) 021303.
[5] K.Abe et al., Jour. of Phys.: Conf. Ser.**120** (2008) 042022.
[6] J.Gascon, arXiv:0906.4232v1[astro-ph.HE].
[7] K.Fushimi et al., Jour. of the Phys. Soc. Japan **74** (2005) 3117-3121.

Topics in Astroparticle and Underground Physics (TAUP 2009) IOP Publishing
Journal of Physics: Conference Series **203** (2010) 012047 doi:10.1088/1742-6596/203/1/012047

Relic signal produced by the annihilation of dark matter particles

Anton N. Baushev

Bogoliubov Laboratory of Theoretical Physics, Joint Institute for Nuclear Research
141980 Dubna, Moscow Region, Russia

E-mail: baushev@gmail.com

Abstract. We discuss the possibility to observe the products of dark matter annihilation that was going on in the early Universe. Of all the particles that could be generated by this process we consider only photons, as they are both uncharged and easily detectable. The earlier the Universe was, the higher the dark matter concentration n and the annihilation rate (proportional to n^2) were. However, the emission from the very early Universe cannot reach us because of the opacity. The main part of the signal was generated at the moment the Universe had just become transparent for the photons produced by the annihilation. Thus, the dark matter annihilation in the early Universe should have created a sort of relic emission. We obtain its flux and the spectrum.

In the second part of the article we consider in greater detail the instance the dark matter is constituted by Weakly Interacting Massive Particles (WIMPs), which is one of the most popular hypotheses. It is shown that in this case we may expect an extragalactic gamma-ray signal in the energy range 0.5 - 20 MeV with a maximum near 8 MeV and there is evidence that an experimentally observed excess in the gamma-ray background at 0.5 - 20 MeV is created by the relic WIMP annihilation.

1. Introduction

In this report, we consider the dark matter annihilation in the epoch near the relic radiation formation ($z \sim 1000$). At that time, the average dark matter particle concentration $n \propto (z+1)^3$ was nine orders higher than now and four orders higher than in our Galaxy near the Sun system. So we may expect that the annihilation, the rate of which is proportional to n^2, was very intensive in that epoch. Of all the particles that can be generated by the dark matter annihilation we will consider only photons, as they are both uncharged and easily detectable. Uncharged particles do not interact with the magnetic field of the Galaxy, which allows one to measure the extragalactic background reliably enough.

The earlier the Universe was, the higher the DM density and the annihilation rate were. However, the emission from the very early Universe cannot reach us because of the opacity. The main part of the signal was generated at the moment the Universe had just become transparent for the photons produced by the annihilation. Later the dark matter density rapidly dropped, decreasing the signal. The moment (and its redshift) depends, of course, on the characteristic energy of the photons, in other words, on the nature of the dark matter. Thus, the DM annihilation in the early Universe should have produced a sort of relic emission. We obtain its flux and the spectrum.

2. Calculations

In case of the homogeneous isotropic Universe the number of photons of the above-mentioned relic emission that come to the local observer from unit solid angle per unit time per unit of area per unit energy interval can be represented [1] as

$$\tilde{Q}(\varepsilon) = \int Q(\varepsilon, \beta) f(\beta)\, d\beta \tag{1}$$

where

$$Q = \frac{\langle \sigma v \rangle\, n_0^2 c}{8\pi H_0 \sqrt{\Omega_m}} \frac{\beta\sqrt{\beta}}{\varepsilon^2 \sqrt{\varepsilon}} \exp\left(-\frac{2\aleph c n_0^b}{3 H_0 \sqrt{\Omega_m}} \frac{\beta\sqrt{\beta}}{\varepsilon\sqrt{\varepsilon}} \right) \tag{2}$$

Here n_0 and n_0^b are present dark matter particle and baryon concentrations respectively, $\langle \sigma v \rangle$ is the annihilation cross section of dark matter particles, $\aleph$ is the averaged cross-section of the interaction of photons generated by the annihilation with the baryonic matter.

Strictly speaking, the annihilation cross-section and the relative contribution of various channels depend on the energy of the DMPs collision, which can be important, if the annihilation occurs, for instance, near a black hole [3]. However, in the considered case the relative velocity of the DMPs is very small, and we may neglect this effect.

If we make a very natural supposition that the DMPs were in thermal equilibrium with other particles in the early Universe, we can estimate their annihilation cross section [2, 6]:

$$\langle \sigma v \rangle \simeq \frac{2 \cdot 10^{-27}\,(\mathrm{cm}^3/\mathrm{s})}{\Omega_{DM} h^2} \tag{3}$$

For the present value [4] of $\Omega_{DM} h^2 = 0.113$ we obtain $\langle \sigma v \rangle \simeq 2 \cdot 10^{-26}\,(\mathrm{cm}^3/\mathrm{s})$. These are the telling reasons to believe that the dark matter consists of Weakly Interacting Massive Particles (WIMPs). They are expected to annihilate into fermion-antifermion or gauge boson pairs with a large fraction of quark-antiquark pairs. A WIMP pair annihilation finally leads to $30 - 40$ photons generation (in the fragmentation process, mainly from π^0 decays). A greater part of the photons has energy in the range from 2 to 4 GeV [5, 6].

In the interval $1 - 5$ GeV the photon spectrum caused by the main $b\bar{b}$ annihilation channel is well approximated if the distribution function $f(\beta) \propto \beta^{-1} \exp(-0.15\beta)$. We shall use the following f (β is expressed in GeVs):

$$f(\beta) = \begin{cases} 26.7 \cdot \beta^{-1} \exp(-0.15\beta), & \beta \in (1-5)\ \mathrm{GeV} \\ 0, & \beta \notin (1-5)\ \mathrm{GeV} \end{cases} \tag{4}$$

We normalized it considering that one act of annihilation generated on average 30 photons.

We use the following set of cosmological parameters: $\Omega_\Lambda = 0.75$, $\Omega_m = \Omega_{DM} + \Omega_b = 0.25$, $\Omega_b = 0.042$ (of course, $\Omega_\Lambda + \Omega_{DM} + \Omega_b = 1$), the Hubble constant $H_0 = 2.4 \cdot 10^{-18} \mathrm{s}^{-1}$, the relic radiation temperature 2.725 K, the baryon-photon ratio $\eta \equiv n_b/n_{ph} = 6.1 \cdot 10^{-10}$. We obtain the present baryon concentration $n_0^b = 2.5 \cdot 10^{-7}\,\mathrm{cm}^{-3}$. The DMP concentration, with DMP mass taken as $M_{DMP} = 50$ GeV, is $n_0 = 2.5 \cdot 10^{-8}\,\mathrm{cm}^{-3}$. Equation (3) gives $\langle \sigma v \rangle \simeq 2 \cdot 10^{-26}\,(\mathrm{cm}^3/\mathrm{s})$. We take $\aleph = \aleph(2\mathrm{GeV}) = 1.63 \cdot 10^{-26}\,\mathrm{cm}^2$ [7].

The resulting spectrum of the relic gamma-ray background, obtained with the aid of (1), (2), grows up to ~ 8 MeV, and the bulk of the signal lies in the range from 0.5 to 20 MeV. Characteristic redshift of the relic gamma-rays is $z \simeq 300$.

3. Comparison with observations and discussion

The cosmic gamma-ray background reportedly (see, for instance, [8, 9, 10, 11, 12]) has a peculiarity in the energy range 0.5 - 20 MeV. A ledge-like feature is visible in the extragalactic gamma-ray spectrum (Fig. 3 in [13]). The photon index here is markedly distinct from those of the softer or harder parts of the spectrum [14, 13, 15], indicating its different origin.

Moreover, this spectral band can be formed neither by too soft emission of normal active galactic nuclei, nor by too hard blazar-type AGNs contribution (see [8] and references therein). Attempts to consider the nuclear-decay gamma rays from Type Ia supernovae as the source have not been successful: the flux expected from the supernovae is several times weaker than the observed [16, 11, 12]. It might be well to point out that the precise determination of the excess boundaries and intensity is model-dependent, and the literature values vary considerably [8, 9, 10, 11]. In any case, however, the excess becomes apparent near 0.5 MeV and disappears at the energies $\gtrsim$ 20 MeV. One can see that the energy range of the feature corresponds closely to the interval characteristic for the relic gamma emission from the WIMPs annihilation. This coincidence looks promising when it is considered that the WIMP is now one of the most probable dark matter candidates.

At the same time, the relic gamma emission predicted by equations (1) and (2), is approximately five orders fainter than the observed feature. This discrepancy might result from inapplicability of the assumption of homogeneous dark matter distribution. The presence of density inhomogeneities does not affect the spectrum of the annihilation signal but increases its intensity. The modern structure of the Universe appeared from some initial perturbations that had already existed, beyond any doubt, in the epoch $z \sim 300$. Let us suppose that the spectrum of primordial fluctuations is not exactly flat, and the intensity of the fluctuations slightly builds up as their scale decreases. In the matter-dominated stage the perturbations grow as $\delta\rho/\rho \propto t^{2/3}$ and $a(t) \propto t^{2/3}$, therefore $\delta\rho/\rho \propto a$ [17]. If we assume that the small-scale perturbations left the radiation-dominated stage with amplitudes more than expected from the Harrison-Zeldovich spectrum by a factor of $\frac{300}{80} \simeq 4$, they collapsed by the moment $z = 300$ (in so doing we imply that the amplitudes of large-scale perturbations are fixed in such a way as they reproduce the observed large-scale structure of the Universe). Considering that the largest and the smallest clump mass scales differ by more than 20 orders, even a negligible tilt of the spectrum of primordial fluctuations with respect to the Harrison-Zeldovich shape is enough to gain the factor of four.

[1] A. N. Baushev, *Monthly Notices of the Royal Astronomical Society*, **398**, Issue 2, pp. 783-789, (2009).
[2] E. Kolb and M. S. Turner, The Early Universe (Addison Wesley) (1990).
[3] A. N. Baushev, *International Journal of Modern Physics D*,Volume 18, Issue 8, pp. 1195-1203, (2009), arXiv:0805.0124
[4] D. N. Spergel, L. Verde, H. V. Peiris et al., *ApJS*, **148**, 175, (2003).
[5] G. Jungman, M. Kamionkowski, and K. Griest, *Phys. Rep.*, **267**, 195, (1996).
[6] W. de Boer, C. Sander, V. Zhukov et al., *A&A*, **444**, 51, (2005).
[7] A.A. Zdziarski and R. Svensson, *Ap.J.*, **344**, 551, (1989).
[8] Y. Inoue, T. Totani, and Y. Ueda, *Ap.J. L.*, **672**, 5, (2008).
[9] K. Ahn, and E. Komatsu, *Phys. Rev. D*, **71**, 021303(R), (2005).
[10] K. Ahn, and E. Komatsu, *Phys. Rev. D*, **72**, 061301(R), (2005).
[11] Y. Rasera, R. Teyssier, P. Sizun, M. Cassé, P. Fayet, B. Cordier, and J. Paul, *Phys. Rev. D*, **73**, 103518, (2006).
[12] L. E. Strigari, J. F. Beacom, T. P. Walker, and P. Zhang, *J. Cosmol. Astropart. Phys.*, **4**, 17, (2005).
[13] A. W. Strong, I. V. Moskalenko, and O. Reimer, *Ap.J.*, **613**, 956, (2004), arXiv:astro-ph/0405441
[14] P. Sreekumar, D. L. Bertsch, B. L. Dingus et al., *Ap.J.*, **494**, 523, (1998), arXiv:astro-ph/9709257
[15] G. Weidenspointner, M. Varendorff, S. C. Kappadath et al., in AIP Conf. Proc., 5th Compton Symposium, ed. M. L. McConnell & J. M. Ryan (New York: AIP), **510**, 467, (2000).
[16] K. Ahn, E. Komatsu, and P. Höflich, *Phys. Rev. D*, **71**, 121301(R), (2005).
[17] A.V. Gurevich, K.P. Zybin, and V.A. Sirota, *Sov. Phys. Usp.*, **40**, 869 , (1997).

Journal of Physics: Conference Series **203** (2010) 012048 doi:10.1088/1742-6596/203/1/012048

Right-handed sneutrino WIMPs in the NMSSM

David G. Cerdeño

Departamento de Física Teórica C-XI, and Instituto de Física Teórica UAM-CSIC, Universidad Autónoma de Madrid, Cantoblanco, E-28049 Madrid, Spain

E-mail: davidg.cerdeno@uam.es

Abstract. Within the context of the Next-to-MSSM the right-handed sneutrino becomes a viable WIMP candidate for dark matter. In this construction a new singlet superfield is included with direct coupling to the singlet Higgs. The right-handed sneutrino can reproduce the correct dark matter relic abundance while not being excluded by current direct searches within a mass-range of 5-200 GeV and for natural values of the input parameters.

The (left-handed) sneutrino [1] is a potentially interesting candidate for dark matter in the Minimal Supersymmetric extension of the Standard Model (MSSM), since it is neutral and has weak-scale interactions. However, its sizable coupling to the Z boson implies that it either annihilates too rapidly (resulting in a very small relic abundance) or gives rise to a too large scattering cross-section off nuclei (being excluded by direct searches of DM) [2]. A reduction of the sneutrino coupling to the Z boson can be achieved by introducing a mixture of left- and right-handed sneutrino states [4, 5, 6] (although a significant mixture is only possible if some particular supersymmetry breaking with an unnaturally large trilinear term [4] is adopted). Another possibility is to consider a pure right-handed sneutrino [7, 8, 9, 10], which can be thermally produced if they are somehow coupled to the observable sector. This is possible through an extension of the gauge [10, 11] or Higgs [13, 14, 15, 16, 17] sectors. Extending the MSSM Higgs sector has also been proposed to address the so-called "μ problem" [12]. The next-to-minimal supersymmetric standard model (NMSSM) offers a simple solution by introducing a singlet superfield, S, and promoting the bilinear $\mu H_1 H_2$ term to a trilinear coupling $\lambda S H_1 H_2$. After radiative electroweak symmetry-breaking (REWSB), S develops a vacuum expectation value (VEV) and provides an effective term, $\mu = \lambda\langle S\rangle$.

The above reasons motivated the study of an extension of the NMSSM in which a singlet scalar right-handed neutrino superfield, N, is included to obtain non-vanishing neutrino Majorana masses with a low scale see-saw mechanism [15, 17]. The superpotential contains that of the NMSSM with the addition of a new trilinear coupling between the singlet superfields S and N and the corresponding Yukawa terms, $W \supset \lambda_N SNN + y_N L \cdot H_2 N$. After REWSB the Higgs fields take non-vanishing VEVs, $(v_{1,2}, v_s) = (\langle H_{1,2}\rangle, \langle S\rangle)$, and a Majorana mass term is generated, $M_N = 2\lambda_N v_s$. Masses for left-handed neutrinos are obtained via a see-saw mechanism, $m_{\nu_L} = y_N^2 v_2^2/M_N$, implying small Yukawa couplings, $y_N < \mathcal{O}(10^{-6})$.

This model contains three new inputs with respect to the NMSSM parameter space, namely the coupling λ_N, the associated soft trilinear term, A_{λ_N}, and the soft right-handed sneutrino mass, $m_{\tilde{N}}$. A crucial feature of this scenario is the existence of direct couplings of the right-handed sneutrino to Higgses and neutralinos. These emerge through the term $\lambda_N SNN$ in the superpotential, since the singlet and singlino components of S mix with the CP-even Higgs bosons and neutralinos, respectively. The strength of the interaction is therefore dependent on the values of λ_N and A_{λ_N}. The coupling between the lightest right-handed sneutrino, $\tilde{N}_1$ and a Higgs boson, H_i^0, determines most of the sneutrino phenomenological properties, and reads

$$C_{H_i^0\tilde{N}_1\tilde{N}_1} = \frac{2\lambda\lambda_N M_W}{\sqrt{2}g}\left(\sin\beta S^1_{H_i^0} + \cos\beta S^2_{H_i^0}\right) + \left[\left(4{\lambda_N}^2 + 2\kappa\lambda_N\right)v_s + \frac{\lambda_N A_{\lambda_N}}{\sqrt{2}}\right] S^3_{H_i^0} \,,$$

where $S^j_{H_i^0}$ $(j = 1, 2, 3)$ are the elements of the Higgs diagonalisation matrix.

The right-handed sneutrino coupling to the Higgs sector provides tree-level interactions with ordinary matter. For adequate values of λ_N (and A_{λ_N}) these couplings would be of electroweak scale, thereby making it a WIMP candidate. The sneutrino annihilation cross section is extremely sensitive to the structure of the Higgs sector, but there are general features which are easy to understand.

The coupling λ_N determines the overall scale of the annihilation cross section. Notice that λ_N affects the right-handed neutrino and sneutrino masses but does not alter the rest of the NMSSM spectrum. Annihilation into a pair of scalar or a pair of pseudoscalar Higgs bosons are among the most effective channels. The flexibility of the Higgs sector of the NMSSM make this possibility very versatile. Very light CP-even and CP-odd Higgses are possible (as long as they are singlet-like), making these channels available for a wide range of sneutrino masses. In particular, very light sneutrinos can thus reproduce the correct relic abundance. Another relevant contribution to the total annihilation cross section is due to the annihilation into a pair of right-handed neutrinos, NN. Finally, for all the possible annihilation products there is always a contribution coming from s-channel CP-even Higgs exchange. This implies that all of them are subject to a resonant effect when $2m_{\tilde{N}_1} \approx m_{H_i^0}$, for $i = 1, 2, 3$.

The direct detection of right-handed sneutrinos could take place through their elastic scattering with nuclei inside a DM detector. There is only one diagram contributing to this process, namely, the t-channel exchange of neutral Higgses. The total spin-independent sneutrino-proton scattering cross section yields

$$\sigma^{\rm SI}_{\tilde{N}_1-p} = \frac{m_p^4}{\pi(m_p + m_{\tilde{N}_1})^2}\left(\sum_{q_i=u,d,s} f^p_{Tq_i}\sum_{j=1}^{3}\frac{C_{H_j^0\tilde{N}_1\tilde{N}_1}Y_{q_i}}{m^2_{H_j^0}m_{q_i}} + \frac{2f^p_{TG}}{27}\sum_{q_i=c,b,t}\sum_{j=1}^{3}\frac{C_{H_j^0\tilde{N}_1\tilde{N}_1}Y_{q_i}}{m^2_{H_j^0}m_{q_i}}\right)^2 ,$$

where Y_{q_i} is the corresponding quark Yukawa coupling and $f^p_{Tq_i}$ and f^p_{TG} are the hadronic matrix elements. The sneutrino detection cross section is extremely dependent on the features of the Higgs sector of the model. In particular, $\sigma^{\rm SI}_{\tilde{N}_1-p}$ becomes larger when the sneutrino-sneutrino-Higgs coupling increases (which can be achieved by enhancing λ_N or with large values for $|A_{\lambda_N}|$), and when the Higgs mass becomes smaller. Once more, this is an interesting possibility, given the flexibility of the Higgs sector in the NMSSM

The theoretical predictions for $\sigma^{\rm SI}_{\tilde{N}_1-p}$ are represented as a function of the sneutrino mass on the right hand-side of Fig. 1 for a wide scan in the NMSSM parameter space as described in Ref. [17]. The right-handed sneutrino in our model is not yet excluded by direct searches for DM, and the predicted $\sigma^{\rm SI}_{\tilde{N}_1-p}$ lies within the reach of projected experiments for some regions in the parameter space. It should be pointed out that the resulting area is contained within the region predicted for neutralinos in the NMSSM [18, 19] (as well as for neutralinos in the MSSM except for the region for very light WIMPs). One may therefore wonder whether these two candidates can be discriminated in the case of a hypothetical detection of dark matter. However, since the sneutrino has no spin-dependent couplings the combination of data from two experiments whose targets are sensitive to spin-dependent and -independent couplings could be used to disentangle between sneutrino and neutralino dark matter in the same way that it was shown for Kaluza-Klein dark matter and neutralinos in Ref. [20].

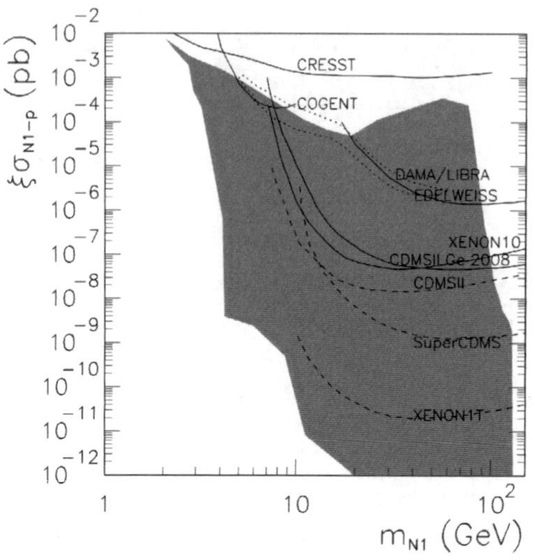

Figure 1. Predicted $\sigma^{\rm SI}_{\tilde{N}_1-p}$ as a function of $m_{\tilde{N}_1}$ for a scan in the NMSSM parameter space.

To conclude, the right-handed sneutrino is a viable thermal WIMP candidate for DM in an extended NMSSM with a new singlet superfield, N. A direct coupling between N and the Higgs provides a sufficiently large annihilation cross section sneutrino and a scattering cross-section off nuclei which is not excluded by current direct DM searches. Sneutrinos within a mass-range of 5-200 GeV can reproduce the correct relic abundance and the prospects for their direct detection can be within the range of future detectors.

References

[1] L.E. Ibáñez, Phys. Lett. B **137** 160 (1984); J.S. Hagelin, G.L. Kane, S. Raby, Nucl. Phys. B **241** 638 (1994).
[2] T. Falk, K. A. Olive and M. Srednicki, Phys. Lett. B **339** 248 (1994).
[3] L. J. Hall, T. Moroi and H. Murayama, Phys. Lett. B **424** 305 (1998).
[4] N. Arkani-Hamed *et al.*, Phys. Rev. D **64** 115011 (2001); D. Hooper, J. March-Russell and S. M. West, Phys. Lett. B **605** 228 (2005).
[5] C. Arina and N. Fornengo, JHEP **0711** (2007) 029.
[6] C. Arina *et al.*, Phys. Rev. Lett. **101** (2008) 161802.
[7] T. Asaka, K. Ishiwata and T. Moroi, Phys. Rev. D **73** 051301 (2006); Phys. Rev. D **75** 065001 (2007).
[8] S. Gopalakrishna, A. de Gouvea and W. Porod, JCAP **0605** (2006) 005.
[9] J. McDonald, JCAP **0701** (2007) 001;
[10] H. S. Lee, K. T. Matchev and S. Nasri, Phys. Rev. D **76** 041302 (2007).
[11] R. Allahverdi, B. Dutta and A. Mazumdar, Phys. Rev. Lett. **99**, 261301 (2007).
[12] J. E. Kim and H. P. Nilles, Phys. Lett. B **138** 150 (1984).
[13] R. Kitano and K. y. Oda, Phys. Rev. D **61** 113001 (2000).
[14] B. Garbrecht *et al.*, JHEP **0612**, 038 (2006).
[15] D. G. Cerdeño, C. Muñoz and O. Seto, Phys. Rev. D **79**, 023510 (2009).
[16] F. Deppisch and A. Pilaftsis, JHEP **0810** (2008) 080.
[17] D. G. Cerdeno and O. Seto, JCAP **0908** (2009) 032 [arXiv:0903.4677 [hep-ph]].
[18] D. G. Cerdeño, C. Hugonie, D. E. Lopez-Fogliani, C. Muñoz and A. M. Teixeira, JHEP **0412** (2004) 048; D. G. Cerdeño, E. Gabrielli, D. E. Lopez-Fogliani, C. Muñoz and A. M. Teixeira, JCAP **0706** (2007) 008.
[19] C. E. Aalseth *et al.* [CoGeNT Collaboration], Phys. Rev. Lett. **101**, 251301 (2008).
[20] G. Bertone, D. G. Cerdeño, J. I. Collar and B. C. Odom, Phys. Rev. Lett. **99** (2007) 151301.

Calculable neutrino masses in an Inverse See-Saw scenario and DM candidates

F. Bazzocchi

Department of Physics and Astronomy, Vrije Universiteit Amsterdam, 1081 HV Amsterdam, The Netherlands

E-mail: f.bazzoc@few.vu.nl

Abstract. Within the supersymmetric inverse seesaw mechanism we provide a scenario where naturally small and calculable neutrino masses arise from a supersymmetry breaking renormalization-group-induced vacuum expectation value. In such a scenario the sneutrino can be the lightest supersymmetric particle potentially viable as WIMP dark matter.

1. Introduction

The solid experimental evidence for neutrino masses and oscillations [1], provide strong evidence for physics beyond the Standard Model. At the same time cosmological studies clearly show that a large fraction of the mass of the Universe is dark and must be non–baryonic.

The idea that the generation of neutrino masses may provide new insight on the nature of the dark matter is quite appealing.

Theory has no clue as to what causes the smallness of neutrino masses. It has become popular to ascribe it as an effect of a very high new physics scale in particular in the context of the so-called minimal type-I seesaw [2].

An alternative approach is the inverse seesaw mechanism [4, 5], which avoids introducing new states above the TeV scale. Neutrino masses arise well below the weak scale, thanks to a very small singlet mass term in whose presence lepton number is violated. In absence of the lepton number violating mass terms the symmetry of the theory increases and neutrinos are massless.

It has been shown that in a minimal supergravity (mSUGRA) [3] scheme where the smallness of neutrino masses is accounted for within the inverse seesaw mechanism the lightest supersymmetric particle is likely to be represented by the corresponding neutrino superpartner (sneutrino), instead of the lightest neutralino. Such a model naturally reconciles the small neutrino masses with the correct relic abundance of sneutrino dark matter and experimentally accessible direct detection rates.

The inverse see-saw scheme has not become as popular as the high-scale seesaw due to some discomfort in assuming by hand the smallness of an $SU(3) \times SU(2) \times U(1)$ invariant mass term.

Here we provide a plausible mechanism where the origin of such small scale would, in addition, find a natural dynamical explanation.

2. The model

The model is defined by the supermultiplets given in table 1, where L denotes the global continuous lepton number and R denotes the R-charge. We also impose a discrete Z_3 symmetry

	$\hat{Q}$	$\hat{U}^c$	$\hat{D}^c$	$\hat{L}$	$\hat{E}^c$	$\hat{\nu^c}$	$\hat{S}$	$\hat{H}^u$	$\hat{H}^d$	$\hat{\Phi}$	$\hat{\Delta}$	$\hat{\tilde{\Delta}}$
$SU(2)$	2	1	1	2	1	1	1	2	2	1	1	1
L	0	0	0	1	-1	-1	1	0	0	0	-2	1
R	1	1	1	1	1	1	1	0	0	0	0	0
Z_3	ω	ω	ω	ω	ω	ω	ω	ω	ω	ω	ω	ω

Table 1. Multiplet content of the model.

that forbids bilinear couplings. The mechanism we develop requires the existence of a singlet sector, perhaps of stringy origin [6]. The required magnitude of the supersymmetric Higgs mass parameter arises from the expectation value of the extra singlet field present in the next-to-minimal supersymmetric standard model (NMSSM) [7], avoiding the so-called μ problem [8, 9], as recently advocated in Ref. [10]. The superpotential is given as

$$\begin{aligned} W &= y^e_{ij}\hat{L}_i\hat{H}^d\hat{E}^c_j + y^u_{ij}\hat{Q}_i\hat{H}^u\hat{U}^c_j + y^d_{ij}\hat{Q}_i\hat{H}^d\hat{D}^c_j + \lambda_1\,\hat{\Phi}\,\hat{H}^u\hat{H}^d + \frac{1}{3!}\lambda_2\,\hat{\Phi}\hat{\Phi}\hat{\Phi} \\ &+ y^\nu_{ij}\hat{L}i\hat{H}^u\hat{\nu}^c_j + \eta_{ij}\hat{\Phi}\hat{\nu}^c_i\hat{S}_j + \frac{1}{2}\xi_{ij}\hat{\Delta}\hat{S}_i\hat{S}_j + \frac{1}{2}\tilde{\xi}\hat{\Delta}\hat{\tilde{\Delta}}\hat{\tilde{\Delta}}\,, \end{aligned} \tag{1}$$

The corresponding soft supersymmetry breaking potential reads,

$$\begin{aligned} V_{soft} &= a^u_{ij}\tilde{Q}_iH^u\tilde{u^c}_j + a^d_{ij}\tilde{Q}_iH^d\tilde{d^c}_j + a^e_{ij}\tilde{L}_iH^d\tilde{e^c}_j + a_{\Phi H}\,\Phi\,H^u\,H^d + \frac{1}{3!}a_\Phi\Phi^3 + a^\nu_{ij}\tilde{L}_iH^u\tilde{\nu^c}_j \\ &+ a^\eta_{ij}\Phi\tilde{\nu^c}_i\,S_j + \frac{1}{2}a^S_{ij}\Delta\tilde{S}_i\tilde{S}_j + \frac{1}{2}a_\Delta\Delta\tilde{\Delta}^2 + H.c. + \Sigma_i m_i^2|\varphi_i|^2 + \frac{1}{2}\Sigma_i M_i\lambda_i\lambda_i\,. \end{aligned} \tag{2}$$

From eq. (1) we see that neutrino masses are induced once the scalar singlet Δ acquires a vacuum expectation value (vev) v_Δ and their expression is given by the inverse see-saw formula

$$m_\nu \sim v_\Delta \frac{(v^u)^2}{v_\Phi^2}\, y^\nu\,\eta^{-1}\,\xi\,\eta^{T-1}\,y^{\nu T}\,. \tag{3}$$

The minimization of the scalar potential determines the values of the vevs of the NMSSM scalar Higgs fields, Φ, H^u, H^d and of the extra guage singlet scalar fields, Δ and $\tilde{\Delta}$. Due to their different charges with respect to $U(1)_L$ Δ and $\tilde{\Delta}$ behave differently. As a result of minimizing the tree-level Higgs potential one finds that Δ develops a vev only in the presence of the trilinear soft term a_Δ. Thus, in the supersymmetric limit, neutrinos are massless. Assuming that all trilinear soft breaking terms vanish at the GUT scale M_0 implies that neutrinos are still massless even in the tree-level limit. However the breaking of supersymmetry due to the gaugino mass induces a small but non-vanishing a_Δ through the RGEs. The singlet nature of the secluded sector protects a_Δ that turns out to be naturally small.

3. The DM candidate

One can see that in most the cases the lightest sneutrino is a combination of the two singlet states $\tilde{\nu}^c$ and $\tilde{s}$. Since all trilinear couplings vanish at the GUT scale, the trilinears involving only gauge singlet fields run very slowly so v_Δ is very small compared to the other vevs. Thus for the above choice the sneutrino mass matrix can be approximated as

$$M^2_{\tilde{\nu}_{ri}} \sim \begin{pmatrix} m_L^2 & 0 & 0 \\ 0 & m_0^2 + \alpha_{ri}v^2 & \pm\delta v^2 \\ 0 & \pm\delta v^2 & m_0^2 + \beta_{ri}v^2 \end{pmatrix}.$$

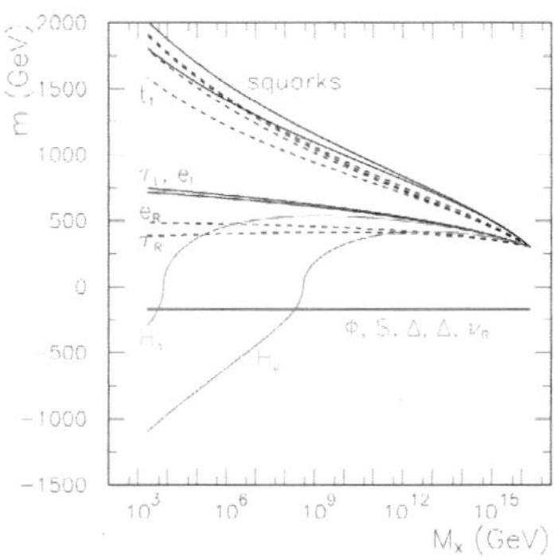

Figure 1. Renormalization group evolution of the scalar masses (with $m \equiv \text{sign}(m_o)\sqrt{|m_0^2|}$) for a representative choice of parameters.

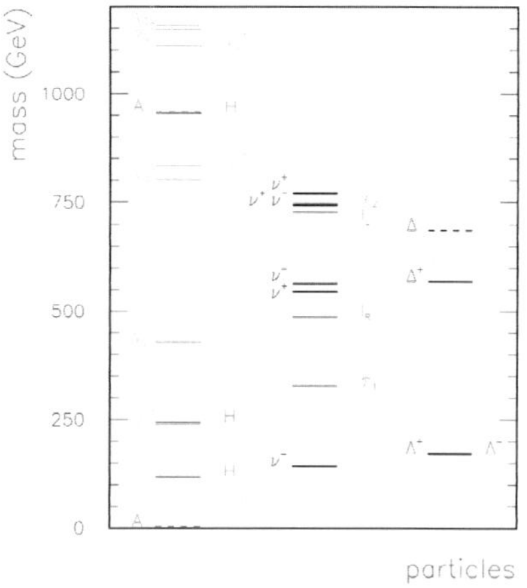

Figure 2. Supersymmetric spectrum for the same choice of parameters as in Fig. 1. Gluino and squark masses are larger than 1.5 TeV and not shown.

with $0 < \alpha_{ri}, \beta_{ri} \sim \mathcal{O}(1)$ while $\delta \sim \mathcal{O}(0.1)$ and where we have used $m_0^2 < 0$. Given that $m_L^2 > 0$, the natural lightest sneutrino is the CP even or CP odd combination of the singlet states.

As illustrated in Fig. 2, due to these mixing effects it is likely that the lightest supersymmetric particle is mainly a mixture of the singlet scalars $\tilde{\nu}^c$ and $\tilde{s}$, instead of the neutralino. In this sense, this model is similar to the scenario of Refs. [10, 11], since in both the right-handed sneutrino component couples directly to the NMSSM Higgs sector through the singlet Φ. As shown there, this makes it possible to fulfil the WMAP result, thereby making the sneutrino a viable WIMP. A similar effect is expected in the present model, this time through the η coupling in (1). Thus, the scheme proposed here opens yet new alternative ways to understand supersymmetric dark matter.

References

[1] For a review see, e. g. M. Maltoni, T. Schwetz, M. A. Tortola and J. W. F. Valle, New J. Phys. **6**, 122 (2004); updated neutrino oscillation results are given in T. Schwetz, M. Tortola and J. W. F. Valle, New J. Phys. **10**, 113011 (2008), [0808.2016].
[2] P. Minkowski, Phys. Lett. **B67**, 421 (1977); R. N. Mohapatra and G. Senjanovic, Phys. Rev. Lett. **44**, 91 (1980); J. Schechter and J. W. F. Valle, Phys. Rev. **D22**, 2227 (1980); Phys. Rev. **D25**, 774 (1982);
[3] C. Arina, F. Bazzocchi, N. Fornengo, J. C. Romao and J. W. F. Valle, Phys. Rev. Lett. **101** (2008) 161802 [arXiv:0806.3225 [hep-ph]].
[4] R. N. Mohapatra and J. W. F. Valle, Phys. Rev. **D34**, 1642 (1986).
[5] M. C. Gonzalez-Garcia and J. W. F. Valle, Phys. Lett. **B216**, 360 (1989); for a recent paper see E. Ma, arXiv:0904.4450 [hep-ph].
[6] E. Witten, Nucl. Phys. **B258**, 75 (1985).
[7] U. Ellwanger, M. Rausch de Traubenberg and C. A. Savoy, Phys. Lett. **B315**, 331 (1993), [hep-ph/9307322].
[8] J. Kim and H. Nilles, Phys. Lett. B **138**, 150 (1984).
[9] G. F. Giudice and A. Masiero, Phys. Lett. **B206**, 480 (1988).
[10] D. G. Cerdeño, C. Muñoz and O. Seto, Phys. Rev. **D79**, 023510 (2009), [0807.3029].
[11] D. G. Cerdeño and O. Seto, arXiv:0903.4677 [hep-ph].

Journal of Physics: Conference Series **203** (2010) 012050 doi:10.1088/1742-6596/203/1/012050

Astrophysical boost factor and Dark Matter indirect detection

Timur Delahaye[1,2], Pierre Brun[3], Jürg Diemand[4], Stefano Profumo[5] and Pierre Salati[2]

[1]Dipartimento di Fisica Teorica, Università di Torino Istituto Nazionale di Fisica Nucleare, via P. Giuria 1, I–10125 Torino, Italy

[2]LAPTH, Université de Savoie, CNRS, BP110, F-74941 Annecy-le-Vieux Cedex, France

[3]CEA, Irfu, Service de Physique des Particules, Centre de Saclay, F-91191 Gif-sur-Yvette, France

[4]University of California, Department of Astronomy and Astrophysics, 1156 High Street, Santa Cruz CA 95064, USA

[5]Santa Cruz Institute for Particle Physics and Department of Physics, University of California, Santa Cruz, CA 95064, USA

E-mail: delahaye@lapp.in2p3.fr

LAPTH-Conf-1340/09

Abstract. The PAMELA [1], ATIC[2] and Fermi[3] collaborations have recently reported an excess in the cosmic ray positron and electron fluxes. These lepton anomalies might be related to cold dark matter particles annihilating within a nearby dark matter sub-halo. We outline regions of the parameter space for both the dark matter sub-halo and particle model, where data from the different experiments are reproduced. We then confront this interpretation of the data with the results of the cosmological N-body simulation Via Lactea 2. Having a sizeable sub-halo at a distance of only 1.2 kpc could explain the PAMELA excess, but such a configuration has a probability of only 0.37 percent. Reproducing also the ATIC bump would require a very large, nearby sub-halo, which is extremely unlikely (p $\sim 3 \times 10^{-5}$). It is even less probable for the smaller Fermi bump to be caused by the presence of such an object. In either case, we predict Fermi will detect the gamma-ray emission from the sub-halo. We conclude that under canonical assumptions, the cosmic ray lepton anomalies are unlikely to originate from a nearby cold Dark Matter sub-halo.

1. Introduction

The positron fraction (that is the positron flux divided by the flux of both positrons and electrons) recently measured by PAMELA is higher than most estimations[1]. This excess starts around 50 GeV and may be related to the one seen at much higher energies ($\sim$ 500 GeV) by ATIC[2], Fermi[3] and HESS[8] in positron plus electron flux. It is possible that theses features have astrophysical explanation but it has also been suggested that annihilating Dark Matter of the Galactic halo may produce these unexpected electrons and positrons.

Relic density considerations require the annihilation cross-section of Dark Matter to be $\langle\sigma v\rangle \simeq 3 \times 10^{-26}\text{cm}^3.\text{s}^{-1}$ and Galactical Dark Matter distribution is fixed thanks to N-body simulations such as Via Lactea 2 [6]. When one applies these conditions and uses propagation models[9] in agreement with various cosmic ray data, it appears that the positron flux is far too low to produce an excess above the astrophysical expectation[4].

2. Boost factors

Two kinds of process have been suggested to "boost" the exotic component of electron/positron cosmic rays. The first one comes from particle physics and consists in finding mechanisms that can increase the annihilation cross-section without changing the thermal density. Here we have not considered particle physics boost. The other mechanism comes from the consideration that the Dark Matter halo of our Galaxy, as it is computed by Via Lactea 2 is not smooth but extremely clumpy. In these sub-haloes the Dark Matter density is much higher and hence the annihilation rate too.

It has been shown[7] that the astrophysical boost, on average, cannot be extremely large. Moreover, as Pamela has seen no excess in anti-proton to proton ratio [3], one should be careful not to overproduce anti-proton while boosting the positrons.

Unlike heavier cosmic ray species, positrons and electrons, when propagating in the Galaxy, loose a lot of energy. Therefore, the ones we detect at the Earth cannot come from much further than 1 kpc away from us. The question we have tried to answer is, how likely is it, considering the sub-halo distribution computed by Via Lactea 2, that the features we are dealing with are produced by a nearby sub-halo ?

3. The method

Considering a few model-independent annihilation channels and masses for the Dark Matter particle, we have computed the positron fraction (or the electron plus positron flux) as in [4] and added the exotic component from the big Galactic halo and from a nearby sub-halo. For each candidate of Dark Matter, we have found the position and size of sub-halo which gives the best fit to the data (either PAMELA or Fermi). Of course this result depends a lot on the electron flux for which we cannot rely on any recent measurement. The choice of the propagation model (even within Boron/Carbon constraints) can also affect the result, but one should not expect extremely different conclusions.

Once we have found the best distance to the Sun and luminosity of the sub-halo that matches observations, we have looked how often this kind of sub-halo actually appears in the results of the many runs of the Via Lactea 2 simulation. From this, we can estimate the probability that the PAMELA (or the Fermi) feature is due to a nearby sub-structure of Dark Matter.

4. Results

Results are summed up in Figure 1. What appears is that for light candidates of Dark Matter (a few hundred of GeV), to fit the PAMELA data, one requires a quite heavy sub-structure situated at less than 1 kpc away, this kind of sub-structures, according to Via Lactea 2, is expected to be much further ($\sim$ 7 kpc) therefore the associated probability is very low (0.37%). If we go to heavier Dark Matter particle candidates (1 TeV), then one needs a sub-structure almost three hundred times heavier and even closer to the Sun. These substructures weight a non-negligible fraction of the total Galactic halo and are extremely unlikely to exist ($\sim 3 \times 10^{-3}$%).

If we consider the cases of the ATIC and Fermi excesses, because the features appears at much higher energy, we need heavy Dark Matter particle ($\geq$1 TeV). Because the annihilation rate of Dark Matter is suppressed by the mass of the particle to the square, we need very bright substructures, so heavy that they can even be of the same mass than the whole Galactic halo. This is absolutely excluded. However, for these very heavy Dark Mater candidates, one should keep in mind that the particle physics boost factor may come into play. If the annihilation cross section is indeed higher at low velocity, it may be possible to release these constraints.

In all the cases we have considered here, the annihilation of Dark Matter should also create gamma rays at a rate high enough for the sensitivity of the Fermi telescope. Therefore, one should expect a clear answer to this problem very soon.

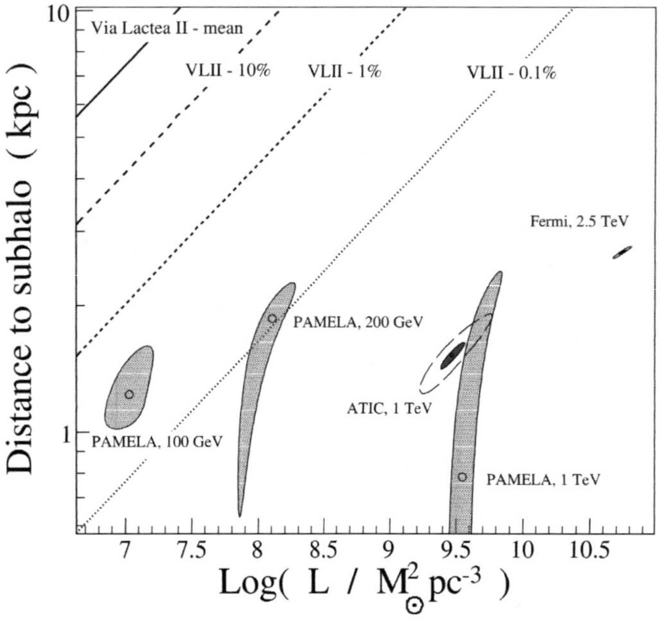

Figure 1. In this plot, the luminosity of the nearest sub-halo is plotted against its distance from the Sun. The full straight line corresponds to the average realisation to the Via Lactea 2 simulation, the dashed and dotted lines delimit the phase space that is reached by less than 10, 1 or 0.1 % of the realisations of the simulation. The elliptic contours correspond to the sub-halo required for the best fit to cosmic ray data, for masses growing from 100 GeV to 2.5 TeV. Only one experiment has been fitted at once.

5. Conclusion

Though there are some way-outs due to the uncertainties on the electron flux and the propagation of cosmic rays, it is extremely unlikely that the ATIC/Fermi/HESS feature can be caused by a nearby sub-halo of Dark Matter. If one tries to explain only the Pamela spectrum, the required sub-halo is a little bit more in agreement with the computations of the N-body simulation Via Lactea 2 but not enough to be seriously considered as a solution to the Pamela puzzle. Therefore it seems extremely difficult to explain the present cosmic ray data with a relic cold Dark Matter even if it is annihilating only into light leptons.

Acknowledgments

T.D. is grateful for financial support from the Rhône-Alpes region (Explora'Doc) and from the International Doctorate on AstroParticle Physics (IDAPP). He would also like to thank the organizers of TAUP for their very good organisation in such difficult conditions.

References

[1] Adriani, O., et al. Nature, **458**, 607 (2009).
[2] J. Chang *et al.*, Nature **456**, 362 (2008).
[3] Abdo, A. A., et al. 2009, Physical Review Letters, **102**, 181101
[4] Delahaye, T., Lineros, R., Donato, F., Fornengo, N., Lavalle, J., Salati, P., & Taillet, R. 2009, Astron. & Astroph., **501**, 821
[5] Brun, P., Delahaye, T., Diemand, J., Profumo, S., & Salati, P. 2009, Physical Review D, **80**, 035023
[6] Diemand, J., Kuhlen, M., Madau, P., Zemp, M., Moore, B., Potter, D., & Stadel, J. Nature **454**, 735 (2008)
[7] Lavalle, J., Yuan, Q., Maurin, D., & Bi, X.-J. 2008, Astron. & Astroph., **479**, 427
[8] Aharonian, F., et al. 2008, Physical Review Letters, 101, 261104
[9] Maurin, D., Donato, F., Taillet, R., & Salati, P. 2001, Astroph. J., 555, 585

Galactic positrons and electrons from dark matter and astrophysical sources

Roberto Lineros

Dipartimento di Fisica Teorica, Università di Torino and INFN – Sezione Torino, via P. Giuria 1, I–10125 Torino, Italy.

E-mail: lineros@to.infn.it

Abstract. A very interesting puzzle about the origin of electron and positron cosmic rays is deduced from the latests experimental results. We model the propagation of such cosmic rays in terms of a successfully tested two–zone propagation model. Several theoretical uncertainties – like ones related to propagation – are considered to study different types of electron and positron sources: dark matter annihilation, secondary production, and supernova remnants.

1. Introduction

During last decades, many efforts have been done to understand the nature of cosmic rays. Recently, the experimental results for the positron and electron signals obtained by PAMELA [1], FERMI [2], and HESS [3] have presented and confirmed a very interesting puzzle related to the origin of such cosmic rays.

The positron fraction, the ratio among number of positrons and number of electrons plus positrons, presents an increment for energies above ∼50 GeV. Similar phenomenon appears in the total flux of electron and positron, which presents a change of behavior. For energies below 100 GeV, the total flux is similar to a power–law with a power index of ∼ 3.3. Above 100 GeV this limit, it changes to a power index of ∼ 3.

In this proceeding, we discuss and study the electron and positron propagation model, different cosmic rays sources as dark matter annihilation and astrophysical ones. We focus to study the supernova remnants as a feasible source of electrons.

2. Electron and positron propagation

The propagation is modeled using a two–zone propagation model [4], in which, cosmic rays number density per unit of energy (ψ) is governed by the following transport equation:

$$\frac{\partial \psi}{\partial t} + \nabla \cdot \left(-K_0 \epsilon^{\delta} \nabla \psi + \mathbf{V}_c \psi \right) + \frac{\partial J_\epsilon}{\partial \epsilon} = q_{\rm src} \,. \tag{1}$$

For pedagogical purpose, the transport equation (eq. 1) is divided into four main parts. The first term represents the temporal evolution which is essential to describe propagation when transient sources are present. The second one describes processes as diffusion due to random magnetic fields – parameterized by K_0 and δ – and drift by the galactic wind $\mathbf{V}_c$. The third one rules the cosmic rays energy losses and gains (J_ϵ). Electrons and positron at energies larger than 10 GeV are dominated mainly by energy losses related to inverse Compton scattering with the interstellar radiation field and synchrotron radiation with galactic magnetic fields. The fourth term is the source term which is directly related to electron and positron production processes. The propagation takes place inside a cylinder – centered in the Galactic Center and oriented like the Galactic Plane – with a radius set in 20 kpc and half–thickness that typically varies in the range of 1–20 kpc.

2.1. Propagation uncertainties

Different propagation models, described by different propagation parameter sets, are compatible with cosmic rays observations [5]; All these models size the propagation uncertainties. The analysis of the ratio secondary cosmic rays/primary cosmic rays – like boron/carbon – are sensitive to different propagation parameter sets. Actually, some nuclei cosmic rays are produced mainly by the spallation of primary cosmic rays on the interstellar gas and not by acceleration in supernova remnants.

For the study of electron and positron cosmic rays, we use all the boron/carbon compatible parameter sets [5]. The compatible sets serve to size the propagation uncertainties in the electron and positron channel [6; 7].

3. Sources and fluxes

Electron and positron sources can be classified by their origin. A first studied case is the production from galactic dark matter annihilation [6; 8]. Electrons, positrons, and other cosmic rays species are not exclusively produced by the dark matter annihilation. Generally, those are the result of decays and/or hadronization processes of unstable particles – like quarks and gauge bosons – likewise created in the annihilation. As well, the cosmic rays production depends of the dark matter distribution; Regions where dark matter is denser are regions where annihilation events are more probable. Another point to consider is that dark matter annihilation corresponds to a type of source which is not localized in the galactic plane, as difference to most of astrophysical sources. Each discussed point reflects its influence on the propagated electron and positron fluxes when different propagation models are considered [6].

Another mechanism to produce electrons and positrons is the interaction between nuclei cosmic rays and the interstellar gas. We study the secondary production of positron from spallations of proton and alpha particle cosmic rays on hydrogen and helium present in the interstellar medium [7; 8]. The propagated secondary positron flux is more dependent on the propagation uncertainties than other factor, for instance, nuclear cross sections. Even though, our estimations are encompassed with current observations. Showing there is consistency among propagation models for nuclei cosmic rays and for electrons and positrons.

A third type of sources are supernova remnants and pulsars. Both types are transients sources, in which, most of electrons and positrons are accelerated and injected into the interstellar space in a very short time compared to cosmic rays propagation scales. A difference with previously discussed sources is that those are rather inhomogeneous in the nearby region to the solar system. Usually these kind of source are taken as a smooth distribution across the whole galaxy due to the diffusive propagation of cosmic rays, but when sources are closer enough – on time and distance – this approximation is no longer valid [9]. In figure 1, we present different examples of supernova remnants at different distances and ages. We show how closer remnants contribute more intense to the electron flux. Nevertheless, younger ones dominates the high energy part of the flux as consequence of electrons have been affected less time by energy losses.

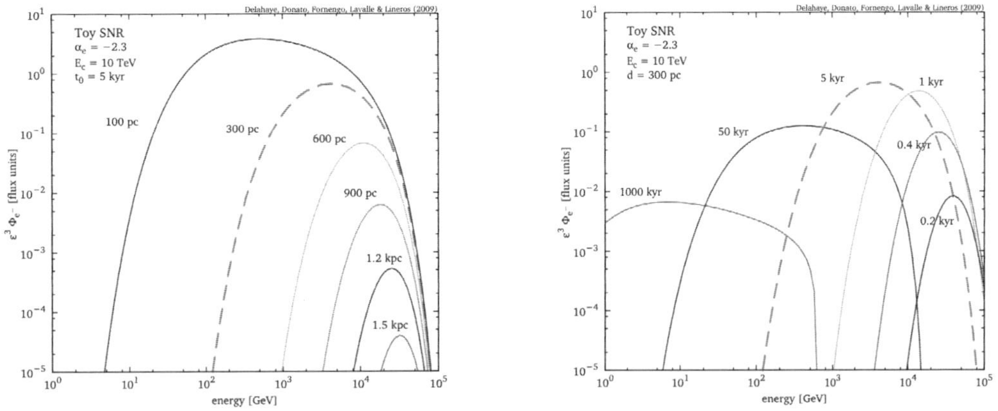

Figure 1. Electron flux ($\varepsilon^3\,\Phi$) versus energy. Left panel shows fluxes produced by supernova remnants with same age but at different distances. Right panel similarly shows cases of remnants with different birth ages and fixed distance to the solar system. Taking as reference the flux from the remnant at 5 kyr and 300 pc, we observe closer sources contributes more than farther ones. On the other hand, older ones contributes at lower energies because particles have lost more energy.

4. Conclusions

The improvement in current electron and positron measurements and data has revealed a very interesting puzzle. Many solution have been proposed during the years. We present the importance of theoretical uncertainties related to propagation and production of electron and positron cosmic rays [6–8]. We stress the necessity to re-estimate secondary and primary component, and to consider already known sources in order to discard/confirm possible presence of an undiscovered component, like the case of dark matter annihilations.

Acknowledgments

Work supported by research grants funded jointly by Ministero dell'Istruzione, dell'Università e della Ricerca (MIUR), by Università di Torino (UniTO), by Istituto Nazionale di Fisica Nucleare (INFN) within the Astroparticle Physics Project and by the Italian Space Agency (ASI) under contract N I/088/06/0. R.L. acknowledges to T. Delahaye, F. Donato, N. Fornengo, J. Lavalle, P. Salati, and R. Taillet for unvaluable comments and suggestions.

References

[1] Adriani O *et al.* 2009 *Nature* **458** 607–609 (*Preprint* `arXiv:0810.4995`)
[2] Abdo A A *et al.* 2009 *Physical Review Letters* **102** 181101–+ (*Preprint* `arXiv:0905.0025`)
[3] Aharonian F *et al.* 2008 *Physical Review Letters* **101** 261104–+ (*Preprint* `arXiv:0811.3894`)
[4] Ginzburg V L *et al.* 1980 *ApSS* **68** 295–314
[5] Maurin D *et al.* 2001 *ApJ* **555** 585–596 (*Preprint* `arXiv:astro-ph/0101231`)
[6] Delahaye T *et al.* 2008 *PRD* **77** 063527–+ (*Preprint* `arXiv:0712.2312`)
[7] Delahaye T *et al.* 2009 *A&A* **501** 821–833 (*Preprint* `arXiv:0809.5268`)
[8] Lineros R A 2008 Ph.D. thesis Università degli Studi di Torino (*Preprint* `arXiv:0812.4272`)
[9] Delahaye T *et al.* 2009 *Preprint DFTT 51/2009 and LAPTH 1339/09*

Gravitational reheating in quintessential inflation

Stefano Scopel

School of Physics and Astronomy, Seoul National University, Gwanakro 599, Gwangak-gu, Seoul 151-749, Korea

E-mail: scopel@phya.snu.ac.kr

Abstract. Quintessential inflation assumes a common origin for inflation and the dark energy of the Universe. In this scenario reheating can occur through gravitational particle production during the inflation-kination transition. We provide a detailed study of gravitational reheating in quintessential inflation and determine the temperature T_* at which radiation starts dominating over kination, generalizing previous analyses only available for the standard case when inflation is followed by an era dominated by the energy density of radiation, The value of T_* is controlled by the Hubble parameter H_0 during inflation and the transition time Δt, scaling as $H_0^2\,[\ln(1/H_0\Delta t)]^{3/4}$ for $H_0\Delta t \ll 1$ and $H_0^2(H_0\Delta t)^{-c}$ for $H_0\Delta t \gg 1$. The model-dependent parameter c is found to be around 0.5 in two different parametrizations for the transition between inflation and kination.

1. Introduction

It is somehow intriguing that both periods of accelerated expansion of the Universe we are aware of, i.e. inflation and Dark Energy, can be explained by some scalar field, the inflaton and quintessence respectively, whose potential energy dominates the energy density at some time of its evolution, so it is natural to imagine to unify the two scenarios, identifying the inflaton and quintessence with the same field ϕ. As a consequence of this, various models of quintessential inflation have been discussed in the literature (see references in [1]). In the quintessential inflation scenario two main qualitative properties emerge: i) The potential $V(\phi)$ needs to account for the large mismatch between the inflationary plateau V_0 at the beginning of the ϕ field evolution and the tiny scale V_f of the quintessential tail eventually accounting for the cosmological constant today, so that at the end of its slow–roll phase the field ϕ experiences a strong acceleration as it "deep-dives" from V_0 toward V_F. Thus the Universe undergoes a period of "kination" expansion when its energy density is dominated by the kinetic energy of ϕ. ii) In this scenario the standard reheating mechanism which is usually assumed to create the initial plasma by the decay of ϕ is not at work, but the mechanism of gravitational particle production is able to reheat the Universe. In this talk I will summarize the results of [1], where a quantitative analysis for the amount of particle production through gravitational reheating is provided, generalizing the analysis of Ref. [2], where the same discussion was restricted to the case of the inflation-radiation transition. The actual value of T_* has important phenomenological consequences. For instance it may change the predictions for the relic density of a thermal cold dark matter candidate or the thermal leptogenesis induced by the CP-violating decay of a right-handed neutrino.

2. Gravitational particle creation

Using the formalism of Bogoliubov coefficients the energy density produced at the end of inflation by gravitational reheating is given by:

$$\rho = \frac{1}{(2\pi a)^3 a} \int d^3\vec{k}k|\beta_k|^2. \tag{1}$$

with:

$$|\beta_k|^2 = \frac{k^2 r_k^2 + (r_k')^2 + \frac{1}{4r_k^2} - k}{2k}.. \tag{2}$$

In the above expression r_k is the solution of the differential equation:

$$r_k'' + \left(\omega_k^2(\eta) - \frac{1}{4r_k^4}\right) r_k = 0, \tag{3}$$

where η is the conformal time, and a prime indicates a derivative with respect to η, and has to be calculated at $\eta \to \infty$ with boundary condition $r_k(\eta \to -\infty) = \frac{1}{\sqrt{2k}}$.

The function $\omega_k(\eta) = k^2+V = k^2 - \frac{a''}{a}$ depends on the scale factor $a(\eta)$ and is uniquely defined during the inflation and radiation/kination periods, while a certain degree of arbitrariness arises from the modelization of the transition period. For this reason in our analysis we will adopt two different parametrizations for $a(\eta)$ during the transition from inflation to kination. The first one is polynomial parametrization joining smoothly the two asymptotic behaviours at early and late times:

$$f(x) = \begin{cases} 1/x^2 & \text{for } x < -1 \quad \text{(inflation)} \\ a_0 + a_1 x + a_2 x^2 + a_3 x^3 & \text{for } -1 < x < x_0 - 1 \quad \text{(transition)} \\ b_0 + b_1 x & \text{for } x > x_0 - 1 \quad \text{(kination)}, \end{cases} \tag{4}$$

with $a_0 = -\frac{1}{x_0} + 6$, $a_1 = -\frac{3}{x_0} + 8$, $a_2 = -\frac{3}{x_0} + 3$, $a_3 = -\frac{1}{x_0}$, $b_0 = 3 + 3x_0 - x_0^2$, $b_1 = 2 + 3x_0$. Here $x = H_0\eta$, H_0 is the value of the Hubble constant during inflation, and $x_0 \equiv H_0 \Delta\eta$ parametrizes the duration $\Delta\eta$ in conformal time of the transition. The second parametrization is a smooth transition applied to the equation of state:

$$w(y) = \tanh\left(\frac{2y}{y_0}\right), \tag{5}$$

where $y \equiv \log(a) + \text{const}$, and y_0 parametrizes the duration of the transition in e–foldings.

3. Numerical Results

A crucial element for the phenomenology of quintessential inflation is the epoch when the standard cosmology with radiation domination starts. The amount of radiation produced by gravitational reheating is very small, so that at the end of inflation the energy of the Universe is dominated by the kinetic energy of the quintessence field. However, the energy density of kination ρ_K is red-shifted away much faster than that of radiation ρ_R as $\rho_K \propto a^{-6}$ and $\rho_R \propto a^{-4}$, thus the latter can eventually dominate at some temperature T_*, given by:

$$T_* = \frac{2 \times 30^{\frac{1}{4}}}{\sqrt{3\pi}} \frac{g_I}{g_*^{\frac{1}{2}}} \frac{I^{\frac{3}{4}}}{b_1} \frac{H_0^2}{m_P}. \tag{6}$$

much below the initial (reheat) temperature T_I. In the above expression m_P is the Plank mass, $g_{I,*}$ are the numbers of relativistic degrees of freedom at $T_{I,*}$ while the combination $I^{\frac{3}{4}}/b_1$, whose

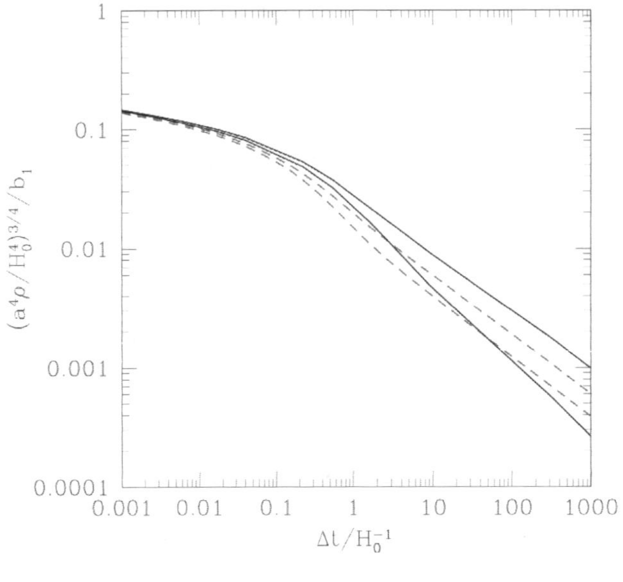

Figure 1. The combination $I^{\frac{3}{4}}/b_1$, entering into the kination–radiation equality temperature T_* given in Eq.(6), as a function of the transition time $\Delta t/H_0^{-1}$ expressed in absolute time. Solid lines are for the transition given in Eq. (5), and dashed lines are for the transition in Eq. (4). In both cases the upper line is the result of a numerical integration of Eq. (3) and the lower one is obtained by making use of a semi–analytic approximation (see text).

expression can be found in [1], is plotted in Fig. 1 in terms of the transition time $H_0\Delta t$ for the two cases when the transition is modeled according to Eqs.(4) and (5). In both cases the upper line is the result of a numerical integration of Eq. (3) and the lower one is obtained using of the approximation $\beta_k \simeq -i/(2k)\int_{-\infty}^{+\infty} e^{-2ik\eta}(-a''/a)\,d\eta$, valid when $\omega(\eta)$ varies slowly with η.

From the above analysis we find the following numerical value of T_*:

$$T_* \sim \frac{g_I}{\sqrt{g_*}}\frac{H_0^2}{m_P}\begin{cases} 0.09\,[\ln \frac{1}{H_0\Delta t}]^{\frac{3}{4}} & \text{for} \quad \Delta t \to 0 \\ {}^{0.03}_{0.039}\,(H_0\Delta t)^{-0.5}_{-0.48} & \text{for} \quad \Delta t \to \infty, \end{cases} \tag{7}$$

which depends on the the inflation scale H_0 and the transition time Δt. In particular we find that T_* can be anywhere above the Big Bang Nucleosynthesis temperature, with potentially interesting consequences for phenomenology.

Acknowledgments

This talk is based on work done in collaboration with E.J. Chun and I. Zaballa.

References

[1] E. J. Chun, S. Scopel and I. Zaballa, *Gravitational reheating in quintessential inflation*, 2009 *JCAP* **0907**, 022 [arXiv:0904.0675 [hep-ph]]

[2] Ford L H, *Gravitational Particle Creation and Inflation*, 1987 *Phys. Rev. D* **35** 2955

Journal of Physics: Conference Series **203** (2010) 012053 doi:10.1088/1742-6596/203/1/012053

Relic density of neutrinos with primordial asymmetries

Sergio Pastor

IFIC (CSIC-Universitat de València), Ed. Instituts, Apt. 22085, 46071 València, Spain

E-mail: `sergio.pastor@ific.uv.es`

Abstract. We study flavor neutrino oscillations in the early universe, assuming primordial ν-$\bar{\nu}$ asymmetries. Including collisions and pair processes in the kinetic equations, we estimate the degree of flavor equilibration and, for the first time, kinetic equilibration among neutrinos and with the ambient plasma. Typically the restrictive BBN bound on the $\nu_e\bar{\nu}_e$ asymmetry indeed applies to all flavors as previously claimed, but fine-tuned initial asymmetries always allow for a large surviving neutrino excess radiation that may show up in precision cosmological data.

The state of the hot medium in the early universe before and around big-bang nucleosynthesis (BBN) is fixed by a small number of cosmological parameters, most importantly the baryon abundance that nowadays is measured most accurately with precision cosmological data. In addition there are three unknown flavor neutrino asymmetries, that modify the cosmic neutrino density and, for the case of the $\nu_e\bar{\nu}_e$ asymmetry, influence $n \leftrightarrow p$ reactions and thus the primordial ^{4}He abundance (see [1] for a review and previous references). It is often assumed that sphalerons equilibrate the lepton and baryon asymmetries, implying that all neutrino asymmetries are of order 10^{-9} or smaller, leaving no observable trace. On the other hand, sphalerons have never been observed, so a non-zero cosmic neutrino asymmetry would raise fundamental questions about both electroweak physics and the early universe.

Nowadays flavor neutrino oscillations driven by the "solar" and "atmospheric" mass differences are well established (see e.g. [2]). In the early universe, flavor conversions are suppressed by large matter effects. Oscillations driven by $\Delta m^2_{\rm atm}$ begin only at $T \sim 10$ MeV whereas those driven by $\Delta m^2_{\rm sol}$ begin as late as 3 MeV, still early enough to achieve strong flavor conversions before BBN [3, 4, 5, 6]. Therefore, all flavors must have similar asymmetries at BBN and the restrictive limit on the ν_e degeneracy parameter ($\xi_{\nu_e} \equiv \mu_{\nu_e}/T$), conservatively $-0.04 < \xi_{\nu_e} < 0.07$ [7], seemingly applies to all flavors. Based on this it was often concluded that primordial asymmetries could not cause a significant increase of the cosmic neutrino density without violating the BBN bound on ξ_{ν_e}. Assuming that neutrinos indeed reach perfect kinetic and chemical equilibrium, the possible increase of the cosmic radiation density is small, $\Delta N_{\rm eff} = \frac{30}{7}\,(\xi_\nu/\pi)^2 + \frac{15}{7}\,(\xi_\nu/\pi)^4 < 0.006$. Here, the usual "effective number of neutrino families" measures the cosmic radiation density, after e^+e^- annihilation, by $\rho_{\rm rad}/\rho_\gamma = 1 + (7/8)\,(4/11)^{4/3}\,N_{\rm eff}$.

We stress, however, that Refs. [4, 5, 6] do not claim that neutrinos after flavor conversions follow Fermi-Dirac distributions in terms of a common degeneracy parameter and the temperature of the ambient plasma. In particular, the transfer of entropy to the electromagnetic plasma was not considered and the final neutrino distributions not studied.

If all initial asymmetries have the same sign, oscillations alone are enough to achieve

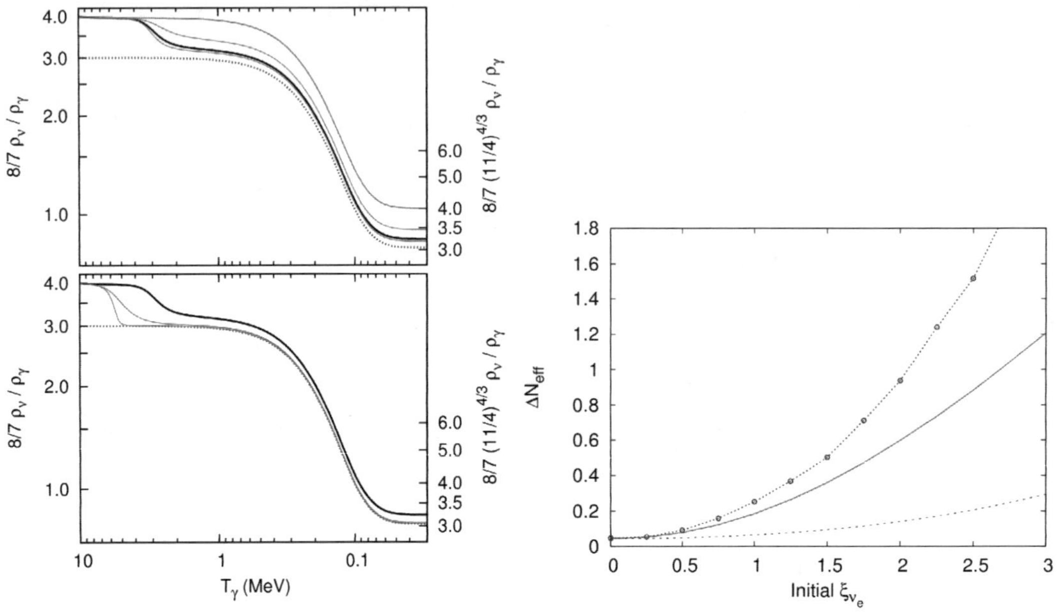

Figure 1. (Left) Evolution of the neutrino energy density. The vertical axis is marked with N_{eff}, left before e^+e^- annihilation, right afterwards. Dotted lines: Baseline case without asymmetries. *Top panel*: $\theta_{13} = 0$, solid lines from top to bottom: $\sin^2\theta_{12} = 0,\, 0.1,\, 0.3$, and 0.5. *Bottom panel:* $\sin^2\theta_{12} = 0.3$, the top solid curve is $\sin^2\theta_{13} = 0$, while the other solid curves correspond to $\sin^2\theta_{13} = 0.04$ with normal (top) and inverted (bottom) hierarchy. **(Right)** Final ΔN_{eff} as a function of the initial ξ_{ν_e} (solid: $\theta_{13} = 0$, dotted: $\sin^2\theta_{13} = 0.04$) for a zero global asymmetry. The upper dotted line is the surviving ΔN_{eff} when the final ξ_{ν_e} is in the range allowed by BBN.

approximate flavor equipartition and the BBN limits on the $\nu_e\bar{\nu}_e$ asymmetry will apply to all flavors. A dangerous case arises, however, if one begins with opposite asymmetries in two flavors. Oscillations, with or without collisions, will achieve approximate flavor equipartition separately among neutrinos and antineutrinos, so in the end all asymmetries vanish. On the other hand, the original excess radiation remains. The efficiency of $\nu\bar{\nu} \to e^+e^-$ entropy transfer determines if the BBN limits on the $\nu_e\bar{\nu}_e$ asymmetry still preclude a large ΔN_{eff}. The main point of our recent paper [8] is to study this case that was not covered in previous treatments [4, 5, 6]. Here we summarize our main results, while we refer the reader to ref. [8] for a description of the corresponding neutrino kinetic equations and how we solved them.

Our main example is the "dangerous case" of two large but opposite initial flavor asymmetries, defined in analogy to the baryon asymmetry as $\eta_{\nu_\alpha} = (n_{\nu_\alpha} - n_{\bar{\nu}_\alpha})/n_\gamma = (\pi^2\xi_{\nu_\alpha} + \xi^3_{\nu_\alpha})/12\zeta(3)$. To be specific, we first consider an exactly vanishing global lepton asymmetry, which requires $2\,(\pi^2\xi_{\nu_x} + \xi^3_{\nu_x}) + (\pi^2\xi_{\nu_e} + \xi^3_{\nu_e}) = 0$, where $\xi_{\nu_x} = \xi_{\nu_\mu} = \xi_{\nu_\tau}$. In the left plot of Fig. 1 we show the evolution of ρ_ν/ρ_γ, properly normalized so that it corresponds to N_{eff} at early and late times, first for the baseline case without asymmetries (dotted lines). The fast drop of ρ_ν/ρ_γ at $T \sim 0.2$ MeV represents photon heating by e^+e^- annihilation. At late times the dotted lines end at $N_{\text{eff}} = 3.046$ instead of 3 because of residual neutrino heating [9].

Next we show (solid lines) the evolution for our main example where $\xi_{\nu_x} = -0.64$ and $\xi_{\nu_e} = 1.17$, corresponding to an initial value $N_{\text{eff}} = 4.0$. The neutrino mass differences and θ_{23} are fixed to their best-fit values [2]. In the upper panel we assume $\theta_{13} = 0$ and for the upper curve also $\theta_{12} = 0$ (no oscillations), leading to an unchanged final $N_{\text{eff}} = 4.0$. The bottom curve is for $\sin^2\theta_{12} = 0.5$ (maximal mixing), showing that at $T_\gamma \sim 3$ MeV the neutrino density depletes relative to photons, when oscillations driven by $\Delta m^2_{\rm sol}$ begin and the flavour asymmetries are reduced. The main effect of collisions is to keep the neutrino spectra closer to their equilibrium form. Physically the excess of entropy is transferred from neutrinos to the electromagnetic plasma, cooling the former and heating the latter. Practically the same behavior is found for the next curve representing $\sin^2\theta_{12} = 0.3$, the experimental best fit, where in the end $N_{\rm eff} = 3.24$. A further curve is shown for the value $\sin^2\theta_{12} = 0.1$, remaining far from equilibrium until the end. In the bottom panel we use the measured $\sin^2\theta_{12} = 0.3$ and repeat the case $\theta_{13} = 0$ (top curve). The other two curves are for $\sin^2\theta_{13} = 0.04$, near the experimental limit, for the normal and inverted hierarchies. As expected, a nonvanishing θ_{13} together with $\Delta m^2_{\rm atm}$ drives the system towards equilibrium at a time when collisions are still efficient. In our numerical example, a significant deviation from equilibrium survives only for a very small θ_{13}. In this case the outcome depends on θ_{12} as found in the original studies [4, 5, 6].

Next we study the final deviation from equilibrium for a range of initial asymmetries $0 \le \xi_{\nu_e} \le 3$, assuming a vanishing global asymmetry as before. In the right plot of Fig. 1 we show the final outcome in terms of the residual excess radiation density $\Delta N_{\rm eff}$, for the experimental best-fit value $\sin^2\theta_{12} = 0.3$ and either $\theta_{13} = 0$ or $\sin^2\theta_{13} = 0.04$. We find that the value of θ_{13} is crucial in determining how close the system comes to equilibrium, which however is never perfect for any value of θ_{13}. If $\sin^2\theta_{13} < 10^{-3}$, the outcome is very close to $\theta_{13} = 0$.

We note that for large initial ξ_{ν_e} the final electron neutrino degeneracy can be so large that it is excluded by BBN. However, we are not bound to the assumption of an initially vanishing global asymmetry, but we can use slightly more negative initial ξ_{ν_x} values such that the final ξ_{ν_e} is pulled into the range allowed by BBN, yet a significant excess radiation density survives as shown in Fig. 1 (right plot), being quite similar to $\Delta N_{\rm eff}$ for the simple case of a vanishing global asymmetry. Thus, up to small adjustments of initial conditions, the simple case of a vanishing global asymmetry gives us a fair idea of a possible surviving residual radiation density.

In summary, the established neutrino mixing parameters together with BBN constraints on ξ_{ν_e} typically assure that the surviving excess radiation density is immeasurably small, in agreement with the previous literature. However, imperfect kinetic and chemical equilibrium achieved by the solar mixing parameters allows for exceptions. In particular, large but opposite primordial asymmetries of two flavors can be fine–tuned such that the ξ_{ν_e} limit is respected, yet a large excess radiation density survives with $\Delta N_{\rm eff}$ of order unity or larger that may be detectable in future cosmological precision data [10, 11, 12, 13]. If precision cosmology were to turn up a nonvanishing $\Delta N_{\rm eff}$, it could be a remnant of a primordial neutrino asymmetry, although the exact interpretation would strongly depend on θ_{13}. In this sense, future measurements of θ_{13} are important for the interpretation of cosmological precision data.

Acknowledgments

This work was partly supported by the Spanish grants FPA2008-00319 (MICINN) and PROMETEO/2009/091 (Generalitat Valenciana).

[1] Iocco F, Mangano G, Miele G, Pisanti O and Serpico P D 2009 *Phys. Rep.* **472** 1
[2] Schwetz T, Tórtola M and Valle J W F 2008 *New J. Phys.* **10** 113011
[3] Lunardini C and Smirnov A Yu 2001 *Phys. Rev.* D **64** 073006
[4] Dolgov A D, Hansen S H, Pastor S, Petcov S T, Raffelt G G and Semikoz D V 2002 *Nucl. Phys.* B **632** 363
[5] Wong Y Y Y 2002 *Phys. Rev.* D **66** 025015
[6] Abazajian K N, Beacom J F and Bell N F 2002 *Phys. Rev.* D **66** 013008
[7] Serpico P D and Raffelt G G 2005 *Phys. Rev.* D **71** 127301
[8] Pastor S, Pinto T and Raffelt G G 2009 *Phys. Rev. Lett.* **102** 241302
[9] Mangano G, Miele G, Pastor S, Pinto T, Pisanti O and Serpico P D 2005 *Nucl. Phys* B **729** 221
[10] Bashinsky S and Seljak U 2004 *Phys. Rev.* D **69** 083002
[11] Hannestad S, Tu H and Wong Y Y Y 2006 *J. Cosmol. Astropart. Phys.* JCAP06(2006)025
[12] Perotto L *et al* 2006 *J. Cosmol. Astropart. Phys.* JCAP10(2006)013
[13] Hamann J, Lesgourgues J and Mangano G 2008 *J. Cosmol. Astropart. Phys.* JCAP03(2008)004

Constraining super-critical string/brane cosmologies with astrophysical data

Vasiliki A Mitsou

Instituto de Física Corpuscular (IFIC), CSIC – Universitat de València,
Apartado de Correos 22085, E-46071 Valencia, Spain

E-mail: vasiliki.mitsou@ific.uv.es

Abstract. We discuss fits of unconventional dark energy models to the available data from high-redshift supernovae, distant galaxies and baryon oscillations. The models are based either on brane cosmologies or on Liouville strings in which a relaxation dark energy is provided by a rolling dilaton field (Q-cosmology). Such cosmologies feature the possibility of effective four-dimensional negative-energy dust and/or exotic scaling of dark matter. We find evidence for a negative-energy dust at the current era, as well as for exotic-scaling ($a^{-\delta}$) contributions to the energy density, with $\delta \simeq 4$, which could be due to dark matter coupling with the dilaton in Q-cosmology models. We conclude that Q-cosmology fits the data equally well with the ΛCDM model for a range of parameters that are in general expected from theoretical considerations.

1. Introduction

There is a plethora of astrophysical evidence today, from supernovae measurements [1, 2], the cosmic microwave background [3], baryon oscillations [4] and other cosmological data, indicating that the expansion of the Universe is currently accelerating. The energy budget of the Universe seems to be dominated at the present epoch by a mysterious dark energy component. Many theoretical models provide possible explanations for the latter, ranging from a cosmological constant [5] to super-horizon perturbations [6] and time-varying quintessence scenarios [7], in which the dark energy is due to a smoothly varying scalar field dominating cosmology in the present era. In the context of string theory, such a time-dependent 'quintessence' field is provided by the scalar dilaton field of the gravitational string multiplet [8–10].

2. Dissipative Q-cosmology basics

Most of the astrophysical analyses so far are based on effective four-dimensional Robertson-Walker Universes, satisfying on-shell dynamical equations of motion of the Einstein-Friedman form. Even in modern approaches to brane cosmology, described by equations deviating during early eras of the Universe from the standard Friedman equation, the underlying dynamics is assumed to be of classical equilibrium (on-shell) nature.

However, cosmology may not be an entirely classical equilibrium situation. The initial Big Bang or other catastrophic cosmic event, such as a collision of two brane worlds in the modern approach to strings, which led to the initial rapid expansion of the Universe, may have caused a significant departure from classical equilibrium dynamics in the early Universe, whose signatures may still be present at later epochs including the present era. Q-cosmology is a specific dark energy model, being associated with a rolling dilaton field that is a remnant of this non-equilibrium phase, that was formulated [10, 11] in the framework of non-critical string theory [8, 12].

3. Data analysis

We highlight the results [13, 14] of the confrontation of cosmological models with data on high-redshift supernovae [2], differential galaxy ages [15] and baryon acoustic oscillation [4]. The predictions for the Hubble rate $H(z)$ of the following three cosmological models are investigated: (a) ΛCDM, a CDM model with a cosmological constant [5]; (b) the super-horizon model [6], where the Universe is assumed to be filled with non-relativistic matter only and there is no dark energy of any sort; and (c) Q-cosmology [10, 11].

We assume a parametrisation for $H(z)$ in the Q-cosmology framework at late eras, such as the ones pertinent to the data ($0 < z < 2$), where some analytic approximations are allowed [13]:

$$\frac{H(z)}{H_0} = \sqrt{\Omega_3(1+z)^3 + \Omega_\delta(1+z)^\delta + \Omega_2(1+z)^2}\ , \quad \Omega_3 + \Omega_\delta + \Omega_2 = 1, \tag{1}$$

with the densities $\Omega_{2,3,\delta}$ corresponding to present-day values ($z = 0$). The a^{-2}-scaling contribution is a feature of the dilaton relaxation and should not be confused with the spatial curvature contribution. A complete analysis of the non-critical and dilaton effects, which turn out to be important in the present era after the inclusion of matter, requires a numerical treatment [11]. In general, the three parameters to be determined by the fit are Ω_3, Ω_δ and δ. In earlier studies [13], a fixed value of $\delta = 4$ was assumed for simplicity, whilst a more complete analysis was performed in ref. [14].

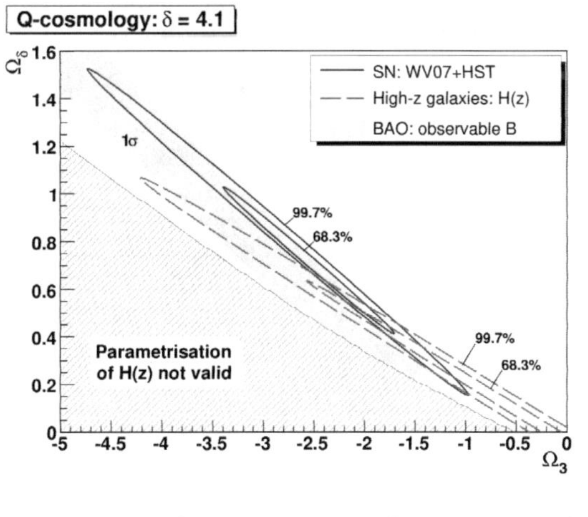

Figure 1. Constraints on Q-cosmology for $\delta = 4.1$: (a) 68.3% and 99.7% C.L. contours for supernova data (blue solid line); (b) 68.3% and 99.7% C.L. contours for $H(z)$ (red dashed line); and (c) 1σ region from the baryon acoustic oscillations (yellow area).

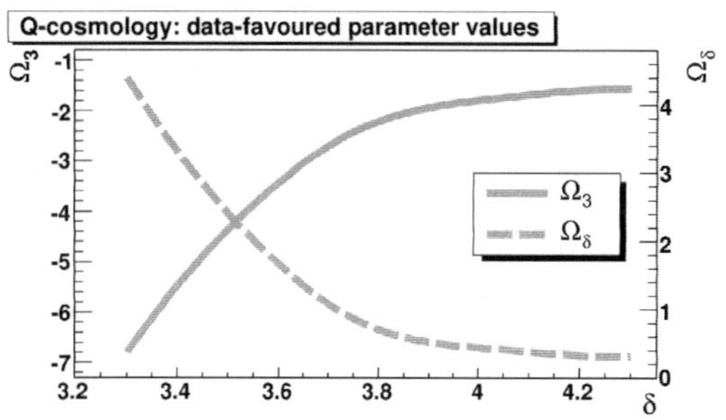

Figure 2. Astrophysical data-favoured values of Q-cosmology parameters as defined from the overlap of the 1σ regions from supernovae, differential galaxy ages and baryon oscillations data.

The ΛCDM model, as expected, fits the data very well and is mainly constrained by the supernova and BAO data. The super-horizon model, on the other hand, is excluded by the BAO-measured matter density and yields contradictory (at 2σ level) parameter determinations by SN and galactic data [14].

The spatially-flat Q-cosmology predictions, as parametrised in (1) for $z \lesssim 2$, fits the data very well providing an alternative scenario to account for the dark energy component of the Universe. The best-fit region for $\delta = 4.1$ is shown in fig. 1. The three model parameters are not univocally determined; their allowed range and the correlation between them is rather well defined. The parameter δ, associated to the exotic scaling of dark matter in the context of Q-cosmology, is restricted in the range $3.3 \lesssim \delta \lesssim 4.3$, whilst the present values of (exotic) matter densities, Ω_3 and Ω_δ, vary with δ as shown in fig. 2.

The fact that the value of $\delta = 4$ is included in the allowed range, also points towards dark radiation terms in brane models [16]. The data also point towards negative-energy dust contributions, which could be due to either dark-energy dilaton terms in Q-cosmologies [10, 13], or Kaluza-Klein compactification (massive) graviton modes in brane-inspired modes [17].

4. Other probes of Supercritical String Cosmology (SSC)

In SSC [18], D-particle defects in space-time lead to Quantum Gravity 'foam. These point-like stringy defects and the interactions of strings with them lead to non-critical strings, giving rise thus to contributions to Q-cosmology-like situations. The modified dispersion relations in turn produce delays in the propagation of photons. Possible probes of these effects are high energy γ-rays from distant sources, such as gamma-ray bursts [19], active galactic nuclei and pulsars. Relevant observations and analyses have been performed by MAGIC [20], HESS [21] and Fermi [22], with the latter setting the stringent limit from GRB090510 [23]. The role of source effects has to be unfolded in the interpretation of possible time lags observed.

The presence of the time-dependent dilaton also affects the dark matter relic density calculation, since it modifies the Boltzmann equations. A dilution in the neutralino density of $\mathcal{O}(10)$ is predicted [24], widening the allowed parameter space of Supersymmetry at collider searches. The LHC signatures themselves are also affected, favouring final states such as Higgs+jets+MET, Z+jets+MET and 2τ+jets+MET, where MET is the missing transverse energy [25].

5. Conclusions and outlook

We confirmed [14] the results of earlier studies [13] on high-redshift supernovae data, establishing stringent constraints on cosmological models by analysing, in addition, data from distant galaxies and baryon acoustic oscillations. We demonstrated that cosmological models with no dark energy, such as Q-cosmology, may be viable alternatives to the Standard ΛCDM model. The data point towards negative-energy dust contributions, which could be due to either dark-energy dilaton terms in Q-cosmologies [10, 13], or Kaluza-Klein compactification (massive) graviton modes in brane-inspired modes [17].

Further detailed studies, such as the theoretical determination of the position of the acoustic peaks in the CMB spectrum, using the underlying formalism of the above models, or the measurement of the pertinent (complicated, in general, z-dependent) equation of state are certainly required in order to provide further evidence that might discriminate among the various models. In addition to the CMB spectrum, further constraints on Q-cosmology may be established by data on look-back times of galaxy clusters and by GRBs.

Acknowledgments

This work was conducted in collaboration with Nick Mavromatos, whom I thank for insightful discussions on the theoretical model. The author acknowledges support from the Ramón y Cajal contract RYC-2007-00631 of the Spanish Ministry of Science and Innovation (MICINN) and the CSIC. This work was supported in part by MICINN under the projects FPA2006-13238-C02-01 and FPA2006-03081

References

[1] Riess A G *et al* 2007 *Astrophys. J.* **659** 98 [arXiv:astro-ph/0611572] and references therein
[2] Davis T M *et al* 2007 *Astrophys. J.* **666** 716 [arXiv:astro-ph/0701510]
[3] Dunkley J *et al* [WMAP Collaboration] 2009 *Astrophys. J. Suppl.* **180** 306 [arXiv:0803.0586 [astro-ph]]
[4] Eisenstein D J *et al* 2005 *Astrophys. J.* **633** 560 [arXiv:astro-ph/0501171]
[5] Bahcall N A *et al* 1999 *Science* **284** 1481
[6] Kolb E W, Matarrese S, Notari A and Riotto A 2005 *Preprint* arXiv:hep-th/0503117; Kolb E W, Matarrese S and Riotto A 2006 *New J. Phys.* **8** 322
[7] Upadhye A, Ishak M and Steinhardt P J 2005 *Phys. Rev.* D **72** 063501 [arXiv:astro-ph/0411803]
[8] Antoniadis I, Bachas C, Ellis J R and Nanopoulos D V 1989 *Nucl. Phys.* B **328** 117
[9] Gasperini M, Piazza F and Veneziano G 2002 *Phys. Rev.* D **65** 023508 [arXiv:gr-qc/0108016]
[10] Ellis J R, Mavromatos N E and Nanopoulos D V 2005 *Phys. Lett.* B **619** 17 [arXiv:hep-th/0412240]; Ellis J R, Mavromatos N E, Nanopoulos D V and Westmuckett M 2006 *Int. J. Mod. Phys.* A **21** 1379 [arXiv:gr-qc/0508105]
[11] Diamandis G A, Georgalas B C, Mavromatos N E and Papantonopoulos E 2002 *Int. J. Mod. Phys.* A **17** 4567; Diamandis G A, Georgalas B C, Mavromatos N E, Papantonopoulos E and Pappa I 2002 *Int. J. Mod. Phys.* A **17** 2241; Diamandis G A, Georgalas B C, Lahanas A B, Mavromatos N E and Nanopoulos D V 2006 *Phys. Lett.* B **642** 179
[12] Ellis J R, Mavromatos N E and Nanopoulos D V 1992 *Phys. Lett.* B **293** 37; 1995 *Mod. Phys. Lett.* A **10** 1685; 1999 *Chaos Solitons Fractals* **10** 345; David F 1988 *Mod. Phys. Lett.* A **3** 1651; Distler J and Kawai H 1989 *Nucl. Phys.* B **321** 509; Distler J, Hlousek Z and Kawai H 1990 *Int. J. Mod. Phys.* A **5** 391; Mavromatos N E and Miramontes J L 1989 *Mod. Phys. Lett.* A **4** 1847; D'Hoker E and Kurzepa P S 1990 *Mod. Phys. Lett.* A **5** 1411
[13] Ellis J R, Mavromatos N E, Mitsou V A and Nanopoulos D V 2007 *Astropart. Phys.* **27** 185 [arXiv:astro-ph/0604272]; Mavromatos N E and Mitsou V A 2006 *Preprint* arXiv:astro-ph/0611788
[14] Mavromatos N E and Mitsou V A 2008 *Astropart. Phys.* **29** 442 [arXiv:0707.4671 [astro-ph]]; Mitsou V A 2007 *Preprint* arXiv:0708.0113 [astro-ph]
[15] Jimenez R, Verde L, Treu T and Stern D 2003 *Astrophys. J.* **593** 622; Simon J, Verde L and Jimenez R 2005 *Phys. Rev.* D **71** 123001
[16] Brax P and van de Bruck C 2003 *Class. Quant. Grav.* **20** R201 [arXiv:hep-th/0303095] and references therein
[17] Minamitsuji M, Sasaki M and Langlois D 2005 *Phys. Rev.* D **71** 084019 [arXiv:gr-qc/0501086]
[18] Mavromatos N E and Nanopoulos D V 2009 *AIP Conf. Proc.* **1166** 26 [arXiv:0902.1507 [hep-ph]] and references therein
[19] Amelino-Camelia G, Ellis J R, Mavromatos N E, Nanopoulos D V and Sarkar S 1998 *Nature* **393** 763 [arXiv:astro-ph/9712103]; Ellis J R, Farakos K, Mavromatos N E, Mitsou V A and Nanopoulos D V 2000 *Astrophys. J.* **535** 139 [arXiv:astro-ph/9907340]; Ellis J R, Mavromatos N E and Nanopoulos D V 2008 *Phys. Lett.* B **665** 412 [arXiv:0804.3566 [hep-th]]; Ellis J R, Mavromatos N E and Nanopoulos D V 2009 *Phys. Lett.* B **674** 83 [arXiv:0901.4052 [astro-ph.HE]]; Mavromatos N E 2009 *J. Phys. Conf. Ser.* **174** 012016 [arXiv:0903.0318 [astro-ph.HE]]
[20] Albert J *et al* [MAGIC Collaboration and Other Contributors] 2008 *Phys. Lett.* B **668** 253 [arXiv:0708.2889 [astro-ph]]
[21] Aharonian F *et al* 2008 *Phys. Rev. Lett.* **101** 170402 [arXiv:0810.3475 [astro-ph]]
[22] Abdo A A *et al* [Fermi LAT and Fermi GBM Collaborations] 2009 *Science* **323** 1688
[23] Abdo A A *et al* [Fermi LAT and Fermi GBM Collaborations] 2009 *Preprint* arXiv:0908.1832 [astro-ph.HE]
[24] Lahanas A B, Mavromatos N E and Nanopoulos D V 2007 *Phys. Lett.* B **649** 83 [arXiv:hep-ph/0612152]; Lahanas A B, Mavromatos N E and Nanopoulos D V 2007 *PMC Phys.* A **1** 2 [arXiv:hep-ph/0608153]
[25] Dutta B, Gurrola A, Kamon T, Krislock A, Lahanas A B, Mavromatos N E and Nanopoulos D V 2009 *Phys. Rev.* D **79** 055002 [arXiv:0808.1372 [hep-ph]]

Topics in Astroparticle and Underground Physics (TAUP 2009) IOP Publishing
Journal of Physics: Conference Series **203** (2010) 012055 doi:10.1088/1742-6596/203/1/012055

Phantom dark energy and the Steady State 'on the average' universe.

Anton N. Baushev
Bogoliubov Laboratory of Theoretical Physics, Joint Institute for Nuclear Research
141980 Dubna, Moscow Region, Russia
E-mail: baushev@gmail.com

Abstract. We consider the hypothesis that the dark energy pervading our universe is phantom ($-p > \rho$). The density of such a substance grows with the cosmological expansion and may become infinite in a finite time producing a Big Rip. In this report we analyze the late stages of the universe evolution and demonstrate that the presence of the phantom energy in the universe is not enough in itself to produce the Big Rip. This singularity occurrence requires the fulfillment of some additional, rather strong conditions. A more probable outcome of the cosmological evolution is the decay of the phantom field into 'normal' matter. The second, more intriguing consequence of the presence of the phantom field is the possibility to introduce a cosmological scenario that does not contain a Big Bang. In the framework of this model the universe eternally expands, while its density and other physical parameters oscillate over a wide range, never reaching the Plank values. Thus, the universe evolution has no singularities at all.

1. Introduction

Experimental data [1, 2, 3] indicate convincingly an accelerating cosmological expansion. The substance responsible for this effect is usually called dark energy. If we speculate that the universe is filled with a perfect fluid with equation of state $p = \alpha\rho$, the accelerating expansion appears if $\alpha < -1/3$. The case of the dark energy with $\alpha < -1$ (it is usually called phantom energy) has currently come to the attention. Physical properties of a dark energy with $\alpha < -1$ differ markedly from the case when $\alpha \geq -1$. The difference most important for the cosmology becomes apparent when we consider expansion of the universe containing dark energy. If $\alpha \geq -1$ the dark energy density is not increasing (or even decreasing) as the universe expands. All bound systems (such as the solar one) remain bound forever; the dark energy presence manifests itself only on cosmological scales. If we allow $\alpha < -1$, the dark energy density grows and becomes infinite in a finite time. Increasing gravitational repulsion produced by the phantom energy will first destroy Galaxies and then any bound systems including elementary particles [4, 5]. This phenomenon is usually called Big Rip. However, as we will see below, realization of such a catastrophic scenario requires the fulfilment of several supplementary conditions. Even if the phantom energy prevails in the universe, the Big Rip does not necessarily occurs.

There is another less evident consequence of $\alpha < -1$: if dark energy is really the phantom one, the universe evolution needs not to contain a Big Bang.

2. The phantom energy-dominated universe evolution

Presence of the phantom energy in the universe is not enough in itself to produce the Big Rip. This phenomenon appears, for instance, if the coefficient averaged over the universe contents $\alpha = const < -1$. However, it is beyond reason to believe that α remains constant during the cosmological evolution. In order to illustrate this let us at first consider a very simple situation when a universe contains only uniformly distributed phantom energy with an ordinary Lagrangian

$$\mathcal{L} = -\frac{\partial_\xi \phi\, \partial^\xi \phi}{2} - V(\phi) \quad (1)$$

For such a field

$$\rho = -\dot{\phi}^2/2 + V(\phi), \quad p = -\dot{\phi}^2/2 - V(\phi) \quad (2)$$

One can see that α is not a constant: it depends on the amplitude of the field and on the potential energy density $V(\phi)$. This question has been discussed in close detail in [6]. To summarize briefly the results: if the potential is not very steep (grows slower than $V(\phi) \propto \phi^4$) then α tends to -1, and the density becomes infinite only when $t \to \infty$. So, no Big Rip appears in this case, though the universe reaches the Planck density in a finite time. For steeper potentials a big rip singularity appears even if $\alpha \to -1$. Even the parameter α can tend to $-\infty$ for a very steep $V(\phi)$. Finally, if $V(\phi)$ has a maximum, $\alpha \to -1$ and $\rho \to const$ [7, 10]. It is important to emphasize that a very steep potential is necessary to provide a constant $\alpha < -1$: for any polynomial potential, for instance, α tends to -1.

The second effect that is able to prevent the Big Rip is a possible phantom energy interaction generating 'normal' particles. The normal substance gives positive contribution to the pressure that decreases the α value averaged over the universe contents. Eventually the Big Rip may be prevented.

It is generally believed that the phantom energy does not interact with 'normal matter', but it cannot be absolutely true, if for no other reason than the gravitational interaction. Let us consider a toy illustration for a start. As noted above, in the phantom scenario increasing gravitational repulsion produced by the phantom energy will finally destroy even elementary particles. We consider a hadron (more precisely, a quark-antiquark pair $q\bar{q}$) ripped by the gravitational forces. When the distance between the quark and antiquark increases, the color field lines of force between them are pressed together into a string-like region. The gravitational repulsion stretches the color lines of force until the increasing potential energy becomes sufficient to create another $q\bar{q}$ pair. It divides the string on two strongly-coupled $q\bar{q}$ pairs, but they are also ripped up by the gravitational repulsion. So, this leads to intensive hadron production in the moments just before the Big Rip.

Another mechanism of 'normal' matter generation in the phantom-dominated universe is particle production in the cosmological gravitational field. In [11] the process was considered under assumption that the universe is filled with a perfect fluid with $\alpha = const < -1$. The generated 'normal' matter density can be written as

$$\rho_{norm} = C\eta^\beta, \quad \text{where} \quad \beta = \frac{4}{1+3\alpha} \quad (3)$$

In order to examine the problem in more detail we consider a universe, initially filled with a perfect fluid with $\alpha = -4/3$. Integration of the system of cosmological equations [8] shows that the Hubble constant grows progressively slower, reaches its maximum H_{max}, and begins to decrease. On the contrary, α constantly increases from the initial value that is close to $-4/3$. Finally it becomes positive. Of course, both the effects result from the high growth fraction of the 'normal' matter in the universe that makes a positive contribution to the pressure. It seems reasonable to contend that in the course of the phantom-dominated universe evolution the quantity of 'normal' matter becomes comparable with the phantom one, and the value of α grows hindering from the Big Rip. To this we can add that we have considered the case of $\alpha = const$, which claims an extremely steep phantom field potential $V(\phi)$, as we could see above. For a more realistic case of the polynomial potential α tends to zero even in the absence of 'normal matter'. This makes the occurrence of the Big Rip highly unlikely.

3. Discussion

Of course, the suggested model is ingenuous. Intensive phantom field may interact in some other way decaying into 'normal' matter. Whatever the mechanism of the transformation may be, the universe after it has the following properties:

(i) It has just passed through the stage of very rapid (at least, exponential) expansion.
(ii) It is flat and homogeneous.
(iii) It is 'normal' matter-dominated.
(iv) The Hubble constant is very large.

It is precisely these physical conditions that existed in our universe ~ 13.7 milliard years ago, just after the inflation. If we assume that the inflation was caused by the phantom field (the equations of the phantom cosmology can be closely analogous to those of the chaotic inflation [8]), this makes possible to construct a model of universe evolution that does not contain a Big Bang. The universe leaves inflation being matter-dominated. As it expands the density of matter decreases, while the phantom density grows: eventually the universe passes into the phantom-dominated stage. As this takes place, the total density and the Hubble constant stop diminishing and begin to increase. Gradually the expansion becomes very fast, leading to an inflation-lake stage. Finally the phantom field decays into 'normal' matter, and the cycle repeats. Thus, the universe eternally expands, while its density and other physical parameters oscillate over a wide range, never reaching the Plank values. The universe evolution has no singularities like a Big Bang or a Big Rip. By analogy with the Steady State theory [12, 13, 14, 9], the considering model may be called the Steady State *on the average* universe.

If the phantom energy does exist, the considering model is an interesting instance of cosmological solutions containing no singular points like a Big Bang or a Big Rip. It is worthy of note that, though in our case the universe density never reaches infinity (or, more precisely, Planck density), all the physical parameters vary over a very wide range and amount up to extremal values. So, the universe passes through the stages that can be called physically critical.

[1] D. N. Spergel, L. Verde, H. V. Peiris, et al., *ApJS*, **148**, 175, (2003)
[2] A. G. Riess, A. V. Filippenko, P. Challis, et al., *Astron. J.*, **116**, 1009, (1998)
[3] S. Perlmutter, G. Aldering, G. Goldhaber, et al., *ApJ*, **517**, 565, (1999)
[4] R. R. Caldwell, *Phys. Lett. B*, **545**, 23, (2002), arXiv:astro-ph/9908168
[5] R. R. Caldwell, M. Kamionkowski, and N. N. Weinberg, *Phys. Rev. Lett.*, **91**, 071301 (2003)
[6] M. Sami, A. Toporensky, *Modern Physics Letters A*, **19**, Issue 20, 1509, (2004), arXiv:gr-qc/0312009
[7] S. M. Carroll, M. Hoffman, M Trodden, *Phys. Rev. D*, **68**, 023509, (2003)
[8] A. N. Baushev, *in press*, (2009), arXiv:0809.0235
[9] G. W. Gibbons, *Grav. Cosmol.*, **8**, 2, (2002), arXiv:hep-th/0104015
[10] P. Singh, M. Sami, N. Dadhich, *Phys. Rev. D*, **68**, 023522, (2003), [hep-th/0305110]
[11] A.B. Batista, J.C. Fabris, and S.J.M. Houndjo, arXiv:0710.0999
[12] F. Hoyle and J. V. Narlikar, *Proc. Roy. Soc.*, **273**, 1, (1963)
[13] F. Hoyle and J. V. Narlikar, *Proc. Roy. Soc.*, **278**, 465, (1964)
[14] F. Hoyle and J. V. Narlikar, *Proc. Roy. Soc.*, **282**, 178, (1964)

Dirac-Born-Infeld and k-inflation: the CMB anisotropies from string theory

Christophe Ringeval

Theoretical and Mathematical Physics Group, Centre for Particle Physics and Phenomenology, Louvain University, 2 Chemin du Cyclotron, 1348 Louvain-la-Neuve, Belgium

E-mail: `christophe.ringeval@uclouvain.be`

Abstract. Inflationary models within string theory exhibit unusual scalar field dynamics involving non-minimal kinetic terms and generically referred to as k-inflation. In this situation, the standard slow-roll approach used to determine the behavior of the primordial cosmological perturbations cannot longer be used. We present a generic method, based on the uniform approximation, to analytically derive the primordial power spectra of scalar and tensor perturbations. At leading order, the scalar spectral index, its running and the tensor-to-scalar ratio are modified by the new dynamics. We provide their new expression, correct previous results at next-to-leading order and clarify the definition of what is the tensor-to-scalar ratio when the sound horizon and Hubble radius are not the same. Finally, we discuss the constraints the parameters encoding the non-minimal kinetic terms have to satisfy, such as the sound speed and the energy scale of k-inflation, in view of the fifth year Wilkinson Microwave Anisotropy Probe (WMAP5) data.

1. Introduction

In the context of string theory, cosmic inflation can be achieved through the motion D-branes in higher dimensional warped and compact spacetimes [1]. There, the inflaton appears as the scalar degree of freedom associated with the position of a brane in these extra-dimensions. From a four-dimensional point of view, Lorentz invariance along a D-brane necessarily leads to four-dimensional scalar fields exhibiting non-standard kinetic terms, and more precisely of the Dirac–Born–Infeld (DBI) type [2, 3, 4]. In fact, DBI-inflation belongs to the class of k-inflationary models in which the accelerated expansion of the universe can be driven by the scalar field's kinetic terms instead of its potential energy [5]. Assuming the gravity sector to be described by General Relativity, the action of the "k-inflaton" $\varphi(x^\mu)$ reads

$$S = \frac{1}{2\kappa}\int \mathrm{d}^4x\sqrt{-g}\left[R + 2\kappa P\left(X,\varphi\right)\right], \tag{1}$$

where $\kappa \equiv 8\pi/m_{\mathrm{Pl}}^2$ and $X \equiv -g^{\mu\nu}\partial_\mu\varphi\partial_\nu\varphi/2$. The quantity $P(X,\varphi)$ is any *acceptable* functional of X and φ (see Ref. [6]). All kinetically modified inflationary models have a "speed of sound"

$$c_{\mathrm{s}}^2 \equiv \frac{P_{,X}}{P_{,X} + 2XP_{,XX}}, \tag{2}$$

which is generically different from the speed of light. In a flat Friedmann-Lemaître–Robertson–Walker (FLRW) universe, it was shown in Ref. [7], that the Mukhanov–Sasaki mode function

$\mu_{\mathrm{S}}(k,\eta) = \zeta\sqrt{2\kappa\epsilon_1}/c_{\mathrm{S}}$ (ζ being the comoving curvature perturbation) satisfies the modified equation of motion

$$\mu_{\mathrm{S}}'' + \left[c_{\mathrm{S}}^2(\eta)k^2 - \frac{\nu^2(\eta) - 1/4}{\eta^2}\right]\mu_{\mathrm{S}} = 0. \tag{3}$$

The effective potential is $(\nu^2 - 1/4)/\eta^2 = [\ln(a\sqrt{\epsilon_1}/c_{\mathrm{S}})]''$, and all derivatives are with respect to conformal time η. The quantity $a(\eta)$ stands for the FLRW scale factor while $\epsilon_1 = -\mathrm{d}\ln H/\mathrm{d}\ln a$ is the first Hubble flow function (H being the Hubble parameter). The standard form of this equation is recovered by setting $c_{\mathrm{S}} = 1$ and can be solved by defining a hierarchy of Hubble flow functions encoding the rate of change of the Hubble parameter and its higher order logarithmic derivatives: $\epsilon_{n+1} \equiv \mathrm{d}\ln|\epsilon_n|/\mathrm{d}\ln a$. Assuming slow-roll, i.e. $\epsilon_n \ll 1$, one can expand the effective potential $(\nu^2 - 1/4)/\eta^2$ and solve order by order Eq. (3) along the lines of Refs. [8, 9, 10]. Generalising this method to the k-inflationary case in which $c_{\mathrm{S}}(\eta)$ is not constant requires some care. Indeed, both the effective potential and the propagation speed are modified. In the following, we use the uniform approximation to solve Eq. (3) order by order to predict the shape of the tensor and scalar primordial fluctuations at the origin of the CMB anisotropies.

2. K-inflationary perturbations

For k-inflation, we can define a new hierarchy encoding the rate of change of the sound speed, the sound flow functions δ_i, defined by

$$\delta_{n+1} = \frac{\mathrm{d}\ln|\delta_n|}{\mathrm{d}\ln a}, \qquad \delta_0 \equiv \frac{c_{\mathrm{sin}}}{c_{\mathrm{S}}}. \tag{4}$$

Expanding both the sound speed and the effective potential in terms of the Hubble and sound flow functions around a particular conformal time η_* gives [11]

$$\nu^2(\eta) = \frac{9}{4} + 3\epsilon_{1*} + \frac{3}{2}\epsilon_{2*} + 3\delta_{1*} + \mathcal{O}(\epsilon\delta) = \nu_*^2 + \mathcal{O}(\epsilon\delta), \quad c_{\mathrm{S}}(\eta) = c_{\mathrm{S}*}\left(1 + \delta_{1*}\ln\frac{\eta}{\eta_*}\right) + \mathcal{O}(\epsilon\delta), \tag{5}$$

where all stars mean that the corresponding function is evaluated at η_*. From these expressions, one can solve Eq. (3) at first order in the flow functions ϵ_i and δ_i by using the uniform approximation [16, 17].

The scalar primordial power spectrum, at first order in Hubble and sound flow functions, then reads [11]

$$\mathcal{P}_\zeta = \frac{H_\diamond^2}{\pi m_{\mathrm{Pl}}^2\epsilon_{1\diamond}c_{\mathrm{S}\diamond}}\left[1 - 2(D+1)\epsilon_{1\diamond} - D\epsilon_{2\diamond} + (D+2)\delta_{1\diamond} - (2\epsilon_{1\diamond} + \epsilon_{2\diamond} - \delta_{1\diamond})\ln\frac{k}{k_\diamond}\right], \tag{6}$$

where $D = 1/3 - \ln 3$. All diamond indexed quantities are evaluated at the time $\eta_\diamond$ defined to be the time at which a chosen pivot wavenumber $k_\diamond$ crossed the sound horizon during inflation, i.e. the solution of $-k_\diamond\eta_\diamond = 1/c_{\mathrm{S}\diamond}$. The constant factor $18\mathrm{e}^{-3}$ typical of WKB methods has also been absorbed in the definition of $H_\diamond$ [18].

Concerning the tensor modes, their evolution is not affected by the non-standard kinetic terms and their power spectrum remains the same as in the standard case $c_{\mathrm{S}} = 1$. However, an important, and so far overlooked, difference is that the standard tensor power spectrum is evaluated at the time at which the perturbations crossed the Hubble radius during inflation. As a result, it is expressed in terms of the Hubble flow functions evaluated at a different time than $\eta_\diamond$ and one cannot evaluate a tensor-to-scalar ratio by simply dividing both power spectra. Using the Hubble and sound flow expansion, one can nevertheless express the tensor power spectrum at $\eta_\diamond$. After some algebra, one obtains

$$\mathcal{P}_h(k) = \frac{16H_\diamond^2}{\pi m_{\rm Pl}^2}\left[1 - 2\left(D + 1 - \ln c_{\rm s\diamond}\right)\epsilon_{1\diamond} - 2\epsilon_{1\diamond}\ln\frac{k}{k_\diamond}\right]. \tag{7}$$

We immediately see that the speed of sound influences $\mathcal{P}_h$, and this effect becomes all the more so important than $c_{\rm s}$ is small. The above expression explains the numerical results on the tensor-to-scalar ratio discussed in Ref. [19].

From Eqs. (6) and (7), one can deduce the scalar spectral index $n_{\rm S}$, its running $\alpha_{\rm S}$ and the tensor-to-scalar ratio at next-to-leading order [10, 11]

$$\begin{aligned}
n_{\rm S} - 1 \equiv \left(\frac{{\rm d}\ln\mathcal{P}_\zeta}{{\rm d}\ln k}\right)_{k=k_\diamond} &= -2\epsilon_{1\diamond} - \epsilon_{2\diamond} + \delta_{1\diamond} - 2\epsilon_{1\diamond}^2 - (2D+3)\epsilon_{1\diamond}\epsilon_{2\diamond} + 3\epsilon_{1\diamond}\delta_{1\diamond} + \epsilon_{2\diamond}\delta_{1\diamond} \\
&\quad - D\epsilon_{2\diamond}\epsilon_{3\diamond} - \delta_{1\diamond}^2 + (D+2)\delta_{1\diamond}\delta_{2\diamond}, \\
\alpha_{\rm S} \equiv \left(\frac{{\rm d}^2\ln\mathcal{P}_\zeta}{{\rm d}\ln^2 k}\right)_{k=k_\diamond} &= -2\epsilon_{1\diamond}\epsilon_{2\diamond} - \epsilon_{2\diamond}\epsilon_{3\diamond} + \delta_{1\diamond}\delta_{2\diamond}, \\
r \equiv \left.\frac{\mathcal{P}_h}{\mathcal{P}_\zeta}\right|_{k=k_\diamond} &= 16c_{\rm s\diamond}\epsilon_{1\diamond}\left[1 + 2\epsilon_{1\diamond}\ln c_{\rm s\diamond} + D\epsilon_{2\diamond} - (D+2)\delta_{1\diamond}\right].
\end{aligned} \tag{8}$$

The spectral index and running correct previous results at next-to-leading order [12, 13, 14], which were assuming $c_{\rm s}$ constant, and match with another method proposed by Kinney and Tzirakis in Ref. [15]. The term in $\ln c_{\rm s\diamond}$ in the tensor-to-scalar ratio has to be considered as soon as $c_{\rm s\diamond}$ becomes small enough.

3. Conclusion

From the scalar and tensor power spectra given above, one can compare the predicted CMB anisotropies with the current data. In Ref. [20], we have performed a Monte–Carlo–Markov–Chains analysis of the WMAP5 data [21] against the k-inflationary power spectra. At 95% of confidence, the flow parameters and the energy scale of k-inflation have to verify

$$0.003 \le 2(\epsilon_{1\diamond} - \delta_{1\diamond}) + (\epsilon_{2\diamond} + \delta_{1\diamond}) \le 0.075, \quad \ln\left(10^5\frac{H_\diamond}{m_{\rm Pl}}\right) \le -0.59. \tag{9}$$

Notably, due to the new degree of freedom introduced by $c_{\rm s}$, we do not longer find any bound on $\epsilon_{1\diamond}$ alone. The class of k-inflationary models is thus weakly constrained by CMB data. Let us however mention that the subclass of DBI models generate a large amount of non-Gaussianity when $c_{\rm s}$ becomes small [22]. In this later case, one can show that the current WMAP5 bounds on non-Gaussianity imposes that $\log\epsilon_{1\diamond} \le -1.1$, at two-sigma [20].

Acknowledgments

It's a pleasure to thank L. Lorenz and J. Martin for their comments on the manuscript. This work is supported by the Belgian Federal Office for Science, Technical and Cultural Affairs, under the Inter-university Attraction Pole grant P6/11.

References

[1] Kachru S *et al.* 2003 *JCAP* **0310** 013 (*Preprint* `hep-th/0308055`)
[2] Dvali G R and Tye S H H 1999 *Phys. Lett.* **B450** 72–82 (*Preprint* `hep-ph/9812483`)
[3] Kachru S, Kallosh R, Linde A and Trivedi S P 2003 *Phys. Rev.* **D68** 046005 (*Preprint* `hep-th/0301240`)

[4] Langlois D, Renaux-Petel S, Steer D A and Tanaka T 2008 *Phys. Rev. Lett.* **101** 061301 (*Preprint* 0804.3139)
[5] Armendariz-Picon C, Damour T and Mukhanov V F 1999 *Phys. Lett.* **B458** 209–218 (*Preprint* hep-th/9904075)
[6] Bruneton J P and Esposito-Farese G 2007 *Phys. Rev.* **D76** 124012 (*Preprint* 0705.4043)
[7] Garriga J and Mukhanov V F 1999 *Phys. Lett.* **B458** 219–225 (*Preprint* hep-th/9904176)
[8] Schwarz D J, Terrero-Escalante C A and Garcia A A 2001 *Phys. Lett.* **B517** 243–249 (*Preprint* astro-ph/0106020)
[9] Leach S M, Liddle A R, Martin J and Schwarz D J 2002 *Phys. Rev.* **D66** 023515 (*Preprint* astro-ph/0202094)
[10] Schwarz D J and Terrero-Escalante C A 2004 *JCAP* **0408** 003 (*Preprint* hep-ph/0403129)
[11] Lorenz L, Martin J and Ringeval C 2008 *Phys. Rev.* **D78** 083513 (*Preprint* 0807.3037)
[12] Peiris H V, Baumann D, Friedman B and Cooray A 2007 *Phys. Rev.* **D76** 103517 (*Preprint* 0706.1240)
[13] Bean R, Shandera S E, Henry Tye S H and Xu J 2007 *JCAP* **0705** 004 (*Preprint* hep-th/0702107)
[14] Henry Tye S H 2008 *Lect. Notes Phys.* **737** 949–974 (*Preprint* hep-th/0610221)
[15] Kinney W H and Tzirakis K 2008 *Phys. Rev.* **D77** 103517 (*Preprint* 0712.2043)
[16] Habib S, Heitmann K, Jungman G and Molina-Paris C 2002 *Phys. Rev. Lett.* **89** 281301 (*Preprint* astro-ph/0208443)
[17] Habib S, Heinen A, Heitmann K, Jungman G and Molina-Paris C 2004 *Phys. Rev.* **D70** 083507 (*Preprint* astro-ph/0406134)
[18] Martin J and Schwarz D J 2003 *Phys. Rev.* **D67** 083512 (*Preprint* astro-ph/0210090)
[19] Agarwal N and Bean R 2009 *Phys. Rev.* **D79** 023503 (*Preprint* 0809.2798)
[20] Lorenz L, Martin J and Ringeval C 2008 *Phys. Rev.* **D78** 063543 (*Preprint* 0807.2414)
[21] Komatsu E *et al.* (WMAP) 2008 (*Preprint* 0803.0547)
[22] Chen X, Huang M x, Kachru S and Shiu G 2007 *JCAP* **0701** 002 (*Preprint* hep-th/0605045)

Journal of Physics: Conference Series **203** (2010) 012057 doi:10.1088/1742-6596/203/1/012057

The MAJORANA Project

C.E. Aalseth[10], M. Amman[6], J.F. Amsbaugh[20], F.T. Avignone III[8,17], H.O. Back[7,11], A.S. Barabash[3], P.S. Barbeau[15], J.R. Beene[17], M. Bergevin[6], F.E. Bertrand[8], M. Boswell[5], V. Brudanin[4], W. Bugg[19], T.H. Burritt[20], M. Busch[2,11], G. Capps[8], Y-D. Chan[6], J.I. Collar[15], R.J. Cooper[8], R. Creswick[17], J.A. Detwiler[6], P.J. Doe[20], Yu. Efremenko[19], V. Egorov[4], H. Ejiri[9], S.R. Elliott[5], J. Ely[10], J. Esterline[2,11], H. Farach[17], J.E. Fast[10], N. Fields[15], P. Finnerty[11,16], B. Fujikawa[6], E. Fuller[10], V.M. Gehman[5], G.K. Giovanetti[11,16], V.E. Guiseppe[5], K. Gusey[4], A.L. Hallin[12], R. Hazama[9], R. Henning[11,16], A. Hime[5], E.W. Hoppe[10], T.W. Hossbach[10], M.A. Howe[11,16], R.A. Johnson[20], K.J. Keeter[1], M. Keillor[10], C. Keller[18], J.D. Kephart[7,10,11], M.F. Kidd[2,11], O. Kochetov[4], S.I. Konovalov[3], R.T. Kouzes[10], L. Leviner[7,11], J.C. Loach[6], P.N. Luke[6], S. MacMullin[11,16], M.G. Marino[20], R.D. Martin[6], D.-M. Mei[18], H.S. Miley[10], M.L. Miller[20], L. Mizouni[10,17], A. Montoya[5], A.W. Myers[10], M. Nomachi[9], J.L. Orrell[10], D.G. Phillips II[11,16], A.W.P. Poon[6], G. Prior[6], J. Qian[6], D.C. Radford[8], K. Rielage[5], R.G.H. Robertson[20], L. Rodriguez[5], K.P. Rykaczewski[8], A.G. Schubert[20], T. Shima[9], M. Shirchenko[4], D. Steele[5], J. Strain[11,16], G. Swift[2,11], K. Thomas[18], R. Thompson[10], V. Timkin[4], W. Tornow[2,11], T.D. Van Wechel[20], I. Vanyushin[3], K. Vetter[6,13], R. Warner[10], J.F. Wilkerson[11,16], J.M. Wouters[5], E. Yakushev[4], A.R. Young[7,11], C.-H. Yu[8], V. Yumatov[3], C. Zhang[18], S. Zimmerman[6]

[1] Department of Physics, Black Hills State University, Spearfish, SD, USA
[2] Department of Physics, Duke University, Durham, NC, USA
[3] Institute for Theoretical and Experimental Physics, Moscow, Russia
[4] Joint Institute for Nuclear Research, Dubna, Russia
[5] Los Alamos National Laboratory, Los Alamos, NM, USA
[6] Lawrence Berkeley National Laboratory, Berkeley, CA, USA
[7] Department of Physics, North Carolina State University, Raleigh, NC, USA
[8] Oak Ridge National Laboratory, Oak Ridge, TN, USA
[9] Research Center for Nuclear Physics and Department of Physics, Osaka University, Ibaraki, Osaka, Japan
[10] Pacific Northwest National Laboratory, Richland, WA, USA
[11] Triangle Universities Nuclear Laboratory, Durham, NC, USA
[12] Centre for Particle Physics, University of Alberta, Edmonton, AB, Canada
[13] Department of Nuclear Engineering, University of California, Berkeley, CA, USA
[14] Department of Physics, University of California, Berkeley, CA, USA
[15] Department of Physics, University of Chicago, Chicago, IL, USA
[16] Department of Physics, University of North Carolina, Chapel Hill, NC, USA
[17] Department of Physics and Astronomy, University of South Carolina, Columbia, SC, USA
[18] Department of Earth Science and Physics, University of South Dakota, Vermillion, SD, USA

[19] Department of Physics and Astronomy, University of Tennessee, Knoxville, TN, USA
[20] Center for Experimental Nuclear Physics and Astrophysics, and Department of Physics, University of Washington, Seattle, WA, USA

E-mail: `elliotts@lanl.gov`

Abstract. The MAJORANA Project, a neutrinoless double-beta decay experiment is described with an emphasis on the choice of Ge-detector configuration.

Experimental evidence of $0\nu\beta\beta$ would establish the Majorana nature of neutrinos. The science of $\beta\beta$ and the large number of experimental and theoretical efforts in the field are described in detail, not only in this volume, but also in many recent reviews [1, 2, 3, 4, 5, 6].

The objective of the first experimental phase of MAJORANA is to build a 60-kg module of high-purity Ge, of which 30 kg will be enriched to 86% in ^{76}Ge, to search $0\nu\beta\beta$. This module is referred to by the collaboration as the DEMONSTRATOR. The physics goals for this first phase are to: probe the neutrino mass region above 100 meV; demonstrate that backgrounds at or below 1 count/tonne-year in the $0\nu\beta\beta$-decay region of interest can be achieved that would justify scaling up to a 1 tonne or larger mass detector; and definitively test the recent claim [7] of an observation of $0\nu\beta\beta$ decay in ^{76}Ge in the mass region around 400 meV.

The DEMONSTRATOR will consist of ^{76}Ge detectors, deployed in multi-crystal modules, located deep underground within a low-background shielding environment. The technique will be augmented with recent improvements in signal processing and detector design, and advances in controlling intrinsic and external backgrounds. Recently, R&D by the MAJORANA and GERDA [8] collaborations have shown the utility of p-type, point-contact (P-PC) Ge detectors for $\beta\beta$. MAJORANA has been described elsewhere [9, 10, 11], so this short note will summarize the efforts developing these detectors.

The efficient commercial production of Ge detectors depends strongly on the yields for producing high-purity single crystals of Ge and for producing p-n junction diodes from these crystals. Single-crystal boules naturally favor right circular cylindrical geometries, making such an obvious choice to make efficient use the precious enriched Ge material.

Previous-generation $\beta\beta$ decay searches in ^{76}Ge favored detectors made of p-type crystals that can be grown more efficiently to larger dimensions than n-type crystals. In addition, charge-trapping effects are reduced, and generally slightly better energy resolution is obtained. P-type detectors are typically fabricated with a p+ B-implanted contact on an inner bore hole and an n+ Li-drifted contact on the outside electrode to obtain efficient and full depletion from the outside. While the inside B contact is very thin (typically <1 μm) the Li-drifted contact can be as thick as 1 mm due to the high mobility of Li in Ge. The thickness of the outside-Li contact provides an advantage since this dead layer absorbs α radiation from surface contamination that might generate background.

While the pulse-shape obtained at the central contact in coaxial detectors can provide radial separation of multiple interactions in any implementation, pulse-shapes obtained in segmented detectors provide improved sensitivity in the radial separation and, more importantly, in complementary directions as well. A high degree of segmentation, such as a 6x6-fold segmentation, enables the full reconstruction of γ-ray interactions within the detectors. Even without absolute event vertex reconstruction, events with multiple energy depositions can be identified and rejected. Preliminary results from GRETINA [12] and other highly-segmented HPGe arrays indicate that a minimum separation of 4 mm will be achievable. However, these advanced capabilities come at the price of a proportionally larger number of small parts such as cables and FETs, with their associated additional sources of background.

An alternative cylindrical detector design has been developed in which the bore hole is removed and replaced with a point-contact in the center of the passivated detector face [13]. The changes in the electrode structure result in a drop in capacitance to ~1 pF, reducing the electronic noise component and enabling sub-keV energy thresholds. This configuration also has lower electric fields throughout the bulk of the crystal and a weighting potential that is sharply peaked near the point contact, resulting in an increased range of drift times and a distinct electric signal marking the arrival of the charge cloud at the central electrode. These features result in an improved pulse-shape analysis (PSA) over semi-coax detectors that rivals the performance of segmented detectors. Hence, p-type versions of this configuration are our favored choice. Instrumented with modern FETs, P-PC detectors have recently been demonstrated by MAJORANA collaborators to provide very low noise and energy threshold, as well as excellent pulse-shape capabilities for distinguishing multiple interactions from single interactions [14].

In our R&D efforts, we have studied a large number of detector designs. Many of these detectors were custom made to our specifications either by vendors or in-house to explore various performance aspects. We studied segmented detectors with a commercial CLOVER detector, a custom ORTEC-fabricated [15] enriched-Ge segmented detector (SEGA), a Canberra-fabricated highly segmented GRETINA prototype, and a LBNL-fabricated segmented P-PC. We have studied passivation issues with a LBNL-fabricated P-PC, and two detectors with various ditch geometries fabricated at PHDs [16]. We also modified an n-type detector for surface α-emitter studies. Recently, a novel variation on the P-PC detector design has been proposed [17] which makes use of a coaxial hole similar in dimensions to that of a coaxial HPGe detector. In this case, however, the hole is part of the outer Li contact. In these coaxial-style P-PC detectors, the hole serves to reduce the depletion voltage of a large detector by about a factor of two, allowing the use of much longer crystals, and crystals with higher concentrations of electrical impurities.

Broad Energy Ge (BEGe) detectors [18] are commercially available. These detectors share many of the features of P-PC detectors, and generally meet the requirements for the DEMONSTRATOR. They have been produced with diameters up to 90 mm, but have a shorter maximum length of about 30 mm. We have studied a variety of sizes of these detectors. We have chosen a BEGe design that is 70-mm diameter, 30-mm high to be the primary detector configuration for the DEMONSTRATOR due to its cost effectiveness, PSA-background rejection capabilities, and simplicity of the contacts. GERDA has also recently reported a BEGe study [19]. We have purchased an initial batch of 18 of these detectors. They have an average leakage current of 3 pA, capacitance 1.3 pF, and a resolution of 1.8 keV at 1332 keV.

References

[1] Elliott S R and Vogel P 2002 *Ann. Rev. Nucl. Part. Sci.* **52** 115
[2] Elliott S R and Engel J 2004 *J. Phys. G: Nucl. Part. Phys.* **30** R 183
[3] Barabash A S, Hubert F, Huber P and Umatov V I 2004 *Phys. At. Nucl.* **67** 438
[4] Ejiri H 2005 *J. Phys. Soc. Jap.* **74** 2101
[5] Avignone F T III, King G S III and Zdesenko Y 2005 *New Journal of Physics* **7** 6
[6] Avignone F T III, Elliott S and Engel J 2008 *Rev. Mod. Phys.* **80** 481 (*Preprint* arXiv:0708.1033)
[7] Klapdor-Kleingrothaus H V and Krivosheina I V 2006 *Mod. Phys. Lett. A* **21** 1547
[8] Schönert S *et al.* 2005 *Nucl. Phys. Proc. Suppl.* **145** 242
[9] Elliott S *et al.* 2008 (*Preprint* arXiv:0807.1741)
[10] Guiseppe V *et al.* 2008 (*Preprint* arXiv:0811.2446)
[11] Henning R *et al.* 2009 (*Preprint* arXiv:0907.1581)
[12] Vetter K *et al.* 2000 *Nuclear Instruments and Methods in Physics Research A* **452** 105–114
[13] Luke P, Goulding F, Madden N and Pehl R 1989 *IEEE trans. Nucl. Sci.* **36** 926
[14] Barbeau P, Collar J and Tench O 2007 *JCAP* **09** 009
[15] 2009 ORTEC, 801 South Illinois Avenue Oak Ridge, TN 37830, USA
[16] 2009 PHDs Co. 777 Emory Valley Road, Suite B, Oak Ridge, TN 37830, USA
[17] Radford D private communication
[18] 2009 Broad Energy Ge Detectors, Canberra Industries, Inc., 800 Research Pkwy, Meriden, CT 06450, USA
[19] Budjáš D, Heider M B, Chkvorets O, Khanbekov N and Schönert S 2009 (*Preprint* arXiv:0909.4044)

Topics in Astroparticle and Underground Physics (TAUP 2009) IOP Publishing
Journal of Physics: Conference Series **203** (2010) 012058 doi:10.1088/1742-6596/203/1/012058

Theoretical status of the double beta decay

Amand Faessler

University of Tuebingen, Institute for Theoretical Physics, Auf der Morgenstelle 14, 72076 Tuebingen, Germany

E-mail: amand.faessler@uni-tuebingen.de

Abstract. The detection of the neutrinoless double beta decay would establish that the neutrino is a Majorana particle and it would allow to measure the absolute neutrino mass in the meV region. To extract from the neutrinoless double beta decay halflife the absolute neutrino mass, the transition matrix element is as important as the experimental data. The dramatic progress in the last years is reviewed to determine these transition matrix elements for the neutrinoless double beta decay. It is shown, that the values calculated in the Quasi-particle Random Phase Approach (QRPA) and the recent results in the interacting boson model agree with each other completely within the uncertainties of the theories, while the shell model yields values which in the region of ^{76}Ge are by a factor $\frac{1}{2}$ smaller than QRPA. Possible reasons for the differences and an experimental way of measuring the Fermi part for the neutrinoless double beta decay transition matrix elements are discussed. This would allow to determine experimentally which of the two approaches is the correct one. Finally we show that the renormalization parameter g_{pp}, which multiplies the nucleon-nucleon interaction matrix elements can be overconstraint together with the axial charge g_A in three nuclei, where the ground state of the intermediate nucleus in the double beta decay system is a 1^+ state and apart of the $2\nu\beta\beta$ also the β^- decay for the second leg and the electron capture (EC) in the first leg is measured. It is shown, that this overconstraint determination of g_{pp} and g_A leads to the same result for the transition matrix elements as when one adjusts g_{pp} to the $2\nu\beta\beta$ alone.

1. Introduction

The neutrinoless double beta decay is presently the only method to determine an absolute scale for the neutrino mass in the meV region. For this the transition matrix elements for the neutrinoless double beta decay are as important as the data.

$$T = \int dE_k(\nu) \sum_k \frac{< f \mid \hat{H}_W \mid k >< k \mid \hat{H}_W \mid i >}{E_{0^+}(^{76}Ge) - [E_k(e_1^-) + E_k(\nu) + E_{Kern}(^{76}As; k)]} \quad (1)$$

$$T = M_m < m_\nu > + M_\theta < tg\vartheta > + M_{WR} < (\frac{M_1}{M_2})^2 > + M_{SUSY}\lambda^{'2}_{111} + M_{VR} < \frac{m_p}{M_{VR}} > + ... \quad (2)$$

$$w = \frac{2\pi}{\hbar} \mid T \mid^2 \rho_f \leq 4.4 \cdot 10^{33}[sec^{-1}] \quad (3)$$

The transition matrix element is calculated with Fermi's golden rule in second order. In equation (1) the expression for the transition from ^{76}Ge is given. The transition matrix element in equation (2) shows on the right first the neutrino mass term, which is proportional to the Majorana double beta decay mass $< m_\nu >$. The matrix element contains additional terms.

Journal of Physics: Conference Series **203** (2010) 012058 doi:10.1088/1742-6596/203/1/012058

$< tg\vartheta >$ depends on the mixing angle ϑ of the vector bosons mediating the left-handed and vector bosons mediating the right-handed weak interaction. The term depending on the heavy vector meson mediating mainly the right-handed weak interaction with the mass M_2 originates from a left-right symmetric Grand unification, while the term with the coupling constant λ'_{111} comes from R-parity violating supersymmetry. The last term in equation (2) describes the exchange of a heavy mainly right-handed neutrino. Equation (3) shows the transition probability according to Fermi's golden rule and as long as we have no confirmed data for the neutrinoless double beta decay we have upper limits, like the one for ^{128}Te in equation (3).

The first term on the right-hand side in equation (2) is the so-called "gold plated term", which is proportional to the absolute scale of the Majorana neutrino mass for the double beta decay. It can be expressed by the three mass eigenvalues of the neutrinos and the transformation coefficients from the mass to the flavor eigenstates. So if one determines the transition matrix element T experimentally and one calculates the matrix elements M_m theoretically and one can assume that this is the leading term, one is able to determine the absolute scale of the Majorana double beta decay neutrino mass.

2. The transition matrix elements

Figure 1 shows as the points with the error bars, the theoretical transition matrix elements calculated in Tuebingen in the QRPA [1, 2]

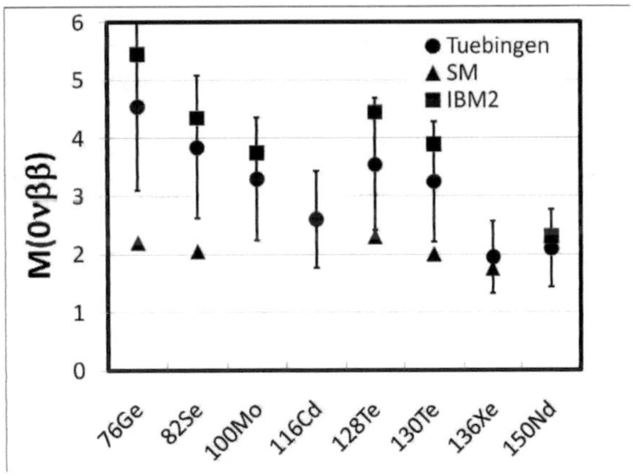

Figure 1. Dimensionless transition matrix elements of the neutrinoless double beta decay. The circles with error bars are calculated [1] in the QRPA. The error bars indicate the highest and the lowest value from a series of theoretical calculations with three different sizes of basis sets, two different short range correlations: Jastrow [3] and UCOM [4], for an unquenched axial charge $g_A = 1.25$ and the quenched value $g_A = 1.00$ and for QRPA and RQRPA for the realistic force Bonn CD. The nucleon-nucleon matrix elements are determined by solving a relativistic version of the Bethe-Goldstone equation. The error bar indicates the highest and the lowest of these 24 calculations, while the circle is put into the middle. The triangles are the result of shell model calculations from Strassburg and Madrid [5], while the squares indicate values calculated in the interacting boson model, treating protons and neutrons separately IBM 2 [6]. The point for ^{150}Nd for QRPA (Tuebingen) is calculated spherically. But the reduction factor 0.6 due to different deformations $\beta(^{150}Nd) = 0.183$ and $\beta(^{150}Sm) = 0.114$ fitted to the measured quadrupole momenta is included.

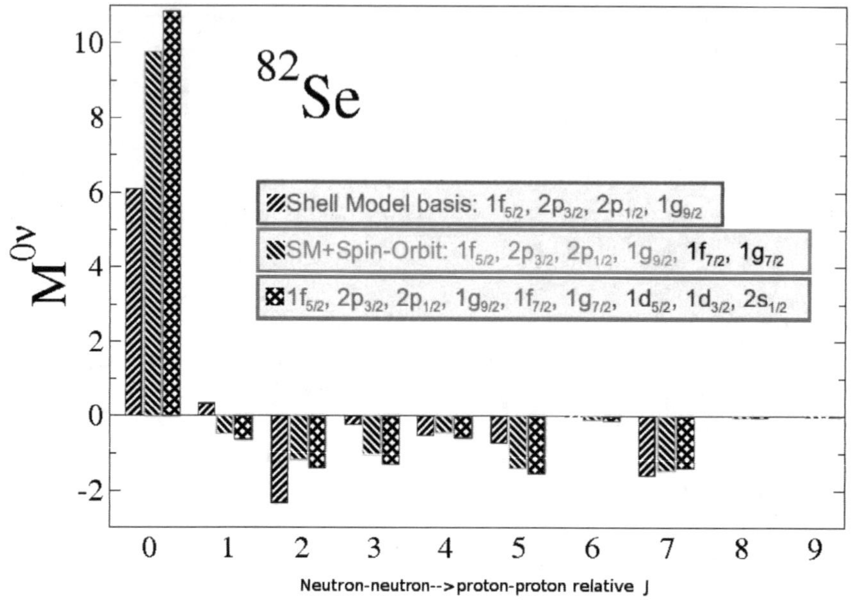

Figure 2. Contributions to the neutrinoless double beta decay matrix element in the decay ^{82}Se to ^{82}Kr for different angular momenta of the pairs of two neutrons which change into pairs of two protons. The left bar is calculated for the small shell model basis: $1f_{5/2}, 2p_{3/2}, 2p_{1/2}, 1g_{9/2}$. The second bar includes in addition the spin-orbit partners $1f_{7/2}$ and $1g_{7/2}$. This guarantees that the Ikeda sum rule is fulfilled. The addition of the spin-orbit partners changes the results appreciably. The addition of the whole sd-shell does not change much the final results. The small single particle basis for the shell model calculations seems to be the cause for the smaller values for the SM than in the QRPA approach.

The shell model is handicapped by the limitation on the number of single particle states which can be included. Figure 2 shows the results for the ^{82}Se decay. The first bar is calculated with QRPA in the shell model basis: $1f_{5/2}, 2p_{3/2}, 2p_{1/2}, 1g_{9/2}$. The second bar is calculated including the spin orbit partners $1f_{7/2}$ and $1g_{7/2}$. This guarantees that the Ikeda sum rule is fulfilled. Inclusion of the full s, d shell does not change too much anymore. Figure 2 indicates the shell model yields to small values, at least in ^{76}Ge and in ^{82}Se due to the small single particle basis. But figure 2 gives also an information on the result in the interacting boson model [6]. This model contains only S and D pairs of two neutrons, which change into two protons. Figure 2 shows that also higher multipole pairs are involved. The very good agreement between QRPA [1] and the IBM 2 comes from the fact that the contribution of the s-bosons is smaller than for the S pairs of neutrons in QRPA. But QRPA has a stronger cancellation due to higher neutron angular momentum pairs which change into proton-proton pairs. In view of this the agreement between QRPA and IBM2 shown in figure 1 is surprisingly good.

In [2] for the first time the transition matrix elements are calculated fully consistently: A realistic force Bonn CD and Argonne V18 is used to solve a relativistic approach to the Bethe-Goldstone equation in the so-called coupled-cluster method (CCM) or the exponential S approach. For the first time one uses the same realistic NN force to calculate the NN interaction matrix elements, the pairing correlations, the nuclear structure problem by QRPA and the short range correlation [2]. Under the assumption that Klapdor and coworkers [7] observed the $0\nu\beta\beta$ decay in ^{76}Ge the CCM yields for the Bonn CD a double beta decay Majorana neutrino mass $< m_\nu >= 0.22 \pm 02(exp) \pm 0.01(theor)[eV]$ and for the Argonne V18 potential $< M_\nu >= 0.24 \pm 0.02(exp) \pm 0.01(theor)[eV]$.

3. Overconstraint determination of g_{pp} and g_A

Due to the finite size of the single particle basis set, the strength of the nucleon-nucleon (NN) interaction must be slightly renormalized by a factor g_{pp}. This factor g_{pp} is usually adjusted to the two neutrino double beta decay transition probability. But in three nuclei $^{100}Mo,^{116}Cd$ and ^{128}Te we know in addition to the $2\nu\beta\beta$ halflives also the β^- decay probability from the intermediate 1^+ state to the final ground state and the electron capture (EC) from the intermediate 1^+ ground state to the initial 0^+ ground state nucleus. With this three data we can overconstrain two parameters, the renormalization constant g_{pp} and the axial charge, which has in nuclei not his vacuum value $g_A = 1.25$ but is quenched already in the pf-shell to about $g_A = 1.00$ and to smaller values in medium heavy and heavy nuclei.

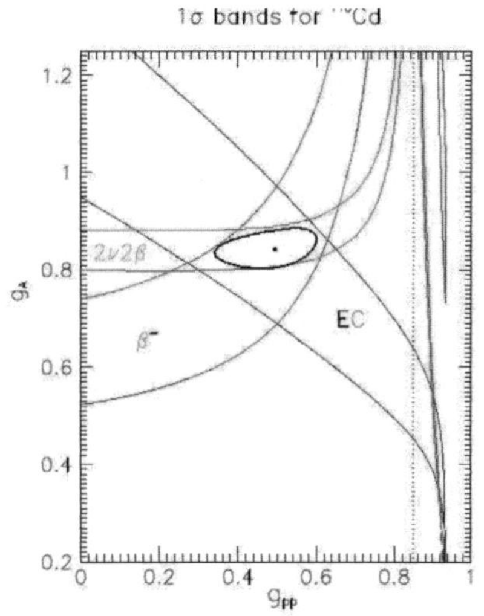

Figure 3. Overconstraint determination of the renormalization factor g_{pp} and the axial charge g_A from the $2\nu\beta\beta$ decay and the β^- transition from the 1+ ground state of the intermediate nucleus for the double beta decay system ^{116}cd to ^{116}Sn.

The overconstraint renormalization factor of the NN-interaction g_{pp} and the axial charge g_A in the three nuclei $^{100}Mo,^{116}Cd$ and ^{128}Te are used in figure 4 to calculate the neutrinoless double beta decay transition matrix element and are compared with the usual adjustment of g_{pp} to the $2\nu\beta\beta$ transition probability. The final results do not change within the uncertainties of the calculations.

4. Summary

The QRPA calculations in Tuebingen [1,2] and the calculations in Jyväskylä [9] obtain now in QRPA the same result. These results are also confirmed by Barea and Iachello [6] within the uncertainty of the calculation (see figure 1), while the shell model calculation gives in the ^{76}Ge region results, which are by a factor 2 smaller. Recently the Tuebingen group [10] proposed a method, how one can at least for the Fermi part of the $0\nu\beta\beta$ transition probability determine experimentally which result is correct. At the moment the score is 2 : 1 in favor of the QRPA result.

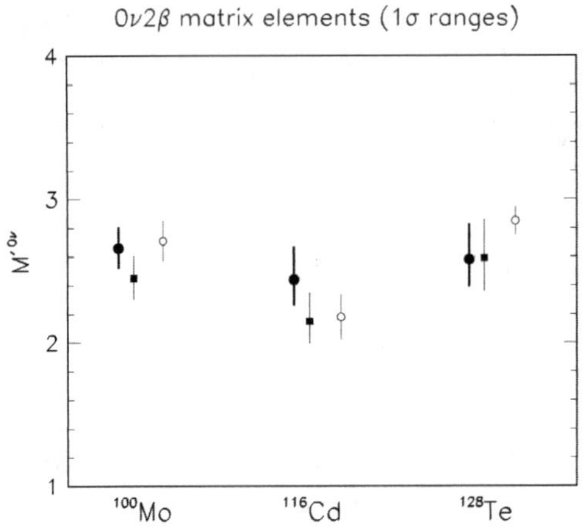

Figure 4. Neutrinoless double beta decay matrix elements for ^{100}Mo,^{116}Cd and ^{128}Te calculated with the overconstraint renormalization factor g_{pp} and axial charge g_A. The black circle is the overconstraint calculation in a large basis, while the black square gives the same results in a smaller basis. The open circle are the old results, where only g_{pp} is adjusted to $2\nu\beta\beta$ transitions. The final results did not change.

I would like to thank Dr. Vadim Rodin and Prof. Fedor Simkovic for the collaboration. The work in Tuebingen was supported by the Sonderforschungsbereich Transregio 27: "Neutrinos and beyond", by the International Graduiertenkolleg GRK 683 Basel - Graz - Tuebingen and the EU project ILIAS N6, WP1: "Double beta decay and related topics".

References

[1] F. Simkovic, A. Faessler, V. Rodin, P. Vogel, Jonathan Engel, *Phys. Rev.***C77**, 045503 (2008)
[2] F. Simkovic, A. Faessler, H. Muether, V. Rodin, M. Stauf, *Phys. Rev.***C79**, 055501 (2009)
[3] G. A. Miller, J. E. Spencer, *Arm. Phys. (N. Y)* **100**, 562 (1976)
[4] R. Roth, H. Hergert, P. Papakoutantinou, T. Neff, H. Feldmeier, *Phys. Rev.***C72**, 034002(2005)
[5] E. Courier, J. Menendez, F. Nowacki, A. Poves, *Phys. Rev. Lett.* **100**, 052503 (2008)
[6] J. Barea, F. Iachello, *Phys. Rev.* **C79**, 044301 (2009)
[7] Klapdor et al. *Mod. Phys. Lett.***A21**, 1547 (2006)
[8] A. Faessler, G. L. Fogli, E. Lisi, V. Rodin, A. M. Rotunno, F. Simkovic, *J. Phys.***G 35**, 075104 (2008)
[9] J. Suhonen, O. Civitarese, *Phys. Lett.***B668**, 277 (2008)
[10] V. Rodin, A. Faessler, http://Arxiv.org/abs/0906.1759

Development and installation of the GERDA experiment

A Smolnikov [1,2] **for the GERDA collaboration**
[1]Joint Institute for Nuclear Research, Dubna 141980, Russia
[2]Max Planck Institute for Nuclear Physics, Heidelberg 69117, Germany

E-mail: smoln@jinr.ru

Abstract. The progress in the development of the GERDA (GErmanium Detector Array) experiment is presented. The goal of the experiment is the search for neutrinoless double beta decay of ^{76}Ge with considerable reduction of background in comparison with predecessor experiments. GERDA will operate bare germanium semiconductor detectors (enriched in ^{76}Ge) submerged in high purity liquid argon supplemented by a water shield. The experimental set up is currently under construction in the underground facility of LNGS, Italy. The results of various R&D efforts and the main steps of the GERDA set up design and installation are given as well as several novel methods for background reduction are described.

1. Introduction

The goal of the GERDA (GErmanium Detector Array) experiment [1] is to search for neutrinoless double beta decay (0νββ) of ^{76}Ge. The 0νββ decay is one of the most sensitive probes of still unknown neutrino properties such as neutrino type and their mass scale. So far the best limits on 0νββ decay half-life 1.9×10^{25} y and 1.6×10^{25} y have been obtained with HPGe detectors enriched in ^{76}Ge in the predecessor Heidelberg-Moscow [2] and IGEX [3] experiments at the background ~ 10^{-1} counts/keV·kg·y . Moreover, the part of the Heidelberg-Moscow collaboration, after additional data treatment, claimed the presence of an excess of events in ROI, which they interpreted as the evidence of 0νββ observation with the best fit $T_{1/2} = 1.2\times10^{25}$ y, $\langle m_{\beta\beta}\rangle = 0.44$ eV at the value of a nuclear matrix element NME = 4.22 [4]. The aim of GERDA is to achieve the considerable reduction of background and thus better sensitivity in comparison with previous experiments.

GERDA will operate bare germanium detectors (enriched in ^{76}Ge) submerged in liquid argon (LAr) for shielding against external radiation. By removing most of the cladding and contact materials and immersing the detectors in an ultra-pure environment, one can reduce the detector background dramatically. Provided that the background can be reduced to 10^{-3} counts/keV·kg·y, it will be possible to operate detectors free of background within the planned exposure. The experimental GERDA strategy is based on three phases. In Phase I all 8 existing and reprocessed enriched HPGe detectors (18 kg of ^{76}Ge in total) from the Heidelberg-Moscow and IGEX experiments and 6 reprocessed HPGe detectors made from natural Ge (15 kg of ^{Nat}Ge) from Genius TF [5] will be used. After 1 year of data taking (an exposure of 15 kg·y with the background level 10^{-2} cts/ keV·kg·y) GERDA can scrutinize the existing claim at the high statistical level without problems with uncertainties of NME or improve the limit on the half life up to $T_{1/2} > 3\times10^{25}$ y , translated into an effective neutrino mass $\langle m_{\beta\beta}\rangle < 0.2$ – 0.5 eV for the range of NME values from 5.46 to 2.22 according to the set of more recent calculations (see Table IV from [6]). In Phase II new segmented or BEGe detectors (>20kg of ^{76}Ge) will be added. In total 40 kg of ^{76}Ge + 15 kg of ^{Nat}Ge will be used in this phase. In addition several detectors from material depleted in ^{76}Ge will be incorporated. After exposure of 100 kg·y with the background reduced up to 10^{-3} cts/ keV·kg·y, the limit on $T_{1/2}$ would improve to 2×10^{26} y leading to $\langle m_{\beta\beta}\rangle < 0.08$ - 0.21 eV and Phase II will cover the degenerate neutrino mass hierarchy.

If no signal for the 0νββ decay is found, a ton scale ^{76}Ge experiment (Phase III) with further background reduction to ~10^{-4} cts/ keV·kg·y undertaken in a worldwide collaboration GERDA-MAJORANA will be required to cover the inverted hierarchy region.

2. R&D performed for Phase I

Due to the main GERDA conceptual design aimed to operate bare HPGe detectors in LAr the success of the experiment depends strongly on the long-term stability of detector parameters in a cryogenic liquid. To verify the ability of a bare detector operation in LN_2 and LAr, three p-type non-enriched HPGe detectors (with masses 1.6 kg, 2.5 kg, and 2.5 kg of ^{Nat}Ge) were produced as the prototypes for Phase I with different designs, namely, 1) with a passivation layer covering all the borehole side, 2) with a passivation layer limited to a groove, 3) without a passivation layer at all. The detectors are mounted in the low-mass holders made of selected low-activity copper, PTFE and silicon (80 g in total per each detector) [7]. For three years of R&D and testing in the underground GERDA Detector Laboratory (GDL) at LNGS the detector handling and mounting procedures have been defined and the Phase I detector technology has been tested successfully [8].

In preparation for GERDA Phase I the IGEX and HdM detectors were removed from thier parent copper cryostats and all detectors were reprocessed by Canberra Semiconductor NV, Olen [9] without passivation layer.The detectors were stored underground during reprocessing with less than 1 week exposure above ground. The main results achieved during modification and testing of bare HPGe detectors are the following: 1) it was shown that bare Ge detectors can work directly in LN_2 and LAr with the leakage current (< 20 pA) and energy resolution (< 3 keV at 1.3 MeV) corresponding to their standard values in traditional cryostats; 2) their parameters are stable after more than 50 cooling/warming cycles in LAr; 3) the detector parameters do not deteriorate after one year of continuous operation in LAr even after irradiation with intensive gamma sources (modification without passivation layer). This shows the feasibility of the overall GERDA project and closes the question about "limited long-term stability of naked detectors in liquid nitrogen" arisen as a result of the GeniusTF investigations [10].

The other important part of the Ge detector assembly is the front-end electronics which should operate perfectly inside LAr in a distance 30 – 60 cm from the bare detectors. Several versions of ASIC charge sensitive preamplifiers were produced and investigated with the prototype detectors inside test bunches filled with LAr. The PZ-0 circuit is built in the AMS HV 0.8 μm CZX process and has discrete input FET and feed-back components. It fulfils all requirements including a bandwith of 30 MHz, an equivalent noise charge of less than 150e^- r.m.s for C_{det} = 30 pF and a rise time of 15 ns after a coaxial driving cable of 10m length. With one of the latest PZ-0 versions totally equipped for Phase I and coupled with 37 cm cable to the bare HPGe detector placed in the center of the test cryostat filled with 100 kg of LAr, an energy resolution of 2.9 keV at 1.3 MeV has been achieved.

Several MC programs have been developed for background analysis [11]. It was shown that the proposed set up allows to suppress the background from external (rocks) and internal (cryostat and crystal support elements) gamma activity to the level required for Phase I.

The radioactivity of a large fraction of the construction materials was measured by several low-background Ge gamma-spectrometers situated in different underground laboratories [12] as well as by the high sensitive systems for Rn emanation measurements [13]. The most important result is that samples of stainless steel with a ^{228}Th activity of less than 1 mBq/kg were found. This gave the possibility to build a low background stainless-steel cryostat which satisfies all GERDA requirements. The main part of construction development and mechanical engineering has been devoted to the design and manufacture of the cryostat which is realized as a double-walled super-insulated pressure vessel from stainless steel 1.4571 [14].

The performance of the water Cherenkov veto was studied and the encapsulated photodetectors were developed. The muon veto modules on the base of plastic scintillator have been developed with a light collection non-uniformity less than 15 % for the 200 x 50 x 3 cm^3 dimensions. The effective approach to discriminate multi-site deposits in the Ge detectors by using a pulse shape analysis of the signal from Phase I detectors as well as to use anticoincidence between nearby detectors assembled in several strings have been carefully investigated and quantified.

3. Installation of the GERDA setup

The construction and assembling of the experimental set up is practically completed in Hall A of the Gran Sasso National Laboratory (3800 m w.e.), Italy. A schematic drawing of the complex GERDA setup is shown in figure 1, left. A stainless steel (U/Th $\leq$ 5 mBq/kg) cryostat of 4.2 m diameter and total height (including the neck) of 8.9 m with the inner vessel volume of 64 m^3 contains 90 tons of LAr. The inner cylindrical shell of the cryostat is covered by a OFE Cu (U/Th $\leq$ 16 μBq/kg) shield with maximum thickness of 6 cm and total mass of about 20 tons (see figure 1, middle). During various acceptance tests the cryostat has been certified as a pressure vessel with a thermal loss of about 200 W. The cryostat is immersed in a water tank with a diameter of 10 m and a total height of 9 m. For a successful long term operation of the cryostat in a tank filled with water, special safety measures have been performed. The cryostat is equipped with an active cooling system using LN_2 which due to sub-cooling of LAr provides minimal losses of evaporated argon (< 0.1 m^3 / 10 days). The ultra pure water buffer (565 tons) serves as a gamma and neutron shield and, instrumented with 66 photomultipliers, as Cherenkov detector for efficiently vetoing cosmic muons. Performed MC simulations show that an efficiency of more than 99 % can be achieved reducing the muon induced background to a level of 10^{-5} cts/ keV·kg·y. Plastic scintillator panels on top of the detector with the total area of 20 m^2 will tag muons which enter the dewar through the neck with the vetoing efficiency of about 98 %.

Figure 1. Left: Schematic view of the GERDA setup. Middle: Stainless steel cryostat with copper shield and active cooling system. Right: The GERDA facility installed in the Hall A of LNGS.

The Ge detector array is made up of individual detector strings and is situated in the central part of the cryostat. A cleanroom and a radon tight lock on top of the vessel assembly allow insert and remove individual detector strings without contaminating the cryogenic volume. Radon tightness throughout the experimental volume is achieved by the exclusive use of metal seals in the components.

The main parts of the "nested type" assembly are already installed in the underground facility, see figure 1, right. Commissioning of the completed set up will start by filling the cryostat with LAr which is scheduled for November 2009.

4. R&D activities for Phase II

To reach the background level required for Phase II (<10^{-3} cts/keV·kg·y) novel methods are developed to suppress as much as possible the intrinsic background of the detectors which is mostly due to cosmogenically produced isotopes such as ^{60}Co and ^{68}Ge.

R&D are carried out to investigate performances of segmented and BEGe types of germanium detectors which can resolve multi-site energy deposits due to background events. The first version of the Phase II detector design foresees 18-fold segmented detectors. The segmentation helps to identify multiple Compton-scattering background events in the region of interest. The 3x6 fold segmented n-type true-coaxial detector has been produced from natural Ge and tested with a novel "snap contact" scheme with very small amount of extra material (19g Cu, 7g PTFE, 2.5g Kapton) [15]. The detector was running successfully in LN_2 for 5 months without deterioration of the performance with a leakage current 30 ± 5 pA. A core resolution of 4.1 keV and segment resolutions between 3.6 keV and 5.7 keV at 1.3 MeV were achieved. Detailed studies have established its power for discriminating single- and multi-site events. For internal ^{60}Co and ^{68}Ge impurities, the suppression factor is of about 10 or even better depending of its location [16]. A long term test of this detector in LAr started recently.

A small-electrode Broad Energy Ge (BEGe) detector [17] is considered as the second candidate for the Phase II detectors. The largest BEGe detector (878 g, ∅80 ×32 mm) commercially available from Canberra Semiconductor, N.V. Olen, has the energy resolution 1.6 keV at 1.3 MeV and is used now for careful investigation of its performances. It was shown that such type of Ge detectors with the applied novel single-parameter PSD method can exhibit very good pulse shape discrimination properties [18] without any segmentation and might be a cost-effective alternative to a segmented detector. On the other hand, BEGe detectors are limited in the currently achievable mass to about 1 kg and a full 3D event reconstruction like in segmented detectors can not be realized.

For the Phase II detectors additional Ge material (37.5 kg) has been enriched (> 86% of ^{76}Ge) in Russia, transported in a shielded container to Europe and stored underground in order to minimize the cosmogenic production of radionuclides.

The novel concept to use LAr scintillation light as anti-coincidence signal for further background suppression was developed. The pilot setup Mini-LArGe on the base of LAr scintillator (19 kg of LAr active volume) was developed and successfully operated. A long-term stability (about 2 years) with light yield of 1800 pe/MeV was achieved. About 95% of the background Compton events in the Ge detector are vetoed in this setup [19]. Suppression factors of 99% are expected for larger active argon volumes. The pulse shape analysis methods were developed, which allow to perform gamma / alpha / neutron selection with a strong discrimination factor ($>10^5$) for background suppression [20]. The results obtained with the Mini-LArGe have successfully proven the efficiency of the LAr scintillation as a powerful tool for background suppression and diagnostics. These investigations will be continued with the up-scaled (1.4 tons of LAr) low background LArGe setup recently installed in the GDL underground facility. A cluster up to 9 bare HPGe detectors can be operated in the LArGe setup. To detect the scintillation light of LAr nine low temperature PMTs are used. A vacuum insulated copper cryostat of 90 cm inner diameter and 205 cm height is lined with the TPB wavelength shifter and VM2000 reflector foil. The setup is surrounded by a graded shield consisting of 20 cm polyethylene, 23 cm steel, 10 cm lead and 15 cm electrolytic copper. A gas-tight lock serves as the access port for insertion of Ge detectors. After construction the cryostat was filled with LAr and the active cooling system was tested. LN_2 at an overpressure of 1 barg is transferred into the heat-exchanger integrated into the top of the cryostat. At a LN_2 flow rate of 2 kg per hour no mass loss was observed within one week of continuous operations. The commissioning of the LArGe setup is planned for autumn 2009.

5. Conclusion

In the framework of the extensive R&D program the main GERDA experimental concepts were proven and the methods of further background reduction were developed and tested. This gave a possibility to start a construction of the GERDA set up which is now practically completed. Commissioning of the set up will start with filling the cryostat with LA scheduled for November 2009.

Acknowledgments

Scientists of the GERDA collaboration acknowledge all involved technicians and engineers from the participating institutions, especially from INFN-LNGS, for their great assistance during construction and installation of the GERDA facility. This work was supported in part by INTAS (Nr. 05-1000008-7996), DFG within the SFB Transregio 27 'Neutrinos and Beyond', RFBR (grant 07-02-01050a).

6. References

[1] The Gerda Collaboration Letter of Intent 2004 [hep-ex/0404039]
Schonert S for GERDA collaboration 2006 *Prog. Part. and Nucl. Phys.* **57** 241

[2] Klapdor-Kleingrothaus H V et al 2001 *Eur. Phys. J.* A**12** 147

[3] Alseth C E et al 2002 *Phys. Rev.* D**65** 092007

[4] Klapdor-Kleingrothaus H V et al 2004 *Phys. Lett.* B **586** 198

[5] Klapdor-Kleingrothaus H V et al 2002 *Nucl. Instr. Methods* A **481** 149

[6] J.Barea and F.Iachello 2009 *Phys.Rev.* C 79044301 Table IV

[7] Barnabe Heider M et al 2008 *Proc. XIV Int. School "Particles and Cosmology" (Baksan Valley)* (2007 April 16-21 editors Demidov S et al.) pp 239-245

[8] Barnabe Heider M et al 2008 *arXiv:0812.1907v1* 10 Dec 2008

[9] Canberra Semiconductor NV, Lammerdries 25, B-2430 Olen, Belgium

[10] Krivosheina I V and Klapdor-Kleingrothaus H V 2006 *Phys. Scr.* T**127** 52

[11] Bauer M et al 2006 *J.Phys.: Conf. Ser.* **39** 1
Pandola L et al 2007 *Nucl. Instr. Methods* A **570** 149
Barabanov I et al 2009 *Nucl. Instr. Methods* A **606** 790

[12] Budjas D et al 2007 *AIP Conference Proc.* **897** 26

[13] Heusser G Simgen H and Zuzel G 2006 *J. Phys.: Conf. Ser.* 39

[14] Maneschg W et al 2008 *Nucl. Instr.Methods* A **593** 448

[15] Abt I et al 2007 *Nucl. Instr.Meth.* A **577** 574

[16] Abt I et al 2008 *Nucl. Instr.Meth.* A **583** 332

[17] Canberra Industries, Inc., http://www.canberra.com/products/485.asp

[18] Budjáš D Marik Barnabé HeiderM Chkvorets O Khanbekov N and Schönert S 2009 *JINST* **4** P10007

[19] Peiffer P et al 2005 *Nucl. Phys. B. Proc. Supp.* **143** 511
Di Marco M et al 2007 *Nucl. Phys. B. Proc. Supp.* **172** 45

[20] Peiffer P Pollmann T Schönert S Smolnikov A and Vasiliev S 2008 *JINST* **3** P08007

Journal of Physics: Conference Series **203** (2010) 012060 doi:10.1088/1742-6596/203/1/012060

Probing the Majorana nature of the neutrino with neutrinoless double beta decay

Stefano Morisi

Institut de Física Corpuscular – C.S.I.C./Universitat de València
Edificio Institutos de Paterna, Apt 22085, E–46071 Valencia, Spain

E-mail: morisi@ific.uv.es

Abstract. Neutrinoless double beta decay ($0\nu\beta\beta$) is the only experiment that could probe the Majorana nature of the neutrino. Here we study the theoretical implications of $0\nu\beta\beta$ for models yielding tri-bimaximal lepton mixing like A_4 and S_4.

1. Introduction

Neutrinoless double beta decay ($0\nu\beta\beta$) is the only experiment that can probe the Majorana Nature of the neutrino. $0\nu\beta\beta$ is a $\Delta L = 2$ lepton number violating decay that can occur if the neutrino is a Majorana particle. For an introduction see for instance [1]. The generic effective Majorana mass term can be write as

$$\mathcal{L}_{d5} = \frac{\lambda^{\nu}_{ij}}{\Lambda} L_i \phi L_j \phi, \tag{1}$$

where ϕ is the standard model higgs doublet and Λ is an effective scale. When ϕ takes a vev $\langle \phi \rangle = v$, the operator in eq. (1) gives the Majorana mass term $m^{\nu}_{ij} = \lambda^{\nu}_{ij} v^2 \Lambda$. In general m^{ν} is an arbitrary 3 by 3 symmetric complex matrix. It has been observed [2] that neutrino data are in very good agreement with zero a rector angle $\sin\theta_{13} = 0$, maximal atmospheric angle $\sin^2\theta_{23} = 1/2$ and large but not maximal solar angle $\sin^2\theta_{12} = 1/3$. A Leptonic mixing mixing matrix with such a values is called tri-bimaximal (TBM). The most generic neutrino mass matrix yielding to TBM mixing is invariant under exchanged the second family with the third family ($\mu \leftrightarrow \tau$ symmetry) and the sum of the elements each row of m_ν is constant, namely $\sum_j m^{\nu}_{1j} = \sum_j m^{\nu}_{2j} = \sum_j m^{\nu}_{3j}$. We call a neutrino mass matrix of such type tri-bimaximal form mass matrix.

We can extend the standard model by means of a flavor symmetry G_f that yields a tri-bimaximal form mass matrix. Examples are the group of even permutation of four objects A_4 [3, 4] or the group of all the permutation of four objects S_4 [5, 6]. We observe that A_4 is the smallest non-Abelian group with a triplet representation allowing to arrange in one multiplet the three families. For now on we focus on A_4 flavor symmetry. We assume three Higgs doublets transforming as a triplet $(\phi_1, \phi_2, \phi_3) \sim 3$ and the lepton doublets L_e, L_μ, L_τ, transforming as a triplet of A_4. The most generic A_4 invariant dimension five operators are given as

$$\mathcal{L}_{5d} = \frac{c_1}{\Lambda}(L_i\phi)_1(L_j\phi)_1 + \frac{c_2}{\Lambda}(L_i\phi)_{1'}(L_j\phi)_{1''} + \frac{c_3}{\Lambda}(L_i\phi)_3(L_j\phi)_3 +$$

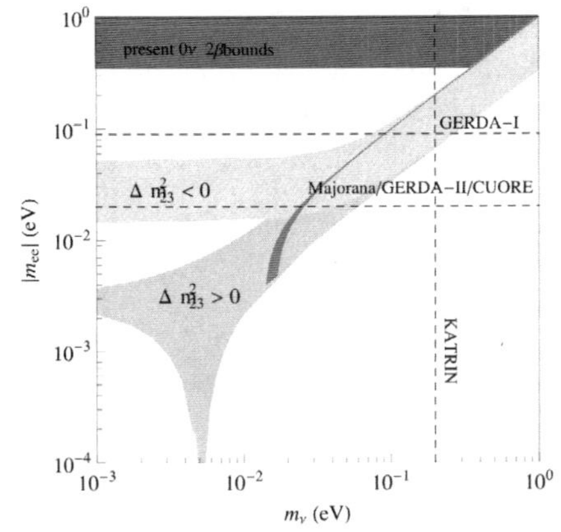

Figure 1. Allowed range of m_{ee} as a function of the lightest neutrino mass for A_4 based model.

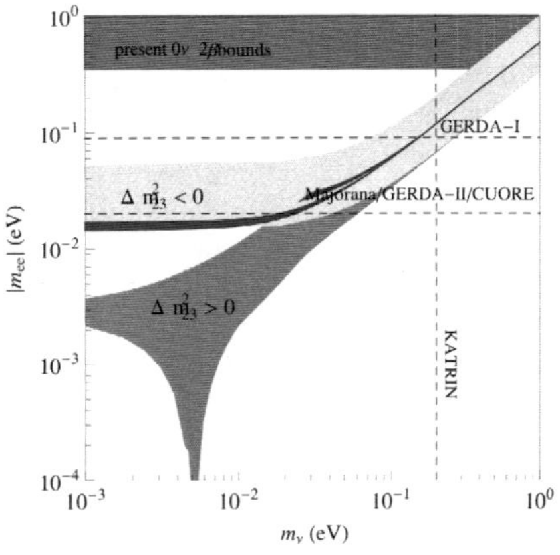

Figure 2. Allowed range of m_{ee} as a function of the lightest neutrino mass for S_4 based model.

$$+ \quad \frac{c_4}{\Lambda}(L_iL_j)_1(\phi\phi)_1 + \frac{c_5}{\Lambda}(L_iL_j)_{1'}(\phi\phi)_{1''} + \frac{c_5'}{\Lambda}(L_iL_j)_{1''}(\phi\phi)_{1'} + \frac{c_6}{\Lambda}(L_iL_j)_3(\phi\phi)_3. \quad (2)$$

The resulting neutrino mass matrix is of tri-bimaximal form if and only if $c_5 = c_5' = 0$. Such a condition can naturally be explained by adding Higgs doublets or flavons carrying extra Abelian symmetries, see for instance [4].

Differently the most generic S_4 invariant dimension five operators

$$\begin{aligned}\mathcal{L}_{5d} \quad &= \quad \frac{\lambda_1}{\Lambda}(L_i\phi)_1(L_j\phi)_1 + \frac{\lambda_2}{\Lambda}(L_i\phi)_2(L_j\phi)_2 + \frac{\lambda_3}{\Lambda}(L_i\phi)_{3_1}(L_j\phi)_{3_1} + \frac{\lambda_3'}{\Lambda}(L_i\phi)_{3_2}(L_j\phi)_{3_2} + \\ &+ \quad \frac{\lambda_4}{\Lambda}(L_iL_j)_1(\phi\phi)_1 + \frac{\lambda_5}{\Lambda}(L_iL_j)_2(\phi\phi)_2 + \frac{\lambda_6}{\Lambda}(L_iL_j)_{3_1}(\phi\phi)_{3_1}, \quad (3)\end{aligned}$$

give neutrino mass matrix of tri-bimaximal form. In S_4 we do not need to introduce extra scalars or add symmetries in order to have tri-bimaximal in the neutrino sector. The resulting neutrino mass matrix is diagonalized by the tri-bimaximal matrix.

Now we study the theoretical implications for $0\nu\beta\beta$ in models based on the two flavor groups A_4 and S_4. The allowed ranges of m_{ee} as a function of the lightest neutrino are presented in Figures (1) and (2). The two bands correspond to the normal hierarchy and the inverse hierarchy with $\sin^2\theta_{12} = 1/3$, $\sin^2\theta_{23} = 1/2$, $\sin^2\theta_{13} = 0$ (tri-bimaximal). We call this bands the *tri-bimaximal region* for $0\nu\beta\beta$.

In order to simplify the discussion and to show the difference between A_4 and S_4 based models, we assume $c_1 = c_2 = c_3 = 0$ and $\lambda_1 = \lambda_2 = \lambda_3 = \lambda_3' = 0$. This could be the case if the dimension five operators arise from type-II seesaw mechanism only. We have already said that A_4 gives tri-bimaximal mixing if $c_5 = c_5' = 0$. If $\lambda_5 = 0$ A_4 and S_4 are phenomenologically equivalent since the operators proportional to c_4 and c_6 correspond to the operators proportional to λ_4 and λ_6 respectively. If $\lambda_5 \neq 0$, then A_4 and S_4 are not phenomenologically equivalent. For instance the S_4 allowed range for m_{ee} is the full tri-bimaximal region. When $\lambda_4 = 0$ the allowed m_{ee} ranges for A_4 and S_4 are given respectively in Figures (1) and (2).

In particular, in Figure (1) we see that A_4 is compatible only with normal hierarchy. The A_4 case has a lower bound with $m_{ee} > 3 \cdot 10^{-3}$, see [7]. In Figure (2) we see that S_4 is compatible only with inverse hierarchy.

[8] and [9] have studied $0\nu\beta\beta$ implications assuming respectively type-I and inverse/linear seesaw mechanisms with A_4 flavor symmetry. An effective neutrino mass model with S_4 flavor symmetry has been studied in [10]. In [11] a model with type-I seesaw with S_4 flavor symmetry has been studied.

Note that for values of the lightest mass bigger than 10^{-1} eV, namely when neutrino are quasi degenerate, A_4 and S_4 allowed ranges are well separated. Therefore it could be in principle possible to distinguish A_4 from S_4 in the quasi degenerate region.

In case of inverse hierarchy there is a lower limit for m_{ee}. In [1] the dependence of such a limit as a function of the solar angle has been studied. However, in case of normal hierarchy in general there is no lower limit. If $m_{\nu_1} = m_{ee} = 0$ we have that the lower limit depends from the values of the solar and reactor angles,

$$\sin^2\theta_{12}\cos^2\theta_{13}e^{i\alpha}\sqrt{\Delta m^2_{12}} + \sin^2\theta_{13}e^{i\beta}\sqrt{\Delta m^2_{13}} = 0. \tag{4}$$

This equation gives a relation between the solar and the reactor angles and the allowed region is represented in Figure (3) with the blue band. The small red region is the range allowed by the data at $3\,\sigma$ and compatible with $m_{\nu_1} = m_{ee} = 0$, namely eq. (4).

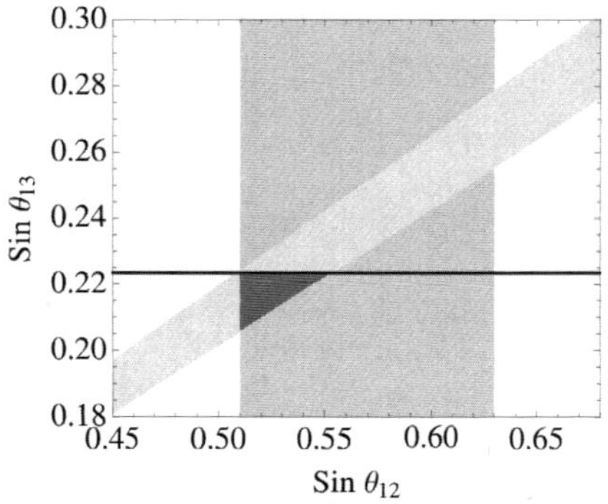

Figure 3. Relation between solar and reactor angles for normal hierarchy when $m_{\nu_1} = m_{ee} = 0$.

Acknowledgments

We thank Martin Hirsch for useful comments. This work was supported by the Spanish grant FPA2008-00319/FPA and PROMETEO/2009/091.

References

[1] M. Hirsch, arXiv:hep-ph/0609146.
[2] P. F. Harrison, D. H. Perkins and W. G. Scott, Phys. Lett. B **530**, 167 (2002) [arXiv:hep-ph/0202074].
[3] K. S. Babu, E. Ma and J. W. F. Valle, Phys. Lett. B **552**, 207 (2003) [arXiv:hep-ph/0206292];
[4] G. Altarelli and F. Feruglio, Nucl. Phys. B **720**, 64 (2005) [arXiv:hep-ph/0504165].
[5] C. S. Lam, Phys. Rev. Lett. **101**, 121602 (2008) [arXiv:0804.2622 [hep-ph]].
[6] F. Bazzocchi and S. Morisi, arXiv:0811.0345 [hep-ph].
[7] G. Altarelli, arXiv:0711.0161 [hep-ph].
[8] M. Hirsch, S. Morisi and J. W. F. Valle, Phys. Rev. D **78**, 093007 (2008) [arXiv:0804.1521 [hep-ph]].
[9] M. Hirsch, S. Morisi and J. W. F. Valle, Phys. Lett. B **679**, 454 (2009) [arXiv:0905.3056 [hep-ph]].
[10] F. Bazzocchi, L. Merlo and S. Morisi, Nucl. Phys. B **816**, 204 (2009) [arXiv:0901.2086 [hep-ph]].
[11] F. Bazzocchi, L. Merlo and S. Morisi, Phys. Rev. D **80**, 053003 (2009) [arXiv:0902.2849 [hep-ph]].

Journal of Physics: Conference Series **203** (2010) 012061 doi:10.1088/1742-6596/203/1/012061

Covariances of nuclear matrix elements for $0\nu\beta\beta$ decay

G.L. Fogli[1,2], **E. Lisi**[2], **A.M. Rotunno**[1,2]

[1] Dipartimento Interateneo di Fisica "Michelangelo Merlin," Via Orabona 4, 70126 Bari, Italy
[2] Istituto Nazionale di Fisica Nucleare, Sezione di Bari, Via Orabona 4, 70126 Bari, Italy

E-mail: annamaria.rotunno@ba.infn.it

Abstract. Estimates of nuclear matrix elements (NME) for neutrinoless double beta decay ($0\nu\beta\beta$) based on the quasiparticle random phase approximations (QRPA) are affected by theoretical uncertainties, which may play a dominant role in comparison with projected experimental errors of future $0\nu\beta\beta$ experiments. We discuss the estimated variances and covariances of NME of several candidate nuclei within the QRPA, focusing on the following aspects: 1) the comparison of $0\nu\beta\beta$ signals, or limits, in different nuclei; 2) the prospects for testing nonstandard $0\nu2\beta$ mechanisms in future experiments.

1. Introduction

The search for the neutrinoless double beta decay ($0\nu\beta\beta$) is being vigorously pursued by several experiments using different nuclei, in order to clarify the nature of neutrinos (Dirac or Majorana) and their absolute mass scale [1]. In a given candidate nucleus, light Majorana neutrinos can induce $0\nu\beta\beta$ decay with half-life T_i depending on the $0\nu\beta\beta$ nuclear matrix element (NME), here denoted as M'_i, and on the "effective Majorana neutrino mass" $m_{\beta\beta}$.

NME's depend: on the effective axial coupling in nuclear matter, g_A (not necessarily equal to its "bare" free-nucleon value $g_A \simeq 1.25$, and on the particle-particle interaction strength parameter, g_{pp}. NME's must be evaluated in a nuclear model, together with their uncertainties. Correlations among uncertainties in different nuclei are also important. In fact, NME positive correlations would lead to a degeneracy effect between theoretical errors.

Within the quasiparticle random phase approximation (QRPA), we discuss the strong covariances between NME in different nuclei and their role in the phenomenology of $0\nu\beta\beta$ decay. Moreover, we show that, assuming prospective half-life data in four promising nuclei (^{76}Ge, ^{82}Se, ^{130}Te and ^{136}Xe), the "standard scenario" characterized by light Majorana neutrino exchange can be distinguished from a few nonstandard physics models.

2. NME uncertainties and their correlations

We consider a set of eight $0\nu\beta\beta$ candidate nuclei: ^{76}Ge, ^{82}Se, ^{96}Zr, ^{100}Mo, ^{116}Cd, ^{128}Te, ^{130}Te and ^{136}Xe. Within QRPA, we consider [2]: (*i*) two values for the axial coupling: $g_A = 1.25$ (bare) and $g_A = 1.00$ (quenched); (*ii*) two approaches to short range correlations (s.r.c.): the so-called Jastrow-type s.r.c., and the unitary correlation operator method (UCOM); (*iii*) three sizes for the model basis; (*iv*) two many-body models: QRPA and its renormalized version (RQRPA). We obtain $2 \times 2 \times 3 \times 2 = 24$ variants, supplemented by errors induced by g_{pp} uncertainties.

Figure 1 shows a scatter plot of the theoretical results for each couple of different nuclei, in the plane charted by the coordinates (η_i, η_j) (where $\eta_i = \log_{10}|M'_i|$), as taken from [3]. Strong positive correlations among theoretical estimates emerge. See [3] for numerical values and details.

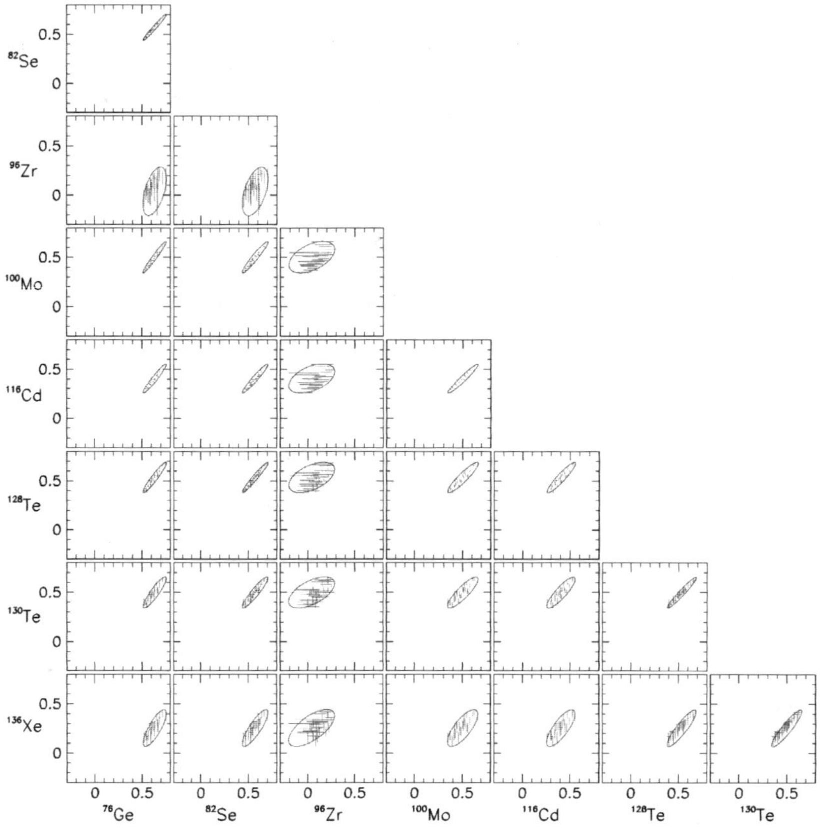

Figure 1. Estimated 1σ error ellipses for any couple of NME's among the indicated $0\nu\beta\beta$ candidate nuclei, as derived from recent QRPA calculations. Scales are logarithmic. Figure taken from [3].

3. Application 1: comparison of half-lives ($\tau_i = \log_{10}(T_i/\text{y})$) in couples of nuclei

We consider two different nuclei i and j, characterized by nuclear matrix elements $\eta_i = \eta_i^0 \pm \sigma_i$ and $\eta_j = \eta_j^0 \pm \sigma_j$ with correlation ρ_{ij}. A positive $0\nu\beta\beta$ signal in the first nucleus ($\tau_i = \tau_i^0 \pm s_i$) translates into a favored range for the second nucleus ($\tau_j = \tau_j^0 \pm s_j$). We apply the procedure in [3] to $i = {}^{76}$Ge and $j = {}^{130}$Te. Then the claim by Klapdor *et al.* [4], $\tau_i = 25.355 \pm 0.118$ (90% C.L.), implies for ^{130}Te that $\tau_j = 24.786 \pm 0.161$ (90% C.L.), with error correlation $\rho_{ij} = 0.447$. Figure 2 shows the corresponding error ellipse at 1.64σ (90% C.L.), in the plane charted by (τ_i, τ_j). The horizontal band rapresents the range preferred by the $0\nu\beta\beta$ claim of [4]. The slanted band is the constraint placed by our QRPA estimates. The combination of the two constraints provides the shaded ellipse. The projection of the ellipse on the x-axis provides the range preferred at 90% C.L. for the ^{130}Te half-life. Projections for other nuclei can be similarly derived, as reported in [3].

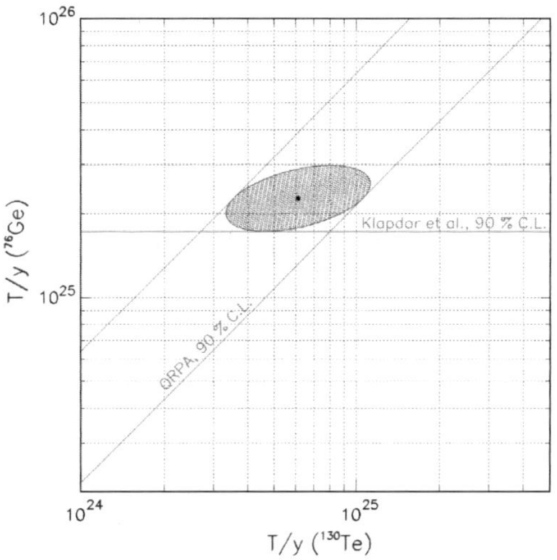

Figure 2. Theoretical and experimental constraints at 90% C.L., in the plane charted by the $0\nu\beta\beta$ half-lives of ^{76}Ge and ^{130}Te. Figure taken from [3].

4. Application 2: probing nonstandard particle physics models

In [5] we have assumed prospective data for four candidate nuclei ^{76}Ge, ^{82}Se, ^{130}Te and ^{136}Xe. Within the standard scenario (light Majorana neutrino exchange and QRPA), we have obtained a set of predictions for the half-lives and their 1σ errors. We have shown that prospective data can distinguish the above "null hypothesis" from other nonstandard mechanism for $0\nu\beta\beta$ decay (see [5] for details). In particular, at least two non standard mechanism, left-right symmetric models with right-handed leptonic currents coupled to right handed hadronic currents (RHC λ) and Kaluza-Klein neutrino exchange with brane shift parameter $a = 10^{+1}$ GeV^{-1} (KK+1) can be distinguished from the standard one at $> 95\%$ C.L. We have shown that correlations of theoretical errors are crucial for the statistical power of the test: in fact, without correlations, significant discrimination is never obtained [5].

5. Conclusions

NME for $0\nu\beta\beta$ decay are affected by relatively large theoretical uncertainties. We have made a first attempt to quantify the covariance matrix of the NME. We have shown that correlations are important in comparing signals or limits from different nuclei. We have also shown that light Majorana neutrino exchange could be distinguished from at least two alternative nonstandard mechanism at $> 95\%$ C.L., using prospective data from four nuclei. Correlations appear to be crucial ingredients also in this context.

References

[1] F.T. Avignone III, S.R. Elliot and J. Engel, "Double Beta Decay, Majorana Neutrinos, and Neutrino Mass", Rev. Mod. Phys. **80**, 481 (2008) [arXiv:0708.1033 [nucl-ex]].

[2] F. Simkovic, A. Faessler, V. Rodin, P. Vogel and J. Engel, "Anatomy of nuclear matrix elements for neutrinoless double-beta decay," Phys. Rev. C **77**, 045503 (2008) [arXiv:0710.2055 [nucl-th]].

[3] A. Faessler, G.L. Fogli, E. Lisi, V. Rodin, A.M. Rotunno and F. Simkovic, "QRPA uncertainties and their correlations in the analysis of neutrinoless double beta decay," Phys. Rev. D **79**, 053001 (2009) [arXiv:0810.5733 [hep-ph]].

[4] H.V. Klapdor-Kleingrothaus and I.V. Krivosheina, "Evidence for the observation of $0\nu\beta\beta$ decay: Identification of $0\nu\beta\beta$ events from the full spectra," Mod. Phys. Lett. A **21**, 1547 (2006).

[5] G.L. Fogli, E. Lisi and A.M. Rotunno, "Probing particle and nuclear physics models of $0\nu\beta\beta$ decay with different nuclei," Phys. Rev. D **80**, 015024 (2009) [arXiv:0905.1832 [hep-ph]].

Journal of Physics: Conference Series **203** (2010) 012062 doi:10.1088/1742-6596/203/1/012062

New Concepts for a Gaseous Xenon Detector for Double Beta Decay

D Sinclair[1] **for the EXO Collaboration**
Department of Physics, Carleton University and TRIUMF

E-mail: sinclair@physics.carleton.ca

Abstract. Xenon gas is an attractive medium for the search for neutrinoless double beta decay because it offers the possibility of reasonable energy resolution, event topology reconstruction, very high intrinsic purity and background rejection through the identification of the daughter barium ion. This talk explores recent developments in the conceptual design of such a detector.

Introduction

To push the sensitivity for searches for neutrino-less double beta decay to increasingly high sensitivities the experimenter is challenged to design a large mass detector with very low backgrounds. Xenon is one of the most attractive among the possible candidates because it is relatively easy to separate the isotope of interest, ^{136}Xe, noble substances such as xenon can be made extremely pure, xenon has no long lived radioactive isotopes, it has a high Q-value for the decay process (2.4 MeV) and high energy resolution has been demonstrated for detectors using xenon as a working substance. The EXO collaboration is developing both liquid and gaseous concepts for such a detector. The liquid detector work is the most advanced and is covered in the talk of L. Kaufman in these proceedings. A liquid detector has the advantage of being compact and this in turn means that the mass of materials making up the detector including the shield can be much smaller which makes tractable the resulting radioactivity constraints. In this talk some of the new concepts for such a gaseous detector will be presented.

A gaseous detector has a number of potential advantages over a liquid detector. The tracks are much longer and this allows information about the event topology to be studied. For example, in the Gottard experiment it was found that a criterion that two Bragg peaks were observed reduced the background by a factor of 25. There is also the chance to reject multi-site events with very high precision. The energy resolution of the gas detector is expected to be at least as good as a liquid detector and may be better as the correlation between ionization fluctuations and the scintillation light are much smaller. Perhaps the most exciting prospect for a gas counter is that it opens new ways to try to identify the daughter barium ion and thus greatly reduce the backgrounds. Some such improvement is likely to be essential for detecting this process if the effective Majorana mass is in the 'Normal Hierarchy' range.

---

[1] To whom any correspondence should be addressed.

The first concept for a gaseous xenon detector with barium identification was published by M. Danilov et al [1] with a design based on the Gotthard laboratory work of J.-C. Vuilleumier et al [2]. The concept was for a gas TPC with a central cathode plane, phototubes to detect the scintillation flash and a fast laser steering scheme to bring the required red and blue lasers to the vertex location to pick up the barium. The ionization would be detected at the anodes using micropattern gas amplification. On further investigation, there appear to be problems with this concept. It has not yet been possible to find a gas composition which meets all of the requirements. These requirements include: (i) The gas must be extremely pure to allow drift of the electrons to the anodes (the use of any additives in the gas makes this more difficult); (ii) the gas must be transparent to the scintillation light at 174 nm. Most quench gases absorb this light. Indeed, to stabilize the gas amplification, one wants a gas that absorbs this uv light.

The barium ion created in the decay process is expected to be Ba++ (higher charge states may be produced due to shake off but in xenon, the ion will relax to 2+). The laser detection techniques require a singly charged ion. Xenon cannot give up an electron to the barium because the ionization potential to too high. Once can find additives that will permit the conversion (such as TEA) but such additives will make the gas purification very difficult and they are likely to form molecules with the barium ion making the tagging impossible.

The New Detector Concept

The new concept for a xenon detector is based on the idea that one would make a detector filled with pure xenon, and detect the barium ion by extracting it from the gas using techniques developed for radioactive ion beams and subsequently identifying it by one of a number of possible techniques.

The motivation for trying to extract the barium ions came from a serendipitous observation by M. Facina [3] who was trying to produce and extract ^{71}Ni ions from proton induced fission fragments stopped in argon gas. His apparatus consisted of a gas cell with flowing argon at 0.5 bar with a nozzle of about 0.5 mm diameter through which lasers tuned to resonantly ionize nickel atoms were focused. Gas flowing out of the nozzle was pumped away with a large roots blower and turbo pumps. The resulting ions were extracted using a sextupole ion guide, accelerated through 40 kV, mass analysed, and stopped on a foil. The foil was then counted using a gamma counter to identify the ions produced. What was observed, in addition to the ^{71}Ni was a strong signal from ^{142}Ba. This signal was independent of the laser and could only come from the production of Ba++ ions in the fission which remained stable in the argon for times of ~0.5 seconds.

The detector would consist of a gaseous time projection chamber filled with xenon. If an ionizing event occurs in the gas, there will be a scintillation flash. This will be detected by light sensors at the ends of the detector to be described later to act as the trigger and time zero signal. Electrons will drift towards the anode under a strong electric field. Once they are close to the anode they will enter a strong field region where they will give rise to light through the electroluminescence process. D Nygren [4] has recently proposed such a concept to improve the energy resolution of a xenon detector. In Nygren's design the light would be detected in a set of phototubes. Because we wish to achieve a fine tracking we will use a fine grained readout described below. A second readout plane is located behind the cathode grids to give a total energy reading. The electron drift time (measured from the scintillation flash) together with the ionization track information, will allow a determination of the barium ion position. The accuracy may not be high because the electron tracks in xenon are so complex but at least the location of the ionization cloud will be known and, at 10 bar this limits the uncertainty in the vertex to about 10 cm. The barium ion is slowly drifting towards the cathode.

The cathode is made using a pair of grids with programmable voltages on each wire. At the point where the barium is expected to be between the wires, the field is changed so that the ions drift to one side. When they reach the edge of the detector, the fields are again changed to drift the ions out along a channel to an extraction nozzle.

After extraction the ions may be identified in a number of ways. One could accelerate them with an electric field and measure the energy and charge/mass to determine that a mass 136 ion was detected in a 2+ charge state. This is almost uniquely barium as any Xe++ ions will quickly convert to Xe+. One can also consider changing the charge state to Ba+ and using the well known laser techniques to form a unique identification.

Our present concept for the light collection is based on a design developed for the Fermilab RICH counter [5]. The electroluminescence region is separated from a parallel plate proportional couonter filled with low pressure methane by a quartz plate. Light of 175 nm passing through the quartz strikes a CsI layer evaporated onto copper pads to produce electrons and hence the signal. There are many aspects of this new concept that need to be demonstrated before a full scale detector could be considered. We need to show that good energy resolution can be achieved in such a chamber. We need to learn what tracking information can be extracted. We wish to explore the use of mixed noble gases such as Ne-Xe which may offer improved tracking and barium location. We need to establish the efficiency for detecting barium ions. Finally, we need to establish the dependence of the measurements on parameters such as the pressure to optimize the detector design. To this end, two substantial prototype systems are being developed. One will study the detection of electrons in the MeV range using a detector similar to the concept discussed above at pressures up to 10 bar. The second will attempt to demonstrate the efficiency with which barium ions can be extracted and detected starting with ions produced in high pressure xenon.

References

[1] Danilov M, DeVoe R, Dolgolenko A, Giannini G, Gratta G, Picchi P, Piepke A, Pietropaolo A F, Vogel P, Vuilleumier J-L, Y-F. Wang Y-F, Zeldovich O 2000 Phys Lett **480** 12-18.

[2] Vuilleumier J-L et al. 1993 Phys. Rev. **D48** 1009-20.

[3] Facina M 2004 Thesis Leuven.

[4] Nygren D 2007 NIM **A581** 632-42.

[5] Anderson DF, Kwan S, Peskov V 1992 Fermilab-conf-92/135.

Journal of Physics: Conference Series **203** (2010) 012063 doi:10.1088/1742-6596/203/1/012063

Test of the single state dominance hypothesis for the two-neutrino double beta decay

O Moreno[1], R Álvarez-Rodríguez[1], E Moya de Guerra[1], P Sarriguren[2], F Šimkovic[3], Amand Faessler[4]

[1] Dpto. Física Atóm., Mol. y Nuclear, Univ. Complutense de Madrid, E-28040 Madrid, Spain
[2] Instituto de Estructura de la Materia, CSIC, Serrano 123, E-28006 Madrid, Spain
[3] Department of Nuclear Physics, Comenius University, SK-842 15 Bratislava, Slovakia
[4] Institut für Theoretische Physik, Universität Tübingen, D-72076 Tübingen, Germany

E-mail: oscar.moreno@iem.cfmac.csic.es

Abstract. The single state dominance hypothesis for the two-neutrino double-beta decay matrix elements is tested in this work for the double-beta decaying nuclei ^{100}Mo, ^{116}Cd, and ^{128}Te. In addition to this, we analyze the contribution to the double-beta matrix elements from the low-lying intermediate states and from the whole set of intermediate states. We use a proton-neutron QRPA calculation based on a deformed Skyrme Hartree-Fock mean field with pairing correlations, and we compare these results with the half-lives of the double-beta emitters for which we have experimental information.

Neutrinoless double-beta ($0\nu\beta\beta$) decay [1] is a nuclear process, not yet experimentally confirmed, that could provide information on the absolute mass of the neutrino. The dependence of this process on the nuclear structure of the decaying nucleus gives rise to a lack of accuracy in the matrix element and consequently in the neutrino properties that could be extracted from the experiments.

On the other hand, the double-beta decay with the emission of two (anti)neutrinos ($2\nu\beta\beta$) proceeds as a perturbative process in the Standard Model and has already been observed in some nuclei. A successful description of the $2\nu\beta\beta$ decay mode is a requirement for a reliable calculation of the matrix element corresponding to the $0\nu\beta\beta$ decay.

An important question in theoretical $2\nu\beta\beta$ decay studies is whether the contributions of higher-lying intermediate states to the $2\nu\beta\beta$ decay amplitude play an important role. According to the single-state dominance (SSD) hypothesis [2], when the ground state of the odd-odd intermediate nucleus is 1^+, the double-beta transition through this intermediate state accounts for a large part of the double-beta decay matrix element (DBDME). If SSD hypothesis is confirmed, the half-lives of $2\nu\beta\beta$ decays could be determined from single β^- and electron capture measurements. The realization of the SSD hypothesis would also lead to a simplification in the theoretical description of the intermediate nucleus, because only the lowest 1^+ wave function is to be considered.

In this work we check the validity of the SSD hypothesis by considering only the lowest-energy term in the summation in the cases of ^{100}Mo, ^{116}Cd, and ^{128}Te, which are emitters with 1^+ intermediate ground states. Then, we analyze the contributions to the $2\nu\beta\beta$ nuclear DBDME from terms with increasing excitation energies, with special attention to the low-energy region and to the full set of excited states.

The expressions for DBDME can be found in Refs. [3, 4]. They are computed within a formalism based on a pnQRPA approach with a selfconsistent quasiparticle basis obtained from deformed Skyrme Hartree-Fock mean field with pairing correlations (HF+BCS+QRPA). For each (N, Z) nucleus, a BCS ground state is obtained from deformed Skyrme (SLy4) HF calculations on a large axially-symmetric harmonic oscillator basis, using phenomenological pairing-gap parameters and solving the BCS equations after each HF iteration. The quadrupole deformation of the nucleus in its ground state is obtained selfconsistently as the shape that minimizes the nuclear energy. The pnQRPA equations are solved using the quasiparticle mean field basis for separable residual spin-isospin interactions in both particle-hole and particle-particle channels [5].

We plot in Fig. 1 the 2ν DBDME for a set of confirmed double-beta emitters as a function of the excitation energy of the intermediate nucleus taken into account in the calculation. The shadowed area in the plots indicates the experimental range of the 2ν DBDME, the two limiting horizontal lines being deduced from experimental half-lives [6] using two different values of the constant g_A, namely g_A=1.25 and g_A=1.00. The vertical lines indicate the experimental energy positions of the Gamow-Teller giant resonances to show the correlation that we find between them and the sudden increase in the value of the DBDME.

A pure SSD analysis would focus on the first contribution to the DBDME for ^{100}Mo, ^{116}Cd and ^{128}Te. From Fig. 1 we see that no evidence to support SSD is found in any of these nuclei, since the first transition (the one at the lowest excitation energy) does not account for a remarkable part of the total DBDME value. One has to consider a few MeV of excitation energy (containing the strength of several transitions) to reach a large percentage (around 60%) of the final DBDME value. This increase in the low energy region, which can be referred to as low-lying state dominance (LLSD) hypothesis, is highlighted in the figure by circles, appearing in ^{48}Ca, ^{96}Zr, ^{100}Mo and ^{116}Cd.

Apart from the SSD or LLSD hypotheses, the figure shows the theoretical prediction of the total DBDME, which is the value reached for a large-enough energy range (as the 30 MeV shown in the figure). In many cases the theoretical value lies within the experimental range, and in all of them the order of magnitude is correct. It is worth mentioning that there are two ingredients that strongly influence the computed value of the DBDME. One is the value of the residual interaction coupling constants, especially for the particle-particle channel, which in the calculation shown here has been assigned the value $\kappa_{pp} = 3/A$ MeV. The other is the different quadrupole deformations of the double-beta partners, whose effect can be seen in the figure in the plots for ^{48}Ca and ^{76}Ge. In both cases when the ground states of the double-beta partners have the same deformations (spherical in the case of ^{48}Ca and ^{48}Ti and prolate for ^{76}Ge and ^{76}Se), the DBDME is considerably larger than in the case when they have different deformations.

To summarize, we have shown that our deformed Skyrme HF+BCS+QRPA calculations reproduce the experimental $2\nu\beta\beta$ DBDME when the whole energy range of excitation energy in the intermediate nucleus is considered, with only two exceptions ($A = 130, 136$). We find important contributions to the 2ν DBDME at relatively high excitation energies in several cases, and consequently dominance of a single or a few low-lying states is not generally found and is at best fulfilled at the 60% level in our theoretical scheme, and only for some nuclei ^{48}Ca, ^{96}Zr, ^{100}Mo, ^{116}Cd. Other important conclusion of the analysis performed here is the verification of shape dependence of DBDME in $A = 48$ and $A = 76$ cases.

[1] Suhonen J and Civitarese O, *Phys. Rep.* **300**, 123 (1998);
Faessler A and Šimkovic F, *J. Phys. G: Nucl. Part. Phys.* **24**, 2139 (1998).
[2] Abad J *et al.*, *An. Fis. A* **80**, 9 (1984).
[3] Moreno O *et al.*, *J. Phys. G: Nucl. Part. Phys.* **36**, 015106 (2009).
[4] Álvarez-Rodríguez R *et al.*, *Phys. Rev. C* **70**, 064309 (2004);

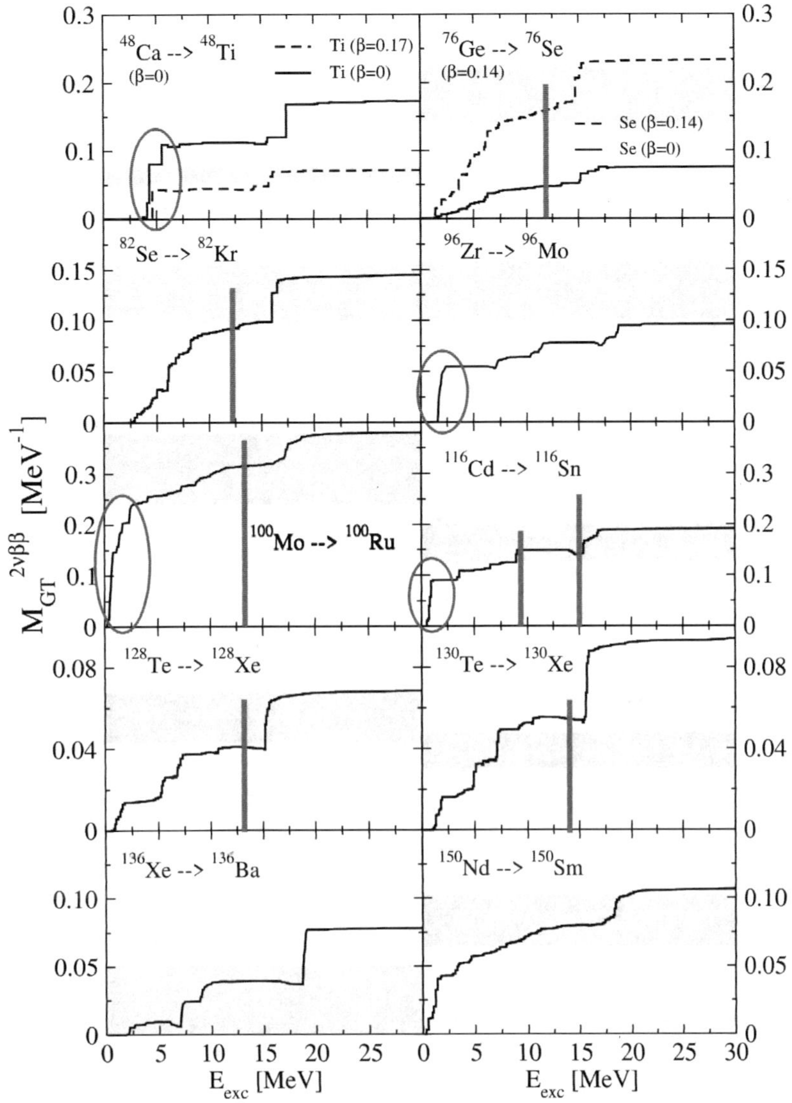

Figure 1. DBDME as a function of the intermediate nucleus excitation energy considered in the calculation.

Šimkovic F *et al.*, *Nucl. Phys. A* **733**, 321 (2004).
[5] Sarriguren P *et al.*, *Nucl. Phys. A* **635**, 55 (1998); *Nucl. Phys. A* **58**, 13 (1999); Sarriguren P *et al.*, *Nucl. Phys. A* **691**, 631 (2001); Sarriguren P *et al.*, *Phys. Rev. C* **64**, 064306 (2001).
[6] Barabash AS, *Czech. J. Phys.* **52**, 567 (2002); Barabash AS, arXiv:nuclex/0602009v2.

Topics in Astroparticle and Underground Physics (TAUP 2009) IOP Publishing
Journal of Physics: Conference Series **203** (2010) 012064 doi:10.1088/1742-6596/203/1/012064

MOON for Double-Beta Decays and Neutrino Nuclear Responses

K.Fushimi[1], Y.Kameda[1], K.Harada[1], S.Nakayama[1], H.Ejiri[2], T.Shima[2], K.Yasuda[3], R.Hazama[4] and K.Imagawa[5]

1. Institute for Socio Arts and Sciences, The University of Tokushima, Tokushima 770-8502, Japan
2. Research Center for Nuclear Physics (RCNP), Osaka University, Ibaraki, Osaka 567-0047, Japan
3. Department of Physics, Osaka University, Toyonaka, Osaka 560-0047, Japan
4. Department of Engineering, Hiroshima University, Higashi Hiroshima, Hiroshima 739-8527 , Japan
5. Horiba Ltd. Kyoto city, Kyoto 601-8510, Japan

E-mail: `kfushimi@ias.tokushima-u.ac.jp`

Abstract. Thin and wide area inorganic crystal was tested for double beta decay experiment. The thin NaI(Tl) whose dimension of 18cm×18cm×0.5cm was developed. The energy resolution at $Q-$value of ^{100}Mo was obtained less than 3% in full-width-half-maximum. Although the backscattering of electrons suffers the detection efficiency, the NaI(Tl) has the advantage for double beta decay experiment.

1. MOON detector for double beta decay

Neutrino-less double beta decays ($0\nu\beta\beta$) are quite realistic and sensitive probes for the Majorana nature of the neutrino. $0\nu\beta\beta$ experiments provides the absolute mass and the CP phase which are the important parameters in the unified theories beyond the standard model. Recent experimental and theoretical studies are given in review articles and references [1, 2, 3]

The interesting range of the effective neutrino mass is between 30meV (quasi-degenerate:QD) and 100meV (inverted-hierarchy:IH). The live time of 0.1 $\sim$ 1ton·yr and the favorable nuclei such as ^{82}Se, ^{100}Mo and ^{150}Nd are needed to search for the $0\nu\beta\beta$ events. Moreover, the good energy resolution is the key factor for the $0\nu\beta\beta$ decay experiments, since it is necessary to reject the most serious background due to the double beta decays with two-neutrino emission ($2\nu\beta\beta$).

The MOON (Mo/Majorana Observatory Of Neutrinos) detector is a multi-layer detector modules for double beta decay experiment [4, 5]. We aimed to develop the detector with the energy resolution about 3% in FWHM at the Q-value ($\sim$ 3MeV). We employed NaI(Tl) detectors whose energy resolutions are quite good. Since the backscattering effect of NaI(Tl) on low-energy electrons is not small, we have to use a higher threshold level for detector signals, and the total efficiency for $0\nu\beta\beta$ may be reduced. We studied the efficiency for $0\nu\beta\beta$ with Monte Carlo simulations, and the loss of the efficiency was found to be about 30$, which is still acceptable for our purpose.

2. Performance of thin and large area NaI(Tl) scintillator

The dimension of the thin and wide area NaI(Tl) crystal is 18cm×18cm×0.5cm. The crystal was encapsulated with thin aluminum sheet and quartz light guides to avoid the humidity. Scintillation photons are guided to the edge of the NaI(Tl) crystal by total reflection and a reflector with high reflective efficiency. Two thin NaI(Tl) modules were piled up, the enriched ^{100}Mo sheet was inserted between two NaI(Tl) modules. The enrichment of ^{100}Mo was 85%. The thickness of the source was 40mg/cm^2 and the total effective mass of ^{100}Mo was 11.02g.

Two types of photomultiplier tubes (PMTs) were contacted at the edge of quartz light guides by optical grease. Two modules of 3 inches diameter PMTs were contacted on each edges in order to collect scintillation photon. One module of 0.75 inches square PMT were contacted on each edges in order to distinguish the layer in which the event occurred.

2.1. Energy resolution

The energy resolution was measured by irradiating several mono-energy gamma-rays from standard gamma-ray sources. The energy dependence of the energy resolution was shown in Fig.1. The energy resolution depends on inverse square of the energy. The energy resolution at the Q-value of ^{100}Mo (3.034MeV) was extrapolated from the fit. About 3% of FWHM was obtained, which is good enough for $0\nu\beta\beta$ experiment.

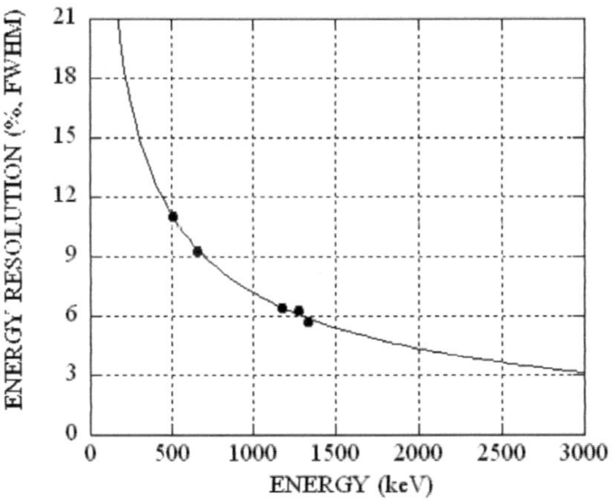

Figure 1. The energy dependence of the energy resolution in full-width-half-maximum (FWHM).

2.2. Low background test

The test experiment was performed at the second laboratory in OTO Cosmo Observatory. OTO Cosmo Observatory is located 100km south from Osaka and 150km east from Tokushima. It is easily accessible site and enough depth (800mwe) for small scale experiment [6]. The thin NaI(Tl) detector system was installed into a shield which was used for ELEGANT-V [6]. Seven modules of large volume NaI(Tl) scintillator whose dimension was 10.1×10.1×101cm^3 were placed below and above the thin NaI(Tl) detectors. The whole detector was covered with an air-tight container and 10cm thick pure copper shield and 15cm thick pure lead shield. Pure nitrogen gas was flushed into the air-tight container to remove radioactive Rn gas. The radioactivity in the air-tight container was kept as low as 150mBq/m^3.

Test measurement for double beta decay was performed for half month with and without Mo-source film. Data analyses to distinguish two layer was successfully performed. Tagging of the layer was performed by using small PMTs. Due to large contamination of ^{40}K and U-chain in aluminum flame of the detector, the background was not sufficiently reduced in the energy region of interest. However, the detector system worked well with good selectivity and the energy resolution.

3. Future prospect

The thin and wide area NaI(Tl) crystal was developed for studying nuclear rare processes. The energy resolution was enough good for not only neutrino-less double beta decay but also for dark matter search. A plastic scintillator is extensively applied to double beta decay experiments [4, 5, 7, 8] because of small backscattering ratio and low cost. However, it has been difficult to develop the plastic scintillator with the energy resolution below 4% in FWHM. The NaI(Tl) scintillator was shown that it is suitable for double beta decay experiment because of the good energy resolution. Figure 2 shows the dependence between the sensitivity for neutrino-less double beta decay and the live time. The sensitivity to the effective neutrino mass reaches to 20meV, it is important to explore the inverted hierarchy of neutrino. One need to carry out the experiment with more than 10 ton·year of live time. In the case of long live time, the NaI(Tl) has better sensitivity for neutrino mass.

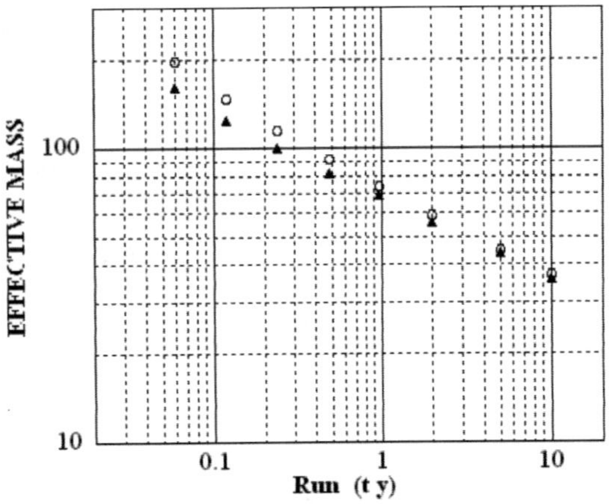

Figure 2. The sensitivity for effective neutrino mass as a function of the experimental live time.
Open circle: The case for NaI(Tl).
Triangle: The case for plastic scintillator (PL).

This work was supported by Yamada Science Foundation and by Grant-in-Aid of Scientific Research of Japan Society for the Promotion of Science No.20540294.

References

[1] H.Ejiri, Jour. of the Phys. Soc. Japan **74** 2101.
[2] F.T.Avignone, S.Elliott and J.Engel, Rev. Mod. Phys. **80** (2008) 481.
[3] J.Vergados, Phys. Rep. **361** (2002) 1.
[4] H.Ejiri, Mod. Phys. Lett. **A22** (2006) 1277.
[5] H.Ejiri et al., Euro Phys. J. **162** (2008) 239.
[6] H.Ejiri et al., Nucl. Instr. & Meth. in Phys. Res. **A302** (1991) 304.
[7] R.Arnold et al., Phys. Rev. Lett. **95** (2005) 182302.
[8] F.Mauger et al., In this proceedings.

Topics in Astroparticle and Underground Physics (TAUP 2009) IOP Publishing
Journal of Physics: Conference Series **203** (2010) 012065 doi:10.1088/1742-6596/203/1/012065

Searching for leptonic number non-conservation with NEMO-3 and SuperNEMO

F. Mauger, on behalf of the NEMO-3/SuperNEMO collaboration
LPC Caen, ENSICAEN, Université de Caen, CNRS/IN2P3, Caen, France
E-mail: mauger@lpccaen.in2p3.fr

Abstract.
The NEMO 3 experiment is devoted to the search for the neutrinoless double beta decay, a leptonic number violating second order weak process which is still to be discovered. The detector has been taking data in the LSM laboratory since 2003. Latest NEMO 3 results for several double beta decay emitters are presented here. The next generation SuperNEMO project, which aims to improve sensitivity to double beta decay by two orders of magnitude using NEMO 3 technique is also briefly described.

1. Introduction

Neutrinoless double beta decay ($0\nu\beta\beta$) (A,Z) $\rightarrow$(A,Z+2) + $2e^-$ is a leptonic number violating process. It has been intensely searched for the last three decades but has not been discovered yet. The observation of such a process, interpreted through the exchange of a light Majorana neutrino (the *mass mechanism* [1]), would indicate the nature of the neutrino and give access to its absolute mass scale.

The NEMO collaboration aims to investigate the $0\nu\beta\beta$ process using a specific experimental technique briefly described in this paper. The NEMO 3 detector is currently running in the Fréjus underground laboratory (LSM, Modane, France). The SuperNEMO project is the next generation experiment that will improve the NEMO 3 sensitivity by two orders of magnitude.

2. The NEMO 3 detector

The NEMO 3 detector has been designed for the direct detection of electrons from neutrinoless double beta decay at the sensitivity of $T_{1/2}^{0\nu\beta\beta} \simeq 10^{24}$ y. It accomodates a 20 m^2 thin source foil made of 7 kg of ^{100}Mo, 1 kg of ^{82}Se for $0\nu\beta\beta$ search, together with smaller amounts of ^{116}Cd, ^{150}Nd, ^{96}Zr, ^{48}Ca and ^{130}Te used to measure two-neutrino double beta decay (A,Z) $\rightarrow$(A,Z+2) + $2e^-$ + $2\bar{\nu}_e$ ($2\nu\beta\beta$). The $\beta\beta$ source is surrounded by a tracking chamber composed of 6400 drift cells working in the Geiger regime and a calorimeter made of 1940 plastic scintillator blocks coupled with low radioactive photomultiplier tubes. A full description of the detector can be found in [2]. The use of such a technique enables the measurement of the total electron energy, the energy of each electron and angular distribution between the electrons. Thanks to its ability to reconstruct the full kinematics of $2\nu\beta\beta$ decays, the NEMO 3 detector also gives very valuable constraints for double-beta decay nuclear models. More, the reconstruction of the topology of beta decays and gamma emission from natural radioactivity is possible and allows to model all sources of background events that would mimic $0\nu\beta\beta$ decays [3]. This feature makes the NEMO 3 detector an unique tool to investigate very rare $\beta\beta$ decays in a ultra low radiactivity environment.

Two-neutrino double beta decay measurements for several isotopes after 3.85 years of data collection are presented in table 1. ^{100}Mo $2\nu\beta\beta$ decays to 0^+ excited state of ^{100}Ru has been measured thanks to NEMO 3 hability to reconstruct $2e2\gamma$ topology: the half-life is $T_{1/2} = [5.7^{+1.3}_{-0.9}$ (stat)$\pm$ 0.8 (syst)] 10^{20} y [4]. More, the analysis of the shape of the single electron energy spectrum in ^{100}Mo $2\nu\beta\beta$ decay strongly supports the single state dominance mechanism hypothesis (SSD) [5].

Isotope	$T_{1/2}^{2\nu\beta\beta}$ (y)
^{100}Mo	$[7.11 \pm 0.02(stat) \pm 0.54(syst)] \times 10^{18}$ (SSD favored)
^{100}Mo(0_1^+)	$[5.7^{+1.3}_{-0.9}(stat) \pm 0.8(syst)] \times 10^{20}$
^{82}Se	$[9.6 \pm 0.3(stat) \pm 1.0(syst)] \times 10^{19}$
^{116}Cd	$[2.8 \pm 0.1(stat) \pm 0.3(syst)] \times 10^{19}$
^{130}Te	$[6.9 \pm 0.9(stat) \pm 1.0(syst)] \times 10^{20}$ *
^{150}Nd	$[9.20^{+0.25}_{-0.22}(stat) \pm 0.73(syst)] \times 10^{18}$ *
^{96}Zr	$[2.35 \pm 0.14(stat) \pm 0.19(syst)] \times 10^{19}$ *
^{48}Ca	$[4.4^{+0.5}_{-0.4}(stat) \pm 0.4(syst)] \times 10^{19}$ *

Table 1. Two-neutrino double beta decay measurements after 3.85 years of data collection with NEMO 3 (* preliminary results).

Concerning the neutrinoless double beta decay, data up to the end of 2008 show no evidence for leptonic number violation; 90% CL lower limits on the $0\nu\beta\beta$ decays half-live for ^{100}Mo [6] and ^{82}Se are set and upper limits on the Majorana neutrino effective mass are derived using recent calculations of nuclear matrix elements [7, 8, 9]:

^{100}Mo: $T_{1/2}^{0\nu\beta\beta} > 1.1\ 10^{24}$ y and $< m_\nu > < 0.45$–0.93 eV

^{82}Se: $T_{1/2}^{0\nu\beta\beta} > 3.6\ 10^{23}$ y and $< m_\nu > < 0.89$–1.61 eV

No evidence for $0\nu\beta\beta$ processes via V+A current or Majoron [10] emission have been proved and the corresponding derived limits are shown on table 2. More details can be found in [11, 12].

	V+A	Majoron(s) emission			
Isotope	$T_{1/2}^{0\nu\beta\beta}$ (y)	n=1	n=2	n=3	n=7
^{100}Mo	$> 5.7\ 10^{23}$ $\lambda < 1.4\ 10^{-6}$	$> 2.7\ 10^{22}$ $g_{ee} < (0.4 - 1.8)10^{-4}$	$> 1.7\ 10^{22}$	$> 1.0\ 10^{22}$	$> 7\ 10^{19}$
^{82}Se	$> 2.4\ 10^{23}$ $\lambda < 2.0\ 10^{-6}$	$> 1.5\ 10^{22}$ $g_{ee} < (0.7 - 1.9)10^{-4}$	$> 6\ 10^{21}$	$> 3.1\ 10^{21}$	$> 5\ 10^{20}$

Table 2. Limits on neutrinoless double beta decay throug V+A current and Majoron emission mechanisms (nuclear matrix elements has been taken from [13, 14]).

3. The SuperNEMO project

The SuperNEMO experiment is being designed to search for $0\nu\beta\beta$ at the sensitivity for the half-life $T_{1/2}^{0\nu\beta\beta} \simeq 2\ 10^{26}$ years, corresponding to $< m_\nu > \sim 50$ meV, for a typical ^{82}Se 500 kg.y exposure. Following NEMO 3 experimental technique, it will accomodate $\sim$ 100 kg of a ^{82}Se source. A three-year R&D phase started in 2006 and a technical design report is due in 2009. A first demonstrator module with 5 kg of ^{82}Se will be constructed in 2010. After validation, 19 more similar modules will be completed to start data collection in 2012. The SuperNEMO R&D program concentrates on the following critical points [15, 16, 17]:

- the making of 100 kg of ultra pure enriched source of ^{82}Se, including experimental validation of the required radiopurity level of the final thin source foils thanks to a dedicated BiPo detector [18],
- the selection and control of the materials used in the construction of the detector to ensure a very low radioactivity level,
- the design of a radon-free highly optimized tracking chamber, including the technology for automated mass production of drift cells,
- the design of the PVT plastic scintillator based calorimeter with required energy, time and spatial resolution, together with a high quality calibration survey procedure.

References

[1] T. Tomoda, *Rep. Prog. Phys.* **54** (1991) 53-126
[2] R. Arnold *et al*, *Nucl. Instrum. Method.* A **536** (2005) 79
[3] J. Argyriades *et al*, *Nucl. Instrum. Method.* A **606** (2009) 449
[4] J. Argyriades *et al*, *Nucl. Phys.* A **781** (2007) 209
[5] F. Simkovic *et al*, *J. Phys.* G **27** (2001) 2233
[6] R. Arnold *et al*, *Phys. Rev. Lett.* **95** (2005) 182302
[7] M. Kortelainen and J.Suhonen, *Phys.Rev.* C **75** (2007) 051303(R)
[8] M. Kortelainen and J.Suhonen, *Phys.Rev.* C **76** (2007) 024315
[9] F. Simkovic *et al*, *Phys.Rev.* C **77** (2008) 045503
[10] R. Arnold *et al,Nucl. Phys.* A **765** (2006) 483
[11] V. Tretyak, *Results from double beta decay experiment NEMO 3*, MEDEX 2009 conference, June 2009, Prague
[12] V. Beillet-Kovalenko, *Search for double beta decay with the NEMO 3 experiment*, Europhysics Conference on High Energy Physics, July 16-22, 2009, Krakow
[13] M. Aunola and J. Suhonen, *Nucl. Phys.* A **643** (1998) 207
[14] J. Suhonen, *Nucl. Phys.* A **700** (2002) 649
[15] L. Vala, *Results of NEMO 3 and Status of SuperNEMO, Proceedings of the Neutrino Oscillation Workshop, Nuclear Physics B - Proceedings Supplements* Volume **188** (2009) 62
[16] E. Chauveau, *SuperNEMO Project Status*, MEDEX 2009 conference, June 2009, Prague
[17] I. Nasteva, *SuperNEMO – the next generation double beta decay experiment*, arXiv:0909.3167v1 [hep-ex] (2009)
[18] S. Cebrian, *The BiPo detector*, TAUP 2009 proceedings (2009)

Journal of Physics: Conference Series **203** (2010) 012066 doi:10.1088/1742-6596/203/1/012066

Neutrino Mass, Neutrinoless Double Electron Capture and Rare Beta Decays

M T Mustonen and J Suhonen

Department of Physics, P.O. Box 35 (YFL), FI-40014 University of Jyväskylä, Finland

E-mail: `jouni.suhonen@phys.jyu.fi`

Abstract. We present results of our theoretical calculations on three nuclei of interest from the neutrino-physics point of view: Firstly, we present the second-forbidden decay branch of ^{115}In with the ultra-low Q value and theoretical open questions related to such decays. Secondly, we have calculated estimates for the half-lives of the single-beta decay channels of ^{96}Zr and concluded that the possible contamination from those to the geochemical measurements of ^{96}Zr double-beta-decay half-life is rather small. Thirdly, we have taken a look at the neutrinoless resonance double-electron-capture decay of ^{112}Sn in the light of recent JYFLTRAP Q value measurements and discovered that the badly fulfilled resonance condition renders the decay unobservable.

1. The ultra-low-Q-value decay of ^{115}In

The decay of ^{115}In to the first excited state of ^{115}Sn was first experimentally discovered and recognized to possibly have the lowest observed beta-decay Q value by Cattadori *et al* [1]. This discovery raised hopes to use this decay as a sensitive neutrino-mass detector. The same decay was recently studied in more detail as a joint effort of one theoretical and two experimental groups [2]: First, a half-life measurement was conducted at the HADES underground laboratory to confirm the first observation and to reduce the uncertainties in both the partial half-life and the branching ratio. Second, the mass difference of the mother and daughter isotopes was measured in the JYFLTRAP Penning trap to evaluate the Q value previously known to be 2(4) keV. Third, we calculated theoretically the decay half-life using our model (the proton-neutron microscopic quasiparticle-phonon model, pnMQPM) that had previous success [3] in describing the ground-state-to-ground-state decay of the same nucleus.

The JYFLTRAP measurement revealed the Q value to be record-setting low 0.35(17) keV, which is an order of magnitude smaller than 2.469(4) keV, the Q value of ^{187}Re beta decay [4]. The HADES measurements for the half-life and branching ratio, $4.1(6) \times 10^{20}$ y and $1.07(17) \times 10^{-6}$, were in perfect agreement with the results of Cattadori *et al* [1], $3.7(10) \times 10^{20}$ y and $1.2(3) \times 10^{-6}$, respectively. From the theoretical point of view this decay is second-forbidden unique, which simplifies the calculation remarkably: The dependence of the decay on nuclear structure is contained in the square of a single nuclear matrix element. However, the difference between the calculated rate and the experimental rate (with one-sigma uncertainties) was roughly two orders of magnitude when using the experimentally derived Q value in the calculations.

J_f	Forbiddeness	Q value (keV)	$T_{1/2}$ [a]
6^+	6th non-unique	161	4.9×10^{29}
5^+	4th unique	117	2.6×10^{20}
4^+	4th non-unique	15	2.3×10^{23}
Total			2.6×10^{20}

Table 1. Theoretical half-lives based on experimental Q values for yet unobserved ^{96}Zr single-beta decay channels.

The initial and final state can both be explained as simple one-quasiparticle states, and taking the small three-quasiparticle contributions into account using the pnMQPM has only a small effect on the nuclear matrix element. We find it therefore very unlikely that the discrepancy could be explained solely by inaccuracies in our nuclear matrix element. However, bearing in mind that we have crossed to the previously unexplored Q value regime, we have realized that there are several atomic effects that are usually neglected but possibly have a dramatic consequences for the decays with such ultra-low Q values. These effects are electron screening, atomic overlap effects, exchange effects and final-state interactions.

None of these effects have been estimated for Q values with order of magnitude this low, but it is known that they are generally the more important the lower the Q value is. The electron screening has been only studied for allowed decays [5]. The existing formulas (Rose prescription [6] and the fully relativistic formula by Lopez and Durand [7]) for taking the screening into account break down for ultra-low Q values. The work on exchange effects [8, 9] has yielded contradictory results. The final-state interactions have only been studied in tritium decay [10]. There is hence a vast gap in our theoretical understanding of nuclear beta decay when we move past the traditional low-Q-value regime.

2. The beta-decay channels of ^{96}Zr

The three single-beta-decay channels of ^{96}Zr are highly forbidden and have low Q values (see Table 1 for details). Due to these two reasons the single beta decays are hindered so strongly that the double beta decay actually has higher transition rate than this first-order weak process. Our first calculation for these single beta decays was published in Ref. [11]. The values of Table 1 have been updated to use more recent experimental data as input parameters. The conclusions remain the same: The total single-beta-decay rate, dominated by the 4th-forbidden unique channel, is much slower than the double beta decay. Therefore the contamination from the single beta decay to geochemical double-beta-decay experiments [12, 13] is well within the experimental uncertainties.

3. The neutrinoless double-EC decay of ^{112}Sn

The neutrinoless double electron-capture (0νECEC) decay with a resonance condition has attracted attention recently [14, 15]. The resonance condition - close degeneracy of the initial and final (excited) atomic states - can enhance the decay rate by a factor as large as 10^6. Fulfillment of the resonance condition [14] depends on the so-called degeneracy parameter $D = Q - E$, where E is the excitation energy of the final atomic state and Q is the difference between the initial and final atomic masses. Possible candidates for the resonant decays are many [14, 15], among them the decay $^{112}\text{Sn} \rightarrow {}^{112}\text{Cd}(0^+)$, depicted in Fig. 1. For a long time this 0νECEC decay of the ^{112}Sn ground state to the fourth 0^+ state in ^{112}Cd was considered to be the flagship of neutrinoless double electron capture.

In [16] the degeneracy parameter was determined by using the JYFLTRAP Penning trap.

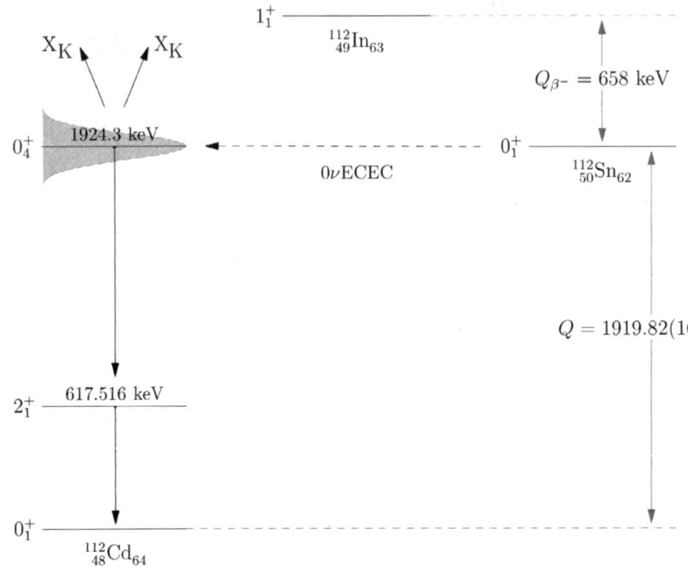

Figure 1. The resonance double-electron-capture decay of ^{112}Sn.

The results $D = -4.5\,\mathrm{keV}$ for KK capture, $D = 18.2\,\mathrm{keV}$ for KL capture and $D = 40.9\,\mathrm{keV}$ for LL capture were derived. This yield for the decay half-life the estimate

$$T_{1/2} > \frac{5.9 \times 10^{29}}{(m_\nu[\mathrm{eV}])^2}\,\mathrm{years}\ . \tag{1}$$

This result indicates that experimental sensitivities of $T_{1/2} > 10^{30}$ years are required to detect the ^{112}Sn decay. This is impossible in the foreseeable future.

Acknowledgments

This work was supported by the Academy of Finland under the Finnish Center of Excellence Program 2006-2011 (Nuclear and Accelerator Based Program at JYFL).

References

[1] Cattadori C M, De Deo M, Laubenstein M, Pandola L and Tretyak V I 2005 *Nucl. Phys.* A **748** 333
[2] Wieslander J S E, Suhonen J, Eronen T, Hult M, Elomaa V V, Jokinen A, Marissens G, Misiaszek M, Mustonen M T, Rahaman S, Weber C and Äystö J 2009 *Phys. Rev. Lett.* **103** 122501
[3] Mustonen M T and Suhonen J 2007 *Phys. Lett.* B **657** 38
[4] Basunia M S 2009 *Nucl. Data Sheets* **110** 999
[5] Matese J J and Johnson W R 1966 *Phys. Rev.* **150** 846
[6] Rose M E 1936 *Phys. Rev.* **49** 727
[7] Lopez J L and Durand L 1988 *Phys. Rev.* C **37** 535
[8] Bahcall J N 1963 *Phys. Rev.* **129** 2683
[9] Harston M R and Pyper N C 1992 *Phys. Rev.* A **45** 6282
[10] Saenz A and Froelich P 1997 *Phys. Rev.* C **56** 2132
[11] Heiskanen H, Mustonen M T and Suhonen J 2007 *J. Phys. G: Nucl. Part. Phys.* **34** 837
[12] Kawashima A, Takahashi K and Masuda A 1993 *Phys. Rev.* C **47** R2452
[13] Wieser M E and De Laeter J R 2001 *Phys. Rev.* C **64** 024308
[14] Bernabeu J, De Rujula A and Jarlskog C 1983 *Nucl. Phys.* B **223** 15
[15] Sujkowski Z and Wycech S 2004 *Phys. Rev.* C **70** 052501
[16] Rahaman S, Elomaa V-V, Eronen T, Hakala J, Jokinen A, Kankainen A, Rissanen J, Suhonen J, Weber C and Äystö J 2009 *Phys. Rev. Lett.* **103** 042501

Searching for Double Beta Decay with the Enriched Xenon Observatory

Lisa J. Kaufman
Physics Department, University of Maryland, College Park, MD 20742, USA
E-mail: lkaufman@umd.edu

Abstract. The Enriched Xenon Observatory (EXO) will search for neutrinoless double beta decay in ^{136}Xe. The first phase of the experiment, EXO-200, uses 200 kg of liquid xenon enriched to 80% in ^{136}Xe in an ultra-low background time projection chamber (TPC). EXO-200 is in the final stages of assembly at the WIPP site in Carlsbad, NM and will begin taking data in 2009 with two-year sensitivity to the half-life for neutrinoless double beta decay of 6.4×10^{25} years. According to nuclear matrix element calculations, this corresponds to an effective Majorana neutrino mass of 0.13 to 0.19 eV. The EXO collaboration is also performing R&D for simultaneous detection of the decay electrons and emerging Ba ion allowing essentially background free detection in a future, ton-scale detector. The status of EXO-200 and of the ion tagging technology in liquid xenon is described.

1. Introduction

Neutrinos have long been a difficult particle to study, yet recent oscillation experiments have answered many questions about neutrinos and their properties. Still, many questions remain unanswered and can be addressed through other similarly challenging techniques. In particular, double beta decay can shed light on the absolute mass scale of the neutrino and the Majorana nature of the neutrino. The Enriched Xenon Observatory (EXO) for double beta decay is a program of experiments, working toward the ton-scale, to look for neutrinoless double beta decay with promising discovery potential.

2. Double Beta Decay

Double beta decay is a second-order standard model process whereby two neutrons in the nucleus of an atom simultaneously decay to protons producing two electrons and two electron anti-neutrinos in the process. This process can occur in particular nuclei where single beta decay is not energetically allowed. These nuclei are also used to look for the neutrinoless double beta decay process ($0\nu\beta\beta$) which is forbidden by the standard model because it requires the violation of lepton number. In addition, neutrinoless double beta decay requires that neutrinos be massive and that they are there own anti-particle classifying them as Majorana particles.

The two-neutrino mode of double beta decay ($2\nu\beta\beta$) has been observed in many isotopes but has not yet been observed in ^{136}Xe. The probability of decay for the $2\nu\beta\beta$ mode is very small and exceedingly smaller for the $0\nu\beta\beta$ mode making these measurements extremely difficult. The measured lifetime for the $0\nu\beta\beta$ mode is proportional to the inverse of the effective Majorana neutrino mass (a coherent superposition of the electron neutrino projections on the mass eigenstates) making the absolute mass scale of the neutrino accessible with such a measurement. The EXO collaboration will search for both the $2\nu\beta\beta$ and $0\nu\beta\beta$ decay modes in ^{136}Xe.

The two decay modes can be differentiated by measuring the energy distribution of the electron spectrum produced in the decay. The neutrinoless signal will appear as a peak at the endpoint of the $2\nu\beta\beta$ spectrum. The width of the peak is determined by the energy resolution of the detector making this a critical component of the experiment design while also eliminating possible backgrounds in the region of interest.

3. EXO-200

EXO-200 is the first phase of the EXO experiment, using 200 kg of xenon enriched to 80% in the ^{136}Xe isotope. EXO-200 will use a time projection chamber (TPC) filled with liquid xenon that will serve as the target and detection medium. When the xenon nucleus decays, it will produce both scintillation light and ionization in the detector. These light is collected by a plane of large-area avalanche photo-diodes (LAAPDs) and the charge is collected on a grid of anode wires at each end of the detector. The energy resolution of the detector is greatly imporoved by collecting both the scintillation and ionization signals in the detector [1][2]. The monolithic detector minimizes surface contamination and the gaseous or liquid volume allows for continuous circulation and purification of the xenon. A high Q-value places the region of interest for $0\nu\beta\beta$ above decay lines, and ^{136}Xe is adavantageous with a $Q = 2.48$ MeV [3].

^{136}Xe offers a unique opportunity to tag the daughter barium ion. Identification of the barium ion for each double beta decay event will eliminate all radioactive backgrounds in the $0\nu\beta\beta$ region of interest leaving the $2\nu\beta\beta$ events the only source of background [4]. The barium tagging technique is in the R&D phase and will not be employed in EXO-200.

EXO-200 will look for $0\nu\beta\beta$ in ^{136}Xe with a sensitivity to the half-life of 6.4×10^{25} years which is an order of magnitude better than the current limit. In addition, EXO-200 will measure the half-life for the standard $2\nu\beta\beta$. Based on the current best limits for the half-life of the $2\nu\beta\beta$ mode, it is expected that EXO-200 will observe anywhere from 23000 to 1.3 million events in one year.

The TPC was designed to maximize fiducial volume and energy resolution and to minimize radioactive contamination. The limits on radioactivity imply significant constraints on the types and quantities of materials that could be used for the TPC construction [5]. A figure of the completed TPC is shown in Figure 1. The xenon vessel, which contains the TPC, is only 1.5 mm thick because of radioactivity requirements. The read-out cables are flex-circuits made of copper on kapton. The field-shaping rings are copper while the anode grid wires and cathode are all phosphor-bronze. The cathode is located in the center of the cylindrical TPC which is 40 cm in length and diameter. The plane of anode grid wires and APD plane of 234 LAAPDs (see [6] for more information on the APDs) are located at each end of the detector. Teflon is used inside the cylindrical volume as a reflective surface to increase light collection.

The TPC is kept at a temperature of 170 K to keep the xenon liquid at a pressure of 1.5 atmospheres. The xenon vessel is kept cold while in thermal contact with a cryogenic fluid made by 3M called HFE-7000. The HFE is contained in a copper cryostat that is cooled by three heat exchangers, each of which is connected to a cryogenic refrigerator. The cryostat containing the HFE is kept cold by a volume of insulating vacuum. This outer cryostat is shielded by a 25 cm thick Pb wall. The entire experimental set-up is contained within a class-100 clean room. Additionally, the clean room is surrounded by a muon veto detector.

EXO-200 is installed at the Waste Isolation Pilot Plant (WIPP) in Carlsbad, New Mexico at a depth of 655 m and $\sim$ 1600 mwe. The assembly of the TPC was completed at Stanford University, and the TPC was shipped to WIPP in October this year and is ready for installation. Cryogenic commissioning of the xenon, HFE, and control systems has just been completed at WIPP. Once the TPC is installed, the experiment will perform an engineering run with natural

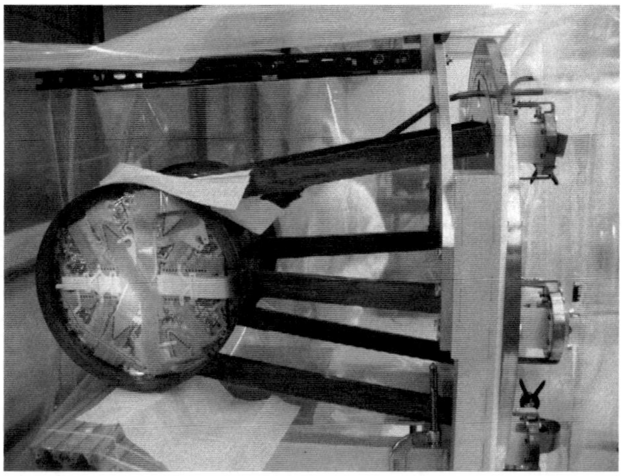

Figure 1. The TPC inserted in the xenon vessel wit the attached cryostat door. The legs contain the readout flex-circuits and are the conduits for xenon circulation.

xenon soon followed by physics data-taking with the enriched xenon.

4. Barium Ion Tagging in Liquid Xenon

The $0\nu\beta\beta$ decay process produces a daughter Ba^{++} ion which is expected to quickly convert to Ba^{+} in a liquid xenon environment. The barium ion produced can be optically identified because of its specific shelving signature. By optically pumping the two transitions with red and blue lasers, a fluorescence rate of $\sim 10^8$ Hz can be measured in vacuum.

The barium ion tagging scheme would be to build a probe that could electrically attract the barium ion out of the liquid xenon and would then be transported to a quadrupole linear ion trap where the ion will be optically pumped and the fluorescence identified by a CCD camera. The ion trap has been built and has successfully shown identification of single Ba ions while a buffer gas is present in the trap [7].

Currently there are two types of probes being developed for the barium tagging in liquid xenon. One probe uses a cryogenic capacitive sensor on a tip to freeze a layer of xenon on the tip after first attracting the ion to the probe tip with a bias voltage [8]. The layer of xenon ice on the tip is well-controlled by the capacitance sensor on the tip. Research is underway to install this probe on the top of linear ion trap to measure efficiency of the probe and trap system. A second technique using resonant ion spectroscopy is being developed.

In order to test the efficiency of the entire tagging system, it is important to have a reliable single-ion barium source. One method is to use ^{137}Cs, which will single beta decay to $^{137}Ba^{+}$ and an electron with a half-life of 30 years. ^{137}Cs comes in the form of CsCl salt and needs to be chemically isolated. This process has been successfully completed in vacuum and and is proceeding in the gaseous and liquid xenon environments.

5. Future Plans

EXO-200 will begin data-taking in 2010 and will provide the first results of the half-life for $2\nu\beta\beta$ decay in ^{136}Xe and set competitive limits on the half-life of the $0\nu\beta\beta$ decay mode. The understanding of the liquid xenon TPC and radioactive background measurements will provide essential knowledge for design work for a future ton-scale EXO experiment. The predicted sensitivities for the the EXO-200 and ton-scale EXO experiments are shown in Table 1. It is assumed that the ton-scale EXO will use the Ba tagging technique. A limit on the effective

Table 1. The predicted EXO sensitivities to the $0\nu\beta\beta$ decay based on nominal experimental parameters and recent nuclear matrix element calculations.

Case	Mass (ton)	Eff (%)	Run Time (yr)	σ_E/E (%) 2.5 MeV	Radioactive BG (events)	$T_{1/2}^{0\nu}$ (yr) 90% CL	Majorana Mass (meV) QRPA[9]	NSM[10]
EXO-200	0.2	70	2	1.6	40	6.4×10^{25}	133	186
Conservative	1	70	5	1.6	0.5 (1) ($2\nu\beta\beta$)	2×10^{27}	24	33
Aggressive	10	70	10	1	0.7 (1) ($2\nu\beta\beta$)	4.1×10^{28}	5.3	7.3

neutrino mass at the tens to few meV scale will exclude the inverted hierarchy for the Majorana neutrino.

Acknowledgments

EXO is supported by the DoE and the NSF in the United States, the NSERC in Canada, FNS in Switzerland, and the Ministry for Science and Education of the Russian Federation in Russia.

References

[1] Conti E *et al.* 2003 *Phys. Rev.* B **68** 054201
[2] Aprile E, Giboni K L, Majewski P, Ni K and Yamashita M 2007 *Phys. Rev.* B **76** 014115
[3] Redshaw M, Wingfield E, j McDaniel and Myers E G 2007 *Phys. Rev. Lett* **98** 053003
[4] Danilov M *et al.* 2000 *Phys. Rev. Lett.* B **480** 12
[5] Leonard D *et al.* 2008 *Nucl. Inst. Meth.* A **591** 490
[6] Neilson R *et al.* 2009 *Nucl. Inst. Meth.* A **608** 68
[7] Flatt B *et al.* 2007 *Nucl. Inst. Meth.* A **578** 399
[8] Fierlinger P *et al.* 2008 *Review of Scientific Instruments* **79** 045101
[9] Rodin V, Faessler A, Simkovic F and Vogel P 2007 *Nuc. Phys.* A **793** 213
[10] Caurier E, Nowacki F, Poves A and Retamosa J 2008 *Phys. Rev. Lett.* **100** 052503

Journal of Physics: Conference Series **203** (2010) 012068 doi:10.1088/1742-6596/203/1/012068

NEXT: searching for the neutrinoless double beta decay with a gas-xenon TPC

P. Novella for the NEXT collaboration

CIEMAT, Av. Complutense 22, 28040 Madrid, SPAIN

E-mail: pau.novella@ciemat.es

Abstract. Although different techniques are used to search for the neutrinoless double beta decay, the common challenges for all the existing or planned experiments are to achieve a good energy resolution and large background rejection factors. The NEXT collaboration addresses these two challenges with a high-pressure gas-Xenon TPC. Natural Xenon consists of almost 9% of ^{136}Xe, a $\beta\beta^{0\nu}$ candidate emitter, and can be easily enriched. When used as a calorimeter, ^{136}Xe yields an excellent energy resolution. This fact, combined with the expected long life of the $\beta\beta^{2\nu}$ mode, accounts for negligible intrinsic backgrounds up to masses of 1 ton. Furthermore, external backgrounds can be rejected with high efficiency by means of the electron tracking capabilities of the TPC. A detector containing about 100 kg of enriched Xenon is expected to be installed at Canfranc Underground Laboratory (LSC) within the next 5 years, with the twofold aim of exploring the degenerated hierarchy of the neutrino mass and providing deep understanding of the experimental techniques which allow extrapolation to larger detectors.

1. Introduction

Results of oscillation experiments in the past few years have demonstrated that neutrinos are massive particles and that individual lepton numbers are not conserved. A direct consequence is the renewed interest in the double beta ($\beta\beta$) decay experiments. $\beta\beta$ decay is a nuclear transition in which two neutrons bound in a nucleus are simultaneously transformed into two protons plus two electrons. The decay mode in which two neutrinos are emitted ($\beta\beta^{2\nu}$) has been observed in many nuclei. However, if neutrinos are massive Majorana particles, the neutrinoless $\beta\beta$ ($\beta\beta^{0\nu}$) decay may be also possible. The detection of the $\beta\beta^{0\nu}$ process would prove the Majorana nature of neutrinos, providing also a measurement of the so-called *neutrino effective mass* $\langle m_\nu \rangle$.

In the search for the $\beta\beta^{0\nu}$ process two main detector capabilities are requested. The first one is to achieve a good energy resolution; the second one is to provide extra handles to reject backgrounds in the energy region around $Q_{\beta\beta}$. The NEXT collaboration aims at finding a good compromise between these two requirements by using a gas ^{136}Xe TPC. So far, there is no clear signal of the $\beta\beta^{0\nu}$ process, although a group of the Heidelberg-Moscow [1] experiment has claimed to observe this decay in ^{76}Ge. To confirm or refute this result, next generation of $\beta\beta$ experiments is meant to reach sensitives around $\langle m_\nu \rangle$=50 meV and to develop the techniques that will allow further improvement in case no signal is detected. These are the two main goals of the NEXT collaboration, along with the measurement of the $\beta\beta^{2\nu}$ decay lifetime, which is not known yet.

2. The NEXT TPC

NEXT stands for Neutrino Experiment with a Xenon TPC. The TPC is defined to operate with gaseous xenon at high pressure (HPGXe), as done in the Gotthard experiment [2]. Xenon is the only noble gas that has a $\beta\beta$-decaying isotope (^{136}Xe) and no other long-lived radioactive isotope. ^{136}Xe natural abundance is almost 9% and further enrichment can be easily achieved by centrifugation. Its $Q_{\beta\beta}$ value is relatively high (2480 keV) and the $\beta\beta^{2\nu}$ decay mode, not measured yet, may be as long as 10^{22} - 10^{23} years. All these characteristics make ^{136}Xe to be a suitable isotope for the search of the $\beta\beta^{0\nu}$ process. A xenon TPC provides both primary scintillation light and ionization electrons when charged particles interact in the TPC volume. The scintillation determines the start-of-event t_0 time when detected with a photosensor device. Although a liquid xenon TPC is a much more compact device than a gas xenon TPC, its tracking capabilities are quite limited. Given the density of the liquid xenon (3 g/cm^3), electrons from a $\beta\beta$ decay deposit all their energy in a single blob. On the contrary, electrons can be easily tracked in the gas: a gas xenon TPC operating at 10 bar is able to track $\beta\beta$ electrons with path lenghts of about 15 cm. In addition, the ionization of liquified noble gases is accompanied by larger fluctuations, thus leading to a worse energy resolution [3].

The NEXT TPC is being designed according to the *SOFT* (Separated-Optimized Function for tracking) concept, based on ideas by D. Nygren [4]: separated and optimized technical solutions for energy and tracking measurements are being explored. Electroluminescence (EL) will be used to amplify the ionization signal, allowing to measure both scintillation and ionization signals with the same photosensors. The readout plane devoted to perform the energy function (and t_0) will be an array of Hamamatsu R8520-06SEL PMTs, currently under development. According to [3], the target for the energy resolution has been set to 1% FWHM at 2480 keV. On the other hand, three main approaches are under study to implement the tracking function. Multi-pixel photon counters (MPPC) and large area avalanche photodiodes (LAPD) could provide tracking information by reading the same EL light used for the energy measurement. Alternatively, Micromegas [5] could perform a charge amplification of the primary ionization and read out the charge.

3. Physics case

The physics case of any $\beta\beta$ experiment can be described in terms of the background rates populating the energy region around the $Q_{\beta\beta}$ value of the isotope under study. In the NEXT experiment, background coming from the tail of the $\beta\beta^{2\nu}$ spectrum is expected to be negligible given the long life of the $\beta\beta^{2\nu}$ decay mode of ^{136}Xe. Furthermore, external backgrounds can be also suppressed down to a negligible value by means of shielding. Therefore, the dominant background comes from ^{214}Bi and ^{208}Tl decays, from the ^{238}U and ^{232}Th chains respectively, taking place in detector materials. A photon of 2614.5 keV is emitted in the ^{208}Tl decay with 100% intensity, along with the β particle and other photons. In addition, ^{214}Bi has two β decays with endpoints above 2480 keV (2663 keV with 1.7% intensity and 3272 keV with 18.2% intensity) and a γ of 2447.8 keV (1.5% intensity). All these decay channels can lead to secondary electrons via Compton and photoelectric interactions.

$\beta\beta$ events have a distinctive topological signature in a HPGXe TPC: an ionization track, of about 30 cm length at 10 bar, tortuous because of multiple scattering, and with larger depositions or *blobs* in both ends corresponding to the two stopping electrons (see Figure 1). This signature is different from that one of a single electron, as only one energy blob is produced. Thereby, the tracking capabilities of the TPC provide a major handle to reject background events when searching for $\beta\beta$ decays. In addition, the specific signature of each physics process allows the measurement of the backgrounds operating the detector on site.

In order to reject backgrounds, three main selection cuts can be applied in the analysis of the data, all of them taking advantage of the topology of the events. First, events with energy

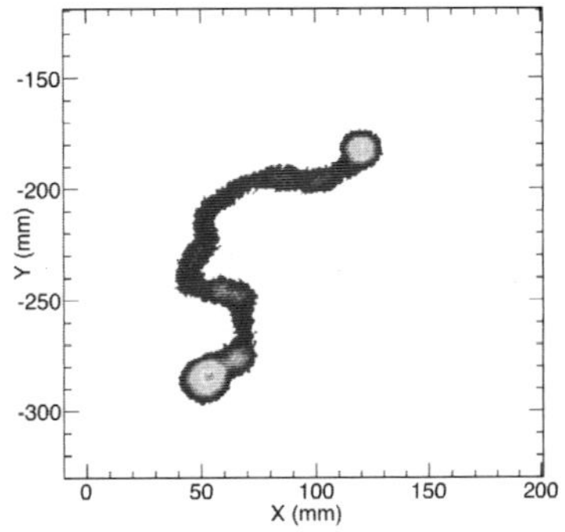

Figure 1. XY projection of a 3D $\beta\beta$ track simulated in ^{136}Xe at 10 bar.

depositions close to the vessel are rejected. Second, only events with one reconstructed track fully contained in the fiducial volume are selected. Finally, the reconstructed track is requested to end up in two blobs of high energy. A full Monte Carlo study has been developed to estimate the background suppression factors according to the above selection cuts. It has been found that a rejection factor of the order of 10^{-7} (10^{-6}) is achieved for ^{214}Bi (^{208}Tl) background events, when only the two first selection cuts are taken into account. According to the results of the Gotthard experiment [2], an additional reduction in the background levels of at least one order of magnitude is expected due to the topology signature. Assuming typical radiopurity levels for the detector materials, it has been concluded that background rates will be less than 10^{-3} count/keV/kg/year, leading to a sensitivity to the $\beta\beta^{0\nu}$ lifetime around 2×10^{23} years at 90% C.L. (below 100 meV in terms of $\langle m_\nu \rangle$) for an exposure of 500 kg $\cdot$ year.

4. The NEXT project

NEXT is an international collaboration already partially funded by the Spanish Ministry of Science and Innovation. NEXT aims at the operation of a 100 kg fiducial volume HPGXe TPC at LSC, by 2018. A Letter of Intent has been already submitted to the LSC Scientific Committee [6] and the project has been approved. The project is divided into three well defined stages. In the first stage, several small prototypes are being built and operated to prove the feasibility of the technologies (energy resolution with PMTs, tracking with photosensors and operation of Micromegas in xenon at high pressure). From 2011 to 2013, a second stage will be devoted to the construction and commissioning of a radioupure HPGXe TPC containing 10 kg of xenon (NEXT-10), to be installed at LSC. Finally, from 2013 to 2018, the ultimate NEXT detector (NEXT-100) containing 100 kg of ^{136}Xe will be built and operated at LSC.

5. Acknowledgments

The NEXT collaboration wishes to acknowledge the support from the CONSOLIDER-INGENIO project CSD2008-0037 (CUP).

References

[1] H. V. Klapdor-Kleingrothaus et al., Phys. Lett. B586 (2004), 198.
[2] R. Luscher et al., Phys. Lett. B434 (1998), 407
[3] A. Bolotnikov et al., Nucl. Ins. Meth A396 (1997), 360
[4] D. Nygren, Nucl. Ins. Meth A581 (2007), 632
[5] Y. Giomataris et al., Nucl. Ins. Meth. A419 (1996), 239
[6] F. Granena *et al.* [The NEXT Collaboration], arXiv:0907.4054 [hep-ex].

From Cuoricino to CUORE: investigating neutrino properties with double beta decay

Monica Sisti, on behalf of the CUORE collaboration
Università di Milano-Bicocca and INFN, Piazza della Scienza 3, 20126 Milano, Italy
E-mail: monica.sisti@mib.infn.it

Abstract. CUORE is the only fully approved 1-ton size neutrinoless double beta decay experiment to date, with the goal of scrutinazing the inverted hierachy region for the effective Majorana mass. CUORE, presently being built in Gran Sasso Underground Laboratory, is an array of 988 TeO_2 cryogenic detectors containing 200 kg of ^{130}Te, the neutrinoless double beta decay candidate. CUORE is due to start data taking in 2012. The feasibility of the project has been proved by Cuoricino, the pilot experiment that took data until 2008, for about five years, with 62 TeO2 cryogenic detectors. Cuoricino will be superseded in 2010 by CUORE-0, the first CUORE tower to be installed in Cuoricino cryogenic facility, which will take data until CUORE start. In this paper the final results of Cuoricino are reported, and CUORE-0 and CUORE potential and state of the art are discussed.

1. Introduction

Neutrinoless double beta decay ($\beta\beta0\nu$) is a key tool for investigating neutrino properties: if observed it would undoubtedly imply that neutrinos are massive Majorana particles. Expected sensitivities of future planned experiments should be able to probe the inverted hierarchy allowed region for the effective Majorana mass [1].

Experimental techniques based on the calorimetric approach with cryogenic detectors offer several advantages like high efficiency, high energy resolution, large masses, and a wide choice of the absorbing material (i.e. of the $\beta\beta0\nu$ source). Cuoricino and CUORE, described in the following sections, are calorimetric experiments searching for the $\beta\beta0\nu$ of ^{130}Te with cryogenic detectors.

2. Cuoricino

The Cuoricino experiment was running from February 2003 until June 2008 in Hall A of the Gran Sasso Underground Laboratory (AQ, Italy) at a depth of $\sim$ 3500 m.w.e. It was by far the largest cryogenic detector in the world, using an array of 62 TeO_2 crystals, for a total active mass of 40.7 kg. The array, in a tower like structure, was composed by 11 modules of 4 crystals $5\times5\times5\,cm^3$ each, and 2 modules of 9 crystals $3\times3\times6\,cm^3$ each. Every module is composed by two copper frames holding the TeO_2 crystals via teflon springs. All the TeO_2 crystals had natural ^{130}Te isotopic abudance ($\sim$34%) except for 4 of the smaller type, two of which enriched to 82.3% in ^{128}Te and the other two to 75% in ^{130}Te, corresponding to a total ^{130}Te mass of 11 kg. The array was operated at a temperature of about 10 mK inside a dilution refrigerator, shielded from environmental radioactivity and neutrons. Further details about the experimental setup can be found in [2]. The performance of the detectors were very good: the average FWHM energy resolution of the entire array was about 8 keV at the ^{130}Te transition energy (2527 keV). The total analyzed exposure adds up to 18 y×kg of ^{130}Te and the final sum spectrum is obtained with the detectors operated in anticoincidence, in order to reduce background contributions from Th and U on the crystal surfaces and from external γs presumably due to Th contamination of the refrigerator. The achieved background level in the $\beta\beta0\nu$ energy region is ~ 0.18 counts/keV/kg/y, which translates into a 90% C.L. lower limit on the half life of ^{130}Te $\beta\beta0\nu$ of $\tau^{0\nu}_{1/2} > 2.94 \times 10^{24}$ y. The corresponding upper limit for the effective Majorana mass is $m_{\beta\beta} < 0.21 - 0.70$ eV, with the spread due to the uncertainties in nuclear matrix element calculations, as given by [3]. This result does not completely rule out the claim of evidence by H.V. Klapdor-Kleingrothaus and his co-workers [4].

A careful analysis of the background level above the highest natural γ line of ^{208}Tl shows a continuum up to about 4 MeV which clearly extends in the region of interest and accounts for most ($\sim$70%) of the count rate around ^{130}Te $\beta\beta0\nu$ transition energy. On the basis of MonteCarlo simulations and of *ad hoc* measurements in the CUORE R&D facility - located in Hall C of Gran Sasso Laboratory - we could disentangle the different sources contributing to the background in the crucial energy region: $\sim$30% of the count rate is due to multi-Compton events of ^{208}Tl γ transition, $\sim$20% is ascribed to degraded alphas coming from U and Th surface contaminations of the TeO_2 crystals, and the remaining $\sim$50% seems to originate from U and Th surface contaminations of the inert materials surrounding the detectors, most probably the copper of the detector holders.

3. From Cuoricino to CUORE

From statistical considerations, the sensitivity $\Sigma(\tau^{0\nu}_{1/2}) \propto \epsilon\, i.a.\, (M t_{meas}/(\Delta E\, bkg))^{1/2}$, where ϵ, $i.a.$, M, t_{meas}, ΔE, and bkg are the detector efficiency, the active isotope abundance, the detector mass, the measuring time, the energy resolution, and the background level at the $\beta\beta0\nu$ transition energy, respectively. Therefore, to improve the sensitivity reached by Cuoricino one could in principle act on several parameters. Since i) the natural isotopic abundance of ^{130}Te is already quite high and isotopic enrichment is expensive and needs R&D effort to maintain a high radiopurity level, and ii) the single crystal mass of TeO_2 is close to the size limit which allows to meet the required performance - especially in terms of the energy resolution ΔE, key parameters to work on remain the number of detectors and the background level. Also an improvement of the experimental live time - the duty cicle of Cuoricino was only 30% - is extremely important.

In order to probe the inverted hierarchy region for the effective neutrino mass Cuoricino upper limit must be improved by a factor of 10, which in turn means to improve $\Sigma(\tau^{0\nu}_{1/2})$ by a factor of 100. By increasing the total mass M by a factor $\sim$20, and the measuring time by $\sim$5, the challenging step to meet the goal will be a background reduction factor of $\sim$100. This could be achieved by a heavier shielding, to reduce the environmental γ contribution, by a better surface cleaning, to reduce the degraded α continuum, and by an improvement of the design, to reduce the amount of material facing the detectors.

4. CUORE

CUORE (Cryogenic Underground Observatory for Rare Events) [5] is an array of 988 natural TeO_2 cryogenic detectors arranged in a cylindrical compact and granular structure of 19 towers, each of them made by 13 planes of 4 detectors 5×5×5 cm^3 each. The total active mass will be of 741 kg, for a ^{130}Te mass of 203 kg. This huge mass, together with its holding structure, must be cooled and steadly operated at $\sim$10 mK. Therefore one of the technological challenges of CUORE will be the cryogenic system, which has to host additionally about 10 ton of internal lead shielding and guarantee at the same time a low mechanical vibration input on the detector.

The CUORE experiment is presently being built in Hall A of Gran Sasso Laboratory, besides Cuoricino facility. The experimental hut is almost completed, the next steps will be the installation of the utilities, of the clean room and of the external shielding. Then the custom-made cryostat will be set up. Assembly of the detector in a controlled radioclean environment is foreseen for 2011 and the data taking start is expected in 2012.

CUORE $\beta\beta0\nu$ sensitivity will strongly depend on the background level, as already discussed. In 5 years of measuring time with a FWHM energy resolution of $\sim$5 keV, a conservative assumption of 0.01 counts/keV/kg/y in the energy region of the ^{130}Te $\beta\beta0\nu$ transition translates into a $1\,\sigma$ sensitivity on $\tau^{0\nu}_{1/2}$ of about 2.1×10^{26} y ($m_{\beta\beta} \leq 24-83$ meV). If all background sources were controlled at the level of 0.001 counts/keV/kg/y, then the sensitivity could be improved to 6.5×10^{26} y ($m_{\beta\beta} \leq 14-47$ meV), deep inside the inverted hierarchy region of the neutrino mass pattern.

A MonteCarlo projection of presently measured radioactive contaminations onto CUORE background around 2.5 MeV shows that almost all identified sources are controlled at the level of 10^{-3} counts/keV/kg/y but surface U and Th contents, which settle the current background extrapolation at 0.04 counts/keV/kg/y. Therefore special efforts are devoted to crystal production and copper surface cleaning. A dedicated radioclean facility has been set-up by the chinese company in charge of making the TeO_2 crystals, with the aim of keeping every step of the crystal growth under control, from the selection of the starting materials to the final surface polishing and packaging of the crystals. The delivery to Gran Sasso is made by ship to reduce cosmogenic activation and the underground storage area, that will also host all other detector components, is constantly flushed with nitrogen. The copper producer has been carefully selected according to the quality and technical characteristics of the raw material (radiopurity, thermal conductivity, mechanical properties) as well as the guaranteed exposure to cosmic rays of the material from casting to delivery. Almost all copper components of the detector mounting structure are being manufactured by electrical discharge machining, thus avoiding possible contamination sources like lubricants, and kept above ground strictly for the time needed for production. The final surface cleaning procedure of the machined copper pieces has been worked out in a facility at the Laboratori Nazionali di Legnaro (PD, Italy), and includes tumbling, electrochemical and chemical etching, and magnetron sputtering.

A final proof of the background achievements will be obtained by operating CUORE-0, the first complete CUORE tower, inside Cuoricino experimental facility. Moreover, from the physics point of view, this measurement with 52 TeO_2 detectors 750 g each, will be a self-consistent experiment by itself, soon overtaking Cuoricino $\beta\beta0\nu$ sensitivity. CUORE-0 assembly is foreseen in 2010 and the data taking will continue until CUORE start.

5. Conclusions

CUORE will be the first next generation $\beta\beta0\nu$ experiment. CUORE has the potentiality to probe deeply the inverted hierarchy region of the neutrino mass pattern. The feasibility of the experiment has been demonstrated by Cuoricino, the pilot experiment taking data until 2008 in Gran Sasso Underground Laboratory, that is the second most sensitive $\beta\beta0\nu$ experiment worldwide. CUORE is presently being built at Gran Sasso Laboratory and the data taking is foreseen by 2012.

References

[1] Schönert S, these proceedings.
[2] Arnaboldi C et al 2008 *Phys. Rev.* C **78** 035502
[3] Rodin V et al 2006 *Nucl. Phys.* A **766** 107; Rodin V et al 2006 Erratum arXiv:nucl-th/0706.430v1
[4] Klapdor-Kleingrothaus HV and Krivosheina IV 2006 *Mod. Phys. Lett.* A **21** 1547
[5] Arnaboldi C et al 2004 *Nucl. Instrum. Methods* A **518** 775-798

Journal of Physics: Conference Series **203** (2010) 012070 doi:10.1088/1742-6596/203/1/012070

Status of the COBRA double beta decay experiment

Kai Zuber

Inst. für Kern- und Teilchenphysik, Technische Universität Dresden, 01069 Dresden, Germany

E-mail: zuber@physik.tu-dresden.de

Abstract. The current status of the COBRA experiment is described. Results on the 4-fold forbidden beta decay of ^{113}Cd and a variety of double beta decay limits of Cd, Zn and Te isotopes are presented based on 18 kg× days of exposure with an array of sixteen CdZnTe semiconductor detectors. A short description on the activities with pixelated detectors for tracking is given.

1. Introduction

A new approach to take advantage of the good energy resolution of semiconductors is COBRA [1] located in the Gran Sasso Underground Laboratory (LNGS). The idea is to use a large amount of CdZnTe (CZT) room-temperature semiconductors. In total, there are nine double beta emitters within the detector including those of $\beta^+\beta^+$ decay. COBRA will mainly explore ^{116}Cd and ^{130}Te for $\beta^-\beta^-$ decay and ^{106}Cd for positron decay also in combination with electron capture. The smallness of the detectors makes a search for coincidences powerful and reduces γ-background. The practical handling is simplified as these detectors are room temperature detectors. In case of pixelated detectors it offers tracking possibilities and even further background reduction by particle identification. Activities are split into two main directions, first of all identifying background components with the aim of removing them and secondly the development of high resolution pixel detectors. The background study is performed at LNGS with commercially available 1 cm^3 CZT detectors. Recently an installation containing 64 detectors was performed, corresponding to about 0.42 kg of CdZnTe. This is the world-largest array of these detectors. The first layer of 16 detectors has been installed in summer 2006, with the full 64 array starting data taking in autumn 2008. The shielding consists of 10 cm of electroformed copper, 10 cm of lead all located within a Faraday cage. The cage is surrounded by a neutron shield made of 5 cm boron loaded polyethylene plates. As one of the major results the red passivation paint on the surface of the detectors and the radon in the air could be identified as dominant background sources. In collaboration with the company an alternative passivation solution was found which, according to material screening results, should have at least three orders of magnitude less contamination. Four detectors with the new passivation were installed at Gran Sasso in 2008. In addition, a nitrogen atmosphere was established to keep the radon out. The combined effect is a background reduction of about an order of magnitude for a wide range of energies. Both data sets obtained at LNGS are shown in Fig. 1. It is obvious, that no gain in knowledge is possible by continuing data taking with the red passivation, hence the array of 64 detectors was dismantled and experimental options are investigated to replace the red paint with the new one.

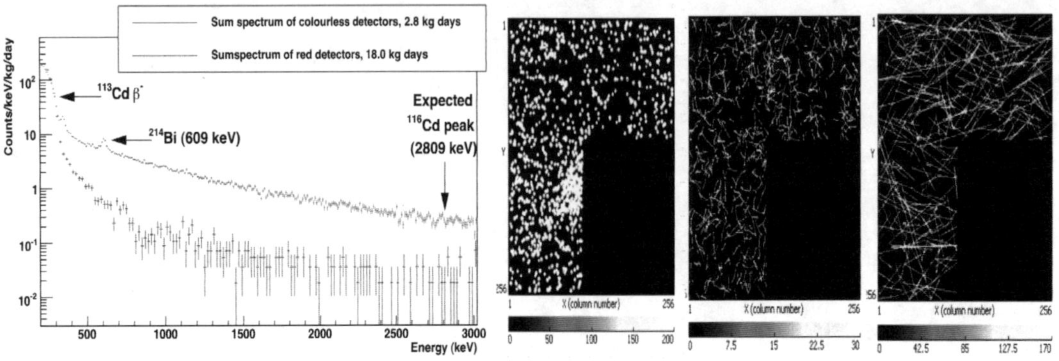

Figure 1. Left: Background spectra obtained at LNGS showing the 16 detector layer with red passivations on it and the spectrum of the four colourless detectors with nitrogen flushing. A massive improvement of about an order of magnitde is seen for a wide range in energy. The shoulder in the low energy range is the beta spectrum of the 4-fold forbidden beta decay of ^{113}Cd with a half-life of about 8×10^{15} years [2, 3]. Right: Preselected samples of particles measured with a 55μm pitch Si pixel-detector. Clearly visible is the difference between alpha particles (left), electrons (middle) and muons (right). With this additional feature of particle identification a major reduction of background can be expected.

2. Results from LNGS measurements

With the first layer of 16 CZT detectors after cleaning cuts a total amount of 18 kg$\times$days of exposure have been accumulated. Besides the background contaminations the most striking feature in the spectrum is a shoulder in the low energy range. It is due to the four-fold forbidden non-unique beta decay of ^{113}Cd to ^{113}In being a $(1/2^+) \rightarrow (9/2^+)$ transition. The statistics is high enough, that a half-life determination could be done for each detector individually. The combined results is a half-life of $(8.00 \pm 0.11(stat.) \pm 0.24(sys.) \times 10^{15})$ years [2]in agreement with our previous measurements [2]. However, it is the first time that a half-life well beyond 10^{10} years has been determined in a statistical representative sample of independent measurements. Using the Q-value as a fit parameter it could be deduced to be 322.2 $\pm$ 0.3 (stat.) $\pm$ 0.9 (sys.) keV in excellent agreement with the current AME value of 320 $\pm$ 3 keV [4]. In the meantime, the noise threshold of the detectors could be reduced significantly and with the installation of a slight cooling the energy resolution is improving significantly [5], thus even more precise data can be expected in the near future.

A whole variety of double beta decay modes have been explored. Some limits obtained by analysing single detector events only are compiled in Tables 1 and 2. For more detailed information see [6], results from former measurements with a four detector array can be found in [7].

3. Pixelated CdZnTe detectors

An independent research within COBRA is the exploration of pixelated CZT detectors. This will add tracking capabilities to the pure energy measurements. Detailed simulations have shown that with $\approx$ 200 μm pixel size, background can be reduced by three orders of magnitude due to particle identification [8]. Double beta decay events will hit several pixels (about 10) for a ^{116}Cd decay, while α particles will be mostly restricted to one pixel and gammas will hit several disconnected pixels. Using the number of pixels hit, the energy deposition in pixels and the energy loss characteristics of particles a strong discrimination seems feasible. Two options are currently explored. First of all, custom made $2 \times 2 \times 0.5$ cm^3 detectors have been fabricated with

Isotope	Decay	Fit range (MeV)	COBRA	World Best
^{116}Cd	to g.s	2.2-3.2	9.4×10^{19}	1.7×10^{23}
^{130}Te	to g.s	2.2-3.2	5.0×10^{20}	3.0×10^{24}
^{130}Te	to 536 keV	1.7-2.3	3.5×10^{20}	9.7×10^{22}
^{114}Cd	to g.s	0.4-1.0	2.0×10^{20}	1.1×10^{21}
^{128}Te	to g.s	0.6-1.3	1.7×10^{20}	1.1×10^{23}

Table 1. Selected half-life results from fits for $\beta^-\beta^-$decay peaks. The 90% confidence limits have conservative systematic uncertainties applied and are compared to the world best limits. For a full list see [6].

Isotope	Decay	Fit range (MeV)	COBRA	World Best
^{64}Zn	$0\nu\beta^+$EC to g.s.	0.5–1.1	1.1×10^{18}	4.3×10^{20}
^{64}Zn	0ν2EC to g.s.	0.5–1.3	$3.3\ \times10^{17}$	1.1×10^{20}
^{120}Te	$0\nu\beta^+$EC to g.s.	1.0–2.0	$\mathbf{4.1\times10^{17}}$	1.9×10^{17}
^{120}Te	0ν2EC to g.s.	0.8–2.0	2.4×10^{16}	6×10^{17}
^{106}Cd	$0\nu\beta^+\beta^+$ to g.s.	0.5–2.0	2.7×10^{18}	2.4×10^{20}
^{106}Cd	$0\nu\beta^+$EC to g.s.	1.4–3.0	4.7×10^{18}	3.7×10^{20}
^{106}Cd	0ν2EC to g.s.	1.4–3.0	1.6×10^{17}	3.5×10^{18}
^{106}Cd	$0\nu\beta^+$EC to 512keV	0.8–2.0	4.6×10^{18}	2.6×10^{20}

Table 2. 90% confidence limits obtained for some $\beta^+\beta^+$ decays analysed in [6] with conservative systematic uncertainties applied, compared to the world best limits. New world best values from this work are shown in bold. The energy range used for each fit is also included.

650 μm (32×32 pixel) and 200 μm (100×100 pixel) and are currently under testing with newly designed ASICS. An alternative is the usage of the Timepix chip with 65000 readout channels, developed by the Medipix2 collaboration, allowing for 55 μm, 110 μm and 220 μm pixel size. First measurements with a 300 μm thick Si-detector have been successfully performed, clearly showing the different patterns for electrons, alpha particles and muons (Fig. 1). In a short test run underground it could be confirmed that the system is reacting on the environment by observing less events and no muons anymore. A new 1mm thick CdTe system is now running underground and first results are expected soon. It is planned to install a first pixel system for background measurements in LNGS later in 2009.

References

[1] K. Zuber, 2001, *Phys. Lett.* B 519, **1**
[2] J. V. Dawson et al., 2009, *Nucl. Phys.* A 818, **264**
[3] C. Goessling et al., 2005, *Phys. Rev.* C 72, **064328**
[4] G. Audi, A. H. Wapstra, C. Thibault, 2007, *Nucl. Phys.* A 729, **337**
[5] J. V. Dawson et al., 2009, *Nucl. Instrum. Methods* A 599, **209**
[6] J. V. Dawson et al., 2009, *Phys. Rev.* C 80, **025502**
[7] T. Bloxham et al., 2007, *Phys. Rev.* C 76, **025501**
[8] T. Bloxham, M. Freer, 2007, *Nucl. Instrum. Methods* A 572, **722**

The DCBA experiment for studying neutrinoless double beta-decay

N Ishihara[1], G Iwai[1], H Iwase[1], Y Kato[1], M Kawai[1], Y Kondou[1], T Haruyama[1], T Inagaki[1], Y Makida[1], T Ohama[1], K Takahashi[1], S Takeda[1], Y Yamada[1], H Igarashi[2], T Ishikawa[2], T Sumiyoshi[2], E Tashiro[3], T Ishizuka[3], S Kitamura[4], Y Teramoto[5], I Nakano[6], Y Sakamoto[7], Y. Nagasaka[8], N Tamura[9], K Tanaka[10] and R Ito[11]

[1] High Energy Accel. Research Organization (KEK), Tsukuba, Ibaraki 305-0801 Japan
[2] Tokyo Metropolitan University, Hachioji, Tokyo 192-0398 Japan
[3] Shizuoka University, Naka, Hamamatsu, Shizuoka 432-8011 Japan
[4] Nihon Institute of Medical Science, Iruma-gun, Saitama 350-0435 Japan
[5] Osaka City University, Sumiyoshi, Osaka 558-8585 Japan
[6] Okayama University, Okayama 700-8530 Japan
[7] Tohoku Gakuin University, Sendai, Miyagi 981-3193 Japan
[8] Hiroshima Institute of Technology, Saeki, Hiroshima 731-5193 Japan
[9] Niigata University, Niigata 950-2181 Japan
[10] Saitama Science Institute, Inarimachi-kita, Fukaya, Saitama 366-0028 Japan
[11] FutureScope, Maruyama-cho, Shibuya, Tokyo 150-0044 Japan

E-mail: nobuhiro.ishihara@kek.jp

Abstract. In order to search for neutrinoless double beta-decay, the DCBA (Drift Chamber Beta-ray Analyzer) experiment uses a momentum analyzer, which mainly consists of drift chamber and a uniform magnetic field. A beta ray from a source plate make helical trajectory in the drift chamber owing to the magnetic field. Momentum of each beta ray is obtained by three-dimensional track reconstruction. A test prototype called DCBA-T2 has been constructed and operated at KEK. Another prototype DCBA-T3 is also under construction to improve the energy resolution. The results of DCBA-T2 engineering run are described together with the status of DCBA-T3.

1. Introduction

Searching for neutrinoless double beta-decay ($0\nu\beta\beta$) is one of the most exciting issues of the present neutrino physics. The observation of $0\nu\beta\beta$ means the strong support of the so-called leptogenesis [1], which is the scenario of the advantage of matter over anti-matter in the early universe. In the leptogenesis, heavy Majorana neutrinos play important roles. According to the so-called seesaw mechanism, both heavy and light Majorana neutrinos must exist, and light Majorana neutrinos produce $0\nu\beta\beta$. If $0\nu\beta\beta$ events are accumulated, the half-life of decay nucleus is measurable, and then the absolute mass scale of light Majorana neutrino is obtained with the help of the calculation of nuclear matrix element.

In order to measure the momentum and kinetic energy of each beta-ray from double beta-decay, momentum analyzers called DCBA (Drift Chamber Beta-ray Analyzer) have been developed at KEK

[2][3]. DCBA is sensitive to only charged particles, thus gamma-rays are automatically eliminated. Electrons and positrons are easily separated because of the opposite directions of their curves in a uniform magnetic field. A background alpha particle is also easily distinguished because of its high momentum and its high pulse signals. The excellent particle identification ability and good spatial resolution of DCBA make it easy to find $0\nu\beta\beta$ events against many particle backgrounds. Only two-neutrino double beta-decay ($2\nu\beta\beta$) events will become backgrounds at the Q-value of decay nucleus. Therefore, the required energy resolution is as less than 5% at Q-value in order to search for $0\nu\beta\beta$ down to about 50 meV, which is predicted in the so-called inverted hierarchy mass spectrum model.

2. DCBA-T2

In order to study the spatial and energy resolutions and to seek operation problems, a test prototype called DCBA-T2 has been constructed and operated. Detection principle and the configuration of DCBA-T2 are described elsewhere [4][5].

2.1. Spatial and energy resolutions

Spatial resolution has been studied with the straight tracks using high-energy cosmic rays. Experimental results in X, Y and Z directions have been obtained as shown in figure 1, where σ_X stands for the resolution in X, which is obtained from the moving time information of drift electrons, taken by Flash Analogue-to-Digital Converter (FADC). Resolutions σ_Y and σ_Z are obtained from the information of anode and pickup wires, respectively. For the study of energy resolution, a point source of ^{207}Bi was installed in the middle point of source plate. The source of ^{207}Bi emitted internal conversion electrons (ICE's) of mainly 4 kinds of monochromatic energies. The obtained energy spectra of ICE's including backgrounds are shown in figure 2. One can see a clear peak with about 0.15 MeV (FWHM) around 1 MeV. This peak consists of 0.98 MeV and 1.05 MeV electrons and backgrounds. The electron of 0.98 MeV is the most intensive with the emission rate of 7% of disintegration, while 1.05 MeV electron has the rate of 2.4%. From figure 2, it seems that DCBA-T2 has the energy resolution of about 0.15 MeV (FWHM), which is in good agreement with simulation studies.

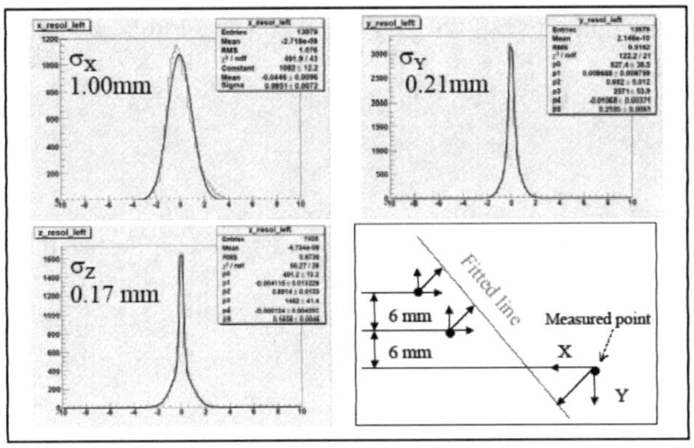

Figure 1. Spatial resolution of DCBA-T2.

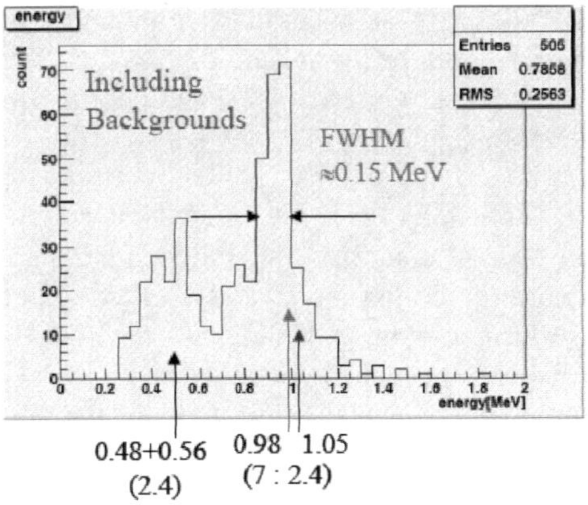

Figure 2. Energy spectra of ICE's from ^{207}Bi in DCBA-T2.

2.2. Half-life measurement of $2\nu\beta\beta$ for ^{100}Mo

Operation problems have been studied through the engineering run of half-life measurement using natural Mo thin plates, which includes ^{100}Mo with the abundance of 9.6%. Since the source area is 728 cm^2 and its thickness is 45 mg/cm^2, the number of ^{100}Mo is 0.03 mol. An example of a $2\nu\beta\beta$ event candidate is as in figure 3. In the figure, upper part is X-Y projection plane and lower part is the X-Z plane. Decay vertex points are shown with LY (RY) and LZ (RZ) in X-Y and X-Z, respectively.

Though LY and RY look like different points in time-count space, they are actually the same point in the actual space. The absolute time count difference 245-239=6 corresponds to 0.3 mm in the actual space, and it is within the spatial resolution in X direction. Of course, source plate is located at the vertex point. From the information in figure 3, three-dimensional reconstruction gives us the momentum and the pitch angle (λ) of each electron. And then we can calculate the kinetic energies of both electrons (T's) and the angular correlation (cosθ).

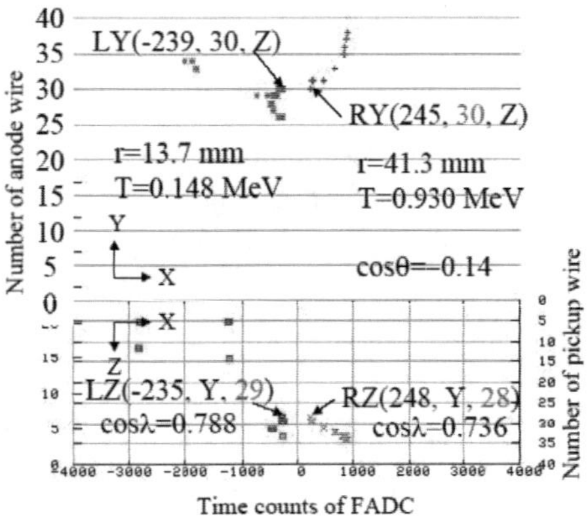

Figure 3. Example of a 2νββ event candidate from natural Mo plate.

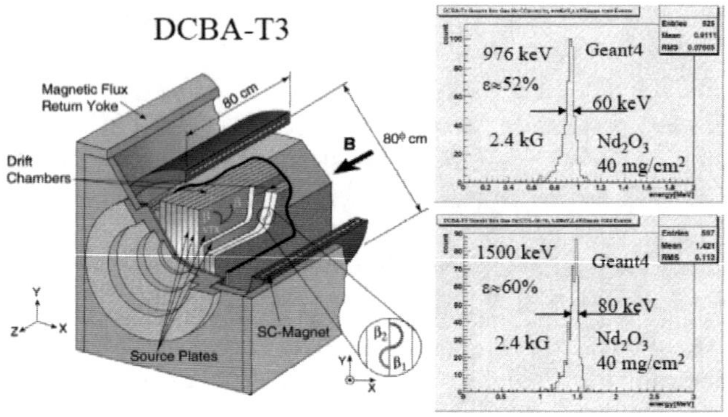

Figure 4. Illustration of DCBA-T3 and the energy resolutions simulated by Geant4.

3. DCBA-T3

The energy resolution obtained in DCBA-T2 is not enough to search for 0νββ down to about 50 meV. Since the energy resolution becomes worse due to the multiple scattering of beta ray in the chamber gas, it is useful to reduce the track length required in the momentum measurement. Figure 4 shows DCBA-T3, which is now under construction at KEK. It has 2 kG magnetic flux density, which is more than twice of that in T2: thus the track length produced in T3 is about half of T2. The sampling pitch is 3 mm for T3, being half of T2: thus the number of sampling points for one track in T3 is the almost same as T2. The energy resolutions obtained by simulation studies are better than those of T2 as shown in figure 4.

4. Conclusions

The background rejection ability of DCBA is excellent especially for gamma rays and alpha particles. According to the simulation studies, the energy resolution is also reasonable for investigating effective neutrino mass down to about 50 meV. DCBA-T3 is under construction at KEK in order to confirm the energy resolution. It will be able to measure the half-life of 2νββ for ^{150}Nd using natural Nd source in T3, while ^{100}Mo has been used in T2 for engineering run.

Acknowledgment

This research was supported in part by the Yamada Science Foundation and in part by Grant-in-Aid for Science Research from Japan Society for the Promotion of Science.

References

[1] Buchmuller W, Peccei R D and Yanagida T 2005 *Annu. Rev. Nucl. Part. Sci.* **55** 311
[2] Ishihara N, Ohama T and Yamada Y 1996 *Nucl. Instr. Meth.* A **373** 325
[3] Ishihara N *et al.* (DCBA) 2000 *Nucl. Instr. Meth.* A **443** 101
[4] Kato Y *et al.* (DCBA) 2003 *Nucl. Instr. Meth.* A **498** 430
[5] Ishihara N *et al.* (DCBA) 2008 *J. Phys.: Conf. Ser.* **120** 052049

Journal of Physics: Conference Series **203** (2010) 012072 doi:10.1088/1742-6596/203/1/012072

Search for double beta decay of ^{106}Cd in TGV-2 experiment

N.I. Rukhadze[1], Ch. Briançon[2], V.B. Brudanin[1], P. Čermák[3], V.G. Egorov[1], A.A. Klimenko[1], A. Kovalik[1], Yu.A Shitov[1], I. Štekl[3], V.V. Timkin[1], Ts. Vylov[1]

[1]Joint Institute for Nuclear research, Joliot-Curie, 6, 141980 Dubna, Russia;
[2]Centre de Spectrométrie Nucléaire et de Spectrométrie de Masse, Orsay, France;
[3]Institute of Experimental and Applied Physics, CTU, Prague, Czech Republic

E-mail: rukhadze@nusun.jinr.ru

Abstract. Search for double beta decay ($\beta^+\beta^+$, β^+/EC, EC/EC) of ^{106}Cd was performed at the Modane underground laboratory (4800 m w.e.) using a spectrometer TGV-2 with 32 HPGe detectors. New limits on the half-lives of 0νEC/EC resonant decay - $T_{1/2} \geq 1.6 \times 10^{20}$ y, and on 2νEC/EC decay of ^{106}Cd - $T_{1/2} \geq 4.1 \times 10^{20}$ y (at 90% CL) were obtained from preliminary calculations of experimental data accumulated for 12900 h of measurement of ~13.6 g of ^{106}Cd with enrichment of 75%. The limits on 2νEC/EC decay of ^{106}Cd to the 2^+,512 keV and $0^+{}_1$,1334 keV excited states of ^{106}Pd and on 2ν$\beta^+\beta^+$ and 2νβ^+/EC decay of ^{106}Cd were improved

1. Introduction

Investigation of double beta decay processes ($\beta^-\beta^-$, $\beta^+\beta^+$, β^+/EC , EC/EC) are of great importance for particle and nuclear physics as a sensitive tool for testing of lepton number conservation and studying neutrino properties and nuclear structure. Up to now, more attention has been given to $\beta^-\beta^-$ channels (two-neutrino and neutrinoless modes). There are also other channels of double beta decay, in particular the double capture of two bound atomic electrons (EC/EC), capture of the bound electron with emission of a positron (β^+/EC) and decay with emission of two positrons ($\beta^+\beta^+$). In comparison with $\beta^-\beta^-$ - decay these channels are disfavored by smaller available kinetic energy, by Coulomb repulsion on the positron or due to a small overlap of the bound electron wave function with nucleus. But the study of neutrinoless 0ν$\beta^+\beta^+$, 0νβ^+/EC, 0νEC/EC decays can yield the same information as the 0ν$\beta^-\beta^-$ - process. In contrast to the $\beta^-\beta^-$ - decay, where the two-neutrino mode of the process was experimentally investigated in the decay of several isotopes, other double beta processes has not been yet observed in direct experiments. The main purpose of our studies is to search for 2νEC/EC decay of ^{106}Cd ($Q_{EC/EC}$ = 2770±7.2 keV) with the transition to the ground state of ^{106}Pd and 0νEC/EC resonant decay of ^{106}Cd with the transition to the 2741 keV excited state of ^{106}Pd. The first decay is characterized by the emission of only two quanta of characteristic KX radiation of palladium (Pd) with the energy of ~21 keV. The second one will have minimum (almost zero) energy of decay and this decay will be accompanied by emission of additional particle(s) - γ2741 keV or γ2229 keV+γ512 keV.

2. Experiment TGV-2

The experiment was performed at the Modane Underground Laboratory (depth of 4800 m w.e.) using the spectrometer TGV-2 (Telescope Germanium Vertical) [1]. The detector part of TGV-2 (Figure 1) consists of 32 HPGe planar type detectors mounted one over another in a low-background cryostat.

The sensitive area of each detector is about 2040 mm^2 and its thickness is 6 mm. The total sensitive volume and the total mass of TGV-2 detectors are about 400 cm^3 and 3 kg respectively. Double beta emitters were the foils of ^{106}Cd (enrichment 75%) with a diameter of 52 mm and thickness of ~50 μm. The foils were inserted between the entrance windows of the neighboring detectors in the TGV-2 cryostat. The distance between detectors and emitters were about 1.5 mm. The energy resolution of the detectors ranged from 3.0 to 4.0 keV at 1332 keV (^{60}Co). The total efficiency of the TGV-2 spectrometer is 50-70% depending on the energy threshold. The detector part of the TGV-2 (Figure 2) was surrounded by copper shielding (>20 cm), a steel airtight box against radon, a led shielding (>10 cm), and an antineutron shielding made of borated polyethylene (16 cm). To suppress both the electronic and microphone noise in the low energy region, a method of comparison of two signals generated by processing with two different shaping times (2 and 8 μs) is used in TGV-2 [1].

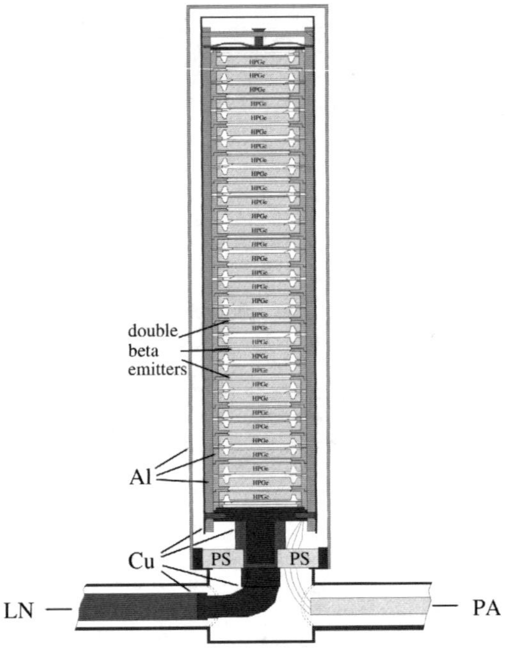

Figure 1. The detector part of TGV-2 spectrometer.

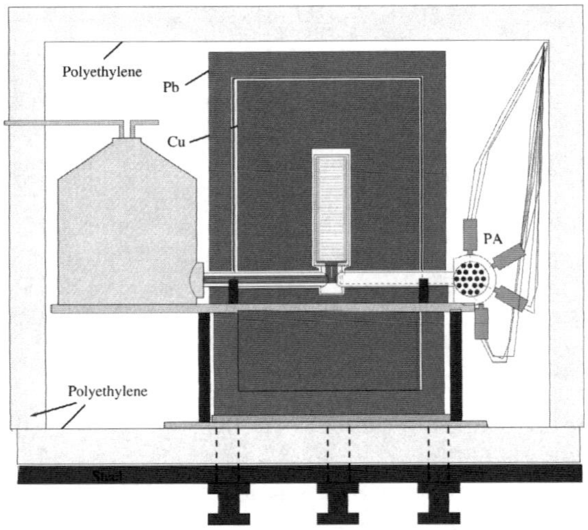

Figure 2. The detector part in passive shielding.

Two experimental runs were performed on the TGV-2 spectrometer to measure the foils of ^{106}Cd with enrichment of 75%. In the first one (February 2005 – February 2006) 12 samples of ^{106}Cd with a total mass of ~10 g and 4 foils of natural Cd with a total mass of ~3.2 g were studied during 8687 h [2]. In the second run (December 2007 – July 2009) 16 samples of ^{106}Cd with a total mass of ~13.6 g were studied during 12900 h. The main differences of the second run in comparison with the first one are: a) the level of TGV-2 background (it was decreased in several times [3]); b) .enlarged energy region in one of spectroscopy channels of each detector from ~800 keV to ~ 3 MeV.

3. Results

The preliminary calculations of experimental data accumulated for 12900h of measurements of the second run were performed. The coincidences between two characteristic KX- rays of Pd detected in neighboring detectors were analyzed to search for 2νEC/EC decay of ^{106}Cd to the ground 0^+ state of ^{106}Pd (Figure 3)(Figure 4). The search for 0νEC/EC resonance decay of ^{106}Cd to the 2741 keV excited state of ^{106}Pd was based on the analysis of KX(Pd) - γ2741 keV and KX(Pd) – γ2229 keV- γ512 keV coincidences. Investigation of other branches of ^{106}Cd decay - 2νEC/EC decay to the 2^+,511.9 keV and

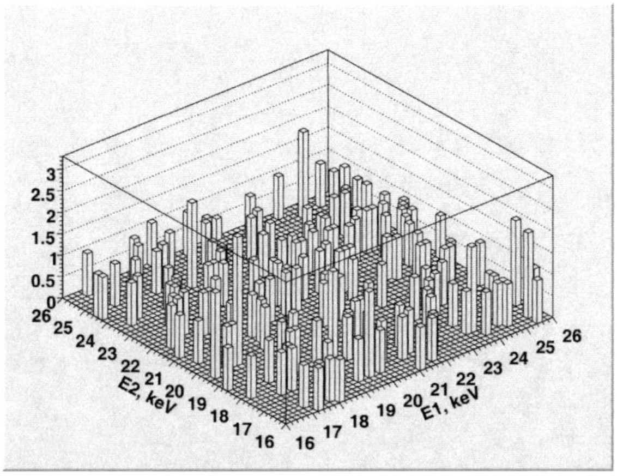

Figure 3. Double coincidence events (DC).

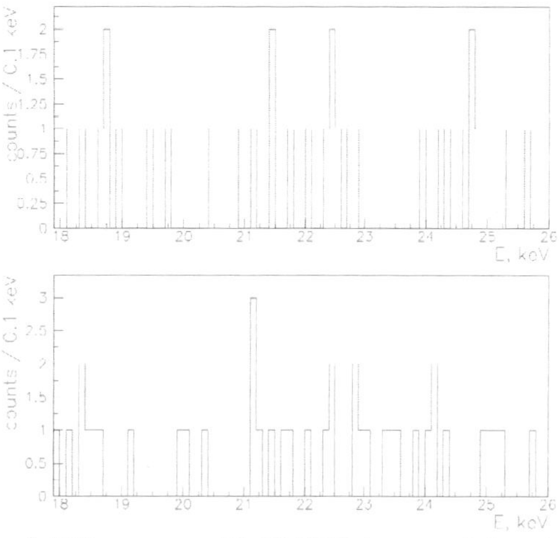

Figure 4. DC spectra with KX(Pd) in one of detectors

0^+_1, 1334 keV excited states of ^{106}Pd, 2νβ⁺/EC and 2νβ⁺β⁺ decays to the ground and excited states of ^{106}Pd were based on the analysis of KX-511 keV, KX-622 keV, 511 keV-511 keV and 511 keV – 622 keV coincidences. Our limits on half-lives of double beta decay of ^{106}Cd are presented in a Table.1.

Table 1. The limits on half-lives of double beta decay of ^{106}Cd

Decay mode	Transition to	$T_{1/2}$, 10^{20} y (C.L. 90%) run 1 [2]	run 2 [this work]
2νEC/EC	0^+, ground state	3.0	4.1
	2^+,511.9 keV	0.42	0.63
	0^+_1, 1334 keV	0.31	0.52
0νEC/EC	1,2^+, 2741 keV	-	1.6
2νβ^+/EC	0^+, ground state	0.59	1.1
	2^+,511.9 keV	0.59	1.1
	0^+_1, 1334 keV	-	1.6
2ν$\beta^+\beta^+$	0^+, ground state	0.6	1.4
	2^+,511.9 keV	0.57	1.7

Our results for 2νEC/EC decay of ^{106}Cd to the ground 0^+ state of ^{106}Pd are more than 2 orders of magnitude higher than those obtained in resent experiment [4] and reaches theoretical predictions ranging [5] from 1.0×10^{20} to 5.5×10^{21} y.

4. Acnowledgements

This work was performed under collaboration agreement No.04-66 between IN2P3 and JINR. and partly supported by grants MSM6840770029, LC07050, LA305 and by Russian Foundation for Basic Research under grant No.08-02-00790. We are grateful to the LSN laboratory that provided service for work underground and highly valuable technical assistance.

References

[1] Beneš P et al 2006 Nucl. Instr. Meth. in Phys. Res. A **569** 737
[2] Rukhadze N I et al 2008 Bulletin of the Russian Academy of Sciences: Physics v.**72** 731
[3] Rukhadze N I et al 2009 Bulletin of the Russian Academy of Sciences: Physics v.**73** 741
[4] Kiel H et al 2003 Nucl. Phys. A **723** 499
[5] Tretyak V I and Zdesenko Yu G 2002 At. Data and Nucl. Data Tables **80** 83

Study of ^{48}Ca double beta decay with CANDLES

I Ogawa[1], T Kishimoto[1,2], S Umehara[1], G Ito[1], K Yasuda[1], M Wada[1], K Matsuoka[1], M Nomachi[1], K Fushimi[3], R Hazama[4], H Ohsumi[5], K Okada[6], Y Tamagawa[7] and S Yoshida[8]

[1] Graduate School of Science, Osaka University, Toyonaka, Osaka 560-0043, Japan
[2] Research Center for Nuclear Physics, Osaka University, Ibaraki, Osaka 567-0047, Japan
[3] Faculty of Integrated Arts and Science, The University of Tokushima, Tokushima 770-8502, Japan
[4] Graduate School of Engineering, Hiroshima University, Higashi-Hiroshima, Hiroshima 739-8527, Japan
[5] Faculty of Culture and Education, Saga University, Saga 840-8502, Japan
[6] Department of Computer Science and Engineering, Kyoto San-gyo University, Kyoto 603-8555, Japan
[7] Graduate School of Engineering, University of Fukui, Fukui 910-8507, Japan
[8] Research Center for Neutrino Science, Tohoku University, Sendai, Miyagi 980-8578, Japan

E-mail: ogawa@phys.sci.osaka-u.ac.jp

Abstract. CANDLES is the project to search for neutrinoless double beta ($0\nu\beta\beta$) decay of ^{48}Ca by using CaF_2 scintillators. The observation of $0\nu\beta\beta$ decay will prove the existence of massive Majorana neutrinos. Expected performances and current status of the CANDLES system are described.

It has been demonstrated that a neutrino has a mass by the observation of neutrino oscillation. The existence of the neutrino mass opens the new possibility that it could be a Majorana particle of which mass term violates lepton number conservation. This fact becomes a key to answer the question why we are living in a matter dominated universe. Leptogenesis is a scenario to realize the matter dominated universe which requires CP violation and lepton number non-conservation. It is only a neutrino-less double beta decay that is known to demonstrate that the neutrino is the Majorana particle and violates lepton number conservation.

We have been studying the double beta decay of ^{48}Ca. The ^{48}Ca has the highest $Q_{\beta\beta}$ value (4.27 MeV) among all double beta decay nuclei, which ensures the least background and the large phase space factor for the decay rate. The least background should be of particular importance for the future study where large amount of target nuclear is employed. We gave the best lower limit on the half-life of neutrino-less double beta decay of ^{48}Ca by using CaF_2(Eu) detector system[1, 2] though further development is highly desirable to reach the mass region of current interest.

CANDLES is the project to search for double beta decay of ^{48}Ca by using undoped CaF_2 (CaF_2(pure)) scintillators[3]. We have constructed CANDLES III detector (figure 1) at sea-level laboratory, OULNS (Lepton Nuclear Laboratory, Osaka University), which consists of CaF_2 modules, liquid scintillator, water buffer and PMTs. 60 CaF_2 modules are immersed in the liquid scintillator which is used as an active veto (veto phase) by utilizing difference of decay time for the CaF_2 ($\sim 1\mu$sec) and the liquid scintillator ($\sim$ 10 nsec) signals. The CaF_2 module consists of a CaF_2(pure) crystal with a dimension of 10 cm cube (3.18 kg), a mineral oil of 5 mm thickness (conversion phase) and an acrylic container. Since both the CaF_2(pure) and the liquid scintillator have long attenuation length (~10 m), the CANDLES has essentially no limit on its size. The conversion phase contains wavelength shifter (Bis-MSB) which converts the emission light of CaF_2(pure) which has a peak in the UV region to the visible one where the quantum efficiency of the PMTs is high enough (maximum at ~ 400 nm) and materials at the optical path have good transparencies[4]. These modules are suspended by wires from the ceiling of a liquid scintillator vessel of 1 m in diameter and 1 m in height. The outside of the liquid scintillator vessel was filled with pure water (water buffer) which also acts as a passive shield against the backgrounds from outside. Scintillation lights from both the CaF_2 modules and the liquid scintillator are viewed by 40 large PMTs and their charge and pulse shapes are recorded using CSADC (partial and full gate width) and high-speed (500 MHz) flash ADCs, respectively. Both data are used for rejecting the events which fire the liquid scintillator and reconstructing the event position. All the detector system described above are contained in an outer tank which is 2.8 m in diameter and 2.6 m in height. The expected energy resolution at the $Q_{\beta\beta}$ region is less than 4 %.

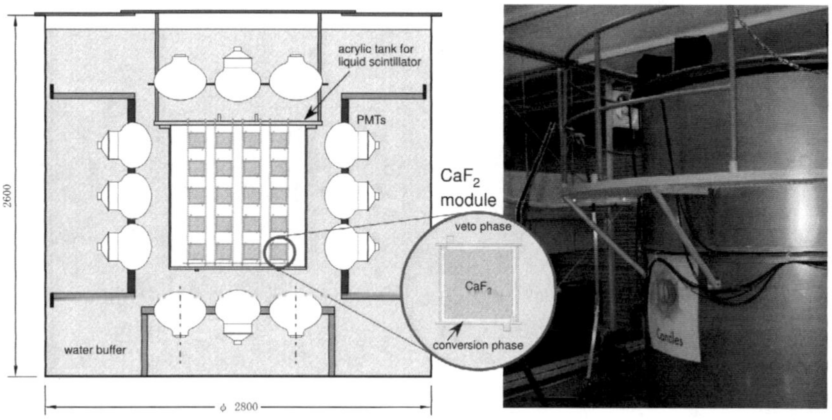

Figure 1. Schematic cross sectional view and a picture of the CANDLES III detector at OULNS. A size of the detector is 2.8 m in diameter and 2.6 m in height.

The background events originate from outside the CaF_2 module are effectively rejected by using 4π active veto system. Since the $Q_{\beta\beta}$ value is higher than typical energies of natural radiations, for instance, 2.6 MeV for γ rays and 3.3 MeV for β rays, backgrounds conceivable in the $Q_{\beta\beta}$ value region is only successive decays of radioactive nuclei in the CaF_2(pure) crystal. The candidates are pile-up 'Bi-Po' events of which delay time is less than about 5 nsec, in the U- and Th-chains;

$$^{214}\text{Bi} \xrightarrow{Q_\beta = 3.27\text{ MeV}} {}^{214}\text{Po}(T_{1/2} = 164.3\ \mu\text{sec}) \xrightarrow{Q_\alpha = 7.83\text{ MeV}} {}^{210}\text{Pb},$$
$$^{212}\text{Bi} \xrightarrow{Q_\beta = 2.25\text{ MeV}} {}^{212}\text{Po}(T_{1/2} = 0.299\ \mu\text{sec}) \xrightarrow{Q_\alpha = 8.95\text{ MeV}} {}^{208}\text{Pb}.$$

The maximum energies deposited are above the $Q_{\beta\beta}$-value of ^{48}Ca where the quenching factor of the CaF_2 scintillator for α particle is about 0.3. We had developed high-purity crystals and confirmed that their measured levels of radioactive contaminations (36(ave.), 14±5(best) μBq/kg for uranium and 28(ave.), 6 ± 1(best) μBq/kg for thorium) within almost 100 crystals are good enough to make the sensitivity of the detector being 0.5 eV with further background rejection using pulse shape discrimination. Figure 3 shows the results of 'shape indicator' analysis[5]

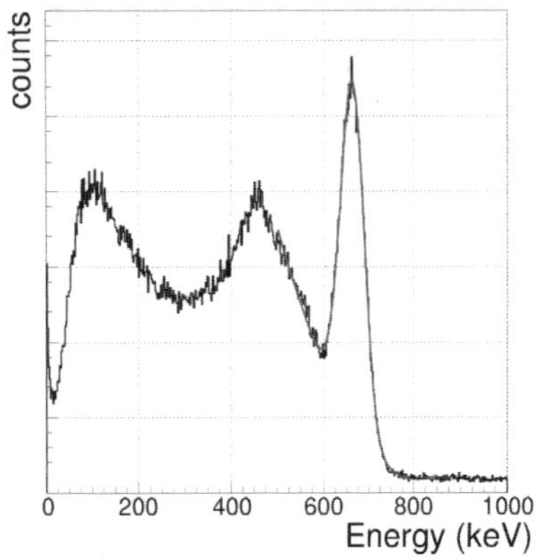

Figure 2. Energy spectrum of the CaF_2 module using ^{137}Cs γ-ray source. Obtained energy resolution at 662 keV is 9.1% (FWHM).

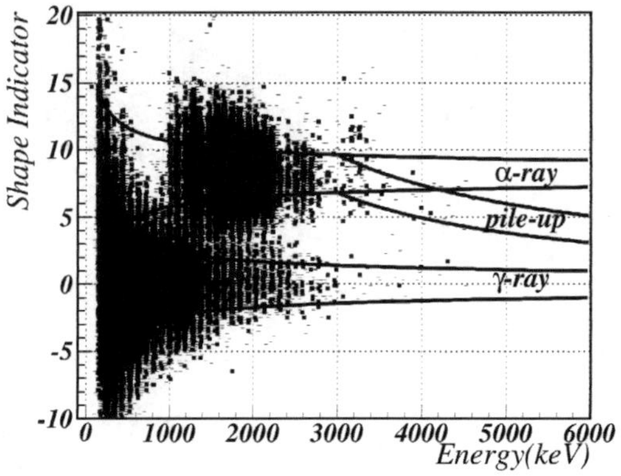

Figure 3. Scattered plot for the pulses corresponding to α-, pile-up and $\beta(\gamma)$-events with 'shape indicator' analysis. Solid lines show the $\pm 1\sigma$ region for each events.

using the difference in the pulse shape between α and $\beta(\gamma)$ events. At the $Q_{\beta\beta}$ region, we can discriminate the pile-up background event which is mainly from α particle. As a consequence, the pile-up events will be reduced by 2 orders of magnitude. We found that background level is low enough when the CANDLES III is installed at the Kamioka underground laboratory.

We are now moving the detector system to new experimental room (room D) at Kamioka underground laboratory (2700 m.w.e.) to avoid large background originated from cosmic rays. At the same time, we increased the total mass of the ^{48}Ca. 96 CaF_2 modules which contain 350 g of ^{48}Ca are located inside a liquid scintillator vessel of 1.4 m in diameter and 1.4 m in height where they are divided into 6 layers in vertical direction and each layer contains 16 modules which are located with 20 cm distance. A new water tank (3 m in diameter and 4 m in height) which contains 62 PMTs (48 × 13" and 14 × 17" tubes) and the liquid scintillator vessel was installed at room D. A purification system of the liquid scintillator together with storage tanks were also installed. The purification system removes the radioactive impurities especially U and Th using the techniques of water-extraction and N_2 purge. Other components including the CaF_2 modules, the PMTs, the liquid scintillator vessel and DAQ system will be installed soon. In addition, we are challenging the enrichment of ^{48}Ca using chemical separation[6] which is relatively low cost . On the bases of experiences in CANDLES III, the CANDLES project will be scaled up to several tons of calcium to have the sensitivity to the mass region of interest.

References

[1] Ogawa I *et al* 2004 *Nucl. Phys.* A **730** 215
[2] Umehara S, *et al* 2008 *Phy. Rev.* C **78** 058501
[3] Umehara S *et al* 2003 *Proc. of the Symp. on Neutrino and Dark Matter in Nuclear Physics (NDM03)* XII-23
Kishimoto T *et al* 2003 *Proc. of the Workshop on Neutrino Oscillations and their Origin (NOON2003)* 338
Yoshida S *et al* 2005 *Nuclear Physics* B (*Proc. Suppl.*) **138** 214
Ogawa I *et al* 2005 *Proc. of the Workshop on Neutrino Oscillations and their Origin (NOON2004)* 260
Umehara S *et al* 2006 *J. of Phys.: Conf. Series* **39** 356
Ogawa I *et al* 2008 *Proc. of the Int. Nuclear Physics Conf. (INPC2007)* vol 2 24
Hirano Y *et al* 2008 *J. of Phys.: Conf. Series* **120** 052053
[4] Yoshida S *et al* 2009 *Nucl. Instr. Meth.* A **601** 282
[5] Fazzini T *et al* 1998 *Nucl. Instr. Meth.* A **410** 213
[6] Hazama R *et al* 2006 *Proc. of the 6th Recontres du Vietnam, Hanoi*

Journal of Physics: Conference Series **203** (2010) 012074 doi:10.1088/1742-6596/203/1/012074

Status and perspectives of the Virgo gravitational wave detector

T. Accadia[12], F. Acernese[12], F. Antonucci[9a], S. Aoudia[15a], K.G. Arun[11], P. Astone[9a], G. Ballardin[2], F. Barone[6ac], M. Barsuglia[1], Th.S. Bauer[14a], M.G. Beker[14a], A. Belletoile[12], S. Bigotta[8ab], S. Birindelli[15a], M.A. Bizouard[11], M. Blom[14a], C. Boccara[3], F. Bondu[15b], L. Bonelli[8ab], R. Bonnand[13], L. Bosi[7a], S. Braccini[8a], C. Bradaschia[8a], A. Brillet[15a], V. Brisson[11], R. Budzyński[17b], T. Bulik[17cd], H.J. Bulten[14ab], D. Buskulic[12], C. Buy[1], G. Cagnoli[4a], E. Calloni[6ab], E. Campagna[4ac], B. Canuel[2], F. Carbognani[2], F. Cavalier[11], R. Cavalieri[2], G. Cella[8a], E. Cesarini[4c], E. Chassande-Mottin[1], A. Chincarini[5], F. Cleva[15a], E. Coccia[10ab], C. N. Colacino[8a], J. Colas[2], A. Colla[9ab], M. Colombini[9b], A. Corsi[9a], J.-P. Coulon[15a], E. Cuoco[2], S. D'Antonio[10a], A. Dari[7ab], V. Dattilo[2], M. Davier[11], R. Day[2], R. De Rosa[6ab], M. del Prete[8ac], L. Di Fiore[6a], A. Di Lieto[8ab], M. Di Paolo Emilio[10ad], A. Di Virgilio[8a], A. Dietz[12], M. Drago[16cd], V. Fafone[10ab], I. Ferrante[8ab], F. Fidecaro[8ab], I. Fiori[2], R. Flaminio[13], J.-D. Fournier[15a], J. Franc[13], S. Frasca[9ab], F. Frasconi[8a], A. Freise*, M. Galimberti[13], L. Gammaitoni[7ab], F. Garufi[6ab], G. Gemme[5], E. Genin[2], A. Gennai[8a], A. Giazotto[8a], R. Gouaty[12], M. Granata[1], C. Greverie[15a], G. Guidi[4ac], H. Heitmann[15], P. Hello[11], S. Hild*, D. Huet[2], P. Jaranowski[17e], I. Kowalska[17c], A. Królak[17af], P. La Penna[2], N. Leroy[11], N. Letendre[12], T.G.F. Li[14a], M. Lorenzini[4a], V. Loriette[3], G. Losurdo[4a], J.-M. Mackowski[13], E. Majorana[9a], N. Man[15a], M. Mantovani[8c], F. Marchesoni[7a], F. Marion[12], J. Marque[2], F. Martelli[4ac], A. Masserot[12], F. Menzinger[2], C. Michel[13], L. Milano[6ab], Y. Minenkov[10a], M. Mohan[2], J. Moreau[3], N. Morgado[13], A. Morgia[10ab], S. Mosca[6ab], V. Moscatelli[9a], B. Mours[12], I. Neri[7ab], F. Nocera[2], G. Pagliaroli[10ab], C. Palomba[9a], F. Paoletti[8a,2], S. Pardi[6ab], M. Parisi[6b], A. Pasqualetti[2], R. Passaquieti[8ab], D. Passuello[8a], G. Persichetti[6ab], M. Pichot[15a], F. Piergiovanni[4ac], M. Pietka[17e], L. Pinard[13], R. Poggiani[8ab], M. Prato[5], G.A. Prodi[16ab], M. Punturo[7a], P. Puppo[9a], O. Rabaste[1], D.S. Rabeling[14ab], P. Rapagnani[9ab], V. Re[16ab], T. Regimbau[15a], F. Ricci[9ab], F. Robinet[11], A. Rocchi[10a], L. Rolland[12], R. Romano[6ac], D. Rosińska[17g], P. Ruggi[2], B. Sassolas[13], D. Sentenac[2], R. Sturani[4ac], B. Swinkels[2], A. Toncelli[8ab], M. Tonelli[8ab], E. Tournefier[12], F. Travasso[7ab], J. Trummer[12], G. Vajente[8ab], J.F.J. van den Brand[14ab], S. van der Putten[14a], M. Vavoulidis[11], G. Vedovato[16c], D. Verkindt[12], F. Vetrano[4ac], A. Viceré[4ac], J.-Y. Vinet[15a], H. Vocca[7a], M. Was[11], M. Yvert[12]

[1]AstroParticule et Cosmologie (APC), CNRS: UMR7164-IN2P3-Observatoire de

Paris-Université Denis Diderot-Paris VII - CEA : DSM/IRFU
[2]European Gravitational Observatory (EGO), I-56021 Cascina (Pi), Italy
[3]ESPCI, CNRS, F-75005 Paris, France
[4]INFN, Sezione di Firenze, I-50019 Sesto Fiorentino[a]; Università degli Studi di Firenze, I-50121[b], Firenze; Università degli Studi di Urbino 'Carlo Bo', I-61029 Urbino[c], Italy
[5]INFN, Sezione di Genova; I-16146 Genova, Italy
[6]INFN, sezione di Napoli [a]; Università di Napoli 'Federico II'[b] Complesso Universitario di Monte S.Angelo, I-80126 Napoli; Università di Salerno, Fisciano, I-84084 Salerno[c], Italy
[7]INFN, Sezione di Perugia[a]; Università di Perugia[b], I-6123 Perugia,Italy
[8]INFN, Sezione di Pisa[a]; Università di Pisa[b]; I-56127 Pisa; Università di Siena, I-53100 Siena[c], Italy
[9]INFN, Sezione di Roma[a]; Università 'La Sapienza'[b], I-00185 Roma, Italy
[10]INFN, Sezione di Roma Tor Vergata[a]; Università di Roma Tor Vergata[b], Istituto di Fisica dello Spazio Interplanetario (IFSI) INAF[c], I-00133 Roma; Università dell'Aquila, I-67100 L'Aquila[d], Italy
[11]LAL, Université Paris-Sud, IN2P3/CNRS, F-91898 Orsay, France
[12]Laboratoire d'Annecy-le-Vieux de Physique des Particules (LAPP), IN2P3/CNRS, Université de Savoie, F-74941 Annecy-le-Vieux, France
[13]Laboratoire des Matériaux Avancés (LMA), IN2P3/CNRS, F-69622 Villeurbanne, Lyon, France
[14]National institute for Subatomic Physics, P.O. Box 41882, 1009 DB [a]; VU University Amsterdam, De Boelelaan 1081, 1081 HV [b], Amsterdam, The Netherlands
[15]Université Nice-Sophia-Antipolis, CNRS, Observatoire de la Côte d'Azur, F-06304 Nice [a]; Institut de Physique de Rennes, CNRS, Université de Rennes 1, 35042 Rennes [b]; France
[16]INFN, Gruppo Collegato di Trento[a] and Università di Trento[b], I-38050 Povo, Trento, Italy; INFN, Sezione di Padova[c] and Università di Padova[d], I-35131 Padova, Italy
[17]IM-PAN 00-956 Warsaw[a]; F.of Phys. Warsaw Univ. 00-681Warsaw[b]; Astro. Obs. Warsaw Univ. 00-478 Warsaw[c]; CAMK-PAN 00-716 Warsaw[d]; Bialystok Univ. 15-424 Bialystok[e]; IPJ 05-400 Swierk-Otwock[f]; Inst. of Astronomy 65-265 Zielona Gora[g], PL
[*]University of Birmingham, Birmingham, B15 2TT, United Kingdom

E-mail: `viviana.fafone@roma2.infn.it`

Abstract. Virgo is designed to detect gravitational waves of both astrophysical and cosmological origin in the frequency range from a few Hz to a few kHz. After the end of the first science run, partially overlapped with the LIGO fifth science run, the detector underwent several upgrades to improve its sensitivity. The second Virgo science run started at the beginning of July 2009 in coincidence with LIGO. A further upgrade is planned at beginning of 2010 with the installation of new suspensions for the test masses and of new mirrors. This will lead to a considerable improvement in the sensitivity and represents an intermediate step toward the development of the advanced detectors.

1. Introduction

Virgo [1] is a 3km power recycled Michelson interferometer where each arm is a Fabry-Perot cavity. It is located close to Pisa (Italy) at the European Gravitational Observatory (EGO). The detector is sensitive to gravitational waves (GW) emitted by spinning neutron stars, coalescing binaries, supernovae and stochastic background of cosmological origin. The commissioning of the Virgo interferometer begun in 2003. These efforts led, in 2007, to the first scientific run VSR1 [2] in coincidence with LIGO (Laser Interferometer Gravitational-wave Observatory) [3]. After VSR1, during 2008 and half of 2009, the commissioning activity restarted and the interferometer underwent several upgrades to further improve its sensitivity. In July 2009, Virgo started its second science run (VSR2) in coincidence with the sixth science run (S6) of the two 4 km long LIGO interferometers L1 and H1 (located in Livingston Parish -Louisiana- and Hanford -Washington State- respectively). At the beginning of 2010, monolithic suspensions will be installed and all the test masses will be replaced. These upgrades will increase the NS-NS coalescing binaries horizon by a factor of 5.

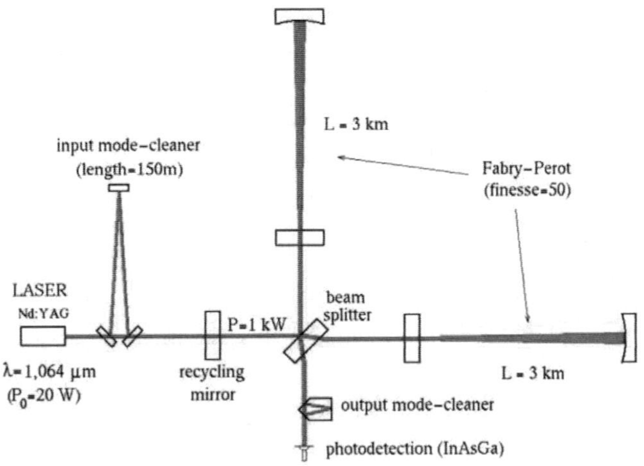

Figure 1. Virgo optical layout.

2. Virgo commissioning in 2008-2009 and the second science run

At the end of VSR1, in October 2007, an intense program of upgrades and commissioning started to lead the detector towards the Virgo+ configuration. Figure 1 shows the Virgo optical layout.

The main upgrades regarded the installation of a more powerful and less noisy read-out and control electronics, the installation of a laser amplifier that can deliver up to 25 W at the input of the interferometer and some upgrades of the injection bench, necessary to cope with the higher laser power and to provide a better seismic noise rejection. Moreover, a thermal compensation system (TCS) has been installed, to reduce thermal effects in the input mirrors. In fact, the absorption of a few ppm of the laser power on the input mirror coatings results in a decrease of the matching between the carrier and the sidebands, thus preventing lock acquisition at input powers above 14 W. TCS uses a CO_2 annular beam that heats the peripheral of the input test masses in order to flatten the optical path length inside the mirror substrates. The commissioning activity allowed to increase the input power from 8 W, used during VSR1, to 17 W at the beginning of the VSR2.

Thanks to these upgrades and the good outcome of the commissioning activity, the measured Virgo sensitivity is now very close to the design curve (see figure 2).

Due to the seismic attenuation performance of the Superattenuator [4], Virgo has a much better sensitivity than LIGO at frequencies below 40 Hz (three orders of magnitude better at 20 Hz). The difficulty to reach a good sensitivity at these low frequencies also depends on the presence of many technical noises (such as control noise, magnetic noise, diffused light noise). A good understanding of these noise sources has been reached, as shown again in figure 2: in fact the pink curve, that is the sum of all the known contributions, superposes very well with the experimental curve. A big effort is devoted to the understanding and reduction of the technical noise contributions. For example, a large fraction of the commissioning work has been devoted to the mitigation of diffused light. In fact, the optics located on the external benches can retro-diffuse light into the interferometer. Since these optics are not seismically isolated, the diffused light introduces phase noise that masks the gravitational wave signal. To reduce this noise, some activities have been performed: mitigation of environmental noise, acoustic isolation of the benches and improvement of the optical set-up.

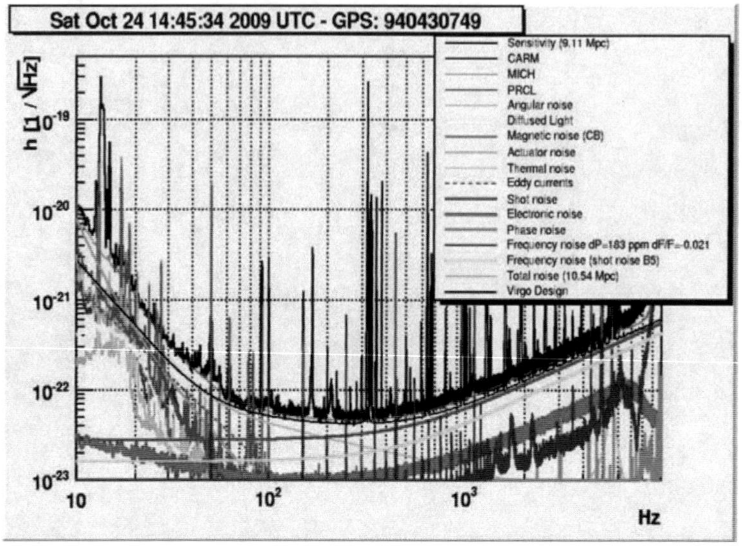

Figure 2. Noise budget: the black curve represents the measured sensitivity. The coloured curves show all the known and modelled noise sources and the thin purple curve is the sum of all these contributions.

All these actions led to an enhancement of the NS-NS averaged horizon from 4 Mpc in VSR1 to more than 9 Mpc during VSR2. After four months of run, the integrated science mode duty cycle is higher than 80%.

3. Virgo+

Virgo+ is a technologically enhanced configuration aimed at improving the detector sensitivity: in particular, when all the upgrades will be implemented, the averaged horizon distance for coalescing binary neutron star systems will increase from 9 to 45 Mpc, corresponding to a 2 orders of magnitude improvement in the observable volume. Both technical and fundamental noises, such as shot noise, mirror and suspension thermal noise, must be reduced.

Shot noise, affecting the high frequency part of the sensitivity, can be reduced by increasing the input power. The present laser amplifier can deliver up to 25 W, 30% more than during VSR2.

Low frequency sensitivity is limited by the thermal noise of the last stage of the suspension system, i.e. the pendulum to which the mirror is suspended. This noise is due to the friction between the mirror substrate and the suspension at the attachment point. The installation of monolithic suspensions will reduce this noise: the input test masses will be suspended with fused silica fibers instead of metallic wires. The technique for fibers production and selection is now well established and the robustness of the new set up has been checked.

The sensitivity around 100 Hz is limited by the mirror thermal noise. Thus, new mirrors with better substrates and coatings are being prepared. A further improvement of the sensitivity in this frequency range comes from the fact that the new test masses will allow to increase the arm cavity finesse from the present 50 to 150. The installation of monolithic suspensions and new test masses will take place at the beginning of 2010. Figure 3 shows the expected Virgo+ sensitivity (black curve) compared to the present one (red). A further reduction of technical noises (such as the beam jitter noise and the noise induced by the magnetic actuators on the test masses) will be necessary to reach the design sensitivity.

Besides representing a relevant scientific step thanks to the increase of sensitivity, these upgrades will give the unique opportunity to explore a region of lower noise before the construction of the advanced detectors (Advanced Virgo and Advanced LIGO) expected to enter in operation in 2014.

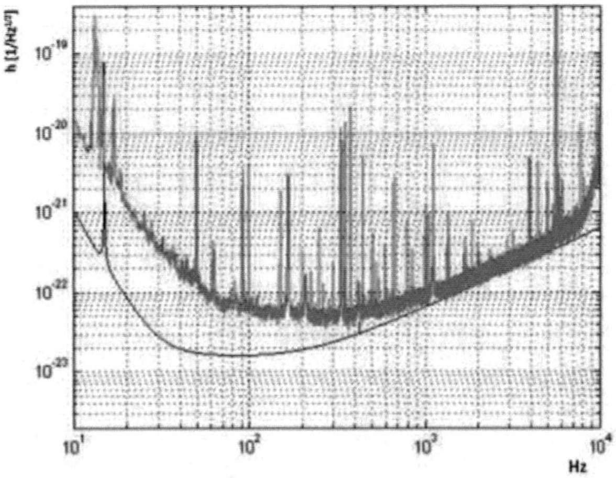

Figure 3. Virgo+ design sensitivity with monolithic suspensions (black), compared to the present sensitivity (red).

4. Conclusions

The Virgo detector is now running its second science run with a sensitivity close to design and a science mode duty cycle higher than 80%. NS-NS coalescing binaries horizon has increased by more than a factor of two with respect to VSR1. Early in 2010, monolithic suspensions and new test masses will be installed, completing the upgrade to Virgo+. The expected sensitivity corresponding to the Virgo+ design curve, expressed in terms of NS-NS coalescing binaries horizon should increase up to of a factor of 5 compared to the present one.

References

[1] www.virgo.infn.it
[2] Frasconi F *et al* (Virgo Collaboration) 2008 *J. Phys.: Conf. Ser.* **120** 032007
[3] www.ligo.caltech.edu
[4] The Virgo Collaboration 2005 *Astrop. Phys.* **23** 557-565

Underground Cryogenic Laser Interferometer CLIO

Shinji Miyoki and CLIO and LCGT Collaborators

5-1-5 Kashiwanoha, Kashiwa, Chiba, 277-8582 JAPAN.

miyoki@icrr.u-tokyo.ac.jp

Abstract. The displacement sensitivity of Cryogenic Laser Interferometer Observatory (CLIO) has been comparable with suspension thermal noise for the frequency range from 20Hz to 80Hz, and sapphire mirror thermo-elastic noise for the frequency range from 80Hz to 200Hz. The best displacement sensitivity is 3×10^{-19} $[\mathrm{m}/\sqrt{\mathrm{Hz}}]$ around 250Hz. In order to verify displacement sensitivity enhancement by cooling suspension wires and sapphire mirrors, we have started modifying mirror suspension systems.

1. Introduction

The final noise barrier to obtain the targeted displacement sensitivity of 1×10^{-20} $[\mathrm{m}/\sqrt{\mathrm{Hz}}]$ around 100Hz for interferometric gravitational wave detectors such as LIGO[1], VIROG[2] is thermal noise caused by coating films on mirrors. One approach to reduce it is to enlarge beam spots on the mirrors and to use SiO_2 mirrors that have high mechanical quality factors. Another approach is to cool sapphire mirrors and their sapphire suspension wires. Large-scale Cryogenic laser interferometer Gravitational wave Telescope (LCGT) project [3], which is proposed in Japan, challenges to utilize the latter. In order to establish several technical bases for cooling mirrors without any additional noise introduction, a 100 meter base-line Cryogenic Laser Interferometer Observatory (CLIO [4]) proto-type is being developed at 1000 meter underground in the Kamioka mine that is the same site for LCGT. We planed firstly to obtain thermal noise limited displacement sensitivity without cooling, then secondly to verify the displacement noise sensitivity enhancement by cooling mirrors and suspension wires. In this letter, we will report about the progress of CLIO displacement noise sensitivity enhancement and about preparation for the cooling.

2. Displacement sensitivity of CLIO operated at the room temperature

The CLIO optical configuration, vacuum configuration, sapphire mirror suspension and seismic noise isolation system, cryostat system for housing sapphire mirrors and ultra-low vibration cryocooler system are detailed in the following references [5,6,7]. A laser beam (1064nm, 2W, MEPHISTO, INNOLIGHT GmbH) is introduced into a triangle-type 9.5 meters base-line Fabry-Perot (FP) cavity as a mode cleaner (MC) after twice phase modulations at 12MHz for the MC length control and 15.804MHz for the 100 meter FP arm cavity control. About 70% transmitted laser beam from MC is bounced on three mode matching telescope (MMT) mirrors, and then is injected into the orthogonally oriented FP arm cavities after split into a half power at a beam splitter (BS). The arm FP cavities have critical coupling mirrors whose reflectivity is 99.9%, which corresponds to a finesse of 3140. One FP arm cavity is used for the further frequency stabilization after the pre-frequency stabilization using MC, while the other FP arm is used to extract GWs signal. The Pound-Drever-Hall method is used to keep MC and two FP arm cavities on resonance. All MC mirrors, MMT mirrors, BS and optical circulator components are isolated from seismic noise by double stages pendulum systems, while the four main sapphire mirrors that composed of both FP arm cavities are by six stages pendulum systems.

Figure.1 shows the best displacement sensitivity of CLIO obtained in November 11th 2008. After removing the suspension thermal noise caused by eddy current in aluminum coil holders for coil-magnet force actuation to control one mirror position, the displacement curve finally touched (a) the suspension thermal noise level that is estimated from the structure damping model accounting for measured mechanical quality factors of 10^5 of the pendulum wires' violin modes and (b) sapphire mirrors thermo-elastic noise level that is also calculated from their physical parameters such as thermal conductivity and so on. For the frequency range below 20 Hz, seismic noise dominates the sensitivity, while shot noise and some unknown noise dominate it above 250Hz.

3. Suspension system modification for mirror cooling

In order to cool sapphire mirrors and their suspension wires, several heat link wires that are pure aluminium (99.999% purity, 0.5mm diameter, Nilaco Co., Ltd.) are connected between (a) an inner radiation shield and a copper cryo-base mass (see reference [6] to understand the mirror suspension system) (b) the cryo-base mass and a copper upper-mass that is suspended from the cryo-base mass by four C-shape type blade springs. The final sapphire mirror mass is suspended by two loops pure aluminium wires, which are same kind of heat link wires and 0.4 meter in length, from the upper mass. Although the original design for the mirror suspension is to use sapphire fibres, which are difficult to be assembled, as suspension wires, we will avoid them in CLIO. The main reason is because even the suspension thermal noise of the pure aluminium wire around 20 K is estimated to be well below the present suspension thermal noise level, accounting for the pendulum mode mechanical quality factor of the pure aluminium wire of 10^4 that is estimated from the measured mechanical quality factor of its violin mode. It will take seven days to cool the sapphire mirrors around 20 K.

4. Summary

The CLIO displacement sensitivity has finally been comparable with the suspension thermal noise and the sapphire mirror thermo-elastic noise level. Especially, the wide frequency range measurement of the mirror suspension noise from the high quality factor pendulum mode is demonstrated. CLIO mirror suspension system is now being modified for cooling mirrors and their suspension fibers.

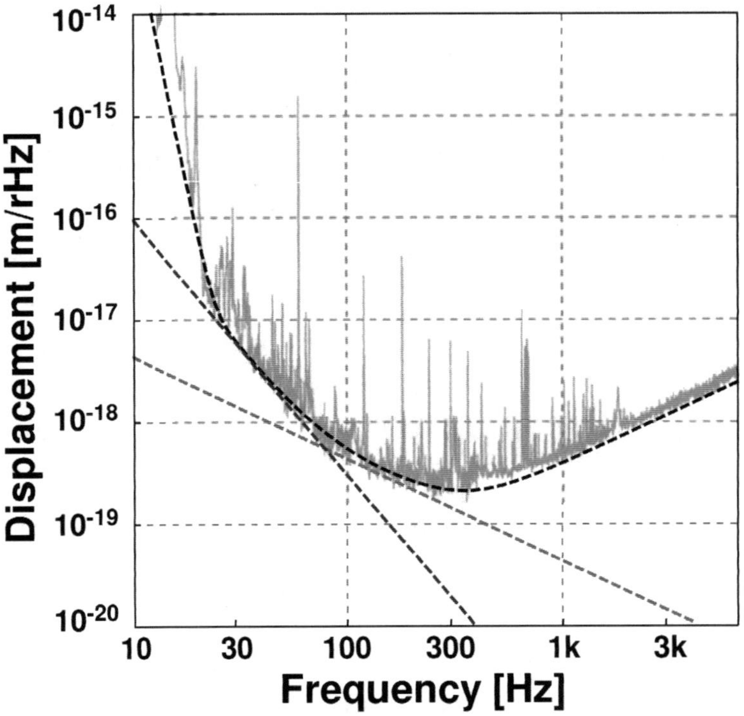

Figure 1. (color online) CLIO best displacement noise denoted by an orage rigid line. The blue and red dotted lines are estimated suspension thermal noise and mirror thermo-elastic noise, respectively.

References

[1] Abramovici A *et al* 1992 *Science* **256** 325
[2] Bradaschia C *et al* 1990 *Nucl. Instrum. Methods Phys.* Sect A **289** 518
[3] Kuroda K *et al* 2006 *Prog. Theor. Phys. Suppl.* **163** 54
[4] Miyoki S *et al* 2006 *Class. Quantum Grav.* **23** S231
[5] Miyoki S *et al* 2004 *Class. Quantum Grav.* **21** S1173
[6] Uchiyama T *et al* 2006 *J. Phys.: Conf. Ser.* **163** 54
[7] Ikushima U *et al* 2008 *Cryogenics* **48** 406

Journal of Physics: Conference Series **203** (2010) 012076 doi:10.1088/1742-6596/203/1/012076

Einstein Telescope site selection: seismic and gravity gradient noise

J.F.J. van den Brand, M.G. Beker, M. Doets, E. Hennes, D.S. Rabeling

Nikhef, National Institute for Subatomic Physics, Science Park 105, 1098 XG Amsterdam, The Netherlands

E-mail: `jo@nikhef.nl`

Abstract. Gravity gradient noise generated by seismic displacements may be the limiting factor for the sensitivity of third-generation gravitational wave detectors at frequencies below 10 Hz. A finite element framework has been developed to calculate the soil response to various excitations. The accompanying gravity gradients as a result of the seismic displacement field can then be evaluated. The results of the gravity gradient noise are in good agreement with previous analytical results. Finally results of gravity gradient noise from a single pulse excitation of a homogenous medium are discussed for an underground detector.

1. Introduction

At this moment the first generation of interferometric gravitational wave (GW) detectors have reached their design sensitivities and second generation detectors are poised to upgrade these sensitivities by a factor of 10 within the next decade. The Einstein Telescope (ET) design study [1] initiates research to explore possibilities beyond the second generation detectors. The third generation of detectors will improve sensitivities by a factor of 10 over a broad frequency spectrum.

A particular challenge will be to detect signals at frequencies below 10 Hz. It is expected that sensitivity at these frequencies will be limited by fluctuations in the local gravitational field as a result of fluctuating density variations in the surrounding soil. These field fluctuations couple directly to the interferometer test masses and are known as gravity gradient (or Newtonian) noise (GGN). A full understanding of seismic activity, seismic wave characteristics and how these affect GW detectors through gravity gradients is imperative to achieve the desired sensitivity. These studies have been the prime focus of working package 1 of the ET design effort in the past eighteen months.

Sources of seismic noise include earthquakes, atmospheric and oceanic disturbances, and cultural noise. The latter is a result of human activity such as footsteps, traffic and machinery. Analytical descriptions of seismic activity fall short when considering a non-homogeneous medium or complex geology and a time domain finite element (FE) framework to model seismic activity has been developed that allows to study the impact of GGN on GW detectors.

2. Studies of seismic noise

Noise studies differentiate noise sources according to frequency f and for Einstein Telescope the critical frequencies f are in the range 1 - 10 Hz, where the response is most variable mainly due to cultural noise. It is therefore important to choose a site location far from human activities both at present and in future. In Europe low noise sites are located in the center of the continent. For example in various locations in Germany displacement noise densities have been measured as low as 1 nm/rtHz at 1 Hz that approximately drop as $1/f^2$. For Einstein Telescope additional suppression of GGN (by a factor 10 – 100) is needed and it is imperative that an underground infrastructure is realized.

2.1 Seismic noise finite element modeling

As a result of seismic disturbances a displacement wave field is produced that is governed by the elasto-dynamic equations. Wave solutions include pressure (P) and shear (S) body waves, and Rayleigh and Love surface waves. Simulation of the ground displacement fields was done with a FE software package.

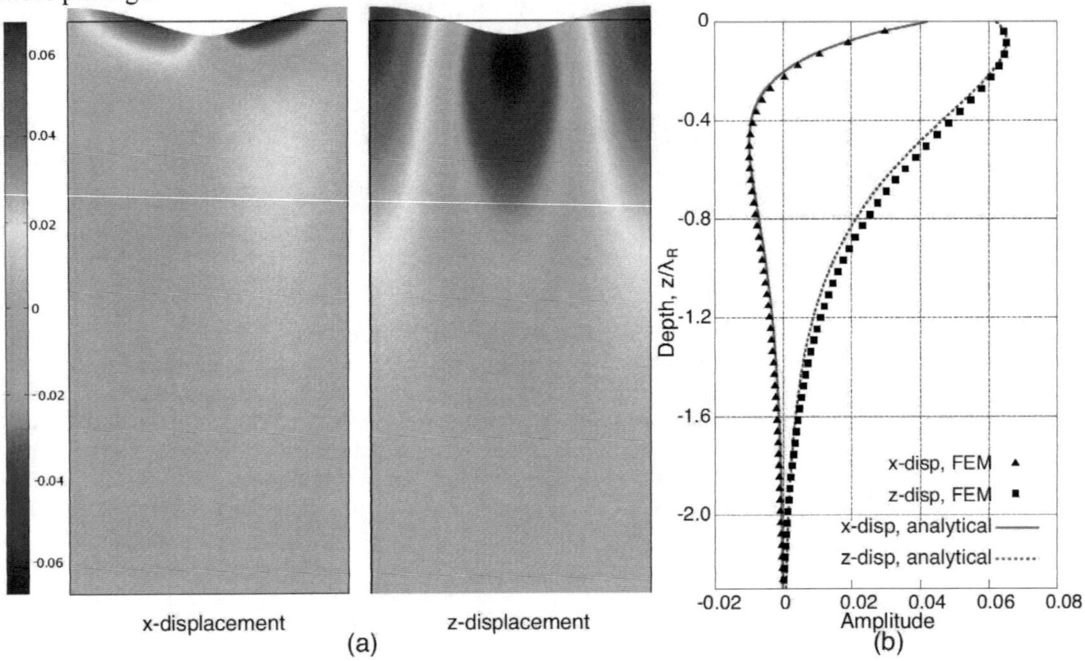

Figure 1. (a) Rayleigh wave displacement fields for a wave propagating in the x direction. (b) FE and analytical calculations of Rayleigh wave displacements (λ_R denotes the Rayleigh wavelength).

The ground motion response calculated by the FE model is in good agreement with analytical solutions for Rayleigh waves (see Fig.1). The x and z components attenuate differently with depth and the x component changes sign at a depth of about 0.2 λ_R.

3. Studies of gravity gradient noise

In our FE analysis the acceleration is given by the summation of the contributions $\mathbf{a}_i$ from each node i with mass m_i. When a seismic displacement field is present, the nodes suffer a displacement denoted by $\xi_i(\mathbf{r}, t)$. The gravity gradient acceleration due to these displacements is given by

$$\mathbf{a}^{NN}(\mathbf{y}, t) = \sum_i (\nabla \otimes \mathbf{a}_i)^T \xi_i(\mathbf{r}, t).$$

Note that in the gravity gradient calculations presented here, the mass associated with each node is assumed to be constant.

3.1 Finite element modeling of gravity gradient noise for a surface detector

The FE results are compared with the analytic results of Saulson [2] and Hughes and Thorne [3] for a surface detector. To facilitate comparison, an integral cut-off radius equal to that used in Saulson's analysis ($\lambda/4$) was employed in the summation process. Fig. 2 shows that good agreement is obtained. In the limit that the cut-off radius decreases to zero, GGN increases by about a factor 2.

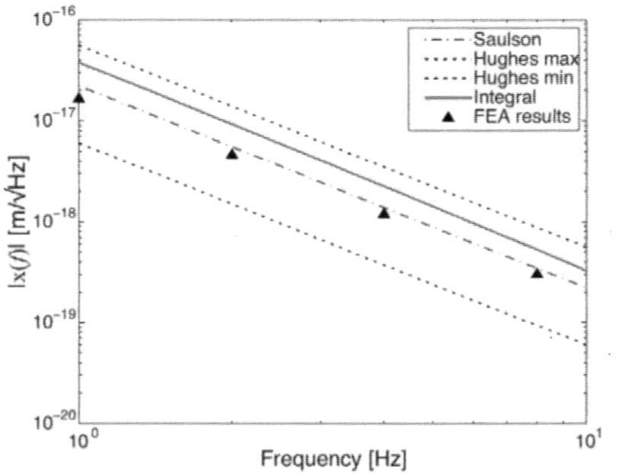

Figure 2. FE results of GGN displacement noise for a surface detector. For comparison the results of Saulson, Hughes and Thorne are shown (the red line represents the limit that the cut-off radius decreases to zero).

3.2 FE modeling of gravity gradient noise for an underground detector

Pulse excitations at the centre of a half-sphere model were used to investigate FE GGN modeling for an underground detector (see Fig. 3). The nodal displacements were recorded as a function of time and the GGN accelerations were calculated at various depths at 800 m from the z-axis.

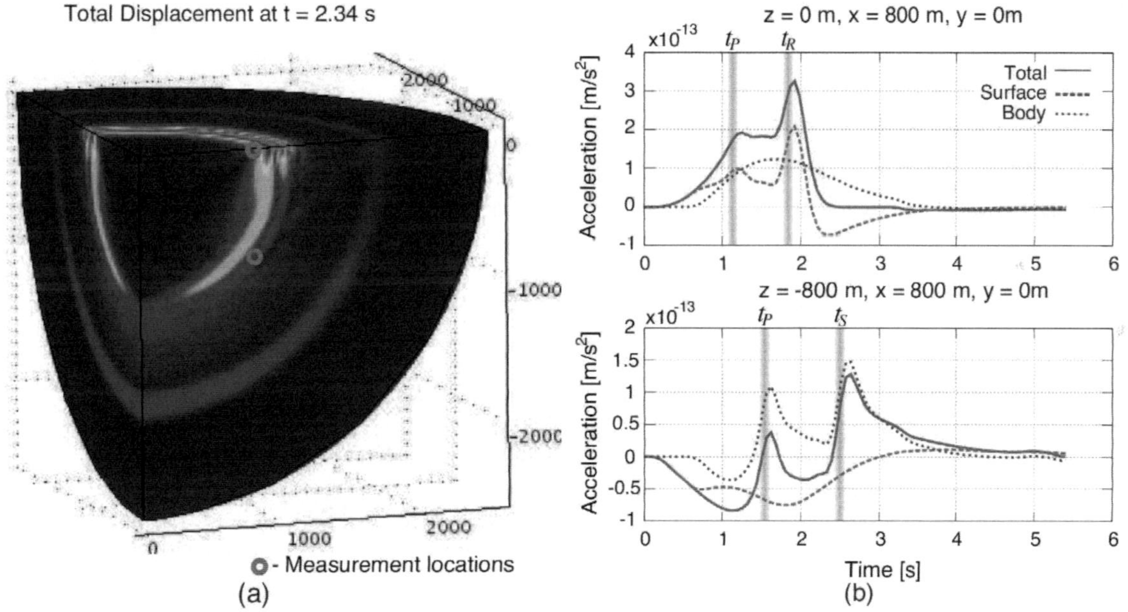

Figure 3. (a) Pulse excitation of a homogeneous half-sphere. (b) Time domain evolution of GGN acceleration (horizontal) at a surface (z=0 m) and underground (z=-800 m) test mass. Arrival times of Rayleigh, S and P-waves are also indicated.

The FE analysis indicates strong contributions from Rayleigh waves for a surface detector and body waves for a subterranean detector. This FE framework can be an effective tool in investigating gravity gradients for subterranean gravitational wave detectors. This allows further research into quantitative effects of seismic motion on GGN and implications of geology and infrastructure.

References

[1] Einstein Telescope design study (Grant Agreement 211743). See http://www.et-gw.eu/
[2] Saulson P 1984 Phys. Rev. **D** 30 732–736
[3] Hughes S and Thorne K 1998 Phys. Rev. **D** 58 122002

Topics in Astroparticle and Underground Physics (TAUP 2009) IOP Publishing
Journal of Physics: Conference Series **203** (2010) 012077 doi:10.1088/1742-6596/203/1/012077

Status of supernova neutrino detectors

Walter Fulgione
Istituto di Fisica dello Spazio Interplanetario and INFN, Torino, Italy
E-mail: walter.fulgione@gmail.com

Abstract. Supernova neutrino observatories currently running are water Cherenkov or scintillator detectors. They differ in technique and performance. These differences have been analyzed and main characteristics of each detector outlined: energy threshold and resolution, pointing capability, duty cycle and number of events expected from a core collapse supernova. Some of these detectors are involved in the Supernova Early Warning System, a network of neutrino observatories dedicated to the prompt recognition of a core collapse supernova in the Galaxy.

Supernova neutrino observatories, presently running, are water Cherenkov and scintillator detectors. Inside these two families there are important differences. Scintillator detectors can be divided into: single volumes (like KamLAND, Borexino and, in the next future, SNO+) and modulars (like Baksan and LVD). They differ in performance, sensitivity and finally interaction channels. Water Cherenkov detectors show, among them, even more evident differences. Super-K and IceCube represent two different ways of detecting SN neutrinos observing Cherenkov light in water. Most of these detectors participate to the Supernovae Early Warning System (SNEWS) [1], the network of SN neutrino observatories whose main goal is to provide the astronomical community with a prompt alert for the next galactic core collapse supernova explosion. To participate to the network, each detector must be able to disentangle a SN neutrino candidate signal by itself, i.e., independently from any other observation. This forces different experiments to develop peculiar strategies for achieving this goal.
In this work I will summarize the main characteristics of detectors concerning SN neutrinos, i.e., number of expected interactions, duty cycle, sensitivity to different neutrino flavors, energy resolution, pointing capability.
The sensitivity of a SN neutrino detector, for example in terms of maximum observable distance, is proportional to the number of detected interactions over the background fluctuations. The expected signal, and the corresponding background, both depend on astrophysical parameters and neutrino properties which are only partially known. All the running SN neutrino detectors, having hydrogenate targets, exploit the inverse beta decay (IBD) interaction of $\bar{\nu}_e$ with free protons, $\bar{\nu}_e + p \rightarrow n + e^+$, to detect SN neutrinos. IBD is definitely the favorite interaction channel in this kind of detectors and in all SN models which predict the emission of neutrinos of all flavors. Moreover the detectors active in occasion of SN1987A were water and scintillator detectors and almost all the interactions detected in that occasion are thought to be IBD. See [2] and references therein for a detailed analysis.
The importance to have different kind of detectors, based on liquid Argon target [3] which has an excellent potential for ν_e tagging, or on Lead [4], to have a good sensitivity to neutrino flavors other than $\bar{\nu}_e$ is well known. The decommissioning of SNO in 2006 [5], with its unique sensitivity

to ν_e, $\bar{\nu}_e$ and neutral currents, left an unrecouped lack.
The number of IBD interactions detected during an interval of duration δt from the beginning of the collapse can be expressed as:

$$N_{\rm IBD} = N_{\rm H} \int_0^{\delta t} dt \int_0^{\infty} dE_{\bar{\nu}_e} \, \Phi(E_{\bar{\nu}_e}, t) \cdot \sigma(E_{\bar{\nu}_e}) \cdot \epsilon(E_{\bar{\nu}_e}) \qquad (1)$$

where $N_{\rm H}$ is the number of free protons, Φ is the flux of electron antineutrinos, σ the IBD cross section, and ϵ the detection efficiency. The number of events expected from a core collapse SN calculated by different models can differ up to a factor two, nevertheless the ratio among the number of events detected by different experiment, through the IBD channel, can be established with a much smaller uncertainty. The detection of a big discrepancy from the expected ratio would indicate a problem in our understanding of the physics of the process. Detection efficiency $\epsilon(E_{\bar{\nu}_e})$, and target mass, $N_{\rm H}$, are known characteristics of each detector.
Table 1 lists running SN neutrino detectors and their characteristics. The number of IBD events expected from a SN (1987A-like) at 10 kpc is calculated assuming 200 positrons events per kiloton of water target mass, disregarding effects due to different energy thresholds because all modern detectors have a nearly saturated efficiency at $E_{e+} = 10$ MeV and because the contribution due to e^+ with lower energies is expected to be small[1] (a different approach is needed for Icecube).

Table 1. Supernova neutrino detectors and their capabilities.

detector	*technique*	*mass kton*	N_{ev}^{IBD} *at 10kpc*	E_{th} *MeV*	σ_E/E *10MeV*	*point*	*duty cycle*
IceCube[6]	**long string**	**120** $\frac{E_{e+}}{MeV}$	**300000**	—	—	**n**	—
SK[7]	**H_2O Ch.**	**22.5(50)**	**4500(10000)**	**7.0**	**15%**	**y**	**89%**
LVD[8]	**Sc. mod.**	**1.0**	**255**	**4.0**	**15%**	**n**	**99%**
KamLAND[9]	Sc. vol.	1.0	255	0.35	2%	n	79%
Borexino[10]	**Sc. vol.**	**0.3**	**75**	**0.2**	**1.5%**	**n**	**75%**
Baksan[11]	Sc. mod.	0.13(0.33)	35(85)	8.0	25%	n	85%

It is important to notice that:

(i) The only reason to have different Φ at different detectors is the Earth matter oscillation effect. It produces a decrease in the number of detected neutrinos for particular neutrino energy, it is quite weak (depending on the value of $\Delta m^2_{\rm sol}$ and on the ratio $T_{\nu_x}/T_{\bar{\nu}_e}$) and, if the position of the source is known, can be calculated (see for example [12]).

(ii) SuperK is the only one with pointing capabilities, through the $\overset{(-)}{\nu}_i + e^- \rightarrow \overset{(-)}{\nu}_i + e^-$ scattering which represents $\sim 5\%$ of IBD interactions.

(iii) For IceCube, whose effective volume is a function of E_{e+}, I assumed $\bar{E}_{\bar{\nu}_e} = 14$MeV. The huge number of interactions must be compared with an important background counting rate. This implies a sensitivity of "only" 5σ at 50 kpc. IceCube cannot distinguish among neutrino flavors nor measuring positron energy, nevertheless, for a Galactic SN, it will furnish a light curve out of reach for the other detector.

[1] The visible energy is $E_{\rm v} \sim E_{\bar{\nu}_e} - \Delta$ in Cherenkov detectors and $E_{\rm v} \sim E_{\bar{\nu}_e} - \Delta + 2m_e c^2$ in scintillator ones. The ratio between the number of free protons per ton is $N_{\rm H_2O}/N_{\rm C_nH_{2n}} = 0.78$.

(iv) Borexino and KamLAND have a high energy resolution and can reach very low energy thresholds because of their very good light collection. They can easily recognize different interactions and, above all, they are sensitive to the $\overset{(-)}{\nu}_i + p \rightarrow \overset{(-)}{\nu}_i + p$ scattering whose number, mainly sensitive to the emission temperature of ν_μ and ν_τ, can be comparable with the number of IBD interactions [13].

(v) Baksan and LVD started their data taking in 1980 and 1992 respectively and they give the most stringent limits to the core collapse SN rate in the Galaxy, even if the capability of Baksan to observe the entire Galaxy, in the present configuration and in the absence of any other signal, is controversial. Modular detectors can be easily operated for a long time with duty cycles near 100%. Independent modules can be treated as independent detectors and the frequency of their coincidences, during a relatively large time window (10-20 sec) can be described by Poisson statistics.

Experiments in bold type in the table are, at present, involved in the SNEWS [14]. Each of them has elaborated its own strategy to disentangle a SN neutrino candidate from the background. To be admitted to the network each detector must keep its average alarm rate lower than a threshold value that is, at present, 1 alarm every 10 days. This relatively high allowed frequency maximizes the detector sensitivity to weak signals, nevertheless it is low enough to keep the accidental two-fold coincidence rate, in a time window of 10 s, lower than one every 100 years. This threshold rate will be modified when the number of detectors involved in the network will increase. SuperK, LVD and SNO have been involved since the very beginning and performed a high rate test in 2001 [1]. IceCube now supersedes AMANDA that had been participating to the network since 2004 and Borexino is in a test phase. LVD and Borexino, because they are host in the same underground laboratory, are not considered, by SNEWS, as completely independent. The neutrino signal, in a core collapse supernova event, precedes the electromagnetic one by hours therefore present detectors can behave as trigger for electromagnetic and gravitational wave detectors [15]. A few galactic supernova are expected per century and SNEWS can provide an alert within minutes. We are patient.

Acknowledgments
I am indebted to M.Nakahata, P.Decowski, A.Chavarria, A.Bouchta, A.Ianni and F.Vissani for precious information and useful discussions.
This work is dedicated to the memory of my teacher Gianni Navarra

References
[1] P. Antonioli *et al.*, New Journal of Physics **6** (2004) 114
[2] G. Pagliaroli, F. Vissani, M. L. Costantini, and A. Ianni, Astropart. Phys. **31**, 163 (2009)
[3] A.Bueno, I.Gil-Botella and A.Rubbia, arXiv:hep-ph/0307222
[4] C A Duba *et al.* 2008 J. Phys. Conf. Ser. **136** 042077
[5] C.J.Virtue 2001, Nucl. Phys. Proc. Suppl. 100, 326
[6] J.Ahrens *el al.* Astrop. Phys. **16** (2002) 345; T.Kowarik *et al.*, Proc. of 31st ICRC, Lodz (2009)
[7] M.Ikeda *et al.*, ApJ, **669** (2007) 519
[8] N.Yu.Agafonova *et al.*, Astrop. Phys., **28**, 6 (2008) 516; Proc. of 31st ICRC, Lodz (2009)
[9] P.Vogel for the KamLAND Collaboration Twenty years after SN1987A Hawaii, February 25, 2007
[10] L.Cadonati, F.P.Calaprice and M.C.Chen 2002 Astrop. Phys. **16**, 361
[11] E.N. Alekseev and L.N. Alekseeva, Journal of Experimental and Theoretical Physics, **95**, 1 (2002) 5-10 and R.V. Novoseltseva *et al.*, Proceedings of the 31st ICRC, Lodz (2009)
[12] N.Yu. Agafonova *et al.*, Astrop. Phys. **27** (2007) 254
[13] J. Beacom, W.Farr and P.Vogel, Phys. Rev. D **66**, 033001 (2002)
[14] K.Scholberg, Astron. Nachr. AN**329**, No. 3,337339 (2008)
[15] G.Pagliaroli, F.Vissani, E.Coccia and W.Fulgione, Phys. Rev. Lett. **103**, 031102 (2009)

Journal of Physics: Conference Series **203** (2010) 012078 doi:10.1088/1742-6596/203/1/012078

Search for neutrinos from core-collapse supernova from the global network of detectors

Alec Habig, for the SNEWS working group

University of Minnesota Duluth, Physics Department, 10 University Dr., Duluth, MN 55812, USA

E-mail: `ahabig@umn.edu`

Abstract. The Supernova Early Warning System (SNEWS) is a cooperative effort between the world's neutrino detection experiments to spread the news that a star in our galaxy has just experienced a core-collapse event and is about to become a Type II Supernova. This project exploits the ~hours time difference between neutrinos promptly escaping the nascent supernova and photons which originate when the shock wave breaks through the stellar photosphere, to give the world a chance to get ready to observe such an exciting event at the earliest possible time. A coincidence trigger between experiments is used to eliminate potential local false alarms, allowing a rapid, automated alert.

1. Introduction

A number of existing neutrino detectors are sensitive to the burst of neutrinos produced when a core collapse supernova ("SN") happens in our Galaxy. The Supernova Neutrino E ystem ("SNEWS") uses an online, real-time coincidence of these experiments' SN triggers to provide an early warning of such a supernova. This alert exploits the neutrino signal's ability to promptly escape the core of the collapsing star, giving observers a several hour lead time over the photospheric shock breakout that produces the first optical signs of the explosion. This early warning will maximize the chances of getting good electromagnetic observations of a previously unobserved time regime in the evolution of a supernova. It also could serve as a trigger for experiments such as gravitational wave detectors which might not trigger on a supernova themselves. Additionally, the SNEWS collaboration aims to optimize global sensitivity to supernova neutrinos via such activities as downtime coordination amongst the participating neutrino experiments.

2. Supernovae

SN1987A provided the first astrophysical neutrinos detected from outside our solar system, with two water Cherenkov proton decay detectors, Kamiokande II [4] and IMB [5] detecting 20 neutrinos of between 10 and 20 MeV between them. Two scintillator detectors also reported neutrino observations (Baksan [6] and LSD [7]). These observations confirmed the gravitational core-collapse model used to explain type II SN. As a large star runs out of fusible material in its core (nuclei heavier than Iron not providing an exothermic fusion reaction), the lack of an energy source to counter gravity's inexorable pull foces the core of that star to collapse past the Chandrasekhar limit to form a proto-neutron star. While the intial neutronization burst of inverse-beta-decay neutrinos is very luminous, it is also very short, with thermally produced neutrinos providing a bright, steady neutrino flux that radiates 99% of the $O(10^{53})$ ergs of gravitational binding energy away over tens of seconds. Average neutrino energies are

Table 1. Summary of neutrino detectors with supernova sensitivity. Neutrino event estimates are approximate.

Detector	Type	Mass (kton)	Location	Events at 8.5 kpc	Live period
Baksan	C_nH_{2n}	0.33	Caucasus	50	1980-present
Super-K	H_2O	32	Japan	8000	1996-present
SNO	D_2O	1 (D_2O)	Canada	400	1999-2006
		1.4 (H_2O)		450	
LVD	C_nH_{2n}	1	Italy	300	1992-present
KamLAND	C_nH_{2n}	1	Japan	300	2002-present
MiniBooNE	C_nH_{2n}	0.7	USA	200	2002-present
Borexino	C_nH_{2n}	0.3	Italy	100	2005-present
SNO+	C_nH_{2n}	0.8	Canada	300	Future
AMANDA	Long string	0.4/PMT	South Pole	N/A	1998-2009
IceCube	Long string	0.4/PMT	South Pole	N/A	2007-present
HALO	Pb	0.07	Canada	80	Future

expected to be about 12 MeV for ν_e, 15 MeV for $\bar{\nu}_e$, and 18 MeV for all other flavors. See [8–10] for many more details of the process and expected signal.

Modern neutrino experiments are orders of magnitude larger than they were two decades ago, with better resolution and reconstruction abilities (Tab. 1. Any given experiment is well over threshold for a solid SN neutrino detection on its own for a SN somewhere in our galaxy. Lack of stars for many MPc outside our galaxy render even the largest insensitive beyond our immediate neighborhood. Given that a star is opaque to electromagnetic radiation but transparent to neutrinos, the neutrino signal precedes the electromagnetic signal by the crossing time of the SN shock front traveling the radius of the star and breaking out of the stellar photosphere. However, in 1987, the neutrino experiments heard about the fortuitous optical discovery of SN1987A and went back days later to examine their data tapes to see what might have happened. Given today's neutrino capabilities, we wish to make use of the several hours leadtime a neutrino signal provides to get the electromagnetic world ready to make observations from as close to $t = 0$ as possible.

3. The Coincidence Network

The SNEWS project is an international collaboration of neutrino experimenters representing supernova neutrino detectors. Currently operational and contributing to the alert are Super-K, LVD, IceCube and Borexino. If any of the participating experiments finds a burst of events which its online systems flag as a possible SN neutrino signal, it sends a datagram to the central SNEWS computer located at Brookhaven National Laboratory (for robustness, there is a duplicate server at BNL, and two more at INFN Bologna). While a SN somewhere in our galaxy will produce a high neutrino flux everywhere on earth, more mundane things which might produce signal bursts in one detector (electronic noise, hardware leaks, spallation from very high energy cosmic rays, human error) are unlikely to happen at the same time in different locations. Thus, if the server sees a coincidence within 10 seconds, an automated supernova alert will be sent out. This eliminates the human-time-scale delay inherent in any one experiment's shifters and supernova experts verifying that what their DAQ just beeped about is really a SN. SNEWS has been operational since 1998, and has been running in a fully-automated mode since 2005 with near-100% up-time.

4. Conclusions

Acknowledgments

We would like to thank the members of the experimental collaborations, without whose hard work the SNEWS project would not be possible. We thank Brett Viren and Brookhaven National Laboratory, and Andrea Paolucci and INFN Bologna, for providing secure homes for the servers. SNEWS is funded by National Science Foundation awards PHY-0303196 and PHY-0302166.

References

[1] Antonioli P *et al.* 2004 *New J. Phys.* **6** 114 (*Preprint* astro-ph/0406214)
[2] Scholberg K 2008 *Astron. Nachr.* **329** 337–339 (*Preprint* 0803.0531)
[3] Pagliaroli G, Vissani F, Coccia E and Fulgione W 2009 *Phys. Rev. Lett.* **103** 031102 (*Preprint* 0903.1191)
[4] Hirata K *et al.* (KAMIOKANDE-II) 1987 *Phys. Rev. Lett.* **58** 1490–1493
[5] Bionta R M *et al.* 1987 *Phys. Rev. Lett.* **58** 1494
[6] Alekseev E N, Alekseeva L N, Volchenko V I and Krivosheina I V 1987 *JETP Lett.* **45** 589–592
[7] Aglietta M *et al.* 1987 *Europhys. Lett.* **3** 1315–1320
[8] Burrows A, Klein D and Gandhi R 1992 *Phys. Rev.* **D45** 3361–3385
[9] Fischer T, Whitehouse S C, Mezzacappa A, Thielemann F K and Liebendorfer M 2008 (*Preprint* 0809.5129)
[10] Budge K G, Fryer C L and Hungerford A L 2008 *J. Phys. Conf. Ser.* **136** 022040

Topics in Astroparticle and Underground Physics (TAUP 2009) IOP Publishing
Journal of Physics: Conference Series **203** (2010) 012079 doi:10.1088/1742-6596/203/1/012079

Future Supernova Neutrino Detectors

Kate Scholberg

Department of Physics, Duke University, Durham, NC, 27708

E-mail: schol@phy.duke.edu

Abstract.
This joint neutrino-gravitational wave session talk will describe current, near future and farther future supernova neutrino detectors. I will comment on the potential of future supernova neutrino-gravitational wave correlation searches.

1. Current and Future Supernova Neutrino Detectors

A core collapse supernova will produce a burst of neutrinos of all flavors with a few tens of MeV energy, over a period of a few tens of seconds. Reference [1] and references therein give an overview of supernova neutrino detection; reference [2] is a recent summary what may be learned from the observation of a supernova neutrino burst. So far the only supernova neutrino observation is from SN1987A [3, 4, 5, 6], and we expect enormously enhanced information from the next nearby observation.

Table 1 summarizes existing and planned detectors. Currently running detectors have sensitivity primarily to the $\bar{\nu}_e$ component of the signal, via inverse beta decay $\bar{\nu}_e + p \rightarrow n + e^+$, which is the dominant component in for water Cherenkov and scintillator. For supernova burst detection, one wants not only statistics but also diversity of flavor sensitivity: neutral current sensitivity, which gives access to the ν_μ and ν_τ components of the flux, and ν_e sensitivity are particularly valuable. Near future (within next five years or so) detectors with supernova neutrino sensitivity should enrich the sample: these are highlighted below.

- The SNO+ experiment [7] in Sudbury comprises 0.77 ktons of liquid scintillator filling the SNO acrylic vessel. It should have supernova sensitivity comparable to that of KamLAND.
- Another future detector now under construction is the HALO experiment [8]: this makes use of 76 tons of lead instrumented with the unused SNO NCD counters to record neutrons and electromagnetic signals. Although this detector is small, it will have unique sensitivity.
- The ICARUS detector at Gran Sasso is a 600 ton liquid argon time projection chamber. It will have excellent electron neutrino sensitivity [9] via $\nu_e +^{40}\mathrm{Ar} \rightarrow e^- +^{40}\mathrm{K}^*$, for which de-excitation gammas will be visible. In principle one may tag modes using gamma spectrum information.
- In spite of large cosmic ray background, new surface detectors like NOνA may observe a statistically significant spike of supernova neutrino events above background [10].

Other future ideas include exploitation of neutral current coherent elastic neutrino-nucleus scattering [11, 12] in dark matter or novel detectors.

Table 1. Summary of neutrino detectors with supernova sensitivity. Neutrino event estimates are approximate and have a fairly large uncertainty. See reference [1] for individual detector references. Not included are are smaller detectors (*e.g.* reactor neutrino scintillator experiments) and detectors primarily sensitive to coherent elastic neutrino nucleus scattering.

Detector	Type	Mass (kton)	Location	Events at 8.5 kpc	Live period
Baksan	C_nH_{2n}	0.33	Caucasus	50	1980-present
Super-K	H_2O	32	Japan	8000	1996-present
LVD	C_nH_{2n}	1	Italy	300	1992-present
KamLAND	C_nH_{2n}	1	Japan	300	2002-present
MiniBooNE	C_nH_{2n}	0.7	USA	200	2002-present
Borexino	C_nH_{2n}	0.3	Italy	100	2005-present
IceCube	Long string	0.4/PMT	South Pole	N/A	2007-present
SNO+	C_nH_{2n}	0.8	Canada	300	Near future
HALO	Pb	0.07	Canada	80	Near future
Icarus	Ar	0.6	Italy	230	Near future
NOνA	C_nH_{2n}	15	USA	3000	Near future
LBNE LAr	Liquid argon	5	USA	1900	Future
LBNE WC	H_2O	300	USA	78,000	Future
MEMPHYS	H_2O	440	Europe	120,000	Future
Hyper-K	H_2O	500	Japan	130,000	Future
LENA	C_nH_{2n}	50	Europe	15,000	Future
GLACIER	Ar	100	Europe	38,000	Future

Very promising for the future are a number of planned mega-detectors, which aim to deploy very large volumes of water, scintillator, or liquid argon: see the table and references [13, 14, 15, 16, 17]. Some such detectors can hope to collect individual neutrino events every few years from beyond the Local Group of galaxies (few Mpc) [18], assuming that background can be reduced sufficiently. In such a regime, some kind of external (non-neutrino) trigger will be essential to distinguish supernova neutrino-induced events from background.

2. What a Gravitational Wave Coincidence Could Add

Gravitational waves (GW) will be produced by core collapses [19], although there are currently significant uncertainties in the scale and nature of the signal. Requiring a coincidence between neutrino and GW signals has the potential to improve the sensitivity of both channels by allowing relaxation of the criteria for detection. For example, Super-K's recent "distant" burst search [20] requires two neutrino events (with energy threshold 17 MeV) within 20 seconds, which corresponds to approximately 8% probability of detecting a supernova in Andromeda. The accidental fake rate for this coincidence criterion is less than one per year; the single event rate at this threshold is about 1 per day. If one could achieve an acceptable accidental coincidence rate by requiring coincidence of a single neutrino event with a GW signal, then the probability of core collapse in Andromeda satisfying the search criterion is about 35%. See Figure 1. Distant burst search parameters could be re-optimized with respect to current ones; the neutrino event energy threshold could potentially be reduced, further improving sensitivity. References [21, 22] explore the possibilities of a neutrino trigger for a GW search.

We note that coordination of gravitational-wave and neutrino experiments will become even more important in the future as the sensitivity of experiments improve. The Advanced LIGO and Virgo detectors [23, 24] are expected to come online starting in 2014, and are expected to be more sensitive than the current generation of detectors by a factor of $\sim$10. Because gravitational

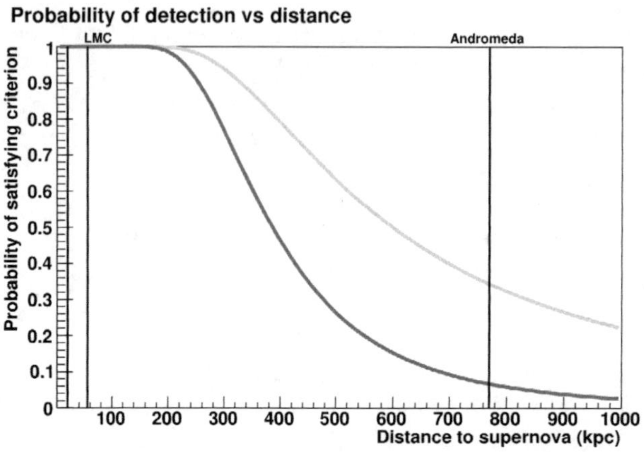

Figure 1. Estimated probability to satisfy Super-K burst search criterion as a function of distance. Red: standard search parameters [20]. Green: probability if only a single neutrino event is required, in coincidence with a GW signal.

wave signals are prompt (in contrast with optical detection, which will have a time window of hours), coincident detection with gravitational waves is very promising. Note also that some supernovae may not produce substantial (or any) optical fireworks, and gravitational waves may be the only way to tag "optically silent" supernovae.

In summary, correlation analysis of gravitational wave and neutrino experiments in a search for core collapse events presents an interesting prospect for the future.

Acknowledgments

The author wishes to thank the GWNU working group.

[1] Scholberg K 2007 (*Preprint* astro-ph/0701081)
[2] Dighe A 2008 *J. Phys. Conf. Ser.* **136** 022041 (*Preprint* 0809.2977)
[3] Bionta R M *et al.* 1987 *Phys. Rev. Lett.* **58** 1494
[4] Hirata K *et al.* (KAMIOKANDE-II) 1987 *Phys. Rev. Lett.* **58** 1490–1493
[5] Alekseev E N, Alekseeva L N, Volchenko V I and Krivosheina I V 1987 *JETP Lett.* **45** 589–592
[6] Aglietta M *et al.* 1987 *Europhys. Lett.* **3** 1315–1320
[7] Chen M C (SNO+) 2008 (*Preprint* 0810.3694)
[8] Virtue C 2009 http://snolab2009.snolab.ca/snolab-workshop-09/HALO20aug09.ppt
[9] Bueno A, Gil Botella I and Rubbia A 2003 (*Preprint* hep-ph/0307222)
[10] Arms K 2009 Proceedings of CIPANP 2009
[11] Horowitz C J, Coakley K J and McKinsey D N 2003 *Phys. Rev.* **D68** 023005 (*Preprint* astro-ph/0302071)
[12] Giomataris Y and Vergados J D 2006 *Phys. Lett.* **B634** 23–29 (*Preprint* hep-ex/0503029)
[13] Nakamura K 2003 *Int. J. Mod. Phys.* **A18** 4053–4063
[14] Kistler M D, Yuksel H, Ando S, Beacom J F and Suzuki Y 2008 ArXiv:0810.1959 (*Preprint* 0810.1959)
[15] Marrodan Undagoitia T *et al.* 2008 *J. Phys. Conf. Ser.* **120** 052018
[16] de Bellefon A *et al.* 2006 ArXiv:hep-ex/0607026 (*Preprint* hep-ex/0607026)
[17] Autiero D *et al.* 2007 *JCAP* **0711** 011 (*Preprint* 0705.0116)
[18] Ando S, Beacom J F and Yuksel H 2005 *Phys. Rev. Lett.* **95** 171101 (*Preprint* astro-ph/0503321)
[19] Ott C D 2009 *Class. Quant. Grav.* **26** 063001 (*Preprint* 0809.0695)
[20] Ikeda M *et al.* (Super-Kamiokande) 2007 *Astrophys. J.* **669** 519–524 (*Preprint* 0706.2283)
[21] Pagliaroli G, Vissani F, Coccia E and Fulgione W 2009 *Phys. Rev. Lett.* **103** 031102 (*Preprint* 0903.1191)
[22] Cadonati L *et al.* 2009 Proceedings of 8th Edoardo Amaldi Conference on Gravitational Waves
[23] Fritschel P 2003 (*SPIE Proc.* vol 4856) p 282
[24] http://www.ligo.caltech.edu/docs/M/M060056-10.pdf

Journal of Physics: Conference Series **203** (2010) 012080 doi:10.1088/1742-6596/203/1/012080

Using neutrinos to search for gravity waves

G. Pagliaroli

INFN, Laboratori Nazionali del Gran Sasso, Assergi (AQ), Italy

E-mail: giulia.pagliaroli@lngs.infn.it

Abstract. We outline the conceptual steps that permit us to use the $\bar{\nu}_e$ signal from the explosion of a galactic core-collapse supernova as an helpful instrument to increase the confidence for the detection of gravitational signal.

Gravitational Waves (GW) predicted by the General Relativity have not been observed directly yet, despite the experimental effort. The main reason is that their small amplitude hides under the background. The general methodology, used by the gravitational community to increase the detection probability, is the search of temporal coincidences between different GW detectors. Obviously the noise can produce accidental coincidences and the number of such accidental coincidences decreases with the size of the search time window. The use of an external trigger, that constrains the temporal window, greatly helps the search for GW signals.

Core Collapse Supernovae (SN) are expected to emit impulsive GW and, in spite of the difficulty to describe the several mechanisms that can contribute to such a GW emission, it is possible to define a generic GW waveform expected during the collapse of the massive star. In this signal the time of the maximum amplitude is strongly correlated to the time of the "bounce", namely the instant when the outer region of the collapsing iron core bounces on the inner compact region. The duration of this signal is about 10 ms. So that, to really help the search of such a GW we need to identify the time of the bounce with an error of the same order.

As discussed in [1], the use of the electronic antineutrinos emitted from the core collapse SN permits us to reach this goal. In fact, considering a gravitational detector and a neutrino detector with clocks synchronized in universal time (U.T.) this relation holds:

$$T_{\text{bounce}} = T_{\text{1st}} - (t_{\text{GW}} + t_{\text{mass}} \pm t_{\text{fly}} + t_{\text{resp}}). \tag{1}$$

Here times in uppercase are absolute times, in U.T., whereas times in lowercase are relative intervals of time. T_{bounce} and T_{1st} are the absolute times of the bounce expected in the gravitational detector and of the first neutrino event detected by the neutrino detector, respectively. The errors δT_{bounce} and δT_{1st} of the detector clocks are lower than $1\mu s$, so they can be neglected for our purposes.

The term t_{GW} is the mean interval time between the starting point of antineutrino luminosity and the bounce of the outer core on the inner core. It is reliably known and ranges within few milliseconds, $\delta t_{\text{GW}} \sim 1.5$ ms.

t_{mass} is the delay of neutrino signal to respect GW one due to the presence of the neutrino mass. However, using cosmological bound on the sum of the neutrino masses this term is negligible.

The time interval t_{fly} is the time of fly between the two detectors separated by the distance $\vec{d}$ is $\pm t_{\text{fly}} = \vec{d} \cdot \hat{n}$, where $\hat{n}$ is the direction pointing to SN. The error on this term is negligible for an astronomically identified SN. The same when we consider the distance between Large Volume Detector (LVD) and Virgo, $d < 1$ ms. But the distances between Super-Kamiokande (SK) or IceCUBE and the GW detectors LIGO or Virgo are such that could imply an error as large as ~ 60 ms. We demonstrated[1] that to reach $\delta t_{\rm fly} \leq 5$ ms, we need to determine the angle of the SN direction with a precision of 20°. This is possible using the elastic scattering (ES) events of SK. Thus, even in absence of an astronomical identification, it should be possible to determine the direction of the SN precisely enough to reduce the error $\delta t_{\rm fly}$ to the desired level.

Finally the non-negative parameter $t_{\rm resp}$ is a "response time", namely the delay related to the detection processes. It is zero for an ideal neutrinos detector but, unfortunately, most neutrinos will be missed. Thus, there is a delay between the time of first neutrino arrived and the first event detected, that we denote by $t_{\rm resp}$. Its value and its uncertainty have to be extracted from the antineutrinos data and both depend on the behavior of the neutrino luminosity.

We suppose that the expected flux of $\bar{\nu}_e$ from a standard core collapse SN explosion can be described by a parameterized model with several free parameters, that take into account the astrophysical uncertainties on the emission processes. We use a model previously developed for SN1987A data analysis [3].

The total $\bar{\nu}_e$ flux includes two phases for neutrino emission. The accretion phase and the corresponding flux Φ_a is generated by the interactions between the neutrons and the positrons above the shock and is described by 3 parameters: the initial accreting mass (M_a), the time scale of the phase (τ_a), and the initial temperature of the e^+ (T_a). The cooling phase, with flux Φ_c, comes from the thermal emission of the new born proto-neutron star and it is proportional to the radius of the neutrino sphere (R_c), with a time scale (τ_c), and the initial temperature of the emitted antineutrinos (T_c). In addition we insert a function describing the initial rise of the signal. This is characterized by a new crucial parameter, the time scale (τ_r), that depends strongly on the velocity of the shock wave.

The dominant detection process for $\bar{\nu}_e$ is the Inverse Beta Decay (IBD). Therefore the expected events rate is $R(t, E_\nu, D) = N_p \sigma_{\bar{\nu}_e p}(E_\nu) \Phi_{\bar{\nu}_e}(t, E_\nu, D) \epsilon(E_{e^+})$, where D is the SN distance, N_p is the number of target protons within the detector, $\sigma_{\bar{\nu}_e p}$ is the process cross section, ϵ is the detector efficiency function and $\Phi_{\bar{\nu}_e}(t, E_\nu, D)$ is the electronic antineutrino flux. The total number of detected SN events is the integral of the rate R over time and energy.

Summarizing, the inputs of our model are the theoretical expected behavior of the electronic antineutrinos flux and the known neutrinos detector features. The outputs, obtained analyzing the data with the maximum likelihood method[2], are the best fit values of the 7 astrophysical parameters that regulate neutrino emission and, finally, the response time, $t_{\rm resp}$, that is the parameter of interest for GW search.

In order to test the overall procedure and its potential, we construct a Monte Carlo simulation for a future SN event adopting the values of the parameters found from SN1987A data analysis [3], that agree with the theoretical expectations within the errors. For the rise-time scale we choose the value $\tau_r = 100$ms. So that, we expect that among the $\sim 2.6 \times 10^{57}$ $\bar{\nu}_e (\sim 5 \times 10^{52}$ erg) leave the SN in a time window of 30 seconds and about the 50% (46% of the energy) are emitted during the first second. For a supernova distance of 20 kpc about 5.5×10^{10}cm^{-2} of these antineutrinos reach the Earth and the number of expected IBD events in a detector with the same mass of SK (*i.e.*, 22.5 kton of water) and efficiency $\epsilon = 0.98$, is on average 1060.

We extract 10 sets of data from the rate function $R(t, E_\nu, 20)$. Each event is characterized by the relative detection time t_i (namely the time elapsed from the first detected event) and by the positron energy E_i; the error on this energy is given by the SK collaboration. For each data set the analysis provides the best-fit values of the unknown parameters, including the one in which we are interested, $t_{\rm resp}$.

The results for the astrophysical parameters show a very good agreement with the true values, proving that the procedure of analysis properly works. This comparison demonstrates that the current uncertainty in astrophysics will be overruled by the data. In other words, if our understanding of neutrino emission (behavior of the $\bar{\nu}_e$ luminosity) is qualitatively correct, we can fix its quantitative aspects (parameters) by the data of a future supernova.

In particular, we found [1] that it is possible to estimate the true value of the response time with an average uncertainty time window of $\langle 2\delta t^{\mathsf{Fit}}_{\mathsf{resp}} \rangle = 10.5$ ms. So, summing in quadrature the errors of the terms in Eq. 1, the U.T. of the bounce can be identified within a temporal window of about 15 ms for a SN at 20 kpc.

For a closer SN event the number of expected events will increase and the statistical error on the response time (that scales approximatively with the inverse square root of the events number) will decrease to few milliseconds. A simplified analysis along these lines has considered the case of a SN distance of 10 kpc and the IceCUBE detector[4].

The knowledge of the GW arrival time in this small time window permits to increase signal-to-noise ratio in the gravitational detectors by decreasing the background. In order to discuss the improvement in sensitivity, let us assume the following idealized situation. We consider a search time window of 30 ms (this is not optimized but it is presumed to contain the 10 ms GW burst) and we suppose that the background can be treated as a gaussian-distributed, random variable (this is certainly an oversimplification). Now we compare the search of the expected burst in a specific triggered search window and in a interval of time when a galactic supernova could presumably occur: say, 10 years of data taking. It is easy to see that a 3 σ fluctuation in the first case has the same chance to occur as a 7.3 σ fluctuation in the second one, for 10 years correspond to about 10 billion search windows. We conclude that the triggered search is sensitive to 2.4 weaker signals.

Let us stress that a search of a signal in a period of 10 years is possible only if we know the extreme tail of the background distribution, which would be possible only after a much longer time of data taking. This knowledge is not needed, instead, for the triggered search: when we know a priori the time window where we should to search, even a 3 σ fluctuation could suffice for a successful detection.

References

[1] G. Pagliaroli, F. Vissani, E. Coccia and W. Fulgione, Phys. Rev. Lett. **103** (2009) 031102 [arXiv:0903.1191 [hep-ph]].
[2] A. Ianni, G. Pagliaroli, A. Strumia, F. R. Torres, F. L. Villante and F. Vissani, Phys. Rev. D **80** (2009) 043007 [arXiv:0907.1891 [hep-ph]].
[3] G. Pagliaroli *et al.*, Astropart. Phys. **31** (2009) 163.
[4] F. Halzen and G. G. Raffelt, arXiv:0908.2317 [astro-ph.HE], accepted to PRD.

Journal of Physics: Conference Series **203** (2010) 012081 doi:10.1088/1742-6596/203/1/012081

Results from the Solar Neutrino Experiment BOREXINO

Lothar Oberauer, on behalf of the BOREXINO collaboration
Physikdepartment E15, TU München, James-Franck-Str 1, 85748 Garching, Germany
E-mail: oberauer@ph.tum.de

Abstract.
We report about recent results from the solar neutrino experiment BOREXINO. With BOREXINO solar neutrino spectroscopy in the sub-MeV range became possible. In particular we demonstrate the measured counting rate of solar ^{7}Be neutrinos. Additionally we show data on solar ^{8}B neutrinos with an energy threshold of 2.8 MeV and discuss the implications for solar and neutrino physics. Finally further prospects of BOREXINO are discussed.

1. Experimental set-up and background suppression

Solar neutrinos are detected in BOREXINO via elastic scattering off electrons: $\nu+e \rightarrow \nu+e$. As a target around 300t of a liquid scintillator (solvent pseudocumene with PPO as wavelength shifter) is used. The scintillator is contained in a spherical nylon vessel. Outside a non-scintillating buffer liquid acts as passive shielding. The scintillation light is registered by more than 2200 photomultipliers (PMs) mounted on the inner surface of a stainless steel sphere. Most of the PMs are equipped with light concentrators. The steel sphere is placed inside a steel dome with a height of about 18m and a diameter with 18m, too. The volume between stell dome and sphere is filled with very pure water. Additional 205 PMs on the outside surface of the sphere and at the floor of the dome are mounted. Hence, the water volume acts as shielding against external gamma and neutron radiation and as an active muon veto. The detector is located in hall C of the deep underground laboratory at Gran Sasso, Italy.

Due to the high light yield of the detector system BOREXINO achieves a hardware energy threshold of well below 100 keV and an energy resolution of about 5% at 1 MeV energy deposition. Recoil electrons of solar ^{7}Be neutrinos deposit a maximal energy of 660 keV. Hence sub-MeV solar neutrino spectroscopy is possible with BOREXINO. However, special care has to be taken concerning background reactions. The muon veto of BOREXINO has a leak rate below 0.5%. Pulse shape techniques applied on the signal of the inner detector, when the muon traverses the steel sphere, allows to suppress this parameter below 0.01%. Hence, muon background in the energy window ($\approx$ 250 keV to $\approx$ 800 keV) of solar ^{7}Be neutrinos is negligible. However, muon induced radioactive nuclei can be a serious background for solar neutrino detection at higher energies.

Very high levels of the purity of the liquid scintillator have to be achieved. In BOREXINO several purification processes before and during the filling procedure have been applied. Degassing with ultra pure N_2, water extraction for the PPO solution as well as distillation of the solvent was used. The contamination level of radon in the scintillator was measured

Journal of Physics: Conference Series **203** (2010) 012081 doi:10.1088/1742-6596/203/1/012081

online during the whole filling period. So the development of air leaks in the filling system could be observed and cured immediately. Hence, dangerous contamination due to radioactive gaseous ^{85}Kr atoms could be avoided.

The measured delayed coincidence signals in the corresponding Bi-Po decays allow an online monitoring of the background due to decays in the Uranium and Thorium chains. Assuming radioactive equilibrium we obtain mass concentration levels of $1.6 \cdot 10^{-17}$ for Uranium and $6.9 \cdot 10^{-18}$ for Thorium, respectively. At the low energy side we observe the beta decays of ^{14}C at a level of $2 \cdot 10^{-18}$. This defines the lower analysis threshold for solar ^{7}Be neutrinos to be around 0.2 MeV. We do observe a peak in the ^{7}Be energy window due to alpha decays of ^{210}Po. However, pulse shape discrimination and/or the good energy resolution allows to separate this background from the solar ^{7}Be signal. Fortunately the ^{210}Bi beta background is clearly not in equilibrium with ^{210}Po, but much smaller. A detailed description of the experimental set-up of BOREXINO can be found in[1].

2. Data reduction and results on solar ^{7}Be neutrinos

Here we report on the results obtained in 192 live days of data taking. In order to reject muon events the signal must be in anti-coincidence to the veto. The position of each single event is reconstructed. In this way we can apply a fiducial volume (ca. 100t) cut for rejecting external gamma background. The delayed coincidence method described above is used to reject background from radon daughters. Finally pulse shape discrimination (PSD) may be applied to reject alpha events from ^{210}Po. Finally a common fit to the remaining spectrum is performed. The characteristic signature to solar ^{7}Be neutrinos is the Compton equivalent shoulder at 660 keV. Two analysis to the data, one with and one without PSD were done. The results of both techniques agree very well as it is shown in fig. 1 and 2. The final result on the counting rate is $R_{Be} = (49 \pm 3_{stat} \pm 4_{syst})\, d^{-1}$ in 100t fiducial mass[2].

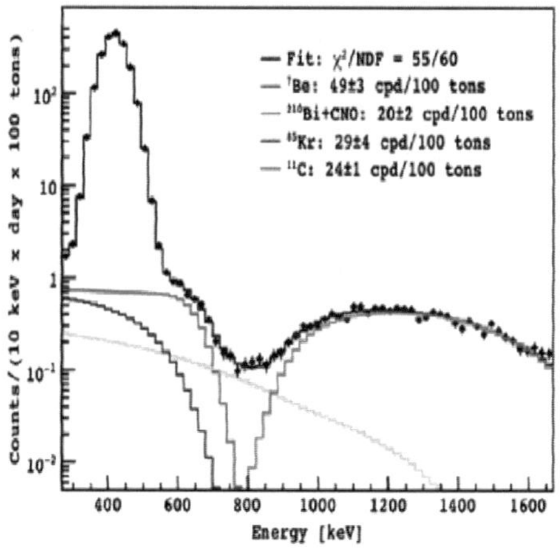

Figure 1. Energy spectrum including solar ^{7}Be neutrino recoil electrons after cuts without PSD

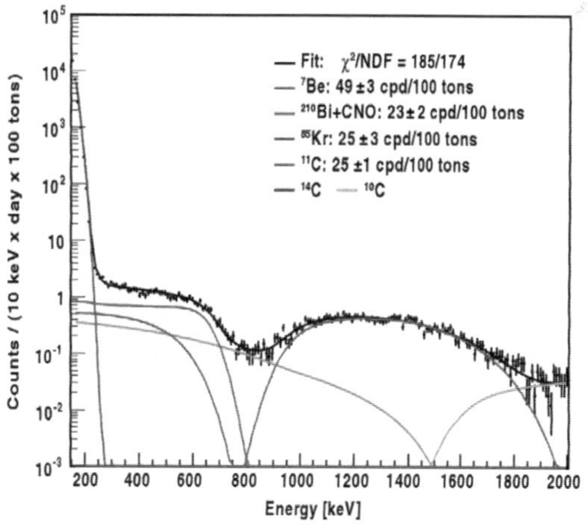

Figure 2. Energy spectrum including solar ^{7}Be neutrino recoil electrons after cuts with PSD

Without neutrino masses and neutrino mixing the expected rate would be $(74 \pm 4)\, d^{-1}$ in the same volume. Our result clearly excludes this scenario. However, taking the actual neutrino mass splitting $\Delta m^2_{12} \approx 8 \cdot 10^{-5}\, eV^2$ and mixing parameter $\theta_{12} \approx 32^0$ including solar matter effects our result is fully compatible with the predictions. For a low inner solar metallicity we expect $44 \pm 4\,d^{-1}$ and for high metallicity a value of $48 \pm 4\,d^{-1}$ in 100t for BOREXINO is expected.

A search for a day/night effect was performed. Our signal is stable in time and no significant difference in the counting rate during day and night was observed. We can report a preliminary limit on the parameter $a = (n-d)/(n+d) = 0.02 \pm 0.04_{stat}$, where n and d denote the averaged rates during night and day, respectively.

With our result new constraints on astrophysical parameters can be obtained. Taking the solar luminosity, the data of all solar neutrino experiments and the actual uncertainties on the neutrino mixing parameters we achieve a ratio f between measured and expected flux on pp-neutrinos of $f = 1.004^{+0.008}_{-0.020}$. Another analysis limits the contribution of the CNO-cycle to the solar energy generation to an amount of 5.4% (90% cl).

The systematic uncertainty is still dominated by the detector response function and the fiducial volume determination. However, due to a calibration campaign in 2009 with radioactive sources at dedicated locations inside the scintillator sphere these uncertainties will be much reduced and we aim to reach a ca. 3% precision on our solar ^{7}Be neutrino measurement.

3. Data reduction and results on solar ^{8}B neutrinos

Again the muon veto and the fiducial volume cut and are applied to the raw data. Radon daughter signals are rejected also. However the long lived background contributions due to radioactive nuclei of ^{10}C and ^{11}Be, created in spallation processes of cosmic muons on Carbon in the scintillator are very important. A 5 second long veto after each muon event in the inner detector rejects short lived cosmogenic background very efficiently. The production of the cosmogenic nuclei often is connected to the production of a neutron which can be detected with high efficiency after it has slowed down in the liquid scintillator. In order to reject the long lived nuclei a three fold coincidence using the muon signal, the neutron event and the cosmogenic generated nucleus is applied. The data from one year were used for this analysis which translates into a a live time period of 246 days. With an energy threshold of 2.8 MeV we measure a counting rate of 0.33 per day in 100t fiducial mass[3]. A spectral fit to the remaining data after cuts is performed.

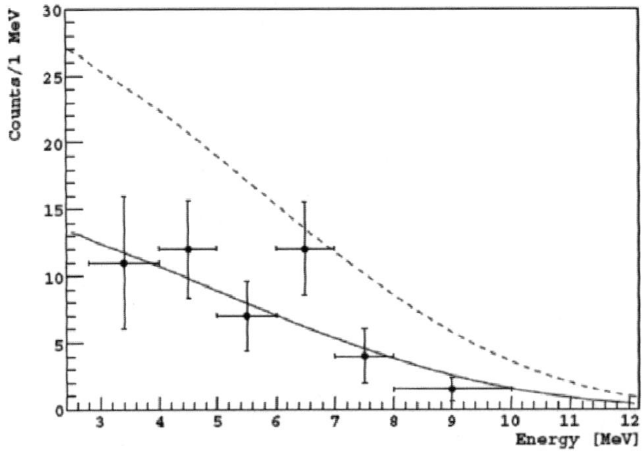

Figure 3. Spectral fit to solar ^{8}B neutrino candidates after cuts. The dashed line shows the expected spectrum in case of no neutrino masses and mixing, the solid line the expected MSW-curve.

In fig. 3 this spectrum of neutrino candidates after all cuts in comparison with the expected curve for the case of no neutrino mixing (dashed line) and in case of the MSW large angle solution is shown. Again our data are in very good agreement with the standard scenario of neutrino masses and mixing. Note, that for the first time data of solar ^{8}B neutrinos below 5 MeV are available.

Combining all available data on solar neutrino experiments it is possible to show the survival probability for solar electron neutrinos to stay in this state as function of the neutrino energy. The precision of the data points in the low energy region have been improved significantly after BOREXINO. In fig. 4[4] the good agreement between experimental values and the MSW prediction is demonstrated.

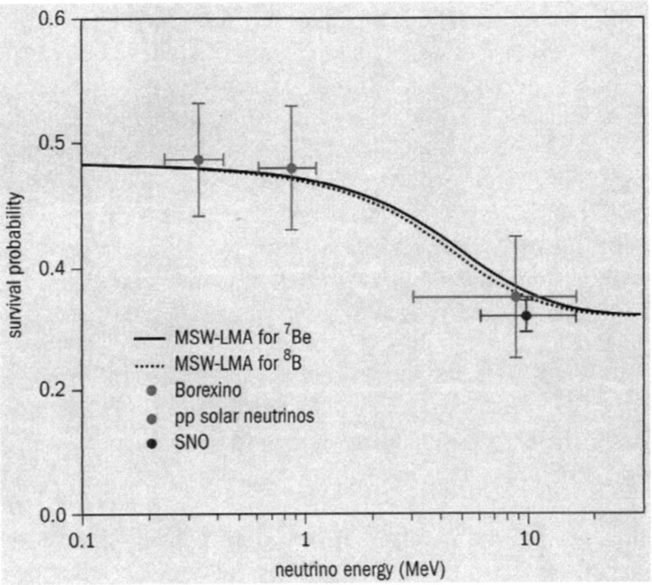

Figure 4. Survival probability of solar ν_e as function of its energy. Data points from solar neutrino experiments are compared with the MSW expected curve.

4. Neutrino magnetic moment and rare event search

With the 192 days sample of data a search for a non zero neutrino magnetic moment was performed. In case of a non vanishing neutrino magnetic moment, the electroweak cross section for neutrino electron scattering is modified by the addition of an electromagnetic term proportional to 1/T, where T is electron recoil kinetic energy. The best limit of $\mu_\nu < 1.1 \cdot 10^{-10}\, \mu_B$ obtained so far using solar neutrino data comes from the SuperKamiokaNDE detector above 5 MeV. As BOREXINO is achieving a significant lower energy threshold we obtain better constraints by investigating the recoil shape of our ^{7}Be signal. An upper limit of $\mu_\nu < 5.4 \cdot 10^{-11}\, \mu_B$ at 90% cl is reached[2]. This result is independent from parameters which usually contribute dominantly to the systematic uncertainty, like fiducial volume or the solar neutrino flux. The currently best limit on the neutrino magnetic moment $\mu_\nu < 3.2 \cdot 10^{-11}\, \mu_B$ was reported recently from an experiment at a nuclear power reactor[5].

Shortly we want to mention the possibility of rare event search in BOREXINO. It is based on the very high level of radiopurity in the liquid scintillator. For example the Pauli exclusion principle is going to be tested with unprecedented precision by searching for the emission of gamma-, neutron-, proton-, and electron/positron radiation emitted in a Pauli blocked transition from the $1P_{3/2}$ to the filled $1S_{1/2}$ state.

5. Prospects

Is it possible to measure solar neutrinos from the pep-reaction and from the sub-dominant CNO cycle? The pep-neutrino flux is predicted in solar models very accurately, as it is strongly connected to the luminosity of the Sun. Of course this is based on the assumption, that its luminosity hasn't change since the last couple of 10^5 years. But if this is the case a pep-neutrino measurement would be very interesting, as these monoenergetic neutrinos with $E_\nu = 1.44 MeV$ are inside the transition region between vacuum and matter dominated flavor transitions (compare with fig. 4). Deviations of the measured survival probability from the MSW expectation could reveal non-standard features like flavor changing neutral current interactions. A measurement of the CNO-neutrino flux at Earth would reveal the solar metallicity in the solar center, which is of astrophysical relevance. Recent data from helioseismology show a discrepancy to existing solar models and deliver hints, that the metallicity in the Sun could be higher as assumed. Common for both solar branches is the difficulty to measure them. Besides ultra high levels in radiopurity of the scintillator (here especially for ^{210}Bi and ^{40}K) the rejection of cosmogenic produced ^{11}C background is necessary. Work is in progress to separate these muon induced events by means of a muon tracking routine and search for coincidences with neutron signals. Probably the CNO-neutrinos are even more difficult to detect as the sum of their recoil spectra is not as distinct as that from the monoenergetic pep-neutrinos, where a distinct Compton like edge should be observed.

Low energy electron anti-neutrinos can be detected in BOREXINO by the inverse beta decay on free protons $\bar{\nu}_e + p \to e^+ + n$. The strong correlation in space and time of the prompt e^+- and the delayed n-signal delivers a tool to separate these events from background with very high efficiency. Sources for $\bar{\nu}_e$ are European nuclear power reactors and neutrinos from the Earth, both emitted in beta decays. As the energy spectra from the two sources are different it is possible to separate them. Only few events per year in the detector are expected. A measurement of terrestrial neutrinos could be used to probe geophysical models, the measurement of reactor neutrinos could be used as a test of the KamLAND result on neutrino oscillation, however at lower statistics. Background can origin from external fast neutrons or from beta-neutron cascade decays, like from ^{9}Li. The former should be identified by our muon veto, the latter can be discriminated by a veto of a few seconds duration after a muon traversed the inner detector.

References

[1] G. Alimonti et al., Borexino coll. *NIM A* **600** 568
[2] C. Arpesella et al., Borexino coll. *Phys. Rev. Lett.* **101** 091302
[3] G. Bellini et al., Borexino coll. *arXiv:0808.2868v1 [astro-ph]*
[4] G. Bellini and A. Ianni *CERN Courier* **Vol 49** Nr 7 (2009)
[5] A.G. Beda et al., *arXiv:0906.1926v1*

Low Energy Neutrino Physics at Super-Kamiokande

Michael B Smy for the Super-Kamiokande Collaboration
University of California, Irvine, 3117 Frederick Reines Hall, Irvine, California 92697-4575
E-mail: smy@michael.ps.uci.edu

Abstract. In spite of intensive studies of solar 8B neutrinos – which resulted in the first evidence that the flavour of solar neutrinos oscillates – the characteristic, expected transition between vacuum oscillations below and matter-dominated oscillations above 3 MeV remains thus far elusive. Also, the only direct demonstration of matter effects on solar neutrino oscillations, the day/night asymmetry, has not been observed yet. After a review of past measurements, Super-Kamiokande's future solar and supernova neutrino physics sensitivity is discussed, in particular in light of the possibility of Gadolinium doping.

1. Introduction

Super-Kamiokande, a 50kton cylindrical water Cherenkov detector located about 3 km w.e. underground, observes solar neutrinos via elastic scattering off electrons. About 11,000 20" photomultiplier tubes view the innermost 32 kton and yield about six detected photo-electrons per MeV of electron energy. As a consequence only the high energy ^{8}B and *hep* neutrinos can be observed. However, the direction and energy of the recoiling electrons are reconstructed, so neutrino interactions are recorded in real-time and point back to the sun. Also, the large fiducial mass of the innermost 22.5kton of the detector results in a large solar neutrino interaction rate of about 15 events/day above 5 MeV after event selection. Consequently, the first phase of the experiment, Super-Kamiokande-I collected about 22,000 solar neutrino events in about five years – by far the largest sample of solar neutrinos in the world. Since all solar neutrino flavors undergo elastic scattering with electrons – although the cross section for the electron flavor is enhanced – a comparison of the elastic scattering rate at Super-Kamiokande with the charged-current interaction rate of purely electron-flavored solar neutrinos with deuterons at the Sudbury neutrino observatory [1] provided the first evidence of solar neutrino flavor transformation.

2. Super-Kamiokande-I Results

Super-Kamiokande-I measured the ^{8}B neutrino flux to be $\Phi = 2.35\pm0.02(\text{stat})\pm0.08(\text{syst})\times10^6/\text{cm}^2\text{sec}$ [2]. The (total) energy threshold for the recoil electrons was 5 MeV. There appears to be no significant distortion in the recoil electron spectrum (see Figure 1). The solar neutrino interaction rate was searched for various time variations; the only significant time variation found is caused by the 1.7% orbital eccentricity of the Earth producing a 7% variation simply from the inverse square radius law. From the solar neutrino data, the eccentricity was measured to be $2.1 \pm 0.3\%(\text{syst})$, the perihelion shift seen in the neutrino data with respect to the true perihelion is 13 ± 18 days. Among the time variations the so-called day/night asymmetry $A_{DN} = \frac{\Phi_D-\Phi_N}{0.5(\Phi_D+\Phi_N)}$ was most carefully studied; a non-zero value would be a sign of Earth matter effects on solar neutrino

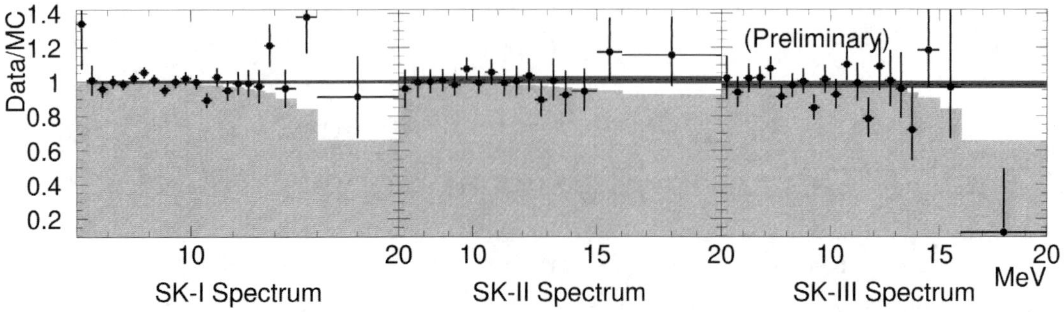

Figure 1. Recoil Electron Spectral Distortion for Super-Kamiokande-I (Left), Super-Kamiokande-II (Middle), and Super-Kamiokande-III (Right). The ratio of the measured and the expected solar neutrino elastic scattering rate from a ^{8}B neutrino flux of $2.33 \times 10^6/\text{cm}^2$sec and a *hep* neutrino flux of $15 \times 10^3/\text{cm}^2$sec is displayed. The shaded area is the contribution from just the ^{8}B neutrino flux. The dark-colored band represents the value and statistical uncertainty of the combined rate 5-20 MeV (7-20 MeV).

oscillation (usually a regeneration of electron-flavor neutrinos). The straight day/night asymmetry was measured to be $A_{DN} = -0.021 \pm 0.020(\text{stat}) \pm 0.012(\text{syst})$. Also, a fit to the amplitude of the expected day/night variation using the solar neutrino oscillation parameters in the Large Mixing Angle region yields $-0.017 \pm 0.016(\text{stat}) \pm 0.012(\text{syst}) \pm 0.0004(\text{osc})$ when expressed as a day/night asymmetry.

3. Super-Kamiokande-II Results

The second phase, Super-Kamiokande-II, confirmed the results of Super-Kamiokande-I. Solar neutrinos were detected for recoil electrons above 7 MeV of (total) energy and the ^{8}B neutrino flux of $\Phi = 2.38 \pm 0.05(\text{stat}) \pm 0.015(\text{syst}) \times 10^6/\text{cm}^2$sec [3] agrees with Super-Kamiokande-I much better than expected even from statistical uncertainties alone (see Figure 1). This is particularly remarkable since Super-Kamiokande-II has only 46.5% of the photomultiplier tubes of Super-Kamiokande-I, those photo detectors are enclosed in blast shields which remove a few percent of the light and add radioactive background. Furthermore, the entire analysis had to be rebuild including event reconstruction, event selection and detector calibration. Like Super-Kamiokande-I, the second phase found no distortions (see Figure 1) in the recoil electron spectrum. The day/night asymmetry was measured to be $A_{DN} = -0.063 \pm 0.042(\text{stat}) \pm 0.037(\text{syst})$ – also in good agreement with Super-Kamiokande-I. There is no significant time variation of the elastic scattering rate of any kind, except the seasonal variation due to the inverse square radius law. Super-Kamiokande-I and II together span an entire 11 year solar cycle, there's no significant dependence of the ^{8}B solar neutrino flux on this cycle (see Figure 2).

4. Super-Kamiokande-III

The third phase of Super-Kamiokande restored the full number of photomultiplier tubes to the experiment. The preliminary Super-Kamiokande-III flux ($\Phi = 2.31 \pm 0.05(\text{stat}) \times 10^6/\text{cm}^2$) and spectrum shown in Figure 1 agrees well with the previous phase within statisical uncertainty only. In spite of the radioactivity introduced by the blast shields, the background level near 5 MeV is significantly lower than Super-Kamiokande-I in the center of the detector due to an improved

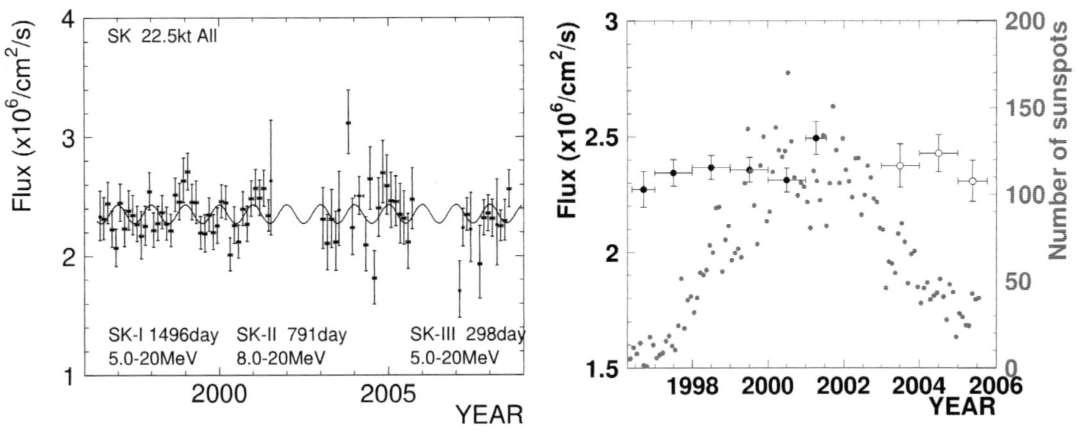

Figure 2. Solar Neutrino Flux Time Dependence (Left) Compared To Sunspot Number (Right).

water circulation pattern which suppresses the transport of Radon to the center. Consequently, the statistical uncertainty of the event rate between 5 and 5.5 MeV of Super-Kamiokande-III is comparable to Super-Kamiokande-I in spite of a factor of about six less exposure! While the trigger efficiency at 5 MeV is 100% (indeed at the very end of Super-Kamiokande-III it was 100% at 4.5 MeV) the hardware trigger threshold does not allow a significant lower analysis threshold than SK-I.

5. Super-Kamiokande-IV Trigger

Last September, the Super Kamiokande electronics and data acquisition was modernized and redone. To fully take advantage of the lower radioactive background of the center of Super-Kamiokande-III with respect to Super-Kamiokande-I there is no event trigger anymore. Instead, data is acquired continuously and events are extracted by a software trigger system from this data stream. This system allows for a secondary data stream apart from the simple coincidence trigger which was previously constructed by hardware and is now emulated by software. The new Wide-band Intelligent Trigger system (WIT) is a computer farm which simultaneously selects and reconstructs very low energy events by applying the timing coincidence after subtraction of the time of flight from the vertex to the PMTs. It consists currently out of four 3GHz quad-core CPUs fed directly via 10GBit/sec fast ethernet lines by the DAQ computers

The WIT system was tested with a Ni-Cf source inserted at the positions (0.4,-0.7,-17.1)m,(0.4,-0.7,-16.1)m, (0.4,-0.7,-15.1)m, (0.4,-0.7,-12)m, (0.4,-0.7,0)m, and (0.4,-0.7,+15.1m). These positions were selected to probe near the edge of the fiducial volume ($\pm$16.1m). The trigger efficiency at 3 MeV is around 85%, the trigger efficency at 4 MeV around 93%.

6. Day/Night Asymmetry

The solar neutrino oscillation parameters predict a fairly small day/night asymmetry (about 1.5%). To eventually measure such a small asymmetry requires excellent control of systematic effects and a very large data sample. The well-calibrated Super-Kamiokande detector with its large fiducial mass of 22.5kton is at present the only detector that can probe such small asymmetries. The fit to the amplitude of the day/night variation was only performed for the Super-Kamiokande-I data (1.6% statistical uncertainty); we expect the Super-Kamiokande-II data yield a (statistical) precision of 3.4% with this method. The A_{DN} uncertainty of the Super-Kamiokande-III data should be 2.9%, so Super-Kamiokande-I/II/III combined can determine

Journal of Physics: Conference Series **203** (2010) 012082 doi:10.1088/1742-6596/203/1/012082

A_{DN} to ±1.3%. The present estimate of the systematic uncertainty for Super-Kamiokande-I is 1.3%, however, it is a rather conservative estimate based on the idea of simply splitting the data into a day and a night sample. The dominant source of systematic uncertainty to day/night variation is the directional dependence of the energy scale. Multiple Coulomb scattering smears the direction of the recoil electrons by 20 to 30 degrees, so a varying energy scale can only explain slow day/night variations while at least some of the flavor conversion variation from the matter effects of the earth happens much faster. Using calibration data taken with a deuterium-tritium (DT) neutron generator (which produces ^{16}N *in situ* via a charge-exchange reaction with ^{16}O) [4] this systematic effect can be verified and understood. If this is indeed a correct description of the effect, the amplitude fit to the day/night variation would be more robust than the estimated 1.3%. So a total uncertainty of 1.4% might be reached assuming the systematic uncertainty can be reduced by 50%.

7. Spectral Distortion

The predicted spectral distortion in the elastic scattering rate (due to the transition of matter-driven oscillations to vacuum-driven oscillations at around 3 MeV) is on the order of 10% in the energy region 4 to 15 MeV. Assuming the observed background level in the central 13.3kton of Super-Kamiokande-III, a fiducial mass of 13.3 kton from 4 to 5.5 MeV, and a fiducial mass of 22.5kton for energies above 5.5 MeV, Super-Kamiokande can resolve this 10% effect within twelve years at three standard deviation significance. If we can further improve the water circulation pattern to suppress Radon transport to take advantage of the full fiducial mass of 22.5ktons for all energies, the same significance will be achieved in ten years. The energy-bin correlated systematic uncertainties are rather important for the spectral sensitivity: if they can be reduced by 50%, the same significance is reached within eight years (assuming full fiducial mass at all energies). This systematics is mostly limited by position dependence of the Super-Kamiokande energy scale. This energy scale is defined by injecting single, downward-going electrons into Super-Kamiokande with an electron linear accelerator [5]. The energy can be tuned between 4.5 and 16.5 Mev. In addition, the DT neutron generator data also defines the energy scale at a single energy, but uniform in direction, with higher statistics, and at more positions. Furthermore, various laser light injectors at different wavelength help with understanding the optical properties of the detector to better control energy-scale related systematic uncertainties.

8. Conclusion

Even in the thirteenth year of measuring solar neutrinos, Super-Kamiokande still impacts solar neutrino physics. The new, threshold-less electronics and data acquisition system in conjunction with better calibration and control of systematic effects will help reveal the role of solar and terrestrial matter effects in solar neutrino oscillations.

References

[1] Q. R. Ahmad et al., **Phys. Rev. Lett.87: 071301**, (2001)
[2] J. Hosaka et al., **Phys.Rev.D73: 112001**, (2006)
[3] J. P. Cravens et al., **Phys.Rev.D78: 032002**, (2008)
[4] E. Blaufuss et al., **Nucl.Instrum.Meth.A458: 638**, (2001)
[5] M. Nakahata et al., **Nucl.Instrum.Meth.A421: 113**, (1999)

Topics in Astroparticle and Underground Physics (TAUP 2009) IOP Publishing
Journal of Physics: Conference Series **203** (2010) 012083 doi:10.1088/1742-6596/203/1/012083

InP solid state detector for a measurement of low energy solar neutrinos

Yoshiyuki Fukuda
Faculty of Education, Miyagi University of Education,
149, Aobaku-aza-aoba, Sendai, Miyagi 980-0845, Japan

E-mail: fukuda@staff.miyakyo-u.ac.jp

Abstract. A large volume radiation detectors using a semi-insulating Indium Phosphide (InP) wafer have been developed for Indium Project on Neutrino Observation for Solar interior (IPNOS) experiment. The volume has achieved to $20mm^3$, and this is world largest size among InP detector observed γs at one hundred keV region. Most of charge are induced by moving carriers, and the charge collection efficiency and the energy resolution are obtained by 60% and 25%, respectively. We measured actual backgrounds related to ^{115}In β decay, and no significant background was found.

1. Introduction

Super-Kamiokande and Sudbery Neutrino Observatory experiment has established ν_e oscillation in their solar neutrino data in 2001[1, 2], and a long way problem so called Solar Neutrino Problem in past 30 years was almost solved by the LMA oscillation. Independently, KamLAND experiment confirmed the oscillation using reactor $\overline{\nu_e}$ in sense of Δm^2 [4]. Next step of neutrino physics should measure both precise oscillation parameters and CP phase in the MNSP matrix elements. For the future solar neutrino experiment, a precise θ_{12} measurement with 1% accuracy will help the determination of θ_{13} [3] and the direct investigation of solar interior on the stellar evolution theory, respectively. In this point of view, new experiment Indium Project on Neutrino Observation for Solar interior (IPNOS) should measure full scale of energy spectrum of solar neutrinos including pp neutrinos.

2. Low energy solar neutrino experiment using ^{115}In

In 1976, R.Raghavan proposed new technique for the measurement of low energy pp/^{7}Be solar neutrinos [5] via following reaction;

$$^{115}\text{In} + \nu_e \rightarrow ^{115}\text{Sn}^* + e^-. \tag{1}$$

The prompt electron has an energy with $E_\nu - 114$ keV, here E_ν is an energy of incident neutrinos. Therefore the spectroscopy will be realized in the solar neutrinos. An excited state of ^{115}Sn shown in Eq.(1) decays into the ground state with a lifetime of 4.76μs, and emits two γs (116 keV and 497 keV). This signature will be also able to use for a triple-coincidence to extract neutrino signal from huge backgrounds. However, ^{115}In itself has natural β decay into the ground state of ^{115}Sn with a lifetime of 4.4×10^{14} years and the end point energy is 499keV. The radiative

Bremsstrahlung could produce fake coincidence for the neutrino signal. In order to avoid these backgrounds, a fine segmented with well energy resolution detector is necessary [6].

We have chosen semi-insulating wafer produced by Sumitomo Electrical Industry Co. LTD with the method of Vapor pressure Controlled Czochralski (VCZ). Hamamatsu Co. LTD developed InP detector with $10mm \times 10mm$ in surface, and $0.2mm$ in thickness. The electrodes consist of Cr-Au with $0.03 \sim 1.0\mu m$ thickness for top and Au-Ge/Ni/Au with 0.13/0.015/0.5 μm thickness for bottom. The junction between electrode and InP are ohmic contact in room temperature, however actually a Schottky barrier has been formed at -79 °C because of the rectification in the measurement of Hall effect.

3. Performance of InP detector

The performance of InP detector was measured by using gammas emitted by usual radio active sources. Carriers generated by the energy deposit of electrons via photoelectric process or Compton scattering are drifted along to electric field. The expected charge could be evaluated by Hecht formula; $Q[C] = Q_0\{\frac{L_e}{d}(1-e^{-\frac{x}{L_e}}) + \frac{L_h}{d}(1-e^{-\frac{d-x}{L_h}})\}$ Here, $L_{e,h}$ is drift length of electron/hole, d is thickness of the detector, and x is distance from the electrode. Generally speaking, the carrier drift length is expressed by $L_{e,h} \equiv \mu_{e,h}\tau_{e,h}\frac{V_0}{d}$, here $\mu_{e,h}$ is the carrier mobility and $\tau_{e,h}$ is life-time of the carrier trapping for electron/hole, and V_0 is the bias voltage.

Figure.1 shows that the observed charge distribution. There found two peaks in each spectra. For instance, the peak for 122keV γ-ray in ^{57}Co appears around 0.3×10^{-14} C and 0.55×10^{-14} C. Higher peak is produced by full carrier collection due to reach the electrode, and it is consistent with other γs assuming by 3.5eV of an average energy for electron/hole pair production. The collection efficiency for lower peak is obtained by the 60% in case of 122keV γs, and this peak was formed by the induced charge as explained by Hecht formula. According to a simulation, the spectral shape could be reproduced assuming both $L_e = 200\mu m$ and $L_h = 30\mu m$ as shown in bottom of Figure.1.

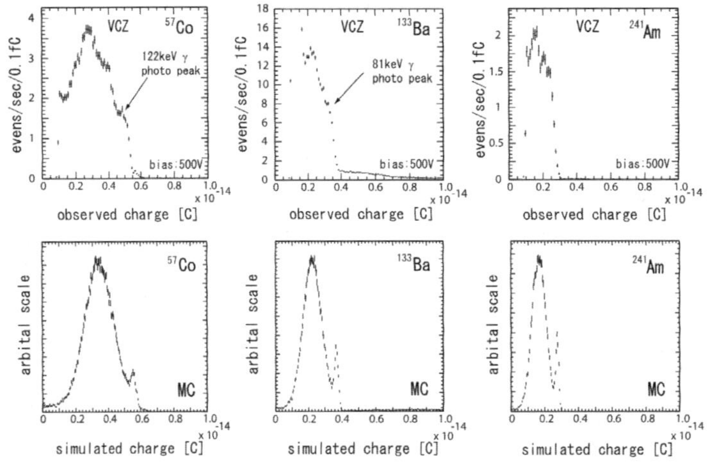

Figure 1. Observed and simulated charge distributions for gamma-rays are shown.

4. Measurement of backgrounds

As described in section 2, ^{115}In decays naturally with β emission, and the radiative Bremsstrahlung due to ^{115}In nuclei might be possible background in the observation of solar neutrinos. For the measurement of those effect, we used InP detector for measurement of β

event and CsI(Tl) scintillator for the measurement of radiative Bremsstrahlung. The CsI size was $50mm \times 50mm \times 20mm$. These two detectors are located by face to face and set inside of radio-active shield which consist of the lead in 5cm thickness and the oxygen free copper in 1cm thickness. The 4-π active veto plastic counter surrounded the shield was used for rejection of cosmic-ray backgrounds. The energy threshold for InP and CsI detector was 100keV and 50keV, respectively.

In order to detect the radiative Bremsstrahlung, we took coincidence between InP and CsI detector within 10 μsec. In case of 17.3 hours measurement, 46 events were observed. Before concerning of the coincidence events, it is necessary to take into account U/Th natural backgrounds. According to measurement of U/Th activity using ultra-low background germanium detector located in Kamioka mine, the InP wafer contains them as order of 10^{-11}g/g. Also right figure of Fig.2 shows the energy spectrum of CsI detector obtained by self trigger. It is clearly seen that the photo-electric peak due to some nuclei in the U/Th decay chain such as ^{214}Bi. The amount corresponds to order of 10^{-10}g/g in the CsI scintillator or surrounded material. According to left of Fig.2, most of coincidence events between InP and CsI detector

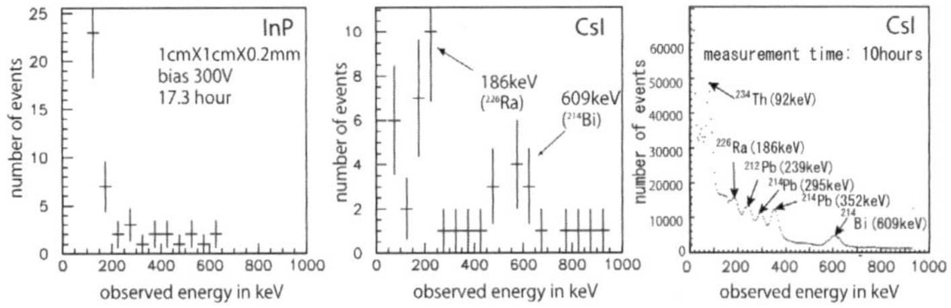

Figure 2. Energy spectrum of coincidence events for CsI and InP detector for 17 hour data, and U/Th background observed in CsI detector by self trigger.

looks consistent with β-γ coincidence of U/Th backgrounds in those detectors as indicated by ^{214}Bi γ observed in CsI, and no clear evidence for the effect of radiative Bremsstrahlung was found. An accidental triple coincidence between InP detector and the scintillator should be evaluated by 5×10^{-6}events/day/unit detector, which corresponds to 10 events/day/4ton detector in case of IPNOS detector. Therefore, we have to reduce the amount of U/Th contamination order of 0.1, however it is not so hard to control the production of materials.

5. Conclusion

An InP detector has been developed, and obtained good performance. Observed charge spectra could be explained by carrier drift with $250\mu m$ and collection efficiency is achieved by 60%. No clear evidence of radiative Bremsstrahlung was found.

Acknowledgments

This work was supported by the Grant-in-Aid Scientific Research (B) 17340065 of Japanese Society for the Promotion of Science, Inamori Foundation, and The Asahi Glass Foundation.

References

[1] Super-Kamiokande collaboration, Phys.Rev.Lett. **86** (2001) 5651.
[2] SNO Collaboration, Phys.Rev.Lett. **87** (2001) 071301.
[3] G.L.Fogli, E.Lisi, A.Marrone, A.Palazzo and A.M.Rotunno, Phys.Rev.Lett. **110** (2008) 141801
[4] KamLAND collaboration, Phys.Rev.Lett. **90** (2003) 021802.
[5] R.S.Raghavan, Phys. Rev. Lett. **37** (1976) 259.
[6] R.S.Raghavan, hep-ex/0106054.

Journal of Physics: Conference Series **203** (2010) 012084 doi:10.1088/1742-6596/203/1/012084

Linear Solar Models: a simple tool to investigate the properties of solar interior

F.L. Villante

Università di L'Aquila and INFN-LNGS, L'Aquila, Italy

E-mail: villante@lngs.infn.it

Abstract. We describe a simple method to study the dependence of the solar properties on a generic (small) modification the physical inputs adopted in standard solar models calculations.

1. Introduction

A new solar problem has emerged recently. The determinations of the photospheric heavy element abundances [1] indicate that the sun metallicity is lower than previously assumed [2]. Solar models that incorporate these lower abundances are no more able to reproduce the helioseismic results. Detailed studies have been done to resolve this controversy (see [3] for a review), but the origin of the "solar composition problem" has still to be identified.

In this situation, it is useful to perform a detailed analysis of the role of physical inputs and of the standard assumptions for standard solar model (SSM) calculations. This task is not always possible in simple terms, since the construction of SSM relys on (time-consuming) numerical integration of a non linear system of partial differential equations. In order to overcome this problem, we provide a tool which is at same time simple and accurate enough to describe the effects of a generic (small) modification of the physical inputs on the solar properties . The starting point is the idea that, despite the present disagreement with helioseismic data, the SSM is a rather good approximation of the real sun. We can thus assume small variations of the physical and chemical properties of the sun with respect to the SSM predictions and use a linear theory to relate them to the properties of the solar plasma.

2. Linear Solar Models

We consider the effects of a generic (small) modification of the input parameters of the sun. Namely, we indicate with $\delta\kappa(r)$ and $\delta\epsilon(r)$ the relative variations of radiative opacity $\kappa(r)$ and of the energy generation $\epsilon(r)$ *along the SSM profile*, i.e.:

$$I(\overline{T}(r),\overline{\rho}(r),\overline{Y}(r),\overline{X}_i(r)) = \overline{I}(\overline{T}(r),\overline{\rho}(r),\overline{Y}(r),\overline{X}_i(r))\;[1+\delta I(r)] \tag{1}$$

where $I = \kappa, \epsilon$ and r indicates the distance from the center of the sun. The notation $\overline{Q}$ indicates, here and in the following, the SSM value for the generic quantity Q. As a result of this modification, we obtain a solar model which deviates from SSM. We indicate with $\delta h(r)$ the relative variation of the physical quantities $h = l,\ m,\ T,\ P$ with respect to SSM predictions, i.e.:

$$h(r) = \overline{h}(r)\;[1+\delta h(r)] \tag{2}$$

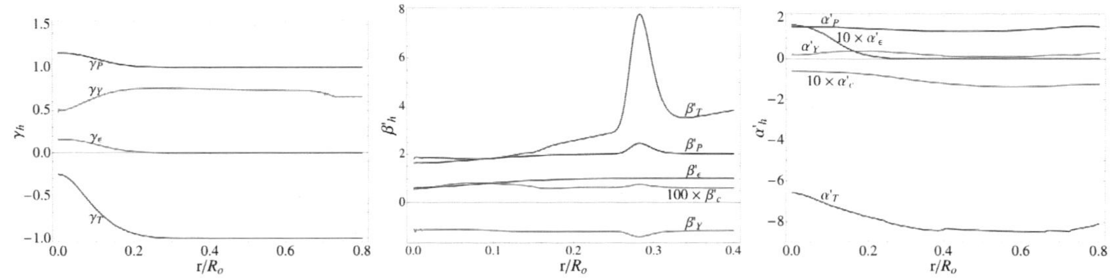

Figure 1. The coefficients $\gamma_h(r)$ (left panel), $\alpha'_h(r)$ (middle) and $\beta'_h(r)$ (right). See [4] for details.

In the radiative region of the sun ($r \leq \overline{R}_{\rm b} = 0.730 R_\odot$), the functions $\delta h(r)$ are described with good accuracy by the solutions of the linear system of ordinary differential equations:

$$\begin{aligned}
\frac{d\delta m}{dr} &= \frac{1}{l_m}\left[\gamma_P\,\delta P + \gamma_T\,\delta T - \delta m + \gamma_Y\,\Delta Y_{\rm ini} + \gamma_\epsilon\,\delta\epsilon\right] \\
\frac{d\delta P}{dr} &= \frac{1}{l_P}\left[(\gamma_P - 1)\,\delta P + \gamma_T\,\delta T + \delta m + \gamma_Y\,\Delta Y_{\rm ini} + \gamma_\epsilon\,\delta\epsilon\right] \\
\frac{d\delta l}{dr} &= \frac{1}{l_l}\left[\beta'_P\,\delta P + \beta'_T\,\delta T - \delta l + \beta'_Y\,\Delta Y_{\rm ini} + \beta'_C\,\delta C + \beta'_\epsilon\,\delta\epsilon\right] \\
\frac{d\delta T}{dr} &= \frac{1}{l_T}\left[\alpha'_P\,\delta P + \alpha'_T\,\delta T + \delta l + \alpha'_Y\,\Delta Y_{\rm ini} + \alpha'_C\,\delta C + \delta\kappa + \alpha'_\epsilon\,\delta\epsilon\right]
\end{aligned} \tag{3}$$

The coefficients γ_h, β'_h and α'_h and the scale heights $l_h \equiv \left[d\ln(\overline{h})/dr\right]^{-1}$ have been calculated in [4] and are shown in fig.1 and fig.2 (left panel). The parameters $\Delta Y_{\rm ini}$ and δC represent the absolute variation of the initial helium abundance and the relative variation of pressure at the bottom of the convective enevelope and can be univocally determined by imposing the appropriate integration conditions. At the center of the sun ($r = 0$) we have:

$$\begin{aligned}
\delta m &= \gamma_{P,0}\,\delta P_0 + \gamma_{T,0}\,\delta T_0 + \gamma_{Y,0}\,\Delta Y_{\rm ini} + \gamma_{\epsilon,0}\,\delta\epsilon_0 & \delta P &= \delta P_0 \\
\delta l &= \beta'_{P,0}\,\delta P_0 + \beta'_{T,0}\,\delta T_0 + \beta'_{Y,0}\,\Delta Y_{\rm ini} + \beta'_{C,0}\,\delta C + \beta'_{\epsilon,0}\,\delta\epsilon_0 & \delta T &= \delta T_0
\end{aligned}$$

where the subscript "0" indicates that a given quantity is evaluated at $r = 0$. At the bottom of the convective envelope ($r = \overline{R}_{\rm b}$), we have instead:

$$\begin{aligned}
\delta m &= -m_{\rm c}\,\delta C & \delta P &= \delta C \\
\delta l &= 0 & \delta T &= A'_Y\,\Delta Y_{\rm ini} + A'_C\,\delta C
\end{aligned}$$

where $A'_Y = 0.626$ and $A'_C = 0.025$ and $m_{\rm c} = M_{\rm conv}/M_\odot = 0.0192$ is the fraction of solar mass contained in the convective region.

The above equations completely determine the physical and chemical properties of the "perturbed" solar models. They have been obtained in [4] by expanding to linear order the structure equations of the present sun close to the SSM solution. We, thus, refer to the models calculated by using this approach as Linear Solar Models (LSM). In the derivation of eqs.(4) we also assumed that:

(i) The relative variation of density $\delta\rho(r)$ can be described by:

$$\delta\rho(r) = \delta P(r) - \delta T(r) - P_Y \Delta Y(r) \tag{4}$$

where $P_Y(r) \simeq -\partial \ln\mu/\partial Y \simeq -5/[8 - 5Y(r)]$ and μ represents the mean molecular weight.

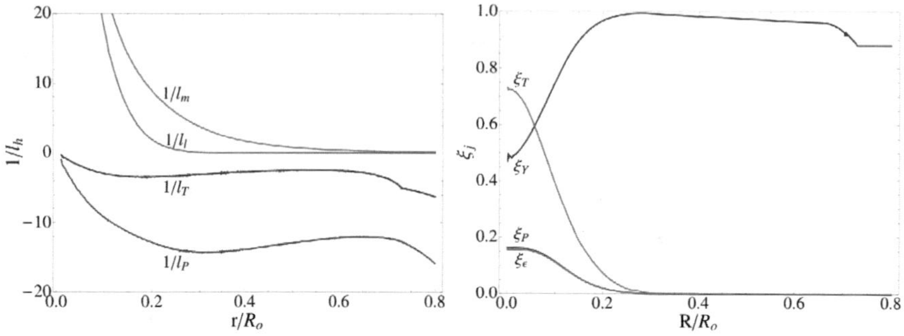

Figure 2. The scale heights $l_h(r)$ (left panel) and the coefficients $\xi_h(r)$ (right panel). See [4] for details.

(ii) The helium produced by nuclear reactions, $Y_{\rm nuc}(r)$, is roughly proportional to the energy generation coefficient in the present sun , i.e. $Y_{\rm nuc}(r) \propto \epsilon(r)$. In this hypothesis (and neglecting the variation of elemental diffusion efficiency in the radiative region), the absolute variation of helium $\Delta Y(r)$ is given by:

$$\Delta Y(r) = \xi_Y(r)\,\Delta Y_{\rm ini} + \xi_p(r)\,\delta P(r) + \xi_T(r)\,\delta T(r) + \xi_\epsilon(r)\,\delta\epsilon(r) \qquad (5)$$

The coefficients $\xi_h(r)$ have been calculated in [4] and are shown in fig.2 (right panel).

(iii) The global effect of elemental diffusion (i.e. integrated over the sun history) on the solar surface abundances is roughly proportional to diffusion efficiency at the bottom of the convective envelope in the present sun. In this assumption, one has:

$$\Delta Y_{\rm b} = 0.838\,\Delta Y_{\rm ini} + 0.033\,\delta C \qquad (6)$$

where the subscript "b" indicates that a given quantity is evaluated at the bottom of the convective region. By taking into account that $(Z_{\rm b}/X_{\rm b})$ is observationally fixed, we can also calculate the relative variation of metals in the convective region:

$$\delta Z_{\rm b} = -1.088\,\Delta Y_{\rm ini} - 0.043\,\delta C \qquad (7)$$

and in the radiative region (where we neglect the variation of diffusion efficiency):

$$\delta Z(r) \simeq \delta Z_{\rm ini} = -0.887\,\Delta Y_{\rm ini} - 0.164\,\delta C \qquad (8)$$

(iv) The relative variation $\delta R_{\rm b}$ of the convective radius $R_{\rm b}$ can be calculated "a-posteriori". By applying the Swartzchild criterion to the solution of eqs.(4) and by expanding to first order (see [4] for details), we obtain:

$$\delta R_{\rm b} = 0.449\,\Delta Y_{\rm ini} - 0.117\,\delta C - 0.085\,\delta\kappa_{\rm b} \qquad (9)$$

In ref.[4] the predictions of LSM for helioseismic observables and neutrinos fluxes have been compared with the results of full non-linear SSM calculations for several selected cases, showing that a very good agreement is achieved.

References

[1] Asplund M., Grevesse N., Sauval A. J. 2005, *Cosmic Abundances as Records of Stellar Evolution and Nucleosynthesis* **25** 336
Asplund M., Grevesse N. , Sauval A. J. and Scott P. 2009, *Ann. Rev. Astron. Astrophys.* **47** 481
[2] Grevesse N., Sauval A .J. 1998, *Space Science Reviews* **85** 161
[3] Basu S. and Antia H. M. 2008, *Phys. Rept.* **457** 217
[4] Villante F. L. 2009, in preparation

Standard and non-standard analysis of solar and reactor neutrino data

Mariam Tórtola

II. Institut für theoretische Physik, Universität Hamburg,
Luruper Chausee 149, 22761 Hamburg, Germany

E-mail: `mariam.tortola@desy.de`

Abstract.
We present an updated analysis of solar and reactor neutrino data in the standard framework of neutrino oscillations. We also consider the presence of non-standard neutrino interactions with matter and perform an alternative analysis of neutrino data studying the non-standard effects both on the neutrino detection and propagation through matter.

1. Introduction

Current solar neutrino data in conjunction with reactor data from the KamLAND experiment show that the neutrino oscillation mechanism is the correct picture to explain the solar neutrino physics. The combination between solar and KamLAND determines a unique solution in the mass-mixing parameter space, the so-called Large Mixing Angle solution [1]. However, while constrained by the solar and KamLAND data in an important way, neutrino non-standard interactions (NSI) still provide an important exception to the robustness of the neutrino oscillation interpretation. Indeed, with oscillations still being the underlying mechanism, an additional degenerate oscillation solution in neutrino oscillation parameters can appear for sufficiently intense non-standard interactions [2].

2. Standard oscillation analysis of solar and reactor neutrino data

Here we present the most important results of our global three–neutrino oscillation analysis [1] in the solar sector. The determination of the leading solar oscillation parameters θ_{12} and Δm^2_{21} emerges from the complementarity of solar and reactor neutrinos, as illustrated at Fig. 1. From our analysis of solar and KamLAND data we find the following best fit points and 1σ errors:

$$\sin^2\theta_{12} = 0.304^{+0.022}_{-0.016}\,, \qquad \Delta m^2_{21} = 7.65^{+0.23}_{-0.20} \times 10^{-5}\,\text{eV}^2\,. \tag{1}$$

Spectral information from KamLAND data leads to an accurate determination of Δm^2_{21} with a precision of 8% at 3σ. KamLAND data start also to contribute to the lower bound on $\sin^2\theta_{12}$, whereas the upper bound is dominated by solar data, most importantly by the CC/NC solar neutrino rate measured by SNO.

The determination of θ_{13}, the mixing angle characterizing the magnitude of CP violation in neutrino oscillations, is one of the main challenges for the future neutrino research. The bound on θ_{13} emerges from the interplay of different data sets, the most important contribution

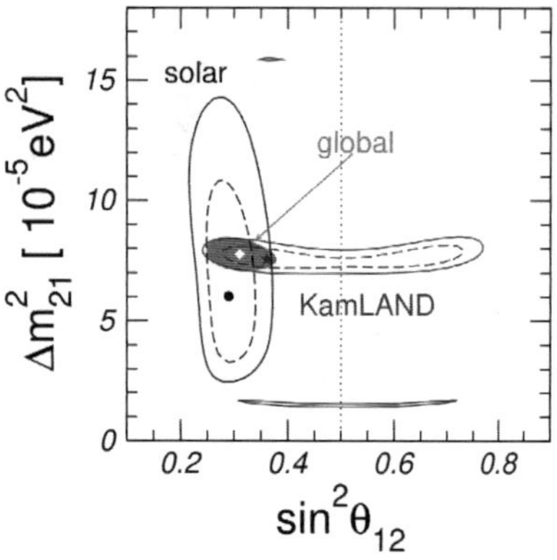

Figure 1. Determination of the leading solar oscillation parameters at 90% CL and 3σ for solar and KamLAND, as well as the 3σ region for the combined analysis.

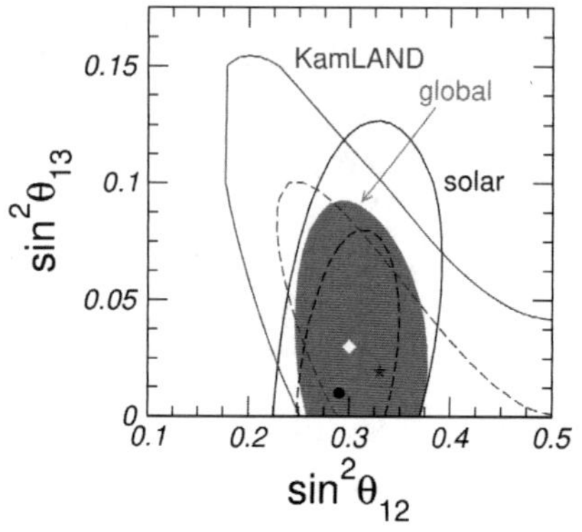

Figure 2. Allowed regions in the $(\theta_{12}-\theta_{13})$ plane at 90% CL and 3σ for solar and KamLAND, together with the 3σ region for the combined analysis.

coming from the CHOOZ reactor experiment combined with the determination of $|\Delta m^2_{31}|$ from atmospheric and long-baseline experiments. Due to the different correlation sign between $\sin^2\theta_{13}$ and $\sin^2\theta_{12}$ for low and high energy solar neutrino data, as well as for solar and KamLAND data, one finds that also solar+KamLAND provide a non-trivial constraint on θ_{13}. The interplay of solar and KamLAND data in constraining θ_{13} is illustrated at Fig. 2. It is also shown the preference of the experimental data for a non-zero value of θ_{13} ($\sin^2\theta_{13} = 0.03$) with an statistical significance of 1.5σ. We obtain at 90% CL (3σ) the following limits:

$$\sin^2\theta_{13} \leq \begin{cases} 0.060\ (0.089) & \text{(solar+KamLAND)} \\ 0.035\ (0.056) & \text{(global data)} \end{cases} \tag{2}$$

3. Non-standard interactions at solar and reactor neutrino data

Most neutrino mass extensions of the standard electroweak model entail non-standard interactions which, in the low energy limit, can be parametrized through the effective four fermion Lagrangian,

$$-\mathcal{L}^{eff}_{\rm NSI} = \varepsilon^{fP}_{\alpha\beta} 2\sqrt{2} G_F (\bar\nu_\alpha \gamma_\rho L \nu_\beta)(\bar f \gamma^\rho P f)\,, \tag{3}$$

where G_F is the Fermi constant and $\varepsilon^{fP}_{\alpha\beta}$ parametrize the strength of the NSI. The chiral projectors P denote $\{R, L = (1 \pm \gamma^5)/2\}$, while α and β denote the three neutrino flavors, and f is a first generation fermion (e, u or d). Here we will focus on the analysis of non-universal (NU) flavor-conserving interactions with electrons [3]. For a recent study on the phenomenology of NSI with quarks see Ref. [2], where an additional degenerate oscillation solution with $\sin^2\theta_{12} > 0.5$ appears for sufficiently intense NSI.

We have analysed the impact of neutrino-electron NSI on the phenomenology of solar neutrinos, taking into account their effect both at the level of propagation where they modify the standard MSW behavior, and at the level of detection. Non-standard couplings of neutrinos with electrons affect the elastic scattering ($\nu_a e \to \nu_a e$) process modifying the number of events and their spectral distribution expected in the solar neutrino detectors. The standard differential cross section for the scattering processes is given by:

$$\frac{d\sigma_a}{dT}(E_\nu, T_e) = \frac{2G_F^2 m_e}{\pi}\left[(g_L^a)^2 + (g_R^a)^2\left(1 - \frac{T_e}{E_\nu}\right)^2 - g_L^a g_R^a \frac{m_e T_e}{E_\nu^2}\right], \tag{4}$$

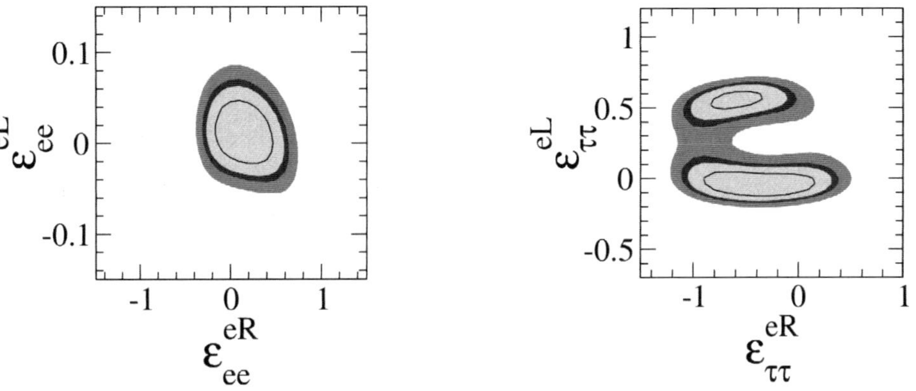

Figure 3. Constraints on the electron and tau neutrino NSI couplings at 68%, 90% 95% and 99% for 2 d.o.f.. The mass-mixing parameters have been marginalized away.

where m_e is the electron mass, E_ν is the incident neutrino energy, T_e is the electron recoil energy. In the presence of NU NSI, the standard left and right SM couplings g_L and g_R get an extra contribution, being replaced at Eq. (4) by the effective non-standard couplings $\tilde{g}_L^a = g_L^a + \varepsilon_{aa}^{eL}$ and $\tilde{g}_R^a = g_R^a + \varepsilon_{aa}^{eR}$. Given the strong limits on ν_μ NSI, in our analysis we focus on possible non-standard couplings of ν_e and ν_τ.

Let's consider now the non-standard effects on the neutrino propagation. The Hamiltonian describing solar neutrino evolution in the presence of NSI contains, in addition to the standard oscillations term, a new term accounting for an effective potential induced by the NSI with matter:

$$H_{\rm NSI} = \sqrt{2} G_F N_e \begin{pmatrix} 0 & \varepsilon \\ \varepsilon & \varepsilon' \end{pmatrix}, \tag{5}$$

where the effective parameters ε and ε', neglecting $\varepsilon_{\alpha\mu}^{fP}$, are given by $\varepsilon = -\sin\theta_{23}\,\varepsilon_{e\tau}^{eV}$ and $\varepsilon' = \sin^2\theta_{23}\,\varepsilon_{\tau\tau}^{eV} - \varepsilon_{ee}^{eV}$. Since we focus on the flavor conserving NU couplings, the flavor-changing off-diagonal coupling ε is set to zero, and in the treatment of solar neutrino propagation we consider only the NU coupling ε'. The vectorial couplings above are related with the left and right couplings through: $\varepsilon_{\alpha\alpha}^{eV} = \varepsilon_{\alpha\alpha}^{eL} + \varepsilon_{\alpha\alpha}^{eR}$.

The main results of our joint analysis of solar and KamLAND data in the $(\Delta m_{21}^2, \sin^2\theta_{12}, \varepsilon_{ee}^{eL}, \varepsilon_{ee}^{eR})$ parameter space are presented at Fig. 3, where the allowed regions for the effective NSI couplings are plotted. The limits on the effective couplings we find (at 90% CL):

$$-0.04 < \varepsilon_{ee}^{eL} < 0.06 \qquad -0.27 < \varepsilon_{ee}^{eR} < 0.59 \tag{6}$$

$$-0.16 < \varepsilon_{\tau\tau}^{eL} < 0.11 \qquad -1.05 < \varepsilon_{\tau\tau}^{eR} < 0.31 \tag{7}$$

are comparable with those found by laboratory experiments. See Ref. [3] for more details.

Acknowledgments
I would like to thank my collaborators A. Bolaños, O. Miranda, A. Palazzo, T. Schwetz and J.W.F. Valle. This work has been supported by the DFG (Germany) under grant SFB-676.

References

[1] Schwetz T, Tortola M A and Valle J W F 2008 *New J. Phys.* **10** 113011 (*Preprint* 0808.2016)
[2] Escrihuela F J, Miranda O G, Tortola M A and Valle J W F 2009 (*Preprint* 0907.2630)
[3] Bolanos A, Miranda O G, Palazzo A, Tortola M A and Valle J W F 2009 *Phys. Rev.* **D79** 113012 (*Preprint* 0812.4417)

Remaining inconsistencies with solar neutrinos: can spin flavour precession provide a clue?

João Pulido[1], C R Das[1], Marco Picariello[2]

[1] Centro de Física Teórica das Partículas, IST, Av. Rovisco Pais 1049-001 Lisboa, Portugal
[2] INFN, Lecce and Dipartimento di Fisica, Universitá di Lecce, Via Arnesano, I-73100, Lecce, Italia

E-mail: pulido@cftp.ist.utl.pt, crdas@cftp.ist.utl.pt, Marco.Picariello@le.infn.it

Abstract. A few inconsistencies remain after it has been ascertained that LMA is the dominant solution to the solar neutrino problem: why is the SuperKamiokande spectrum flat and why is the Chlorine rate prediction over two standard deviations above the data. There also remains the ananswered and important question of whether the active neutrino flux is constant or time varying. We propose a scenario involving spin flavour precession to sterile neutrinos with three active flavours that predicts a flat SuperK spectrum and a Chlorine rate prediction more consistent with data. We also argue that running the Borexino experiment during the next few years may provide a very important clue as to the possible variability of the solar neutrino flux.

Ever since the first KamLAND results were released [1], the large mixing angle solution (LMA) [2] has been generally accepted to be the dominant one for the solar neutrino problem. Nevertheless a few intriguing questions remain open in this area. In fact it is customary to assume that the active solar neutrino flux is constant, although the possibility of its time variation is still a challenging issue to be investigated [3], [4] both experimentally and theoretically. Moreover, the observed flatness of the SuperKamiokande spectrum [5] and the Chlorine rate prediction lying at more than 2σ above the data [6] are disturbing facts providing evidence that LMA may not be the ultimate solution.

Based on these motivations, we developed a model which includes a light sterile neutrino in addition to the three active ones [7] which communicates with these by means of transition magnetic moments. Somewhere inside the sun, in a location determined by the balance between the solar density and the mass square difference between the sterile neutrino and the active ones, a resonant transition may occur converting the latter into the sterile in the presence of a large magnetic field. Hence, besides the solar and atmospheric mass differences, a third one is introduced implying another resonance in addition to the LMA one. For simplicity no extra vacuum mixing angle is assumed, so that the matter Hamiltonian which includes vacuum propagation is [7]

$$H_M = \begin{pmatrix} \frac{\Delta m^2_{01}}{2E} & \tilde{\mu}_1 B & \tilde{\mu}_2 B & \tilde{\mu}_3 B \\ \tilde{\mu}_1 B & V_n + V_c u^2_{e_1} & V_c u_{e_1} u_{e_2} & V_c u_{e_1} u_{e_3} \\ \tilde{\mu}_2 B & V_c u_{e_1} u_{e_2} & \frac{\Delta m^2_{21}}{2E} + V_n + V_c u^2_{e_2} & V_c u_{e_2} u_{e_3} \\ \tilde{\mu}_3 B & V_c u_{e_1} u_{e_3} & V_c u_{e_2} u_{e_3} & \frac{\Delta m^2_{31}}{2E} + V_n + V_c u^2_{e_3} \end{pmatrix}. \quad (1)$$

Here V_c, V_n are the charged and neutral current interaction potentials, $\tilde{\mu}_{1,2,3}$ are transition magnetic moments between mass eigenstates 0 and 1, 2, 3 and u_{e_i} are the first row entries of the (3x3) PMNS matrix. The newly introduced mass squared difference Δm^2_{01} is smaller than Δm^2_{21} and so the spin flavour precession resonance lies closer to the solar surface than the LMA one.

We have used the numerical integration of the Hamiltonian equation and considered the following two field profiles which are approximately complementary to each other [7]

Profile 1

$$B_1 = B_0/\cosh[6(x-0.71)] \quad (0<x<0.71), \qquad B_1 = B_0/\cosh[15(x-0.71)] \quad (0.71<x<1)$$

Profile 2

$$B_2 = B_0/[1+\exp[10(2x-1)] \quad 0<x<1.$$

The experimental data on all rates (Cl, Ga, SK rate and spectrum, SNO rates and spectrum, Borexino [8]) were used to assess the quality of the fits with the following χ^2 standard definition

$$\chi^2 = \sum_{j_1,j_2} (R^{th}_{j1} - R_{j_1}{}^{\exp}) \left[\sigma^2(tot)\right]^{-1}_{j_1 j_2} (R^{th}_{j2} - R_{j_2}{}^{\exp}).$$

Using the experimental values of Δm^2_{21}, Δm^2_{atm}, θ_{12}, θ_{atm} and B_0 (field at the peak), Δm^2_{01} as free parameters, we obtained the fits for profile 1 shown in table 1.

$B_0(kG)$	$\sin\theta_{13}$	Ga	Cl	SK	SNO_{NC}	SNO_{CC}	SNO_{ES}	χ^2_{rates}	$\chi^2_{SK_{sp}}$	χ^2_{SNO}	χ^2_{gl}
	0	67.2	2.99	2.51	5.62	1.90	2.49	0.07	42.7	57.2	99.9
0	0.1	66.0	2.94	2.49	5.62	1.87	2.46	0.30	42.1	55.2	97.6
	0.13	65.0	2.90	2.46	5.62	1.84	2.44	0.62	41.7	53.7	96.0
	0	66.4	2.82	2.32	5.37	1.76	2.31	0.20	37.6	46.0	83.8
140	0.1	65.3	2.77	2.29	5.37	1.73	2.28	0.53	37.9	44.9	83.3
	0.13	64.3	2.72	2.27	5.37	1.70	2.25	0.95	38.4	44.1	83.4

Table I. χ^2's and rates (in SNU for Cl and Ga experiments, in $10^6 cm^{-2}s^{-1}$ for SK and SNO).

For the best fit we found B_0=140 kG, $\Delta m^2_{01} = 1.25\times10^{-7}eV^2$ with transition moments between flavour eigenstates $\mu_{(\mu,\tau)s} = 1.4\times10^{-12}\mu_B$ and $\mu_{es} \le \mu_{(\mu,\tau)s}$.

The corresponding results for the SuperKamiokande and Borexino spectra in the case of profile 1 are shown in figs. 1 and 2. For profile 2 the quality of the fits and the results are much the same, except that the best fit is obtained for $B_0 = 0.75MG$, $\Delta m^2_{01} = 2.7\times10^{-6}eV^2$.

From table 1 and figs. 1, 2 a clear preference of the data for a sizable magnetic field is apparent. In fact

- the SuperKamiokande spectrum becomes flat
- the rate prediction for the Chlorine experiment strongly improves (instead of a 2σ discrepancy, a prediction within 1σ of the data is obtained).

Furthermore no conclusion can be drawn on the value of θ_{13}. As for the Ga rate, vanishing and sizable fields are equivalent, as both classes of predictions lie within 1σ of the central value. Regarding Borexino it is unclear whether it can favour either a negligible or a sizable solar magnetic field owing to the size of the experimental errors. Likewise no conclusion is obtained regarding the magnitude of θ_{13} in this case. So as this stage Borexino results are consistent with both a vanishing and a large field.

Since the intensity of the solar magnetic field is possibly related to the solar activity, the time dependence of the data is an open possibility consistent with this model. So we examine the periods in which the data were taken. We see that

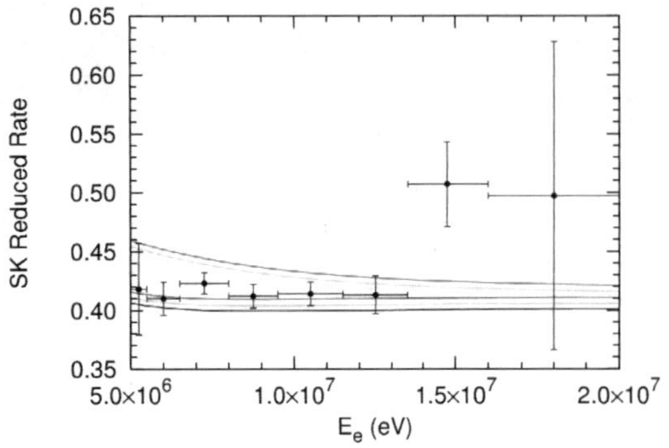

Figure 1. SuperKamiokande spectrum: data points and predictions for vanishing field (upper curves) and $B_0 = 140kG$ (lower curves). From top to bottom and for each set $sin\theta_{13} = 0, 0.1, 0.13$.

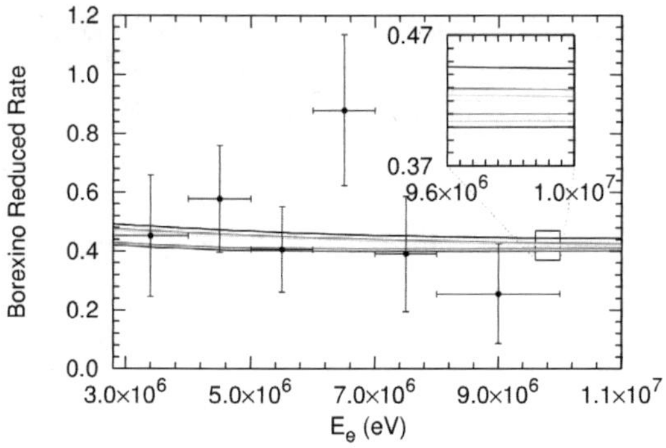

Figure 2. The same as fig.1 for Borexino spectrum and ^8B neutrinos: all experimental errors are larger here and no conclusion can be drawn as for the preference of the data.

- the SuperKamiokande spectrum refers to the period from May 31, 1996 to July 15, 2001 during which the average sunspot number was 65
- the Borexino ^8B spectrum refers to the period July 15, 2007 to June 21, 2008 when the average sunspot number was 4
- in most of the former period the solar magnetic activity increased and reached an 11 year peak in the Summer of 2000, whereas in the latter the activity was continuously at its minimum.

Therefore in the light of this model one expects the Borexino ^8B spectrum to coincide with the LMA prediction and the SuperKamiokande one to reflect a moderately active sun. If and when Borexino are able to reduce their errors to 1/3 (2/3 reduction), solar activity will probably have increased. Depending on whether this will be reflected in the increase of the solar field strength and a consequent decrease in the event rate, an oportunity may arise to clearly test the model.

Signatures of profiles 1 and 2 may also be distinguishable. In fact for profile 1 and Borexino phase 1 (low energy fluxes, $E < 1.7MeV$) all $\nu's$ have their resonances at $x < 0.5$. Here the field is weak and the density is large so that the event rate modulation will be too small ($\simeq 1\%$) to ever be seen. For profile 2 and Borexino phase 1, resonances lie at $x < 0.23$ where the field is close to maximal ($O(1MG)$), so that a field modulation in connection with solar activity accounts for as much as ($\simeq 9\%$) variability in the Borexino event rate. Hence low energy fluxes can detect modulation for profile 2 but not for profile 1. Finally the ^8B flux (seen by SK and Borexino) can detect modulation for both classes of profiles, so it cannot tell profile 1 from 2.

Hence we believe it essential to keep Borexino taking data for all neutrino fluxes during at least the first half of the present solar cycle expected to peak in 2011 or 2012 and present their data in time bins.

[1] Eguchi K *et al.* [KamLAND Collaboration] 2003
[2] Anselmann P *et al.* [GALLEX Collaboration] 1992 *Phys. Lett.* B **285** 390
[3] Pandola L 2004 *Astropart. Phys.* **22** 219
[4] Sturrock P A and Caldwell D O 2006 *Astropart. Phys.* **26** 174
[5] Fukuda S *et al.* [Super-Kamiokande Collaboration] 2002 Phys. Lett. B **539** 179
[6] Cleveland B T *et al.* 1998 *Astrophys. J.* **496** 505
[7] Das C R, Pulido J and Picariello M 2009 *Phys. Rev.* D **79** 073010
[8] Bellini G *et al.* [Borexino Collaboration] arXiv:0808.2868.

Journal of Physics: Conference Series **203** (2010) 012087 doi:10.1088/1742-6596/203/1/012087

Geo-neutrino flux calculation

S.T. Dye[1,2]

[1]Hawaii Pacific University, Kaneohe, HI 96744, USA
[2]University of Hawaii at Manoa, Honolulu, HI 96822, USA

sdye@hpu.edu

Abstract. Observation of terrestrial antineutrinos provides information on the origin, composition, and thermal evolution of the earth. Comparing the measured flux of terrestrial antineutrinos with the modeled flux constrains the quantity and distribution of uranium and thorium, the main heat-producing elements within the earth. Because the flux is dominated by nearby crustal material, detailed studies of the local geology are essential to a refined understanding of the bulk earth. This contribution outlines the calculation of the terrestrial antineutrino flux.

1. Introduction

Geo-neutrinos result from the beta decay of terrestrial nuclei. They have maximum energy of a few MeV, allowing them to pass freely through the earth to infrequently interact in large detectors near the surface. A dominant flux of solar *neutrinos* currently restricts geo-neutrino observations to terrestrial *antineutrinos* energetic enough to initiate inverse neutron decay. Only four nuclei in the decay series of uranium-238 and thorium-232 produce sufficiently energetic antineutrinos. An initial observation of these antineutrinos agrees with the prediction within the reported measurement uncertainty[1]. Radiation from these decay series is primarily responsible for the radiogenic heating of the earth. A comparison of rate of heating with the rate of heat flowing from the surface of the earth determines the thermal evolution of the planet. The distribution of heat-producing elements within the earth influences mantle convection, plate tectonics, hot-spot volcanism, and possibly the geo-magnetic field. Geo-neutrino observation remotely senses terrestrial heat-producing elements, enabling novel investigations of the earth's chemical composition and dynamic processes.

According to earth models, the crust and mantle are the main reservoirs of the principal heat-producing elements, with concentrations greater in continental crust than in mantle and oceanic crust[2]. This distribution predicts a surface flux of geo-neutrinos significantly higher on continents than in ocean basins, although uncertainties are difficult to quantify[3]. Moreover, a dominant portion of the continental flux originates from the crust within several hundred kilometers of the detector, making detailed studies of the local geology extremely useful[4]. As exposure of the two detectors currently recording the interactions of terrestrial antineutrinos accumulates[1,5], and construction of a third detector nears completion[6], flux modeling becomes increasingly important for constraining proposed distributions of uranium and thorium.

2. Geo-neutrino flux calculation

Geo-neutrino flux is a scalar quantity, diminishing as the inverse square of the distance from the source point to the measuring point. The rate of geo-neutrinos emerging from a source region depends on the concentrations and activities of the parent nuclei. Where the decay of daughter nuclei is involved, secular equilibrium is assumed. Flux calculation is a straightforward integration over all source regions. Assuming radial symmetry the flux contribution $d\varphi$ at a surface detector due to a source region at a given radius r and zenith angle θ is approximately

$$d\varphi \propto \rho a r^2 dr \sin\theta d\theta \left(R^2 - 2rR\cos\theta + r^2\right)^{-1} \quad (1)$$

where a is the concentration of parent nuclei in the region, ρ is the density of the region, and R is the radius of the earth. The main uncertainty in the flux calculation is the unmeasured concentration of parent nuclei in the inaccessible regions of the deep earth. This is, of course, a principal reason for geo-neutrino studies.

2.1. Mantle

The calculation of the geo-neutrino flux from the mantle assumes radial symmetry[7], which produces a spatially invariant surface flux. Specification of the physical parameters of the concentric spherical shells follows from seismology[8]. In addition to density, the relevant parameters for each shell include the inner and outer radii. Although uniform in this study, it is a simple matter to vary the concentrations of parent nuclei. The seismic data[8] extend to the center of the earth, allowing for inclusion of the core should models predict appreciable parent nuclei concentrations in this potential reservoir.

2.2. Crust

The predicted geo-neutrino flux from the crust varies significantly over the surface of the earth, being strongly correlated with the thickness of continental crust. A physical description of the crust in two-degree pixels specifies density, thickness, and elevation of seven layers[9]. Of primary importance for calculation of the geo-neutrino flux are the upper, middle, and lower crust layers of 360 possible types. A particular challenge is the assignment of parent nuclei concentrations to these myriad layers.

2.3. Model Earth

An investigation of the dependence of the flux from the mantle and crust as a function of distance from the detector results from a numerical integration over spherically symmetric shells with density and thickness determined by seismology[8]. Integration begins at the core-mantle boundary and extends to a uniform crustal shell at the surface, in which parent nuclei are 125 times more concentrated than in the interior mantle shells. One-half of the flux from the crustal shell originates from source regions within 390 km of the detector. In contrast, one-half of the flux from the mantle shells originates from source regions within 3300 km of the detector. Together these reservoirs form a model earth from which one-half of the flux originates from source regions within 1200 km of the detector. The crustal shell contributes 68% of the total flux in this model. Figure 1 displays the results of these numerical integrations of the geo-neutrino flux without neutrino oscillation effects[1].

3. Local geology

The results of the numerical integration of the flux from the model earth demonstrate the importance of detailed knowledge of the geology local to the detector. This is particularly relevant to continental locations where the crustal flux dominates. Moreover, predicted parent nuclei concentrations are higher in the upper continental crust than in the less accessible middle and lower continental crust[10]. Geological surveys to determine uranium and thorium concentrations in surface deposits within several hundred kilometers of the location of a continental geo-neutrino detection site are crucial for constraining the quantity and distribution of heat-producing elements within the earth.

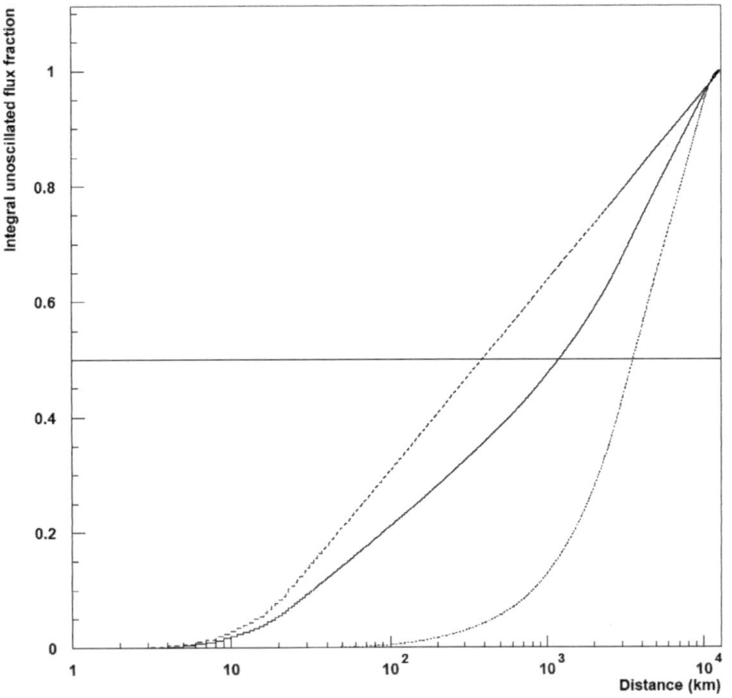

Figure 1. The integral, unoscillated flux fraction as a function of distance from the detector for a crust layer (dashed), mantle layers with seismically determined densities (dots), and model earth with crust layer containing parent nuclei concentrations 125 times greater than in the mantle layers (solid). Horizontal line indicates 50% of the integral flux.

4. Conclusions

Calculation of the surface flux of terrestrial antineutrinos is essential for refining understanding of the origin, composition, and thermal evolution of the earth. Comparison of the calculated flux with the measured flux constrains the quantity and distribution of uranium and thorium within the earth. Detailed knowledge of the geology local to the geo-neutrino detection site leads to tighter constraints, especially for continental locations.

References

[1] Araki T *et al.* 2005 *Nature* **436** 499-503
Abe S *et al.* 2008 *Phys Rev Lett* **100** 221803
[2] McDonough WF and Sun S-s 1995 *Chem Geol* **120** 223-253
[3] Mantovani F, Carmignani L, Fiorentini G and Lissia M 2004 *Phys Rev D* **69** 013001
[4] Enomoto S, Ohtani E, Inoue K and Suzuki A 2007 *Earth Planet Sci Lett* **258** 147-59
[5] Arpesella C *et al.* 2008 *Phys Lett B* **658** 101-8
Giammarchi MG and Miramonti L 2006 *Earth Moon Planets* **99** 207-20
[6] Chen MC 2006 *Earth Moon Planets* **99** 221-8
[7] Krauss LR, Glashow SL and Schramm DL 1984 *Nature* **310** 191-8
[8] Dziewonski AM and Anderson DL 1981 *Phys Earth Planet Inter* **25** 297-356
[9] Bassin C, Laske G and Masters G 2000 *EOS Trans AGU* **81** F897
[10] Rudnick R and Fountain DM 1995 *Rev Geophys* **33** 267-309

Topics in Astroparticle and Underground Physics (TAUP 2009) IOP Publishing
Journal of Physics: Conference Series **203** (2010) 012088 doi:10.1088/1742-6596/203/1/012088

A Search for Supernova Relic Neutrinos at Super-Kamiokande

Takashi Iida and Kirk Bays for the Super-Kamiokande Collaboration
ICRR, University of Tokyo, Mozumi, Hida, Gifu, Japan
E-mail: iida@suketto.icrr.u-tokyo.ac.jp

Abstract. Supernova relic neutrinos (SRN) are the diffuse supernova neutrino background from all past supernovae. No experiment has succeeded in detecting SRN yet. Currently, the Super-Kamiokande experiment has the world's best flux upper limit of 1.2 $\bar{\nu_e}/cm^2/sec$ for $E_\nu > 19.3MeV$. We have worked to improve this value by improving the data analysis. We have achieved better reduction efficiency and lowered the analysis energy threshold by developing a new spallation cut as well as optimizing other cuts.

1. Introduction

In 1987, Kamiokande detected neutrinos originating from SN1987A, a core collapse supernova in the Large Magellanic Cloud[1]. This proved that type II supernova emit an energy of about 10^{53} ergs in the form of neutrinos. Since the beginning of the universe, innumerable supernovae have occured. Neutrinos from such supernovae should be observable now as a diffuse background: the Supernova Relic Neutrinos (SRN). Measurement of SRN will enable us to investigate the history of past supernova. For example, the flux of SRN would show the star formation rate and supernova rate in galaxies. Figure 1 shows the predicted SRN flux in [3].

The current world's best limit for SRN comes from Super-Kamiokande(SK), a large water Cherenkov detector located in the Kamioka mine in Japan. SK is a high-performance neutrino detector consisting of 11,146 20-inch PMTs and 50,000 tons of pure water, 22,500 tons of which are the fiducial volume for the SRN search. SRN includes all six types of neutrinos, but the main interaction in the SK detector is the charged current quasi-elastic interaction of inverse β decay ($\bar{\nu_e}$ + p $\rightarrow$ e^+ + n). A search for SRN using SK's first phase (April, 1996 - July, 2001) is reported in [2] and a flux upper limit of $< 1.2/cm^2/sec$ was obtained (spectrum is shown in Figure 2). This limit is not far above theoretically predicted SRN fluxes.

So SRN may be detectable in the near future if we achive better sensitivity and collect more statistics. In this report, analysis improvements for SRN using SK first phase data are described.

2. Analysis improvements

In order to select candidates of SRN events, several cuts are applied to raw data to remove cosmic ray induced backgrounds, solar neutrinos, and external gamma-ray backgrounds. We are improving these cuts to increase sensitivity of the measurement.

- Spallation cut

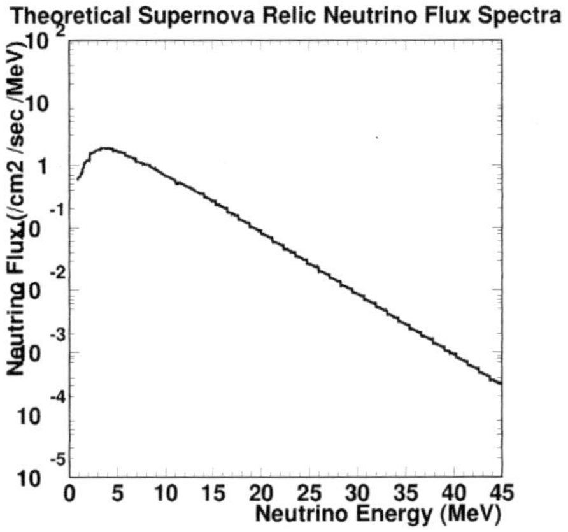

Figure 1. Predixted SRN flux in [3].

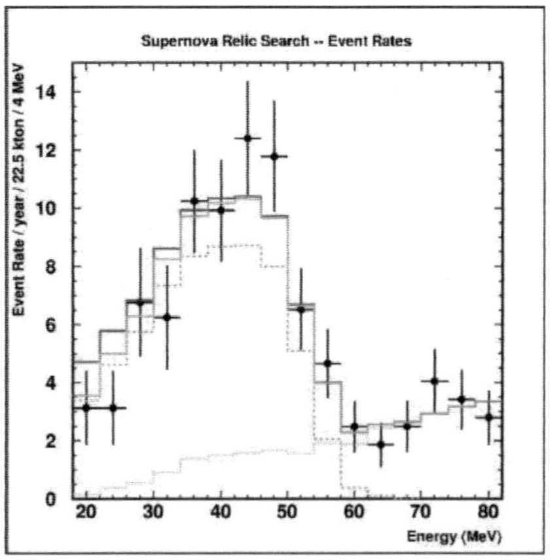

Figure 2. Energy spectrum of data and expected BG in SK-I. The dotted and dash-dot histgrams are the fitted background from invisible muons and atmospheric ν_e.

Cosmic ray muons spall oxygen and produce unstable nuclei called spallation products. Their decays provide one of the most abundant backgrounds in the SRN energy region; the ability to remove this background determines the lower threshold of our SRN search.
The standard spallation cut is applied in two steps. First is a cut based on a three variable likelihood method which was developed and optimized for SK's solar ν analysis. A second cut rejects events whose time difference from the last muon is shorter than 0.15 sec. The combined inefficiency of these cuts is 36% - the dominant inefficiency of the SRN analysis. A new spallation cut for SRN has been developed to achieve better efficiency and lower threshold. It uses a likelihood method based on four variables. As before, the four variables include the time difference between the preceding muons and SRN candidate event, and the distance from muon track to reconstructed relic candidate position. However, for the new cut we also make a histogram like Figure 3 for every muon. The horizontal axis corresponds to the muon track, and the vertical axis is the charge produced from each segment of muon track (extracted from the observed charge in every PMT and their geometrical relations to the muon track). When we find a strong peak this indicates where spallation occurs along the muon track. So the third variable for our new likelihood method is the distance from the reconstructed position of the relic candidate to the position where we expect the spallation event along the muon track. The fourth variable is the total charge which is emitted within +/-5 m of the peak position on the muon track.
This new method halves the inefficiency of the spallation cut from 36% to 18.5% in the 18 - 24 MeV region. It also allows us to lower the energy threshold below 18 MeV with 23% inefficiency. We eventually expect to lower the threshold down to 16 MeV.

- Solar neutrino cut

  Solar ^8B and *hep* neutrinos have energies up to $\sim$ 18MeV, high enough to be a background to our analysis. In order to remove solar neutrino events, events coming from the direction of the sun are removed.
  In the old SRN analysis, a cut tuned for the solar analysis was used. This old solar cut removed events whose $cos\theta_{sun}$ was less than 0.866 and had energies below 34 MeV. This time we (re)tuned it for SRN analysis using solar B8 and hep Monte Carlo. Significance (simply signal over square root background) is calculated and solar cut criteria are tuned

Journal of Physics: Conference Series **203** (2010) 012088 doi:10.1088/1742-6596/203/1/012088

for each energy. According to this study, we do not need to apply any solar cut above 19 MeV, and its efficiency below that is better than the old cut criteria.

- Gamma-ray cut
  Events originating from outside of the fiducial volume have the possibility of being reconstructed within the fiducial volume of SK. Gamma-rays from the materials of the detector structure and surrounding rock are the source of these events. We removed those backgrounds using the expected travel distance of gamma rays through the detector (the distance projected backwards from vertex). Under the old criteria, we applied a cut at 450 cm for all energies. In order to reduce inefficiency, energy dependent cut criteria were adopted. The 7% inefficiency of the old gamma cut decreased to 2.5% above 19 MeV, and is now lower than 7% (with the exact values depending on energy) below 19 MeV.
- Pion-like event cut
  Sometimes atmospheric neutrinos create a pion in the detector and it can be a background for SRN analysis. In the old analysis, these events remained in the final sample. A new cut to remove pion-like events was developed. Cherenkov rings from electrons tend to be broad due to multiple scattering, while rings from pions are sharper. Using this sharpness, signal events and background events are separated. Inefficiency of this cut is less than 1%.

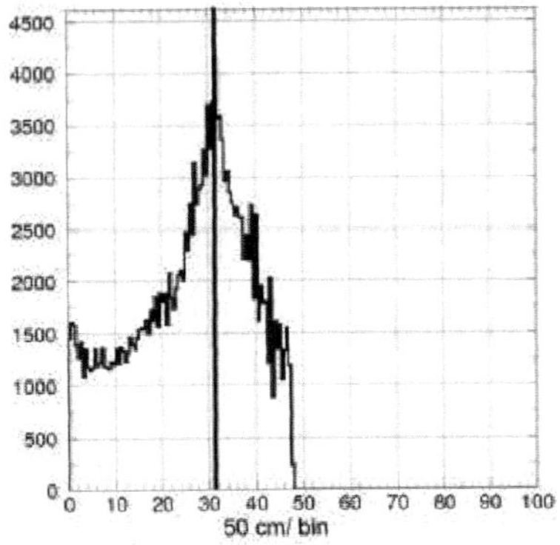

Figure 3. Muon charge projected onto muon track.

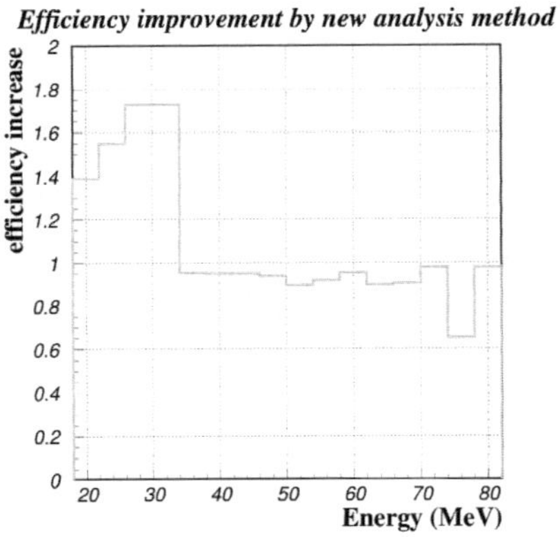

Figure 4. Efficiency increase as a function of energy.

3. Conclusion

We developed new cuts for rejecting spallation and pion-like backgrounds. Other cuts were optimized to improve their efficiencies. These analysis improvements increase the efficiency for SRN signal detection, especially in the lower energy region. Figure 4 shows the efficiency increase as a function of the reconstructed energy. In addition, the energy threshold was lowered below 18MeV, primarily due to the improvement of the spallation cut. We plan to lower the threshold down to 16MeV. Currently we are reprocessing SK-I data with these new analysis tools; they should give us better sensitivity for the SRN search. We will provide a new flux limit using the new method in the near future.

References

[1] K.S. Hirata *et al.* [The Kamiokande Collaboration], Phys. Rev. Lett. **58**, 1490 (1987).
[2] M. Malek *et al.* [The Super-Kamiokande Collaboration], Phys. Rev. Lett. **90**, 061101 (2003).
[3] C. Lunardini, Phys.Rev. D **75**, 073022 (2007)

Neutrino and antineutrino spectral splits from collective effects in supernovae

Irene Tamborra[1,2]

[1] Dipartimento Interateneo di Fisica "Michelangelo Merlin,"
Via Amendola 173, 70126 Bari, Italy
[2] Istituto Nazionale di Fisica Nucleare, Sezione di Bari,
Via Orabona 4, 70126 Bari, Italy

E-mail: irene.tamborra@ba.infn.it

Abstract. In core-collapse supernovae, $\nu - \nu$ interactions affect appreciably the evolution of flavor. Collective effects are discussed in the case of supernova luminosities which are different for different flavors. The observable spectral split features of ν and $\bar{\nu}$ are highlighted. Detection of such effects could provide a handle on two unknown: the neutrino mass hierarchy and the mixing angle θ_{13}.

1. Introduction and supernova framework

Neutrino flavor states $(\nu_e, \nu_\mu, \nu_\tau)$ are related to the mass states (ν_1, ν_2, ν_3) by means of a unitary matrix that is a function of the mixing angles $(\theta_{12}, \theta_{13}, \theta_{23})$ and of a possible CP-violating phase δ. Although the squared mass differences $(\delta m^2 \ll \Delta m^2)$ are precisely known, the sign of Δm^2 (i.e., the mass hierarchy) is unknown, as well as θ_{13}. Hints about these two unknowns might come from core-collapse supernova (SN) neutrinos.

In the SN matter, the $\nu_e - \nu_{\mu,\tau}$ interaction energy difference is described by the Mikheev-Smirnov-Wolfestein (MSW) matter potential $\lambda(r) = \sqrt{2}G_F N_e(r)$ where $N_e(r)$ is the electron number density (see [1] for a review). Moreover, $\nu - \nu$ interactions induce collective flavor changes. The corresponding self-interaction potential is given by

$$\mu = \sqrt{2}G_F(N + \overline{N}) \ , \tag{1}$$

as a function of the total effective density of neutrinos $(N = N_e + N_\mu + N_\tau)$ and antineutrinos $(\overline{N})$ per unit volume. For sake of simplicity we focus on μ-induced collective effects. Since ν_μ and ν_τ behave similarly at typical SN energies, we call each of them as ν_x. We can reduce the full 3ν evolution to an effective 2ν one, namely: $3\nu = (\nu_e, \nu_x) \oplus \nu_x$, one ν_x acts as "spectator." The relevant 2ν subspace (ν_e, ν_x) is governed by $(\pm\Delta m^2, \theta_{13})$[2]. We choose as default values: $\sin^2\theta_{13} = 10^{-6}$ and $\Delta m^2 = 2 \times 10^{-3}$ eV^2.

The total energy luminosity, $L_{\rm tot}$, is distributed over six $(3\nu + 3\overline{\nu})$ species. One spectator neutrino family ν_x only shares a fraction of luminosity, but does not take part to oscillations. The fractional luminosities ($l_\alpha = L_\alpha/L_{\rm tot}$ for $\alpha = e, \overline{e}, x$, and with $l_x \equiv l_{\overline{x}}$) obey the constraint:

$$1 = l_e + l_{\overline{e}} + 4l_x \ . \tag{2}$$

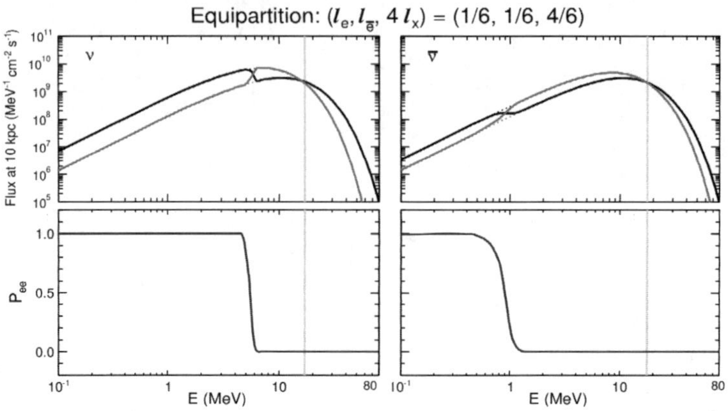

Figure 1. Equipartition case in inverted hierarchy. Left upper panel: flux of ν_e (black, solid) and of ν_x (red, solid) at the end of collective effects; dotted lines indicate initial spectra. Left lower panel: electron flavor survival probability P_{ee}. Right upper and lower panel: as before, but for $\bar{\nu}$. Vertical green lines mark the crossing energies where the ν_e (or $\overline{\nu}_e$) and ν_x fluxes are equal.

Our purpose is to go beyond the usual assumption of "energy equipartition" among flavors, which amounts to take $l_e = l_{\overline{e}} = l_x$.

We assume a total luminosity $L_{\mathrm{tot}} = 10^{53}$ erg/s, and average energies as

$$\langle E_e \rangle = 10 \text{ MeV}, \; \langle E_{\overline{e}} \rangle = 12 \text{ MeV}, \; \langle E_x \rangle = 15 \text{ MeV}. \tag{3}$$

The densities for unit of volume and energy are then [3]:

$$n_\alpha(r, E) = \frac{L_{\mathrm{tot}} l_\alpha}{4\pi R_\nu^2} \frac{\Phi_\alpha(E)}{\langle E_\alpha \rangle} g(r) \;, \tag{4}$$

where $g(r)$ is a damping factor (decreasing as $\sim 1/r^4$), $\Phi_\alpha(E)$ are thermal ν spectra, and $R_\nu = 10$ km is the ν-sphere radius.

In Fig. 1, oscillated spectra for the equipartition case in inverted hierarchy ($-\Delta m^2$) are shown [2, 4, 5, 6]. In this case, above a critical energy neutrinos swap their flavors, and similarly for $\bar{\nu}$ (but at lower energy).

2. Ternary Luminosity Diagram

The constraint in Eq. 2 can be represented in terms of a ternary diagram with fractional luminosities on each side. Each internal point of the triangle corresponds to different initial luminosity distributions. The equipartition case, described in Fig. 1, corresponds to $(l_e, l_{\bar{e}}, 4l_x) = (1/6, 1/6, 4/6)$. Variations from this case induce dramatic spectral changes [7].

Fig. 2 shows the qualitative spectral split patterns emerging from our numerical exploration of the ternary luminosity diagram, in the case of inverted hierarchy [7]. We use different markers for different split patterns, as indicated in the legenda of Fig. 2. In the lower half of the diagram, corresponding to relatively low ν_x luminosity ($4l_x < 0.5$), we always find one ν and one $\overline{\nu}$ split. More precisely, the blue triangles on the right correspond to one dominant ν split at high energy (HE) plus a minor $\overline{\nu}$ split at low energy (LE), qualitatively similar to the equipartition case shown in Fig. 1; for the yellow triangles on the left, the situation is reversed for ν and $\overline{\nu}$. As the ν_x luminosity increases, some 1ν (HE) + $1\overline{\nu}$ (LE) split cases survive (blue triangles), including the equipartition point. However, these cases are now flanked, on the left, by a couple of points where a double ν split occurs (blue squares) and, on the right, by two points with a double $\overline{\nu}$ split (yellow squares) plus two points with a double split for both ν and $\overline{\nu}$ (red circles). One of these two last cases is shown in Fig 3.

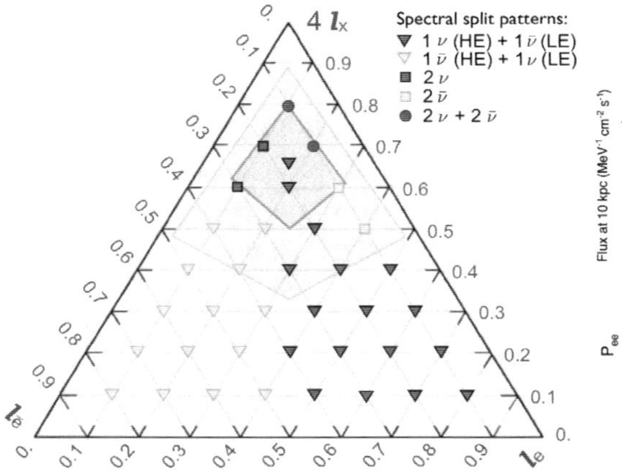

Figure 2. Spectral split patterns in inverted hierarchy. Blue (yellow) triangles: one HE ν split and one LE $\overline{\nu}$ split (one LE ν split and one HE $\overline{\nu}$ split). Blue (yellow) squares: two ν splits and no $\overline{\nu}$ split (two $\overline{\nu}$ splits and no ν split). Red circles: two splits for both ν and $\overline{\nu}$.

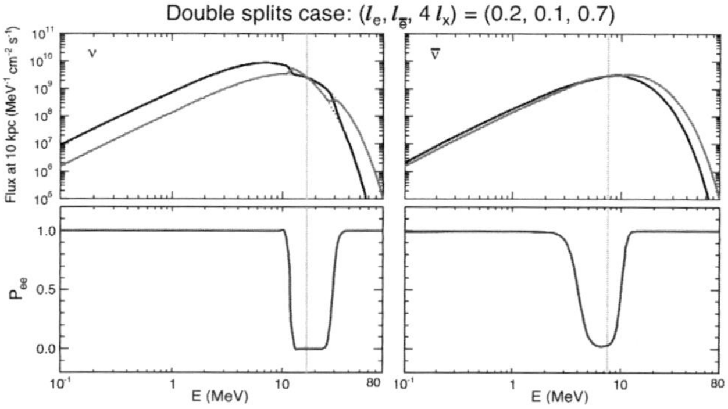

Figure 3. As in Fig.1, but for a case different from energy equipartition, leading to $2\nu + 2\overline{\nu}$ split.

In conclusion, the luminosity equipartition scenario (1ν HE $+1\overline{\nu}$ LE splits) is somewhat "special." In many other cases, the spectral split pattern may be significantly different [7, 8]. This must be taken into account in the analysis of prospective SN signals. For each different luminosity distribution if a double or a single split should take place, could be understood in terms of the initial conditions of the global vectors.

3. Conclusions

In a core-collapse supernova the ν density is high, and $\nu - \nu$ interactions are not negligible. They induce collective flavor conversion, typically leading to spectral splits/swaps. We have analyzed different luminosity distribution scenarios, embedding the equipartition scenario as a particular case. We find that several split patterns emerge in the luminosity diagram.

Acknowledgments

This work is supported in part by the Italian "Istituto Nazionale di Fisica Nucleare" (INFN) and "Ministero dell'Istruzione, dell'Università e della Ricerca" (MIUR) through the "Astroparticle Physics" research project. The results presented here have been obtained in Ref. [7] in collaboration with G.L. Fogli, E. Lisi, A. Marrone. I.T. is grateful to TAUP 2009 organizers for kind hospitality.

References

[1] T. K. Kuo, J. Pantaleone, *Rev. Mod. Phys.* **61**, 937 (1989).
[2] G. L. Fogli, E. Lisi, A. Marrone and I. Tamborra, *JCAP* **0904**, 030 (2009).
[3] H. Duan, G. M. Fuller, J. Carlson and Y. Z. Qian, *Phys. Rev.* D **74**, 105014 (2006).
[4] G. L. Fogli, E. Lisi, A. Marrone, A. Mirizzi and I. Tamborra, *Phys. Rev.* D **78**, 097301 (2008).
[5] S. Hannestad, G. G. Raffelt, G. Sigl and Y. Y. Y. Wong, *Phys. Rev.* D **74**, 105010 (2006) [Erratum-ibid. D **76**, 029901 (2007)].
[6] G. G. Raffelt and A. Y. Smirnov, *Phys. Rev.* D **76**, 125008 (2007).
[7] G. Fogli, E. Lisi, A. Marrone and I. Tamborra, arXiv:0907.5115 [hep-ph], to appear in *JCAP* (2009).
[8] B. Dasgupta, A. Dighe, G. G. Raffelt and A. Y. Smirnov, *Phys. Rev. Lett.* **103** (2009) 051105.

Topics in Astroparticle and Underground Physics (TAUP 2009) IOP Publishing
Journal of Physics: Conference Series **203** (2010) 012090 doi:10.1088/1742-6596/203/1/012090

The nuclear response of terrestrial detectors to low-energy neutrino spectra

T.S. Kosmas and V. Tsakstara

Theoretical Physics Section, University of Ioannina, GR 45110 Ioannina, Greece

E-mail: hkosmas@uoi.gr, vtsaksta@cc.uoi.gr

Abstract. In current probes searching for rare–event processes, appropriate nuclear targets are employed. In this work the response of such detectors to various low-energy neutrino spectra is explored starting from state-by-state calculations of the neutrino-nucleus reactions cross sections obtained by using the quasi particle random phase approximation (QRPA) based on realistic two-body residual interactions.

1. Introduction

During the last few decades, the search of semi-leptonic weak processes involving lepton-nucleus interactions (β-decay modes, neutrino-induced reactions on nuclei, nuclear muon capture, etc.) has deepen our knowledge on the fundamental electro-weak interactions and enriched our understanding on nuclear structure [1, 2, 3]. Such precious information inspired significant probes within and beyond the standard electro-weak theory and provided valuable interpretations to various experiments like those looking for neutrino masses, neutrino oscillations, neutrino detection, etc. [2, 4]. Using nuclei as micro-laboratories in reactor, accelerator and underground neutrino production/detection experiments, neutrinos have been extensively studied as key elementary particles in nuclear weak responses and in new physics beyond the standard model [2, 4, 5].

Recently, terrestrial neutrino experiments (combined with measurements of neutrino telescopes) have provided crucial information about the weak processes taking place in the interior of stars and showed that neutrinos are extremely sensitive probes for studying stellar evolution and astro-nuclear processes [4, 6]. To this end, measurements of solar neutrinos (at KAMLAND, Borexino, etc.) used to test the standard solar model (SSM) while recent probes at SNO (SNO+ experiment) aim to measure low-flux solar neutrinos (pep and CNO-cycle neutrinos) to check the abundances of solar core and clarify if the metalicity in the Sun is homogeneous [5]. Furthermore, recent stellar evolutions models describing the explosion mechanism of type II supernovae have provided us with important information regarding the role of neutrinos in the evolution of massive stars, explosive nucleo-synthesis, etc. [4, 6, 7].

During the collapse and explosion phase neutrino-nucleus reactions play an interesting role for supernova nucleosynthesis. However, uncertainties on astrophysical interactions of neutrinos and supernova physics opened numerous questions that are still unanswered and the degree of this knowledge is limited by our understanding of the neutrino-nucleus cross sections. Also, projected neutrino experiments need good control over systematic errors originating from neutrino-nucleus cross section uncertainties at low and intermediate nuclear excitation energies.

2. Low-energy neutrino-nucleus probes

As is well known, neutrinos excite nuclear modes not accessible to the electromagnetic probes that allow the study of the characteristics of the nuclear dynamics and of the nuclear interaction usually hidden in other processes. Thus, inelastic scattering of neutrinos on nuclei may induce nuclear excitations below or above particle emission thresholds. The latter excitations might subsequently decay via particle emission, thus, contributing to nucleosynthesis in the interior of stars [6, 7]. These processes could be incorporated into nuclear networks of explosive nucleosynthesis when the cross sections for neutrino-induced reactions like

$$(A, Z) + \nu \longrightarrow \nu' + X + a \,, \qquad (1)$$

where (A, Z) and X are the target and daughter nucleus, respectively, and a the emitted light particle (it may be a proton, a neutron or an a-particle), are reliably estimated. Such reactions may be considered as two-step processes in the compound nucleus picture as: (i) the target nucleus is excited via the neutral current neutrino-nucleus reaction

$$(A, Z) + \nu \longrightarrow \nu' + (A, Z)^* \qquad (2)$$

and (ii) the compound nucleus then decays into the particle channels, where the relevant decay rates might be taken from a statistical model (e.g. a conventional Hauser-Feshbach model) [2]. Focusing on the neutrino-nucleus probes, they are not only important in our efforts to better understand the nature of reactions like the above, but also they are significant in the current experimental studies of low and intermediate energy neutrinos. Nowadays, the study of neutrino scattering off nuclei is motivated by the interest to know how the nuclear medium affects the nucleon axial current and probe the structure of the nucleons (e.g. to study the strangeness content of the nucleon) and nuclei [1, 8]. The importance of the neutrino-nucleus interaction probes is not limited to the neutrino detection, but it is also related to the study of the processes generating them, and to their scattering inside matter (earth, stars, supernovae).

Recently, it became feasible to detect low-rate (solar and supernova) neutrinos by measuring the recoiling nucleus with very-low threshold-energy gaseous-detectors. Such studies are in conjunction with the direct-detection of cold dark matter (CDM) events and double beta decay measurements [2, 5]. Among the various ongoing or proposed double beta decay detectors, the molybdenum and the Cd-based semiconductor detectors CdTe and CdZnTe (they are currently employed for various rare processes in several areas of physics) have proposed to be used in the MOON and COBRA experiments, respectively, as detectors of double beta decay processes [5]. Since the response of such materials as neutrino (solar, supernova, etc.) detectors has not been addresses up to the present, a systematic study of neutrino–nucleus reaction cross sections (folded with various neutrino energy spectra) for their stable isotopes is needed [9, 10].

3. Nuclear detector response to low-energy neutrino sources

In general, nuclear responses to neutrinos are crucial for low-energy neutrino studies in nuclei that include the vector and the axial-vector weak interactions. Accordingly, the nuclear responses connected to the charged current neutrino-nucleus interactions are nuclear isospin and spin isospin responses, which reflect the spin isospin structures. Such responses in nuclear medium are modified much by strong nuclear spin and isospin interactions [2]. Isospin and spin isospin giant resonances, which absorb most of isospin and spin isospin strengths, are located at the excitation region of $E_{ex} = 10 - 25$ MeV. Thus, nuclei show large responses for neutrinos in that energy region. In the case of the neutral current neutrino-nucleus reactions, in addition to the other neutrino-induced nuclear excitations, the coherent channel ($g.s. \rightarrow g.s.$ transitions) is also possible and, as has been shown [9], this is the dominant channel for low-energy neutrinos.

From a nuclear theory point of view, up to now neutrino-nucleus cross sections have been investigated for various nuclear isotopes by employing several methods, such as: (i) Closure approximation within the shell model and the random phase approximation (RPA) to calculate the average contribution of the transitions to all excited states of the target nucleus (inclusive processes). (ii) Relativistic or non-relativistic Fermi gas methods utilizing a Lindhard function to compute the sum of all partial rates of the inelastic neutrino-nucleus reaction channels [1]. The incoherent rate is evaluated by integrating over a continuum of the excited states of a local Fermi sea. (iii) State-by-state calculations by using shell model or continuum RPA versions [6] for the explicit construction of the final nuclear states. Such calculations are reliable for low neutrino energies in nuclei where the transitions to definite nuclear states (ground-state or some low-lying excitations) could be dominant.

Recently, an advantageous numerical approach has been constructed for nuclear structure calculations based on analytic expressions of the reduced matrix elements of all the basic tensor operators that allow the systematic calculation of multi-pole matrix elements by separating out the geometrical coefficients from the kinematical parameters (energy, momentum, scattering angle) of the studied reaction. This method improves over other previous methods employing multi-pole expansion procedure in nuclear structure calculations. It is applicable in nuclear systems where the initial and final nuclear states have good quantum numbers of energy (E), angular momentum (J) parity (π) and isospin (T) so as the multipole expansion of the Donnelly-Walecka method is reliable [9].

In the present work, we perform extensive calculations of the differential and integrated cross sections of the neutrino-nucleus reactions for the isotopes contained in the COBRA double beta decay detector mentioned before. The many-body nuclear matrix elements are evaluated in the context of the quasi-particle random phase approximation (QRPA) by utilizing the advantageous numerical approach constructed in Ref. [9]. It is worth mentioning that, our QRPA method is based on realistic two-body forces and has been successfully used in other semi-leptonic nuclear processes. It offers a relatively simple and detailed state-by-state calculation of all the individual low-lying excitations induced by the interaction of neutrinos with nuclei [9].

Since our main purposes is to explore the response of the COBRA isotopes as solar and supernova neutrino detectors, we subsequently convolute the original neutrino-nucleus cross sections with appropriate neutrino-energy distributions (for supernova such distributions are the two-parameter Fermi-Dirac, the Power-Law, etc. distributions [10]) in order to obtain folded differential, cumulative and total cross sections for various values of the parameters entering the aforementioned energy distributions.

Closing, it must be stressed that, from a nuclear theory point of view, it is important to compute the transition matrix elements entering the neutrino-nucleus reactions cross sections and other observables with accurate many-body nuclear wave functions based on realistic two-body forces, a task extremely difficult due to the fact that neutrino-nucleus cross sections are highly suppressed in neutrino-detection experiments [9, 10].

References

[1] E. Kolbe and T.S. Kosmas, *Springer Trac. Mod. Phys.* **163**, 199–225 (2000).
[2] H. Ejiri, *Phys. Rep.* **338**, 265–351 (2000).
[3] T.S. Kosmas, Prog. Part. Nucl. Phys. **48** (2002) 307.
[4] M. T. Keil, G. G. Raffelt, and H.-T. Janka, *Astrophys. J.* **590**, 971-991 (2003).
[5] K. Zuber, *Phys. Lett. B* **519**, 1–7 (2001); *Prog. Part. Nucl. Phys.* **57**, 235–240 (2006).
[6] E. Kolbe, K. Langanke and P. Vogel, *Phys. Rev. D* **66**, 013007 (2002).
[7] N. Jachowicz, G.C.McLaughling and C.Volpe*Phys. Rev. C* **68**, 055502 (2003); **77**, 055501-12 (2008).
[8] R. Arnold, *Phys. Rev. Lett.* **95**, 182302 (2005).
[9] V.C. Chasioti and T.S. Kosmas, *Nucl. Phys. A* **7777** (2009).
[10] T.S. Kosmas and V. Tsakstara, Prog. Part. Nucl. Phys. , in press.

Long-term study of low energy counting rate with the Large Volume Detector

G.Bruno
On behalf of the LVD collaboration
E-mail: bruno@to.infn.it

Abstract. The Large Volume Detector (LVD), located in the hall A of the INFN Gran Sasso National Laboratory, is a ν observatory mainly designed to study low energy neutrinos from the gravitational collapse of galactic objects. The experiment, which is continuosuly taking data and consists of 840 scintillator counters, can monitor the local radioactivity background at energies > 0.5 MeV.
In this paper we study the long-term behavior of this low-energy counting rate since 1997. The Fourier analysis of the period 1997-2002 shows the existence of a seasonal modulation with the maximum occurring at the end of August. The same modulation is found by analyzing data collected by a Radon-meter placed inside the detector: this shows that changes in the concentration of Radon in the experimental hall are responsible for the measured variations of the counting rate in LVD.

1. Introduction
The INFN Gran Sasso National laboratory is located at a depth of about 3600 m w.e. Three main experimental halls, denoted A, B and C are connected to the 10.2 Km long highway tunnel. The laboratory is equipped with a ventilation system working, in normal conditions, at a flow rate of about 40000 m^3/h. This system provides one "lab volume" of fresh air in 3.5 hours [2] and keep in the laboratory a slight over-pressure compared to the outside tunnel.
LVD, installed in the hall A, consists of an array of 840 parallelepiped-shaped stainless steel (4 mm thick) counters (*tank*) of 1.5 m^3 filled with liquid scintillator and viewed, from the top, by 3 PMTs (d=15 cm). Counters are contained inside iron structures (*portatank*), 1.5 cm thick.
Neutrinos interact in scintillator mainly trough inverse beta decay (IBD) reaction: $\bar{\nu}_e p \rightarrow e^+ n$. Signals coming from each pmt are discriminated at two different thresholds (L and H), corresponding to the energies $\mathcal{E}_L \simeq 0.5$MeV and $\mathcal{E}_H \simeq 4$MeV. The H coincidence in any counters represents the scintillator trigger condition and enable the L coincidence channel that allows the detection of the signal due to np,d gamma capture, whose time delay is 185±5 μs [1].
The single counter rate at the low energy threshold (LTCR) is measured by a system of 840 scalers enabled, every 10 min for 10 s, by a dedicated trigger, designed for monitoring purpose. The rate per counter is around 50 Hz, and it is due to gammas from rock radioactivity and decays of Radon products.

2. Data and methods of analysis
To study the long-term behavior of the low-energy counting rate we consider scaler data between 1997 and 2009. We limit the analysis to counters that shows a stable behavior versus time. The

definition of stability is based on the counting rate deviation from its average. We present in fig. 1 the LTCR time series obtained averaging the contribution of a subset of selected counters. The selection is updated every two years and the subset is never smaller than about 50 tanks.

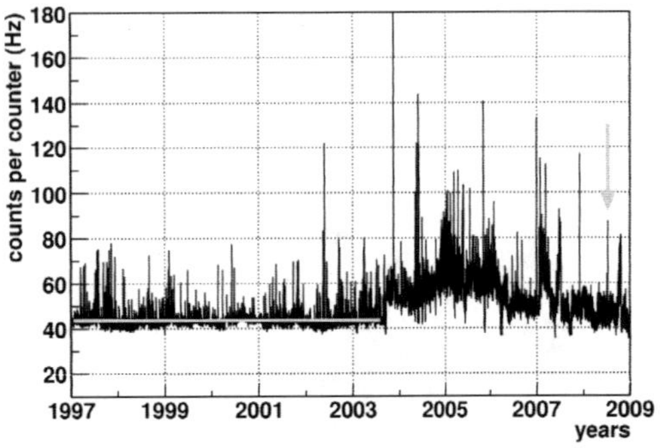

Figure 1. Average LTCR collected in the period 1997-2009.

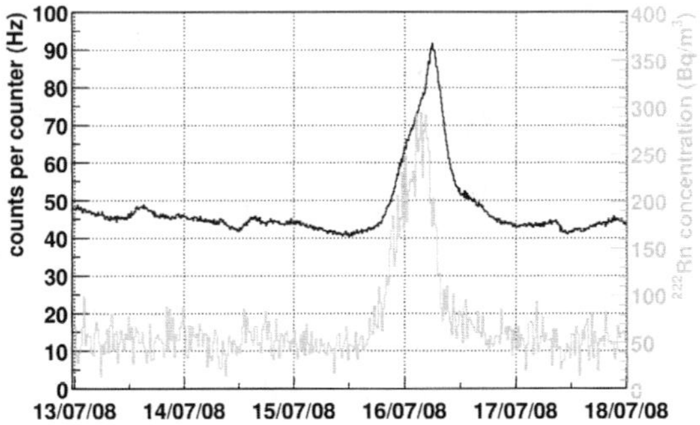

Figure 2. Average LTCR (thick line) since july 13th to 18th, 2008, compared with the radon-meter data (light line).

Each peak of fig. 1 shows a well structured shape if observed in detail. An example is reported in fig. 2 (thick line), where the behavior of the counting rate in 5 days in 2008 is shown, together with data recorded by an alpha radon-meter (light line) located just inside the LVD experiment. In this particular case the increase observed is due to a scheduled switch off of the ventilation system. The LTCR and radon-meter series are clearly correlated as we have already demonstrated in [3]. As highlighted by other authors [4], the radon concentration in the laboratory strongly depends on the conditions of the ventilation system. The change in the average counting rate visible in fig. 1, just before the beginning of the 2004, is related to modifications of the ventilation system. Mean value and weekly behavior of LTCR are influenced

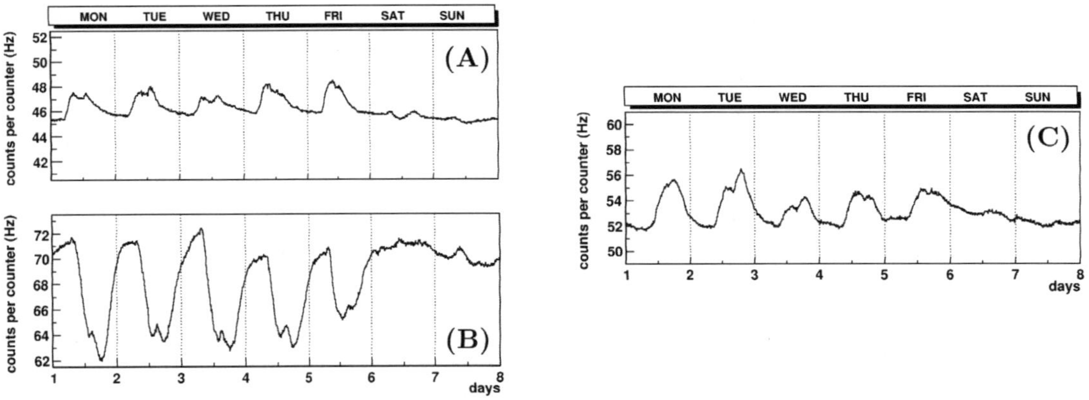

Figure 3. Average counting rate on a weekly time-base during: 2001-2002 (A), 2004-2005 (B), and currently up to april 2009 (C).

by these change as shown in fig. 3A and B.

The pattern visible in figs. 3 is well explained by the anti-correlation between pressure and Radon emanation from rock in caves [5]: in the LVD hall the pressure changes due to the opening/closing of the main entrance door in correlation with the start/end of human working

activity. The opposite behavior between panel A and B of the fig. 3, is due to an inversion of the pressure gradient between inside and outside the hall A. Indeed, during 2004-2005, the ventilation system was configured to maintain hall A de-pressurized assuring air re-circulation between the experimental halls and then ejecting outside the laboratory hot air through an extractor located on the northern side of the hall A [6]. From fig. 3 we can argue that at present the environmental conditions of the Hall A are returned to their original conditions.
To study long-term modulations in the counting rate, we limit ourselves to the period up to beginning of 2003, when the counting rate was more homogeneous. The analysis is carried out applying Discrete Fourier Transform (DFT) algorithm and the obtained power spectrum shows evidence of a seasonal variation. To better visualize the modulation we plot in fig. 4A the yearly rate averaged over 6 years. The seasonal modulation shows the maximum at day 241±32, corresponding to the end of August. The same procedure, applied to the radon-meter data set, is reported in fig. 4B and supports the previous result.

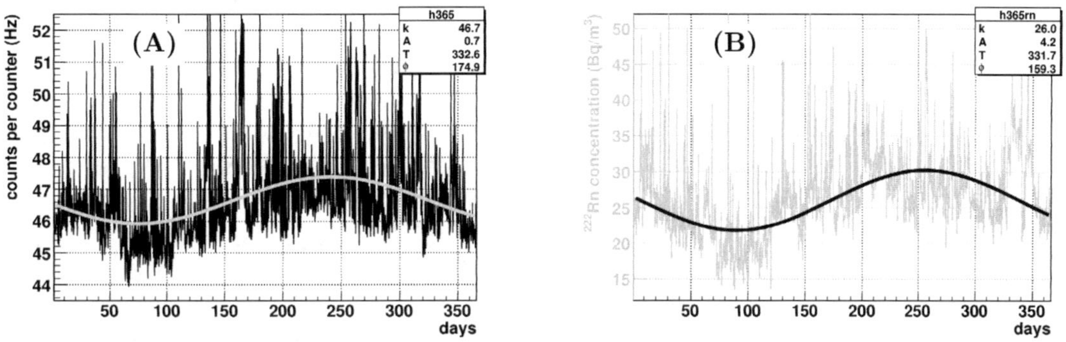

Figure 4. Yearly LTCR (A) and Rn concentration (B) averaged over 6 years (1997-2002) and fitted with the function $k + A \cdot sin(\frac{2\pi}{T}(x + \phi))$.

3. Conclusions

We have studied the behavior of the low-energy (E>500 KeV) counting rate as measured in the LVD detector between 1997 and 2009. We find that the rate variations are correlated with the changes in Radon concentration in the Hall A of LNGS, where LVD is located. The average rate per counter is rather homogeneous up to 2003, where a change is found, which corresponds to an inversion in the laboratory pressure conditions due to a change in the ventilation system of Hall A. We search for possible modulations in the low-energy counting rate, by restricting the analysis to the time period when the rate was homogeneous, i.e., between 1997 and 2003. We find a seasonal modulation, with the maximum at day 241±32. The phase of this modulation is consistent with the one found by the analysis of data from a radon-meter located close to LVD.

Acknowledgements

We wish to thank M. Laubenstein and M. Dedeo for providing the radon-meter data and for helpful discussions.

References

[1] N.Yu.Agafonova *et al.*, Astrop. Phys., **28**, 6 (2008) 516 [arXiv:0710.0259]
[2] A Bettini 2008 *J. Phys.: Conf. Series* **120** 082001
[3] G Bruno and H Menghetti 2006 *J. Phys.: Conf. Series* **39** 278-280
[4] A Bassignani et al. 1995 *Radiation Measurements* **Vol. 25** Nos 1-4, pp.557-560
[5] *Radon Measurements By Etched Track Detectors*, Saeed A Durrani & Radomir Ilic
[6] *Ing. D. Franciotti, Private communication*

Reactor neutrino detection for non proliferation with the Nucifer experiment

A. Porta for the Nucifer collaboration

Laboratoire SUBATECH, Ecole des Mines, 4 rue Alfred Kastler, 44307 NANTES CEDEX 3, FRANCE

e-mail: porta@subatech.in2p3.fr

Abstract. The idea of using a neutrino detector to monitor nuclear reactors was first proposed in 1977 by L.A.Mikaelian. Tremendous progresses have been achieved since then in both detection technology and understanding of fundamental neutrino physics. This allows us to apply reactor antineutrino detection to thermal power measurement and survey of isotopic composition of reactor fuel. Here we present the Nucifer experiment, a 1 ton Gd-doped liquid scintillator detector to be installed a few 10 m from the core for reactor surveillance. The design of such a small surface detector has been focused on high detection efficiency (~50%) and good background rejection. Detailed simulations of reactor emitted antineutrino spectrum and detector response have been developed and used to calculate the Nucifer sensitivity to illicit retrieval of Pu from the core as it will be discussed. These results have been presented in October 2008 to the International Atomic Energy Agency (IAEA) which has expressed its interest in the potentialities of this detector as a new safeguard tool.

1. Introduction

Nuclear reactor power comes from the energy produced by the fission of heavy elements (i.e. U and Pu) into neutron rich nuclei. Reactor neutrinos are produced by the β-decay of these fission products (FPs) into more stable nuclei: ${}^{A}_{Z}X \rightarrow {}^{A}_{Z-1}Y + e^- + \bar{\nu}_e$.

Each fission releases about 200 MeV and 6 antineutrinos, which means that the flux emitted by a 1 GW_{th} reactor is ~1.5 10^{20} antineutrinos/second. Although the interaction cross section between matter and neutrinos is very tiny (~10^{-43} cm^2), the huge emitted flux allows us to detect their signal with a relatively small detector (1 ton scale) placed a few 10 m from the core.

Since the number of emitted neutrinos and their mean energy depend on the fissioning isotopes (^{235}U, ^{238}U, ^{239}Pu and ^{241}Pu), their detection would give a direct image of the core composition.

2. Reactor monitoring by antineutrinos

Reactor antineutrinos are usually detected by using large volumes of liquid scintillator doped with Gadolinium. The detection reaction is the inverse β-decay $\bar{\nu}_e + p \rightarrow e^+ + n$. The e^+ detection produces a prompt signal of energy $E_{prompt} = E_{\bar{\nu}_e} - (M_n - M_p) + m_e$. This first pulse is followed by a delayed signal induced by the radiative capture of the neutron on Gd with the emission of a gamma cascade of total energy ~8MeV. The mass excess of the final state of the inverse β reaction implies a kinematical threshold of $M_n – M_p + m_e$ = 1.8 MeV for the antineutrino energy.

The two main fissile isotopes contained in the fuel of a pressurized water reactor (PWR) are ^{235}U and ^{239}Pu. Fresh Uranium fuel is typically enriched at 3.5% in ^{235}U, while ^{239}Pu is produced by neutron captures on the original ^{238}U followed by two consecutive β-decays: ^{239}U→^{239}Np→^{239}Pu. During a reactor cycle ^{235}U is burned while ^{239}Pu is produced. This means that the relative contribution to the total number of fissions induced by these two isotopes changes over time: it increases for the ^{239}Pu while decreasing for the ^{235}U. This is called the "burn-up" effect.

Because the FPs of these two isotopes have different atomic masses their β-decays result in different neutrino fluxes. The idea of a neutrino probe to monitor the reactor fuel composition is based on the hypothesis that we are able to discriminate between different generated fluxes.

Table 1. Relevant fission parameters of ^{235}U and ^{239}Pu

Fission parameters	^{235}U	^{239}Pu
E/fission[a] (MeV)	193.5	198.9
<Eν> above threshold (MeV)	2.94	2.84
Nν/fission above threshold	1.92	1.45
<σ_{int}> (10-43 cm2)	3.20	2.76

[a] Only the energy contributing to the thermal power is quoted here (i.e. total released energy minus antineutrino energy)

Considering the key parameters for ^{235}U and ^{239}Pu fissions summarized in table 1, we can calculate the ratio of detected antineutrinos in the two extreme cases, where all fissions come from pure ^{235}U or pure ^{239}Pu to produce the same thermal power: $N_\nu^U/N_\nu^{Pu} = [(N_\nu/\text{fission} \times \sigma)^U \times (E/\text{fission})^{Pu}]/[(N_\nu/\text{fission} \times \sigma)^{Pu} \times (E/\text{fission})^U] \approx 1.6$. This large difference suggests the possibility to use the antineutrino rate to monitor changes in the relative amounts of ^{235}U and ^{239}Pu in the core.

3. The Nucifer experiment

The detector design has been performed by following the IAEA requirements for a small, portable, cheap, remote controlled and safe tool. To optimize and validate the detector and shielding geometry we developed a dedicated GEANT4 simulation and we performed many background measurements at the two research reactors where Nucifer will be first tested.

The resulting detector (figure 1) is composed of a steel cylindrical tank (height 1.6 m, diameter 1.2 m) filled with about 0.85 m^3 of Gd-doped liquid scintillator. 16 photomultiplier tubes observe the liquid scintillator from the top through a 25cm thick acrylic disk which ensures light response uniformity and safe requirements. For calibration purpose, small radioactive sources can be deployed along the target central axis inside a vertical tube. The detector is shielded by 10 cm of Pb, 14 cm of bored polyethylene and an active muon-veto made by 5 cm thick plastic scintillators.

Presently we are building the detector and we foresee to install it at 7 m from the Osiris research reactor core (CEA-Saclay) at the beginning of 2010 for the commissioning phase. In 2011 we will move it at the ILL research reactor of Grenoble. Thanks to the almost pure ^{235}U composition of this reactor fuel we will be able to perform a precise calibration of the detector. In 2012 we plan to install Nucifer close to an EDF to make a final validation of the detector concept.

4. Nucifer performance

The detailed GEANT4 simulation of Nucifer developed to optimize its design has been also used to study its performance and its response to reactor neutrinos. The estimated energy resolution for the detection of 4 MeV gammas randomly generated in the target is ~6% while the estimated reactor neutrino detection efficiency is ~50%.

Beside the detector simulation a detailed simulation of reactor neutrino emission has been performed. It consist of a dedicated twofold code: a first package called MURE [1] simulates the time evolution of the core by solving the Bateman equation for the FPs and feeds a second part, called BESTIOLE, which reads in the ENSDF database all the relevant information on nuclear levels of the FPs to build their β and ν spectra.

Merging the results from this simulation to the detector response we have been able to reproduce different scenarios of neutrino detected rate corresponding to different fuel compositions. Figure 2 shows the expected weekly detected rate of neutrinos generated by a 2.9 GWth PWR placed at 25 meters from Nucifer. At the beginning of the cycle the fuel composition in relative number of fissions of ^{235}U and ^{239}Pu is 70/30%, while it is 50/50% at the end of the cycle. We can clearly see the decrease in the neutrino rate induced by the burn-up effect. The hole after week 51 is to simulate the reactor stop to refuel the core. After week 67 the reactor starts again with the initial fuel composition. The relative neutrino rate change is ~7% produced by a 20% relative change in the number of fissions induced by ^{235}U and ^{239}Pu.

Figure 1. The Nucifer detector.

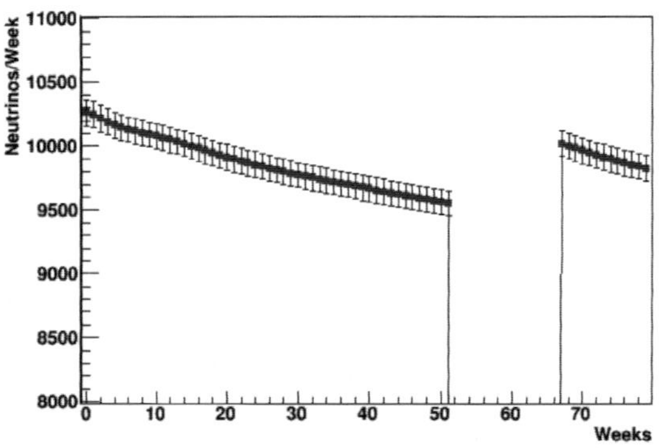

Figure 2. Weekly neutrino rate detected during one cycle by Nucifer installed 25 m away from a 2.9 GWth PWR.

These simulations also allow us to determine the Nucifer sensitivity to illicit Pu retrivals. If we ask for an upper limit on the probability of giving a false alarms of 4% and a lower limit for the probability of valid alarms of 75% Nucifer can be sensitive up to ~80 Kg of extracted Pu when we make two relative neutrino flux measurements of 15 days before and after the reactor stop performed to extract the fuel [2]. This calculation has been made for the detector placed at 25 m from the considered 2.9 GW_{th} power reactor. Different reactor powers, data taking periods and detector distances would lead to different sensitivities.

Reference

[1] O. Meplan et al. 2005 *MURE : MCNP Utility for Reactor Evolution* Proc. European Nuclear Conf., Versailles, France.
L. Giot et al. 2008 *Monte-Carlo simulations in the context of reactor monitoring with antineutrinos* Proc. Inter. Conf. on the physics of Reactors (Physor), Interlaken, Switzerland ref. FP438.

[2] A.Porta et al. 2009 *Reactor Neutrino Detection for Non Proliferation with the Nucifer Experiment* Proc. Inter. Conf. on Advancements in Nuclear Instrumentation, Measurement Method and their Applications (ANIMMA), Marseille, France, ref. 176. Accepted for publication by IEEE (Institute of Electrical and Electronics Engineers).

Topics in Astroparticle and Underground Physics (TAUP 2009) IOP Publishing
Journal of Physics: Conference Series **203** (2010) 012093 doi:10.1088/1742-6596/203/1/012093

Investigating the nuclear response of Te isotopes to SN-neutrinos

V. Tsakstara[1*] , T.S. Kosmas[1] and P.C. Divari[2]

[1] Theoretical Physics Section, University of Ioannina, GR 45110 Ioannina, Greece
[2] Department of Physical Sciences, Hellenic Army Academy, Vari 16673, Attica, Greece

E-mail: * vtsaksta@cc.uoi.gr

Abstract. The response of Te isotopes to the energy-spectra of supernova neutrinos is explored by applying the folding procedure. We present double differential cross sections for ^{128,130}Te isotopes convoluted with a two-parameter Fermi-Dirac (FD) and a Power-Law (PL) neutrino energy distributions. The original cross sections are calculated in the context of the quasi particle random phase approximation (QRPA) by using realistic two-body forces.

1. Introduction

Recently, the detection of low-energy neutrinos is searched in conjunction with double beta and neutrinoless double beta decays by employing very-low threshold-energy detectors in terrestrial experiments [1, 2, 3, 4]. Among the various ongoing or planed double beta decay detectors, the Cd-based semiconductor detectors CdTe and CdZnTe, which are currently used to study various rare processes in several areas of physics (X-ray physics, astrophysics, medical applications, etc.), have been proposed to be used in the COBRA double beta decay experiment [4]. The great advantage of this experiment is related to the fact that the double beta emitters are part of the detectors themselves and, in contrast to scintillators, they have good energy resolution.

It is worth mentioning that the detector of this experiment (it has recently been funded) the stable Te isotopes participate by 40-45% and a systematic investigation of the response of these isotopes as low-energy neutrino detectors has not been addresses up to the present. We mention that CUORE experiment is using a TeO_2 detector. It is the purpose of our present work to study the nuclear response to supernova neutrino energy spectra of all nuclear isotopes of COBRA detector. We, first, focus on the ^{128,130}Te isotopes which have big abundances on the natural Te and about 40% in the matter of COBRA detector. To this aim, the cross sections obtained in the context of the QRPA are folded with a known neutrino-energy distribution (see Sect. 2) [2, 3]. The folding procedure could be subsequently carried out for integrated and total cross section as well [2, 3]. In this work, results of the convoluted double differential cross sections for neutral current reactions of neutrinos with ^{128,130}Te isotopes are presented and discussed.

2. Supernova neutrino distributions

According to predictions of recent numerical simulations [1], the shape of the energy distributions of supernova neutrinos is determined by the conditions under which the neutrinos are emitted from the collapsing star. Even though the use of a thermal spectrum to describe the energy distribution of the supernova neutrinos is intuitively correct, various interactions cause deviations of the spectral shape from a purely thermal one. These deviations are described by the introduction of a chemical potential (degeneracy parameter η), narrowing the width w of the spectrum compared to the purely thermal shape, an effect referred to as pinching. Furthermore, the spectral shape is influenced by the fact that the neutrino cross section scales with the square of the lepton energy, forcing high-energy neutrinos to leave the star promptly.

In addition, the fact that different kinds of neutrinos are involved in different interactions, and that the reaction cross sections of the (anti-) neutrino depends on its energy, flavor, and helicity, modulates this picture. For all (anti-)neutrino flavors, the energies are in the range of a few to a few tens of MeV although calculations of neutrino transport that use different opacities achieve somewhat different spectra [1]. In the present work we employ the following distributions for supernova neutrino energy spectra.

2.1. The Fermi-Dirac distributions

Traditionally, Fermi-Dirac spectra described by the analytic expression

$$\eta_{FD}(\epsilon_\nu) = \frac{N_{FD}(\eta)}{T^3} \frac{\epsilon_\nu^2}{1 + exp(\epsilon_\nu/T - \eta)}, \tag{1}$$

where $N_{FD}(\eta)$ is a normalization factor depending on the degeneracy parameter η, were utilized as the convenient description for the energy distribution of supernova neutrinos [2, 3].

2.2. The power-law distributions

As it also supported by recent core collapse supernova simulations [1], the supernova-neutrino energy spectra can be rather accurately parameterized by a power-law distribution of the form

$$\eta_{PL}(\epsilon_\nu) = N_{PL}(\alpha)\Big(\frac{\epsilon_\nu}{\langle\epsilon_\nu\rangle}\Big)^\alpha exp\Big(-\frac{(\alpha+1)\epsilon_\nu}{\langle\epsilon_\nu\rangle}\Big), \tag{2}$$

where $N_{PL}(\alpha)$ is a normalization factor depending on the pinching parameter α [2, 3].

It is worth mentioning that the above neutrino energy distributions are not very satisfactory in the high energy dail $\epsilon_\nu \approx 60 - 80$ MeV. In this energy region other important effects (like the neutrino-neutrino interaction, neutrino oscillations, neutrino hierarchy problem, etc.) must be taken into account [5, 6]. In the present work we do not consider such effects since the nuclear excitations derived by our QRPA method lie in the energy region $E_x \leq 30 - 40$ MeV [7].

3. Results and discussion

In this work we studied folded double differential cross sections for ^{128}Te and ^{130}Te isotopes. By using the original results of the double differential cross sections obtained with our QRPA [7] and Eqs. (1) and (2) we calculated the corresponding folded cross sections as functions of the incoming neutrino energy ε_ν for a mean energy $\langle E\rangle$ and various values of the width parameters w and temperature T.

The results are illustrated in Fig. 1, for ^{128}Te where a Fermi-Dirac distribution was used, and Fig. 2, for ^{130}Te where a Power-law distribution was employed. More specifically, in Fig. 1 we plot the folded double differential cross section of ^{128}Te for T= 2.57, 3.58 MeV and η= 4.4, 1.1 while in Fig. 2, those of ^{130}Te for pinching parameter values α =5.1, 2.7.

The plots of Figs. 1, and 2 have been obtained by varying the scattering angle from 0° to 180° with a step $\Delta\Phi = 155^o$. For a fixed value of the width w the peak of the folded cross section is shifted towards larger incoming neutrino energies ε_ν as the scattering angle Φ increases.

The peaks lie around ε_ν = 20-30 MeV for each fixed value of the scattering angle Φ. As can also be seen, there is a clear T and w dependent increase of the peak shown for each curve.

In conclusion, our results show a rich response of both isotopes to supernova neutrino spectra. A detailed study of the response of Te isotopes to SN neutrinos is done elsewhere [8].

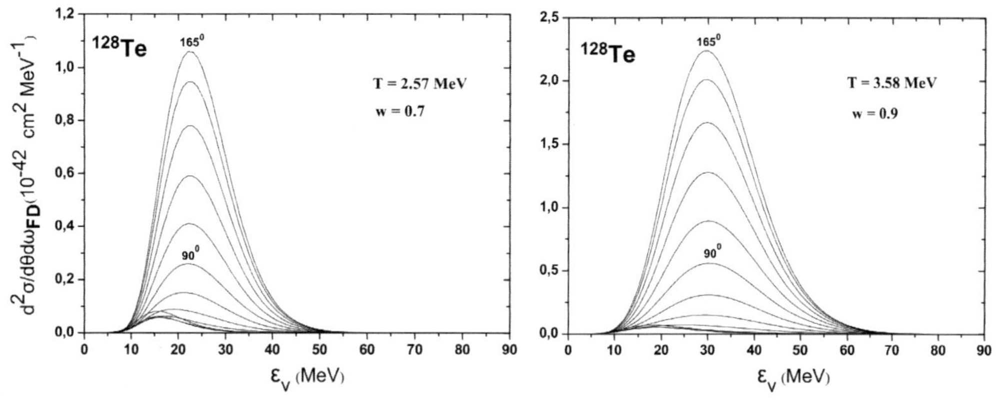

Figure 1. Folded double differential cross sections for ^{128}Te. A Fermi-Dirac (FD) distribution for width parameter values $w=$ 0.7, 0.9 and average neutrino energy $\langle E\rangle = 12$ MeV is used.

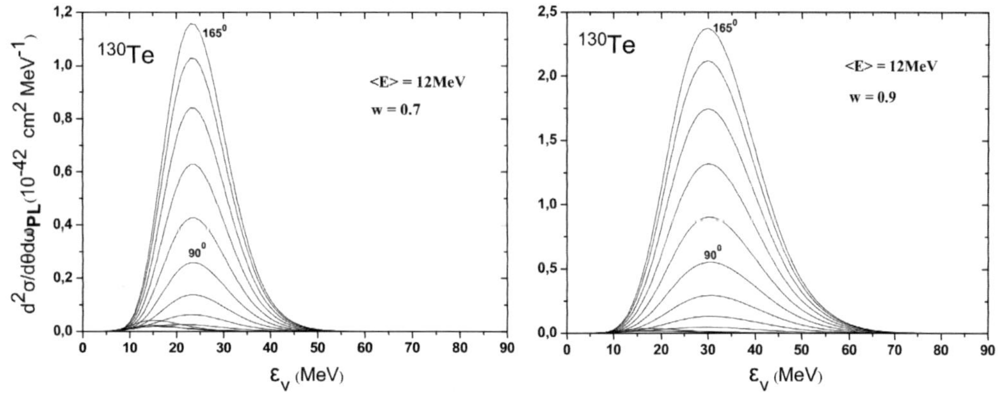

Figure 2. Same as in Fig. 1 but for ^{130}Te with the use of a Power-Law (PL) distribution.

3.1. Acknowledgments

We wish to acknowledge financial support from the ΠΕΝΕΔ/807 project of the Greek General Secretariat for Research and Technology.

References

[1] Keil M.T., Raffelt G.G, and Janka H-T. 2003 *Astrophys. J.* **590** 971–991.
[2] Jachowicz N., McLaughling G.C. and Volpe C. 2003 *Phys. Rev. C* **68** 055502.
[3] Volpe C. 2008 *Phys. Rev. C* **77**, 055501-12; Gava J. and Volpe C. 2008 *Phys. Rev. D* **78** 083007.
[4] Zuber K. 2001 *Phys. Lett. B* **519** 1–7; 2006 *Prog. Part. Nucl. Phys.* **57**, 235–240.
[5] Gava J., Kneller J., Volpe C and McLaughlin G.C. 2009 *Phys. Rev. Lett.* **103** 071101.
[6] Volpe C. 2009 Low energy neutrino scattering, hep-ph/09061476.
[7] Chasioti V.C. and Kosmas T.S. 2009 *Nucl. Phys. A* **829** 234–252.
[8] Tsakstara V., Kosmas T.S. *et al.*, in preparation.

Topics in Astroparticle and Underground Physics (TAUP 2009) IOP Publishing
Journal of Physics: Conference Series **203** (2010) 012094 doi:10.1088/1742-6596/203/1/012094

Lepton-nucleon interactions at the next-to-the-leading order

A. Aleksejevs[1], **S. Barkanova**[2]

SWGC, Memorial University, Division of Science, Corner Brook, NL, A2H 6P9, Canada[1],
Acadia University, Department of Physics, Wolfville, NS, B4P 2B2, Canada[2]

E-mail: `aaleksejevs@swgc.mun.ca`[1], `svetlana.barkanova@acadiau.ca`[2]

Abstract. Next-to-Leading-Order (NLO) effects play a crucial role in tests of the Standard Model (SM), and require careful theoretical evaluation. Neutrino physics, which has just entered the precision age, is an excellent place to search for new physics, but also requires considerable theoretical input. We show how we adopted computational symbolic packages such as FeynArts, FormCalc, Form and LoopTools for the evaluation of one-loop electroweak and hadronic radiative corrections.

1. Introduction

With recent advances in the automatization of the (NLO) calculations, it is reasonable to consider these methods in the applications towards the electro-weak and hadronic processes. Computer packages such as FeynArts [2], FormCalc, LoopTools [3] and Form [4] created a possibility to perform such type of calculations. In the work presented here, we extend FeynArts for the NLO symbolic calculations of amplitude or differential cross section. Using Dirac and Pauli type couplings, we construct the computational model [5] enabling us to deal with electron-nucleon scattering up to NLO level. Using this model we compute parity-violating asymmetries up to NLO level in electron-proton (e-p) scattering. Additionally we have developed an extension named Computational Hadronic Model (CHM) [7] of the FeynArts towards the hadronic sector using the Chiral Perturbation Theory (ChPTh). Later we have applied CHM to extract strong coupling constants arising in the chiral lagrangian from the experimental branching ratios (BR) for the following processes: $\Sigma^{*+} \to \Sigma^0 + \pi^+$, $\Sigma^{*-} \to \Lambda + \pi^+$, $\Xi^{*0} \to \Xi^{*-} + \pi^+$ and $\Delta^{++} \to p + \pi^+$.

2. Electroweak Sector

Electroweak properties of the nucleon can be studied by parity-violating electron-nucleon scattering at low energies. Such experiments can measure the asymmetry coming from the difference between cross sections of left- and right-handed electrons ($A = \frac{d\sigma_+ - d\sigma_-}{d\sigma_+ + d\sigma_-}$). This asymmetry between left- and right-handed particles is clearly predicted in the Standard Model (SM) of particle physics. Extracting the physics of interest from the measured asymmetry requires evaluating NLO contribution to electroweak scattering at very high precision. The dominant contribution normally comes from the Leading Order (LO) correction in perturbation theory. The NLO effects in the physics of the electroweak interactions plays a crucial role in the tests of the Standard Model. In this project, we took into consideration the NLO effects in parity-violating lepton scattering and have computed radiative corrections to the parity violating

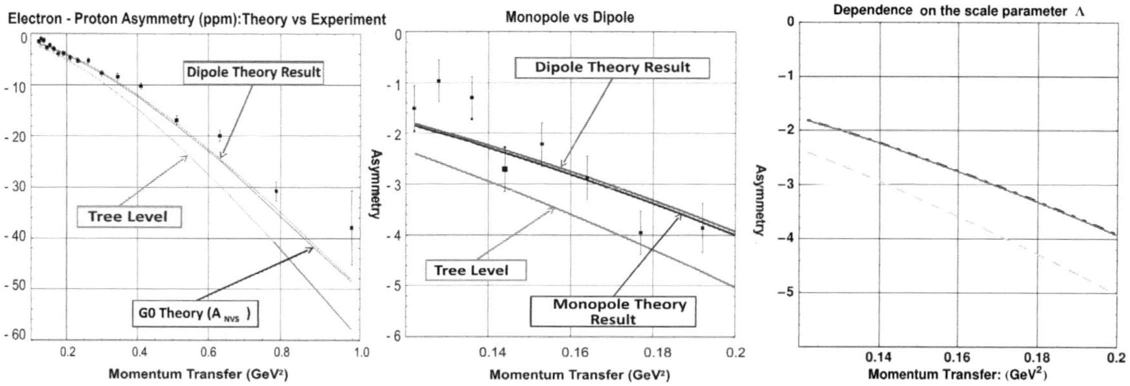

Figure 1. a) Asymmetry for the PV electron-proton scattering is given on the left figure. Dipole theory results are given by LO+NLO contributions, and A_{NVS} is no-vector-strange asymmetry given in [1]. b) Figure in the centre shows the difference in the asymmetries for monopole and dipole formfactors. c) Figure on the right shows the difference in asymmetries for the two limiting values of scale parameter, $\Lambda^2 = 0.4 \cdot m_N^2$(solid red curve) and $\Lambda^2 = 2.0 \cdot m_N^2$(dashed blue curve). Dashed green curve is a leading order contribution.

asymmetries. In general, we have extended FeynArts by including Dirac and Pauli form factors in couplings taken in the dipole/monopole form without strange quark contribution. Calculations were done in the on-shell renormalization scheme using Feynman gauge. Detailed description of this model is given in [5]. To avoid the infrared divergences, we have treated the final asymmetries with both Soft and Hard Photon Bremsstrahlung (SPB+HPB) [8] contributions. Computed results for the asymmetry are given in the Fig.[1a] for the range of the momentum transfers up to $1.0\,GeV^2$ in the forward scattering.

It is clear from Fig.[1a] that NLO PV effects are of the order of 20%. Also we can see that our results are in excellent agreement with theoretical predictions of G0 group [1]. Comparing our predictions with the G0 [1] experimental asymmetry we can conclude that there is an evident discrepancy between theory and experiment. Recalling the fact that our calculations were completed without an account for the sea of the strange quarks in the nucleon, good explanation to this would be a non negligible strange content of the nucleon. Moreover, we have used a model dependent form-factors and it is reasonable to investigate impact of this model dependence on calculated asymmetries. In this case we looked at the difference between monopole and dipole results and dependence on the scale parameter Λ (see Fig.[1b,c]). From Fig.[1b,c] it is evident that calculations of the asymmetries are independent of the choice of the type of form-factors and have virtually no dependence on the scale parameter Λ. Although model dependence will become evident when calculating absolute cross sections, for the asymmetries calculations presented results are model independent. We also reserved the kinematic dependence in all types of our radiative corrections. It will make it easier to adopt our results to the current and future parity-violating experiments for any lepton-hadron scattering processes.

3. Hadronic Sector

Tremendous success of the ChPTh in the description of the hadronic interactions in the non-perturbative regime of the QCD attracted attention of the physics community for decades. To calculate amplitudes or cross sections, we need a theoretical input at the NLO retaining full kinematic dependencies. To date, there are several packages (FeynArts, FormCalc, FeynCalc and Form) allowing us to produce semi-automatic calculations in the high energy physics.

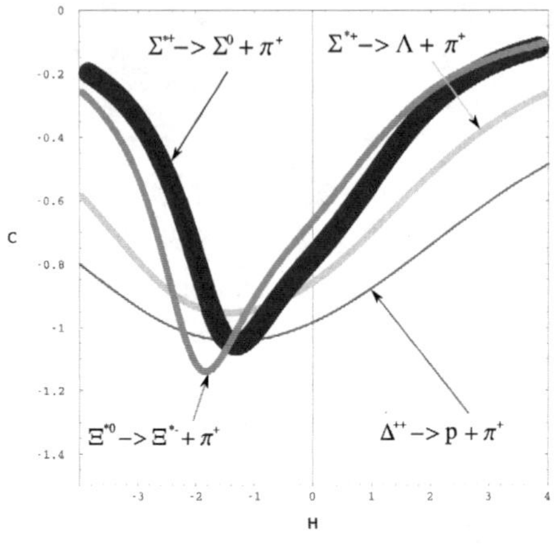

Figure 2. Parametric plots for the strong coupling constants C and H fitted to the BR of $\Sigma^{*+} \to \Sigma^0 + \pi^+$, $\Sigma^{*-} \to \Lambda + \pi^+$, $\Xi^{*0} \to \Xi^{*-} + \pi^+$ and $\Delta^{++} \to p + \pi^+$. Thickness of the curves represents the experimental uncertainty taken from the BR for these decays.

Although FeynArts and FormCalc were originally designed for SM calculations, the flexibility of the programs allows us to extend them to interactions appropriate for the hadronic sector. This was a main reason to use FeynArts and FormCalc as a base languages for the automatization of chiral hadronic calculations. Here we have used a CHM which detailed description reader can find in [7]. As an application and test of this model, we decided to extract strong coupling constants of ChPTh from the experimental values of BR to decays of resonances to baryons, such as: $\Sigma^{*+} \to \Sigma^0 + \pi^+$, $\Sigma^{*-} \to \Lambda + \pi^+$, $\Xi^{*0} \to \Xi^{*-} + \pi^+$ and $\Delta^{++} \to p + \pi^+$. The BR are centered about 1.4, 1.4, 1.3 and 1.5 respectively and as it is expected, leading order calculations are SU(3) conserving and will result in equally valued BR. This fact prompts us to look at the NLO corrections. The NLO calculations were completed with an account of the octet of mesons, baryons and decuplet of resonances participating in the one loop diagrams. If we take experimental BR and fit strong coupling constants C and H to the values of BR we can reproduce results predicted earlier in [6]. For the each decay outlined earlier in this article, parametric plots for the C coupling constant as a function H are shown in Fig.[2].

From Fig.[2] it is clear that all curves indicate the central value of the C coupling constant is around $|C| = 1.0 \pm 0.2$. From this we can derive constrains on the H coupling constant as $H = -1.5 \pm 0.5$. Our predictions are consistent with the results of [6]: $|C| = 1.2 \pm 0.1$ and $H = -2.1 \pm 0.7$. In general, these calculations served us as an excellent test of rather involved computer based calculations. The results of the CHM can be expressed analytically, using FormCalc with an amplitude expressed in the Passarino-Veltman basis as well as numerically with application of the package LoopTools. The future applications of the CHM can be seen in the automatization of the calculations of NLO radiative corrections for the production/decay channels of hadronic physics.

References

[1] G0 Collaboration, arXiv:nucl-ex/0506021v1.
[2] T. Hahn, Comput. Phys. Commun. 140 (2001) 418, hep-ph/0012260.
[3] T. Hahn, M. Perez-Victoria, Comput. Phys. Commun. 118 (1999) 153 , hep-ph/9807565.
[4] J.A.M.Vermaseren, math-ph/0010025.
[5] A Aleksejevs, S. Barkanova and P. Blunden, J. Phys. G: Nucl. Part. Phys. 36, 045101, (2009).
[6] M. Butler, M. Savage, R. Springer, Nuclear Physics B 399, Issue: 1, (1992).
[7] A. Aleksejevs and M. Butler, arXiv:0710.3580 (2007).
[8] A. Aleksejevs, S. Barkanova, P. Blunden and N. Deg, arXiv:0707.0657 (2007).

The Double Chooz Experiment - toward the high precision measurement in the quest for θ_{13}

Jelena Maricic for the Double Chooz collaboration
Drexel University, 3141 Chestnut St. 12-816, Philadelphia, PA, 19104
E-mail: jelena@physics.drexel.edu

Abstract. The Double Chooz experiment will perform a highly sensitive measurement of the neutrino mixing angle $\sin^2 2\theta_{13}$. The Double Chooz will utilize a two detector design where the near detector provides an "unoscillated" neutrino flux measurement (high count rate, but higher background as well), while the far detector provides the "oscillated" neutrino flux characterized by the lower count rate and lower background level. The signal comparison between the two detectors will be necessary to see the subtle sign of θ_{13}. Double CHOOZ will be built at the site of the original CHOOZ experiment which still provides the best world upper limit on the value of the neutrino mixing angle $\sin^2 2\theta_{13} < 0.2$. We will report on the status of the Double Chooz experiment that will contribute to the discovery of θ_{13} if above 0.03 or improve the existing limit on $\sin^2 2\theta_{13}$ from 0.2 to 0.03 in just 3 years of running of the Double Chooz experiment. The Double Chooz is currently undergoing the construction of the far detector and will start taking data in April 2010.

1. Introduction

Recent discoveries and measurements in neutrino physics with experiments like KamLAND [1], SNO [2] [3], Superkamiokande [4], and others, have advanced the knowledge of neutrino physics from mere observation to the point where precision measurements are necessary. While the mixing angles θ_{sol} and θ_{atm} have been measured to some level of precision, the same has not been true for the angle θ_{13}. The measurement of the mixing angle θ_{13} is of great importance for the complete understanding of the neutrino oscillations and observability of CP-violation.

A new two-detector reactor neutrino experiment at the CHOOZ site, the Double Chooz experiment, will be sensitive to $\sin^2 2\theta_{13} > 0.03$ (Figure 1 presents limits on the value of θ_{13} that can be achieved with Double Chooz). In order to successfully measure such a small value of $\sin^2 2\theta_{13}$, the Double Chooz experiment will use near detector to obtain an "unoscillated" flux with high count rate and higher level of backgrounds, and far detector to obtain "oscillated" flux at a lower count rate, but lower background level as well. Although the next generation accelerator experiments will also have the potential to measure θ_{13}, their measurement will be degraded not only by statistical and systematic uncertainties, but also by correlations and degeneracies between θ_{13}, θ_{atm}, $\mathrm{sgn}(\Delta m^2_{31})$, CP-$\delta$ phase and matter effects [7] [5]. The measurement of θ_{13} by Double Chooz is much simpler, and does not suffer from the above-mentioned ambiguities. Thus, the Double Chooz experiment will provide a clean, independent measurement of θ_{13} which will provide necessary robustness to our knowledge of θ_{13} and neutrino mixing properties [8]. The Double Chooz will start operating with the far detector only, in spring of 2010. Double Chooz will be able to probe 90% of the $\sin^2 2\theta_{13}$ range of interest for the future

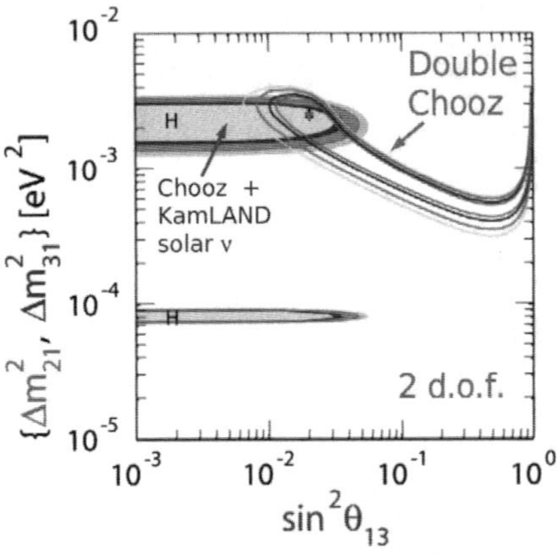

Figure 1. DC sensitivity contours (gray 1 σ, blue 90%, magenta 2σ, cyan 3σ C.L.) in ($\sin^2 2\theta_{13}$, Δm^2_{31}) plane generated for $\sin^2 2\theta_{13}$ = 0.02 and $\Delta m^2_{31} = 2.5 \times 10^{-3}\,\text{eV}^2$ [10].

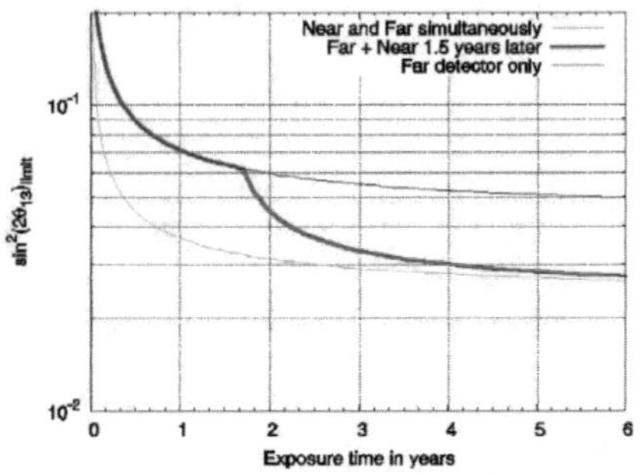

Figure 2. $\sin^2 2\theta_{13}$ sensitivity limit for Double Chooz as a function of exposure time. The expected startup for Double Chooz is in 2010.

CP-violation measurement. This experiment can also improve the $\sin^2 2\theta_{13}$ limit with time, and Figure 2 illustrates this.

2. Measurement of $\sin^2(2\theta_{13})$ with reactor anti-neutrinos in Double Chooz

Successful measurement of $\sin^2 2\theta_{13}$ requires a dramatic lowering of the systematic error from 2.7% in CHOOZ to 0.6% in Double Chooz. This will mainly be achieved by a two-detector design that will cancel part of the systematic errors by comparison of two detectors [9] [10]. The anti-neutrino flux in Double Chooz will be provided by the Chooz pressurized water nuclear reactors. Fuel composition as well as dynamics of the fuel burn-up and replacement are known in detail, providing excellent knowledge of the expected anti-neutrino rates that will be used as a cross-check for the measurement obtained by the nearer of two detectors.

Double Chooz will measures the survival probability $P_{\bar{\nu}_e \to \bar{\nu}_e}$ of the $\bar{\nu}_e$s emitted from the nuclear plant. $\bar{\nu}_e$ are detected in Double Chooz via inverse-beta decay reaction $\bar{\nu}_e + p \to e^+ + n$. The target mass is loaded with gadolinium, which increases energy emitted in neutron capture on gadolinium to 8 MeV. Figure 3 shows an example measurement for a relatively large value of $\sin^2 2\theta_{13} \sim 0.1$.

3. Detector design

The Double Chooz experiment [9] [10] will consist of two identical detectors containing 10.3 m^3 of liquid scintillator target doped with 0.1% of Gadolinium ((see Figure 4). The far detector will be located at the site of the original CHOOZ detector at 1 km and 1.1 km from the two cores of the Chooz nuclear reactors. The far site has shielding of ~300 mwe of 2,800 kg/m^3 rock density. The Chooz near site will be built 400 m from reactor cores with 100 mwe. This will be sufficient to keep the signal to noise ratio above 100 at the near detector site, keeping the associated systematic error much smaller than 1%.

Another important study that has been conducted is the stability of scintillator. As a solvent we will use 20% PXE, 80% Dodecane mixture. The scintillator will be loaded with 0.1% of Gd and it is being developed at Max-Planck-Institut fur Kernphysik (MPIK) at Heidelberg and Gran Sasso National Laboratory (LNGS). Results on the different scintillator formulation are now available on more than 4 years long tests. Even more, heating of the scintillator samples to

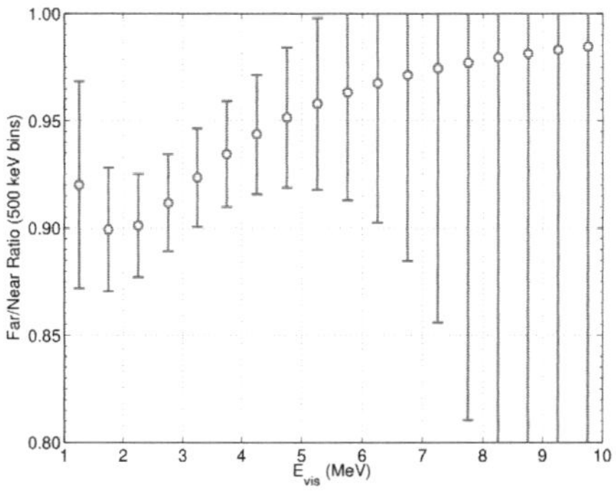

Figure 3. Far to Near detector spectrum ratio for $\sin^2 2\theta_{13} = 0.1$ and $\Delta m^2_{31} = 2.5 \times 10^{-3}\,\mathrm{eV}^2$. Error bars are only statistical [10].

Figure 4. The Double Chooz far detector design at the CHOOZ underground site.

40°C has been used to study the aging process in the accelerated manner. Chosen scintillator mixture is Gd-Dmp Beta Dikitonate. Measured attenuation length is a few meters at 420 nm.

Data acquisition system will use Flash-ADC CAEN N(V)1726 modules developed by APC-Paris + CAEN. The modules have 4 channels, wave-form sampling is performed at 500MHz. They posses 8-bit resolution (few PEs/ch for $\bar{\nu_e}$ events). Modules will continuously digitize with zero deadtime (if DAQ can sustain the trigger rate).

Extensive calibration of the detectors are planned for detector related systematic error evaluation and reduction. Target region will be tested in several different ways: deployment of radioactive sources along the detector central axis and throughout the detector volume (University of Alabama (UA), Drexel University and Argonne National Laboratory), laser and LED systems (UA, Tokyo Tech, Oxford University and Illinois Institute of Technology).

4. Conclusion

The construction of the Double Chooz detector has started in the late spring of 2007. As of October 2009, inner veto detector, buffer and gamma catcher have been installed, with target vessel to follow. Detector filling and outer veto installation are planned for winter 2009/2010. The Double Chooz experiment will start taking data in Spring of 2010 and in just 3 years of running, under the condition of achieving projected systematic uncertainty of 0.6%, reach sensitivity to 0.03 value for θ_{13}, providing an ~85% reduction in the existing limit set by Chooz experiment and therefore having an excellent discovery potential.

References

[1] S. Abe *et al* 2008 *Phys. Rev. Lett.* **100** 221803 *hep-ex/0801.4589*
[2] S. N. Ahmed *et al* 2004 *Phys. Rev. Lett.* **92** 181301 (2004) *nucl-ex/0309004*
[3] B. Aharmim *et al* 2005 *Phys. Rev.* **C72** 055502 *nucl-ex/0502021*
[4] Y. Ashie *et al* 2004 *Phys. Rev. Lett.* **93** 101801 *hep-ex/0404034*
[5] P. Huber, M. Lindner, T. Schwetz and W. Winter 2003 *Nucl. Phys.* **B665** 487
[6] M. Apollonio *et al* 2003 *Eur. Phys. J.* **C27** 331-374 *hep-ex/0301017*
[7] H. Minakata, H. Sugiyama, O. Yasuda, K. Inoue and F. Suekane 2002 *Eur. Phys. J.* **C27** *hep-ph/0211111*
[8] K. Anderson *et al* 2004 *hep-ex/0402041*
[9] S. Berridge *et al* 2004 *hep-ex/0410081*
[10] M. Goodman *et al* 2006 *hep-ex/0606025*

Magnus approximation for neutrino oscillations with three flavors in matter

Alexis A. Aguilar-Arevalo and Juan Carlos D'Olivo

Instituto de Ciencias Nucleares, Universidad Nacional Autónoma de México, Apartado Postal 70-543, 04510, Distrito Federal, México

E-mail: alexis@nucleares.unam.mx, dolivo@nucleares.unam.mx

Abstract. The Magnus expansion of the evolution operator is used to find approximate analytical solutions to the problem of three neutrino oscillations in matter with varying density. Survival probabilities are calculated for the case of solar and supernova neutrinos.

A neutrino propagating through matter can be described by a Schrödinger equation in the basis of definite flavor states $\{|\nu_e\rangle, |\nu_\mu\rangle, |\nu_\tau\rangle\}$, with a Hamiltonian $\mathcal{H}(t)$ that depends on the neutrino interaction potential with the medium $V(t)$, the neutrino mass splittings Δm^2_{ij} between the definite mass states $\{|\nu_1\rangle, |\nu_2\rangle, |\nu_3\rangle\}$, and the neutrino mixing matrix in vacuum U. Diagonalization of this Hamiltonian at any time t defines the *adiabatic* basis $\{|\nu_{1m}(t)\rangle, |\nu_{2m}(t)\rangle, |\nu_{3m}(t)\rangle\}$, in which the time evolution operator $\mathcal{U}(t,t_0)$ satisfies the equation

$$i\hbar\frac{\partial}{\partial t}\mathcal{U}(t,t_0) = \left[\mathcal{H}_D(t) - i\hbar\, U_m^\dagger(t)\frac{\partial U_m(t)}{\partial t}\right]\mathcal{U}(t,t_0), \tag{1}$$

where $\mathcal{H}_D(t) = U_m(t)^\dagger\mathcal{H}(t)U_m(t)$ is the diagonalized Hamiltonian with energy eigenvalues E_1, E_2, E_3, and $U_m(t)$ is a unitary matrix that takes the amplitudes in the adiabatic basis at time t into the corresponding ones in the flavor basis. This is, $U_m(t)$ is the generalization of the mixing matrix in matter. The smallness of the ratio $\Delta m^2_{21}/\Delta m^2_{31} \simeq 1/32$, allows one to find the following approximate expression for $U_m(t)$, valid throughout the material medium:

$$U_m(t) \simeq \mathcal{O}_{23}\,\Gamma\,\mathcal{O}_{13}(\theta^{\rm m}_{13}(\rm t))\,\Gamma^*\mathcal{O}_{12}(\theta^{\rm m}_{12}(\rm t)), \tag{2}$$

where $\mathcal{O}_{12}(\theta^m_{12}(t))$ is a rotation in the 12 plane by an angle $\theta^m_{12}(t)$, $\mathcal{O}_{13}(\theta^m_{13}(t))$ is a rotation in the 13 plane by an angle $\theta^m_{13}(t)$, $\mathcal{O}_{23}$ is a rotation in the 23 plane by the vacuum mixing angle θ_{23}, and $\Gamma = \mathrm{diag}(1, 1, e^{i\delta})$ is a diagonal matrix containing the CP violating phase δ. The matter mixing angles $\theta^m_{12}(t)$ and $\theta^m_{13}(t)$ arise from the diagonalization of $\mathcal{H}(t)$ in a low density and a high density regions, respectively, in each of which, the Hamiltonian reduces to that of a two-state problem. We make the additional change of representation $\mathcal{U}(t,t_0) = \Gamma\; P(t,t_0)\; \mathcal{U}_P(t,t_0)\; \Gamma^*$, where $P(t,t_0) = \mathrm{diag}(e^{-i\alpha_1}, e^{-i\alpha_2}, e^{-i\alpha_3})$, with $\alpha_k(t) = \frac{1}{\hbar}\int_{t_0}^t dt' E_k(t)$ to define the evolution operator $\mathcal{U}_P(t,t_0)$ which satisfies the simple equation $i\hbar\; \partial\mathcal{U}_P/\partial t = \mathcal{H}_P\mathcal{U}_P$ with

$$\mathcal{H}_P(t) = \begin{bmatrix} 0 & h_a & 0 \\ h_a^* & 0 & 0 \\ 0 & 0 & 0 \end{bmatrix} + \begin{bmatrix} 0 & 0 & h_b \\ 0 & 0 & h_c \\ h_b^* & h_c^* & 0 \end{bmatrix} = \mathcal{H}_l(t) + \mathcal{H}_h(t), \tag{3}$$

where

$$\begin{aligned} h_a(t) &= -i\hbar\, \dot{\theta}^m_{12}\, \exp -i[\alpha_2 - \alpha_1], \\ h_b(t) &= -i\hbar\, \dot{\theta}^m_{13}\, c^m_{12}\, \exp -i[\alpha_3 - \alpha_1], \\ h_c(t) &= -i\hbar\, \dot{\theta}^m_{13}\, s^m_{12}\, \exp -i[\alpha_3 - \alpha_2], \end{aligned} \tag{4}$$

and we have used the shorthand notation $s^m_{ij} = \sin\theta^m_{ij}$, $c^m_{ij} = \cos\theta^m_{ij}$.

A factorized solution $\mathcal{U}_P(t,t_0) \simeq \mathcal{U}_l(t,t_0)\ \mathcal{U}_h(t,t_0)$, where $\mathcal{U}_{l,h}(t,t_0)$ is the solution to the problem defined by the Hamiltonian $\mathcal{H}_{l,h}(t)$, is possible given that $[\mathcal{U}_l(t,t_0), \mathcal{H}_h(t)] = 0$. That this condition is approximately satisfied follows from the smallness of Δ_{21}/Δ_{31}, which causes the low and high density regions to be well separated as functions of the neutrino potential. Furthermore, it can be shown that $h_b(t) \simeq 0$ for all evolution, because whenever $\dot{\theta}^m_{13} \neq 0$, we have $\theta^m_{12} \simeq \pi/2$. In turn, this also implies that $s^m_{12} \simeq 1$ in $h_c(t)$.

A solution of the form $\mathcal{U}(t,t_0) = \exp[\Omega(t,t_0)]$ where the operator Ω is given by a series $\sum_{n=1}^{\infty} \Omega_n$ can be constructed following the method outlined by Magnus [1], in which the nth term in the series is calculable as a sum of integrals of n-fold nested commutators of $H(t)$. Each Ω_n is is separately anti-Hermitian, thus truncating the series at any n yields a unitary approximation to $\mathcal{U}(t,t_0)$. For a recent review on the method see [2]. Following the calculation in [3] for the two neutrino case, we use the first order term of the Magnus expansion to write $\mathcal{U}_{l,h}(t,t_0) \simeq \exp\left(-\frac{i}{\hbar}\int_{t_0}^{t} dt'\ \mathcal{H}_{l,h}(t')\right)$, giving the following explicit matrix form for $\mathcal{U}_P$:

$$\mathcal{U}_P(t,t_0) \simeq \begin{bmatrix} C_{K_l} & 0 & -S_{K_l}S_{K_h}\frac{K_lK_h}{|K_l||K_h|} \\ 0 & C_{K_l}C_{K_h} & 0 \\ 0 & 0 & C_{K_h} \end{bmatrix} - i \begin{bmatrix} 0 & S_{K_l}C_{K_h}\frac{K_l}{|K_l|} & 0 \\ S_{K_l}\frac{K_l^*}{|K_l|} & 0 & S_{K_h}C_{K_l}\frac{K_h}{|K_h|} \\ 0 & S_{K_h}\frac{K_h^*}{|K_h|} & 0 \end{bmatrix}, \tag{5}$$

where $K_{l,h}(t) = \int_{t_0}^{t} dt'\ h_{a,c}(t')$, and we have defined $C_{K_{l,h}} \equiv \cos|K_{l,h}|$, $S_{K_{l,h}} \equiv \sin|K_{l,h}|$, It is straightforward to verify that the unitarity condition ${\mathcal{U}_P}^\dagger \mathcal{U}_P = I$ is satisfied.

The survival probability for an electron neutrino produced at the time t_0 and escaping an astrophysical source with decreasing matter density profile, averaged over a large travel distance to the Earth and the production position, can be calculated as $\langle P_{ee}\rangle = \sum_i |U_{ei}|^2 |\psi^m_i(T)|^2$, where $\psi^m_i(T)$ is the amplitude of the state in the adiabatic state vector $|\nu_{im}(T)\rangle$ at the time T when the neutrino reaches the border of the medium. Because of the normalization condition $\sum_i |\psi^m_i(T)|^2 = 1$, it suffices to know $|\psi^m_1(T)|$ and $|\psi^m_3(T)|$, which are given by:

$$\begin{aligned} |\psi^m_1(T)|^2 &\simeq {c^m_{12}}^2(t_0)\, {c^m_{13}}^2(t_0)\, \cos^2|K_l(T)| + {s^m_{12}}^2(t_0)\, {c^m_{13}}^2(t_0)\, \sin^2|K_l(T)|\cos^2|K_h(T)| \\ &\quad + {s^m_{13}}^2(t_0)\, \sin^2|K_l(T)|\sin^2|K_h(T)| \\ |\psi^m_3(T)|^2 &\simeq {s^m_{13}}^2(t_0)\, \cos^2|K_h(T)| + {s^m_{12}}^2(t_0)\, {c^m_{13}}^2(t_0)\, \sin^2|K_h(T)|. \end{aligned} \tag{6}$$

The quantities $|K_l(T)| \simeq [\theta^m_{12} - \theta^m_{12}(t_0)]\ e^{-\kappa_l}$ and $|K_h(T)| \simeq [\theta^m_{13} - \theta^m_{13}(t_0)]\ e^{-\kappa_h}$ are given in terms of the *adiabatic scale factors* $\kappa_{l,h}$, which in our approximation depend on the point where the matter mixing angles have maximum derivative.

Figure 1 shows two applications of the derived solution. In Figure 1(a) we calculate the survival probability for solar neutrinos under the approximation of the exponential matter density profile of [4] as a function of $\Delta m^2_{21}/2E_\nu$, for parameters consistent with current experimental measurements, and compare between direct ($\Delta m^2_{31} > 0$) and inverted ($\Delta m^2_{31} < 0$) hierarchies. The points are the results from the Borexino and SNO experiments, and an expectation for *pp* neutrinos. The shaded area corresponds to the region of interest for neutrinos produced in the Sun (E_ν < 17 MeV). Figure 1(b) shows the iso-probability contours for a ν_e

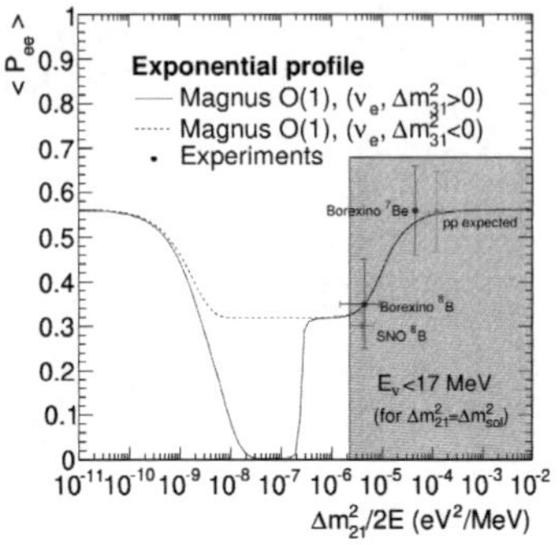

(a) Sun: exponential profile

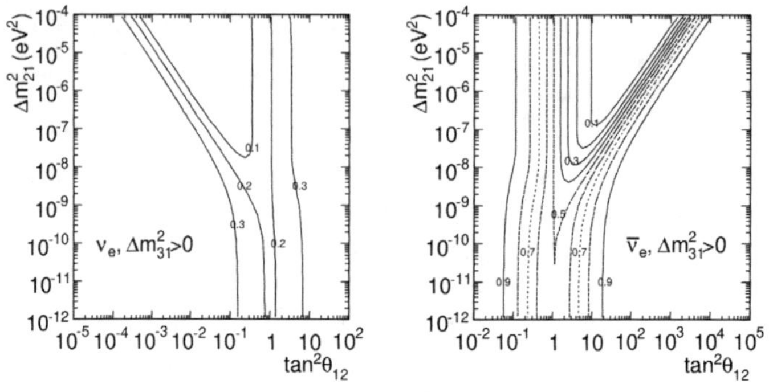

(b) Supernova: power-law profile ($n = 3$)

Figure 1. (a) Survival probability for solar neutrinos for $\Delta m^2_{21} = 7.59\times10^{-5}$ eV^2, $\theta_{12} = 34.4°$, $|\Delta m^2_{31}| = 2.4 \times 10^{-3}$ eV^2, $\sin^2 2\theta_{13} = 0.01$, with direct (solid) and inverted (dashed) hierarchy. (b) Iso-probability contours for supernova ν_e and $\bar{\nu}_e$ with fixed energy $E_\nu = 15MeV$, and $|\Delta m^2_{31}| = 2.4 \times 10^{-3}$ eV^2, $\tan^2\theta_{13} = 4 \times 10^{-4}$.

and a $\bar{\nu}_e$, produced at the core of a supernova in the plane $\Delta m^2_{12}/2E_\nu$ *vs.* $\tan^2\theta_{12}$ with direct hierarchy, assuming a power-law density profile $\rho \propto r^{-n}$, with $n = 3$ [5]. Symmetry properties relating the neutrino, antineutrino, direct and inverse hierarchy cases have been applied as described in [5].

In the case of the Sun, a comparison with the numerical solution given in [6] for the exponential profile has shown that our solution deviates from the exact result by as much as 30% between 10^{-10}eV$^2 < \Delta m^2_{21}/2E_\nu < 10^{-9}$eV^2. In this region $\langle P_{ee}\rangle$ is essentially that of a two neutrino problem with the small mixing angle θ_{13}, and is not accessible to solar neutrinos produced in fusion processes. Outside this region the approximation is better than 1%.

Conclusions

We have calculated the evolution operator of a neutrino propagating through matter with varying density for three neutrino flavors using a first order Magnus approximation, and have applied it to the case of solar and supernova neutrinos. Our solution provides a good account for non-adiabatic effects for neutrinos with arbitrary energies, altough higher order terms are needed to improve the accuracy in this regime.

Acknowledgments

The authors acknowledge the support of PAPIIT-UNAM through grant IN115607 and CONACyT through grant 83534. A.A.-A. also acknowledges the support of CIC-UNAM.

References

[1] W. Magnus, *Commun. Pure Appl. Math.*, **7**, 649 (1954).
[2] S. Blanes, F. Casas, J. Oteo, and J. Ros, Phys. Rep., **470**, 151-238 (2009).
[3] J.C. D'Olivo, Phys.Rev.D **45**, 924 (1992).
[4] J.N. Bahcall and R.K. Ulrich, Rev. Mod. Phys. **60**, 297 (1988).
[5] G.L. Fogli *et. al*, Phys. Rev. D **65**, 073008 (2002).
[6] J.S. Kim and K. Lee, Comp. Phys. Commun. **135**, 176 (2001).

The KATRIN Experiment

Marcus Beck for the KATRIN collaboration
Institut für Kernphysik, Westfälische Wilhelms-Universität Münster, Germany
E-mail: marcusb@uni-muenster.de

Abstract. The KArlsruhe TRitium Neutrino mass experiment, KATRIN, aims to search for the mass of the electron neutrino with a sensitivity of $0.2\,\mathrm{eV/c^2}$ (90% C.L.) and a detection limit of $0.35\,\mathrm{eV/c^2}$ (5σ). Both a positive or a negative result will have far reaching implications for cosmology and the standard model of particle physics and will give new input for astroparticle physics and cosmology. The major components of KATRIN are being set up at the Karlsruhe Institut of Technology in Karlsruhe, Germany, and test measurements of the individual components have started. Data taking with tritium is scheduled to start in 2012.

1. Introduction

After neutrino oscillation experiments have shown that neutrinos possess a non-zero rest mass the question of the absolute mass scale of neutrinos has become important for particle physics and cosmology (see e.g. [1, 2]). The goal of the KArlsruhe TRItium Neutrino experiment (KATRIN, [3]) is to search for the mass of the electron antineutrino with a sensitivity of $0.2\,\mathrm{eV/c^2}$. This will probe most of the mass range in which the three neutrino flavours have nearly degenerate masses and where neutrinos are of cosmological importance.

To this end a precision measurement of the endpoint region of the β-decay of tritium will be performed with KATRIN. The shape of this spectrum depends sensitively on the neutrino mass. This will be done by using a windowless gaseous tritium source for small systematic uncertainties, and an electrostatic energy filter of MAC-E filter type (electrostatic filter with magnetic adiabatic collimation, [4, 5]) for the analysis of the electron energy with high luminosity and high resolution. This is the same method as used at the Mainz [4, 6] and Troitzk neutrino mass experiments [5, 7], which have set the until now best upper limits of $2.3\,\mathrm{eV/c^2}(95\%)$ on the neutrino mass.

2. The set-up of KATRIN

Figure 1 gives an overview of the KATRIN set-up. It consists of a source section with the windowless gaseous tritium source, the WGTS (a), a transport section with a differential- and a cryopumping section (b), a pre-spectrometer (c), the main spectrometer (d) and a detection section with the focal plane detector (e).

The WGTS provides a source of electrons from tritium β-decay with high and uniform intensity. It basically consists of a 10 m long beam tube of 90 mm diameter through which tritium is circulated in a loop with a flow of 4.7 Ci/s and a maximum pressure of $4 \cdot 10^{-3}$ mbar. This results in a column density of $5 \cdot 10^{17}\,\mathrm{T_2/cm^2}$ and $\approx 10^{10}$ β-particles per second accepted by the spectrometer. In order to keep density fluctuations at 0.1% or less the WGTS is temperature stabilized to 30 mK at a temperature of 30 K by two-phase neon cooling. The purity of the T_2 is

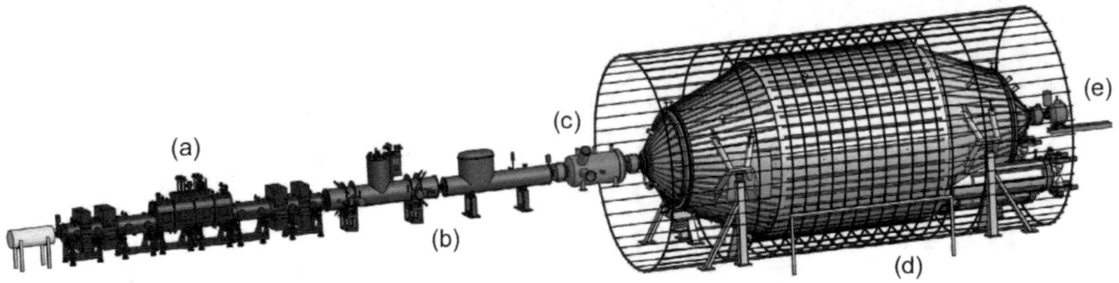

Figure 1. The KATRIN set-up (explanation see text)

maintained at $95 \pm 1\%$ by continous purification of the gas in the tritium loop. These stringent stability requirements result in a highly complex cryostat system [8, 9]. The technologies used will be tested using a demonstrator set-up, which will be available in spring 2010. In the final set-up the properties of the source will be checked regularly with an electron source, which is under development [10], by laser Raman spectroscopy, which already has been deployed successfully for test measurements [11], and a Penning trap in the beamline to the spectrometers using FT-ICR, which has recently been tested successfully [12].

The transport section serves to guide the electrons from tritium β-decay to the spectrometers without any losses and to prevent the transport of tritium at the same time, with a reduction factor for tritium of 10^{14} and a final flow of 10^{-14}mbar-l/s. This is achieved by a differential pumping section, which recently arrived at Karlsruhe for testing and commissioning, followed by a cryogenic pumping section. Here the remaining tritium gets cryosorbed on a layer of Argon frost on the walls of the stainless steel vessel at a temperature of ≈ 3.4 K. This method has been tested with the TRAP experiment successfully [13, 14].

The pre-spectrometer is a MAC-E filter like the main spectrometer. It rejects all electrons with energy < 200 eV below the endpoint, which are of no interest for the neutrino mass measurement and which even may increase the main spectrometer background. It has been successfully tested and is working at design parameters with an intrinsic background of $\mathcal{O}(10^{-2}$ counts per second (cps)). For now it is being used as a test case for the technologies and methods used for the main spectrometer, such as the vacuum system (see e.g. [15, 16]) and the electromagnetic design. For these tests it is presently equipped with a 64-pixel PIN diode [17].

The endpoint region of the β-spectrum is measured with the main spectrometer. This consists of a stainless steel vacuum vessel of 23 m length and 10 m diameter and will reach a vacuum of 10^{-11} mbar using turbomolecular pumps and non-evaporating getter material. The magnetic field at the entrance of this MAC-E filter is 4.5 T and at its exit 6 T while the field in the central analysis plane will be $3 \cdot 10^{-4}$ T resulting in an energy resolution of 0.93 eV. The retardation voltage is applied directly to the spectrometer hull. For fine shaping of the field and background suppression a modular inner wire electrode system will be used. The main spectrometer has been set-up at Karlsruhe and an outgassing rate of $< 10^{-12}$mbar-l/s has been achieved after bakeout. High demands on the quality of the electric potential and therefore for the precision of the machining of the wire modules required stringent quality control (see e.g. [18]). The installation of the wire modules is in progress.

The high voltage (HV) that creates the retardation potential in the analysis plane has to be highly stable. To reach the sensitivity of KATRIN, fluctuations of the HV of 3 ppm, i.e. of 60 mV, have to be detected. For this purpose a precision HV-divider has been developed [19] to measure the applied voltage with this precision. In conjunction with calibration sources based on conversion electrons from the decay of ^{83m}Kr this will ensure the required short and long term stability of the high voltage.

Electrons with sufficient energy to pass the retardation potential are counted with the focal plane detector. This is a 148-fold segmented PIN diode with surroundings made of low-activity materials to reduce the intrinsic background of the detector. Simulations show that a background of $3 \cdot 10^{-3}$ cps can be achieved in the signal region. The detector will be commissioned at Karlsruhe in spring 2010.

The KATRIN background count rate necessary to reach the intended sensitivity is $1 \cdot 10^{-2}$ cps. By shifting the signal in the detector via a DC-voltage the intrinsic detector background can be reduced to $1 \cdot 10^{-3}$ cps. However, other sources of background have to be controlled as well. Two major sources are background from the tank walls due to muon interactions and radioactive decays, and background from incidential Penning traps due to the magnetic and electric fields employed in the MAC-E filter. The former is suppressed by the axial magnetic field, which guides most electrons from the wall back to the wall. The remaining wall electrons will be suppressed by the wire electrode system, which is slightly more negative (O(100V)) than the tank wall. The background due to Penning traps has been investigated in detail at the pre-spectrometer and can be avoided in large part by proper shaping of the electrodes. However, some unavoidable Penning traps still exist, like the trap formed by the pre- and the main spectrometer. Investigation of these traps has started and first results are promising [20].

3. Conclusion

The KATRIN experiment searching for the neutrino mass is under construction. The major components are being assembled and tested, and integration of the components will start soon. Installation of the inner electrode system of the main spectrometer is ongoing. In 2010 the properties of the main spectrometer will be investigated using calibration sources. Data taking using tritium will start in 2012 and will proceed for five years.

4. Acknowledgement

The work by the author is supported by the German Bundesministerium für Bildung und Forschung.

References

[1] E. W. Otten and C. Weinheimer, Rep. Prog. Phys. **71** (2008) 086201
[2] J. Lesgourgues and S. Pastor, Phys. Rep. **429** (2006) 307
[3] The KATRIN collaboration (J. Angrik *et al.*), *KATRIN Design Report 2004*, FZKA Scientific Report 7090, 2005, available online at `http://bibliothek.fzk.de/zb/berichte/FZKA7090.pdf`
[4] A. Picard *et al.*, Nucl. Instr. Meth. **B 63** (1992) 345
[5] V. M. Lobashev and P. E. Spivak, Nucl. Instr. Meth. **A 240** (1985) 305
[6] Ch. Kraus *et al.*, Eur. Phys. J. **C 40** (2005) 447
[7] V. M. Lobashev *et al.*, Phys. Lett. **B 460** (1999) 227
[8] S. Grohmann, Cryogenics **49** (2009) 413
[9] B. Bornschein, Prog. Part. Nucl. Phys. **57** (2006) 38
[10] K. Valerius *et al.*, New Journal of Physics **11** (2009) 063018
[11] M. Sturm *et al.*, to be published in Las. Phys. **20(2)**, 2010
[12] M. Ubieto-Díaz *et al.*, Int. J. Mass Spectr. **288** (2009) 1
[13] O. Kazachenko *et al.*, Nucl Instr. and Meth. **A 587** (2008) 136
[14] F. Eichelhardt *et al.*, Fusion Science and Technology **54** (2008) 615
[15] X. Luo *et al.*, Vacuum **81** (2008) 777
[16] J. Wolf, Journal of the Vacuum Society of Japan **52** (2009) 278
[17] S. Wuestling *et al.*, Nucl. Instr. Meth **A 568** (2006) 382
[18] M. Prall *et al.*, submitted to IEEE Transactions on Nuclear Science, arXiv:0907.5490
[19] Th. Thümmler, R. Marx and Ch. Weinheimer, New Journal of Physics **11** (2009) 103007
[20] M. Beck, submitted to European Physical Journal **A** (2009), arXiv:0909.3337

Topics in Astroparticle and Underground Physics (TAUP 2009) IOP Publishing
Journal of Physics: Conference Series **203** (2010) 012098 doi:10.1088/1742-6596/203/1/012098

Measuring neutrino mass with radioactive ions in a storage ring

Mats Lindroos[1], Bob McElrath[2], Christopher Orme[3] and Thomas Schwetz[4]

[1]AB division, CERN, Geneva, Switzerland,
[2]Theory division, CERN, Geneva, Switzerland,
[3]IPPP, Department of Physics, Durham University, Durham DH1 3LE, United Kingdom and
[4]Max-Planck-Institute for Nuclear Physics, PO Box 103980, 69029 Heidelberg, Germany.

E-mail: c.d.orme@durham.ac.uk

Abstract. We propose a method to measure the neutrino mass kinematically using beams of ions which undergo beta decay. The idea is to tune the ion beam momentum so that in most decays, the electron is forward moving with respect to the beam, and only in decays near the endpoint is the electron moving backwards. Then, by counting the backward moving electrons one can observe the effect of neutrino mass on the beta spectrum close to the endpoint. In order to reach sensitivities for $m_\nu < 0.2$ eV, it is necessary to control the ion momentum with a precision better than $\delta p/p < 10^{-5}$, identify suitable nuclei with low Q-values (in the few to ten keV range), and one must be able to observe at least $\mathcal{O}(10^{18})$ decays.

1. Introduction

The present bound on the neutrino mass from kinematic studies of beta decay endpoints was obtained from the Mainz and Troitsk Tritium beta decay experiments [1, 2, 3]:

$$m_\nu^{\text{eff}} < 2.3 \text{ eV}\,, \tag{1}$$

where m_ν^{eff} is the effective neutrino mass in beta decay. The next iteration of the Tritium and beta decay technologies [4] and Rhenium calorimeters [5] should be able to place a limit $m_\nu^{\text{eff}} < 0.2$ eV; however it is unlikely that these approaches will scale to lower neutrino masses. Thus, new experimental technologies are required to reach the level of hierarchical masses. In this talk we present an idea to measure the neutrino mass by measuring the kinematics of the electrons ejected in a beta decay using a beam of ions in a storage ring.

2. The concept

A low boost radioactive ion beam is sent through an evacuated chamber with a weak magnetic field parallel to the beam line. A detector is set up on the back wall of the chamber to record the number of electrons still traveling backwards after the boost (see Fig. 1). The purpose of the boost is to perform a cut on the electron momenta, only selecting electrons very close to the beta spectrum endpoint where the effect of the neutrino mass manifests itself. In our proposal we cut on the momentum component parallel to the beam. Hence, we use only a slice of the momentum sphere (for small ϵ): $p_{\max} - \epsilon < p_\parallel < p_{\max}$. However, this approach has the disadvantage that many electrons close to the endpoint are lost.

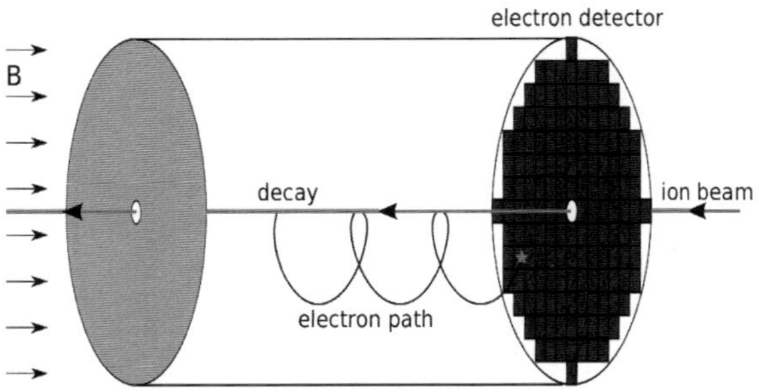

Figure 1. Diagram of the proposed experiment. The ion beam enters an evacuated cavity whose back wall holds an electron detector. Each ejected electron follows a helical trajectory. Electrons moving in the backward direction in the laboratory frame are counted by the detector.

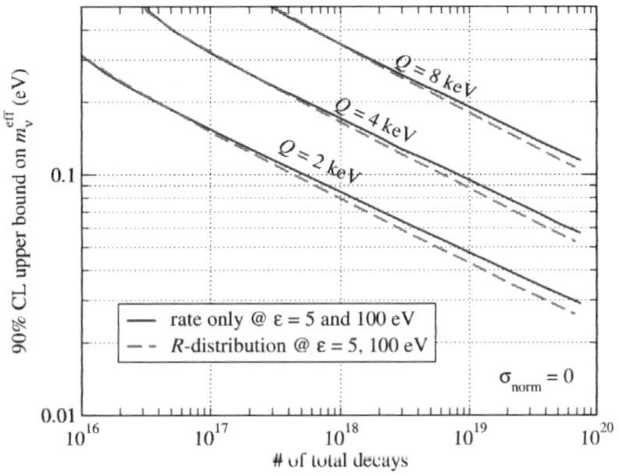

Figure 2. Sensitivity to the neutrino mass at 90% confidence level as a function of useful decays for $Q = 2, 4, 8$ keV. We show the total rate analyses with two ion boosts corresponding $\epsilon = 5$ and 100 eV (solid) and an analysis with using also the radial distribution with 20 bins of equal width in R (dashed).

3. Results

In the simulations carried out in this study, we combine two experimental runs: one with small ϵ close to the endpoint and one with large ϵ. At large ϵ the total count rate is several orders of magnitude larger than the change invoked by a non-zero neutrino mass. Hence, a 2-parameter fit for the neutrino mass and Q-value at large ϵ will be largely independent of the neutrino mass, i.e. one effectively makes a measurement of the Q-value. For small ϵ, however, the total count rate becomes comparable to the reduction for non-zero neutrino mass. There is a strong neutrino mass dependence in this case which, when combined with the Q-value measurement from the run with large ϵ, constrains the neutrino mass.

In Fig. 2, we present the upper bound on $m_\nu^{\rm eff}$ at 90% confidence level, which can be obtained if the true value is $m_\nu^{\rm true} = 0$, as a function of useful decays for Q-values of 2, 4 and 8 keV. Runs of $\epsilon = 5$ eV and $\epsilon = 100$ eV are considered in the ratio 99:1 such that the total useful number of decays sum up to the value shown on the horizontal axis. No backgrounds or systematics have been included, and we neglect the momentum spread of the ions. We also show in Fig. 2 an analysis that takes into account the radial distribution of the backward moving electrons (see [6] for details).

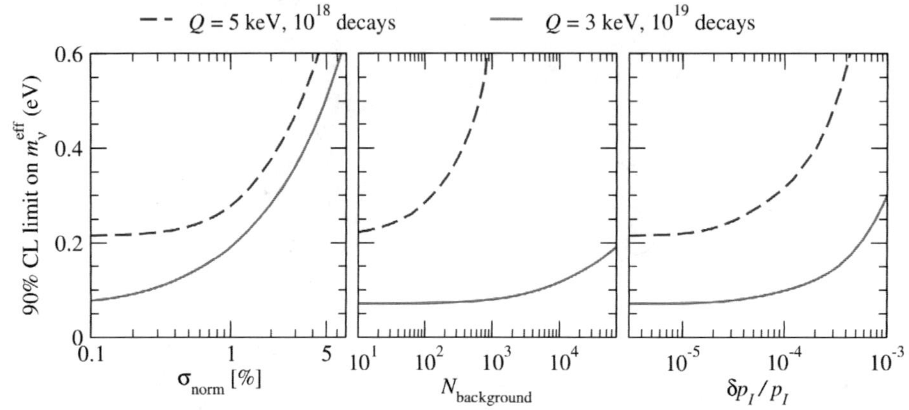

Figure 3. Sensitivity to $m_\nu^{\rm eff}$ at 90% confidence level for $Q = 5$ keV and 10^{18} useful decays (dashed/blue) and $Q = 3$ keV and 10^{19} useful decays (solid/red). In the left panel, the effect of the normalization uncertainty on the flux is considered; background levels are varied in the center panel; whilst the effect of the momentum spread of the initial ion is taken into account in the right panel.

In Fig. 3 we examine the behavior of neutrino mass sensitivity with systematics, backgrounds and the ion beam momentum spread. We consider the cases $Q = 3$ keV with 10^{19} useful decays ("high sensitivity") and $Q = 5$ keV with 10^{18} useful decays ("low sensitivity"), and, as before, in both cases we run at $\epsilon = 5$ and 100 eV with a ratio of 99:1. We include each one of the three above mentioned effects separately, in order to investigate at which level the sensitivity starts to deviate from the nominal sensitivities.

4. Conclusions

The talk discussed the possibility to use radioactive ion beams to study the kinematic effects of a non-zero neutrino mass close to the endpoint of the beta decay spectrum. The crucial questions are whether it will be possible to accelerate enough ions within reasonable time such that of order $10^{18} - 10^{20}$ decays can be observed; the identification of a suitable ion with a low enough Q-value and a small enough half life.

Acknowledgments

The speaker acknowledges the support of a STFC studentship and from CARE (contract RII3-CT-2003-05063950). The authors would like to thank colleagues for many useful comments and suggestions throughout this work. The speaker would also like to thank the organisers for the stimulating scientific atmosphere of the conference.

[1] C. Kraus *et al.*, Eur. Phys. J. C **40**, 447 (2005) [arXiv:hep-ex/0412056].
[2] V. M. Lobashev *et al.*, Phys. Lett. B **460** (1999) 227.
[3] V. M. Lobashev *et al.*, Nucl. Phys. Proc. Suppl. **91** (2001) 280.
[4] A. Osipowicz *et al.* [KATRIN Coll.], arXiv:hep-ex/0109033.
[5] A. Monfardini *et al.*, Prog. Part. Nucl. Phys. **57** (2006) 68 [Nucl. Instrum. Meth. A **559** (2006) 346] [arXiv:hep-ex/0509038].
[6] M. Lindroos, B. McElrath, C. Orme and T. Schwetz, arXiv:0904.1089 [hep-ph].

Measurement of Neutrino-Electron Scattering Cross Section and the EW Parameters at the Kuo-Sheng Reactor Neutrino Laboratory

M. Deniz[1,2], H. T. Wong[2], S. T. Lin[2] (for TEXONO Collaboration)
[1] Department of Physics, Middle East Technical University, Ankara 06531, Turkey.
[2] Institute of Physics, Academia Sinica, Taipei 11529, Taiwan.

E-mail: htwong@phys.sinica.edu.tw

Abstract. The $\bar{\nu}_e - e^-$ elastic scattering cross-section was measured at the Kuo-Sheng Nuclear Power Station with a CsI(Tl) scintillating crystal detector array with a total mass of 187 kg. The detector was exposed to a reactor $\bar{\nu}_e$ flux of 6.4×10^{12} cm^{-2}s^{-1} originated from a core with 2.9 GW thermal power. A cross-section ratio of $R_{expt} = [1.08 \pm 0.21 \text{ (stat)} \pm 0.16 \text{ (sys)}] \times R_{SM}$ was measured. Constraints on the electroweak parameters (g_V, g_A) were placed, corresponding to a weak mixing angle measurement of $\sin^2\theta_W = 0.251 \pm 0.031 \text{ (stat)} \pm 0.024 \text{ (sys)}$. The destructive interference in the SM $\bar{\nu}_e$-e processes was verified with the value of $R^{INT}_{expt} = [0.92 \pm 0.30\ (stat) \pm 0.24\ (sys)] \times R^{INT}_{SM}$. Bounds on neutrino anomalous electromagnetic properties were placed: neutrino magnetic moment at $\mu_{\bar{\nu}_e} < 2.2 \times 10^{-10}\mu_B$ at 90% C. L., and the neutrino charge radius at $\langle r^2_{\bar{\nu}_e} \rangle = [0.61 \pm 1.30 \text{ (stat)} \pm 1.01 \text{ (sys)}] \times 10^{-32}$ cm^2.

Neutrino-electron scatterings ($\nu_e(\bar{\nu}_e) - e^-$) are fundamental electroweak processes which play important roles in neutrino oscillation studies and in probing the electroweak parameters of the Standard Model (SM) as well as in the studies of neutrino properties such as the electromagnetic moments and charge radius[1]. The interactions $\nu_e(\bar{\nu}_e) - e^-$ have the additional unique features of being sensitive to the contributions of charged current (CC), neutral current (NC) and their interference (INT). The differential cross section for $\bar{\nu}_e - e^-$ scattering can be written as[1, 2]:

$$\frac{d\sigma_{SM}}{dT}(\bar{\nu}_e e) = \frac{G_F^2 m_e}{2\pi}\left[(g_V - g_A)^2 + (g_V + g_A + 2)^2\left(1 - \frac{T}{E_\nu}\right)^2 - (g_V - g_A)(g_V + g_A + 2)\frac{m_e T}{E_\nu^2}\right] \quad (1)$$

where T is the kinetic energy of the recoil electron, E_ν is the incident neutrino energy and g_V, g_A are coupling constants which can be expressed as $g_V = -\frac{1}{2} + 2\sin^2\theta_W$ and $g_A = -\frac{1}{2}$. In the low energy neutrino studies we must consider the electron mass dependent term because of its significant contribution to the cross section[2]. The value of weak mixing angle ($\sin^2\theta_W$) was measured precisely at high energy ($Q^2 > GeV^2$) at the accelerators, and at lower energy with Moller scattering and atomic parity violation experiments[1]. The cross-sections of $\nu_e - e^-$ have been measured at accelerators[3]. For reactor $\bar{\nu}_e - e^-$, the existing data are either controversial[4] or with large uncertainties[5]. There is room for improvement and this work is an attempt to bridge this gap. An important component of the TEXONO research program is to study $\bar{\nu}_e - e^-$ elastic scattering at $Q^2 \sim (3 \times 10^{-6})$ GeV^2 reactor neutrino range. The neutrino laboratory is located at the Kuo-Sheng Nuclear Power Plant a distance of 28 m from the reactor core with

2.9 GW of thermal power, having a total flux of about 6.4×10^{12} $cm^{-2}s^{-1}$. The details of neutrino source and neutrino spectrum were discussed in Ref. [6]. The CsI(Tl) scintillation detector array is enclosed by 4π low-activity passive shielding materials with a total mass of 50 tons, as well as a layer of active cosmic-ray (CRV) plastic scintillator panels. The entire target space is covered by a plastic bag flushed with dry nitrogen to suppress background due to the diffusion of the radioactive radon gas. The CsI(Tl) crystals were arranged as a 12 × 9 array matrix inside an

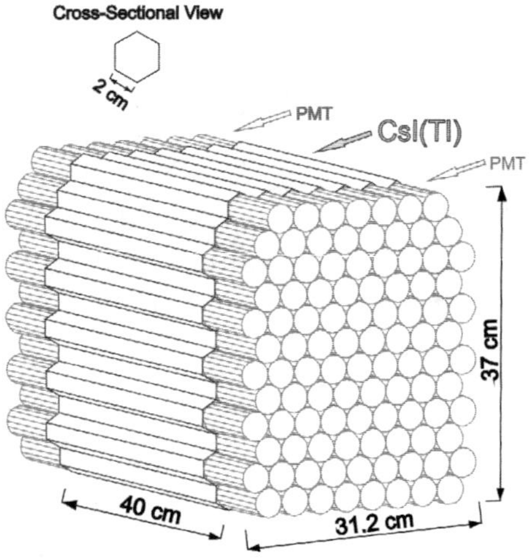

Figure 1. Schematic diagram of the CsI(Tl) crystal scintillator array.

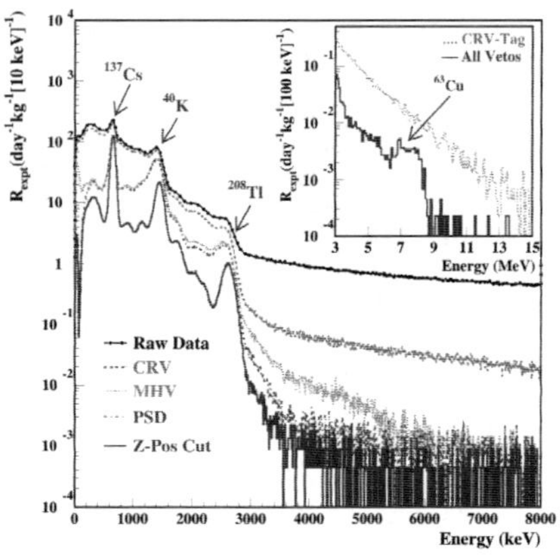

Figure 2. Measured spectra at the various stages of background suppression.

OFHC copper box, as shown schematically in Figure 1. The detector consisted of 100 crystals giving a total mass of 187 kg. Each single crystal module has a hexagonal-shaped cross-section with 2 cm side, 40 cm length and modular mass of 1.87 kg. The properties, advantages and the performance of the prototype modules of CsI(Tl) scintillating crystal detector were documented elsewhere[7]. These properties make crystal scintillator suitable for the study of low energy neutrino experiments. The sum of the two PMT signals gives the energy of the event, while their difference provides information on the longitudinal "Z" position. An energy resolution of 4% and a Z-resolution of <1 cm root-mean-square (RMS) at 660 keV as well as excellent α/γ event identification by pulse shape discrimination (PSD) were demonstrated in prototype studies.

Neutrino-induced candidate events were selected through the suppression of: (a) cosmic-ray and anti-Compton background by CRV and multiplicity cuts (MHV), (b) accidental and α-events by PSD, and (c) external background by Z-position cut. The spectra at the various stages of the background rejection were displayed in Figure 2. In situ calibration was achieved using the measured γ-lines from ^{137}Cs, ^{40}K and ^{208}Tl. The internal contaminations of the ^{238}U and ^{238}Th series were measured[8] and found to be negligible compared to the observed background rates. Residual background at the relevant 3−8 MeV range are cosmic-ray induced and full energy deposition of ambient gamma such as due to neutron capture by ^{63}Cu or due to coincidence of γ-emissions following ^{208}Tl decays. Their intensities were evaluated from the *in situ* multi-hit samples, the ^{208}Tl-2614 keV lines as well as from simulation studies, and the results provide the second background measurement. The background from both methods was subsequently combined (BKG) and subtracted from the candidate Reactor-ON samples.

A total of 29882/7369 kg-day of Reactor ON/OFF data was recorded and the combined ON−BKG residual spectrum is displayed in Figure 3, from which various electroweak parameters were derived. There is an excess in the residual spectrum corresponding to ∼414 neutrino-induced events. The residual spectrum of Figure 3 corresponds to $R_{expt} = [1.08 \pm 0.21 \pm 0.16] \times R_{SM}$ with $\chi^2/dof = 8.7/9$, giving $\sin^2\theta_W = 0.251 \pm 0.031 \pm 0.024$. The allowed region in the $g_V - g_A$ plane is depicted in Figure4. The accuracy is comparable to that achieved in accelerator-based $\nu_e - e$ scattering experiments[3].

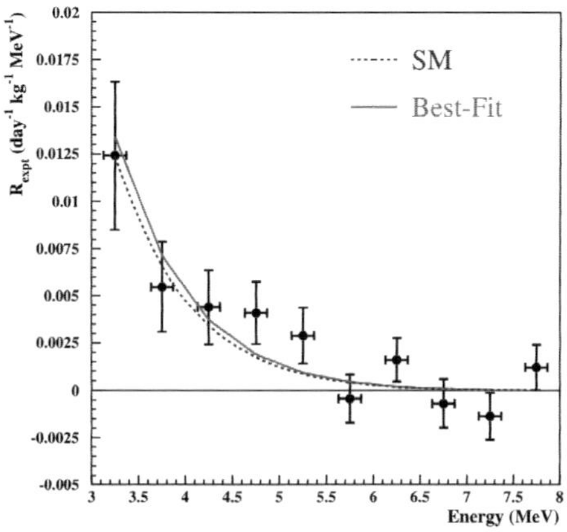

Figure 3. The combined ON-BKG residual spectrum.

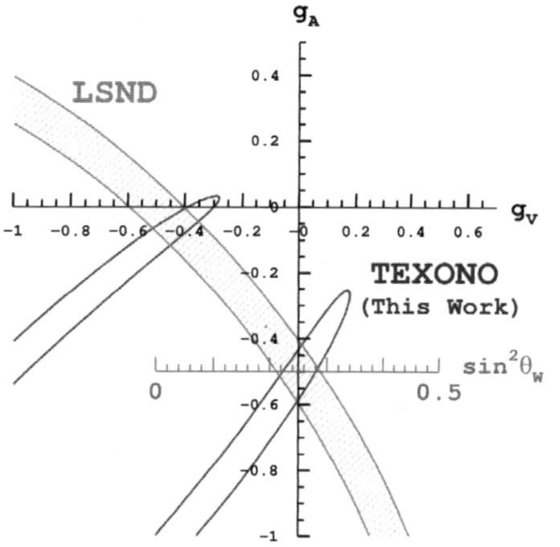

Figure 4. The 1-σ allowed region $g_V - g_A$ space, together with $\sin^2\theta_W$.

The measured value of R_{expt}^{INT} is $[0.92 \pm 0.30 \pm 0.24] \times R_{SM}^{INT}$ which verifies the SM prediction of destructive interference. Existence of neutrino magnetic moment ($\mu_{\bar{\nu}_e}$) would contribute an additional term[4] to the cross-section of Eq. 1:

$$\left(\frac{d\sigma}{dT}\right)_{\mu_\nu} = \frac{\pi\alpha_{em}^2\mu_\nu^2}{m_e^2}\left[\frac{1 - T_e/E_\nu}{T_e}\right] \quad . \tag{2}$$

From the best fit a limit of $\mu_{\bar{\nu}_e} < 2.2 \times 10^{-10} \times \mu_B$ at 90% CL was derived. A finite neutrino charge radius $\langle r_{\bar{\nu}_e}^2 \rangle$ would lead to radiative corrections[4, 10] which modify the electroweak parameters by $g_V \rightarrow -\frac{1}{2} + 2\sin^2\theta_W + (2\sqrt{2}\pi\alpha_{em}/3G_F)\langle r_{\bar{\nu}_e}^2\rangle$; $\sin^2\theta_W \rightarrow \sin^2\theta_W + (\sqrt{2}\pi\alpha_{em}/3G_F)\langle r_{\bar{\nu}_e}^2\rangle$ where α_{em} and G_F are the fine structure and Fermi constants, respectively. The results of $\langle r_{\bar{\nu}_e}^2\rangle = (0.61 \pm 1.30 \pm 1.01) \times 10^{-32}$ cm^2 at $\chi 2/dof = 8.7/9$ were obtained. The complete version of this work is archived at Ref. [9].

References

[1] B. Kayser, *Phys. Lett.* **B 667**, 163 (2008); J. Erler and P. Langacker, *Phys. Lett.* **B 667**, 125 (2008), and references therein.

[2] B. Kayser et al., *Phys. Rev.* **D 20**, 87 (1979).

[3] R. C. Allen et. al., *Phys. Rev.* **D 47**, 11 (1993); L. B. Aurbach et. al., *Phys. Rev.* **D 63**, 112001 (2001).

[4] F. Reines, H.S. Gurr, and H.W. Sobel, *Phys. Rev. Lett.*, **37**, 315 (1976); P. Vogel and J. Engel, *Phys. Rev.* **D 39**, 3378 (1989).

[5] G. S. Vidyakin et al., *JETP Lett.* **55**, 206 (1992); A. I. Derbin et al., *JETP Lett.* **57**, 796 (1993); Z. Daraktchieva et al., *Phys. Lett.* **B 615**, 081601 (2005); A.G. Beda et al., [hep-ex/0705.4576v1].

[6] H. T. Wong et al., *Phys. Rev.* **D 75**, 012001 (2007).

[7] H. T. Wong et al., *Astroparticle Phys.* **14**, 141 (2000); H. B. Li et al., *Nucl. Instrum. Methods* **A 459**, 93 (2001); W. P. Lai et al., *Nucl. Instrum. and Methods* **A 465**, 550 (2001); Y. Liu et al., *Nucl. Instrum. and Methods* **A 482**, 125 (2002).

[8] Y.F. Zhu et al., *Nucl. Instrum. Methods* **A 557**, 490 (2006).

[9] M. Deniz, et al., [hep-ex/0911.1597] (2009)

[10] J. Barranco et al., *Phys. Lett.* **B 662**, 431 (2008).

Journal of Physics: Conference Series **203** (2010) 012100 doi:10.1088/1742-6596/203/1/012100

Electromagnetic properties of neutrinos

Carlo Giunti
INFN, Section of Turin, Turin, Italy
E-mail: giunti@to.infn.it

Alexander Studenikin
Department of Theoretical Physics, Moscow State University, Moscow, Russia
E-mail: studenik@srd.sinp.msu.ru

Abstract. A short review on electromagnetic properties of neutrinos is presented. In spite of many efforts in the theoretical and experimental studies of neutrino electromagnetic properties, they still remain one of the main puzzles related to neutrinos.

Neutrino electromagnetic vertex function. The neutrino electromagnetic properties (see [1] for a recent review on this topic and the corresponding references on original papers) are determined by the neutrino electromagnetic vertex function $\Lambda_\mu(q,l)$ that is related to the matrix element of the electromagnetic current between the neutrino initial state $\psi(p)$ and final state $\psi(p')$, $<\psi(p')|J_\mu^{EM}|\psi(p)>=\bar{u}(p')\Lambda_\mu(q,l)u(p)$, where $q_\mu = p'_\mu - p_\mu$, $l_\mu = p'_\mu + p_\mu$. Lorentz and electromagnetic gauge invariance imply [1] (see also [2, 3, 4]) that the vertex function can be written in terms of four form factors,

$$\Lambda_\mu(q) = f_Q(q^2)\gamma_\mu + f_M(q^2)i\sigma_{\mu\nu}q^\nu + f_E(q^2)\sigma_{\mu\nu}q^\nu\gamma_5 + f_A(q^2)(q^2\gamma_\mu - q_\mu \not{q})\gamma_5, \tag{1}$$

where $f_Q(q^2)$, $f_M(q^2)$, $f_E(q^2)$ and $f_A(q^2)$ are charge, dipole magnetic and electric, and anapole neutrino form factors. The matrix element of the electromagnetic current can be considered also between neutrino initial $\psi_i(p)$ and final $\psi_j(p')$ states with different masses, $p^2 = m_i^2$ and $p'^2 = m_j^2$. The corresponding vertex function can be written in the form

$$\Lambda_\mu(q) = \Big(f_Q(q^2)_{ij} + f_A(q^2)_{ij}\gamma_5\Big)(q^2\gamma_\mu - q_\mu \not{q}) + f_M(q^2)_{ij}i\sigma_{\mu\nu}q^\nu + f_E(q^2)_{ij}\sigma_{\mu\nu}q^\nu\gamma_5, \tag{2}$$

where the form factors are matrices in the space of neutrino mass eigenstates [5].

In the case of Dirac neutrinos, the assumption of CP invariance combined with the hermiticity of the electromagnetic current J_μ^{EM} implies that the electric dipole form factor vanishes. In the case of Majorana neutrinos, regardless of whether CP invariance is violated or not, the charge, dipole magnetic and electric form factors vanish [3, 6], $f_Q = f_M = f_E = 0$ (the anapole moment can be non-vanishing, see also [7], as well as transition magnetic and electric moments). Since Dirac and Majorana neutrinos exhibit quite different electromagnetic properties, the investigation of neutrino electromagnetic interactions provides a tool for specifying the neutrino nature.

Neutrino electric form factor. It is usually believed that the neutrino electric charge is zero. This is often thought to be attributed to gauge invariance and anomaly cancellation constraints imposed in the Standard Model. In the Standard Model of $SU(2)_L \times U(1)_Y$ electroweak interactions it is possible to get [8] a general proof that neutrinos are electrically neutral which is based on the requirement of electric charges quantization. The direct calculations of the neutrino charge in the Standard Model for massless (see, for instance [9] and references therein) and massive neutrino [10] also prove that, at least at the one-loop level, the neutrino electric charge is gauge independent and vanishes. However, if the neutrino has a mass, the statement that a neutrino electric charge is zero is not so evident as it meets the eye. As a result, neutrinos may become electrically millicharged particles [8].

The most severe experimental constraints on the electric charge of the neutrino, $q_\nu \leq 10^{-21}e$, are obtained assuming electric charge conservation in neutron beta decay $n \to p + e^- + \nu_e$, from the neutrality of matter (from the measurements of the total charge $q_p + q_e$) [11]. A detailed discussion of different constraints on the neutrino electric charge can be found in [12].

Even if the electric charge of a neutrino is vanishing, the electric form factor $f_Q(q^2)$ can still contain nontrivial information about neutrino static properties. A neutral particle can be characterized by a superposition of two charge distributions of opposite signs so that the particle's form factor $f_Q(q^2)$ can be non zero for $q^2 \neq 0$. The mean charge radius (in fact, it is the charged radius squared) of an electrically neutral neutrino is given by $< r_\nu^2 > = -6 \frac{df_Q(q^2)}{dq^2}|_{q^2=0}$, which is determined by the second term in the expansion of the neutrino charge form factor $f_Q(q^2) = f_Q(0) + q^2 \frac{df_Q(q^2)}{dq^2}|_{q^2=0}$ in series of powers of q^2.

Note that there is a long standing discussion (see [1] for details) in the literature on the possibility to obtain (calculate) for the neutrino charged radius a gauge independent and finite quantity. In the corresponding calculations, performed in the one-loop approximation including additional terms from the $\gamma - Z$ boson mixing and the box diagrams involving W and Z bosons, the following gauge-invariant result for the neutrino charge radius have been obtained [13]: $< r_{\nu_l}^2 > = \frac{G_F}{4\sqrt{2}\pi^2}\left[3 - 2\log\left(\frac{m_l^2}{m_W^2}\right)\right]$, where m_W and m_l are the W boson and lepton masses $(l = e, \mu, \tau)$[1]. Numerically, for the electron neutrino electroweak radius it yields $< r_{\nu_e}^2 > = 4 \times 10^{-33}\, cm^2$. This theoretical result differs at most by one order of magnitude from the available experimental bounds on $< r_{\nu_i}^2 >$. Therefore, one may expect that the experimental accuracy will soon reach the value needed to probe the neutrino effective charge radius.

Neutrino magnetic and electric moments. The neutrino dipole magnetic moment (along with the electric dipole moment) is the most well studied among neutrino electromagnetic properties. A Dirac neutrino may have non-zero diagonal electric moments in models where CP invariance is violated. For a Majorana neutrino the diagonal magnetic and electric moments are zero.

The explicit evaluation of the one-loop contributions to the Dirac neutrino dipole moments in the leading approximation over the small parameters $b_i = m_i^2/m_W^2$ (m_i are the neutrino masses, $i = 1, 2, 3$), that however exactly accounts for $a_l = m_l^2/m_W^2$, leads to the following result [15]:

$$\left.\begin{array}{c}\mu_{ij}^D \\ \epsilon_{ij}^D\end{array}\right\} = \frac{eG_F m_i}{8\sqrt{2}\pi^2}\left(1 \pm \frac{m_j}{m_i}\right)\sum_l f(a_l)U_{lj}U_{li}^*, \;\; f(a_l) = \frac{3}{4(1-a_l)^3}\left(2 - 7a_l + 6a_l^2 - a_l^3 - 2a_l^2 \ln a_l\right), \quad (3)$$

where U_{li} is the neutrino mixing matrix. From (3) in the limit $a_l \ll 1$, the diagonal magnetic moment of a Dirac neutrino is given by [5] $\mu_{ii}^D = \frac{3eG_F m_i}{8\sqrt{2}\pi^2} \approx 3.2 \times 10^{-19}\left(\frac{m_i}{1\ eV}\right)\mu_B$. On the other hand, the magnetic moment of a hypothetical heavy neutrino ($m_\ell \ll m_W \ll m_\nu$) is given by [10] $\mu_\nu = \frac{eG_F m_\nu}{8\sqrt{2}\pi^2}$. Note that much larger values for the neutrino magnetic moments can be obtained in various extensions of the Standard Model (see, for instance, [1]).

[1] This result, however, revived the discussion [14] on the definition of the neutrino charge radius.

Bounds on neutrino magnetic moments. Constraints on the neutrino magnetic moment have been obtained in laboratory $\nu - e$ scattering experiments from the observed lack of distortions of the recoil electron energy spectra. Upper bounds on the neutrino magnetic moment have been obtained in recent reactor experiments: $\mu_\nu \leq 9.0 \times 10^{-11}\mu_B$ (MUNU collaboration [16]), $\mu_\nu \leq 7.4 \times 10^{-11}\mu_B$ (TEXONO collaboration [17]. The best world limit $\mu_\nu \leq 3.2 \times 10^{-11}\mu_B$ has been recently obtained by the GEMMA collaboration [18]. A stringent limit has also been obtained in the Borexino solar neutrino scattering experiments: $\mu_\nu \leq 5.4 \times 10^{-11}\mu_B$ [19]. Note that the magnetic and electric transition moments can contribute to the effective value of μ_ν (see Section 3.6 of [1]).

Neutrino electromagnetic interaction effects. If a neutrino has the non-trivial electromagnetic properties discussed above, a direct neutrino coupling to photons is possible and several processes important for applications exist [12]. A set of most important neutrino electromagnetic processes is: 1) neutrino radiative decay $\nu_1 \to \nu_2 + \gamma$, neutrino Cherenkov radiation in an external environment (plasma and/or electromagnetic fields), spin light of neutrino, $SL\nu$, in the presence of a medium [20]; 2) photon (plasmon) decay to a neutrino-antineutrino pair in plasma $\gamma \to \nu\bar{\nu}$; 3) neutrino scattering off electrons (or nuclei); 4) neutrino spin (spin-flavor) precession in a magnetic field (see [21]) and resonant neutrino spin-flavour oscillations in matter [22]. The tightest astrophysical bound on neutrino magnetic moments, $(\sum_{i,j} \mid \mu_{ij} \mid^2)^{1/2} \leq 3 \times 10^{-12}\mu_B$, applicable to both Dirac and Majorana neutrinos, has been obtained from the observed lack of anomalous stellar cooling due to plasmon decay [12].

Acknowledgements. One of the authors (A.S.) is thankful to Eugenio Coccia, Nicolao Fornengo and Luciano Pandola for the invitation to attend the TAUP-2009 and to all the conference organizers for their kind hospitality in Rome.

[1] Giunti C and Studenikin A 2009 *Phys.Atom.Nucl.* **72** 2151, arXiv: 0812.3646
[2] Kayser B 1982 *Phys.Rev.* D **26** 1662
[3] Nieves J F 1982 *Phys.Rev.* D **26** 3152
[4] Nowakowski M, Paschos E and Rodriguez J 2005 *Eur.J.Phys.* **26** 545
[5] Marciano W J and Sanda A I 1977 *Phys.Lett.* B **67** 303
Lee B W and Shrock R E 1977 *Phys.Rev.* D **16** 1444
Fujikawa K and Shrock R E 1980 *Phys. Rev. Lett.* **45** 963
[6] Schechter J and Valle J W F 1981 *Phys.Rev.* D **24** 1883
[7] Kobzarev I and Okun L 1972 *Problems of Theoretical Physics* (Moscow: Nauka) p 219
[8] Foot R, Lew H and Volkas R R 1993 *J. Phys.* G **19** 361; *ibid.* 1067 [Erratum]
Babu K S and Mohapatra R N 1989 *Phys.Rev.* D **63** 938
[9] Bardeen W, Gastmans R and Lautrup B 1972 *Nucl.Phys.* B **46** 319
Cabral-Rosetti L, Bernabeu J, Vidal J and Zepeda A 2000 *Eur.Phys.J.* C **12** 633
[10] Dvornikov M and Studenikin A 2004 *Phys.Rev.* D **69** 073001
[11] Marinelli M and Morpurgo G 1984 *Phys.Lett.* B **137** 439
[12] Raffelt G 1996 *Stars as Laboratories for Fundamental Physics* (Univ. of Chicago Press)
[13] Bernabeu J, Papavassiliou J and Vidal J 2004 *Nucl.Phys.* B **680** 450
[14] Fujikawa K and Shrock R 2004 *Phys.Rev.* D **69** 013007
Bernabeu J, Papavassiliou J and Binosi D 2005 Nucl. Phys. B **716** 352
[15] Pal P and Wolfenstein L 1982 *Phys.Rev.* D **25** 766
[16] MUNU Collab. (Darakchieva Z *et al.*) 2005 *Phys.Lett.* B **615** 153
[17] TEXONO Collab. (Wong H T *et al.*) 2007 *Phys.Rev.* D **75** 012001
[18] Beda A G *et al.* 2009 *Particle Physics on the Eve of LHC* ed. by Studenikin A (World Scientific: Singapore) p 112, arXiv:09.06.1926.
[19] Borexino Collab. (Arpesella C *et al.*) 2008 *Phys.Rev.Lett.* **101** 091302
[20] Lobanov A and Studenikin A 2003 *Phys.Lett.* B **564** 27; *ibid.* 2004 **601** 171
Grigoriev A, Studenikin A and Ternov A 2005 *Phys.Lett.* B **622** 199
Studenikin A 2008 *J.Phys.A: Math.Theor.* **41** 164047
[21] Okun L, Voloshin M and Vysotsky M 1986 *Sov.Phys.JETP* **64**, 446
[22] Lim C and Marciano W 1988 *Phys.Rev.* D **37** 1368
Akhmedov E 1988 *Phys.Lett.* B **213** 64

Journal of Physics: Conference Series **203** (2010) 012101 doi:10.1088/1742-6596/203/1/012101

Neutral current reaction cross sections for the stable ^{100}Mo isotope

K.G. Balasi [a], **T.S. Kosmas** [a], **P.C. Divari** [b] **and H. Ejiri** [c]

[a] Theoretical Physics Section, University of Ioannina, GR 45110 Ioannina, Greece
[b] Department of Physical Sciences, Hellenic Army Academy, GR 16673 Vari, Attica, Greece
[c] Research Center for Nuclear Physics, Osaka University, Ibaraki, Osaka 567-0047, Japan

E-mail: hkosmas@uoi.gr

Abstract. Motivated by the ongoing MOON neutrino experiment at Japan aiming to search for double beta and neutrinoless double beta decay events, we investigate inelastic neutrino scattering cross sections for the stable ^{100}Mo isotope by performing state-by-state calculations. The required many body nuclear wave functions are constructed within the context of the quasi-particle random phase approximation (QRPA) tested in the reproducibility of the low-lying spectrum of the ^{100}Mo isotope.

1. Introduction

Current experiments searching for low-energy neutrinos and rare event processes are significant sources of information for the fundamental weak interactions, the nuclear structure, the neutrino properties and the astro-nuclear physics. The MOON (Molybdenum Observatory Of Neutrinos) is a modern next generation neutrino experiment mainly aiming to search for double-beta and neutrinoless double-beta decay events. MOON is also sensitive to low-energy solar neutrinos above the ^{100}Mo beta-decay threshold of 168 keV and it has the capability of addressing multiple physics question within a single detector [1, 2].

The MOON experiment provides real time sensitivity to charged current neutrino-nucleus reactions. Its large reaction rates for all solar neutrino sources are due to the low threshold energy and the large Gamow-Teller strengths for the ground and excited states of the produced ^{100}Tc nucleus.

In the present work, we investigate inelastic cross sections of the neutral current reaction $^{100}Mo(\nu,\nu')^{100}Mo^*$ at low incoming-neutrino energies ($\epsilon_i \leq 100$ MeV). This enables us to explore the role of this isotope as low-energy neutrino detector (solar, supernova neutrinos, etc.).

2. The QRPA method for neutral current reactions

For neutral current neutrino-nucleus reactions, the wave functions for the initial and final nuclear states can be constructed within the quasi-particle RPA where the mth excited state of a multipolarity $|J^\pi_m M\rangle$ (J = total angular momentum, M = projection and π = parity) is created by acting on the QRPA ground state with the phonon operator

$$\hat{Q}^{m\dagger}_{J^\pi M} = \sum_{\substack{k\leq l\\ \tau}} \Big[X^m_\tau(kl,J) A^\dagger_\tau(kl,JM) + Y^m_\tau(kl,J)\tilde{A}_\tau(kl,JM) \Big] . \qquad (1)$$

X and Y are the forward and backward going amplitudes determined by the QRPA matrix equations. The quasi-particle pair creation and annihilation operators $A^\dagger$, A are defined as

$$A^\dagger_\tau(kl, JM) \equiv (1+\delta_{kl})^{-\frac{1}{2}} \left[a^\dagger_{\tau k} a^\dagger_{\tau l}\right]^J_M \qquad \tilde{A}_\tau(kl, JM) = (-1)^{J-M} A_\tau(kl, J-M) \tag{2}$$

where $a^\dagger$ (a) the creation (annihilation) quasi-particle operators. The next step in using QRPA is the construction of the eigenvalue problem (QRPA matrix equations) which reads

$$\begin{pmatrix} \mathcal{A} & \mathcal{B} \\ -\mathcal{B} & -\mathcal{A} \end{pmatrix} \begin{pmatrix} X^m \\ Y^m \end{pmatrix} = \Omega^m_{J^\pi} \begin{pmatrix} X^m \\ Y^m \end{pmatrix}, \tag{3}$$

where $\Omega^m_{J^\pi}$ denotes the excitation energy of the nuclear state $|J^\pi_m\rangle$. The QRPA matrices $\mathcal{A}$, $\mathcal{B}$ contain matrix elements of the double commutators of $A^\dagger$, A with the Hamiltonian operator $\hat{H}$.

3. Results

In the present work we perform realistic state-by-state calculations for all inelastic neutrino-nucleus channels by using the quasi-particle RPA method [3, 4]. We calculated both Fermi and Gamow-Teller type contributions of the polar vector and axial vector operators. Our method has been checked in the reproducibility of the low-lying spectrum of the ^{100}Mo isotope by using: (i) at the BCS level the pairing parameters for proton-pairs g^p_{pair}, and neutron-pairs g^n_{pair}, and (ii) at the QRPA level the fitting parameters for the strength of the residual interaction, i.e. the g_{ph}, for the particle-hole, and the g_{pp}, for the particle-particle channel, respectively. The model space used consists of the following eleven active single particle levels: $0f^{7/2}$, $1p^{3/2}$, $1d^{1/2}$, $0f^{5/2}$, $0g^{9/2}$, $1d^{5/2}$, $2s^{1/2}$, $1d^{3/2}$, $0g^{7/2}$, $0h^{11/2}$, $0h^{9/2}$, for which the corresponding single-particle energies were produced by a Coulomb corrected Woods-Saxon potential.

In Fig. 1 we show the calculated cross sections for neutrino scattering off the ^{100}Mo isotope as a function of the incoming neutrino energy, ϵ_i. More specifically, in Fig. 1 (left side) the differential cross sections for the dominant multipole states 0+, 1+, 2+, 3+, 3-, 4+, 5- and 6+ are shown. These cross sections resulted from the original double differential ones $d^2\sigma/d\Omega d\epsilon_f$ by summing over the individual contributions of each multipole state. The curve labeled "total" corresponds to the sum over the differential cross sections $d\sigma/d\Omega$ originating from all J^π multipole states (essentially up to $J^\pi = 6$). As can be seen, the contribution of the 0+ and 1+ multipolarities dominate the total differential cross section $d\sigma/d\Omega$.

In the right side of Fig. 1, the latter differential cross sections are plotted for various values of the scattering angle θ (step $\Delta\theta = 15^o$, for $0^o \le \theta \le 180^o$) of the reaction $^{100}Mo(\nu, \nu')^{100}Mo^*$. These plots show clearly that, for low neutrino energies ($\epsilon_i \le 10 - 12$ MeV) the differential cross section decreases as the scattering angle increases, but for higher energies the cross section increases with the scattering angle. By using the above differential cross sections and numerical integration techniques we could obtain the total cross section $\sigma_{tot}(\epsilon_i)$ [3, 4].

Our calculated cross sections show a smooth increase similar to that of the cross sections obtained for the charged current reaction

$$\nu_e + {}^{100}Mo \longrightarrow e^- + {}^{100}Tc \tag{4}$$

by the MOON experiment group [1, 2].

The results of the cross sections calculated in the present work are useful for investigating the response of the ^{100}Mo isotope as a neutrino detector. At the neutrino energies of Fig. 1 one could extract information about the response of ^{100}Mo isotope to solar and supernova neutrinos.

Closing we mention that there exists a number of detectors having the ability to study neutrino events and detect either the charged-current processes (through the produced electrons or positrons) or the neutral current reaction (mostly through the emitted neutrons, protons or a-particles). For the charge-exchange process (4) the ^{100}Mo isotope has a threshold (Q value) of only Q= 0.17 MeV [2].

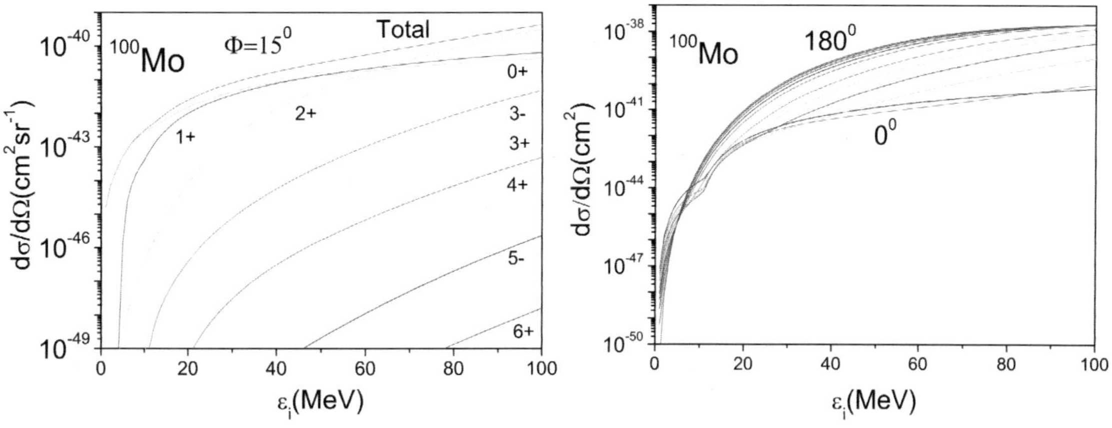

Figure 1. Differential cross sections $d\sigma/d\Omega$ with respect to the initial neutrino energy, ε_i for the reaction $^{100}Mo(\nu,\nu')^{100}Mo^*$ illustrating: (i) the dominant multipolarities for scattering angle $\Phi = 15^o$ (left), and (ii) the angular dependence by increasing the scattering angle by a step $\Delta\theta = 15^o$, for $0^o \leq \theta \leq 180^o$ (right).

4. Summary and Conclusions

^{100}Mo isotope is used for real-time spectroscopic studies of the individual solar neutrino sources, as well as for studying supernova neutrino and double beta decay processes. We have investigated neutral current reaction cross sections at low and intermediate energies of the reaction $^{100}Mo(\nu,\nu')^{100}Mo^*$. We have performed systematic calculations for all the even-even Mo isotopes [5] and we plan to calculate charged current cross sections in order to support the MOON experiment from a theoretical point of view.

5. Acknowledgments

This research was supported by the ΠΕΝΕΔ No 03ΕΔ807 project of the General Secretariat for Research and Technology of the Hellenic Ministry of Development. One of us (TSK) acknowledges support from the ILIAS EU Project.

References

[1] H. Ejiri, 2000 *Phys. Rep.* **338** 265–351.
[2] H. Ejiri, 2003 *Nucl. Phys. A* **721** 517c-520c; 1995 *J.Phys. Soc. Japan* **64**, 339.
[3] K.G. Balasi, T.S. Kosmas, P.C. Divari and V.Ch. Chasioti,2008 *AIP Conf. Proc.* **972** 554.
[4] K.G. Balasi, T.S. Kosmas, P.C. Divari and V.Ch. Chasioti,2008 *Proc. of Hell. Nucl. Phys.* **17** 97-104.
[5] K.G. Balasi, et.al., in preparation.

Current MINOS neutrino oscillation results

Alec Habig, for the MINOS collaboration

University of Minnesota Duluth, Physics Department, 10 University Dr., Duluth, MN 55812, USA

E-mail: ahabig@umn.edu

Abstract. The MINOS experiment is now making precise measurements of the ν_μ disappearance oscillations seen in atmospheric neutrinos, tests possible disappearance to sterile ν by measuring the neutral current flux, and has extended our reach towards the so far unseen θ_{13} by looking for ν_e appearance in the ν_μ beam. It does so by using the intense, well-understood NuMI neutrino beam created at Fermilab and observing it 735km away at the Soudan Mine in Northeast Minnesota. High-statistics studies of the neutrino interactions themselves and the cosmic rays seen by the MINOS detectors have also been made. Results from MINOS' first three years of operations will be presented.

1. Introduction

The MINOS experiment uses two magnetized steel/scintillator calorimeters [1] to investigate the neutrino oscillations previously observed using atmospheric neutrinos [2, 3]. This long-baseline experiment observes the intense and well-understood NuMI beam both near its source at Fermilab with a 0.98 kton "near detector", and again 735 km to the northwest in the Soudan Mine Underground Lab with the 5.4 kton "far detector". This before and after comparison of the neutrinos as seen in the similar detectors greatly reduces the systematic errors associated with comparing the differences in the observed neutrino spectra to various neutrino oscillation scenarios, allowing for a more accurate probe of the physics of neutrino propagation. The NuMI beam is composed of 92.9% ν_μ, 5.8% $\overline{\nu}_\mu$, 1.2% ν_e and 0.1% $\overline{\nu}_e$. The bulk of the data come from the "low energy" beam configuration, peaked at several GeV (see the red line in Fig. 1). This paper summarizes the status of several analyses of the neutrino data acquired over the two year time period starting with the beginning of NuMI operations in May of 2005 and ending during the summer shutdown in June 2007, an integrated exposure of over 3×10^{20} protons on target ("pot") with a neutrino yield on order of one neutrino per proton. The intrinsic divergence of the beam results in a neutrino flux at the far detector which is a factor of 10^6 lower than that at the near detector.

2. Oscillation Analyses

2.1. ν_μ Disappearance Oscillations

The design goal of the MINOS experiment is to use quasi-elastic ν_μ interactions to make a precision measurement of mixing. A ν_μ of energy $E_\nu[GeV]$ observed after traveling some distance $L[km]$ has a probability of being detected as a ν_μ given by $P(\nu_\mu \rightarrow \nu_\mu) = 1 - \sin^2(2\theta)\sin^2\left(1.27\Delta m^2 \frac{L}{E}\right)$, where $\Delta m^2[eV^2]$ is the mass difference between ν_2 and ν_3 and $\sin^2(2\theta)$ is the mixing amplitude. The oscillation minima at the 735 km baseline is less than the τ production threshold, so the oscillatory signature is that of ν_μ disappearance.

An exposure of 3.36×10^{20} pot has been analyzed [4], selecting 848 events as ν_μ with good purity. The observed, unoscillated near detector signal is used to calculate an expectation of 1065±60 far detector events, including a small background of 2.3 external μ, 5.9 neutral current ("NC") induced showers, and 1.5 τ decays, and are shown with the observed data in Fig. 1. Systematic errors (dominated by relative normalization, NC background, and overall hadronic energy scale) are still smaller than statistical errors, so the measurement will improve as data from the current run is added.

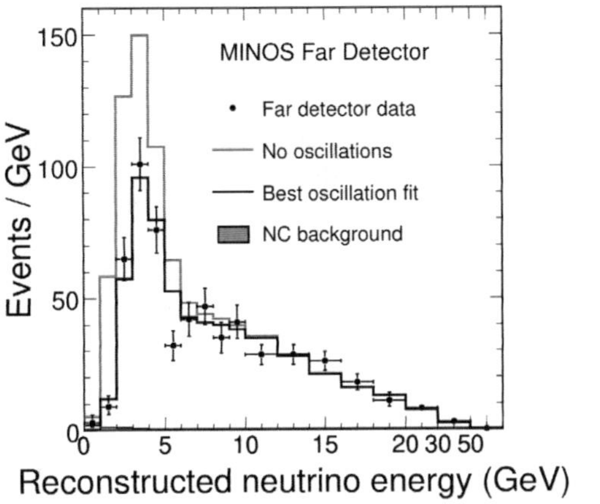

Figure 1. The MINOS far detector $\nu_m u$ spectrum [4]. The data (points with statistical errors) show a significant deficit from the null hypothesis (red line), but well-match a $\nu_\mu \leftrightarrow \nu_\tau$ oscillation scenario (black line), with best fit mass splitting $|\Delta m^2| = (2.43 \pm 0.13) \times 10^{-3}$ eV^2 (68% cl) and mixing angle $\sin^2(2\theta) > 0.90$ (90% cl).

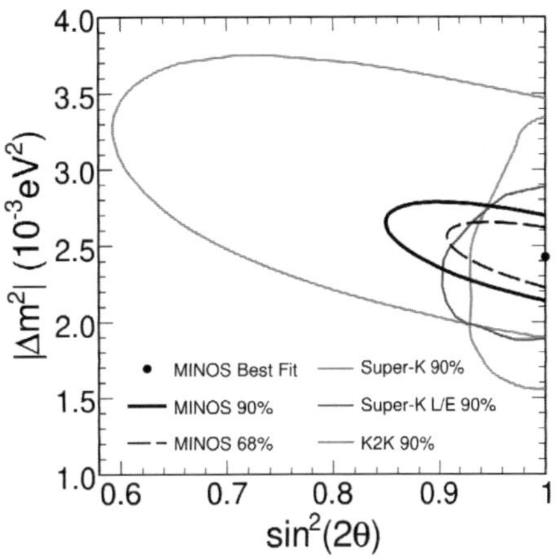

Figure 2. The allowed region in oscillation parameter space for the data in Fig. 1, at 90% (solid black) and 68% (dashed black) confidence levels. Two Super-K atmospheric analyses are shown in red [3] and blue [2], and the K2K long baseline experiment results produce the grey contour [5].

2.2. Anti-neutrinos

The magnetized nature of the MINOS detectors allows the event-by-event determination of the charge sign of muons, and thus the identification of the parent neutrino or anti-neutrino. Selection of wrong-sign muons in the ν_μ beam tests if $\bar{\nu}_\mu$ oscillate in the same fashion as ν_μ in Sec. 2.1, and if some fraction α of disappearing ν_μ reappear as $\bar{\nu}_\mu$ – both tests of CPT conservation. Given that only 6.4% of the neutrino events in a 3.2×10^{20} pot far detector exposure are due to anti-neutrinos, the relative backgrounds are higher and statistics lower. 42 $\bar{\nu}_\mu$ events are seen while $64.6 \pm 8.0_{stat} \pm 3.9_{syst}$ are expected in the no-oscillation case, and $58.3 \pm 7.6_{stat} \pm 3.6_{syst}$ if CPT is conserved given the observed ν_μ oscillation parameters. This places a 90%cl upper limit on $\alpha < 0.026$, and the anti-neutrino oscillation parameters are consistent with the neutrino parameters given these low statistics. The next year of NuMI beam running will be optimized to produce anti-neutrinos to better understand anti-neutrinos.

2.3. Sterile Neutrinos

Another possible explaination of the ν_μ disappearance is oscillation into sterile neutrinos which experience no interactions. This would suppress the rate of NC events in the far detector compared to the traditional explanation of sub-threshold ν_τ, since ν_τ still undergo NC interactions. Thus, NC showers have been selected from an exposure of 3.18×10^{20} pot, following the analysis outlined in [6].

The ratio of observed to expected NC events in the far detector is $R = 1.04 \pm 0.08_{stat} \pm 0.07_{syst} - 0.10_{\nu_e}$, resulting in a limit on the fraction of ν_s participation of $f_s < 0.51$ at 90%cl.

2.4. The Search for ν_e Appearance

MINOS was designed to be a good muon calorimeter for ν_μ disappearance, but is coarse for resolution of ~GeV electromagnetic showers. It retains sensitivity to the ~2% ν_e appearance signal which a θ_{13} near the CHOOZ limit [7] would create, and the first 3.14×10^{20} pot of MINOS data have been examined [8] by a neural network to select electromagnetic shower candidates. When applied to Monte Carlo data this is 41% efficient while rejecting >92% of NC showers (the dominant background) and >99% of ν_μ charged current ("CC") interactions (high-y hadronic showers). Given the small expected signal and large uncertainties in hadronic shower modeling, data-driven methods are used to better estimate the background. At the near detector no oscillation has yet occurred, so with the exception of the well-modeled inherent beam ν_e, all events selected must be examples of such background events. This yields an expected background of 26.6 (18.2 NC, 5.1 CC, and 2.2 beam ν_e) at the far detector, while 35 ν_e like events are seen, a 1.5 σ excess (including 7.3% statistical and 19% systematic errors) (Fig. 3). If fit for oscillations, this is just below the CHOOZ limit and consistent within errors with no ν_e appearance.

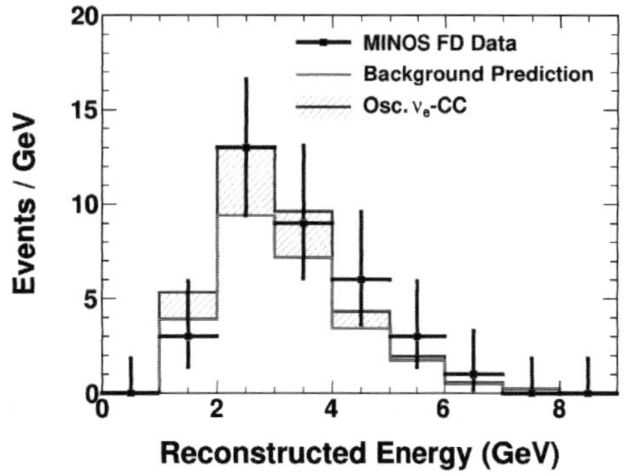

Figure 3. The spectrum of potential ν_e interactions in the MINOS far detector, with statistical plus systematic error bars [8]. The 1.5σ excess is consistent with both the expected large background (red) and a $\sin^2(2\theta_{13})$ comparable to the CHOOZ limit [7] (purple).

3. Conclusions

MINOS has measured neutrino oscillation parameters in the "atmospheric" $\nu_2 \leftrightarrow \nu_3$ sector with high precision, favoring standard $\nu_\mu \leftrightarrow \nu_\tau$ oscillations. Further statistics will be of great interest to investigate the possibility of ν_e appearance just under the CHOOZ limit.

Acknowledgments

This work was supported by the U.S. Department of Energy, the U.K. Science and Technologies Facilities Council, the U.S. National Science Foundation, and the State and University of Minnesota. This researcher is supported by NSF RUI grant #0653016.

References

[1] Michael D G *et al.* (MINOS) 2008 *Nucl. Instrum. Meth.)* **A596** 190–228 (*Preprint* `0805.3170`)
[2] Ashie Y *et al.* (Super-Kamiokande) 2004 *Phys. Rev. Lett.* **93** 101801 (*Preprint* `hep-ex/0404034`)
[3] Ashie Y *et al.* (Super-Kamiokande) 2005 *Phys. Rev.* **D71** 112005 (*Preprint* `hep-ex/0501064`)
[4] Adamson P *et al.* (MINOS) 2008 *Phys. Rev. Lett.* **101** 131802 (*Preprint* `0806.2237`)
[5] Ahn M H *et al.* (K2K) 2006 *Phys. Rev.* **D74** 072003 (*Preprint* `hep-ex/0606032`)
[6] Adamson P *et al.* (MINOS) 2008 *Phys. Rev. Lett.* **101** 221804 (*Preprint* `0807.2424`)
[7] Apollonio M *et al.* (CHOOZ) 1999 *Phys. Lett.* **B466** 415–430 (*Preprint* `hep-ex/9907037`)
[8] Adamson P *et al.* (MINOS) 2009 *submitted to Phys. Rev. Lett.* (*Preprint* `0909.4996`)

Probing θ_{13} with global neutrino data analysis

G.L. Fogli[1,2], E. Lisi[2], A. Marrone[1], A. Palazzo[3] and A.M. Rotunno[1]

[1] Dipartimento di Fisica, Università di Bari,
Via Amendola 173, 70126, Bari, Italy
[2] Istituto Nazionale di Fisica Nucleare (INFN), Sezione di Bari,
Via Orabona 4, 70126 Bari, Italy
[3] AHEP Group, Institut de Física Corpuscular – C.S.I.C./Universitat de València
Edificio Institutos de Paterna, Apt. 22085, E–46071 València, Spain

E-mail: antonio.palazzo@ific.uv.es

Abstract. We discuss the results of an updated global analysis of neutrino oscillation data, focusing on the determination of θ_{13}, the smallest and unknown leptonic mixing angle. We discuss three independent and converging hints of $\theta_{13} > 0$: a first one coming from atmospheric neutrino data; a second one from the combination of solar and long-baseline reactor (KamLAND) neutrino data; and a third one from the latest MINOS measurements in the appearance ($\nu_\mu \to \nu_e$) channel. Their combination provides an indication for $\theta_{13} > 0$ at the 2σ (95% C.L.) level.

1. Introduction

The measurement of the third unknown leptonic mixing angle θ_{13} is a crucial step towards the possible discovery of CP violation in the leptonic sector. The first robust upper bound on this fundamental parameter was established by the CHOOZ reactor experiment [1] in 1998. In the last decade this bound has been corroborated and strengthened by the increasingly refined global neutrino data analyses performed by various groups. The past year (2008), however, for the first time, the world neutrino data have shown an intriguing preference for a non-zero value of θ_{13} at a non-negligible statistical level [2]. Remarkably, an analogous hint has emerged in the first searches of ν_e appearance in MINOS [3], reinforcing the statistical significance of the global indication for non-zero θ_{13}, now at the 95% C.L. In the following we review the current status of these weak, but intriguing, indications.

2. The hint from the "atmospheric sector"

In [4] we pointed out a weak preference of atmospheric neutrino data (in combination with CHOOZ) for a non-zero value of θ_{13}. We traced a possible source for this preference in subleading 3ν oscillation terms driven by the "solar parameters" [5, 6], which could partly explain the observed excess of sub-GeV atmospheric electron-like events. We find such a hint unaltered after combination with current (disappearance) LBL accelerator neutrino data, which are not sufficiently sensitive to θ_{13} [7]. In particular, after inclusion of the latest MINOS (disappearance) data [8], and marginalizing over the leading mass-mixing parameters $(\Delta m^2, \theta_{23})$, we still find a $\sim 0.9\sigma$ hint of $\theta_{13} > 0$ from the combination of atmospheric, LBL accelerator (disappearance), and CHOOZ data,

$$\sin^2\theta_{13} = 0.012 \pm 0.013 \quad [1\sigma,\ \mathrm{Atm} + \mathrm{LBL(disapp.)} + \mathrm{CHOOZ}] \tag{1}$$

where the error scales almost linearly up to $\sim 3\sigma$, within the physical range $\sin^2\theta_{13} \geq 0$. A similar hint has been found in [9, 10], but not in [11, 12]. Given the slight disagreement among equally refined analyses, we deem that only a complete 3-flavor analysis (including the subleading effects induced by the solar parameters) performed by the Super-Kamiokande collaboration can add further relevant information on the issue.

3. The hint from the "solar sector"

As noticed in [13], as well as in [14], the latest KamLAND data [15] prefer values of the mixing angle θ_{12} somewhat higher than those indicated by solar data. As stressed in [13, 14], this tension could be alleviated for $\theta_{13} > 0$, as a result of the different dependence of the survival probability $P_{ee} = P(\nu_e \to \nu_e)$ on the mixing angles in KamLAND and SNO, respectively. In [13] we also remarked that such a preference, although rather small at that time ($\sim 0.5\sigma$), could be potentially corroborated by new solar neutrino data. This trend has indeed been reinforced by the SNO-III data [16], whose inclusion leads to a hint of $\theta_{13} > 0$ at the $\sim 1.2\sigma$ level [2],

$$\sin^2\theta_{13} = 0.021 \pm 0.017 \ (1\sigma,\ \mathrm{Solar} + \mathrm{KamLAND})\ , \tag{2}$$

with errors scaling linearly, to a good approximation, up to $\sim 3\sigma$.

It must be stressed that such a hint arises from a disagreement in the comparison of matter dominated transitions (solar ν's) with vacuum-like oscillations (KamLAND), which, in principle may be due to some unaccounted effect in the propagation of solar neutrinos in the Sun. One such an example is offered by non-standard interactions (NSI), recently investigated in [17], which can mimic the values of $\theta_{13} > 0$ indicated by the fit. This circumstance, underlines the importance of future reactor searches, which are unaffected by matter effects.

4. The hint from MINOS and the global combination

The first MINOS data in the $\nu_\mu \to \nu_e$ appearance channel [3] have shown a weak preference (90% C.L.) for $\theta_{13} > 0$. Although, very prudently, the collaboration did not emphasize this fact, a combination of their results with ours enhances the statistical significance of the global hint. Indeed, from the combination of the solar and atmospheric data, *without* the appearance MINOS results, we find a hint [2], at the 90% C.L.,

$$\sin^2\theta_{13} = 0.016 \pm 0.010 \ \ (1\sigma,\ \text{all data 2008})\ , \tag{3}$$

which, after inclusion of the MINOS result becomes [2],

$$\sin^2\theta_{13} = 0.02 \pm 0.01 \ \ (1\sigma,\ \text{all data 2009})\ , \tag{4}$$

implying an overall preference at the level of $\sim 2\sigma$ (95% C.L). Figure 1 summarizes the hints we obtain using different data sets and their combinations.

5. Conclusions

The latest neutrino data have disclosed the opportunity to probe the unknown mixing angle θ_{13}, showing an interesting preference for a nonzero value of this parameter which lies just below the CHOOZ upper limit. At the time of writing these proceedings, new relevant information has become available from the low energy threshold analysis (LETA) performed by the SNO collaboration [18], whose results confirm the trend for a preference for $\theta_{13} > 0$ in the solar sector.

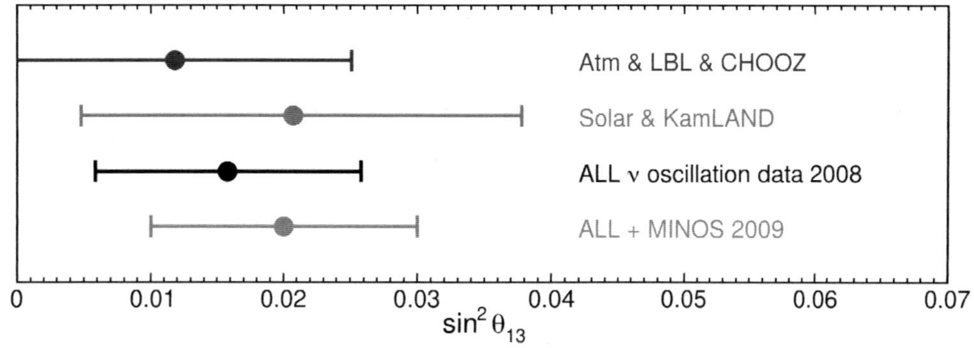

Figure 1. Ranges (at the 1σ level) preferred by different data sets and combinations.

The MINOS appearance results with doubled statistics are expected soon and will add further information. However, it is difficult that they can change substantially the statistical level of the present global indication, which will likely remain with us until the direct (dis-)confirmation by the new dedicated reactor experiments. The strength of these new experiments relies also in the fact that their inference of θ_{13} is unaffected by (standard and/or non-standard) matter effects, which instead can alter the indirect inferences of the global analyses [17].

6. Acknowledgments

G.L.F., E.L., A.M., and A.M.R. acknowledge support by the Italian MIUR and INFN through the "Astroparticle Physics" research project, and by the EU ILIAS through the ENTApP project. A.P. acknowledges support by MEC under the I3P program, by Spanish grants FPA2008-00319/FPA, FPA2008-01935-E/FPA, PROMETEO/2009/091 and ILIAS/N6 Contract RII3-CT-2004-506222.

References

[1] Apollonio M *et al.* [CHOOZ Collaboration] 1998 *Phys. Lett. B* **420** 397
[2] Fogli G L, Lisi E, Marrone A, Palazzo A and Rotunno A M 2008 *Phys. Rev. Lett.* **101** 141801
[3] Sanchez M 2009 *FNAL Wine and Chees Seminar*, February 27
[4] Fogli G L, Lisi E, Marrone A and Palazzo A 2006 *Prog. Part. Nucl. Phys.* **57** 742
[5] Peres O L G and Smirnov A Y 1999 *Phys. Lett. B* **456** 204
[6] Peres O L G and Smirnov A Y 2004 *Nucl. Phys. B* **680** 479
[7] Fogli G L *et al.* A 2007 *Phys. Rev. D* **75** 053001
[8] Adamson P *et al.* [MINOS Collaboration] 2008 *Phys. Rev.Lett.* **101** 131802
[9] Roa J E, Latimer D C and Ernst D J 2009 *Phys. Rev. Lett.* **103** 061804
[10] Roa J E, Latimer D C and Ernst D J 2009 *Preprint* arXiv:0904.3930 [nucl-th]
[11] Schwetz T, Tortola M and Valle J W F 2008 *New J. Phys.* **10**, 113011
[12] Maltoni M and Schwetz T 2008 *Preprint* arXiv:0812.3161 [hep-ph]
[13] Fogli G L, Lisi E, Marrone A, Palazzo A and Rotunno A M 2008 *Proceedings of Workshop on "Neutrino Oscillations in Venice", Venice, Italy, April 15-18, 2008, edited by M. Baldo Ceolin (University of Padova, Papergraf Editions, Padova, Italy, 2008)*, p. 21
[14] Balantekin A B and Yilmaz D 2008 *J. Phys. G* **35**, 075007
[15] Abe S *et al.* [KamLAND Collaboration] 2008 *Phys. Rev. Lett.* **100** 221803
[16] Aharmim B *et al.* [SNO Collaboration] 2008 *Phys. Rev. Lett.* **101**, 111301
[17] Palazzo A and Valle J W F 2009 *Preprint* arXiv:0909.1535 [hep-ph]
[18] Aharmim B *et al.* [SNO Collaboration] 2009 *Preprint* arXiv:0910.2984 [nucl-ex]

Status of the T2K long baseline neutrino oscillation experiment

Atsuko K. Ichikawa
Department of Physics, Kyoto University, Kitashirakawa-Oiwakecho, Sakyo-ku, Kyoto 606-8502, Japan
E-mail: ichikawa@scphys.kyoto-u.ac.jp

, for the T2K collaboration

Abstract. The T2K experiment is a long baseline neutrino oscillation experiment. The main goals of the experiment are a measurement of the neutrino mixing angle θ_{13} by the observation of electron neutrino appearance and a precise measurement of θ_{23} and mass difference by the observation of muon neutrino disappearance. Construction of the neutrino beamline has recently been completed; production of the first neutrino beam was confirmed by observing muons produced by the proton beam. In this paper, the T2K physics goals and the status of preparations toward the start of the experiment in fall 2009 are reported.

1. Physics goals of the T2K experiment

The T2K experiment is a long baseline neutrino oscillation experiment. A muon neutrino beam of sub GeV energy is produced using a high intensity proton accelerator complex, J-PARC at Tokai village, Japan. Then, after traveling 295 km, it is measured at a gigantic water Čerenkov detector, Super-Kamiokande, at Kamioka, Japan. Figure 1 shows the baseline of the experiment.

In the three generations framework, the neutrino oscillation probability for this energy region and this baseline length is written as

$$P_{\mu\to x} \approx 1 - \cos^4\theta_{13}\sin^2\theta_{23}\sin^2(1.27\Delta m^2_{23}L/E_\nu)$$

for ν_μ disappearance, and

$$P_{\mu\to e} \approx \sin^2\theta_{23}\sin^2 2\theta_{13}\sin^2(1.27\Delta m^2_{13}L/E_\nu)$$

for ν_μ to ν_e appearance, where L =295 km and E_ν is a neutrino energy (GeV).

The main physics motivation of the T2K experiment comprises three goals.

(i) Discovery of $\nu_\mu \to \nu_e$
Among three mixing angles, only θ_{13} is unknown and set an upper limit[1]. In the T2K experiment, θ_{13} can be explored by searching for the $\nu_\mu \to \nu_e$ appearance signal. The ν_e signals are identified from the distinctive quantity of observed Čerenkov rings and reconstructed neutrino energy. The sensitivity is $\sin^2 2\theta_{13} = 0.006$ at 90% C.L. This represents an order of magnitude improvement over the current limit.

Journal of Physics: Conference Series **203** (2010) 012104 doi:10.1088/1742-6596/203/1/012104

(ii) Precision measurements of oscillation parameters in ν_μ disappearance.
The goal is an observation of the oscillation minimum, a 1% measurement of the mixing angle and a 10% measurement of $|\Delta m^2|$ ($\delta(\Delta m^2_{23}) = 10^{-4}$ eV^2 and $\delta(\sin^2 2\theta_{23}) = 0.01$).

(iii) Search for sterile components in ν_μ disappearance by detecting the neutral-current events.

In order to achieve these goals, T2K will accumulate data with 3.75 MW.yr integrated beam power of J-PARC, which corresponds to 5 years of running with the design beam power of 0.75 MW.

With the successful achievement of these measurements, the construction of more than 100 kt far detectors, and a possible upgrade of the accelerator from 0.75 MW to a few MW in beam power, further experiments can be envisaged. These include a sensitive search for the CP violation in the lepton sector (CP phase δ down to $10° \sim 20°$).

Figure 1. Baseline of the T2K experiment.

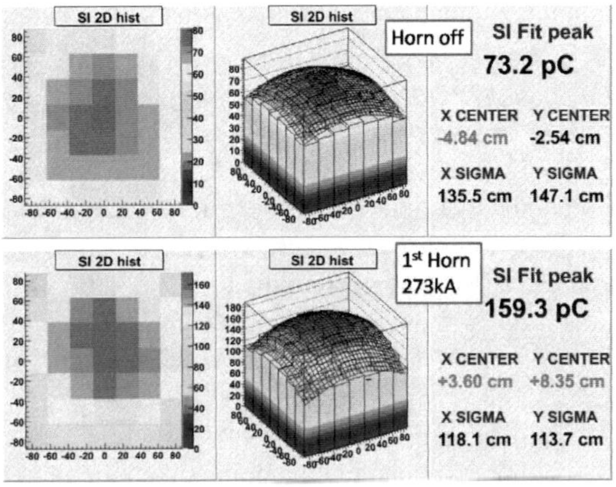

Figure 2. Muon flux measured by MUMON Silicon PIN photodiode. Top figure : muon profile when all the horns are off. Bottom figure : muon profile when the 1st horns are operated with 273 kA.

2. Overview of the Experiment

The far detector, Super-Kamiokande, located in the Kamioka Observatory, Institute for Cosmic Ray Research (ICRR), University of Tokyo, has been successfully taking data since 1996.

It is a 50,000 ton water Čerenkov detector located at a depth of 2,700 meters water equivalent in the Kamioka mine in Japan. Its performance and results in atmospheric neutrinos and solar neutrinos have been well-documented in [3]. The total fiducial mass is 22,500 ton.

Photons from the Čerenkov rings produced by relativistic charged particles are detected by PMT's. Their pulse-height and timing information are fitted to reconstruct the vertex, direction, energy, and particle identification. The Čerenkov ring shape, clear ring for muons and fuzzy ring for electrons, provides good e/μ identification.

A ν_μ beam at J-PARC is produced from the decay of pions produced by protons hitting a target. The pions are focused by magnets called electromagnetic horns into a volume where they decay in flight. The beam consists of almost pure ν_μs.

A key element of the design of the T2K facility is that the neutrino beam is directed so that the beam axis actually misses Super-Kamiokande . This off-axis neutrino beam technique is described in Ref.[2]. Due to the kinematics of the pion decay, pions in a wide momentum range contribute to a narrow energy range of neutrinos. This results in a considerable improvement in the quality of the beam for the ν_e appearance experiment. The peak neutrino energy can be adjusted by choosing the off-axis angle. The J-PARC neutrino facility is capable of producing a beam whose off-axis angle is from 2.0 degree to 2.5 degree. The peak energy is 500 $\sim$ 700 MeV and corresponds to $\Delta m^2_{23} = 2 \sim 3 \times 10^{-3}$ eV^2 for the baseline length of 295 km. Contamination of ν_e in the beam, which is a background for the ν_e appearance search, is expected to be 0.2% at the peak energy.

At J-PARC, primary protons are first accelerated by the LINAC and then by Rapid Cycle Synchrotron up to 3 GeV and by the Main Ring synchrotron (MR) up to 30 GeV. A 30 GeV proton beam is extracted from the MR in a single turn to the neutrino facility. The primary proton beamline of the neutrino facility transfers the proton beam from the MR to the neutrino production target. It consists of normal conducting magnets and super-conducting combined-function type magnets. The neutrino production target is a graphite rod of 26 mm diameter and 90 cm length The 3 magnetic horns are driven by a pulsed current of 320 kA at peak synchronized with the proton beam timing. The decay volume (DV) is a free space down stream of the horns where pions decay in flight into ν_μ's and muons. A beam dump is placed at the end of DV, 110m from the target. Muon monitors (MUMON) is placed just behind the beam dump to measure the muon flux. While almost all hadrons are absorbed by the beam dump, muons of energy $>$ 5 GeV can penetrate the beam dump. MUMON provides pulse-by-pulse information on the intensity and profile (direction) of the beam.

At 280 m from the target, neutrino detectors are placed in order to measure the properties of neutrinos just after their production. Details of these near detectors are described in Ref.[4].

3. Present status

Super-Kamiokande has been running steadily after the renewal of the readout electronics in 2008.

The construction of the J-PARC accelerator complex was completed in 2008 and the beamline of the neutrino facility in Mar. 2009. On 23rd April 2009, the first proton beam was extracted to the neutrino beamline and MUMON observed the muon signal confirming the generation of neutrinos. The beamline magnets and proton beam monitors worked as expected. Only the first horn was excited and its focusing effect was confirmed by the measurement of the muon profile with MUMON as shown in Fig.2. The nominal intensity during this commissioning run was $\sim 4 \times 10^{11}$protons per spill , which is $\sim$ 0.1% of the design intensity. From Autumn 2009, further commissioning is planned to increase the beam intensity and all 3 horns will excited. First physics data taking is expected to begin in 2010.

Acknowledgments

The author acknowledges the support by Grant-in-Aid for Young Scientist(S, no.20674004) from MEXT, Japan.

References

[1] M. Apollonio, *et al.*, Eur. Phys. J. **C27**, 331-374 (2003)
M. Apollonio, *et al.*, Phys. Lett. B **466**, 415-430 (1999).
[2] D. Beavis, A. Carroll, I. Chiang, *et al.*, Proposal of BNL AGS E-889 (1995).
[3] Y. Ashie *et al.* [Super-Kamiokande Collaboration], Phys. Rev. D **71**, 112005 (2005)
Y. Fukuda *et al.*, Nucl. Instrum. Meth. A **501**, 418 (2003).
and citations therein.
[4] M. Besnier, "Proceedings of the TAUP2009 Conference"

The contribution of Near Detectors to the T2K neutrino measurements

Magali Besnier, on behalf of the T2K collaboration
Laboratoire Leprince Ringuet, Route de Saclay, 91241 Palaiseau, France
E-mail: magali.besnier@llr.in2p3.fr

Abstract. The T2K long baseline experiment located in Japan is dedicated mainly to constrain or measure the θ_{13} neutrino mixing angle by searching for the electron neutrino appearance at Super Kamiokande in a pure muon neutrino beam produced at JParc 295 km away. The beam recently started on time in April 2009. Near detectors are situated 280 m from the primary target. One is on axis and a second one is situated off axis with an angle of 2.5 degrees in the direction of Super Kamiokande. They are essential to study the different systematic errors which have to be well known for the θ_{13} determination. This paper reviews the overall status of the near detectors and their role in the understanding of systematics.

1. The T2K experiment

T2K (Toaki to Kamiokande) is a long baseline neutrino experiment in Japan aiming to study several neutrino parameters. A muon neutrino beam is produced at JParc accelerator located in Tokai and sent in the direction of Super Kamiokande (SK) 295 km downstream. The first goal of T2K is to measure or constrain the θ_{13} mixing angle by studying the $\nu_\mu \rightarrow \nu_e$ oscillation (electron neutrino appearance). It will also be possible to measure the "atmospheric parameters" θ_{23} and $|\Delta m^2_{23}|$ with a very good precision by studying the muon neutrino disappearance. The muon neutrino beam is produced by 30 GeV protons sent on a carbon target. The resulting hadrons (pions and kaons) decay mainly into a muon and a muon neutrino. Electron neutrinos can be produced by other decay channels and secondary interactions. The amount of tau neutrino is negligible. A complete description of the beamline can be found in [1].

The muon neutrino beam is sent in the direction of SK with an off axis angle of 2.5 degrees. The off axis angle decreases the neutrino flux but provides several advantages thanks to the kinematics of a 2-body decay:

- The muon neutrino energy is peaked to 700 MeV corresponding to the maximum oscillation probability along the 295 km distance.
- The high energy tail and the ν_e beam component are strongly reduced.

The expected fluxes at SK of muon neutrinos and oscillated $\nu_\mu \rightarrow \nu_e$ strongly depend on the off-axis angle. This angle, in addition to other parameters like fluxes and cross sections, has therefore to be known very precisely in order to limit the systematic errors. The following section presents the needs and means of T2K concerning systematics and their measurement. The Two near detectors located 280 m from the primary target will play a crucial role in this task. A detailed description of the near detectors is given in section 3.

2. Systematic errors and sensitivities to neutrino parameters

Thanks to its features detailed in section 1, T2K can achieve competitive sensitivities after 5 years running [1] :

- $\sin^2(2\theta_{13}) < 6\ 10^{-3}$ at 90% C.L. if systematic errors are less than 10%.

- $\Delta(\sin^2 2\theta_{23}) \approx 10^{-2}$ and $\Delta(\Delta m^2_{23}) \approx 10^{-4}$ at 90% C.L. if systematic errors are kept below statistical errors.

The different sources of systematics can be studied thanks to different means:

-The near/far ratio (calibration of fluxes at SK with respect to near detectors) depends on the off-axis angle but also on hadronic production and hadron decays. This point is studied by the NA61 experiment at CERN [2]. Their goal is to reach 2 or 3 % error for the near/far ratio.

-The energy calibration can be done thanks to SK-IV data analysis with cosmic events and π^0 mass measurements. A 2% precision is presently achieved.

-Off-axis angle, fluxes and cross-section of signal and background events will be studied with the near detectors, described in the next section.

3. The role of near detectors in systematic error measurements

The T2K near detectors, located 280 m from the primary target in a 25 m deep pit, are represented on figure 1. They play the main role in systematic error measurements.

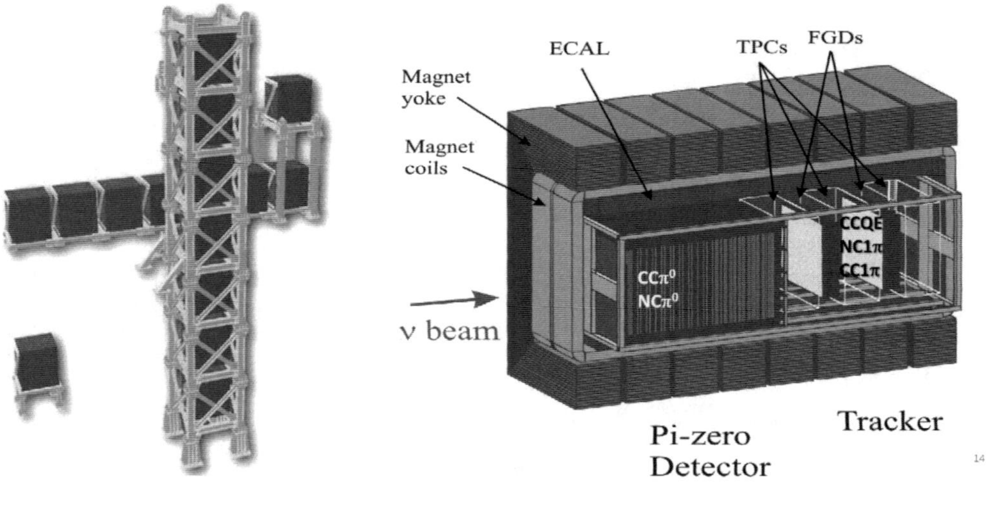

Figure 1. Sketch of T2K near detectors: Left, on axis INGRID. Right, off axis ND280.

INGRID is a cross-shaped detector made with seven horizontal and seven vertical modules composed by iron plates interleaved with X-Y plastic scintillator planes. INGRID is on axis and aims to measure the off axis angle with a precision better than 1 mrd. The complete installation and cabling was finished in August 2009 and the commissioning with neutrino beam will start in October. Moreover, as can be seen on figure 1, two off-axis modules will be added by 2011 in order to estimate the azimuthal symmetry of the beam. Also, a movable module without iron, called proton-module, will be installed in the front of the horizontal modules. It will discriminate between the different hadronic final states and quasi-elastic interactions, and will give a more precise measurement of the vertex positions.

The ND280 detector is located off axis in the direction of SK and is composed of several sub-detectors dedicated to the measurement of neutrino fluxes and different neutrino cross-sections.

- The UA1 magnet is surrounding all sub-detectors. Its C-shape iron elements are instrumented with muon range detectors made of plastic scintillators. It will be possible to measure the muon energy with 10% resolution.

- The first sub-detector located upstream the neutrino beam is called π^0-detector (POD). It is made of lead foils and scintillator planes interleaved with water tanks. It will measure the π^0 production in neutrino interactions on water, which corresponds to the main background of the electron neutrino appearance at SK.

- Downstream the POD is located the tracker. It is composed of three TPC interleaved with two fine grained detectors (FGD) made of plastic scintillator bars. The TPC will separate electrons and muons using dE/dX, also providing energy measurements. The FGD will distinguish quasi-elastic events from multihadronic states which constitute a huge background for the "atmospheric parameters" measurement. Moreover the most upstream FGD is filled with water allowing to measure cross-sections like in SK. Finally the tracker will be able to study several cross-sections (CCQE, NC1π, CC1π) and measure neutrino fluxes.

- The ECAL detector is located downstream and on the top of the POD and the tracker. It is composed of scintillator bars interleaved with lead sheets. The part surounding the POD acts as a γ/μ tagger for large angle tracks escaping the POD. The part around the tracker will reconstruct electromagnetic showers with their energy and improve the electron/muon identification.

Table 1 summarises the near detector contributions to the systematic error measurements in T2K. All the near detectors will be installed and instrumented by the end of the year 2009 in order to take neutrino data early 2010.

Table 1. Near detector contributions to the systematic error measurements in T2K.

Source of systematic error	Required performances	Actual achievable performances	Principal means
off-axis angle	1 mrad	<1mrd	INGRID
CCQE ν_e^{beam} event flux	10%	5%	TPC+ECAL
CCQE ν_μ event flux	5%	4%	TPC+Magnet
QE/Non-QE ratio (ν_μ events)	5-10%	10%	TPC+FGD
Number of π^0 events	10%	10%	POD+ECAL

4. Conclusion

The T2K experiment will measure several neutrino mixing parameters with competitive sensibilities and precisions. To achieve these performances, the near detectors will study and measure systematic errors in order to keep them below 10% for θ_{13} study and under statistical errors for "atmospheric parameter" measurements. The full set of near detectors will be installed and instrumented in the pit by the end of 2009 and precise measurements will be performed during the first year of data taking. First physic results are expected by summer 2010.

References

[1] Hayato Y et al. 2003 *Letter of intent Neutrino Oscillation Experiment at JHF*
[2] Posiadala M 2009 Status of the NA61 (SHINE) experiment at CERN (*preprint* hep-ex/09013332)

Journal of Physics: Conference Series **203** (2010) 012106 doi:10.1088/1742-6596/203/1/012106

Emulsion analysis in the OPERA experiment

N. Naganawa on behalf of the OPERA collaboration

Fundamental Particle Physics Laboratory, Graduate School of Science of Nagoya University, Furo-cho, Chikusa-ku, Nagoya, 464-8602, Japan

Email : naganawa@flab.phys.nagoya-u.ac.jp

Abstract. OPERA is a unique experiment aimed at the first detection of ν_τ appearance in a flux of ν_μ due to the neutrino oscillation from ν_μ to ν_τ. The CERN CNGS beam is the source of ν_μ. The detector is hybrid; it is composed of nuclear emulsion films and electronic detectors. It is located in the LNGS underground laboratory. The target consists of 150,000 Emulsion Cloud Chambers (ECC) bricks, which are stacks of interleaved emulsion films and lead plates. The ν_τ charged current interactions will be detected by identifying the decay topology of the τ in the ECC bricks. The first run started in 2008. The experiment is currently in the phase of data taking and analysis. The experimental methods, the status and the summary of the results from the 2008 run are presented in this paper.

1. Introduction

OPERA [1] is a unique experiment aimed at the first direct observation of the appearance of ν_τ caused by ν_μ to ν_τ oscillation by detecting the decay topology of the τ induced in the ν_τ CC interactions. The CERN ν_μ CNGS beam has an average optimised energy of 17 GeV. The hybrid detector composed of nuclear emulsion films and electronic detectors is located 1400m underground in the LNGS laboratory, 730km away from CERN. The first physics run with a target partly filled started in spring 2008. The experiment is currently in the phase of its second year of data taking and of data analysis.

2. The OPERA detector

The OPERA detector [2] consists of two identical Super-Modules. Each consists of an instrumented target followed by a magnetic muon spectrometer. The targets, for a total mass of 1.25 kton, is composed of 150 000 units called hereafter "bricks" arranged in twice 31 walls. These are interleaved with pairs of planes of orthogonal plastic scintillator strips constituting the Target Tracker (TT) - see Figure 1.

Bricks are fabricated following the so-called Emulsion Cloud Chamber (ECC) technique: 57 nuclear emulsion films and 56 1-mm thick lead plates are stacked alternately (see Figures 2 and 3). This technique is adequate to recognise τ decay topologies as proven by the DONUT experiment [3].

An emulsion film is made of two layers of nuclear emulsion gel 44 μm thick deposited on each side of a 205 μm thick plastic base of 12.5×9.9 cm^2. A track of a minimum ionising particle is shown in Figure 4. The position resolution of the emulsion film is 0.3 micron. A brick weighs 8.3 kg and its thickness, 7.5 cm, corresponds to about 10 radiation lengths. With this structure, it is also possible to measure momentum by Multiple Coulomb Scattering [4] and identify electrons and measure their energy by counting the number of track segments in the electromagnetic showers [5].

A pair of emulsion films called the Changeable Sheets (CS) is packed into a thin box glued on the downstream face of each brick. The films can be removed and if necessary replaced without opening the brick. They serve as interface between the TT and the brick.

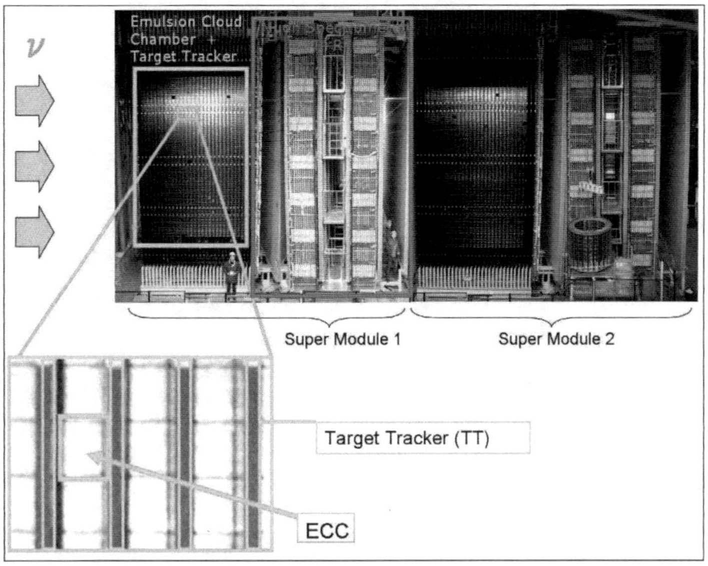

Figure 1. The OPERA detector.

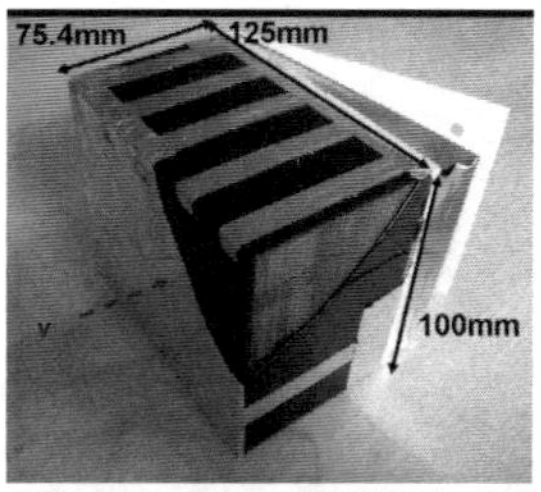

Figure 2. An ECC brick and its structure.

Figure 3. OPERA emulsion films and lead plates.

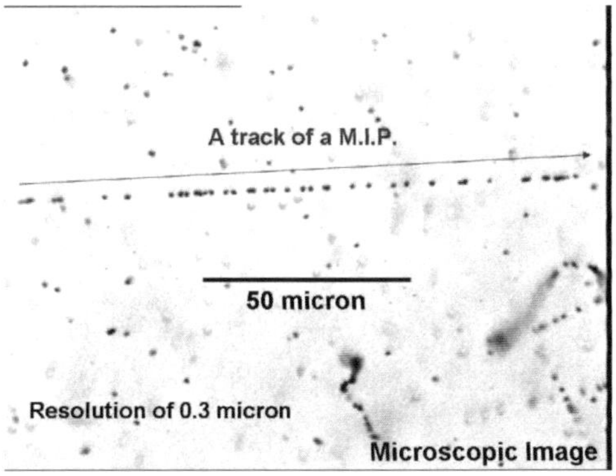

Figure 4. The microscopic image of a track of a M.I.P. The linear sequence of black dots along the blue arrow is the track of a M.I.P. in a developed emulsion film.

3. Localisation of the neutrino interaction vertices in the ECC bricks

From the tracks reconstruction in the TT, the brick with the highest probability to contain the interaction vertex is identified and removed from the target by an automaton. The CS is developed underground and scanned at either LNGS and or in Japan. If tracks from the neutrino interaction are found in the CS, the films of the corresponding brick are developed, scanned and analysed. The tracks in the CS are extrapolated to the most downstream film of the brick and searched for. They are then followed upstream until they disappear. A volume scan of about 1 cm^3 around the point of disappearance is performed. Tracks emerging from a single point are extracted from the data. This localises the position of the vertex inside the brick.

In case tracks belonging to the event are not found in the CS, the brick is equipped with a new pair of CS and reinserted into the target. The next most probable brick is then removed and its CS are analysed.

Journal of Physics: Conference Series **203** (2010) 012106 doi:10.1088/1742-6596/203/1/012106

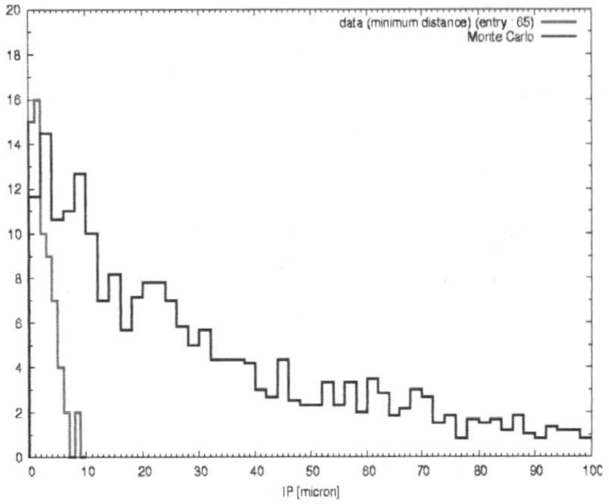

Figure 5. Data of minimum distance of pairs of tracks from neutrino interactions and the IP of the decay daughters of τ generated by MC.

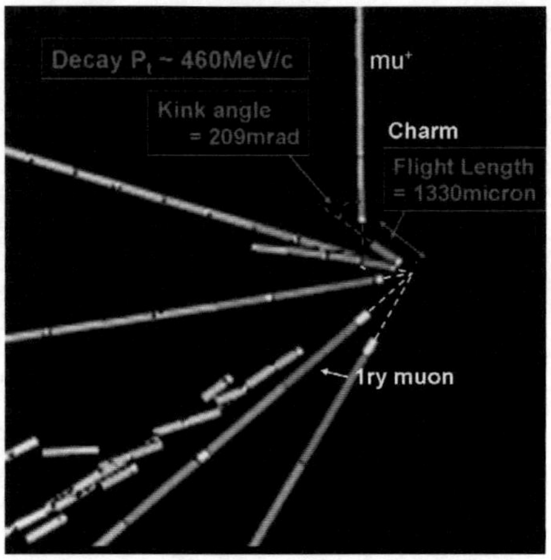

Figure 6. An example of a charm decay candidate.

4. Separation of the vertices of the neutrino interaction and of the τ decay

The histogram in red in Figure 6 shows the minimum distance between pairs of tracks from neutrino interactions in lead plates computed from the data recorded in the emulsion films immediately downstream of the vertex plate. Only tracks with a momentum larger than 1 GeV/c are considered. The uncertainty of the position of vertices in the lead plates results from Multiple Coulomb Scattering, the angular resolution and the position resolution of the tracks. The blue histogram shows the Monte Carlo simulation of the distribution of the impact parameter (IP) of the decay daughters of τ produced in CC ν_τ interactions. The comparison between the two plots demonstrates that the vertices of the neutrino interaction and of the τ decay are separable. The 1 mm thickness of the lead plates does not affect significantly the τ decay search. This is confirmed by the detection of several charm decay candidates with topologies similar to that of τ decays. An example of such event is shown in Figure 6.

5. Summary of the run status and future prospects

During the run of 2008 [6], the CNGS provided 1.782×10^{19} protons on target (p.o.t.) and 1690 neutrino events have been registered in the OPERA target. This corresponds to an expectation of 0.7 detected ν_τ CC interactions for $\Delta m^2 = 2.43\times10^{-3}$ eV^2 and full mixing. Up to now, 746 vertices of neutrino interactions have been localised in the bricks. No τ candidate has been detected yet. However, several charm decay candidates have been observed, in particular charged charm decays into one and three prongs with topologies similar to τ decay topologies.

It is verified from the data that the lead plates are thin enough and the spatial resolution of the films is good enough to allow efficiently separating the vertices of the neutrino interaction and the τ decay.

For the 2009 run which has started on June 1, we requested 4.5×10^{19} p.o.t. About 2.5 ν_τ CC interactions would then be expected to be detected at the end of this run.

References

[1] Guler M et al. 2000 *CERN/SPSC* 2000-028, SPSC/P318, LNGS P25/2000
[2] Acquafredda R et al. 2009 *JINST* 4 P04018
[3] Kodama K et al. 2001 *Phys. Lett.* B **504** 218
[4] Park B D et al. 2007 *Nucl. Instrum. Meth.* A **574**
[5] Toshito T et al. 2003 *Rev. Sci. Instrum.* **74** 53-56
[6] Agafonova N et al. 2009 *JINST* 4 P06020

Status of the ICARUS T600 detector at the LNGS

Alessandro Menegolli, on behalf of the ICARUS Collaboration
INFN, Sezione di Pavia, via Bassi 6, 27100 Pavia
E-mail: alessandro.menegolli@pv.infn.it

Abstract. The installation of the ICARUS T600 Liquid Argon TPC detector at the INFN (Istituto Nazionale di Fisica Nucleare) LNGS underground Laboratory (Assergi, Italy) is almost completed and its commissioning is forthcoming: the detector, the cryogenic plant and the data acquisition are now essentially ready. The main phases of the installation of the T600 plant will be resumed and the first tests on the inner detector constituents will be presented. Finally, the next steps needed to bring the T600 apparatus to the full operation in view of the physics run with the CNGS neutrino beam and the cosmic neutrinos will be explained.

1. The ICARUS T600 plant at the LNGS

ICARUS T600 is the largest detector exploiting the Liquid Argon Time Projection Chamber (LAr TPC) technique [1] ever realized. After the successful surface test run held in Pavia in 2001, with one of the two 300 tons half-modules successfully put into operation [2] [3] [4] [5] [6] [7], the T600 detector was brought to the LNGS underground Laboratory (Assergi, Italy), to be activated in view of the physics run with CNGS neutrino beam[1] and cosmic neutrinos.
A large amount of activities were carried out in the last years in order to bring the detector to the full functionality in Hall B at LNGS, see fig. 1. They concerned the cryogenic plant (Air Liquide[2] company plant), the N_2 liquefaction system (Stirling[3] company plant) and the LNGS infrastructures, responsability of the Laboratory. The ICARUS Collaboration finalized the flange and feed-through mounting for wires, Photo-Multipliers Tubes and Slow Control sensors and the cabling of all feed-throughs to the electronic racks.
The Air Liquide cryogenic plant was completed in summer 2008 with the activation of the vacuum pumping for the insulation panels. The unloading station was also completed and already used (September 2008 and February 2009) to unload the liquid Nitrogen needed to the liquefier system commissioning. The ICARUS Control System has been cabled and almost tested. The interface with the Control System of the Laboratory is almost completed. Final tests will be carried out at the commissioning level.
The liquefier plant has been completed and its commissioning was successful: all the ten units of the system have been switched-on, certifying that the liquid Nitrogen is produced at the required rate. The system is now in continuous operation.
The Laboratory infrastructures (power supply, exhaust ventilation system, water cooling circuit for Stirling system, safety facilities) have been completed and tested.

[1] The CERN CNGS project, http://proj-cngs.web.cern.ch/proj-cngs
[2] Air Liquide Italia Service, Milano (Italy)
[3] Stirling Cryogenics BV, Eindhoven (Netherlands)

Journal of Physics: Conference Series **203** (2010) 012107 doi:10.1088/1742-6596/203/1/012107

Figure 1. ICARUS T600 plant inside Hall B at the LNGS underground Laboratory (July 2009).

2. First tests on the T600 detector at LNGS

In the course of the last year for the first time a number of tests was carried out by the Collaboration on the fully assembled T600 detector, the main results of which are reported in the following:

- *Electronics tests.* The installation of the readout electronics racks on the T600 top has been completed, together with the cabling of clock and trigger distribution. Each rack has been commissioned with test pulse data taking, see fig. 2, in order to identify any broken or malfunctioning boards. The event builder architecture has been deployed including DAQ computers, storage, networking (cabling, switches and fibres to control room and to external labs). Moreover, these data were used to test the new software developed for underground data reading and event reconstruction in view of the physics run.

- *Inner detector sensor tests.* A Slow Control system has been implemented to acquire and monitor the following information: inner wall displacement during vacuum and cooling phases; liquid Argon level during filling; pressure and temperatures inside the cryostat during all phases of commissioning and run. Different types of sensors are installed: cold cathode gauges for pressure monitor; resistors for temperature (Pt1000) and liquid Argon level measurement (carbon resistors); capacitive tubes for inner wall displacement and liquid Argon level measurement. Dedicated boards have been designed, built and calibrated in order to read the signals coming from the capacitive tubes. A MySQL[4] database manages the data coming from all sensors, while a LabView interface allows the online monitor. The Slow Control system has been already tested and it is ready for the commissioning phase.

- *Pressure/vacuum leaking tests.* These tests showed that some of the flanges housing the wire signal feed-throughs were leaking in a sensitive way, at the rate of a few mbar/h, see fig. 3. Already in 2001, during the test in Pavia, this problem was observed. For this reason, all flanges were reworked at CERN to add a layer of Araldite^{TM}. These flanges were tested under vacuum after the reworking and the problem seemed solved. The common assumption is that the aging and the mechanical stress could induce the loss of the tight performance. Because of this, the Collaboration, soon after the 2001 surface test in Pavia, developed a new flat flange that showed higher reliability. A set of 42 new flanges (96 is the total number

[4] Sun Microsystems

of wire signal flanges in T600) has been used to substitute all the bad flanges during the first half of 2009: this allowed a strong reduction of the leak rate, see fig. 3, which is now suitable for the vacuuming of the T600. The Collaboration is currently operating to get enough flanges of the new type in order to substitute all the old flanges, in order to prevent possible future problems.

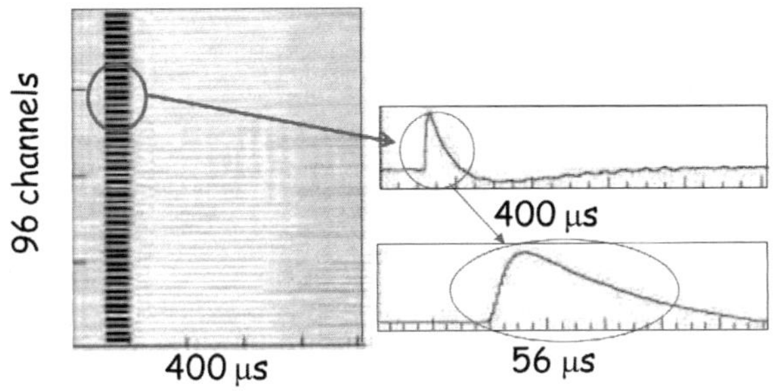

Figure 2. Example of test pulse data acquired with the ICARUS T600 new read-out.

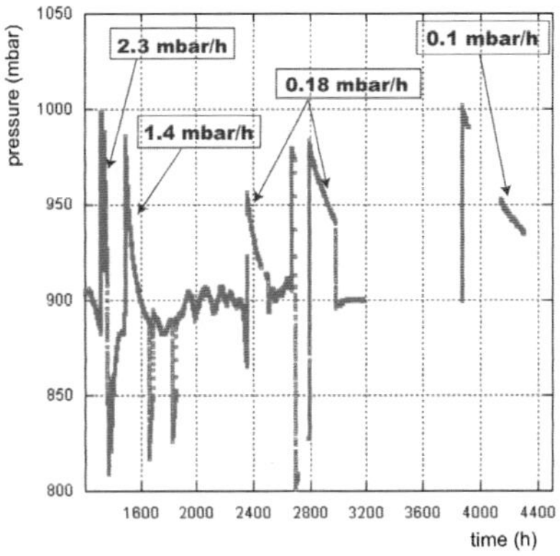

Figure 3. Results of pressure/vacuum tests on one of the T600 half-modules: a strong reduction of the leak rate is achieved with the substitution of the old flanges. The ambient pressure in Hall B is lower than one bar because of the altitude.

3. Forthcoming activities

The installation of ICARUS T600 has progressed in 2008/2009 up to its completion. The detector, the cryogenic plant and the data acquisition are now essentially ready. The Laboratory infrastructures, required for the detector filling, have been completed and the T600 start-up procedures, logistics and authorizations have been almost completely defined. The last step before the detector commissioning concerns the reparation of one of the thermal insulation panels, which partially collapsed during its evacuation. The strategy for the repair of the panel has been set and works for its recovery already began. After this, the ICARUS T600 commissioning will start with evacuation of the two half-modules and their subsequent cooling. The filling of the detector with 600 tons of liquid Argon will be subsequent. Finally, ICARUS T600 activation and the data taking will follow.

References

[1] C. Rubbia, The Liquid-Argon Time Projection Chamber: A New Concept For Neutrino Detector, CERN-EP/77-08 (1977).
[2] S. Amoruso et al., Nucl. Instr. Meth. A **516**, 68 (2004).
[3] S. Amoruso et al., Nucl. Instr. Meth. A **523**, 275 (2004).
[4] S. Amerio et al., Nucl. Instr. Meth. A **527**, 329 (2004).
[5] S. Amoruso et al., Eur. Phys. J. C **33**, 233 (2004).
[6] A. Ankowski et al., Eur. Phys. J. C **48**, 667 (2006).
[7] A. Ankowski et al., Nucl. Instr. Meth. A **556**, 146 (2006).

Journal of Physics: Conference Series **203** (2010) 012108 doi:10.1088/1742-6596/203/1/012108

ArgoNeuT, a liquid argon time projection chamber in a low energy neutrino beam

Joshua Spitz, for the ArgoNeuT Collaboration
Department of Physics, Yale University, New Haven, CT 06520, USA
E-mail: joshua.spitz@yale.edu

Abstract. ArgoNeuT (Argon Neutrino Test), a NSF/DOE project at Fermilab, is the first LArTPC to go in a low energy neutrino beam and just the second LArTPC to go in a neutrino beam ever. ArgoNeuT sits just upstream of the on-axis MINOS near detector in the NuMI beamline, about 1 km from the target station and 100 m underground. The detector features a 47×40×90 cm (169 L) active volume TPC with a fully contained recirculation and purification system. Among other physics, ArgoNeuT will measure the charged-current quasi-elastic (anti-) neutrino cross section on argon at $E_\nu \sim 3$ GeV.

1. Introduction

Liquid Argon Time Projection Chambers (LArTPCs) [1] feature high-resolution three dimensional tracking and calorimetry for efficient and low background neutrino detection. An accelerator-based long baseline neutrino oscillation experiment will require a multi-kiloton scale far detector [2][3][4][5] for sensitivity to θ_{13} and leptonic CP-violation. LArTPCs of up to 600 tons [6] have been successfully built and operated but further research and design is necessary for a multi-kiloton detector.

In a LArTPC, the charged particles emanating from the interaction vertex ionize argon atoms as they traverse the detector. As argon is a noble element, this ionization is free to drift across the medium. Argon, with a comparatively large density, short radiation length, and high scintillation yield, is ideal for neutrino detection and event containment. Krypton and Xenon, although superficially attractive for the same reasons, are far less abundant in the atmosphere and therefore very expensive. An electric field is imposed in the active volume and the ionization tracks are drifted towards readout wire planes that are oriented at an angle with respect to one another. The ionization induces a current on the induction plane(s) and is collected by the collection plane. Combining the positional information from the wire planes and detecting the events in time allows a high-resolution three-dimensional image of the event with calorimetric information[1] to be obtained.

2. ArgoNeuT

The ArgoNeuT LArTPC is currently taking data in the on-axis Neutrinos at the Main Injector (NuMI) beamline (peaking at $E_\nu \sim 3$ GeV) at Fermilab. ArgoNeuT instruments a total of 480 wires on an induction plane and a collection plane, oriented at an angle of 60° with respect to

[1] The charge induced/collected is proportional to the amount of energy the particle deposited in the detector.

one another. The wire spacing on both planes is 4 mm, the planes are separated by 4 mm, and the electric field is 500 V/cm in the 47×40×90 cm TPC. The event trigger and time of the initial neutrino interaction, t_0, are obtained from beam timing information (the NuMI beam spill window is approximately 10 μs). ArgoNeuT purifies its liquid argon through a fully contained recirculation system which includes a cryocooler and oxygen/water filters. The filter system is similar to that which is described in [7].

2.1. Charged-Current Quasi-Elastic Events in ArgoNeuT

The simple two particle event topology and dominant cross section in the relevant energy region make the Charged-Current Quasi-Elastic (CCQE[2]) interaction the most important interaction channel for accelerator-based neutrino oscillation experiments. Current (MiniBooNE [8], MINOS [9]) and future (T2K [10], NOvA [11], wide-band-beam to DUSEL [12], ...) accelerator-based neutrino oscillation experiments search for the appearance of electron neutrinos largely via the ν_e CCQE interaction $\nu_e n \longrightarrow e^- p$. The ν_μ CCQE interaction is used to measure the muon-neutrino flux, constraining the expected ν_e flux at the near and far detectors. The anti-neutrino analog interactions are important in the search for CP violation in the lepton sector (e.g. $P[\nu_\mu \rightarrow \nu_e] \neq P[\overline{\nu}_\mu \rightarrow \overline{\nu}_e]$). The CCQE cross section is known with 20-30% uncertainty over most relevant energies and only one (preliminary) measurement of the ν_μ CCQE cross section on argon has ever been taken (at $< E_\nu >\sim 28$ GeV) [13].

Figure 1. Left: The fully instrumented TPC being inserted into the ArgoNeuT inner cryostat. The cryostat is composed of an inner, liquid argon layer and an outer, vacuum-jacket layer. Right: The ArgoNeuT cryostat in the NuMI beamline.

ArgoNeuT will detect ~1230 anti-neutrino CCQE events/180 days with 54% proton containment inside the active volume. ArgoNeuT will employ the downstream MINOS near detector for sign determination and muon energy reconstruction as it is usually necessary to range out the muon in order to know its energy precisely. Larger LArTPCs can reconstruct the muon's energy via multiple scattering if the muon travels longer than about 1 m [14]. However, a muon will rarely travel this far in ArgoNeuT. Note that in 47% of CCQE events, the proton is contained inside ArgoNeuT and the muon enters the MINOS near detector. ArgoNeuT will measure the CCQE interaction rate as a function of neutrino energy and determine cross section using the well known NuMI neutrino flux[3].

[2] For neutrinos, $\nu_l n \longrightarrow l^- p$. For anti-neutrinos, $\overline{\nu}_l p \longrightarrow l^+ n$.
[3] The NuMI neutrino flux at ArgoNeuT is known with 5-10% uncertainty.

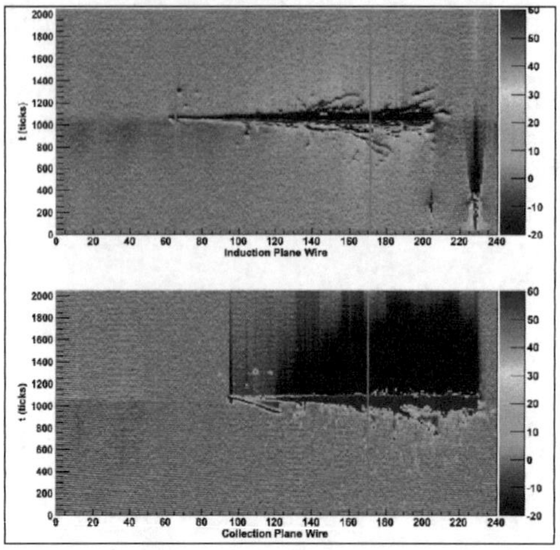

Figure 2. A ν_e charged-current candidate event an seen in the induction and collection views. Please note that this event image is raw data and that the "ν_e charged-current" interpretation is preliminary.

3. Outlook

ArgoNeuT was installed in the NuMI beamline in early 2009 and filled with liquid argon in May 2009. The first neutrino events were observed shortly thereafter. ArgoNeuT ran in neutrino-commissioning-mode for ~1 month up until the planned 3 month beam shutdown on June 14, 2009 and the physics run began on September 14, 2009 (ending on March 1, 2010). A recent neutrino event in ArgoNeuT can be seen in Figure 2.

Acknowledgments

The author would like to acknowledge the support staff at Fermilab for their invaluable contributions to the planning and construction of ArgoNeuT. This work is supported by the Department of Energy and the National Science Foundation.

References

[1] Rubbia C 1977 *CERN-EP/77-08*
[2] Baibussinov B *et al.* 2008 *Astroparticle Physics* **29** 174
[3] Cline D *et al.* 2003 *Nucl. Instr. Meth. A* **503** 136
[4] Rubbia A 2004 Experiments For CP-Violation: A Giant Liquid Argon Scintillation, Cerenkov And Charge Imaging Experiment? *arXiv:hep-ph/0402110*
[5] Bartoszek L *et al.* 2004 FLARE, Fermilab Liquid Argon Experiments *arXiv:hep-ex/0408121*
[6] Amerio S *et al.* 2004 *Nucl. Instr. Meth. A* **527** 329
[7] Curioni A *et al.* 2009 *Nucl. Instr. Meth. A* **605** 306
[8] Aguilar-Arevalo A *et al.* 2007 *Phys. Rev. Lett.* **98** 231801
[9] Michael *et al.* 2006 *Phys. Rev. Lett.* **97** 191801
[10] Itow Y *et al.* 2001 The JHF-Kamioka Neutrino Project *arXiv:hep-ex/0106019*
[11] Ayres D *et al.* 2002 Letter of Intent to build an Off-axis Detector to study numu to nue oscillations with the NuMI Neutrino Beam *arXiv:hep-ex/0210005*
[12] DUSEL Homepage: http://www.lbl.gov/nsd/homestake
[13] de la Ossa A M 2007 Study of accelerator neutrino interactions in a liquid argon TPC *Ph.D. Thesis. University of Granada*
[14] Ankowski A *et al.* 2006 *Eur. Phys. Jour. C* **48** 667

Journal of Physics: Conference Series **203** (2010) 012109 doi:10.1088/1742-6596/203/1/012109

The Long Baseline Neutrino Oscillation Experiment at DUSEL

Jelena Maricic for the LBNE DUSEL collaboration
Drexel University, 3141 Chestnut St. 12-816, Philadelphia, PA, 19104
E-mail: jelena@physics.drexel.edu

Abstract. Rapid progress in neutrino physics in recent years has brought us closer to realization of a massive neutrino detector at the Deep Underground Science and Engineering Laboratory (DUSEL) in the Homestake mine in the town of Lead, South Dakota, USA. The detector is being designed with the following scientific goals in mind: value of CP-violation phase δ in lepton sector, neutrino mass hierarchy and value of the angle θ_{13}. In addition, the Long Baseline Neutrino Experiment (LBNE) will be able to search for proton decay, get a detailed energy spectrum in the case of galactic supernovae, detect solar and atmospheric neutrinos, possibly geoneutrinos and even relic supernovae neutrinos that have never been successfully detected before. The physics goals dictate the minimal size of 300 kton fiducial volume, and the plan is to realize it with three 100 kton fiducial volume modules that will be placed at the depth of 1480 m. LBNE will be able to obtain 3σ C.L. measurement of θ_{13} if the value of $\sin^2 2\theta_{13} \geq 0.005$ for any value of CP-δ phase; 3σ C.L. measurement of the neutrino mass hierarchy if $\sin^2 2\theta_{13} \geq 0.012$ for any value of CP-δ phase and measure 50% of all CP-δ phases with 3σ C.L. for $\sin^2 2\theta_{13} \geq 0.012$. This is all under the assumption that an upgraded neutrino beam is sent to LBNE from Fermilab, 1300 km away and experiment is run for 6 years. This paper describes physics reach, status and current undergoing research and development effort toward construction of the LBNE.

1. Introduction

Recent discoveries in neutrino physics, including oscillations and precision measurements of majority of neutrino mixing parameters, have paved the path for even greater open physics questions: the value of charge parity (CP) violation phase δ in leptons; the ordering of the three neutrino masses; and the value of the last non-measured neutrino mixing angle θ_{13}. The plans are underway for the Long Baseline Neutrino Experiment (LBNE) that will have the potential to address these questions and maintain a broad physics program at the same time - search for the proton decay, galactic supernovae detailed neutrino spectrum measurement, relic supernovae neutrinos, solar and atmospheric neutrinos and possibly geoneutrinos. The LBNE will be located at the Deep Underground Science and Engineering Laboratory (DUSEL) in the Homestake mine in the town of Lead, South Dakota, USA. Strong neutrino beam will be sent from Fermi National Laboratory, 1300 km away, in order to search for CP-δ phase, θ_{13} and mass hierarchy.

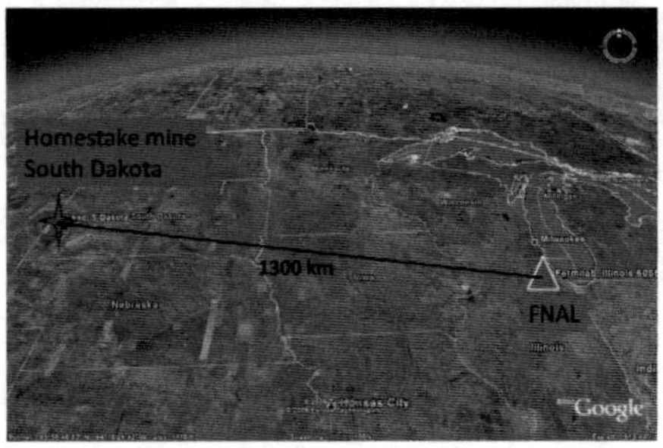

Figure 1. Location of the LBNE in Lead, South Dakota, with respect to Fermi national Laboratory that will send a strong neutrino beam 1300 km away.

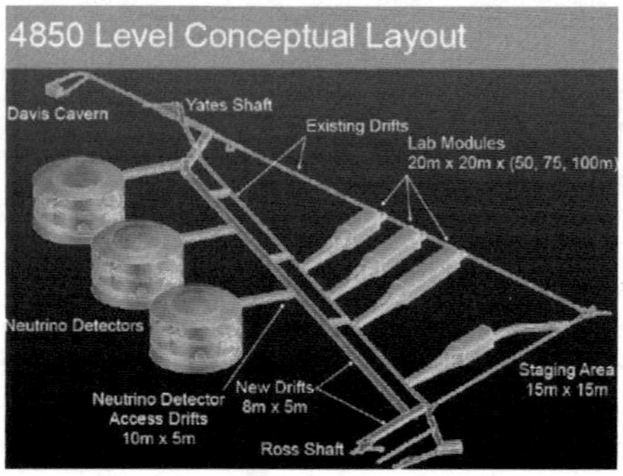

Figure 2. The layout of water Cherenkov modules.

Table 1. The sensitivity of the experiment with a water Cherenkov detector for different integrated luminosity and detector size. For $\sin^2 2\theta_{13}$ and the mass hierarchy, the sensitivity is given as the minimum value of $\sin^2 2\theta_{13}$ at which the experiment achieves a 3σ reach for all δ_{CP}. For CP violation, the sensitivity is given as the minimum vale of $\sin^2 2\theta_{13}$ at which the experiment achieves 3σ reach for 50% of δ_{CP}. This assumes a 5% systematic uncertainty on the background.

Detector Size (kTon)	POT($\times 10^{20}$)@120GeV 1MW 10^{21} POT	Years $\nu_\mu + \bar{\nu}_\mu$	$\sin^2 2\theta_{13} \neq 0$	Mass Hierarchy	CP Violation
100	30 + 30	3 + 3	0.014	0.031	¿ 0.1
300	30 + 30	3 + 3	0.008	0.017	0.025
600	30 + 30	3 + 3	0.005	0.012	0.012
300	60 + 60	3 + 3	0.005	0.012	0.012

2. Physics Potential

Neutrino Physics Program Using a strong muon neutrino beam from Fermi National Laboratory (Fig. 1), appearance search for electron neutrinos will be conducted at the neutrino detector in Homestake mine, to search for the value of CP violation phase, ordering of neutrino masses and value of the the neutrino oscillation angle θ_{13}. Evaluation of the physics prospects has been conducted for two possible wide band beam power levels 1 MW and 2 MW, and initial proton beam energy 60 to 120 GeV that produces neutrinos in 1-10 GeV range. Preliminary studies have been conducted for various detector sizes, assuming 3 years of running in the neutrino mode and additional 3 years of running in the anti-neutrino mode. The summary of results is shown in the Table 1. The conclusion is that reaching science goals requires beam power of at least 2 MW, 300 kton fiducial volume for a water Cherenkov detector (50 kton for the Liquid Argonne detector) and six years of neutrino, anti-neutrino combined running. Under these conditions LBNE will be able to obtain 3σ C.L. measuremet of θ_{13} if the value of $\sin^2 2\theta_{13} \geq 0.005$ for any value of CP-δ phase; 3σ C.L. measurement of the neutrino mass hierarchy if $\sin^2 2\theta_{13} \geq 0.012$ for any value of CP-δ phase and measure 50% of all CP-δ phases with 3σ C.L. for $\sin^2 2\theta_{13} \geq 0.012$. Thus, meeting these requirements will ensure strong neutrino physics potential of LBNE.

Nucleon Decay Proton decay is one of the most important predictions of Grand

Unification, but has not been observed yet. A number of theoretical models predict proton decay with a lifetime in the range of 10^{35-36} years. There are a lot of possible branches for the proton decay with the modes $p \rightarrow e^+ + \pi^0$ and $p \rightarrow K^+ + \nu$ being common benchmarks.

Based on the experience from SuperKamiokande detector, 20 years of running with 300 kton water Cherenkov detector will result in 10^{35} years lifetime limit at 90% C.L. for $p \rightarrow e^+ + \pi^0$ decay mode.

Mode $p \rightarrow K^+ + \nu$, K^+ is affected by atmospheric neutrino background concurrent to kaon decay at rest. As a result, for a 20 year long run, expected sensitivity is in the range $1 - 2 \times 10^{34}$ years long proton lifetime in this decay mode.

Cosmological Supernovae Neutrino remnants from all previous supernovae stream throughout the universe and can be detected on Earth with neutrino detectors. Learning more about this flux and spectrum of relic supernovae neutrinos will provide insight in the star formation mechanism and era when they were formed. Supernovae relic neutrinos are detected in water Cherenkov via inverse beta decay interaction on protons, resulting in positron and neutron capture on proton in the energy range between 10 MeV (above reactor neutrino energy) and below 100 MeV (atmospheric anti-neutrino background). Above 20 MeV, the dominant background is due to atmospheric neutrino interactions. Water Cherenkov detector has no means of tagging γ photons from the neutron capture due to its low energy of 2.2. MeV. However, if the the detector is loaded with Gd, then the γ's emitted in neutron capture are at the level of 8 MeVs and can be tagged successfully. A 300 kton Gd loaded water Cherenkov detector should be able to unambiguously detect relic supernovae neutrino flux.

Galactic Supernovae and Solar neutrinos Galactic supernova will produce a strong signal in the LBNE neutrino detector and provide a detailed temporal and spectral profile of detected neutrinos. This information will lead to better understanding of core collapse, accretion, neutron star cooling and possible transition to quark matter or black hole, tests of exotic physics and measurement of neutrino oscillation parameters including mass hierarchy and θ_{13}. Such a large neutrino detector will collect the largest number of solar and atmospheric neutrinos in the energy range from 5 MeV to 1 TeV.

3. Experimental design

The LBNE consists of several integral components that are required to reach the physics goals and these are: neutrino detector at DUSEL inside the Homestake mine, neutrino beam that will be built at Fermi lab and near detector that will be built adjacent to the neutrino beam to measure detailed beam profile, normalize neutrino flux and monitor neutrino beam stability. Two complementary technologies are planned for neutrino detector in South Dakota. The first one is well understood and proven to work, water Cherenkov detector technology and the second one is Liquid Argonne detector technology that is promising, but requires a lot of research and development before a large scale detector (of the order of 50 kton) can be built.

Water Cherenkov detector (Fig. 2) will consist of three 100 kton fiducial volume modules and they are envisioned as right cylinders 54 m high and 53 m in diameter. Approximately 50,000 ten inch PMTs are needed per module to get more than 20% photocoverage.

4. Status and Prospects

Work is underway to prepare the cavern at 1479 m depth for the water Cherenkov and liquid Argonne detectors. There is a vigorous research and development effort in the USA to produce conceptual design for LBNE, followed by the full technical design. The plan is to start building the detector by 2013 fiscal year. LBNE has a strong discovery potential and will seek answers to a number of open questions in particle and astroparticle physics.

A search for sterile neutrinos at CERN-PS

E Calligarich[1], S Centro[2], D Gibin[2], A Guglielmi[2], F Pietropaolo[2], C Rubbia[3] and P R Sala[4]

[1] INFN sezione di Pavia, via Bassi 6, I-27100 Pavia, Italy
[2] Dipartimento di Fisica dell'Università e INFN Padova, via Marzolo 8, I-35131 Padova, Italy
[3] CERN, European Laboratory for Particle Physics, CH-1211 Geneve 23, Switzerland
[4] INFN sezione di Milano, via Celoria 2, I-20123 Milano, Italy

E-mail: paola.sala@mi.infn.it

Abstract.
The LSND experiment has observed a 3.8 sigma excess of $\bar{\nu}_e$ events from an $\bar{\nu}_\mu$ beam coming from pions at rest. If confirmed, the LSND anomaly would imply new physics beyond the standard model. The MiniBooNE experiment shows a 1.6 σ disagreement with LSND results, but it introduces an unexplained, new 3.0 sigma anomaly at lower energies, down to 200 MeV.

The proposal for a new eperiment based on two liquidArgon TPCs at CERN-PS is presented in this paper. The superior quality of the Liquid Argon imaging TPC and its unique electron - pi-zero discrimination allow full rejection of the NC background, without efficiency loss for electron neutrino detection. The use of two detectors at a near and a far location enables to minimize the effects of systematic uncertainties related to the neutrino beam and to the neutrino cross-sections. This experiment could collect 10^6 charged current events in two years data taking, thus allowing for a high priecision and high statistics study of the LSND and MiniBoone anomalies

1. Introduction

Positive evidence for positron excess of in ν_μ interactions coming from muon decays at rest has been reported by the LSND experiment at LANSCE in Los Alamos [1]. The anomalous $\bar{\nu}_e$ signal (87.9 ± 22.4 ± 6.0 events) represents a 3.8σ effect and it occurs for L/E distances of about 0.5-1.0 m/MeV. To explain this signal with neutrino oscillations requires a mass-squared difference $\Delta m^2 \approx 1eV^2$, much larger and inconsistent with the ones required by the standard atmospheric and long-baseline neutrino experiments. Additional Physics beyond the Standard Model in the neutrino sector will be necessary if more than the two oscillation signals due to Atmospheric and Solar transitions were to be eventually observed. Many theoretical hypotheses (see for instance [2]) have been put forward to explain the signal excess observed by LSND and to accommodate it into a coherent and more general neutrino physics framework. All of them imply the presence of new physics beyond the Standard Model, in most instances with additional sterile neutrino families, whose existence is probed through their coupling with ordinary matter via neutrino oscillation, decay or CPT violation. Other more exotic scenarios have also been suggested. More recently the MiniBooNE experiment[3] at FNAL-Booster has further searched for the LSND anomaly with a horn focused neutrino beam from protons of the FNAL Booster. The LSND expectation in MiniBoone is an additional signal over a background of 375 CC-QE counts, due to a variety of backgrounds mostly calculated with Monte Carlo predictions. For

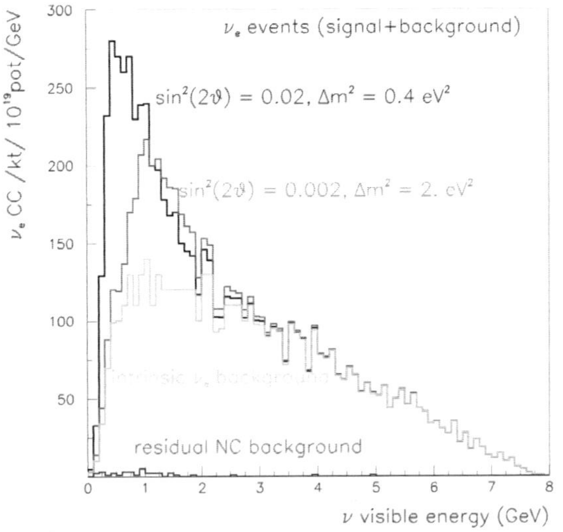

Figure 1. Simulated spectra of ν_e CC-like events in the far detector

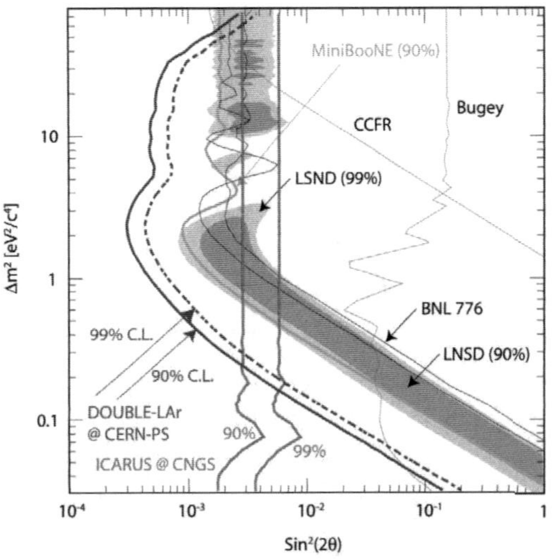

Figure 2. Expected sensitivity for the proposed experiment for 2.5 10^{20} pot.

contamination, which, however, is fully determined by the measure at the near detector. Thanks to the excellent imaging capability of LAr-TPC, and to dE/dx discrimination, the background coming from π^0 misidentification in NC events can be reduced of a factor 10^{-3} maintaining an electron identification efficiency of 90%. The expected sensitivity in a two-flavour oscillaton scheme, shown in fig. 2, extends well over the region explored by LSND and MiniBoone. Even in anti-neutrino mode, 2 years of data taking would be sufficient to cover the whole LNSD region.

4. References

[1] Aguilar A *et al.* (LSND) 2001 *Phys. Rev.* **D 64** 112007
[2] Schwetz T 2009 *Pramana* **72** 119–129
[3] Aguilar-Arevalo A A *et al.* (The MiniBooNE) 2007 *Phys. Rev. Lett.* **98** 231801 (*Preprint* 0704.1500)
[4] Aguilar-Arevalo A A *et al.* (MiniBooNE) 2009 *Phys. Rev. Lett.* **102** 101802 (*Preprint* 0812.2243)
[5] Aguilar-Arevalo A A *et al.* (MiniBooNE) 2009 *Phys. Rev. Lett.* **103** 111801 (*Preprint* 0904.1958)
[6] Rubbia C 2006 CNGS2 (ICARUS) presented at the 78th meeting of the SPSC , CERN, 3 october 2006
[7] Chen H *et al.* (MicroBooNE) Proposal for a New Experiment Using the Booster and NuMI Neutrino Beamlines: MicroBooNE Tech. Rep. PROPOSAL-0974 FERMILAB
[8] Stancu I *et al.* (OscSNS) 2008 The oscsns white paper `http://physics.calumet.purdue.edu/ oscsns`
[9] Baibussinov B *et al.* 2009 (*Preprint* 0909.0355)
[10] Rubbia C 1977 The liquid-argon time projection chamber: A new concept for neutrino detector Tech. Rep. CERN-EP/77-08 CERN
[11] 2001 Icarus initial physics program Tech. Rep. LNGS P28/01 LNGS-EXP 13/89 add.1/01 INFN
[12] 2001 Cloning of T600 modules to reach the design sensitive mass Tech. Rep. LNGS-EXP 13/89 add.2/01. INFN
[13] Guler M *et al.* (I216/P311) 1999 Search for ν_e oscillation at the cern ps (proposal) Tech. Rep. CERN-SPSC/99-26, SPSC/P311 CERN
[14] Amerio S *et al.* (ICARUS) 2004 *Nucl. Instrum. Meth.* **A527** 329–410
[15] Ferrari A, Sala P, Fassò A and Ranft J 2005 Fluka: a multi-particle transport code Tech. Rep. CERN-2005-10, INFN/TC_05/11, SLAC-R-773 CERN
[16] Battistoni G, Muraro S, Sala P, Cerutti F, Ferrari A, Roesler S, Fassò A and Ranft J 2007 *AIP Conference Proceedings* vol 896 ed Albrow M and Raja R pp 31–49
[17] Battistoni G, Ferrari A, Lantz M, Sala P and Smirnov G 2009 Prepared for the 12th International Conference on Nuclear Reaction Mechanisms, june 15-19 2009 (Varenna, Italy)

Journal of Physics: Conference Series **203** (2010) 012111 doi:10.1088/1742-6596/203/1/012111

Non-standard neutrino interactions in the type-II seesaw model

He Zhang[1]

Department of Theoretical Physics, School of Engineering Sciences,
Royal Institute of Technology (KTH) – AlbaNova University Center,
Roslagstullsbacken 21, 106 91 Stockholm, Sweden

E-mail: zhanghe@kth.se

Abstract. Within the framework of the type-II seesaw model, we investigate in detail the non-standard neutrino interactions (NSIs). Non-trivial correlations between NSI parameters and neutrino masses and mixing parameters are established. We show that sizable NSIs can be generated as a consequence of a nearly degenerate neutrino mass spectrum. Significant zero distance effects in the near detector of a future neutrino factory, as well as characteristic decays of the doubly charged Higgs at the Large Hadron Collider are discussed.

1. Introduction

Non-standard neutrino interactions (NSIs) emerge from the effective Lagrangian of the form

$$\mathcal{L}_{\rm NSI} = -2\sqrt{2}G_F \varepsilon^{ff'C}_{\alpha\beta} \left(\overline{\nu_\alpha}\gamma^\mu P_L \nu_\beta\right) \left(\overline{f}\gamma_\mu P_C f'\right) , \tag{1}$$

where f and f' denote charged lepton or quark fields, G_F is the Fermi coupling constant, and P_L and P_C (with $C = L, R$) stand for different chiral projectors. It is generally expected that the effective operator comes out of an underlying theory respecting the Standard Model (SM) gauge symmetry upon integrating out a certain set of "heavy" degrees of freedom. For purely leptonic NSIs, i.e., the two charged-fermions in Eq. (1) are leptons, the heavy mediators can be either charged scalars or vector bosons, which may usually induce interactions involving four charged leptons. For a fermion mediated NSI operator, non-unitarity effects usually arise due to the mixing between the neutral components of heavy fermions and light neutrinos [1, 2]. Generally, the mass scale of seesaw particles is chosen to be much higher than the electroweak scale and close to the Grand Unified scale in order to sufficiently suppress the light neutrino masses. Therefore, no significant NSI effects could be expected. Here, instead, we focus on a particular extension of the SM featuring an extra $SU(2)_L$-triplet Higgs not far away from the electroweak scale, which provides a very popular and simple scheme for accommodating Majorana masses of neutrinos within a renormalizable framework, and indeed, can induce significant NSI effects [3].

[1] In collaboration with Michal Malinský and Tommy Ohlsson.

2. NSIs from the type-II seesaw model

In the simple extension of the SM with an extra Higgs triplet Δ, gauge invariance allows one to add the couplings between the Higgs triplet and two lepton doublets [4, 5, 6]

$$\mathcal{L}_Y = Y_{\alpha\beta}\left[\Delta^0 \nu_\alpha P_L \nu_\beta - \frac{1}{\sqrt{2}}\Delta^+ \left(\overline{\ell^c_\alpha} P_L \nu_\beta + \overline{\nu^c_\alpha} P_L \ell_\beta\right) - \Delta^{++}\overline{\ell^c_\alpha} P_L \ell_\beta\right] + \text{H.c.}\,, \qquad (2)$$

where Y is a 3×3 symmetric matrix in flavor space. Integrating out the heavy triplet field at tree level, one can obtain effective dimension five and six operators, which are responsible for light neutrino masses, NSIs and non-standard four charged-lepton interactions, respectively, [7]

$$\mathcal{L}^m_\nu = \frac{Y_{\alpha\beta}\lambda_\phi v^2}{m^2_\Delta}\left(\nu^c_{L\alpha}\nu_{L\beta}\right) = -\frac{1}{2}(m_\nu)_{\alpha\beta}\nu^c_{L\alpha}\nu_{L\beta}\,, \qquad (3)$$

$$\mathcal{L}_{\rm NSI} = \frac{Y_{\sigma\beta}Y^\dagger_{\alpha\rho}}{m^2_\Delta}\left(\overline{\nu_\alpha}\gamma_\mu P_L \nu_\beta\right)\left(\overline{\ell_\rho}\gamma^\mu P_L \ell_\sigma\right)\,, \qquad (4)$$

$$\mathcal{L}_{4\ell} = \frac{Y_{\sigma\beta}Y^\dagger_{\alpha\rho}}{m^2_\Delta}\left(\overline{\ell_\alpha}\gamma_\mu P_L \ell_\beta\right)\left(\overline{\ell_\rho}\gamma^\mu P_L \ell_\sigma\right)\,, \qquad (5)$$

where $v \simeq 174$ GeV is the vacuum expectation value of the SM Higgs field and λ_ϕ denotes the coupling of the scalar triplet to the SM Higgs doublet. Comparing Eqs. (3) and (4) with Eq. (1), we can establish relations between the light neutrino mass matrix and NSI parameters as

$$\varepsilon^{\rho\sigma}_{\alpha\beta} = -\frac{m^2_\Delta}{8\sqrt{2}G_F v^4 \lambda^2_\phi}(m_\nu)_{\sigma\beta}\left(m^\dagger_\nu\right)_{\alpha\rho}\,. \qquad (6)$$

In order to accommodate sizable NSIs, one naturally expects the triplet Higgs to be rather light, typically at the TeV scale. Notice that, in such a case, the dimensionful parameter λ_ϕ associated with the trilinear Higgs coupling must be small enough to keep the absolute neutrino mass scale proportional to $\lambda_\phi v^2/m^2_\Delta$ in the sub-eV range. However, this could be natural, since the symmetry is enhanced in the zero λ_ϕ limit. It is also easy to see that the NSI parameters are not independent, and in fact, they are strongly tied to the structure of the light neutrino mass matrix m_ν. However, not all the ε's are physically interesting parameters. For the propagation process in long baseline experiments, neutrinos encounter Earth matter effects and only electron type of NSIs $\varepsilon^m_{\alpha\beta} \equiv \varepsilon^{ee}_{\alpha\beta}$ contributes to the matter potential. In addition, Eq. (6) affects neutrino production at neutrino sources, especially for a neutrino factory. More generally, both the processes $\mu^- \to e^-\nu_\mu\overline{\nu_\beta}$ (corresponding to the NSI parameter $\varepsilon^{e\mu}_{\mu\beta}$) and $\mu^- \to e^-\nu_\alpha\overline{\nu_\beta}$ (corresponding to $\varepsilon^{e\mu}_{\alpha\beta}$ with $\alpha \neq \mu$) may occur, and their contributions have to be added to the SM transition rate either coherently or incoherently, depending on the specific situation.

The most stringent experimental bounds on these NSI parameters come from the lepton flavor violating (LFV) decays $\mu \to 3e$ and $\tau \to 3\ell$, the rare radiative lepton decays $\ell_\alpha \to \ell_\beta\gamma$, and the Bhabha scattering and muonium to antimuonium conversion. In order to receive sizable NSIs effects in neutrino experiments and avoid large LFV processes at the same time, one expects the flavor non-diagonal parts of m_ν to be relatively small. Hence, m_ν should take an approximately diagonal form, which is quite favorable in the case of a nearly degenerate (ND) neutrino mass spectrum $m_1 \simeq m_2 \simeq m_3$. Then, m_ν approximates to a unit matrix, and the only relevant NSI parameters are ε^m_{ee} and $\varepsilon^{e\mu}_{e\mu} = (\varepsilon^{\mu e}_{\mu e})^*$. Note that another possible way to evade the LFV constraints is discussed in Ref. [8]. We illustrate in Fig. 1 the generic upper bounds on the NSI parameters with respect to the lightest neutrino mass m_1. One can find that only ε^m_{ee} and $\varepsilon^{e\mu}_{e\mu}$ are significant at large m_1 regions, and they are similar in size. The upper bound $|\varepsilon^{e\mu}_{e\mu}| \sim |\varepsilon^m_{ee}| < 3 \times 10^{-3}$ can be obtained according to Fig. 1. As for the other NSI parameters, the upper bounds are rather strong, which means that there is no hope for them to be discovered in the near-future experiments.

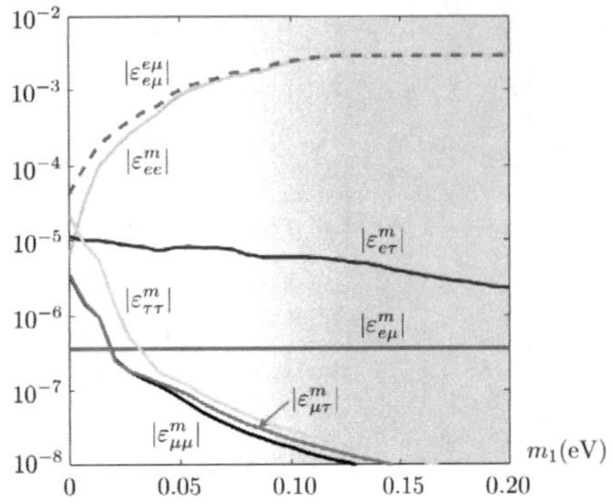

Figure 1. The upper bounds (at 90 % C.L.) on the NSI parameters governed by Eq. (6) with respect to the lightest neutrino mass m_1. The solid curves correspond to constraints on $|\varepsilon^m_{\alpha\beta}|$ parameters, while the dashed curve shows the bound on the $|\varepsilon^{e\mu}_{e\mu}|$ (or $|\varepsilon^{\mu e}_{\mu e}|$) parameter.

3. Signatures at a neutrino factory and the LHC

Wrong sign muons at a near detector: The most striking signal of the type-II seesaw model at a neutrino factory corresponds to the processes $\mu^- \to e^-\nu_e\overline{\nu_\mu}$ and $\mu^+ \to e^+\overline{\nu_e}\nu_\mu$ leading to would-be observable rates of the "wrong sign" muon tracks in a near detector, the so-called zero-distance effect. We find in our numerical analysis that a neutrino factory provides an excellent sensitivity to probe this type of NSI effects. The precision of ε is limited by the baseline length L, especially for a large θ_{13}. Thus, a distance $L < 10$ km would be favorable for the detector [9].

Earth matter density profile uncertainties: Precision measurements of ε^m_{ee} are related to how well the matter density uncertainty can be constrained. Once an accurate geophysical estimate of the Earth matter density uncertainties becomes available, one then may hope to perform a high sensitivity test on ε^m_{ee} in practice [10].

Like-sign di-lepton production at the LHC: According to the above analysis, in the case of sizable NSIs, $\Delta^{\pm\pm}$ should predominantly decay into a pair of identical leptons [11]. Therefore, one can expect significant NSI effects, if the branching ratios of decays $\Delta^{\pm\pm} \to \ell^\pm_\alpha \ell^\pm_\alpha$ are dramatically larger than the other channels at the LHC.

4. conclusions

The NSI effects in the type-II seesaw were discussed in detail. We found upper bounds for the effective couplings like $|\varepsilon| < 3 \times 10^{-3}$. We argued that a combined analysis of the neutrino oscillation experiments and the future LHC results would be very helpful to figure out the underlying physics behind the neutrino masses and mixing as well as non-standard interactions.

Acknowledgments

This work was supported by the Göran Gustafsson Foundation and the Royal Institute of Technology (KTH), project no. SII-56510.

References

[1] Malinský M, Ohlsson T and Zhang H 2009 *Phys. Rev.* **D79** 073009 (*Preprint* arXiv:0903.1961)
[2] Malinský M, Ohlsson T, Xing Z z and Zhang H 2009 *Phys. Lett.* **B679** 242–248 (*Preprint* arXiv:0905.2889)
[3] Malinský M, Ohlsson T and Zhang H 2009 *Phys. Rev.* **D79** 011301 (*Preprint* arXiv:0811.3346)
[4] Schechter J and Valle J W F 1980 *Phys. Rev.* **D22** 2227
[5] Lazarides G, Shafi Q and Wetterich C 1981 *Nucl. Phys.* **B181** 287
[6] Mohapatra R N and Senjanović G 1981 *Phys. Rev.* **D23** 165
[7] Chao W and Zhang H 2007 *Phys. Rev.* **D75** 033003 (*Preprint* hep-ph/0611323)
[8] Ohlsson T, Schwetz T and Zhang H 2009 (*Preprint* arXiv:0909.0455)
[9] Tang J and Winter W 2009 *Phys. Rev.* **D80** 053001 (*Preprint* arXiv:0903.3039)
[10] Kopp J, Ota T and Winter W 2008 *Phys. Rev.* **D78** 053007 (*Preprint* arXiv:0804.2261)
[11] Fileviez Pérez P, Han T, Huang G Y, Li T and Wang K 2008 *Phys. Rev.* **D78** 015018 (*Preprint* arXiv:0805.3536)

Journal of Physics: Conference Series **203** (2010) 012112 doi:10.1088/1742-6596/203/1/012112

The very low energy range of atmospheric neutrinos

Orlando L. G. Peres[1,2,]

[1]The Abdus Salam International Centre for Theoretical Physics, I-34100 Trieste, Italy
[2]Instituto de Física Gleb Wataghin - UNICAMP, 13083-970 Campinas SP, Brazil

E-mail: orlando@ifi.unicamp.br

Abstract. We discuss the oscillation effects of sub-sub GeV atmospheric neutrinos, the sample with energies $E \lesssim 100$ MeV. The energy spectra of the $e-$like events in water Cerenkov detectors are computed and dependence of the spectra on the 2-3 mixing angle, θ_{23}, the 1-3 mixing and CP-violation phase are studied.

1. Introduction

Since 1996, Super-Kamiokande experiment have shown that atmospheric muon neutrinos disappeared [1]. From that we understand the we learn that $\nu_\mu \to \nu_\mu$ and $\nu_\mu \to \nu_\tau$ oscillations, with a difference mass scale of $\Delta m^2_{31} \sim 2.8 \times 10^{-3}\mathrm{eV}^2$ [2], are the leading mechanism to explain the deficit of $E \gtrsim 100$ MeV atmospheric neutrino data.

Later on, it was realized that the scale $\Delta m^2_{21} \sim 7.5 \times 10^{-5}\mathrm{eV}^2$ can have a effect for sub-GeV events of energy $100 \lesssim E \lesssim 1200$ MeV [3]. As realized in this paper, although very stronger conversion probabilities for $\nu_\mu \to \nu_e$ are possible with present parameters, there is a stronger cancellation for maximal mixing, θ_{23}, and the effect is small.

In this talk I will describe the effect of oscillation for the very lowest energy range of atmospheric neutrinos, $E \lesssim 100$ MeV.

2. Oscillation in atmospheric neutrinos

Since the first computation of atmospheric neutrino fluxes [4], we know that atmospheric neutrinos are produced mainly in electron and muon flavor, where we include neutrinos and anti-neutrinos of these flavor. In this article we will use the Bartol fluxes from Ref. [5], where you can find a more detailed discussion of the characteristics of the neutrino fluxes.

Assuming the the PMNS mixing matrix defined through $\nu_f = U\nu_{mass}$ can be parametrized as $U = U_{23}\Gamma_\delta U_{13} U_{12}$. Here $\Gamma_\delta \equiv \mathrm{diag}(1, 1, e^{i\delta})$, $\hat{V} \equiv diag(V, 0, 0)$, and $V = \sqrt{2} G_F n_e$, and U_{ij} is the rotation in the $ij-$plane onto the angle θ_{ij}. The ν_e-flux at the detector with (without) oscillations taken into account, F_e (F_e^0),

$$\frac{F_e}{F_e^0} = 1 + (rc^2_{23} - 1)\tilde{P} - rs_{13}c^2_{13} \sin 2\theta_{23}(\cos\delta\ R_{e\mu} + \sin\delta\ I_{e\mu}) + s^4_{13}(1 - rs^2_{23})(2 - \tilde{P})$$
$$-2s^2_{13}\left[(1 - rs^2_{23}) + \tilde{P}(r - 2)\right] \quad (1)$$

where $r(E, \Theta_\nu) \equiv \frac{F^0_\mu(E,\Theta_\nu)}{F^0_e(E,\Theta_\nu)}$, $\tilde{P} \equiv |\tilde{A}_{\mu e}|^2 = |\tilde{A}_{e\mu}|^2$,, $R_{e\mu} \equiv \mathrm{Re}(\tilde{A}^*_{e\mu}\tilde{A}_{ee})$, $I_{e\mu} \equiv \mathrm{Im}(\tilde{A}^*_{e\mu}\tilde{A}_{ee})$, $R_{\mu e} \equiv \mathrm{Re}(\tilde{A}^*_{\mu e}\tilde{A}_{ee})$, and $I_{\mu e} \equiv \mathrm{Im}(\tilde{A}^*_{\mu e}\tilde{A}_{ee})$. The terms $\tilde{A}_{ee}$, $\tilde{A}_{\mu\mu}$ and $\tilde{A}_{\mu e}$ are the solution for the sub-matrix that have a effective Hamiltonian given by $H_2 \equiv H_2(\Delta m^2_{21}, \theta_{12}, V c^2_{13})$. We use notations $c_{13} \equiv \cos\theta_{13}$, $c_{12} \equiv \cos\theta_{12}$, $s_{12} \equiv \sin\theta_{12}$, etc. For details look the Ref. [6]. Following the same procedure for muons, then the ratio of fluxes with and without oscillations equals

$$\begin{aligned}\frac{F_\mu}{F^0_\mu} &= 1 - \frac{1}{2}\sin^2 2\theta_{23} - c^2_{23}\tilde{P}\left(c^2_{23} - \frac{c^2_{13}}{r}\right) - s_{13}\sin 2\theta_{23}\left\{\cos\delta\left[\frac{R_{\mu e}}{r} - c^2_{23}(R_{e\mu} + R_{\mu e})\right]\right. \\ &- \left.\sin\delta\left[\frac{I_{\mu e}}{r} + c^2_{23}(I_{e\mu} + I_{\mu e})\right]\right\} + O(s^2_{13}). \qquad (2)\end{aligned}$$

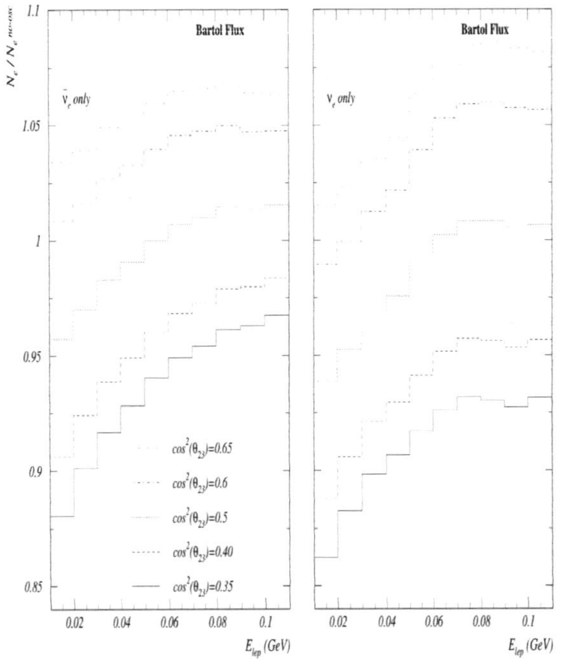

Figure 1. Effect of the 2-3 mixing on the spectra of events induced by the electron neutrinos and anti-neutrinos. We assume $s_{13} = 0$ and the curves are for different values of c^2_{23}, from upper to lower curve, $c^2_{23} = 0.65, 0.6, 0.50, 0.40, 0.35$. *Left panel:* $\bar{\nu}_e$, *right panel:* ν_e.

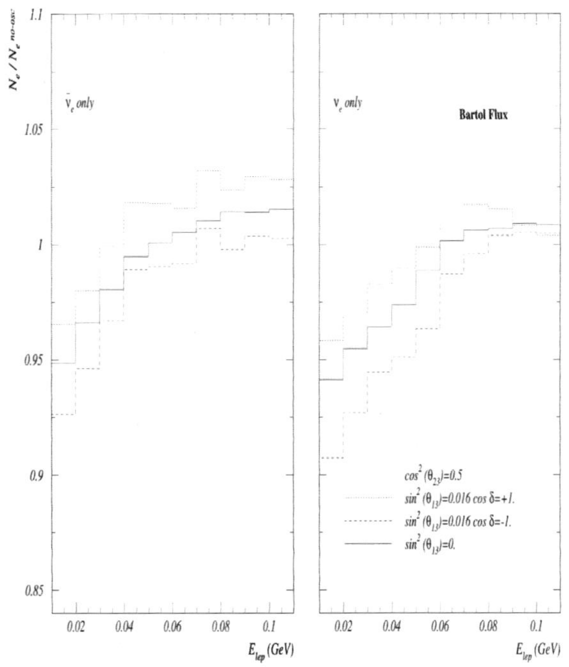

Figure 2. Effect of the 1-3 mixing on the spectra of events induced by the electron neutrinos and anti-neutrinos. We take $c^2_{23} = 0.5$. *Left panel:* $\bar{\nu}_e$, *right panel:* ν_e.

3. Number of events

We will consider here only events that have lepton energies below 120 MeV, because these upper limit is below the threshold for muon production we will only consider events with electrons (or positrons) in the final state.

We can classify the events in two categories:

(i) electron events, are detected via the quasi-elastic scattering $\bar{\nu}_e + p \to e^+ + n$, $\bar{\nu}_e +^{16} O \to e^+ + N$, $\nu_e +^{16} O \to e^- + F$. As described in the previous section, these electron neutrino events, Eq. 1 depend on the mixing parameters.

(ii) invisible muons events, are originated from muon neutrinos , $\nu_\mu + N \to N' + \mu$, that produce muons with kinetic energies below 160 MeV (the threshold of Cerenkov signal in water Cerenkov detectors). They quickly lose energy, stop and decay at rest emitting electron (positron).

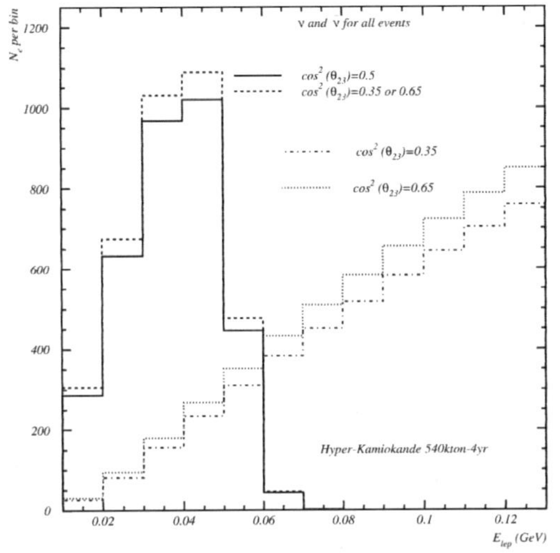

Figure 3. The spectrum of neutrino and anti-neutrino events expected for 4 years of the Hyper-Kamiokande (0.5 Mton) data-taking. The dotted and dot-dashed histograms are for the e-like events for two different values of 2-3 mixing. The solid and short-dashed lines correspond to the $e-$like events from the invisible muon decays for $\cos^2\theta_{23} = 0.35, 0.50$.

First considering electron events, we show the ratio of $N_{\bar{e}}/N_{\bar{e}}^0$, the number of oscillated events over the non-oscillated events in Fig. 1 (Fig. 2) the effects of the 2-3 mixing (1-3 mixing). For both plots we are the mixing parameters $\sin^2\ 2\theta_{12}\ =\ 0.82$, $\Delta m^2_{21} = 7.3\times 10^{-5}$ eV^2. We see the energy distortion is mains signal of these oscillations, and we can be sensitive mainly to 2-3 mixing and slightly worst to 1-3 mixing.

4. Conclusions

We have show that the lowest energy range for atmospheric neutrinos, that have sensitive to the 2-3 mixing, due distortions in the energy spectra. Future experiments like the proposed Hyper-Kamiokande can look for these distortions.

References

[1] Fukuda Y. *et al.* 1998 *Phys. Rev. Lett.* **81**, 1562
[2] T. Kajita in these proceedings.
[3] Peres O. L. G. and Smirnov A. Y. 1999 *Phys. Lett.* B **456** 204 ; Peres O. L. G. and Smirnov A. Y. 2004, *Nucl. Phys.* B **680** 479
[4] Zatsepin G. T. and Kuzmin V. A. 1962 *Sov. Phys. JETP* **14**, 1294
[5] Gaisser T. K. , Stanev T. and Barr G. 1988 *Phys. Rev.* D **38**, 85 ; Honda M. , Kajita T. , Kasahara K. and Midorikawa S. 1995 *Phys. Rev.* D **52**, 4985 ; Battistoni G. , Ferrari A. , Montaruli T. and Sala P. R. 2005 *Astropart. Phys.* **23** 526
[6] Peres O. L. G. and Smirnov A. Y. 2009 *Phys. Rev.* D **79** 113002

Acknowledgments

The author is grateful to G. Battistoni and T. Stanev for discussions of the atmospheric neutrino fluxes at low energies. O. L.G. P. thanks FAPESP, CNPQ and ICTP for support.

Topics in Astroparticle and Underground Physics (TAUP 2009) IOP Publishing
Journal of Physics: Conference Series **203** (2010) 012113 doi:10.1088/1742-6596/203/1/012113

Sensitivity on Earth Core and Mantle densities using Atmospheric Neutrinos

Celio Adrega de Moura[1,2] **and Ofelia Pisanti**[3,1]

[1] INFN - Sezione di Napoli, Complesso Universitario di Monte S.Angelo, Via Cinthia, 80126, Napoli, Italy

[2] Instituto de Física "Gleb Wataghin" - UNICAMP 13083-970 Campinas SP, Brazil

[3] Dipartimento di Scienze Fisiche - Università di Napoli "Federico II", Complesso Universitario di Monte S.Angelo, Via Cinthia, 80126, Napoli, Italy

E-mail: `pisanti@na.infn.it`

Abstract. Neutrino radiography may provide a tool, alternative to conventional seismic studies, to study the very deep structures of the Earth. The aim of this paper is to assess how well the core and mantle averaged densities can be reconstructed through atmospheric neutrino radiography. We find that about a 2% sensitivity for the mantle and 5% for the core could be achieved for a ten year data taking at an underwater km^3 Neutrino Telescope. This result does not take into account systematics related to the details of the experimental apparatus.

1. Introduction

Earth tomography with high–energy neutrinos is based on the fact that cosmic neutrinos with an energy of a few TeV have an interaction length of the order of the Earth radius and thus sample the density profile along the path [1]. Detecting at a km^3 Neutrino Telescope the flux of emerging charged leptons (mainly muons) versus the arrival direction can be therefore a promising approach, complementary to the more conventional seismic studies, for measuring at least some of the features of the Earth density radial profile as recently discussed in Ref. [2]. Until energies of the order of 100 TeV the neutrino flux crossing the Earth is essentially made of Atmospheric Neutrinos (AN) which are produced in collisions of cosmic rays with nuclei in the Earth's atmosphere. On the other hand, at higher energies a considerable prompt neutrino flux from the decay of heavy mesons is expected, as well as an extragalactic neutrino component. We are limited in our analysis by the transparency of the Earth to the neutrinos of the lowest energies, and by their prompt and yet uncertain extragalactic flux components at the highest energies. So, we use the electron and muon (anti)neutrino fluxes calculated in Ref. [3] in the energy range $(10^3 - 10^4)$ GeV and extrapole it as a power-law behavior till 10^5 GeV. We notice that neutrino flavor oscillations can be neglected in the energy range we are interested in, since they are only effective at energies lower than 1 TeV.

The detection perspectives of high energy neutrinos have received a great interest in the past few years, in view of several proposals and R&D projects for Neutrino Telescopes (NT's) in the deep water of the Mediterranean sea [4], leading to the construction of a km^3 telescope as pursued by the `KM3NeT` project [5]. Furthermore, `IceCube` experimental setup, a cubic-kilometer under-ice neutrino detector [6] is now under construction and already taking data.

In Ref. [2] the authors point out the possibility to use the arrival direction distribution of events in ten years of data taking at IceCube, to distinguish, at 3σ level, between the Earth matter density profile of the Preliminary Reference Earth Model (PREM) [7] and a homogeneous Earth toy model. However, in Ref. [2] no indication of the real sensitivity of neutrino Earth radiography to PREM parameters is reported. In this contribution we present a study of the sensitivity of an underwater NT to the Earth interior in the case of a *simplified* PREM (sPREM), where the Earth is divided in three constant density layers (crust, mantle, and core). We assume an Earth radius of 6378 km and a crust with a thickness of 37 km, while the crust density is fixed to be 2.68 g cm^{-3}.

2. Simulation and data analysis

We choose as an example of undersea NT a km^3 detector placed at NEMO site, and generate a large number of tracks crossing the NT fiducial volume (for simplicity a cube of 1 km edge placed at 100 m from seabed) by means of a detailed Digital Elevation Map of the under-water Earth surface (details can be found in [8]). We perform a Monte Carlo simulation, following tracking particles inside the rock with a maximum of 5 zones of 3 different densities, corresponding to the two regions of sPREM, as well the crust thin layer. This leads to three possible kinds of neutrino tracks inside the Earth: those going i) through the core, ii) through mantle and crust, and iii) through the crust only. We inject a number of electron and muon (anti)neutrinos for a given energy at each angular bin according to the flux of AN given in [3]. We do not consider tau neutrino contribution in this range of energy, since we are neglecting neutrino oscillations.

The detectable events can be classified in two categories: the *track* events where the charged lepton is produced outside the fiducial volume, and the *contained* ones, where neutrino converts inside the NT.

Concerning muon neutrinos, which provide the main contribution to the total amount of events, the Monte Carlo simulates their interaction in the Earth and propagates the outgoing lepton. In this respect, we take into account the phenomenon of neutrino regeneration for a neutral current neutrino interaction. In the case of a charged current neutrino interaction, we consider the muon energy loss in matter due to ionization, bremsstrahlung, e^+e^- pair production and nuclear interaction. An energy threshold of 1 TeV is considered in counting the muons detected in the fiducial volume and the condition of a minimal track length of 300 m in the apparatus defines detectable events. This energy threshold value results to be a good compromise between the need of a sufficiently large statistics, thus a not too high lower energy threshold, and the necessity to reduce the neutrino interaction length in order to increase the sensitivity to Earth density profile. Contained events, which take contributions both from electron and muon neutrinos, are treated separately. To be conservative, in analogy to track ones we assume that they are detected if charged lepton energy is larger than 1 TeV. This, of course, does not take into account the amount of energy released in the hadronic channel accompanying the charged current interaction that for contained events could be in principle detectable. Since the contribution of contained processes is in any case subdominant with respect to the track one, we assume this conservative point of view which has the advantage to make our analysis almost independent of the details of the experimental apparatus.

In order to carry out a sensitivity study, we vary the mantle and core densities in a grid of 5×4 values: $\rho_m = \{4.00, 4.25, 4.50, 4.75, 5.00\}$ g cm^{-3}, $\rho_c = \{9.0, 10.0, 11.0, 12.0\}$ g cm^{-3}, and calculate with the Monte Carlo the corresponding number of events in ten years of data taking, $N_i(\rho_m, \rho_c)$, in the five equally spaced angular bins of $\cos\vartheta$ in the interval [0,1] (upgoing events correspond to $\cos\vartheta = 1$). For each pair of chosen values of ρ_m and ρ_c, the radius of the core/mantle boundary R_c is constrained by the mass of the Earth.

We then compare the counts $N_i(\rho_m, \rho_c)$ with the expected counts N_i^0 for the benchmark case, $\rho_m = 4.48$ g cm^{-3}, $\rho_c = 11.0$ g cm^{-3}, $R_c = 3450$ km, by means of a likelihood analysis, in which

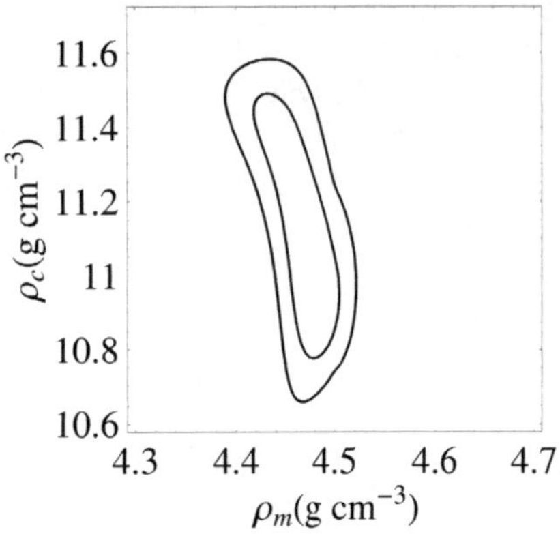

Figure 1. The 68 and 95% C.L. contours of the marginalized likelihood function $\mathcal{L}$ for the measured mantle and core Earth densities, for 10 years of data taking at a NT.

the likelihood function, $\mathcal{L}'(\rho_m, \rho_c, \xi, \eta) \propto \exp(-\chi(\rho_m, \rho_c, \xi, \eta)^2/2)$, is defined using the following expression for the χ^2:

$$\chi(\rho_m, \rho_c, \xi, \eta)^2 = \sum_{i=1}^{5} \frac{[N_i(\rho_m, \rho_c)(1+\xi)(1-\eta\,\langle\cos\vartheta\rangle_i) - N_i^0]^2}{N_i^0} + \left(\frac{\xi}{\Delta\xi}\right)^2 + \left(\frac{\eta}{\Delta\eta}\right)^2, \quad (1)$$

where ξ takes into account an overall uncertainty of the atmospheric neutrino fluxes and neutrino interaction cross-section ($\Delta\xi = 0.25$), while η encodes the uncertainty between horizontal and vertical events ($\Delta\eta = 0.05$) [2].

We show in Fig. 1 the 68 and 95% C.L. contours of the marginalized function with respect to ξ and η, $\mathcal{L}(\rho_m, \rho_c) = \int \mathcal{L}'(\rho_m, \rho_c, \xi, \eta)\, d\xi\, d\eta$. By using $\mathcal{L}(\rho_m, \rho_c)$ we can derive the "measured" values of the densities and the radius of core/mantle boundary at $1\sigma(2\sigma)$: $\rho_m = 4.47\,^{+0.02}_{-0.03}\left(^{+0.04}_{-0.06}\right)$ g cm^{-3}, $\rho_c = 11.0\,^{+0.3}_{-0.1}\left(^{+0.5}_{-0.2}\right)$ g cm^{-3}, $R_c = 3440 \pm 30\left(^{+70}_{-50}\right)$ km.

Our analysis hence suggests that a 2% and 5% uncertainties (at 2 σ level) on the averaged mantle and core densities respectively, can be reached in a neutrino radiography campaign with a ten years of data taking at a typical km^3 NT, placed in the `NEMO` site. Note that these results are obtained in a very simplified PREM model, and this justifies the good level of sensitivity reachable on ρ_m and ρ_c determinations. Notice also that we do not take into account systematics related to the details of the experimental apparatus.

References

[1] De Rujula A, Glashow S L, Wilson R R, Charpak G 1983 *Phys. Rept.* **99** 341
[2] González-García M C, Halzen F, Maltoni M and Tanaka H K M 2008 *Phys. Rev. Lett.* **100** 061802
[3] Honda M, Kajita T, Kasahara K, Midorikawa S and Sanuki T 2007 *Phys. Rev.* D **75** 043006
[4] Spurio M [ANTARES Collaboration] 2006 *Preprint* hep-ph/0611032
Migneco E *et al.* [NEMO Collaboration] 2008 *Nucl. Instrum. Meth.* A **588** 111
Aggouras G *et al.* [NESTOR Collaboration] 2006 *Nucl. Instrum. Meth.* A **567** 452
[5] Katz U F 2006 *Nucl. Instrum. Meth.* A **567** 457
[6] Krasberg M [IceCube Collaboration] 2006 AIP Conf. Proc. **867** 209
[7] Dziewonski A M and Anderson D L 1981 *Phys. Earth Planet. Interiors* **25** 297
[8] Cuoco A *et al.* 2007 *JCAP* **0702** 007
Borriello E *et al.* 2008 *Phys. Rev.* D **77** 045019
Borriello E, Mangano G, Marotta A, Miele G, Migliozzi P, Moura C A, Pastor S, Pisanti O, and Strolin P E 2009 *JCAP* **0906** 030

Journal of Physics: Conference Series **203** (2010) 012114 doi:10.1088/1742-6596/203/1/012114

Cosmic neutrinos at IceCube: θ_{13}, δ and initial flavor composition

Arman Esmaili

Department of Physics, Sharif University of Technology, P.O.Box 11365-8639, Tehran, IRAN
School of Physics, Institute for Research in Fundamental Sciences (IPM), P.O.Box 19395-5531, Tehran, IRAN

E-mail: arman@mail.ipm.ir

Abstract. We discuss the prospect of extracting the values of the mixing parameters δ and θ_{13} through the detection of cosmic neutrinos in the planned and forthcoming neutrino telescopes. We take the ratio of the μ-track to shower-like events, R, as the realistic quantity that can be measured in the neutrino telescopes. We take into account several sources of uncertainties that enter the analysis. We then examine to what extent the deviation of the initial flavor composition from $w_e : w_\mu : w_\tau = 1 : 2 : 0$ can be tested.

The neutrino mixing parameters can be extracted from cosmic neutrinos data according to the following argument [1]: assume that the flavor ratio of neutrinos at the source is $w_e : w_\mu : w_\tau$; after that neutrinos travel the large distances between the astrophysical sources and the Earth the oscillatory terms in the flavor transition probabilities average out such that the flavor ratio at the detector will become

$$\sum_\alpha w_\alpha P_{\alpha e} : \sum_\alpha w_\alpha P_{\alpha\mu} : \sum_\alpha w_\alpha P_{\alpha\tau}, \quad (1)$$

where

$$P_{\alpha\beta} \equiv P(\nu_\alpha \to \nu_\beta) = P(\bar{\nu}_\alpha \to \bar{\nu}_\beta) = \sum_i |U_{\alpha i}|^2 |U_{\beta i}|^2, \quad (2)$$

and $U_{\alpha i}$ are the elements of the neutrino mixing matrix. In a wide range of models the flavor ratios at the source are predicted to be $w_e : w_\mu : w_\tau = 1 : 2 : 0$. Thus, by measuring the flavor ratio at Earth, one can derive the absolute values of the mixing matrix elements which in principle yield information on the yet-unknown neutrino parameters θ_{13} and δ.

IceCube [2] and its Mediterranean counterparts (such as KM3NET [3]) can basically distinguish only two types of events: 1) shower-like events; 2) μ-track events. It is possible to derive information on the flavor composition of neutrinos by studying the ratio R =number of μ-track events/number of shower-like events [4]. In the analysis in this paper we consider neutrinos with energies 100 GeV $< E_\nu <$ 100 TeV, where the upper and lower limits come from the absorption of neutrinos in Earth and the energy threshold of detection in neutrino telescopes, respectively.

Two sources contribute to the μ-track events: (i) Charged Current (CC) interaction of ν_μ or $\bar{\nu}_\mu$ producing μ or $\bar{\mu}$; (ii) CC interaction of ν_τ and $\bar{\nu}_\tau$ producing τ or $\bar{\tau}$ and the subsequent decay of τ and $\bar{\tau}$ into μ and $\bar{\mu}$. In the literature, the contribution of ν_τ (via $\nu_\tau \to \tau \to \mu$) to

μ-track events has been overlooked but to study the effect of θ_{13}, one should take into account such sub-dominant effects (see the Appendix of [5] for details). Three types of events appear as shower: i) the Neutral Current (NC) interactions of all kinds of neutrinos; ii) the CC interactions of ν_e and $\bar{\nu}_e$; iii) the CC interactions of ν_τ ($\bar{\nu_\tau}$) and the subsequent hadronic decay of τ ($\bar{\tau}$). Details of the event rate calculation for each case in both μ-track and shower-like events can be found in [5].

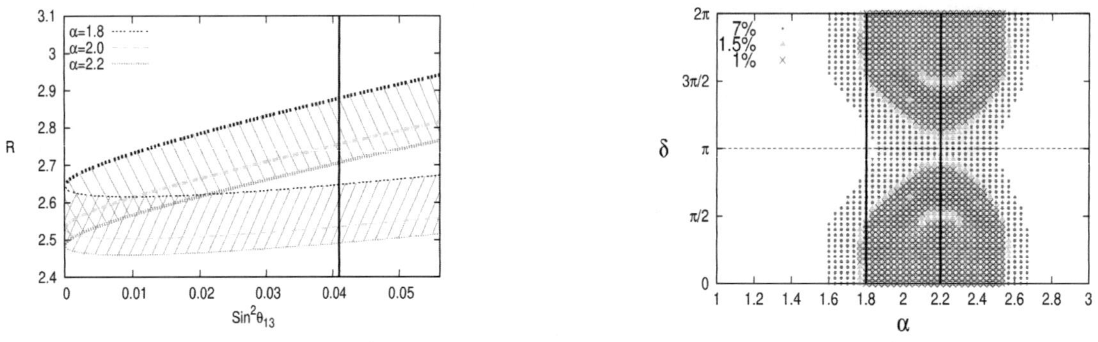

(a) (b)

Figure 1. (a) The dependence of R on $\sin^2\theta_{13}$ for different values of the spectral index, α. The thicker lines correspond to $\delta = \pi$ and the thinner ones correspond to $\delta = 0$. In drawing this figure we have set $\mathcal{N}_{\bar{\nu}_e}/\mathcal{N}_{\nu_e} = 0.5$ and $(\mathcal{N}_{\bar{\nu}_\mu} + \mathcal{N}_{\nu_\mu})/(\mathcal{N}_{\bar{\nu}_e} + \mathcal{N}_{\nu_e}) = 2$. The input for θ_{12} and θ_{23} are set equal to the best fit in [6]. The vertical line at 0.041 shows the present bound at 3σ [6] (b) Points in the (α, δ) space consistent with $R = 2.53 \pm \Delta R$. True values of the (α, δ) pair are $(2, \pi/2)$. Points displayed by dots, triangles and crosses respectively correspond to $\Delta R/\bar{R} = 7\%$, $\Delta R/\bar{R} = 1.5\%$ and $\Delta R/\bar{R} = 1\%$. To draw this figure we have varied $\sin^2\theta_{13} \in (0.028, 0.032)$, $\sin^2\theta_{12} \in (0.30, 0.34)$, $\sin^2\theta_{23} \in (0.47, 0.53)$ and $\mathcal{N}_{\bar{\nu}_e}/\mathcal{N}_{\nu_e} \in (0, 1)$.

Several input parameters enter the calculation of R and their uncertainties induce imprecision in the calculated value of R. We consider the following uncertainties in the calculation of R: i) for the energy spectrum of neutrinos we assume a power-law spectrum $dF_{\nu_\beta}/dE_{\nu_\beta} = \mathcal{N}_{\nu_\beta} E_{\nu_\beta}^{-\alpha}$. Neutrino production through the Fermi acceleration mechanism of particles in the source predict the value of the spectral index equal to $\alpha = 2$, but taking into account non-linear effects in the acceleration mechanism results in spectral index values $\alpha \in (1, 3)$. It is shown in [4] by assuming the flux $E_\nu^2 dF_\nu/dE_\nu = 0.25$ GeV cm^{-2} sr^{-1} yr^{-1}, after one year of data-taking α can be determined with 10% uncertainty. For the normalization factor $\mathcal{N}_{\nu_\beta}$ it can be shown that: $\mathcal{N}_{\bar{\nu}_\mu} = \mathcal{N}_{\nu_\mu}$ and $\mathcal{N}_{\bar{\nu}_e}/\mathcal{N}_{\nu_e} \in (0, 1)$.

Fig. (1-a) shows R versus $\sin^2\theta_{13}$ for $\cos\delta = \pm 1$ and various values of α. As seen from the figure when $\delta = 0$, the sensitivity of R to s_{13}^2 is very mild and less than 2 %. That is while for $\cos\delta = -1$, the sensitivity to s_{13}^2 is about 10 %. As seen from the figure, even for $\cos\delta = -1$, the sensitivity to s_{13}^2 can be obscured by the 10 % uncertainty in α. However, for $s_{13}^2 > 0.02$, the bands between $\alpha = 2.2$ and 1.8 for $\cos\delta = 1$ and $\cos\delta = -1$ have no overlap. This means that for $s_{13}^2 > 0.02$, 10 % precision in α is enough to distinguish $\cos\delta = 1$ from $\cos\delta = -1$.

Fig. (1-b) addresses the question that whether it will be possible to extract the value of δ. Drawing the plot, we have assumed that $\bar{R}$ will be found to have a typical value of 2.53 with an uncertainty of $\Delta R/\bar{R}$. This value of $\bar{R}$ can be obtained by taking maximal CP-violation ($\delta = \pi/2$), $\sin^2\theta_{13} = 0.03$, $w_e : w_\mu : w_\tau = 1 : 2 : 0$, $\alpha = 2$ and $\mathcal{N}_{\bar{\nu}_e}/\mathcal{N}_{\nu_e} = 0.5$. We have looked for solutions in the $\delta - \alpha$ plane for which $R = 2.53(1 \pm \Delta R/\bar{R})$, varying the rest of the

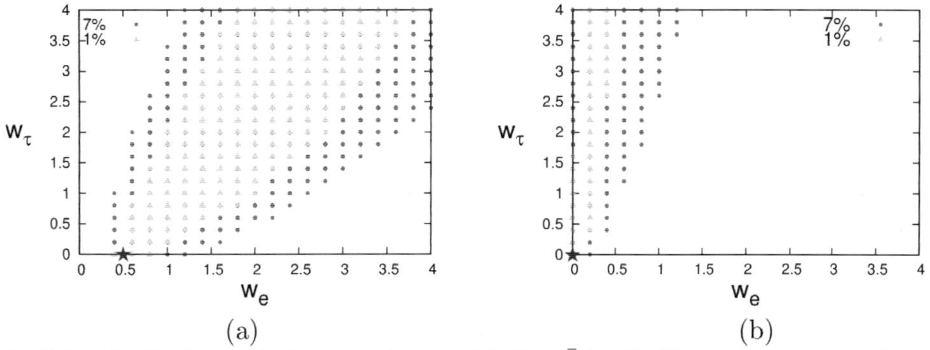

Figure 2. Points in the (w_e, w_τ) plane consistent with $R = \bar{R} \pm \Delta R$. The ratios are normalized such that $w_\mu = 1$. The true values of (w_e, w_τ) are denoted by ★. Points displayed by dots and triangles respectively correspond to $\Delta R/\bar{R} = 7\%$ and $\Delta R/\bar{R} = 1\%$. In drawing this figure we have varied $\sin^2\theta_{13} \in (0, 0.003)$, $\delta \in (0, 2\pi)$, $\alpha \in (1.8, 2.2)$, and $\mathcal{N}_{\bar{\nu}_e}/\mathcal{N}_{\nu_e} \in (0, 1)$. Drawing Fig. (a), we have taken $\bar{R} = 2.53$ which corresponds to the standard picture with $w_e = 1/2$ and $w_\tau = 0$. In case Fig. (b), we have set $\bar{R} = 3.2$ which corresponds to the stopped muon scenario with $w_e = w_\tau = 0$.

relevant parameters in the ranges indicated in the caption of Fig. (1-b). The regions covered with dots, little triangles and crosses respectively correspond to 7%, 1.5% and 1% precision in the measurement of R. As seen from the figure, with $\Delta R/\bar{R} = 7\%$, δ cannot be constrained. In fact, any point between the vertical lines can be a solution. The figure shows that reducing $\Delta R/\bar{R}$ to 1% (but keeping the rest of the uncertainties as before), some parts of the solutions can be excluded. In particular, the region around $\delta = \pi$ will not be a solution anymore. Notice that along with $\delta = \pi/2$, $\delta = 0$ is also a solution. This means that despite maximal CP-violation, the CP-violation cannot still be established.

Taking into account the relevant uncertainties in the input parameters, we look for values of $w_e : w_\mu : w_\tau$ that are consistent with $R = \bar{R} \pm \Delta R$. To perform this analysis, we take $\theta_{13} = 0$. In Fig. (2), we consider two possibilities: (i) the standard case with $w_e : w_\mu : w_\tau = 0.5 : 1 : 0$ leading to $\bar{R} = 2.53$ (see Fig. 2-a); (ii) the case of stopped muons with $w_e : w_\mu : w_\tau = 0 : 1 : 0$ yielding $\bar{R} = 3.20$ (see Fig. 2-b). From these figures we observe that with a precision of $\Delta R/\bar{R} = 7\%$, these two scenarios can be easily discriminated. These two can also be discriminated from the scenario in which the neutrino production mechanism is $n \to pe\bar{\nu}_e$ (*i.e.*, $w_e : w_\mu : w_\tau = 1 : 0 : 0$). When we restrict the analysis to $w_\tau = 0$ (*i.e.*, the case without exotic neutrino properties) from these figures we observe that the measurement of R stringently constrains $w_e : w_\tau$ which in turn sheds light on the production mechanism. However, once the assumption of w_τ is relaxed, a wide range of $w_e : w_\mu : w_\tau$ can be a solution. For example, the exotic case of $w_e : w_\mu : w_\tau = 0 : 0 : 1$ leads to the same value of $\bar{R}$ as the stopped muon scenario.

Acknowledgments
I am grateful Y. Farzan for useful discussions and comments. I would like to thank "Bonyad-e Melli-e Nokhbegan" for partial financial support.

References
[1] J. F. Beacom *et al.* Phys. Rev. D **69** (2004) 017303 [arXiv:hep-ph/0309267];
Y. Farzan and A. Y. Smirnov, Phys. Rev. D **65** (2002) 113001 [arXiv:hep-ph/0201105].
[2] J. Ahrens *et al.* [IceCube Collaboration], Astropart. Phys. **20** (2004) 507 [arXiv:astro-ph/0305196].
[3] U. F. Katz, Nucl. Instrum. Meth. A **567** (2006) 457 [arXiv:astro-ph/0606068].
[4] J. F. Beacom *et al.* Phys. Rev. D **68** (2003) 093005 [Erratum-ibid. D **72** (2005) 019901] [arXiv:hep-ph/0307025].
[5] A. Esmaili and Y. Farzan, Nucl. Phys. B **821** (2009) 197 [arXiv:0905.0259 [hep-ph]].
[6] T. Schwetz, Phys. Scripta **T127** (2006) 1 [arXiv:hep-ph/0606060].

Astronomy's New Messengers: A traveling exhibit on gravitational-wave physics

Marco Cavaglià[1], Martin Hendry[2], Szabolcs Márka[3], David H. Reitze[4], Keith Riles[5]

[1] The University of Mississippi, University, MS 38677, USA
[2] University of Glasgow, Glasgow, G12 8QQ, United Kingdom
[3] Columbia University, New York, NY 10027, USA
[4] University of Florida, Gainesville, FL 32611, USA
[5] University of Michigan, Ann Arbor, MI 48109, USA

E-mail: marco.cavaglia@ligo.org, szabolcs.marka@ligo.org, martin.hendry@ligo.org, david.reitze@ligo.org, keith.riles@ligo.org

Abstract. The Laser Interferometer Gravitational-wave Observatory exhibit *Astronomy's New Messengers: Listening to the Universe with Gravitational Waves* is traveling to colleges, universities, museums and other public institutions throughout the United States. *Astronomy's New Messengers* primarily communicates with an adolescent and young adult audience, potentially inspiring them into the field of science. Acknowledging that this audience is traditionally a difficult one to attract, the exhibit publicly announces itself in a charismatic fashion to reach its principal goals of broadening the community of people interested in science and encouraging interest in science among young people.

As a frontier physics effort, a core mission of the Laser Interferometer Gravitational-wave Observatory (LIGO) [1] Scientific Collaboration is to inspire interest in astronomy and fundamental science among students and to educate the broader community. In this context, *Astronomy's New Messengers: Listening to the Universe with Gravitational Waves* is an innovative outreach project whose aim is to inform young students and the general public of the physics of gravitational waves and the science of the LIGO detectors.

Funded by the National Science Foundation through grant NSF-0852870 from the Informal Science Education program, the EPSCOR program, and the Office of Multidisciplinary Activities, and managed by the LIGO Scientific Collaboration, *Astronomy's New Messengers* consists of an approximately 200 sq. ft. touring exhibit which will be displayed at colleges, universities, museums, and other public institutions through the nation. *Astronomy's New Messengers* includes possibilities for age-appropriate self exploration and a messaging that allows visitors to first discover their own excitement in the subject, and then delve deeper into its complexity. The exhibit's introductory area presents a general overview of the LIGO detectors and its science, while inviting visitors to step inside, explore and find out more. Text panels and a large LCD screen with a looping high-quality video produced by Milde Marketing [2] for the International Year of Astronomy [3] cornerstone project "100 Hours of Astronomy" deliver key informational points.

The main area of *Astronomy's New Messengers* explains how gravitational waves are generated by cataclysmic events in the distant universe, how LIGO "listens" for these events, and

Figure 1. Young students at the University Museum, Oxford MS, watching the Einstein's Messengers video and playing the black hole hunter game. The interferometer kiosk is in the background. (Photos by Elizabeth Herren.)

demonstrates how scientists decode their signals. Three hands-on components engage visitors in discovering how LIGO operates and understanding some of the foundations of gravitational wave astronomy: a working and interactive laser interferometer, a grid-patterned rubber sheet which illustrates the curvature of space-time, and the black hole hunter game [4] kiosk. A second LCD screen with looping video compiled of clips from the NSF movie *Einstein's Messengers* [5] and several panels with photographs and diagrams further encourage visitors' interest in LIGO and the deep universe.

The interferometer model, constructed by the Ann Arbor Hands-On Museum [6] in collaboration with the Michigan LIGO group, intuitively shows visitors how an interferometer operates. A fringe pattern is projected on a screen, illustrating the concept of light interference and the effect of environmental noise. By tapping the interferometer case, visitors can disturb the fringe pattern. A photodiode at the output port of the interferometer measures the fringe variation and speakers produce a sound for an ultimate multi-sensory experience. Each key element of the interferometer is called-out and explained on a LCD screen.

The concepts of space-time and of gravity as space-time curvature are illustrated with the rubber sheet interactive. Visitors can set a heavy steel ball on a grid-patterned sheet, which becomes stretched around the ball. Rolling a second, smaller ball into the depression formed around the larger one, visitors can visually understand how space-time warps around massive objects and affects the motion of bodies in space.

The purpose of the black hole hunter game, developed by Cardiff University, is to give visitors an opportunity to do (in a figurative sense) what LIGO scientists do, i.e. look for gravitational signals in noisy streams of data. Simulated gravitational waves are translated into sound clips. Through a GUI interface on a computer screen, visitors try to detect recorded sounds of black hole events buried in different static noise clips.

Crucial to the design, integration and implementation of these elements was the collaboration between the LIGO Education and Public Outreach group and professional exhibit designers. *Astronomy's New Messengers* was designed by Lee H. Skolnick Architecture + Design Partnership [7], a multi-disciplinary firm of architects, designers, and educators providing award-winning architecture and exhibit design services.

Astronomy's New Messengers premiered at the Street Fair of the World Science Festival [8] in New York City in June 2009. After a month-long stay at the Adler Planetarium in Chicago, the traveling LIGO exhibit will be on display in a network of museums, science museums, public libraries and academic institutions across a cluster of southern states from August 2009 to February 2010.

In 2010, an extended version of this exhibit will appear in a New York City venue that is accessible to a large and diverse cross section of the general public. This second phase of the project envisages the design and construction of a large-scale, interactive exhibit, which will draw upon elements of the touring exhibit (and upon lessons learned from the former's evaluation). The large-scale exhibit will blend LIGO science with high-concept artwork through incorporation of an interpretive 3-D lighting display. This large-scale exhibit is scheduled to debut at a prominent public location at the 2010 World Science Festival in New York City, alongside a signature event on the theme of gravitational-wave astronomy and featuring high-profile keynote speakers and artists. After the 2010 Festival, the exhibit will be transferred to a permanent home in a public institution or science museum.

The large-scale exhibit will incorporate elements from the touring exhibit, but on a grander scale – e.g. using larger flatscreen displays and poster boards and including a larger table-top interactive interferometer. A key additional element of the large-scale exhibit will be the blending of the table-top interferometer with a major, high-concept work of art in the form of an interactive three-dimension lighting display designed and realized by Leni Schwendinger Light Projects [9], a NYC-based art studio which specializes in creating lighting environments for architectural and public spaces all over the world. To highlight and illustrate the grand scale of gravitational-wave astrophysical sources, the output of the interferometer will be linked directly to a large-scale three-dimensional artwork suspended from the ceiling, symbolizing the Universe above the Earth. Changes in the interferometer's fringes will be translated and displayed on a spatial field of strands of diodes, creating a dazzling show of light and sound in real time, with each and every visitor who disturbs the interferometer contributing to the patterns of light, color and sound of the "sky" above them, representing the everlasting link between Earth and the Universe.

Acknowledgments

The authors gratefully acknowledge the support of the United States National Science Foundation for the construction and operation of the LIGO Laboratory and the Science and Technology Facilities Council of the United Kingdom, the Max-Planck-Society, and the State of Niedersachsen/Germany for support of the construction and operation of the GEO600 detector. The authors also gratefully acknowledge the support of the research by these agencies and by the Australian Research Council, the Council of Scientific and Industrial Research of India, the Istituto Nazionale di Fisica Nucleare of Italy, the Spanish Ministerio de Educación y Ciencia, the Conselleria d'Economia, Hisenda i Innovació of the Govern de les Illes Balears, the Royal Society, the Scottish Funding Council, the Scottish Universities Physics Alliance, The National Aeronautics and Space Administration, the Carnegie Trust, the Leverhulme Trust, the David and Lucile Packard Foundation, the Research Corporation, and the Alfred P. Sloan Foundation.

References

[1] http://www.ligo.org
[2] http://www.milde-marketing.de
[3] http://www.astronomy2009.org
[4] http://www.blackholehunter.org
[5] http://einsteinsmessengers.org
[6] http://www.aahom.org/
[7] http://www.skolnick.com
[8] http://www.worldsciencefestival.com
[9] http://www.lightprojectsltd.com

Journal of Physics: Conference Series **203** (2010) 012116 doi:10.1088/1742-6596/203/1/012116

Gran Sasso National Laboratory: Outreach and Communication activities.

R Antolini, A Di Giovanni, M Galeota, S Sebastiani
INFN Laboratori Nazionali del Gran Sasso, SS 17 bis km 18+910, 67010 Assergi(AQ)

E-mail: adriano.digiovanni@lngs.infn.it

Abstract. Due to its fascinating structures, the Gran Sasso National Laboratory (LNGS) offers huge opportunities for communication and outreach activities conceived for students and general public. A great effort is devoted to the organisation of the "OPEN DAY", in which the scientific staff of Gran Sasso introduces non expert people to the main relevant research topics of the laboratory through interactive demonstrations and particle detectors. In particular, a portable cosmic rays telescope has been realized: the detector is used by LNGS team in pubblic events as well as to promote the scientific activities of the Laboratory. In order to point out the importance of the scientific culture for young people, LNGS is involved in the organisation of several training courses for students and teachers focused on the improvement of the knowledge on modern physics topics. Since May 2008 is operating in Teramo the "Galileium", an interactive museum for physics and astrophysics.

1. Introduction

The Gran Sasso National Laboratory (LNGS), located between the towns of L'Aquila and Teramo, about 120 km from Rome is one of the four laboratories of the National Institute for Nuclear Physics. It is the largest underground laboratory in the world for experiments in astroparticle physics, nuclear astrophysics and rare decays. Main research topics of the present programme are neutrino physics, with neutrinos naturally produced in the Sun and in Supernova explosions, neutrino oscillations with an artificial beam from CERN (CNGS program), search for neutrino mass in neutrinoless double beta decay, dark matter search, nuclear reactions of astrophysical interest. The Gran Sasso National Laboratory is located inside the National Park of Gran Sasso and Monti della Laga. Since its first years of operation the LNGS has been involved in the organisation of several events meant to promote the scientific culture in general.

2. Outreach actvities

The LNGS staff is involved in many activities to make the local communities aware of the reasearches and progresses in cosmology and astroparticle physics, taking advantage of the beauty of the surrounding mountains of the Abruzzo Region as well as of the fascinating majesty of the civil engineering underground structures, coupled with the high-tech service plants and the experimental apparata. Our strategy is reported in the following points.

2.1. Visits

Last year the number of people visiting the underground facilities and main experiments hosted was around 8000. It is important to remark that, due to the constraints imposed by the construction of new detectors, visits are allowed just at week ends. Visitors consist mainly of high-school students which come for scientific tours organized by their teachers, who give them introductory lectures in advance.

They do not normally come from sorrounding regions cause, by tradition, programmes for cultural trips are made on a long-distance basis and last a few days. In order to face this matter the labs has been seeking a number of new initiatives to encourage visits at LNGS among local people.
Furthermore a modern Visitor Center devoted to astroparticle Physics is in progress. It will include exhibits, interactive devices and a sample of working particle detectors as a Fog chamber, PhotoTubes, nuclear emulsions and, of course, the Compact Cosmic Ray Telescope, that will be the milestone of the outreach and educational mission of the LNGS.

Figure 1. A group of students visiting the underground laboratory.

2.2. Open Day
Aseries of activities, in collaboration with school teachers and local scientific associations as the Association for Physics Teaching, addressed to students as well as to teachers, they are normally organized.
Although visits are mainly reserved for high school students, recently we have received an increasing number of requests from younger students wishing to visit our underground laboratory. To properly answer the requests of this particular range of people, we are planning special initiatives. As from 2002 we launched an initiative called Open Day. All the editions were a real success, with almost 2000 people attending.

Figure 2. On the left children are learning about the motion laws by playing "scientific toys". On the right a view of the Fermi conference hall during the award ceremony of the schools' competition "Anch'io Scienziato...".

The Open Day is a whole day during which the laboratory is open to the public with a number of events to entertain youngsters and less young; these include a visit to the underground laboratory, the possibility to ask researchers about the experiments in progress, a sort of "Carnival of Physics" in the

external infrastructures, both open-air and in the buildings, in which it is possible to learn physics and to participate in simple experiments.
In the stand "Outdoor Physics" students, mainly from schools in the surrounding area, perform experiments to explain daily life phenomena to the public.
To involve schools directly we usually organise the competition "Anch'io scienziato..." (I'm a scientist too...) dedicated to students (from kindergarden to high school) from all over the region. Students are supposed to conceive, plan and realize a scientific project in any field of science. In the last edition, more than 1179 students from 49 classes all over the region took part in the competition. The winning projects are awarded during the Open Day.

2.3. Gran Sasso – Princeton Summer School

The Gran Sasso-Princeton summer school, supported since 2004 by the Abruzzo Region, stems from the scientific collaboration between the Gran Sasso National Laboratory and the physics department at Princeton University.

Figure 3. On the left students are arranging thermodynamics experience during the laboratory session at Princeton. On the right the Italian students of the 2009 class awarded with the Diploma at the end of school.

The school takes place at Princeton University during the summer and lasts 3 weeks. The program is open to students coming from Abruzzo region and attending the fourth and fifth year of high school. For each edition a number of 20 students are selected on a competitive basis giving preference to fourth year students. The participants in the program are offered plane ticket to and back from Princeton and full accommodation and board in the Princeton campus. The activities in Princeton include physics courses and training in the field of physics and astrophysics, special relativity, fundamental calculus and English classes. Special seminars about the Gran Sasso activities and the main topics on modern physics are generally held. Students have weekends and some afternoons free to explore one of the most beautiful campuses in the United States, several facilities, observatories and museum. Furthemore, guided tours to New York, Philadelphia and Washington are scheduled at week ends. By visioning the results of the periodical survey the service uses to make on students of the previous editions, the feedback seems to be extremely positive. As a matter of fact a good percentage of students is now studying Physics or other scientific subjects at university.

3. Special tools for outreach: the compact cosmic ray telescope

By using the technology adopted for the realization of detectors in particle physics, in collaboration with the electronics service, a portable cosmic ray telescope to introduce non expert people in astroparticle physics through the direct visualization of the particle tracks has been realized.

The telescope consists of 10 planes made of Glass Resistive Plate Chambers equipped with electronics and readout. To display the particle path, the readout has been complemented by two LEDS matrices lightning when a particle crosses the device as shown in fig. 4.

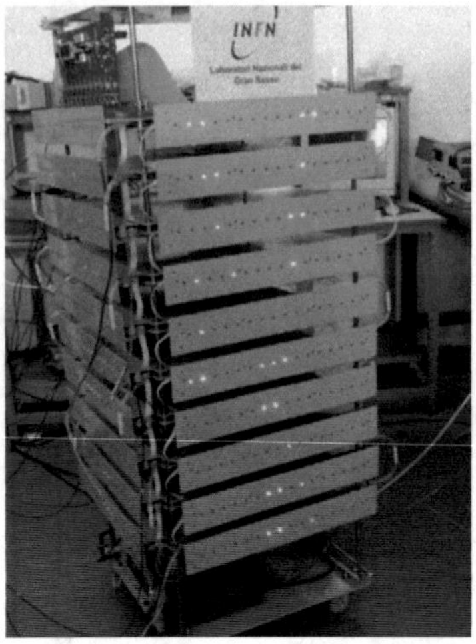

Figure 4. The compact cosmic ray telescope developed at LNGS. A multi-track is visualized by using LED matrices in double view.

The telescope has been already used in outreach and public events with great success. In particular the detector has been running during the exhibition entitled "Eccellenze d'Italia" held in July in L'Aquila during the G8 meeting, the forum for governments of the eight richest countries in the world.

4. Museum of Physics and Astrophysics: the Galileium

The Galileium, the Museum of physics and astrophysics located in Teramo (40 km far the LNGS) and operational since May 2008, is the first example of a Science Center led by a research institution. It hosts permanent and temporary exhibitions with interactive paths, workshops and lessons. The mission for a science center is to connect the territory and its scientific "excellences" knowing how to exploit the different cultural resources, paying attention to the world of school, creating spaces of approach between the citizens and the scientific progress and technological innovation. The Galileium aim at "involving" the visitor, making science approachable to everyone and allowing everyone to interact regardless their level of education. The Galileium hosted public iniziatives in his first year of operations: the performance "Guarda che Luna", the "Solar Day" and several seminars for the community.

5. Conclusion

Since the LNGS is a very attractive place, for its location in the heart of the mountain as well as for the "futuristic" experiments carried out, the main commitment of the Service in charge of Outreach and Communication is to constantly respond to the requests of visitors. The fashinating mistery of the origin and evolution of the Universe as well as the properties of neutrinos and astroparticles feed the curiosity of visitors and their thirst for knowledge making easier the outreach work at LNGS. Therefore, organizing scientific conferences, realizing prototypes, commissioning the construction of new exciting exhibits for the Museum or managing guided tours, is our way to involve people in the world of physics, making them aware of the importance of our scientific research and its applications to daily life.

Journal of Physics: Conference Series **203** (2010) 012117 doi:10.1088/1742-6596/203/1/012117

M 87 and Centaurus A: laboratories for VHE physics of jets and near supermassive black holes

M Raue[1] for the H.E.S.S. Collaboration

[1]Max-Planck-Institut für Kernphysik, Heidelberg, Germany

E-mail: martin.raue@mpi-hd.mpg.de

Abstract. In the recent years radio-galaxies have emerged as a new class of extragalactic very-high energy γ-ray emitters (VHE; $E > 100\,\mathrm{GeV}$). Their proximity and well studied multi-wavelength behavior enable unique studies of the physics of relativistic plasma jets and the immediate surroundings of super-massive black holes. Here, the two most prominent and nearby sources - Centaurus A and M 87 - are presented and their VHE properties are discussed. Special attention is given to the location of the VHE emission, where recent observational results from a joined multi-wavelength campaign on M 87 can offer new insights.

1. Introduction

Up to now, the known extra-galactic very-high energy sky (VHE; $E > 100\,\mathrm{GeV}$) is almost exclusively populated by a single source class: the VHE blazar. VHE blazars are active galactic nuclei (AGN) with their relativistic jets pointing close to the line of sight of the observer, resulting in a boosting of the flux and energy of the emitted photons. This boosting enables to detect these source up to large distances, with the furthest VHE source currently ranging at a redshift > 0.5 [1], and thereby also giving the means to study the intervening diffuse radiation fields (e.g. [2]).

Recent discoveries of VHE emission from the radio-galaxies M 87 and Centaurus A opened up a new source class for studies. Radio-galaxies also belong to the AGN class, but their jets make a larger angle to the line of sight and therefore the emission is less boosted. The different geometry also enables to study the individual components of the AGN system in greater details. Even more importantly: the most nearby AGN are radio galaxies. The detection of VHE emission from these source enable unique studies of the spatial origin of the VHE emission with unprecedented resolution.

2. Centaurus A

Centaurus A (Cen A) is the closest active radio galaxy (for a review see [3]). At radio wavelengths rich jet structures are visible, extending from the core and the inner pc and

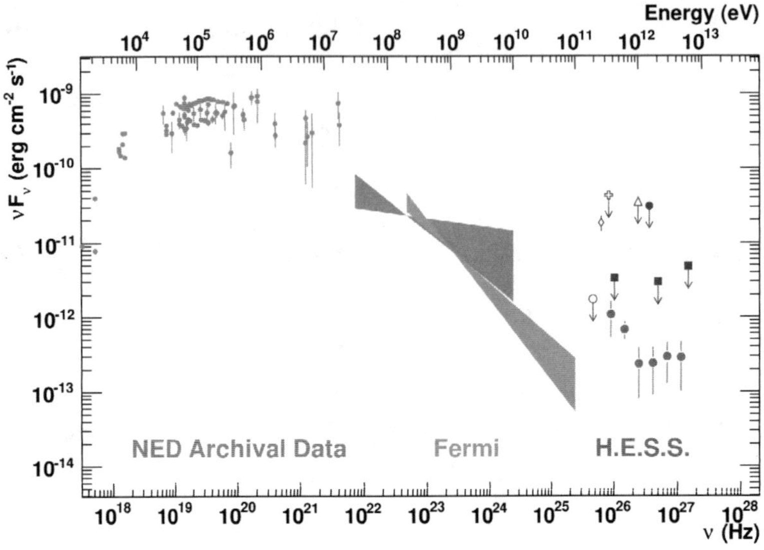

Figure 1. High energy part of the spectral energy distribution of Cen A.

kpc jet to giant outer lobes with an angular extension of $8^\circ \times 4^\circ$. The inner kpc jet has also been detected in X-rays, revealing a complex structure of bright knots and diffuse emission [4]. Recent estimates for the mass of the central supermassive black hole give $(5.5 \pm 3.0) \times 10^7\, M_\odot$ [5]. Cen A has also been proposed as a possible source of ultra-high energy cosmic rays.

Centaurus A has been observed by the H.E.S.S. experiment for over 115 h between 2004 and 2008 resulting in a detection of VHE γ-ray emission with a flux of $\sim$0.8% Crab [6]. The measured differential photon spectrum is well described by a power-law function $dN/dE = \Phi_0 \cdot (E/1\,\mathrm{TeV})^{-\Gamma}$ with normalization $\Phi_0 = (2.45 \pm 0.52_{\mathrm{stat}} \pm 0.49_{\mathrm{sys}}) \times 10^{-13}\,\mathrm{cm}^{-2}\,\mathrm{s}^{-1}\,\mathrm{TeV}^{-1}$ and photon index $\Gamma = 2.73 \pm 0.45_{\mathrm{stat}} \pm 0.2_{\mathrm{sys}}$. No significant variability has been found on time-scales of 28 min, nights and months (moon periods). Fig. 1 shows the spectral energy distribution of Cen A ranging from X-rays to the VHE regime. The flux measured by H.E.S.S. is clearly below all previous upper limits in the VHE regime. Recently, the *Fermi* LAT team reported a detection of Cen A at GeV energies [7] (Fig. 1 orange bow-tie). A simple extrapolation of the reported power law function would result in a too low flux at $\sim$TeV energies.

Several authors have predicted VHE emission from Cen A, and more generally discussed VHE emission from radio galaxies. A first class of models proposed the immediate vicinity of the supermassive black hole as the region of VHE emission, e.g. in pulsar-type scenarios [8, 9]. A second class of models propose a mechanism similar to the mechanism at work in other VHE blazars [10, 11]. [12] discussed a two-flow type model [13, 14], with a fast spine and a slower, mildly relativistic sheath propagating within the jet, which has been successfully applied to M 87 [15]. [16] modeled the VHE emission of Cen A with a multi-blob SSC model. Extended VHE emission may also be expected from Cen A. In this context, [17] proposed that γ-rays emitted in the immediate vicinity of the active nucleus are partly absorbed by the starlight radiation in the host galaxy and initiated pair production cascades. The created $e^{\pm}$ pairs are quickly isotropized and radiate VHE γ-rays by inverse Compton scattering the starlight radiation, resulting in an extended VHE emission. Hadronic models have also been invoked to predict VHE emission from radio galaxies [18]. Given the weakness of the signal for the current generation instruments the H.E.S.S. results do not yet strongly constrain these models.

Recently, [19] reported the detection of non-thermal X-ray synchrotron emission from the shock of the southwest inner radio lobe. They investigated inverse Compton scattering the starlight radiation and the CMB from high energy particles in this lobe and predicted a VHE emission well compatible with the H.E.S.S. measurement. This would imply that the VHE emission from Cen A is produced analogous to a gigantic supernova remnant (SNR). While the position is $\sim 3\sigma$ away from the best fit position of the VHE excess, it is well within the upper limit of the extension.

3. M 87

The giant radio galaxy M 87 is located in the Virgo cluster at a distance of 16 Mpc[21]. The angle between the plasma jet in M 87 and the line of sight is estimated to lie between $20° - 40°$ [22]. With its proximity, its bright and well resolved jet, and its very massive black hole with $6 \times 10^9\,M_{\odot}$ [23], M 87 provides an excellent opportunity to study the inner structures of the jet, which are expected to scale with the gravitational radius of the black hole. Substructures of the jet are resolved in the X-ray, optical and radio wavebands [24] and high-frequency radio very long baseline interferometry (VLBI) observations with sub-milliarcsecond (mas) resolution are starting to probe the collimation region of the jet [25].

A first indication of VHE γ-ray emission ($> 4\,\sigma$) from the direction of M 87 in 1998/99 was reported by HEGRA [26]. The VHE γ-ray emission was confirmed by H.E.S.S. [27], establishing M 87 as the first non-blazar extragalactic VHE γ-ray source. Utilizing causality arguments, the reported day-scale variability strongly constrains the size of the γ-ray emission region. VERITAS detected M 87 in 2007 in a low flux state [28] but did not find variability in their data-set.

The measured flux variability rules out large-scale emission from dark matter annihilation or cosmic-ray interactions as dominant origin of the VHE emission. Leptonic [29, 16] and hadronic [18] jet emission models have also been proposed to explain the VHE emission. The location of the VHE γ-ray emission is still unknown, but the nucleus [8], the inner jet [18, 29, 16, 15] or larger structures in the jet [30] have been suggested as possible sites. The 2005 VHE γ-ray flare was detected during an exceptional, several years lasting X-ray outburst of the innermost knot in the jet 'HST-1' [31].

To further investigate the origin of the VHE emission the three major VHE instruments H.E.S.S., MAGIC and VERITAS joined for a coordinated monitoring campaign on the source in winter/spring of 2008. During the campaign a strong VHE flux outburst of M 87 was detected [32], again showing variability at the time scale of days. At the same time, further observations at X-ray with Chandra and at radio with VLBA have been performed [33, 34]. The high resolution VLBA observations at 43 GHz achieved a direct imaging of the innermost jet region and the immediate surrounding of the central supermassive black hole with a resolution $< 100\,R_s$ (0.2 mas or 0.016 pc

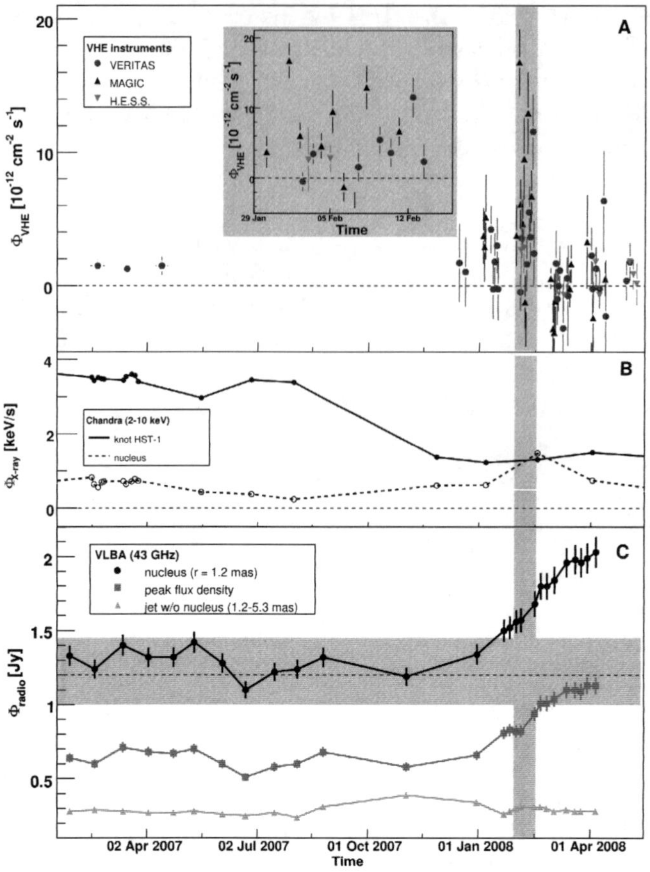

Figure 2. Results of the joined M 87 monitoring campaign in 2008 combining observations at VHE (H.E.S.S./MAGIC/VERITAS; top panel), X-rays (CHANDRA; middle panel), and radio (VLBA; bottom panel). From [20]. Reprinted with permission from AAAS.

at 16 Mpc).[33]. The results of this joined multi-wavelength effort have recently been published [20] and are summarized below:

The combined lightcurve of the 2008 data-set is shown in Fig. 2. At VHE energies (top row) the sources is mostly in a quiet state with the very notable exception of February/March 2008, where several strong outburst of the source were detected, again with variability on time-scales ~day. In X-rays (middle panel) the lightcurve for the innermost resolvable X-ray bright knot in the jet HST-1 and the core are shown. Contrary to the VHE flare in 2005, the X-ray flux from HST-1 is in a low state, while the core shows a minor increase in flux contemporaneous with the VHE flare. The lowest panel shows the radio flux for the core, which is associated with the black hole and its immediate surrounding at scales $< 100\,R_s$. The radio flux of the core shows a significant increase coinciding with the VHE flare. The detected multi-wavelength behavior and its comparison with previous observations leads to the conclusion that there is strong indication for a connection between the radio and the VHE emission - readily explained by a simple model - and that therefore the central black hole and its immediate surrounding are the most likely production site of the VHE emission [20].

These findings lend support to models preferring the immediate vicinity of the central black hole or the jet base as origin of the VHE emission.

4. Conclusion & Outlook

The discovery of VHE emission from the nearby radio galaxies Centaurus A and M 87 enables unique studies of acceleration processes in relativistic jets and in the vicinity of supermassive black holes. Given the proximity of the sources (3.8 and 16 Mpc) and the larger jet angle to the line of sight compared to VHE blazars, the outer and inner kpc jet structures will be resolvable by future VHE experiments like CTA, enabling to further spatially pin down the site of the emission. With the help of simultaneous multi-wavelength observations and (temporal) correlation studies, different section of the jet and the core can be probed, down to the smallest sub-pc (milli-arcsecond) scale, only accessible to VLBA radio observations.

The multi-wavelength behavior observed from M 87 during the 2008 joined campaign points towards the immediate surrounding of the central super-massive black hole as origin of the VHE emission. While this surely needs further observational confirmation, it demonstrates the potential of such joined campaigns to probe VHE emission sites.

Acknowledgments

The support of the Namibian authorities and of the University of Namibia in facilitating the construction and operation of H.E.S.S. is gratefully acknowledged, as is the support by the German Ministry for Education and Research (BMBF), the Max Planck Society, the French Ministry for Research, the CNRS-IN2P3 and the Astroparticle Interdisciplinary Programme of the CNRS, the U.K. Science and Technology Facilities Council (STFC), the IPNP of the Charles University, the Polish Ministry of Science and Higher Education, the South African Department of Science and Technology and National Research Foundation, and by the University of Namibia. We appreciate the excellent work of the technical support staff in Berlin, Durham, Hamburg, Heidelberg, Palaiseau, Paris, Saclay, and in Namibia in the construction and operation of the equipment. The author acknowledges the close and extremely productive collaboration with the VERITAS Collaboration, the VLBA 43 GHz M 87 Monitoring Team, and the MAGIC Collaboration on the M 87 results in particular M. Beilicke, H. Krawczynski, C. Walker, P. Hardee, D. Mazin, and R. Wagner.

[1] Albert J, Aliu E, Anderhub H, Antonelli L A, Antoranz P *et al.* 2008 *Science* **320** 1752– (*Preprint* 0807.2822)
[2] Mazin D and Raue M 2007 *A&A* **471** 439–452 (*Preprint* arXiv:astro-ph/0701694)
[3] Israel F P 1998 *A&A Rev.* **8** 237–278 (*Preprint* arXiv:astro-ph/9811051)
[4] Kraft R P, Forman W R, Jones C, Murray S S, Hardcastle M J and Worrall D M 2002 *ApJ* **569** 54–71 (*Preprint* arXiv:astro-ph/0111340)
[5] Cappellari M, Neumayer N, Reunanen J, van der Werf P P, de Zeeuw P T and Rix H 2009 *MNRAS* **394** 660–674 (*Preprint* 0812.1000)
[6] Aharonian F, Akhperjanian A G, Anton G *et al.* 2009 *ApJ* **695** L40–L44 (*Preprint* 0903.1582)
[7] Abdo A A, Ackermann M, Ajello M, Atwood W B, Axelsson M *et al.* 2009 *ApJ* **700** 597–622 (*Preprint* 0902.1559)
[8] Neronov A and Aharonian F A 2007 *ApJ* **671** 85–96 (*Preprint* arXiv:0704.3282)

[9] Rieger F M and Aharonian F A 2008 *A&A* **479** L5–L8 (*Preprint* 0712.2902)
[10] Bai J M and Lee M G 2001 *ApJ* **549** L173–L177 (*Preprint* arXiv:astro-ph/0102314)
[11] Chiaberge M, Capetti A and Celotti A 2001 *MNRAS* **324** L33–L37 (*Preprint* arXiv:astro-ph/0105159)
[12] Ghisellini G, Tavecchio F and Chiaberge M 2005 *A&A* **432** 401–410 (*Preprint* arXiv:astro-ph/0406093)
[13] Sol H, Pelletier G and Asseo E 1989 *MNRAS* **237** 411–429
[14] Marcowith A, Henri G and Renaud N 1998 *A&A* **331** L57–L60
[15] Tavecchio F and Ghisellini G 2008 *MNRAS* **385** L98–L102 (*Preprint* 0801.0593)
[16] Lenain J P, Boisson C, Sol H and Katarzyński K 2008 *A&A* **478** 111–120
[17] Stawarz Ł, Aharonian F, Wagner S and Ostrowski M 2006 *MNRAS* **371** 1705–1716 (*Preprint* arXiv:astro-ph/0605721)
[18] Reimer A, Protheroe R J and Donea A C 2004 *A&A* **419** 89–98
[19] Croston J H, Kraft R P, Hardcastle M J, Birkinshaw M, Worrall D M *et al.* 2009 *MNRAS* **395** 1999–2012 (*Preprint* 0901.1346)
[20] Acciari V A, Aliu E, Arlen T, Bautista M, Beilicke M *et al.* 2009 *Science* **325** 444–448 (*Preprint* http://www.sciencemag.org/cgi/reprint/325/5939/444.pdf) URL http://www.sciencemag.org/cgi/content/abstract/325/5939/444
[21] Macri L M, Huchra J P, Stetson P B, Silbermann N A, Freedman W L *et al.* 1999 *ApJ* **521** 155–178 (*Preprint* arXiv:astro-ph/9901332)
[22] Gebhardt K and Thomas J 2009 *ApJ* **700** 1690–1701 (*Preprint* 0906.1492)
[23] Macchetto F, Marconi A, Axon D J, Capetti A, Sparks W and Crane P 1997 *ApJ* **489** 579–+ (*Preprint* arXiv:astro-ph/9706252)
[24] Wilson A S and Yang Y 2002 *ApJ* **568** 133–140 (*Preprint* arXiv:astro-ph/0112097)
[25] Junor W, Biretta J A and Livio M 1999 *Nature* **401** 891–892
[26] Aharonian F, Akhperjanian A, Beilicke M *et al.* 2003 *A&A* **403** L1–L5 (*Preprint* arXiv:astro-ph/0302155)
[27] Aharonian F, Akhperjanian A G, Bazer-Bachi A R *et al.* 2006 *Science* **314** 1424 (*Preprint* astro-ph/0612016)
[28] Acciari V A, Beilicke M, Blaylock G, Bradbury S M, Buckley J H *et al.* 2008 *ApJ* **679** 397–403 (*Preprint* 0802.1951)
[29] Georganopoulos M, Perlman E S and Kazanas D 2005 *ApJ* **634** L33–L36 (*Preprint* arXiv:astro-ph/0510783)
[30] Cheung C C, Harris D E and Stawarz L 2007 *ApJ* **663** L65–L68 (*Preprint* arXiv:0705.2448)
[31] Harris D E, Cheung C C, Biretta J A, Sparks W B, Junor W, Perlman E S and Wilson A S 2006 *ApJ* **640** 211–218 (*Preprint* arXiv:astro-ph/0511755)
[32] Albert J, Aliu E, Anderhub H, Antonelli L A, Antoranz P *et al.* 2008 *ApJ* **685** L23–L26 (*Preprint* 0806.0988)
[33] Walker R C, Ly C, Junor W and Hardee P J 2008 *Journal of Physics Conference Series* **131** 012053–+
[34] Harris D E, Cheung C C, Stawarz Ł, Biretta J A and Perlman E S 2009 *ApJ* **699** 305–314 (*Preprint* 0904.3925)

Topics in Astroparticle and Underground Physics (TAUP 2009) IOP Publishing
Journal of Physics: Conference Series **203** (2010) 012118 doi:10.1088/1742-6596/203/1/012118

Recent Results from VERITAS

David Hanna (for the VERITAS Collaboration)[1]
Physics Department , McGill University, Montreal, Canada
E-mail: hanna@physics.mcgill.ca

Abstract. The Very Energetic Radiation Imaging Telescope Array System (VERITAS) [1] is a ground-based gamma-ray observatory used for a variety of observations at energies between 100 GeV and 30 TeV. This contribution reports on detections of two active galactic nuclei, W-Comae and 3C 66A, of the IBL class and discusses some features of their detection within the context of astroparticle physics, one of the central themes of the TAUP conference series.

1. Introduction

TeV gamma-ray astronomy plays an important role in astroparticle physics, one of the two themes of the TAUP conference series. Some gamma-ray observations have direct links to fundamental physics while others are concerned more with astrophysical phenomena that involve subatomic particles. Searching for evidence of Lorentz invariance violation by measuring the arrival times and energies of photons from the flares of distant sources [2, 3, 4] is an example of the first kind of study. An example of the second is the search for the origin of cosmic rays by studying the gamma-ray emission from supernova remnants within our galaxy, or from the more distant active galactic nuclei (AGNs). Dark matter is topical in both astrophysics, where evidence for its existence was first obtained, and particle physics, which supplies one of the leading candidates for dark matter. Searching for high energy gamma rays from annihilation of dark matter particles in places such as dwarf galaxies is one of the activities of TeV gamma-ray groups.

VERITAS [1], as one of the new generation of TeV instruments, is pursuing an active program in particle astrophysics. In this contribution I will highlight some recent results which bear on the broader program, namely discoveries of TeV emission from two AGNs. I will describe the VERITAS instrument and observing details and describe our results. Space limitations preclude a more comprehensive review; readers are invited to refer to the proceedings of the 31st International Cosmic Ray Conference [5] for more details.

2. VERITAS Instrument and Observational Details

The VERITAS detector comprises an array of four imaging atmospheric Cherenkov telescopes (IACTs) located at the Whipple Base Camp on Mount Hopkins in southern Arizona. Each telescope consists of a reflector, 12 metres in diameter, which collects Cherenkov light from extensive air showers and directs it onto a 'camera' made from 499 photomultiplier tubes (PMTs). Each camera has a field of view 3.5° in diameter. The array was built in stages between 2003 and 2007 with four-telescope observations beginning in September, 2007.

[1] see http://veritas.sao.arizona.edu/conferences/authors?taup2009 for a list of collaboration members

VERITAS covers the energy range from approximately 100 GeV to beyond 30 TeV with an energy resolution of between 15 and 20% and single-event angular resolution (68% containment) of 0.1°. It can make a five-sigma detection of a source with a flux of 1% of that of the Crab Nebula in less than 50 hours. More details can be found in [6].

During the first years of observations we have concentrated on four key science projects: blazars, galactic sources, dark matter and a sky survey of the Cygnus region. Additional observing campaigns proposed by members of the collaboration and approved by a time-allocation committee have also been carried out.

3. Active Galactic Nuclei

TeV gamma-rays from beyond our galaxy have been detected from a limited set of astrophysical objects, the majority of which belong to the *blazar* class of AGNs. According to the standard blazar paradigm, gamma rays are produced in relativistic jets powered by accretion onto a super-massive black hole at the centre of the AGN. An AGN with one of its jets oriented towards the Earth is a blazar. The spectral energy distributions of blazars display a double-peaked structure with the low-energy peak likely due to synchrotron radiation from high energy electrons in the jet and the high-energy peak due to inverse-Compton scattering from these electrons, or perhaps arising from other processes like synchrotron radiation from protons. Blazars can be classified according to the mean energy of the synchrotron peak. Low-frequency-peaked BL Lac objects (LBLs) have this peak at infra-red and optical wavelengths and the corresponding high energy peak at X-ray energies. For HBLs the synchrotron peak occurs in X-rays and the high energy peak in the TeV range. Until recently, no intermediate-frequency-peaked BL Lac objects (IBLs) had been detected.

3.1. W-Comae

VERITAS has detected three IBLs so far; the first was W-Comae [7]. A 4.9 σ detection was made using a 40-hour data set acquired between January and April 2008, using all four telescopes. The sky map generated from these observations is shown in figure 1. Approximately 70% of the signal resulted from a four-night flare [8] which occured in mid March, as shown in the light curve displayed in figure 2. The energy spectrum from the two flare nights with the highest flux can be fit with a power law with index of $\Gamma = 3.8 \pm 0.4$. The integral flux above 200 GeV is approximately 9% of the Crab Nebula flux above that energy. The soft energy spectrum is likely due to absorption of high-energy photons in pair-production collisions with the extra-galactic background light (EBL [9]) since the source lies at z=0.102. It is also consistent with the notion that the mean of the high-frequency peak of this source is in the GeV region and we are observing its rapidly falling upper side. The spectral energy distribution (SED) can be modeled using a one-zone synchrotron-self-Compton scenario but this needs an unnaturally low magnetic field value (0.007 G). A more natural fit can be obtained with an external-Compton (EC) model.

3.2. 3C 66A

The second IBL to be detected by VERITAS was 3C 66A. A detection of this variable source at very high flux levels was reported by the Crimean group [11] in 2002 and VERITAS detected it during a flaring episode [12] in 2008. Although data were acquired over the 2007-08 (4.7 hours after quality cuts) and 2008-09 (28.1 hours) observing seasons, 80% of the signal comes from the period from MJD54740 through MJD54749. The sky map made with the entire data set is shown in figure 3 and the corresponding energy spectrum in figure 4. The sky map shows that the TeV source detected by VERITAS is very close to the nominal position of 3C66A and is definitely not (at a significance level of 4.8σ) associated with the nearby radio galaxy 3C 66B. This is an important feature of our measurement since it contrasts with an earlier TeV detection by the MAGIC [13] collaboration which favours 3C 66B as the source. The energy

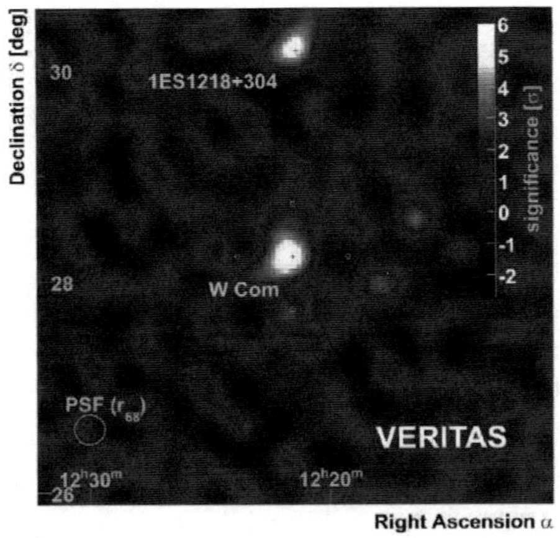

Figure 1. Sky map for the region around W-Comae. Note the presence of a signal from the blazar 1ES 1218+30.4, demonstrating VERITAS' off axis sensitivity.

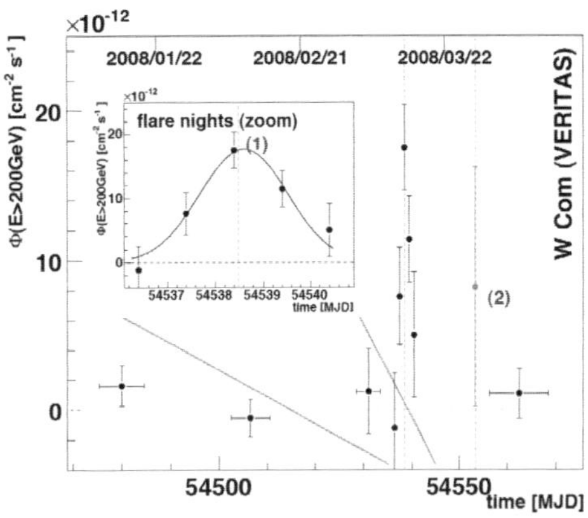

Figure 2. Light curve for the W-Comae measurments made during 2008. The flare centred on March 13 (MJD 5438), which contributed most of the detected flux, is shown in the inset.

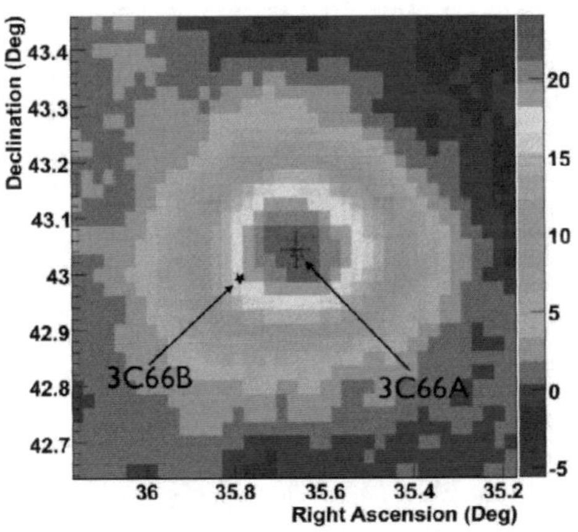

Figure 3. Sky map for the region around 3C 66A indicating that the VERITAS measurement strongly disfavours 3C 66B as the source of the gamma-ray signal.

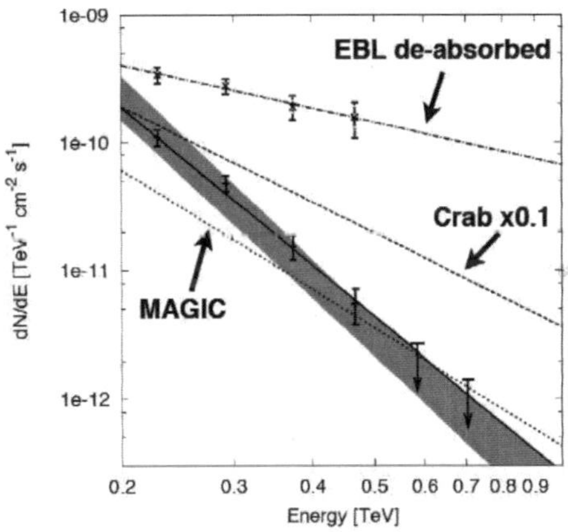

Figure 4. Energy spectrum for 3C 66A as measured by VERITAS, along with its possible intrinsic spectrum as calculated using a model of extra-galactic background light[10] and assuming z=0.444. The MAGIC [13] spectrum, from their detection of 3C 66B, is also shown.

spectrum of 3C 66A is very soft; a power-law fit results in an index of $\Gamma = 4.1 \pm 0.4_{stat} \pm 0.6_{syst}$. One expects absorption by EBL photons to steepen the spectrum of distant blazars but using a leading model [10] to correct for this effect yields an anomalously hard intrinsic spectrum ($\Gamma = 1.1 \pm 0.4_{stat}$). It is possible that this is due to the use of an erroneous value for the distance; the redshift for 3C66A (z = 0.444) is not well measured [16, 17, 18].

4. Discussion and Conclusions

VERITAS has detected TeV gamma-ray emission from two AGNs which are members of the IBL class. The catalog of these objects will surely grow now that the *Fermi* Gamma-ray Space Telescope is in orbit since the high-energy peaks of these objects lie in the range where the *Fermi* Large Area Telescope [19] has good sensitivity. The measurements presented here are very much astronomical but they are part of a body of knowledge that must be increased and consolidated to improve our understanding of the basics of AGNs. Only by understanding the physics of gamma-ray emission from AGNs, especially any energy-dependent timing issues, will we be able to make confident statements about apparent Lorentz-invariance violation effects. The modeling uncertainties associated with the W-Comae observations show that much is left to be done. Similarly, the uncertainties in source red-shifts, highlighted in the 3C 66A observation need to be reduced. Finally, a detailed understanding of AGN particle production in general, accessible through high-quality gamma-ray observations, is critical for determining whether AGNs are the sources of ultra high energy cosmic rays [20], another topic of current interest. High-energy gamma-ray astronomy will continue to have much to offer astroparticle physics in the future.

5. Acknowledgements

VERITAS research is supported by grants from the US Department of Energy, the US National Science Foundation, and the Smithsonian Institution, by NSERC in Canada, by Science Foundation Ireland, and by STFC in the UK. The author would like to commend the TAUP conference organizers on an interesting and enjoyable meeting.

References

[1] T. C. Weekes et al. (VERITAS Collaboration) Astroparticle Physics, 17:221, May 2002.
[2] S. D. Biller et al. (Whipple Collaboration), Phys. Rev. Lett. 83, 2108 (1999)
[3] J. Albert et al. (MAGIC Collaboration) and J. Ellis et al., Phys. Lett. B 668, 253 (2008)
[4] F. Aharonian et al. (H.E.S.S. Collaboration) Phys.Rev.Lett.101:170402,2008
[5] VERITAS Collaboration Contributions to the 31st International Cosmic Ray Conference, arXiv:0908.0150v2
[6] J. Holder et al. (VERITAS Collaboration) Astroparticle Physics, 25:391, July 2006.
[7] V. Acciari et al. (VERITAS Collaboration) Ap.J 684 L73 (2008)
[8] Swordy, S. (VERITAS Collaboration) 2008, ATel-1422 http://www.astronomerstelegram.org/?read=1422
[9] Gould, R. J., & Schréder, G. P. 1967, Physical Review, 155, 1408
[10] A. Franceshini et al. 2008, A&A, 487, 837
[11] A. Stepanyan et al., Astronomy Reports, 46, 634 (2002)
[12] V. Acciari et al. (VERITAS Collaboration) ApJ, 693, L104 (2009)
[13] E. Aliu et al. (MAGIC Collaboration) 2009, ApJ, 692, L29.
[14] F.A. Aharonian et al. (HEGRA Collaboration) 2001, A&A, 370, 112
[15] J. Albert et al. (MAGIC Collaboration) 2009, A&A, 474, 937
[16] D. Bramel et al. (STACEE Collaboration) 2005, ApJ 629:108-114 (2005)
[17] Miller, J. S., French, H. B., & Hawley, S. A. 1978, in BL Lac Objects, ed. A. M. Wolfe, 176
[18] Lanzetta, K. M., Turnshek, D. A., & Sandoval, J. 1993, ApJS, 84, 109
[19] Atwood, W. B., et al. 2009, ApJ, 697, 1071
[20] Pierre Auger Collaboration, *Science* 318, 938-943 (2007)

Journal of Physics: Conference Series **203** (2010) 012119 doi:10.1088/1742-6596/203/1/012119

Gamma-ray astronomy with the ARGO-YBJ experiment

Tristano Di Girolamo on behalf of the ARGO−YBJ collaboration
Università "Federico II" and Istituto Nazionale di Fisica Nucleare, Napoli, Italy
E-mail: tristano@na.infn.it

Abstract. The ARGO-YBJ experiment is an Extensive Air Shower detector which combines high altitude location (Tibet, P.R. China, 4300 m a.s.l.) and full coverage with Resistive Plate Counters, resulting in an energy threshold of a few hundred GeV. The large field of view (~2 sr) and high duty cycle allow the continuous monitoring of the Northern sky, searching for unknown sources and unpredictable events, such as flares in blazar emissions or high energy Gamma Ray Bursts. In this paper I will present some results obtained in γ-ray astronomy with the ARGO-YBJ experiment.

1. The detector

The ARGO−YBJ experiment is located at Yangbajing (Tibet, P.R. China, 4300 m a.s.l.) and consists of a single layer of Resistive Plate Counters (RPCs) on a total area of about ($110 \times 100\ m^2$) [1]. The detector has a modular structure, the basic module being a CLUSTER ($5.7 \times 7.6\ m^2$), made of 12 RPCs. Each RPC is divided into 10 PADs, whose data acquisition is independent and which represent the high granularity pixel of the detector ($56 \times 62\ cm^2$). The digital readout of each PAD is made by means of 8 strips ($6 \times 62\ cm^2$). The detector carpet is connected to two different DAQ systems, which work independently: in shower mode, for each event the location and timing of each detected particle is recorded, allowing the reconstruction of the lateral distribution and of the arrival direction; in scaler mode, the counting rate of each CLUSTER is measured every 0.5 s, with little information on the space distribution and arrival direction of the detected particles. The current trigger of the shower mode is $N_{PAD} \geq 20$ in a time window of 420 ns, with a rate $\sim 3.8\ kHz$. In the scaler mode DAQ, for each CLUSTER four scalers record the rate of counts ≥ 1, ≥ 2, ≥ 3 and ≥ 4 in a time window of 150 ns. The corresponding measured rates are, respectively, $\sim 40\ kHz$, $\sim 2\ kHz$, $\sim 300\ Hz$ and $\sim 120\ Hz$. The use of four different scalers may give an indication of the source spectrum in case of positive detection [2]. The experiment has been taking data with its full layout since November 2007.

The detector angular resolution has been determined studying the deficit in the cosmic ray flux due to the Moon. The result for cosmic ray events is less than 0.6° for $N_{strip} \geq 300$, while an improvement of at least 30% is expected for γ-induced showers because of their better defined time profile [3].

2. Gamma-ray astronomy

Five sky maps for different thresholds of hit pads ($N_{PAD} \geq$40, 60, 100, 200, 300) were obtained analysing the shower data of 424 days. Figure 1 (left) shows the resulting sky map for events with $N_{PAD} \geq 40$, which corresponds to a median primary energy for protons of ~2 TeV. Two large spots appear in the region of the Galactic anticentre, which were already reported by the Milagro experiment [4]. These regions have been interpreted as excesses in the cosmic ray flux (~0.1%), however their origin is still under debate. These hot spots affect the search for point γ-ray sources, overestimating the significances inside them. Therefore, a procedure to renormalize the background was applied [5] and the map after this correction is shown in Figure 1 (right). The Crab Nebula and the blazar Mrk 421 are visible with statistical significances 7.0 and 8.0 standard deviations, respectively. The resulting Crab energy spectrum is $dN/dE = (3.7 \pm 0.8) \times 10^{-11} E^{-2.67 \pm 0.25}$ $ph\ cm^{-2} s^{-1} TeV^{-1}$, in agreement with measurements by other experiments [6].

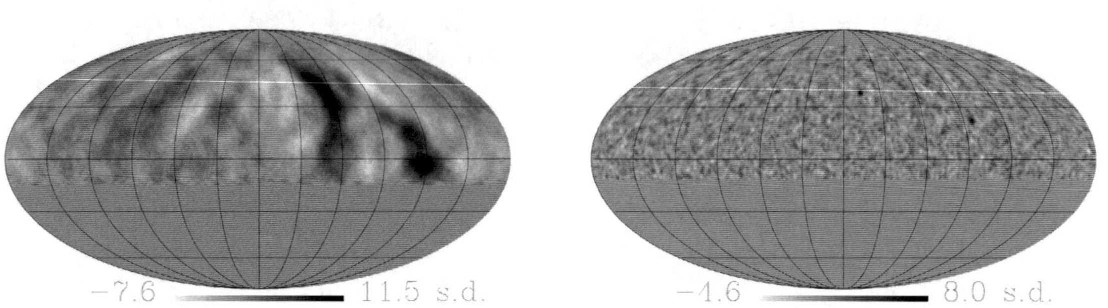

Figure 1. Left: sky map in equatorial coordinates obtained with 424 days of data, for events with $N_{PAD} \geq 40$. The scale gives the statistical significance in standard deviations. Right: sky map after correcting for large size excesses. The γ-ray median energy is $\approx$ 0.6–2.0 TeV, depending on the source spectrum.

Thanks to its continuous sky monitoring, ARGO-YBJ was able to detect the flaring states of Mrk 421 in 2006 and 2008. In particular, the observations from day 41 to 180 of 2008 provided a spectrum $dN/dE = (7.5 \pm 1.7) \times 10^{-11} E^{-2.51 \pm 0.29} e^{-\tau(E)}$ $ph\ cm^{-2} s^{-1} TeV^{-1}$, where the exponential factor $e^{-\tau(E)}$ takes into account the absorption of γ-rays in the Extragalactic Background Light (EBL). The resulting flux above 1 TeV was about twice that of the Crab Nebula, and the emission appeared to be correlated with X-rays [7]. An extraordinary set of simultaneous data, covering twelve decades of energy from optical to γ-rays, was available for the first of two flares in June 2008 [8], while during the second and more intense one observations by Cherenkov telescopes were hampered by the moonlight. Therefore in this case ARGO-YBJ completed the multiwavelength information, detecting a signal with a significance of 4.2 standard deviations and a flux of about 7 Crab units in the period June 11-13. The resulting spectrum is consistent with the prediction of the synchrotron self-Compton model. This is the first time that an air shower array is able to detect a sub-TeV γ-ray flare in a few days period.

3. Search for emission from Gamma Ray Bursts

In scaler mode, the energy threshold for photons is about 1 GeV, lower than the highest energies detected by satellite experiments. Moreover, the modular structure of the ARGO-YBJ detector allowed the collection of data during the different mounting phases. Therefore a search for emission from Gamma Ray Bursts (GRBs) in coincidence with satellite detections started in November 2004, when Swift was launched [9]. Until September 2009 a sample of 83 GRBs was analysed, 14 of them with known redshift z. Since no significant signal was found in the data, the fluence upper limits in the 1-100 GeV energy range were determined at 99% confidence level assuming two different power law spectra: a) extrapolation of the keV-MeV index measured by satellites; b) differential index -2.5. For the set of 14 GRBs with known redshift the ranges of upper limits between the values corresponding to the two spectral assumptions are represented by rectangles in Figure 2, while a simple arrow is shown if the low energy spectrum is a cutoff power law, and thus only case b) is considered. Moreover, an upper limit on the GRB cutoff energy E_{cut} can be set by the intersection of the fluence upper limit, as a function of E_{cut}, with the extrapolation of the fluence measured by satellites. Figure 3 reports the E_{cut} upper limits for the 18 GRBs for which the intersection occurs in the 1-100 GeV range, as a function of the differential spectral index.

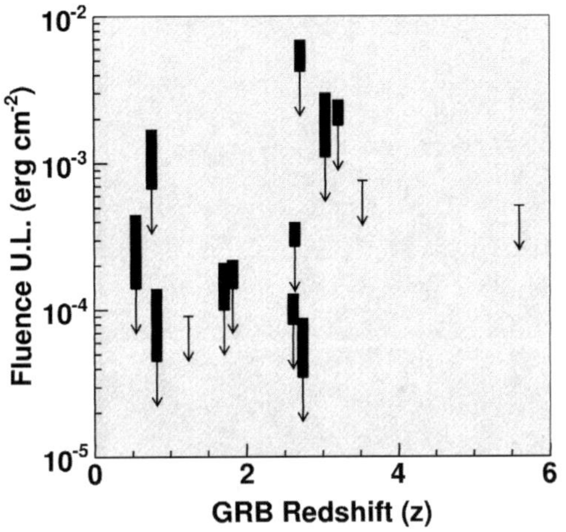

Figure 2. Fluence upper limits in the 1-100 GeV range for the 14 GRBs with known redshift, obtained with differential spectral indexes ranging from the value measured by satellites to -2.5 (only this latter case is considered for cutoff power law spectra).

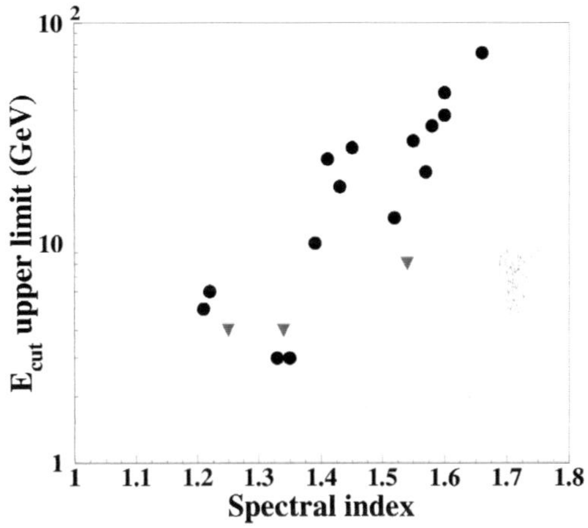

Figure 3. Cutoff energy upper limits as a function of the spectral index obtained extrapolating the measured keV-MeV spectra. The values represented by triangles are obtained taking into account the extragalactic absorption at the known GRB redshift, for the others z=1 is assumed.

Also the shower mode data have been analysed in a search for directional signals from GRBs detected by satellites, and fluence upper limits were set in the two energy ranges 10-100 GeV and 10 GeV-1 TeV [10].

References

[1] Aielli G et al. 2006 *Nucl. Instrum. Methods* A **562** 92
[2] Aielli G et al. 2008 *Astropart. Phys.* **30** 85
[3] Iuppa R et al. 2009 *Proc. 31st Int. Cosmic Ray Conf.* July 7-15, 2009, Łódź, Poland
[4] Abdo A A et al. 2008 *Phys. Rev. Lett.* **101** 221101
[5] Vernetto S, Guglielmotto Z et al. 2009 *Proc. 31st Int. Cosmic Ray Conf.* July 7-15, 2009, Łódź, Poland
[6] Marsella G et al. 2009 *Proc. 31st Int. Cosmic Ray Conf.* July 7-15, 2009, Łódź, Poland
[7] Vernetto S, Martello D et al. 2009 *Proc. 31st Int. Cosmic Ray Conf.* July 7-15, 2009, Łódź, Poland
[8] Donnarumma I et al. 2009 *Astrophysical Journal Lett.* **691** L13
[9] Aielli G et al. 2009 *Astrophysical Journal* **699** 1281
[10] Aielli G et al. 2009 *Astropart. Phys.* **32** 47

Have Cherenkov telescopes detected a new light boson?

Marco Roncadelli
INFN, Sezione di Pavia, via A. Bassi 6, I – 27100 Pavia, Italy
E-mail: marco.roncadelli@pv.infn.it

Alessandro De Angelis
Dipartimento di Fisica, Università di Udine, Via delle Scienze 208, I – 33100 Udine, and INAF and INFN, Sezioni di Trieste, Italy
E-mail: deangelis.alessandro@gmail.com

Oriana Mansutti
Dipartimento di Fisica, Università di Udine, Via delle Scienze 208, I – 33100 Udine, and INFN, Sezione di Trieste, Italy
E-mail: oriana.mansutti@gmail.com

Abstract. Recent observations by H.E.S.S. and MAGIC strongly suggest that the Universe is more transparent to very-high-energy gamma rays than previously thought. We show that this fact can be reconciled with standard blazar emission models provided that photon oscillations into a very light Axion-Like Particle occur in extragalactic magnetic fields. A quantitative estimate of this effect indeed explains the observed data and in particular the spectrum of blazar 3C279.

1. Introduction

So far, Imaging Atmospheric Cherenkov Telescopes (IACTs) have detected about 30 very-high-energy (VHE) blazars above 100 GeV over distances ranging from the parsec scale for Galactic objects up to the Gigaparsec scale for extragalactic ones. By now, the fartest blazar observed by IACTs is 3C279 at redshift $z = 0.536$ detected by MAGIC. Given that these sources extend over a wide range of distances, not only can their intrinsic properties be inferred but also photon propagation over cosmological distances can be probed. This is particularly intriguing because VHE photons from distant sources scatter off soft background photons, thereby disappearing into e^+e^- pairs. Since the cross section $\sigma(\gamma\gamma \to e^+e^-)$ peaks where the VHE photon energy E and the background photon energy ϵ are related by $\epsilon \simeq (500\,\mathrm{GeV}/E)\,\mathrm{eV}$, the horizon of the observable VHE Universe rapidly shrinks above 100 GeV due to the presence of the Extragalactic Background Light (EBL) produced by galaxies during the whole cosmic history. In the presence of such an energy-dependent opacity, photon propagation is controlled by the photon mean free path $\lambda_\gamma(E)$ for $\gamma\gamma \to e^+e^-$, and so the observed photon spectrum $\Phi_{\rm obs}(E, D)$ is related to the emitted one $\Phi_{\rm em}(E)$ by

$$\Phi_{\rm obs}(E,D) = e^{-D/\lambda_\gamma(E)}\,\Phi_{\rm em}(E)\;. \quad (1)$$

It turns out that $\lambda_\gamma(E)$ decreases like a power law from the Hubble radius 4.3 Gpc around 100 GeV to nearly 1 Mpc around 100 TeV [1]. Thus, Eq. (1) implies that the observed flux is *exponentially* suppressed both at high energies and at large distances, so that sufficiently far-away sources become hardly visible in the VHE range and their observed spectrum should be *much steeper* than the emitted one.

Yet, the behaviour predicted by Eq. (1) has *not* been detected by observations. A first indication in this direction was reported by the H.E.S.S. collaboration in connection with the discovery of the two blazars H2356-309 ($z = 0.165$) and 1ES1101-232 ($z = 0.186$) at $E \sim 1$ TeV [2]. Stronger evidence comes from the observation of blazar 3C279 ($z = 0.536$) at $E \sim 0.5$ TeV by the MAGIC collaboration [3]. In particular, the signal from 3C279 collected by MAGIC in the region $E < 220$ GeV has more or less the same statistical significance as the one in the range 220 GeV $< E <$ 600 GeV (6.1σ in the former case, 5.1σ in the latter).

We argue that this circumstance strongly hints at the existence of a very light Axion-like particle (ALP), which is indeed predicted by many extensions of the Standard Model, including compactified Kaluza-Klein theories as well as in superstring theories [4].

2. DARMA scenario

Our proposal – to be referred to as the DARMA scenario – can be sketched as follows [5]. Because ALPs are characterized by a coupling to two photons, in the presence of an external magnetic field **B** the interaction eigenstates differ from the propagation eigenstates, so that photon-ALP oscillations show up (much in the same way as it happens for massive neutrinos). Photons are supposed to be emitted by a blazar in the usual way. In the presence of extragalactic magnetic fields – whose existence is strongly suggested by AUGER observations [6] – some of them can turn into ALPs. Further, some of the produced ALPs can convert back into photons and ultimately be detected. In free space this would obviously produce a flux dimming. Remarkably enough, because of the EBL such a double conversion can make the observed flux *considerably larger* than in the standard situation! This is due to the fact that ALPs do *not* undergo EBL absorption. As a consequence, the observed photons travel a distance in excess of $\lambda_\gamma(E)$ and Eq. (1) becomes

$$\Phi_{\rm obs}(E,D) = e^{-D/\lambda_{\gamma,\rm eff}(E)}\,\Phi_{\rm em}(E)\;, \quad (2)$$

which shows that even a *small* increase of $\lambda_{\gamma,\rm eff}(E)$ gives rise to a *large* enhancement of the observed flux. It turns out that the DARMA mechanism makes $\lambda_{\gamma,\rm eff}(E)$ shallower than $\lambda_\gamma(E)$ although it remains a decreasing function of E. So, the resulting observed spectrum is *much harder* than the one predicted by Eq. (1), thereby ensuring agreement with observations even for a *standard* emission spectrum. As a bonus, we get a natural explanation for the fact that only the most distant blazars would demand $\Phi_{\rm em}(E)$ to substantially depart from the emission spectrum predicted by the standard mechanism.

3. Predicted energy spectrum

Even though the existence of extragalactic magnetic fields **B** has been strongly suggested by AUGER observations [6], their morphology is largely unknown. So, it is usually supposed that they have a domain-like structure. It looks plausible to assume that $0.3\,{\rm nG} < B < 1.0\,{\rm nG}$ and that their coherence length $L_{\rm dom}$ is in the range $1\,{\rm Mpc} < L_{\rm dom} < 10\,{\rm Mpc}$ [7].

We evaluate of the probability $P_{\gamma\to\gamma}(E,D)$ that a photon remains a photon after propagation from the source to us when allowance is made for photon-ALP oscillations as well as for photon absorption from the EBL, so that Eq. (2) gets replaced by

$$\Phi_{\rm obs}(E,D) = P_{\gamma\to\gamma}(E,D)\,\Phi_{\rm em}(E)\;. \quad (3)$$

Our procedure is as follows. We first solve exactly the beam propagation equation over a single magnetic domain, assuming that the EBL is described by the "best-fit model" of Kneiske *et al.* [8]. Starting with an unpolarized photon beam, we next propagate it by iterating the single-domain solution as many times as the number of domains crossed by the beam, taking each time a *random* value for the angle between **B** and a fixed overall fiducial direction. We repeat such a procedure 10 000 times and finally we average over all these realizations of the propagation process.

As far as 3C279 is concerned, we find that about 13% of the photons arrive to the Earth for $E = 500\,\text{GeV}$, representing an enhancement by a factor of about 20 with respect to the expected flux without DARMA mechanism (the comparison is made with the above "best-fit model"). The same calculation gives a fraction of 76% for $E = 100\,\text{GeV}$ (to be compared to 67% without DARMA mechanism) and a fraction of 3.4% for $E = 1\,\text{TeV}$ (to be compared to 0.0045% without DARMA mechanism). The resulting spectrum is exhibited in Fig. 1. These conclusions hold for ALPs lighter than $10^{-10}\,\text{eV}$ and for their two-photon coupling within the experimentally allowed range [5].

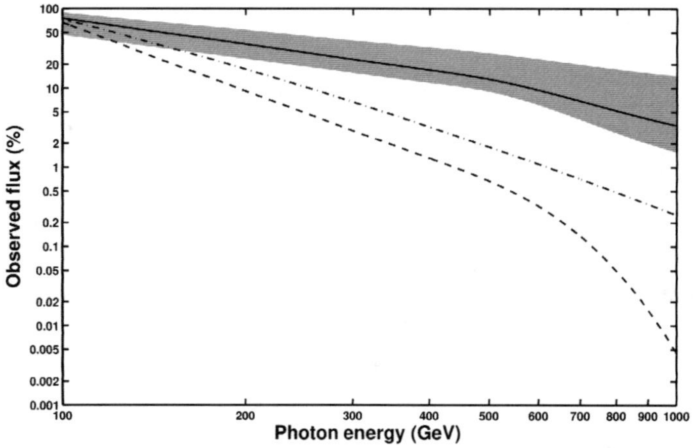

Figure 1. The two lowest lines give the fraction of photons surviving from 3C279 without the DARMA mechanism within the "best-fit model" of EBL (dashed line) and for the minimum EBL density compatible with cosmology (dashed-dotted line). The solid line represents the prediction of the DARMA scenario for $B \simeq 1\,\text{nG}$ and $L_{\text{dom}} \simeq 1\,\text{Mpc}$ and the gray band is the envelope of the results obtained by independently varying **B** and L_{dom} within a factor of 10 about such values.

References

[1] Coppi P and Aharonian F 1997 *Astrophys. J.* **487** L9
[2] Aharonian F *et al.* (H.E.S.S. Collaboration) 2006 *Nature* **440** 1018
[3] Albert J *et al.* (MAGIC Collaboration) 2008 *Science* **320** 1752
[4] Arvanitaki A *et al.* 2009 String Axiverse Preprint arXiv:0905.4720
[5] De Angelis A, Roncadelli M and Mansutti O 2007 *Phys. Rev.* D **76** 121301
[6] Abraham J *et al.* [Pierre Auger Collaboration] 2007 *Science* **318** 939
[7] De Angelis A, Persic M and Roncadelli M 2008 *Mod. Phys. Lett.* A **23** 315
[8] Kneiske T M, Bretz T, Mannheim K and Hartmann D H 2004 *Astron. Astrophys.* **413** 807

Cherenkov Telescope Array – The Future of Ground-Based Gamma-Ray Astronomy

Robert Wagner (for the CTA Consortium)
Max-Planck-Institut für Physik, Föhringer Ring 6, D-80805 München, Germany
E-mail: `robert.wagner@mpp.mpg.de`

Abstract. Very-high energy (VHE; $E \gtrapprox 100$ GeV) γ-rays provide a unique probe for non non-thermal processes in the universe. The ground-based Imaging Air Cherenkov Technique for detecting VHE γ-rays has matured, and a fast assembly of inexpensive and robust telescopes is possible. The goal for the next generation of instruments is to increase their sensitivity by a factor $\gtrsim 10$ compared to current facilities, to extend the accessible γ-ray energies from a few tens of GeV to a hundred TeV, and to improve on other parameters like angular and energy resolution. I discuss the key physics goals and resulting design considerations for the Cherenkov Telescope Array (CTA), a project for a new generation of highly automated telescopes for γ-ray astronomy. The technical solutions chosen for CTA and the status of the project are discussed.

1. Introduction

Very-high energy (VHE) γ-rays are produced in nonthermal processes in the universe, namely in galactic objects like pulsars, pulsar-wind nebulae, supernova remnants (SNR), binary systems containing compact objects, or OB associations. Among the extragalactic VHE γ-ray sources are active galactic nuclei (AGN), particularly blazars and radio-galaxies, and starburst galaxies. Galaxy clusters and gamma-ray bursts are also potential, although not yet discovered, sources of VHE γ rays. Apart from the astrophysics of specific astronomical objects, γ-ray astronomy can be used to search for the annihilation of dark matter particles, and for studying the transparency and history of the universe. Further fundamental physics searches, like for the violation of Lorentz invariance, can be performed. For recent reviews, see, e.g., [1, 2].

Upon reaching the Earth's atmosphere, VHE γ-rays interact with atmospheric nuclei and generate electromagnetic showers. The showers extend over several kilometers in length and few tens to hundreds of meters in width. At VHE, the shower particles are stopped high up in the atmosphere. A sizeable fraction of the charged secondary shower particles, mostly electrons and positrons in the shower core, move with ultra-relativistic speed and emit Cherenkov light. Imaging atmospheric Cherenkov telescopes reflect the Cherenkov light onto multi-pixel cameras that record the shower images.

2. Towards a precision gamma-ray astronomy

Despite the promising achievements of current-generation Cherenkov telescopes [1], there are limitations that future instruments will need to overcome: current instruments are sensitive in an energy range of $\gtrsim 80$ GeV$-$50 TeV. At the low energy end, limitations come from the background from atmospheric hadronic (and electronic) showers. At the high end, the limit is statistics; due to the lack of a calibrated cosmic γ-ray source, spectral reconstruction is limited by systematic bias and statistical uncertainties; the aperture of Cherenkov telescopes is limited to a typical field of view (FOV) of $3-5°$ diameter, as is the angular resolution, currently around a few arcmin. Also, current facilities are rather poorly automatized. From a physics point of view, there are strong arguments to improve in the following aspects: decrease the energy threshold to few tens of GeV; acquire sensitivity beyond 50 TeV; increase sensitivity in the core range (100 GeV$-$50 TeV); improve energy and angular resolution. Cherenkov Telescope Array (CTA) is a next-generation ground-based project which aims at implementing all these improvements.

2.1. Low-energy physics (sub-50 GeV)

MAGIC has opened the field of sub-100 GeV γ-ray astronomy [3]. Observations with higher sensitivity in this region will have important consequences for galactic and extragalactic physics. By studying the sub-50 GeV energy band, CTA could provide the final answer to the acceleration mechanism in pulsars. A high sensitivity in the low energy regime is also vital for studying AGNs, which typically exhibit rather steep power-law spectra (due to the increasing suppression of γ-ray flux with energy by the extragalactic background light). Further, low-energy sensitivity is vital to complement the catalog of *Fermi*-LAT detected γ-ray emitters with a possibly larger significance than obtainable by LAT at its high-energy end. This will provide a unique way to understand the nature of the tens of yet unidentified *Fermi*-LAT detected sources [5].

2.2. High-energy physics (above 50 TeV)

Typical astrophysical γ-ray spectra have bimodal distributions with one peak at lower energies due to synchrotron emission, and a second peak at higher energies due to inverse Compton scattering of VHE electrons on seed infrared/optical photons. For galactic objects, one may expect to observe VHE power-law γ-ray spectra with cutoffs due to intrinsic mechanisms. γ-rays from distant blazars suffer a severe attenuation after pair production with local IR-UV photons of the extragalactic background light (EBL). In all current IACT data, the evidences for spectral cutoffs and steepening are rather poor. There is no simple justification for this. CTA will explore this region with unprecedented sensitivity. This will allow to understand the acceleration mechanism in galactic objects like SNRs, and discriminating hadronic/leptonic models. The high energy region is also important for identifying PeVatrons, i.e. cosmic-ray accelerators to TeV/PeV energies. Detecting outbursts ("flares") from distant AGNs at super-TeV energies would allow to increase the prospects for firm limits on Lorentz invariance violation.

2.3. Core energy region (0.1–50 TeV)

CTA will provide an increase of the sensitivity in this core energy region by at least a factor 10 as compared to the current IACTs, reaching a level of 10^{-3} C.U.[1] sensitivity in this range. This will promote γ-ray astrophysics to a *γ-ray astronomy*. In fact, for the first time, CTA will allow for a full VHE sky survey, with approx. a thousand new VHE γ-ray sources expected to be detected.

An increase in sensitivity will, by reducing the required observation times, allow more follow-up observations and higher time resolution of variable sources. The current telescopes are sensitive enough to detect variations on timescalew of minutes. CTA will allow a sub-minute resolution, and thus to understand the complex phenomena of γ-ray flares, directly connected to the acceleration mechanisms and the local environment. Morphological studies will profit from reduced required observation times. These are of importance to study spatially extended γ-ray emitters, like SNRs. An increase of the angular resolution by a factor $4-5$, down to

[1] The differential photon flux of the Crab nebula between 60 GeV and 9 TeV is [4] $\mathrm{d}F/\mathrm{d}E = (6.0 \pm 0.2)10^{-10}(E/300\,\mathrm{GeV})^{a+b\log_{10}(E/300\,\mathrm{GeV})}\,\mathrm{TeV}^{-1}\mathrm{cm}^{-2}\mathrm{s}^{-1}$ with $a = -2.31 \pm 0.06$, $b = -0.26 \pm 0.07$.

0.02 arcmins (current theoretical limits are discussed in [6]), will reduce source confusion and improve collaboration with instruments observing at other wavelengths.

3. General design ideas

To meet the physical requirements specified in the previous section and to maintain an overall high technical performance, the CTA concept is based on few general ideas: Increase the array from currently 4 (H.E.S.S., VERITAS) to ≈ 100 telescopes; distribute them over a large area ($1 - 10$ km^2); make use of telescopes of $2 - 3$ different sizes; take advantage of well-proven technology of current IACTs; high automatization and remote operation; run array as observatory and open the facility to the astronomy and astrophysics community.

An increase in sensitivity over the full energy range can only be achieved by combining many telescopes distributed over a large area of at least 1 km^2 and using telescopes of $2 - 3$ different sizes: several medium size telescopes (MST) of 12 m, few large size telescopes (LST) of 24 m diameter, and probably several small size telescopes (SST) of 7 m diameter. The number of the telescopes, their size, configuration and the overall performance are under investigation.

The LSTs detect sub-100 GeV photons thanks to their large reflective area. Technologically, they are the most challenging telescope type. A design is currently under development. Several tens of MSTs will perform the γ-ray detection in the core energy region. Those telescopes will be based on the experience gained with the H.E.S.S. and MAGIC telescopes. The main goal is to reduce cost and maintenance efforts. The MSTs constitute the core of the array, and will also perform the fundamental task of vetoing the LST triggers to reduce hadronic background. The MST design is currently studied and the construction of a prototype is expected in few years from now. In case 3 different telescope sizes are required, several tens of SSTs will complete the array to perform the super-TeV search, increasing the effective collection area of the array. Very simple in construction, contributing only a small percentage to the cost of the full array.

The trigger systems will support different operation modes. In the "deep field" mode, all telescopes will be pointed to the same sky position to maximize the sensitivity. In a more flexible mode, parts of the telescopes could point to different positions, with few telescopes making follow-up observation of single sources as, e.g., to monitor blazar activity. Finally, the array can be operated in a "wide-FOV" mode, to perform an all-sky scan in a time-efficient way at a moderate sensitivity.

4. The CTA consortium

CTA is a partnership between the H.E.S.S. and MAGIC collaborations and several European institutes, with recent interests from more institutions world-wide. Activities are coordinated with AGIS (Advanced Gamma-ray Imaging System; cf. [2]), a similar US-based project. The consortium comprises > 70 institutes from 17 countries, involving more than 400 scientists.

For the current design phase, CTA is organized in several work packages: Management, Physics, Monte Carlo, Site, Mirror, Telescope, Focal-Plane Instrumentation, Electronics, Atmospheric Transmission and Calibration, Observatory, Data, and Quality Assurance. The telescope design and component prototyping are expected to be completed in 2011 and 2012, respectively. Array prototypes could exist by 2012/13, and the construction of the full array and partial operation could be started in 2014/15.

References

[1] Aharonian, F., Buckley, J., Kifune, T., Sinnis, G., 2008, Rep. Prog. Phys., 71, 096901
[2] Buckley, J., Byrum, K., Dingus, B. et al., 2008, arXiv:0810.0444 [astro-ph]
[3] Aliu, E. et al. (MAGIC collaboration), 2008, Science, 322, 1221
[4] Albert, J. et al. (MAGIC collaboration), 2008, Astrophys. J., 674, 1037
[5] Abdo, A. A. et al. (Fermi-LAT collaboration), 2009, Astrophys. J. Suppl., 183, 46
[6] Hofmann, W., 2006, arXiv:astro-ph/0603076

Topics in Astroparticle and Underground Physics (TAUP 2009) IOP Publishing
Journal of Physics: Conference Series **203** (2010) 012122 doi:10.1088/1742-6596/203/1/012122

First results of the ANTARES neutrino telescope

J.D. Zornoza for the ANTARES collaboration

IFIC, Ed. Institutos de Investigacion, AC22085, CP46071, Valencia, Spain

E-mail: zornoza@ific.uv.es

Abstract.
The ANTARES neutrino telescope was completed in May 2008 with the installation of its twelfth detector line. The main goal of the experiment is the detection of high-energy neutrinos from extraterrestrial sources. Other searches, like neutralinos, are also possible. Since the connection of the first lines, the detector ANTARES has been providing data smoothly, enabling physics analyses. In this talk we describe the status of the operation of the telescope and its first results.

1. Introduction

Neutrino astronomy is an important tool to complement our knowledge of the universe at high energies. Up to now, most of the information about the cosmos comes from photons and cosmic rays. However, these "traditional" probes have significant drawbacks. On the one hand, they interact with radiation (infrared and cosmic microwave backgrounds) and matter, so their range is limited, in particular at high energies (the mean free range path is $\sim$10 kpc for PeV photons and $\sim$1 Mpc for ZeV protons). Moreover, cosmic rays are also deflected by galactic and extragalactic magnetic fields. On the other hand, neutrinos are neutral and only interact weakly, which means they are a unique probe to study the high-energy universe.

Neutrinos have been predicted in several astrophysical scenarios [1] and could help to explain the origin of cosmic rays. Moreover, they could also be the key to discriminate about the origin of the TeV gamma rays detected from several sources. In some of these sources, the observed gamma rays can be explained by leptonic mechanisms (where no neutrinos are produced) or from neutral pions produced by collisions of accelerated hadrons. In the latter case, neutrinos are expected.

The scientific goals of neutrino astronomy also include the search for dark matter. In particular, if neutralinos are the favored candidate to explain dark matter, they would accumulate in massive objects like the Sun or the Galactic Centre and annihilate. Among the indirect products of these annihilations, there would be neutrinos. This indirect detection is complementary to direct searches, since it would correspond to neutralinos with large masses (above several tens of GeV). Other studies like oscillations or monopoles are also possible.

2. Detector installation and operation

The ANTARES detector consists of a tri-dimensional array of about 900 photomultipliers (encapsulated in glass spheres called optical modules), distributed along 12 lines anchored at 2475 m in the Mediterranean Sea. It is located at (42°48'N, 6°10'E) at 40 km from the French city of Toulon. High energy neutrinos interacting with the water or rock near the detector can produce relativistic muons which would induce the emission of Cherenkov light. The information of the position and time of the hits of this Cherenkov light in the photomultipliers allow the reconstruction of muon track. Above ~10 TeV, the muon and neutrino trajectory are almost collinear.

The first detector line of ANTARES was installed in 2006. By the beginning of 2007, five lines were connected and providing data smoothly. The rest of the lines were installed in the following months, the detector being completed in May 2008 [2].

Several calibration systems are necessary in order to ensure the expected angular resolution (better than 0.3 degrees for neutrino energy above ~10 TeV. First, the time calibration is performed by a high-precision clock which sends a common signal to the front-end electronics in the optical modules. This system is complemented by a set of blue LED beacons which allow to calculate the relative time offsets due to the photomultiplier transit times and the front-end electronics. This system allows a relative time calibration better than 1 ns. Second, an acoustic position system comprised by a set of hydrophones along the lines and in the surroundings of the detector allows to know the position of the optical modules with a precision of 20 cm.

There are two kinds of background in ANTARES. The so-called "optical background" refers to the light produced by the decay of potassium-40 and bioluminescence. The background from potassium-40, which yields a constant rate, is due to the Cherenkov emission induced by the relativistic electrons produced in its decay. There is also a constant contribution from bacteria and other microscopic organisms. In addition to this, there are bursts of bioluminescent produced by macroscopic organisms in the sea.

The other kind of background is produced by cosmic rays. The dominant source of this "physical background" is due to the atmospheric muons interacting with the Earth's atmosphere. There is a second source of physical background which are the atmospheric neutrinos also produced in the cosmic rays interactions in the atmosphere. These neutrinos can traverse the Earth so they are also present when only up-going events are selected and cluster search methods have been developed to look for accumulations.

3. First results

The technical success of ANTARES has made possible the detection of millions of down-going atmospheric muons and hundreds of neutrino candidates in the first months of operation. In figure 3 (left), the distribution of the zenith angles of these data is shown, together with the Monte Carlo predictions. The sky-map with the equatorial coordinates of the upgoing neutrino candidates is shown in figure 3 (right).

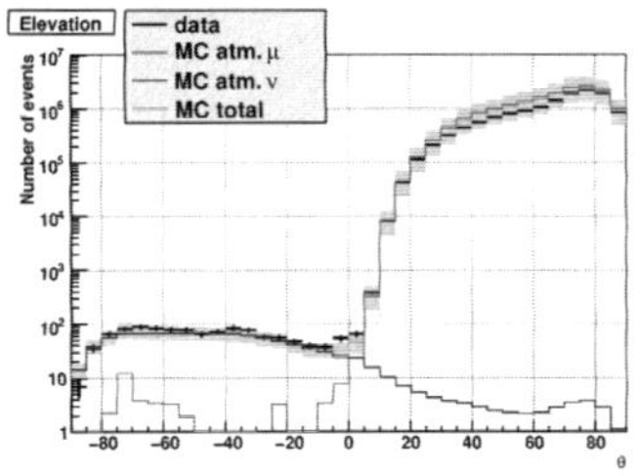

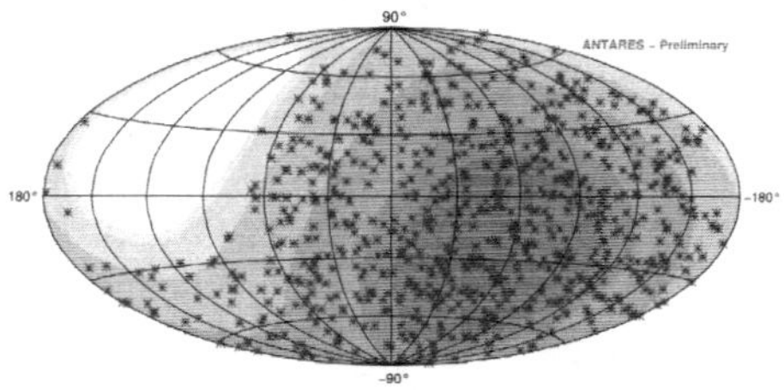

Figure 1. Left: Distribution of zenith angles of selected events in the 2007-2008 data. The plot also shows the expected contribution from atmospheric muons and atmospheric neutrinos. Right: Skymap in equatorial coordinates showing the positions of the upgoing events selected as neutrino candidates among the 2007-2008 sample (the right ascensions have been scrambled to respect the unblinding policy of the collaboration.)

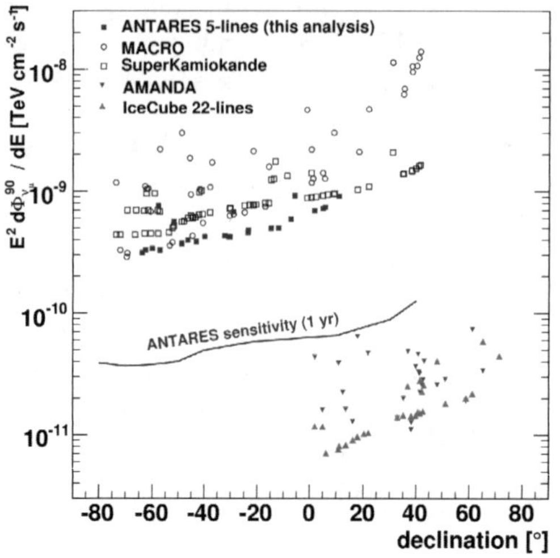

Figure 2. Neutrino flux upper limits at 90% C.L. obtained by ANTARES with data of 2007, compared with the results from other experiments (IceCube, AMANDA, SuperKamiokande and MACRO). The sensitivity of ANTARES for one year with twelve lines is also shown (solid line). The source spectrum assumed in these results is E^{-2}.

The first analysis looking for point sources was made using 2007 data. As mentioned before, five lines were already deployed and connected. The expected angular resolution with this configuration and the selection cuts applied is better than 0.5 degrees (above 10 TeV). The active time corresponding to this selection criteria was 140 days. An cluster search method called "Expectation-Maximization" [3] was used in order to look for accumulation in the sky [4]. Two strategies were followed: a search within a list of 25 potential neutrino sources and a scan in the whole visible sky of ANTARES. None of this searches found any excess above the background. The corresponding flux upper limits are shown in figure 2. These limits are competitive with those set by previous multi-year experiments looking at the Southern Hemisphere.

4. Conclusions

The ANTARES collaboration completed in May 2008 the installation of the largest submarine neutrino telescope. Its final configuration comprises ~900 photomultipliers, but data taking has been taking place since the connection of the first line in 2006. The operation of the detector is successful and has allowed the first physics analysis. With the data corresponding to 2007, when five of the twelve lines were connected, the first search for point sources has been made. With these data, corresponding to 140 active days, flux limits for 25 candidate sources have been set, competitive with the results from other multi-year experiments looking at the Southern Hemisphere.

References

[1] Halzen F and Hooper D, Rep. Prog. Phys. 65 (2002) 1025.
[2] Carr J, for the ANTARES collaboration, J. Phys. Conf. Ser. **136** (2008) 022047.
[3] Aguilar J A and Hernandez-Rey J J, Astrop. Phys. **29** 2 (2008), 117.
[4] Aguilar J A et al. (ANTARES collaboration) *Search for cosmic neutrino point sources with the 5-line ANTARES telescope*, Submitted to Journal of Physics G, preprint arXiv:0909.1262.

Journal of Physics: Conference Series **203** (2010) 012123 doi:10.1088/1742-6596/203/1/012123

The Baikal Experiment - from Megaton to Gigaton

A. Avrorin[a], V. Aynutdinov[a], V. Balkanov[a], I. Belolaptikov[d], D. Bogorodsky[b], N. Budnev[b], I. Danilchenko[a], G. Domogatsky[a], A. Doroshenko[a], A. Dyachok[b], Zh.-A. Dzhilkibaev[a], S. Fialkovsky[f], O. Gaponenko[a], K. Golubkov[d], O. Gress[b], T. Gress[b], O. Grishin[b], A. Klabukov[a], A. Klimov[h], A. Kochanov[b], K. Konischev[d], A. Koshechkin[a], V. Kulepov[f], D. Kuleshov[a], L. Kuzmichev[c], E. Middell[e], S. Mikheyev[a], M. Milenin[f], R. Mirgazov[b], E. Osipova[c], G. Pan'kov[b], L. Pan'kov[b], A. Panfilov[a], D. Petukhov[a], E. Pliskovsky[d], P. Pokhil[a], V. Poleschuk[a], E. Popova[c], V. Prosin[c], M. Rozanov[g], V. Rubtzov[b], A. Sheifler[a], A. Shirokov[c], B. Shoibonov[d], Ch. Spiering[e], O. Suvorova[a], B. Tarashansky[b], R. Wischnewski[e], I. Yashin[c] and V. Zhukov[a]

[a] Institute for Nuclear Research, 60-th October Anniversary pr. 7a, Moscow 117312, Russia
[b] Applied Physics Institute of Irkutsk State University, Gagarin blvd. 20, Irkutsk, Russia
[c] Skobeltsyn Institute of Nuclear Physics MSU, Moscow, Russia
[d] Joint Institute for Nuclear Research, Dubna, Russia
[e] DESY, Zeuthen, Germany
[f] Nizhni Novgorod State Technical University, Nizhni Novgorod, Russia
[g] St.Petersburg State Marine University, St.Petersburg, Russia
[h] Kurchatov Institute, Moscow, Russia

E-mail: bair@nusun.jinr.ru

Abstract. We review the status of the Lake Baikal Neutrino Experiment. Preparation towards a km3-scale Gigaton Volume Detector (GVD) in Lake Baikal is currently a central activity. As an important milestone, a km3-prototype string comprising 6 optical modules and based on a completely new technology, has been installed and was put into operation together with NT200+ in April, 2008. An upgraded version of the prototype string, which comprises 12 optical modules, was put into operation in April 2009. We also present new results from the long-term operation of NT200, including an improved limit on the diffuse astrophysical neutrino flux.

The Baikal Neutrino Telescope NT200 is taking data since 1998 in Lake Baikal. Since 2005, the upgraded 10-Mton scale detector NT200+ is in operation. Detector configuration and performance have been described elsewhere [1, 2, 3, 4, 5]. The most recent milestone of the ongoing km^3-telescope research and development work (R&D) was the installation of a new technology prototype string in spring 2008, as a part of NT200+ [6]. Fig.1 gives a sketch of the current status of the telescope NT200+, including the km^3-prototype string.

In this paper we discuss the R&D activities towards a km^3-scale Baikal telescope and review recent astroparticle physics results from long-term operation of NT200 - an improved limit on a diffuse astrophysical neutrino flux, and upper limit on muon flux from annihilation of WIMPs in the Sun.

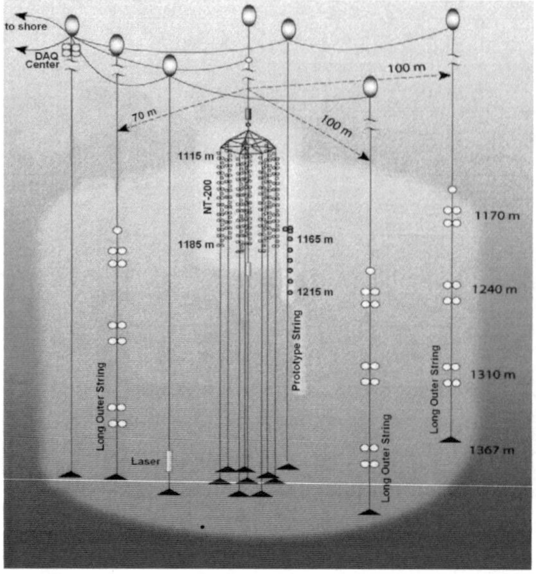

Figure 1. The Lake Baikal neutrino telescope: the compact NT200 (center), 3 long outer strings and the new technology km^3-prototype string deployed in 2008.

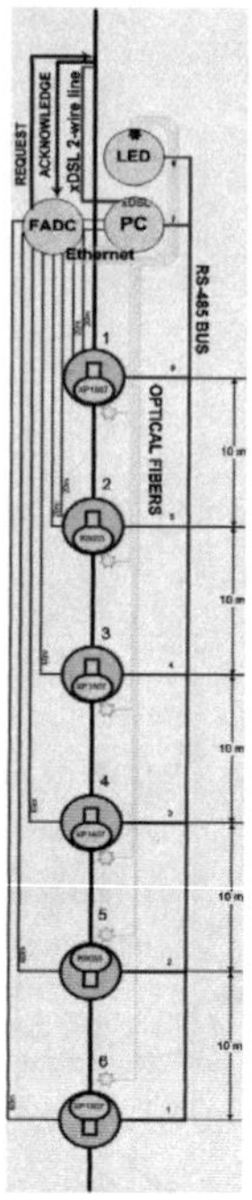

Figure 2. The km^3-prototype string.

1. Towards a km^3 detector in Lake Baikal: the new technology string

The Baikal collaboration follows since several years a R&D program for a km^3-scale neutrino telescope in Lake Baikal. The Baikal km^3-detector will have a relatively flexible structure, which allows for a rearrangement of the main building blocks (clusters), to adapt for requirements of new scientific goals, if necessary. The total number of 2100–2500 optical modules (OMs) will be arranged at 95–100 strings with 22–24 OMs each, and an instrumented length of 350–460 m. Interstring distances will be 60–80 m. The strings are grouped in 12–14 clusters which will form independent arrays. A total volume of 0.4–0.6 km^3 will be instrumented with photo-sensors. The effective area for muons above 10 TeV with an angular resolution of 0.5°–1° is 0.1–1 km^2, and the effective volume for cascade events above 100 TeV with angular resolution $\sim 5°$ is 0.3–0.8 km^3. MC-optimization for the km^3-detector design is going on, as well as studies for optimal trigger technologies.

The basic goals of the prototype string installation are: investigation and in-situ test of basic elements of the future detector (new optical modules, DAQ system and cable system); studies of the basic DAQ/triggering approach for the km^3-detector; comparison of the classical TDC/ADC approach with a FADC-based full pulse shape readout. The design of the prototype string is presented in Fig.2. The string consists of 6 optical modules with photomultipliers R8055 (Hamamatsu) and XP1807 (Photonis). The distance between OMs along the string is 10 m. The preamplified dynode outputs of all 6 PMs are connected through underwater coaxial cables to 200 MHz FADC boards, located in a separate glass sphere. Data from the FADC unit are transmitted via a local Ethernet line to the underwater micro-PC unit. Synchronization of prototype string and telescope data acquisition systems is the same as for the outer strings of NT200+ [3]. Time and amplitude calibration is provided by a string LED flasher located in a separate glass sphere near the FADC and PC units. Light pulses from the flasher are transmitted to each OM via optical fibers with calibrated length. Control and monitoring of OM and LED flasher operation are provided by the string PC unit via a RS-485 underwater bus. Two basic modes of string operation are available: joint operation with NT200+ and standalone operation.

Below, we present some preliminary results of tests of the string response to LED flasher and laser calibration sources in the standalone mode.

The accuracy of the pulse time measurement with the 200 MHz FADC was studied by using two pulses of LED flasher, which were delayed on 497 ns. We find the average delay values, calculated from the FADC data, to be close to the nominal value of 497 ns with average RMS of 1.8 ns. A basic function of the LED flasher is the direct measurement of the relative time shifts and a direct amplitude calibration of the spectrometric channels. Time shifts were determined as the time difference between pulses on different FADC channels for simultaneously detected LED flashes. Two pulses with a delay of about 500 ns are produced by the LED flashers during the amplitude calibration. The first LED provides a light pulse at single electron level. The second LED has a value significantly larger than the PM noise amplitude and is used as a trigger for the phototube dark noise suppression. An average value of a single electron distribution is used as amplitude calibration coefficient for the given FADC channel. The results of the time shift and amplitude calibration were used for the analysis of the prototype string calibration with the NT200+ laser light source. From a preliminary analysis of the laser data we conclude that the pulse arrival times are in good agreement with the expected values.

2. Recent physics results from NT200

A possible signal from dark matter WIMP annihilations in the Sun would reveal as an excess of upward going muons over atmospheric neutrinos arriving from the direction of the Sun. We have used the experimental data of NT200 taken between April, 1998 and March, 2003 (1008 live days). No indication for excess muons was found. The upper limit at 90% c.l. on an additional muon flux from the Sun is obtained as function of the WIMP mass for b anti-b (soft channel) and W^+W^- (hard channel) neutrino energy spectrum. For the WIMP masses greater than 500 GeV the limit depends weakly on the WIMP mass and is about $F < 3 \times 10^3$ km^{-2}yr^{-1} [7].

The BAIKAL survey for high energy neutrinos searches for bright cascades produced at the neutrino interaction vertex in a large volume around the telescope. A full cascade reconstruction algorithm (for vertex, direction, energy) was applied to the 1038 live days data taken with NT200 in 1998-2002 years [8]. Cuts were then placed on this reconstructed cascade energy to select neutrino events. No neutrino events were found. Upper limits on diffuse neutrino fluxes predicted by several theoretical models of AGN-like astrophysical sources were derived [8]. For an E^{-2} neutrino spectrum the 90% c.l. upper limit on the neutrino flux of all flavors obtained with the Baikal neutrino telescope NT200 is: $E^2\Phi < 2.9 \times 10^{-7}\text{cm}^{-2}\text{s}^{-1}\text{sr}^{-1}\text{GeV}$.

Acknowledgments

This work was supported in part by the Russian Ministry of Education and Science, by the German Ministry of Education and Research, by the Russian Foundation for Basic Research (grants 08-02-00432-a, 09-02-10012-k, 07-02-00791, 08-02-00198, 09-02-10001-k, 09-02-00623-a, 09-02-12295), by the grant of the President of Russia NSh-321.2008-2 and by the program "Development of Scientific Potential in Higher Schools" (projects 2.2.1.1/1483, 2.1.1/1539, 2.2.1.1/5901).

References

[1] Belolaptikov I et al 1997 *Astropart. Phys.* **7** 263
[2] Aynutdinov V et al 2006 *Astropart. Phys.* **25** 140
[3] Aynutdinov V et al 2006 *Nucl. Instrum. Methods* A **567** 433
[4] Aynutdinov V et al 2008 *Nucl. Instrum. Methods* A **588** 99
[5] Aynutdinov V et al 2009 *Nucl. Instrum. Methods* A **602** 14
[6] Aynutdinov V et al 2009 *Nucl. Instrum. Methods* A **602** 227
[7] Aynutdinov V et al 2009 *Proc. Int. Cosmic Ray Conf. (Lodz)* in press (*Preprint* astro-ph/0909.5589)
[8] Avrorin A et al 2009 *Astronomy Lett.* **35** 651

Journal of Physics: Conference Series **203** (2010) 012124 doi:10.1088/1742-6596/203/1/012124

KM3NeT: a cubic-kilometre scale deep sea neutrino telescope in the Mediterranean Sea

A. Margiotta on behalf of the KM3NeT Consortium[1]

Dipartimento di Fisica dell'Università and INFN Bologna

Viale C. Berti-Pichat,6/2 – 40127 Bologna (Italy)

e-mail: Annarita.Margiotta@bo.infn.it

Abstract. The European KM3NeT consortium is presently carrying out R&D activities towards the construction of a cubic-kilometre scale deep-sea Cherenkov detector for high-energy astrophysical neutrinos. Located in the Mediterranean Sea, KM3NeT will complement the field of view of the IceCube neutrino telescope currently under construction at the South Pole. KM3NeT will allow for observations of the Galactic Centre and most of the Galactic plane, where H.E.S.S. has discovered several TeV-gamma sources, some of which are candidate neutrino sources. The three Mediterranean pilot projects, ANTARES, NEMO and NESTOR, are actively involved in KM3NeT, which is declared a high-priority project by the astronomy (ASTRONET Roadmap) and astroparticle (ASPERA Roadmap) scientific communities and has been included in the 2006/08 Roadmaps of the European Strategy Forum for Research Infrastructures (ESFRI). Additionally, a number of oceanography, geophysics, biology, environmental sciences and geology institutions are active in the KM3NeT consortium with the aim of helping create a future multidisciplinary deep-sea research infrastructure in the Mediterranean Sea.

1. Introduction

The discovery of cosmic rays dates back more than a century, but the most fundamental questions on their origin and acceleration mechanisms remain unanswered.

Recent observations of high energy gamma rays from several Supernovae Remnants make these objects plausible sources of high energy cosmic rays [1], though the smoking gun evidence is still missing and can be provided by the detection of high energy neutrinos. Thanks to their weakly interacting nature, neutrinos represent the ideal probe to carry information from the core of very distant astrophysical objects. On the other hand the same property, together with the low expected fluxes, makes them difficult to detect and requires huge detectors. A possible solution is the exploitation of sea water and/or polar ice, equipped with a lattice of photo-sensors which are able to detect the Cherenkov radiation emitted by the charged particles, mainly muons, produced in the neutrino interactions, thus allowing for the reconstruction of their trajectories and energies [2].

A cubic-kilometre detector is currently under construction in the deep South Polar ice (IceCube) [3]. A km^3 detector in the Mediterranean Sea would complement the field of view of IceCube, and

[1] A complete list of the authors is available here: http://www.km3net.org

would have the major advantage of looking directly at the Galactic Centre which is considered an extremely interesting region to explore, see Fig. 1.

The construction of this detector requires a joint effort of the national scientific communities. A consortium of 40 institutes from 10 European countries, including all the groups involved in the Mediterranean neutrino telescope projects: ANTARES, NEMO, NESTOR [4], the KM3NeT [5] Design Study (DS), was created in 2006 and funded in the EU 6th framework programme. Its main goal is to develop a cost-effective design for the construction of a neutrino telescope with at least 1 km^3 of instrumented volume. In April 2008 the consortium published the KM3NeT Conceptual Design Report (CDR) [6], which describes the scientific objectives, and the concepts behind design, construction and operation of the KM3NeT Research Infrastructure. The DS will end in October 2009, the completion of the Technical Design Report (TDR) is scheduled for December 2009. The TDR will define the technological solutions for the construction of a km^3 neutrino telescope in the Mediterranean Sea.

In March 2008 the Preparatory Phase (PP) project started. It is funded under the EU 7th framework programme. The PP will address legal, governance, funding and strategic issues of the construction of the KM3NeT in the Mediterranean, including site selection. The PP foresees also prototyping work, with a view to start the telescope construction in 2013.

The KM3NeT research facility will host also an earth and marine sciences observatory.

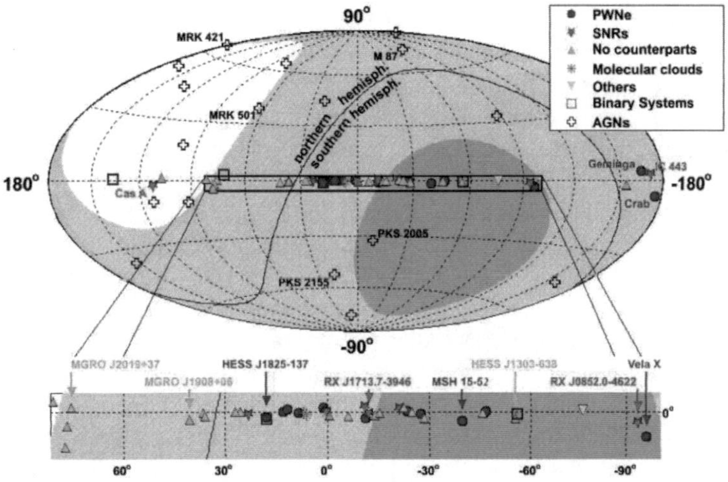

Fig.1 - Sky coverage of a neutrino telescope in the Mediterranean Sea: dark (light) blue regions are visible at least 75%(25%) of the time.

2. Design goals and technical solutions.

The CDR contains the detailed description of all the requirements for a km^3 neutrino telescope in the Mediterranean Sea. This neutrino telescope is planned as a long-term observatory, with a data taking period longer than 10 years. This requires a high reliability of the critical components of the telescope. The infrastructure will be designed to "survive" in the deep sea, under high pressure and in the aggressive salt water environment.

The neutrino telescope effective area increases with energy due to the features of neutrino interactions and the detection technique, while the energy spectra of neutrino fluxes are expected to fall with energy. This requires the optimisation of detector sensitivity in the energy range 1 TeV –1 PeV.

The reduced light scattering in sea water represents a major advantage of a Mediterranean Sea neutrino telescope. To exploit this characteristic good positioning and time calibration systems are required to reduce the detector effects on the angular resolution when reconstructing muons. The goal is an angular resolution of $0.1°$ for neutrino energies $\geq$ 30 TeV.

A full cost of 200-250 M€ is foreseen for KM3NeT.

Journal of Physics: Conference Series **203** (2010) 012124 doi:10.1088/1742-6596/203/1/012124

The technical solutions investigated have been inspired mainly by the pilot projects (ANTARES, NEMO, NESTOR), and by new developments conducted partly in collaboration with industry.

Though there is a general convergence on a unique design, several technical options are still available.

- *Photo-sensor.* The rejection of optical background, consisting mainly of single photoelectrons, requires a local set of photomultipliers (PMTs) for an efficient 1-2 photon separation. This can be realised with a multi-PMT optical module (OM) option, 2 large 8"/10" PMTs or several 3" PMTs, housed in pressure resistant glass spheres, or by local groups of more single-PMT OMs.
- *Data acquisition and transmission.* The concept of "all-data-to-shore" will be applied: all PMT signals above a given threshold will be sent to shore. This implies an overall data rate of ~100-300 Gb/s for a typical threshold of 1/3 of a single photo-electron. On shore, a computer farm will reduce this rate by about 5 orders of magnitude, by online filtering. Optical fibres will be used for the communication between shore and the detection units.
- *Mechanical structures.* Simulations indicate that a horizontal spacing of a few metres between OMs on the same storey improves detector sensitivity. A possible mechanical solution is represented by a tower-like structure, made by 6-metre horizontal bars carrying 3 groups of 2 single-PMT OMs or some single multi-PMT OMs. An alternative solution can be a slender string-like structure with multi-PMT OMs.
- *Deep-sea infrastructure.* The deep-sea infrastructure consists of one or several main electro-optical cables connecting the shore station to a deep-sea network made of junction boxes and secondary cables for electrical power and data distribution. The KM3NeT infrastructure will also host nodes for earth and marine science instrumentation.
- *Calibration.* An acoustic positioning system will measure the OM positions with a precision better than 40 cm using a network of transponders and receivers. The stability of the timing calibration of the OM signals will be monitored by synchronisation signals sent from shore, and using LED/Laser flashers installed in the detector. The absolute orientation of the telescope can be evaluated using acoustic triangulation coupled to GPS positioning and precision pressure probes. An independent way to check the absolute pointing is the measurement of coincident down-going events between a sea surface detector and the deep-sea telescope. Finally, the observation of the Moon shadowing effect on cosmic rays can be used as a cross-check of the absolute orientation of the apparatus, down to the level of a fraction of a degree.
- *Candidate sites.* The Mediterranean Sea offers optimal conditions, on a worldwide scale, to host an underwater neutrino telescope. A few sites have been identified to be suitable for hosting a future KM3NeT detector. Several criteria can help to define an optimal site:
  - proximity to the coast to ease deployment and reduce the expense of the power and signal cable connections to shore;
  - a sufficient depth to reduce background from atmospheric muons;
  - good optical properties of water;
  - low level of bioluminescence;
  - low rates of biofouling and sedimentation on optical surfaces;
  - stable low sea current velocities.

  The ANTARES, NEMO and NESTOR Collaborations have performed extensive studies to characterise their sites. The final decision will take into account also technological and budgetary input. A multi-site option has been proposed as well.

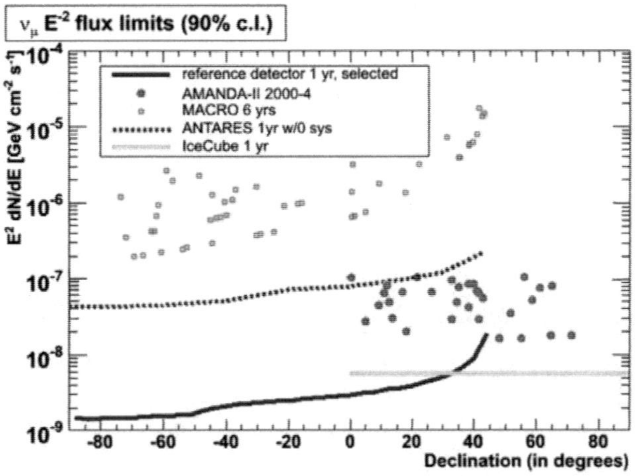

Fig.2 - Expected point source sensitivity for a KM3NeT ν_μ typical detector for one year data taking, assuming an E^{-2} energy spectrum (lower solid line).

3. Earth and marine science observatory

The KM3NeT research infrastructure will serve also as a platform for a wide spectrum of earth and marine sciences research. It will enable continuous data to be collected without interruption over long periods. Permanent connections from the deep sea to the shore for powering and reading out instruments are rare and have a great value. The only cabled sites at present in Europe are those of the neutrino telescope pilot projects, which already feature extensive programmes in these fields.

4. Conclusions

The experience of the pilot projects demonstrated the feasibility of a cubic kilometre scale underwater high-energy neutrino telescope.

By the end of this year the KM3NeT European Consortium will publish the Technical Design Report with the definition of the technological solutions for the construction of such a detector.

In March 2008 KM3NeT entered a Preparatory Phase (PP) which aims at the definition of legal, financial, governance and strategic issues that need to be settled before the start of construction, and foresees also a prototyping activity. The KM3NeT neutrino telescope in the Mediterranean Sea will complement IceCube in its field of view, exceeding its sensitivity (Fig. 2), and looking directly at the Galactic centre. The KM3NeT detector construction is expected to start in 2013.

Acknowledgments

The KM3NeT project is supported by the EU in the 6th framework programme under contract no. 011937 and in the 7th framework programme under contract no. 212525.

References

[1] F. Aharonian et al. (H.E.S.S. Collab.), Astron. Astrophys. 449 (2006) 223; F. Aharonian et al. (H.E.S.S. Collab.), Astrophys. Journal Lett. 695 (2009) L40.
[2] M.A.Markov and I.M.Zheleznykh, Nucl. Phys. 27 (1961) 385
[3] R. Abbasi et al. (IceCube Collab.), Phys. Rev. D79 (2009) 062001.
[4] M. Ageron et al., (ANTARES Collab.), Astropart. Phys. 31(2009) 277; E. Migneco et al. (NEMO Collab.), Nucl. Instrum. Meth. A588(2008)111; G. Aggouras et al. (NESTOR Collab.), Nucl. Instrum. Meth. A 552(2005)420.
[5] http://www.km3net.org/
[6] http://www.km3net.org/CDR/CDR-KM3NeT.pdf

Topics in Astroparticle and Underground Physics (TAUP 2009) IOP Publishing
Journal of Physics: Conference Series **203** (2010) 012125 doi:10.1088/1742-6596/203/1/012125

Limit on the diffuse flux of ultra high energy neutrinos using the Pierre Auger Observatory

Pierre Billoir[1], **for the Auger Collaboration**
[1]LPNHE CNRS/Univ. P. et M. Curie/Univ. D. Diderot, 4 place Jussieu, Paris, France
E-mail: billoir@lpnhe.in2p3.fr

Abstract. The surface array of water-Cherenkov detectors of the Pierre Auger Observatory is sensitive to neutrinos of energies around or above 1 EeV. All flavours interact through charged and neutral current interactions in the atmosphere (down-going), and tau neutrinos may also induce the "earth skimming" mechanism (up-going or crossing mountains). Both types of interaction may produce nearly horizontal atmospheric showers close to the ground, triggering the surface array; they may be distinguished from normal showers thanks to the broad time structure of the signals in the water tanks. Using data collected from january 2004 to february 2009, we use an analysis based on down-going neutrinos to obtain a limit on the all-flavour diffuse neutrino flux. We also update the previous limit for up-going tau neutrinos. Sources of possible backgrounds and systematic uncertainties are discussed.

1. Introduction

Neutrinos of energy in the EeV range are predicted by essentially all models of Ultra High Energy Cosmic Rays (UHECR), as the result of meson decays. In the standard scenario of acceleration mechanism, the mesons are produced in the interaction of the CRs, either in the medium surrounding their source, or along their propagation through the background radiation fields (GZK mechanism) [1, 2]. In the so-called "top-down" mechanisms, they are copiously produced by the decay of very massive exotic objects, with a hard spectrum [1].

The surface detector (SD) of the Pierre Auger Observatory (PAO) [3] is able to detect UHE neutrinos in the EeV range and above. In particular, nearly horizontal tau neutrinos interacting within the Earth produce a τ lepton which can emerge and decay in the lower atmosphere ("earth-skimming" mechanism [4]), producing a shower with a lateral expansion down to the ground. A limit on the diffuse flux of UHE ν_τ was already obtained through this technique using data collected from Jan 04 to Aug 07 [5]. This limit is updated including the new data (up to Feb 09) with the same analysis.

With similar criteria the SD of the PAO may detect nearly horizontal "down-going" neutrinos interacting deeply in the atmosphere [6]. Using the first part of the data (Jan 04 to Oct 07) as a training sample, discriminating variables have been tuned for zenith angles above 75 degrees, and then applied to the rest of the data (Nov 07 to Feb 09) to obtain a limit on the diffuse flux of all flavours.

2. Discriminating neutrinos in data

The background of nucleonic showers is much larger than the expected rate of neutrino events (typically one event per year or less). However, UHE protons or nuclei interact in the upper atmosphere, and at zenith angles larger than 75 degrees the shower crosses more than 2000 g/cm^2 before reaching the ground. At this stage the electromagnetic cascade is fully extinguished, and only high energy muons can hit the ground: in this "old" stage, all particles (including the products of the decay and the radiative interactions of the muons) are concentrated into a thin and flat shower front moving practically at the speed of light. On the contrary, neutrinos interact proportionally to the density, and a shower induced in the lower atmosphere can reach the ground in a "young" stage: the arrivals of the particles into the SD detector, especially the low energy photons and electrons, are spread over a time which is much larger than the sampling time of the FADC traces (25 ns) and the decay time of the Cherenkov light in a tank (70 ns). The first triggered tanks have broader signals, because they see an earlier stage of the electromagnetic cascade, and because the divergence of the photons and electrons favours late arrivals in the tanks under the shower core. Requesting several detectors to have signals significantly broader than the one of a single particle provides a powerful discrimination against the background of nucleonic showers. In [7] several variables describing the time structure of the signals are defined and their discriminating power is discussed.

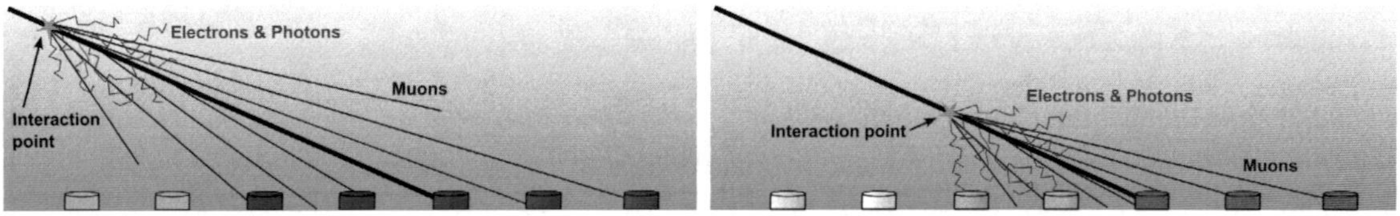

Figure 1. Left panel: sketch of an inclined shower induced by a hadron interacting high in the atmosphere. The EM component is absorbed and only the muons reach the detector. Right panel: deep inclined shower. Its early region has a significant EM component at the detector level.

Detailed simulations of UHE neutrinos of all flavours interacting in the lower atmosphe were produced, in both charged current (CC) and neutral current (NC) modes, using Herwig [8] for the first interaction and AIRES [9] for the shower development. "Double bang" (showers produced by a CC interaction of ν_τ followed by the decay in flight of the τ lepton) were generated using TAUOLA [10] to compute the decay products, to be reinjected in AIRES as additional primary particles. In all modes, a hadronic jet is produced, with, in average, a small fraction of the initial neutrino energy, which induces a mixed shower (hadronic+electromagnetic components). In the CC ν_e mode, the energy of the electron is fully transmitted to the electromagnetic shower. In the CC ν_τ mode, a part of the τ energy is given to a secondary shower if it decays. In NC or CC ν_μ modes, only the energy of the hadronic jet shower is visible.

As the down-going neutrino showers are expected to be seen over a wider range of zenith angle than the up-going ones, the criteria of selection were modified and new discriminating variables were designed. Data collected between Jan 04 and Oct 07, excluding a few periods of unstable data acquisition was used as a training sample (equivalent to $\sim$ 1.2 years of the full SD array) to optimize the discrimination of simulated neutrino showers from real data. In events triggering the SD array with at least 4 stations, the FADC traces are first cleaned to remove parasitic signals produced by accidental particles; ambiguous traces are rejected. This sample is then searched for very inclined events with an elongated pattern on the ground, in the same way as in [12]. The first condition is obtained with a cut on the average of the speeds (distance/difference of starting time) between pairs of triggered stations: $< V > \leq 0.313$ m/ns, as expected for a planar front with $\theta \geq 75°$, allowing for some spread due to fluctuations; the

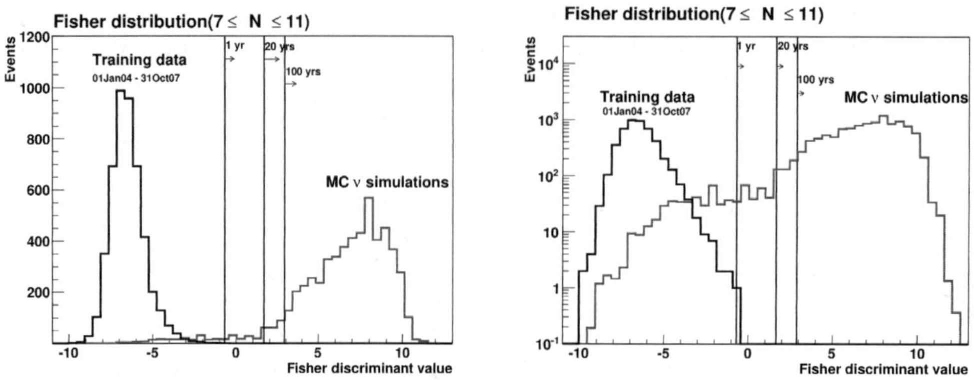

Figure 2. Distribution of the Fisher discriminant (see text for details) in linear (left) and logarithmic (right) scale for real data in the training period (1 Jan 04 - 31 Oct 07) and Monte Carlo simulated down-going neutrinos for events with multiplicity $7 \leq N \leq 11$.

r.m.s. is requested to be less than 0.08 m/ns. The second condition is $L/W > 3$, where L and W are the length and the width of the footprint, weighted by the integrated signals in the tanks (see [12] for the precise definition). The sample of inclined events is searched for "young" showers, using the "Area-over-Peak" ratio (AoP) of the traces (integrated signal over peak value) . A Fisher discriminant method is applied to the AoP's of the earliest 4 tanks, their squares, their product, and a global early-late asymmetry parameter. To account for significant differences between small and large tank multiplicities, the method is applied separately on 3 subsamples: $4 \leq N \leq 6$, $7 \leq N \leq 11$ and $N \geq 12$. Distributions of these observables in real and simulated data are shown in [7]. In each subsample, the Fisher method provides a linear combination f of the variables which maximizes the separation between the real data and the simulated ν showers. Fig. 2 shows that a clear separation is achieved. In each subsample a value f_{cut} is defined such that the exponential extrapolation of the distribution of the training sample predicts less than 1 event per 20 years with $f > f_{cut}$. This cut keeps most of the neutrinos interacting in the lower atmosphere and satisfying the trigger condition.

3. Exposure and limit on the flux of UHE neutrinos

The same selection procedure and cut on f was applied on the data collected from Nov 07 to Feb 09 ($\sim$ 0.8 year of the full SD array). No neutrino candidate was found.

The exposure of the SD array is calculated by folding the geometrical aperture, the interaction probability and the identification efficiency, and integrating over time accounting for the variable array configuration and the instabilities of the data taking. The triggering and selection efficiency is evaluated from simulated data as a function of the flavour of the neutrino, its energy, its zenith angle, the type of interaction, the depth of the interaction, the position in the instantaneous configuration of the array. The exposure is summed over all parameters and all flavours (assuming equipartition).

Several sources of systematic uncertainties have been taken into account and their effect on the exposure evaluated. We tentatively assign a $\sim$ 20% systematic uncertainty due to the neutrino-induced shower simulations and the hadronic model (SIBYLL 2.1 vs QGSJETII.03). Another source of uncertainty comes from the neutrino cross section. Using [15] we estimate a systematic uncertainty of $\sim$ 10%. The topography around the Southern Site of the Pierre Auger Observatory enhances the flux of secondary tau leptons. In this work we neglected this effect. Our current simulations indicate that including it will improve the limit by roughly $\sim 15-20\%$.

Finally, assuming a differential neutrino flux $f(E) = k/E^2$ we obtain a 90% C.L. limit on the single flavour flux using down-going showers: $k < 3.2 \times 10^{-7}$GeV cm^{-2} s^{-1} sr^{-1}. The updated limit based on up-going tau neutrinos using the data up to feb 09 with the same analysis as in [12] is $k < 4.7^{-2.5}_{+2.2} \times 10^{-8}$ GeV cm^{-2} s^{-1} sr^{-1},where the upper/lower values correspond to best/worse scenario of systematics [12]. These limits are plotted on Fig.4, together with the limits in differential format to show how the sensitivity of the Pierre Auger Observatory depends on the initial neutrino energy.

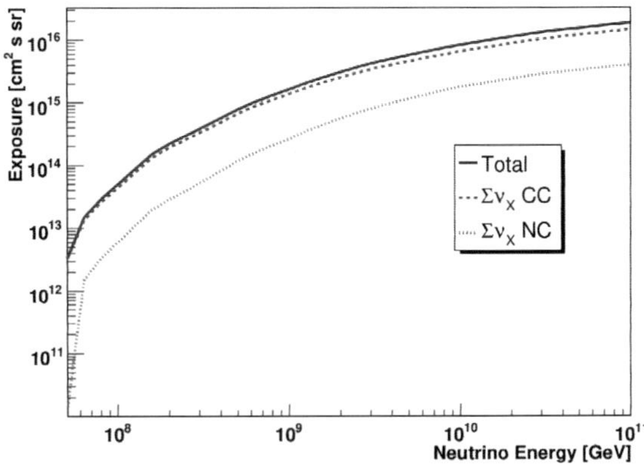

Figure 3. Exposure of the SD array to down-going neutrinos in the search period (1 Nov 07 - 28 Feb 09).

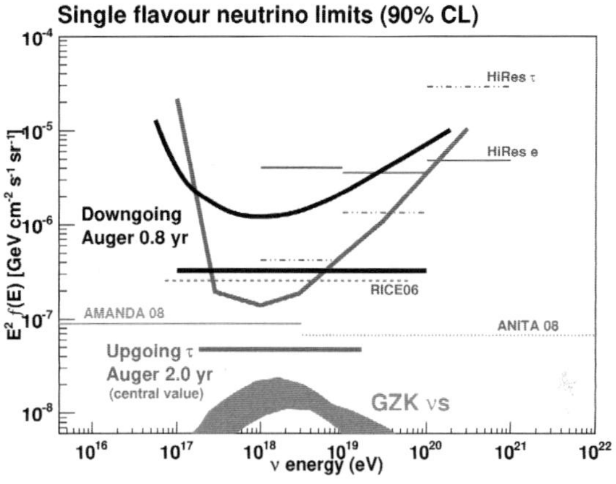

Figure 4. Differential and integrated upper limits (90% C.L.) for a diffuse flux of down-going ν and up-going ν_τ. Limits from other experiments [14] are also plotted. A theoretical flux for GZK neutrinos from [2] is shown.

References

[1] F. Halzen *et al.* Rep. Prog. Phys. **65**, 1025 (2002); P. Bhattacharjee *et al.* Phys. Rep. **327**, 109 (2000).
[2] R. Engel *et al.* Phys. Rev. D **64**, 093010 (2001); D. Allard *et al.*, JCAP **9**, 5 (2006); L. Anchordoqui *et al.*, Phys. Rev. D **76**, 123008 (2007).
[3] J. Abraham *et al.* [Pierre Auger Collaboration] Nucl. Instr. and Meth. A **523**, 50 (2004).
[4] A. Letessier-Selvon, Pro. AIP Conf **566** (2001), 157; D. Fargion, Astrophys. J. **570** (2002), 909; X. Bertou *et al.*, Astropart. Phys **17** (2002), 183.
[5] J. Abraham *et al.* [Pierre Auger Collaboration], Phys. Rev. Lett. **100**, 211101 (2008).
[6] K. S. Capelle *et al.*; *Astropart. Phys.* **8** (1998) 321; P. Billoir *et al.* Procs. of NOW 2006, Lecce, Italy.
[7] D. Gora [Pierre Auger Collaboration], Procs. 31th ICRC 2009, Lódź, Poland.
[8] G. Corcella *et al.*, Herwig 6.5, JHEP 0101 (2001).
[9] S. Sciutto, AIRES. http://www.fisica.unlp.edu.ar/auger/aires
[10] S. Jadach *et al.*, Comput. Phys. Commun. **76**, 361 (1993).
[11] J. Alvarez-Muñiz [Pierre Auger Collaboration], Procs. 30th ICRC 2007, 607, Mérida, México. I. Valiño (PhD thesis) Univ. de Santiago de Compostela, ISBN: 9788497509664, 2008.
[12] J. Abraham *et al.* [Pierre Auger Collaboration], Phys. Rev. D **79**, 102001 (2009).
[13] R. Fisher, Ann of Eugenics **7**, 179 (1936).
[14] Ackermann *et al.* [IceCube], Astrophys. J. 675, 1014-1024 (2008); I. Kravchenko *et al.*, Phys. Rev. D **73**, 082002 (2006) [RICE]; P.W. Gorham *et al.* [ANITA], arXiv:0812.2715v1; R.U. Abbasi *et al.* [HiRes], Astrophys. J. 684, 790-793 (2008); K. Martens [HiRes], arXiv:0707.4417.
[15] A. Cooper-Sarkar and S. Sarkar, JHEP **0801**, 075 (2008).

Journal of Physics: Conference Series **203** (2010) 012126 doi:10.1088/1742-6596/203/1/012126

Evolution of the cosmic ray anisotropy above 10^{14} eV

Piera L. Ghia, for the EAS-TOP Collaboration[1]

IFSI-INAF, Torino, LNGS-INFN, Assergi, Italy and IPN-CNRS, Orsay, France

E-mail: piera.ghia@lngs.infn.it

Abstract. The study of the evolution of the amplitude and phase of the cosmic ray anisotropy into the energy region $10^{14} - 10^{16}$ eV can provide a significant tool for the understanding of the "knee" of the primary spectrum. We extend here the EAS-TOP measurement performed at $E_0 \approx 10^{14}$ eV, to higher energies by using the full data set (8 years). Results derived at about 10^{14} and $4 \cdot 10^{14}$ eV are compared and discussed. Hints of increasing amplitude and change of phase above 10^{14} eV are reported.

1. Introduction

The "knee" observed at $E_0 \approx 3 \cdot 10^{15}$ eV represents a remarkable feature of the energy spectrum of cosmic rays and its characterization is therefore a main tool for the understanding of the galactic radiation. Composition studies have shown that it is related to the steepening of the lightest primaries (protons, helium, CNO) spectra [1, 2].

Such effect can be due, on the one side, to energy limits of the acceleration process at the source, namely diffusive shock acceleration in supernova remnants, generally considered to be the sources of galactic cosmic rays (see e.g. [3]). On the other side, this feature has been possibly explained in terms of a change in the cosmic ray propagation properties inside the Galaxy [4, 5]. Galactic propagation is described through diffusion models whose parameters have been obtained through composition studies (mainly from the ratio of secondary to primary nuclei) at energies well below 1 TeV (see e.g. [6, 7]). The diffusion coefficient, D, is found to increase with magnetic rigidity ($D \propto R^{0.6}$, or $D \propto R^{0.3}$ for models including reacceleration). However, no confirmation, and no information has till now been obtained at higher energies, where the main observable is represented by the large scale anisotropy in the cosmic rays arrival directions, that is known to be strictly related to the diffusion coefficient (see e.g. [8]). The study of the evolution of the anisotropy in the "knee" energy region can therefore provide a

[1] The EAS-TOP Collaboration: M. Aglietta[1,2], V.V. Alekseenko[3], B. Alessandro[2], P. Antonioli[4], F. Arneodo[5], L. Bergamasco[2,6], M. Bertaina[2,6], R. Bonino[1,2], A. Castellina[1,2], A. Chiavassa[2,6], B. D'Ettorre Piazzoli[7], G. Di Sciascio[7,8], W. Fulgione[1,2], P. Galeotti[2,6], P.L. Ghia[1,5,9], M. Iacovacci[7], G. Mannocchi[1,2], C. Morello[1,2], G. Navarra[2,6], O. Saavedra[2,6], A. Stamerra[6,10], G.C. Trinchero[1,2], S. Valchierotti[2,6], P. Vallania[1,2], S. Vernetto[1,2], C. Vigorito[2,6]. [1]Istituto di Fisica dello Spazio Interplanetario, INAF, Torino, Italy, [2]Istituto Nazionale di Fisica Nucleare, Torino, Italy, [3]Institute for Nuclear Research, AS Russia, Baksan Neutrino Observatory, Russia, [4]Istituto Nazionale di Fisica Nucleare, Bologna, Italy, [5]Laboratori Nazionali del Gran Sasso, INFN, Assergi (AQ), Italy, [6]Dipartimento di Fisica Generale dell'Università, Torino, Italy, [7]Dipartimento di Scienze Fisiche dell'Università and INFN, Napoli, Italy, [8]Presently at Istituto Nazionale di Fisica Nucleare, Roma Tor Vergata, Italy, [9]Institut de Physique Nucleaire, CNRS, Orsay, France, [10]Presently at Dipartimento di Fisica dell'Università and INFN, Pisa, Italy.

significant test of the diffusion models, and a valuable insight for the discrimination between the two possible explanations of the spectral steepening.

At $E_0 \approx 10^{14}$ eV the EAS-TOP results [9] demonstrated that the main features of the anisotropy (i.e. of cosmic ray propagation) are similar to the ones measured at lower energies ($10^{11} \div 10^{14}$ eV), both with respect to amplitude ($(3 \div 6) \cdot 10^{-4}$) and phase ($(0 \div 4)$ h LST). At higher energies the limited statistics does not allow to draw any firm conclusion (see [10], for an extended bibliography). We present here the EAS-TOP measurement based on the full data-set and we extend the analysis to about $4 \cdot 10^{14}$ eV.

2. The experiment and the analysis

The EAS-TOP Extensive Air Shower array was located at Campo Imperatore (2005 m a.s.l., lat. 42° 27′ N, long. 13° 34′ E, INFN Gran Sasso National Laboratory). The electromagnetic detector [12] consisted of 35 modules of scintillator counters, 10 m^2 each, distributed over an area of about 10^5 m^2. The trigger was provided by the coincidence of any four neighbouring modules (threshold $n_p \approx 0.3$ m.i.p./module), the event rate being $f \approx 25$ Hz.

The data under discussion have been collected between January 1992 and December 1999 for a total of 1431 full days of operation. To select different primary energies, a cut is applied to the events based on the number of triggered modules. Events with at least 4 (12) triggered modules correspond to an average primary energy of 1.1×10^{14} eV (3.7×10^{14} eV), evaluated for primary protons and QGSJET01 hadron interaction model in CORSIKA [13].

For the analysis of the anisotropy, we adopt a method based on the counting rate differences between East-ward and West-ward directions, that allows to remove counting rate variations of atmospheric origin. The events used in the analysis are the ones with azimuth angle ϕ inside $\pm 45°$ around the East and West directions, and zenith angle $\theta < 40°$. The number of collected events in the East+West sectors is 1.5×10^9 (1.7×10^8) at 1.1×10^{14} eV (3.7×10^{14} eV). The difference between the number of counts measured from the East sector, $C_E(t)$, and from the West one, $C_W(t)$, at time t in a fixed interval ($\Delta t = 20$ min), is related to the first derivative of the intensity $I(t)$ as: $\frac{dI}{dt} \simeq D(t) = \frac{C_E(t) - C_W(t)}{\delta t}$ where δt is the average hour angle between the vertical and each of the two sectors (1.7 h in our case). The harmonic analysis is performed on the differences $D(t)$; the amplitudes and phases of the variation of $I(t)$ are obtained through the integration of the corresponding terms of the Fourier series [14].

3. Results

The harmonic analysis has been performed in solar, sidereal and anti-sidereal time. For the two different primary energies, the reconstructed amplitudes and phases of the first and second harmonics are shown in table 1, together with their Rayleigh imitation probabilities (P).

E_0[eV]	A^I_{sol} 10^4	ϕ^I_{sol}[h]	P^I_{sol}(%)	A^I_{sid} 10^4	ϕ^I_{sid}[h]	P^I_{sid}(%)	A^I_{asid} 10^4	ϕ^I_{asid}[h]	P^I_{asid}(%)
$1.1 \cdot 10^{14}$	2.8 ± 0.8	6.0 ± 1.1	0.2	2.6 ± 0.8	0.4 ± 1.2	0.5	1.2 ± 0.8	23.9 ± 2.8	32.5
$3.7 \cdot 10^{14}$	3.2 ± 2.5	6.0 ± 3.4	44.1	6.4 ± 2.5	13.6 ± 1.5	3.8	3.4 ± 2.5	22.3 ± 3.2	39.7
	A^{II}_{sol} 10^4	ϕ^{II}_{sol}[h]	P^{II}_{sol}(%)	A^{II}_{sid} 10^4	ϕ^{II}_{sid}[h]	P^{II}_{sid}(%)	A^{II}_{asid} 10^4	ϕ^{II}_{asid}[h]	P^{II}_{asid}(%)
$1.1 \cdot 10^{14}$	1.4 ± 0.8	7.0 ± 1.2	21.6	2.3 ± 0.8	6.3 ± 0.7	1.6	0.6 ± 0.8	-	75.5
$3.7 \cdot 10^{14}$	1.7 ± 2.5	-	79.4	1.5 ± 2.5	-	83.5	1.2 ± 2.5	-	89.1

Table 1. *Results of the analysis of the first (amplitude A^I, phase ϕ^I, and Rayleigh imitation probability P^I) and second harmonic (A^{II}, ϕ^{II}, P^{II}) in solar, sidereal and anti-sidereal time. Phases are not defined when amplitudes are smaller than their uncertainties.*

– Concerning the **first harmonic**:

(a) At $1.1 \cdot 10^{14}$ **eV**, from the analysis in solar time, the obtained amplitude and phase ($A^I_{sol} = (2.8 \pm 0.8) \cdot 10^{-4}$, $\phi^I_{sol} = (6.0 \pm 1.1)$ h, $P^I_{sol} = 0.2\%$) are in excellent agreement with the expected ones from the Compton-Getting effect [15] due to the revolution of the Earth around the Sun: at our latitude $A_{sol,CG}$=$3.0 \cdot 10^{-4}$, $\phi_{sol,CG}$=6.0 h. With respect to the sidereal time analysis, the measured amplitude and phase ($A^I_{sid} = (2.6 \pm 0.8) \cdot 10^{-4}$, $\phi^I_{sid} = (0.4 \pm 1.2)$ h LST), with imitation probability $P^I_{sid} = 0.5\%$, confirm the previous EAS-TOP result [9].

(b) At $3.7 \cdot 10^{14}$ **eV** the amplitude and phase of the first harmonic in solar time are still consistent with those expected for the Compton-Getting effect, although, due to the reduced statistics, the chance imitation probability is rather high. Concerning the analysis in sidereal time, we obtain $A^I_{sid} = (6.4 \pm 2.5) \cdot 10^{-4}$, $\phi^I_{sid} = (13.6 \pm 1.5)$ h LST, with an imitation probability of about 3.8%. This indicates therefore a change of phase (from 0.4 to 13.6 h) and an increase of amplitude (by a factor 2.5) with respect to the first harmonic measured at $1.1 \cdot 10^{14}$ eV.

– Concerning the **second harmonic** most significant ($P^{II}_{sid} = 1.6\%$) is the amplitude observed in sidereal time in the lower energy class of events (comparable with the first harmonic one: $A^{II}_{sid} = (2.3 \pm 0.8) \cdot 10^{-4}$, $\phi^{II}_{sid} = (6.3 \pm 0.7)$ h LST) (see also [11]).

Both at $1.1 \cdot 10^{14}$ eV and $3.7 \cdot 10^{14}$ eV, no significant amplitude is observed in anti-sidereal time, showing that no additional correction is required due to residual seasonal effects.

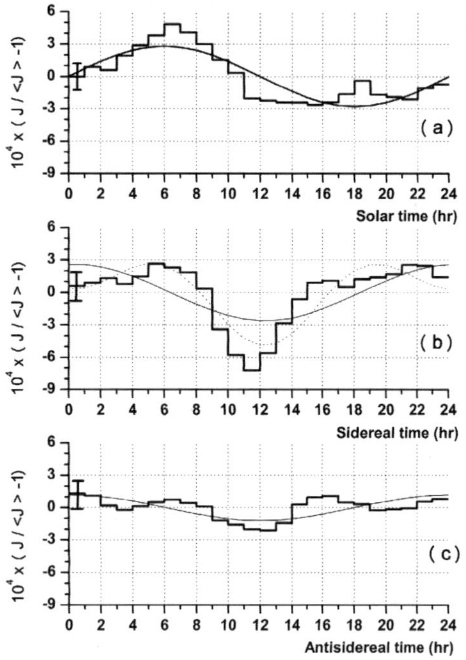

Figure 1. *Thick black lines: counting rate curves in solar (a), sidereal (b) and anti-sidereal (c) time at* $1.1 \cdot 10^{14}$ *TeV. Results from the first harmonic analysis are also shown (light lines); for the sidereal time curve, the combination of I and II harmonics is superimposed as a dotted line.*

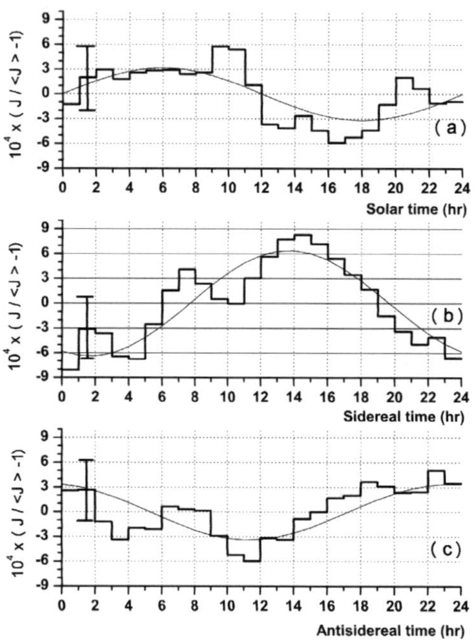

Figure 2. *Thick black lines: counting rate curves in solar (a), sidereal (b) and anti-sidereal (c) time at* $3.7 \cdot 10^{14}$ *TeV. The curves resulting from the first harmonic analysis are also shown (light black lines). The statistical uncertainty for each bin is given in the first one.*

Besides the harmonic analysis, it is interesting to visualize the variations of the cosmic ray intensity versus time, $I(t)$, as reconstructed by integration of the East-West differences, $D(t)$. They are shown in figs. 1 and 2, for the classes of events at $1.1 \cdot 10^{14}$ eV and $3.7 \cdot 10^{14}$ eV, respectively (a,b,c for solar, sidereal, and anti-sidereal time scales). As already shown by the harmonic analysis, at both energies the curves in solar time are dominated by the Compton-Getting effect due to the motion of the Earth, and no modulation is visible in the anti-sidereal time scale. A main difference is observed in the sidereal time curves: while the shape of the curve at $1.1 \cdot 10^{14}$ eV is in remarkable agreement with the EAS and muon measurements reported at and below 100 TeV, the curve related to the highest energy class of events is characterized by a broad excess around 13-16 h LST.

4. Conclusions

High stability data obtained from long time observations (8 years) from the EAS-TOP array confirm the amplitude and phase of the cosmic ray anisotropy already reported at 10^{14} eV: $A^I_{sid} = (2.6 \pm 0.8) \cdot 10^{-4}$, $\phi^I_{sid} = (0.4 \pm 1.2)$ h LST, with Rayleigh imitation probability $P^I_{sid} = 0.5\%$. The result is supported by the observation of the Compton-Getting effect due to the revolution of the Earth around the Sun, and by the absence of anti-sidereal effects. It confirms the homogeneity of the anisotropy data over the energy range 10^{11}-10^{14} eV.

At higher energies (around $4 \cdot 10^{14}$ eV) the observed anisotropy shows a larger amplitude, $A^I_{sid} = (6.4 \pm 2.5) \cdot 10^{-4}$, and a different phase, $\phi^I_{sid} = (13.6 \pm 1.5)$ h LST, with an imitation probability of 3.8%. The statistical significance is still limited, but the measurement has the highest sensitivity with respect to previous experiments at these energies, and it is not in contradiction with any of them.

The dependence of the anisotropy amplitude over primary energy ($A \propto E_0^\delta$) deduced from the present two measurements can be represented by a value of $\delta = 0.74 \pm 0.41$. Therefore, at least in the energy range $(1-4) \cdot 10^{14}$ eV, such dependence is compatible with that of the diffusion coefficient as derived by composition measurements at lower energies.

On another side, the sharp increase of the anisotropy above 10^{14} eV may be indicative of a sharp evolution of the propagation properties, and therefore of the diffusion coefficient just approaching the steepening of the primary spectrum. This opens the problems of obtaining an improved theoretical and experimental description of the whole evolution of the diffusion processes vs primary energy, and understanding how such evolution could affect the energy spectra at the "knee". From the experimental point of view, the extension of the anisotropy measurements with high sensitivity to and above 10^{15} eV will be of crucial significance.

Acknowledgments

This paper is a tribute to Gianni Navarra, the spokeperson of the EAS-TOP experiment and the real heart of it. We would like to remember him not only for his contributions to the growth of cosmic ray physics in Italy and in the world, but also to the growth of our group. We are left with his memory and his teachings.

References

[1] Aglietta, M. et al. 2004, Astrop. Phys., 21, 583
[2] Antoni, T. et al. 2005, Astrop. Phys., 24, 1
[3] Berezhko, E.G., & Volk, H.J. 2007, ApjL, 661, L175
[4] Peters, P. 1960, Proc. 6th ICRC, Vol. 3, 157
[5] Zatsepin, G.T. et al. 1962, Izv. Akad. Nauk USSR S.P., 26, 685
[6] Jones, F.C. et al. 2001, ApJ, 547, 264
[7] Strong, A.W et al, 2007, Annual Review of Nuclear and Particle Science, 50, 1, 285
[8] Berezinsky, V. et al. 1990, Astrophysics of Cosmic Rays, North Holland (V.L. Ginzburg ed.)
[9] Aglietta, M. et al. 1996, ApJ, 470, 501
[10] Aglietta, M. et al, 2009, ApjL, 692, L130
[11] Alekseenko, V.V. et al. 1981, Proc. 17th ICRC, 1, 146
[12] Aglietta, M. et al. 1993, Nucl. Instr. Meth. Phys. Res. A., 336, 310
[13] Heck, D. et al. 1998, FZK Report, 6019
[14] Aglietta, M. et al. 2007, Proc. 30th ICRC
[15] Compton, A.H., & Getting, I. A. 1935, Phys. Rev, 47, 817

Hybrid detection of Ultra High Energy Cosmic Rays with the Pierre Auger Observatory

Mariangela Settimo[a] **for the Pierre Auger Collaboration**[b]
[a]Department of Physics, University of Salento and INFN, 73100 Lecce, Italy
[b]Av. San Martin Norte 304, (5613) Malargue, Mendoza, Argentina
E-mail: mariangela.settimo@le.infn.it

Abstract. The Pierre Auger Observatory has been designed to observe ultra high energy cosmic rays employing two independent techniques. It consists of a surface array overlooked by air fluorescence telescopes which together provide a powerful instrument for air shower reconstruction. The hybrid reconstruction approach and its performance are described and the hybrid detector exposure derived using an accurate simulation which takes into account the exact and time dependent working conditions of the experiment.

1. Introduction

The measurement of the cosmic ray flux is one of the main goals of the Pierre Auger Observatory [1]. In the energy region above 10^{18} eV two important features, the ankle and the GZK cut-off are expected. In particular, the ankle position may be related to the transition between galactic and extragalactic components [2, 3] of the cosmic ray flux. The Pierre Auger Observatory employes a hybrid detection technique, allowing the reconstruction of extensive air showers with two complementary measurements. It consists of 1600 water-Cherenkov detectors deployed over an area of 3000 km^2. The stations are deployed with a spacing of 1500 m yielding full efficiency for extensive air shower detection above 5×10^{18} eV. The Surface Detector (SD) array is overlooked by the Fluorescence Detector (FD) consisting of four buildings, each hosting six fluorescence telescopes. The SD array samples the particle density of the shower at ground with an almost 100% duty cycle. The fluorescence telescopes observe the longitudinal development of the shower in the atmosphere providing a calorimetric measurement of the primary particle energy, only weakly dependent on theoretical models through the "invisible" energy carried out by penetrating particles. The FD operates during clear and moon-less nights allowing a duty cycle of 10-15% [4]. The information of the two detectors are combined together providing a more reliable energy and direction measurement for those events simultaneously observed by FD and at least one SD detector (*"hybrid"* events). Due to the lower energy threshold of the fluorescence detector, the hybrid mode allows to safely extend the measurable energy range down to 10^{18} eV. Indeed, above this energy, each event detected by FD has at least one triggered SD station, independently of mass composition (fig. 1, left panel). Moreover, a sub-sample of events recorded and independently reconstructed by both detectors can be used to calibrate the energy scale of the SD array [5, 6].

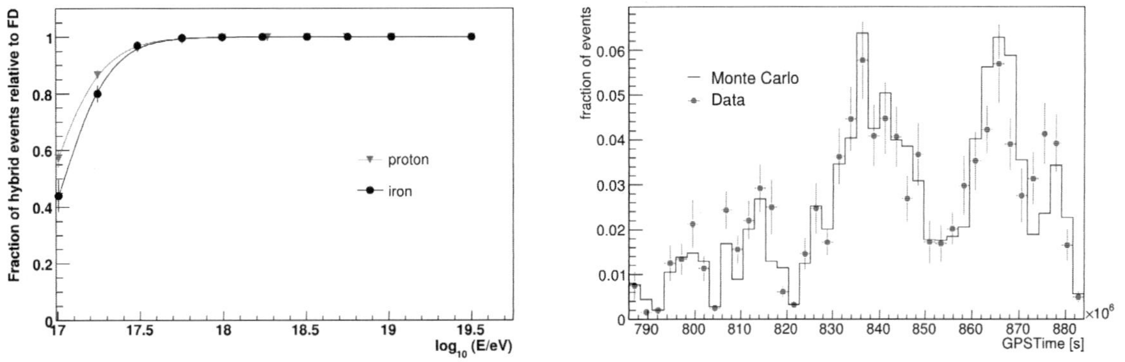

Figure 1. Left: Hybrid trigger efficiency relative to FD derived from CORSIKA showers and full hybrid detector simulation for proton (red triangle), iron (black circles). Right: fraction of hybrid events as a function of time.

2. Performance of the hybrid detector

Hybrid data supplement the traditional FD direction fitting method with the arrival time of the shower at the ground measured by the SD station with the largest signal. The hybrid geometry reconstruction [7] proceeds in two steps: the plane containing the shower axis and the fluorescence telescope is derived from the direction of the triggered pixels. Then, the orientation of the shower axis within the shower detector plane (SDP) is determined from timing information. The hybrid technique assures an accuracy of about 0.5° when reconstructing the arrival direction [8] and the core position is determined within 50 m [7]. Once the geometry is known, the observed energy deposit profile is reconstructed taking into account the absorption of the atmosphere and the fraction of forward-emitted and scattered Cherenkov light [10]. The energy resolution of the hybrid reconstructed events is better than 6% above 10^{18} eV. The total systematic uncertainty in the energy determination is estimated as 22% [5].

3. Hybrid exposure and spectrum measurement

The aperture calculation, data selection and spectrum measurement are discussed in detail in [4, 9]. Data collected between November 2005 and May 2008 are used for these analyses. During the inspected period the site was under construction and the fast growth of the detector introduces a significant dependence of the aperture on time. This effect has been taken into account and simulated using an accurate calculation of the hybrid detector *On-time*. To reproduce the exact data taking conditions, a large sample of Monte Carlo simulations has been performed using longitudinal energy deposit profiles generated with CONEX [11]. The distribution of particles at ground is not provided by CONEX. Nevertheless, the time of the station with the highest signal is a sufficient information for the geometry reconstruction. The SD trigger is simulated according to a parameterisation of the "Lateral Trigger Probability" functions (LTPs) [12] derived from CORSIKA [13] showers processed through a detailed detector simulation [14]. The used method has been validated by comparing the distribution of reconstructed observables to experimental data. In the right panel of Fig. 1, the distribution of selected events is shown as a function of time. In this plot the growth of the hybrid detector and the seasonal trend of data taking are visible. Fig. 2 (left) shows the hybrid exposure for proton (solid circles) and iron (open squares) primaries. The black arrow indicates the region above 10^{18} eV where the exposure is used for the measurement of the hybrid spectrum. At the selection level, the systematic uncertainty due to mass composition is about 8% at 10^{18} eV and

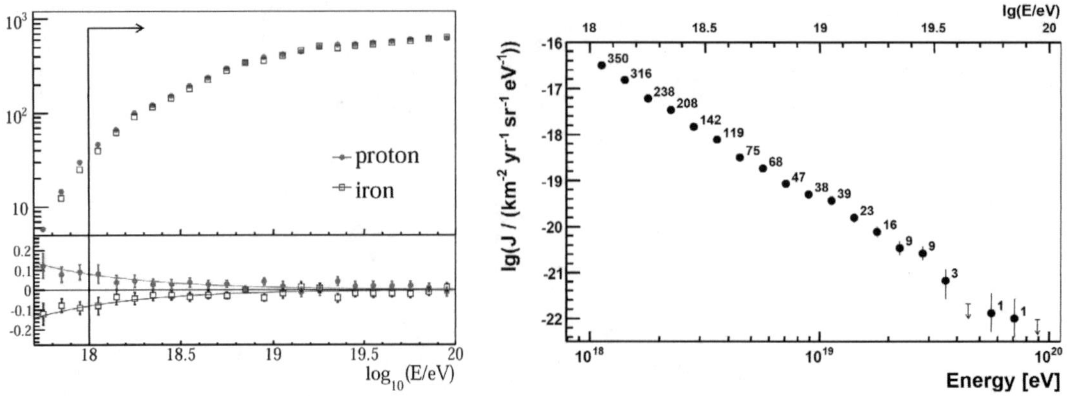

Figure 2. Left: Hybrid exposure for proton (solid circles) and iron (open squares) primaries obtained from Monte Carlo simulations together with the relative differences between the exposures for pure proton (iron) and mixed composition (lower panel). Right: Energy spectrum of cosmic rays derived from hybrid data. Only statistical error bars are shown.

decreases to less than 1% above 10^{19} eV. The dependence on hadronic interaction models has been checked using QGSJet-II and Sibyll-2.1 and found to be negligible. The measured spectrum, assuming a 50% proton - 50% iron mixed composition, is shown in Fig. 2 (right). Details on this calculation and its systematic uncertainties are given in [9].

4. Conclusions

The hybrid detection mode of the Pierre Auger Observatory provides a very precise geometry reconstruction and an almost model independent energy determination. The measurement of the cosmic ray flux using the hybrid detector relies on an accurate determination of the detector exposure. A set of detailed and time dependent Monte Carlo simulations has been performed for reproducing all the data taking conditions. The derived exposure is known with a systematic uncertainty smaller than 10%. The hybrid specturm has been measured above 10^{18} eV being the systematic uncertainty dominated by the overall energy scale (22%). As a next challenge, the sensitivity of the Observatory will be extended towards the lower energy region by the deployment of a dense surface detector array (AMIGA) [15] and of high elevation telescopes (HEAT) [16].

References

[1] J.Abraham et al., Pierre Auger Coll., *Nucl. Instrum. Methods* A, 523, 50-95, 2004.
[2] V.S.Berezinsky, A.Z.Gazizov and S.I.Grigorieva, Phys. Rev.,D74, p.043005, 2006.
[3] D.Allard, E.Parizot and A. Olinto, Astrop.Phys, 27, 61-75, 2007.
[4] F.Salamida for the Pierre Auger Coll., 31th ICRC, Lodz, Poland, 2009, arXiv:0906.2189.
[5] C.Di Giulio for the Pierre Auger Coll., 31th ICRC, Lodz, Poland, 2009, arXiv:0906.2189.
[6] Pierre Auger Collaboration [J. Abraham et al.], Phys.Rev.Lett.101:061101, 2008
[7] Pierre Auger Collaboration [J. Abraham et al.], arXiv:0907.4282
[8] C.Bonifazi for the Pierre Auger Coll., Nucl.Phys.Proc.Supp., 190:20-25, 2009.
[9] F.Schüssler for the Pierre Auger Coll., 31th ICRC, Lodz, Poland, 2009, arXiv:0906.2189.
[10] M.Unger, B.R.Dawson, R.Engel, F.Schüssler, and R.Ulrich, *Nucl. Instrum. Methods* A, 588, 433, 2008
[11] T.Bergmann et al., Astropart.Phys. 26 (2007)
[12] E.Parizot for the Pierre Auger Coll., Proc. 29th ICRC. Pune (2005) 7,71-74.
[13] D.Heck et al, Report FZKA 6019, (1998).
[14] J. Allen *et al.*,J.Phys.Conf.Ser., 119, 032002, 2008
[15] M.Platino for the Pierre Auger Coll., 31th ICRC, Lodz, Poland, 2009, arXiv:0906.2354
[16] M.Kleifges for the Pierre Auger Coll., 31th ICRC, Lodz, Poland, 2009, arXiv:0906.2354

Journal of Physics: Conference Series **203** (2010) 012128 doi:10.1088/1742-6596/203/1/012128

Neutrino oscillations in MHD supernova explosions

S Kawagoe[1], T Takiwaki[2] and K Kotake[1,2]

[1]Division of Theoretical Astronomy, National Astronomical Observatory of Japan, 2-21-1, Osawa, Mitaka, Tokyo, 181-8588, Japan

[2]Center for Computational Astrophysics, National Astronomical Observatory of Japan, 2-21-1, Osawa, Mitaka, Tokyo, 181-8588, Japan

E-mail: `shio.k@nao.ac.jp`

Abstract. We calculate the neutrino oscillations numerically in magnetohydrodynamic (MHD) explosion models to see how asphericity has impacts on neutrino spectra. Magneto-driven explosions are one of the most attracting scenarios for producing large scale departures from spherical symmetric geometry, that are reported by many observational data. We find that the event rates at Super-Kamiokande (SK) seen from the polar direction (e.g., the rotational axis of the supernovae) decrease when the shock wave is propagating through H-resonance. In addition, we find that L-resonance in this situation becomes non-adiabatic, and the effect of L-resonance appears in the neutrino signal, because the MHD shock can propagate to the stellar surface without shock-stall after core bounce, and the shock reaches the L-resonance at earlier stage than the conventional spherical supernova explosion models. Our results suggest that we may obtain the observational signatures of the two resonances in SK for Galactic supernova.

1. Introduction

A large number of neutrinos are produced when a massive star undergoes a core-collapse supernova explosion. Such supernova neutrinos will carry valuable information from deep inside the core (e.g., [1]). Over the last decades, significant progress has been made in many neutrino detectors [2]. The neutrino detections are important to study not only the supernova physics but also the unknown neutrino oscillation parameters (e.g., [3, 4]). The supernova neutrinos interact with electrons when the neutrinos propagate through stellar matter via the Mikheyev-Smirnov-Wolfenstein (MSW) effect. The neutrino conversion efficiency via the MSW effect depends on the density gradients. Recently, such shock effects have been focused (e.g., [5]). The time dependence of the neutrino events monitors the evolution of the density profile thus could provide a powerful test of the mixing angle and the mass hierarchy (e.g., [6, 7, 8]). Most of those phenomenology of supernova neutrinos have been based on the spherically symmetric models of core-collapse supernovae (e.g., [9, 10]). On the other hand, there are observations indicating that core-collapse supernovae are globally aspherical (e.g., [11]). Pushed by them, various mechanisms have been explored thus far by supernova modelers to understand the central engine (e.g., [3]).

We study the neutrino oscillations in the case of the magnetohydrodynamic (MHD) explosions of core-collapse supernovae [12]. Based on such models, we calculate numerically the detection of supernova neutrinos in the highly non-spherical envelope through the MSW effect.

2. Numerical method

We take time-dependent density profiles from our MHD simulations of core-collapse supernovae [12], in which a 25 $M_\odot$ presupernova model [13] was adopted and the magnetic fields is $\sim 10^{12}$G. This progenitor lacks the hydrogen and helium layer during stellar evolution, which is reconciled with observations that the progenitors associated with long-duration gamma-ray bursts (LGRB) are type Ib/c supernovae [14]. In this case, the highly collimated shock pushed by the strong magnetic pressure can blow up the massive stars along the rotational axis [12]. It is noted that to maximize the shock effect a sharpness of the shock [15]. The flavor conversion through the MSW effect occur in two resonances, one is at higher density called H-resonance, and the other is at lower density called L-resonance.

Along the time-dependent density profiles, we solve numerically the time evolution equation for the neutrino wave functions [15]. The neutrino oscillation parameters are taken as $\sin^2 2\theta_{12}$=0.84, $\sin^2 2\theta_{23}$=1.00, Δm^2_{12}=8.1$\times$10^{-5}eV^2 and $|\Delta m^2_{13}|$=2.2$\times$10^{-3}eV^2. We assume the inverted mass hierarchy and $\sin^2 2\theta_{13}$=1.0$\times$10^{-3} in our computations. The neutrino energy spectra at the surface of the star are calculated [4] using the results of a numerical simulation by the Lawrence Livermore group (e.g., [9, 8]). The luminosity is taken to decay exponentially with a timescale of 3 s [15]. We calculate expected event numbers of the supernova neutrinos at the SK [15], and assume that the supernova occurs in our Galactic center (10 kpc). We neglect the effects of neutrino self-interactions, the resonant spin-flavor conversion and do not consider the Earth matter effect in this study.

3. Result and Discussion

We define a ratio of the high-energy to the low-energy event numbers as in [8]: R(High/Low) = (event number of 20 - 60 MeV)/(event number of 5 - 20 MeV). The influence of the shock wave in the H-resonance appears in $\bar{\nu}_e$, and that in the L-resonance appears in ν_e in the inverted mass hierarchy. Left panels of Figure 1 show the time evolution of the event number, and right panels show R(High/Low). Solid (red) line and dotted (blue) line are for the polar and the equatorial direction, respectively. Top panels are $\bar{\nu}_e$, and bottom panels are ν_e. In the left panels, we can see the event number of the polar direction increases slightly, and decreases compared with that of the equatorial direction. This is because the shock reaches to the resonances, and the low-energy neutrinos increase and the high-energy neutrinos decrease. On the other hand, the event number of the equatorial direction does not change sharply, because the shock does not reach to the resonances for the equatorial direction. It is noted that the decrease in the events comes mainly from the decrease of the high-energy neutrinos rather than the increase of the low-energy neutrinos. This is because the cross section of main reaction for detection of $\bar{\nu}_e$ is proportional to the square of the neutrino energy. The decrease of the event number by the shock passage is about 36% of the event number without the shock. Since the expected events are $\sim$ 2500 at the sudden decrease, it seems to be possible to identify such a feature by the SK class detectors. Such a large number imprinting the shock effect is thanks to the mentioned early shock-arrival to the resonances, peculiar for the MHD explosions. The event number of ν_e become much fewer than that of $\bar{\nu}_e$. This is because the cross section of the main reaction for detecting ν_e is about 10^{-2} times smaller than that of $\bar{\nu}_e$.

In right panels, R(High/Low) begins to decrease, which might be an observable signature of the shock entering to the resonances. Along the polar axis, the shock reaches to the H-resonance at $\sim$ 0.5 s, and the L-resonance at $\sim$ 1.2 s. It should be noted that those timescales are very early in comparison with the ones predicted in the neutrino-driven explosion models, typically $\sim$ 5 s and $\sim$ 15 s for the H- and L- resonances, respectively (e.g., [6, 7]). This arises from the fact that the MHD explosion is triggered promptly after core bounce without the shock-stall, which is in sharp contrast to the neutrino-driven explosion models ([6, 7]). The progenitor of the MHD models, possibly linked to LGRB, is more compact due to the mass loss of hydrogen and

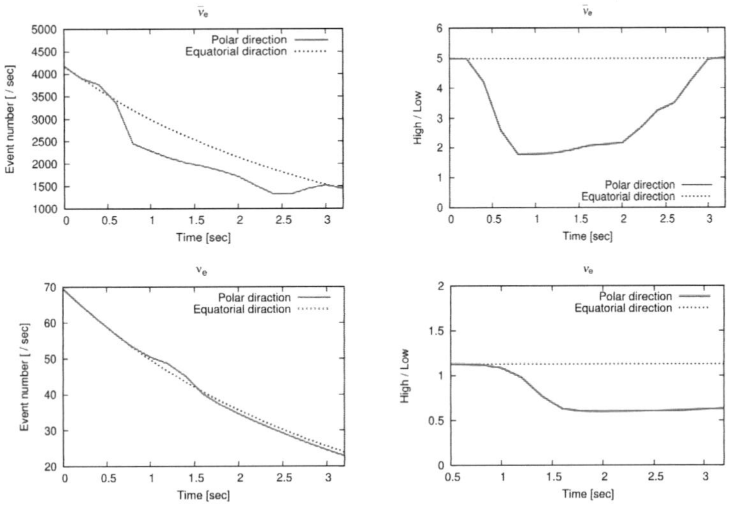

Figure 1. Left panels are the time evolution of the event number, and right panels are the ratio of the high-energy to low-energy event number. Solid (red) lines and dotted (blue) lines are for the polar and the equatorial direction, respectively. Top panels are $\bar{\nu}_e$, and bottom panels are ν_e.

helium envelopes during stellar evolution, which is also the reason for the early shock-arrival to the resonances.

4. Summary

We studied neutrino oscillations from core-collapse supernovae that produce MHD explosions, which are attracting attention recently as a possible relevance to magnetars and/or LGRB. Based on our simulation, we calculated numerically the event number of the supernova neutrinos in the highly non-spherical envelope through the MSW effect, and investigated how the explosion anisotropy could have impacts on the observed neutrinos at the SK for a Galactic supernova. In the case of the inverted mass hierarchy with a relatively large θ_{13}, the event numbers observed from the polar direction show steepest decrease, reflecting the passage of the shock to the resonances. This reflects a unique nature of the MHD explosion featuring a very early shock-arrival to the resonances. Our results suggested that the two features in the $\bar{\nu}_e$ and ν_e signals, if visible to the SK for a Galactic supernova, could be an observational signature of magneto-driven supernovae.

References

[1] Raffelt G G 2002 *Nucl. Phys. B Proc. Suppl.* **110** 254
[2] Totsuka Y 1992 *Rept. Prog. Phys.* **55** 377
[3] Kotake K, Sato K and Takahashi K 2006 *Rept. Prog. Phys.* **69** 971
[4] Dighe A S and Smirnov A Y 2000 *Phys. Rev.* D **62** 3007
[5] Duan H and Kneller J P 2009 *J. Phys. G: Nucl. Part. PHys.* **36** 113201
[6] Fogli G, Lisi E, Mirizzi A and Montanino D 2005 *JCAP* **04** 002
[7] Tomàs R, Kachelrieß M, Raffelt G, Dighe A, Janka H T and Scheck L 2004 *JCAP* **0409** 015
[8] Takahashi K, Sato K, Dalhed H E and Wilson J R 2003 *Astropart. Phys.* **20** 189
[9] Totani T, Sato K, Dalhed H E and Wilson J R 1998 *ApJ* **496** 216
[10] Lunardini C, Müller B and Janka H Th 2008 *Phys. Rev.* D **78** 023016
[11] Wang L, Howell D A, Hoeflich P and Wheeler J C 2001 *ApJ* **550** 1030
[12] Takiwaki T, Kotake K and Sato K 2009 *ApJ* **691** 1360
[13] Heger A, Langer N and Woosley S E 2003 *ApJ* **528** 368.
[14] Piran T 2004 *Rev. Mod. Phys.* **76** 1143
[15] Kawagoe S, Takiwaki T and Kotake K 2009 *JCAP* **09** 033

Topics in Astroparticle and Underground Physics (TAUP 2009) IOP Publishing
Journal of Physics: Conference Series **203** (2010) 012129 doi:10.1088/1742-6596/203/1/012129

Cosmic ray studies with the OPERA detector at Gran Sasso

Nicoletta Mauri for the OPERA Collaboration

Università di Bologna and INFN-Sezione di Bologna, Dipartimento di Fisica, Viale Berti Pichat 6/2, 40127, Bologna, Italy

E-mail: mauri@bo.infn.it

Abstract. OPERA is a long-baseline neutrino experiment located in the Hall C of the Gran Sasso underground Laboratory at an average depth of 3.8 km w.e., corresponding to average muon energies at surface higher than 1.4 TeV. After a brief description of the main scientific goals of the experiment, we focus on the potentialities of OPERA used as a cosmic ray detector. In particular, we report on the measurement of the atmospheric muon charge ratio $R_\mu = N_{\mu^+}/N_{\mu^-}$ and on the analysis of upgoing muons induced by atmospheric neutrinos.

1. Introduction

OPERA [1, 2] is a long baseline neutrino experiment located in the Gran Sasso underground laboratory (LNGS). It is aimed at detecting for the first time the appearance of tau neutrinos from the oscillation of muon neutrinos in the CERN to Gran Sasso beam-line (CNGS). The detector is located under 1400 meters of rock overburden, where only high energy cosmic ray muons can arrive: the minimum surface energy threshold is ~1 TeV (1.4 TeV averaged over all the directions and rock depths). OPERA, as a cosmic ray detector, provides the capability to measure the charge and momentum of cosmic muons with large statistics.

Cosmic ray data are collected in the OPERA detector together with CNGS beam data: CNGS neutrino induced events are removed from the sample using the coincidence with the CERN data timing system ("on time" flag).

After a short description of the OPERA detector, we present the results obtained from the analysis of atmospheric neutrino induced muons and of the atmospheric muon charge ratio.

2. The OPERA detector

The OPERA detector is a hybrid apparatus composed of two identical parts, called supermodules (SM), each consisting of a target section and a magnetic spectrometer. It is equipped with passive detectors, the Emulsion Cloud Chamber (ECC) units, called "bricks", and with electronic detectors. In the target section, the bricks are arranged in 29 vertical "walls", transverse to the beam direction, interleaved with Target Tracker (TT) walls. Each TT wall consists of a double layered plane of 64 long scintillator strips.

The target section is followed by a magnetic spectrometer. A large dipolar iron magnet is instrumented with Resistive Plate Chambers (RPC). The magnetic field intensity is 1.53 T, directed along the vertical axis. The RPC planes are inserted between the iron slabs: they provide the tracking inside the magnet. The deflection of charged particles in the magnet is measured by six stations of vertical drift tubes, the High Precision Trackers (HPT). Each HPT station is formed by four staggered layers of aluminium tubes, 8 m long, with 38 mm outer diameter. The spatial resolution of a HPT station is better than 500 μm in the bending (horizontal) plane.

A muon crossing the spectrometer is deflected in the horizontal plane: the charge and momentum reconstruction is performed for tracks crossing at least one magnet arm using the bending angle information ($\Delta\phi$) coming from HPT stations.

3. Atmospheric neutrinos

Atmospheric neutrino interactions in the rock surronding the LNGS underground laboratory produce muons. In OPERA, the TT timing system measures particle time-of-flights, thus it is used to discriminate between up-going and down-going particles: up-going muons (not within the CNGS spill window) with a restricted speed range are selected as atmospheric neutrinos.

This analysis classifies events in terms of the "speed side", a composition of velocity and direction informations:

$$\text{speed side} = (1/\beta) \times (\text{track angle sign}) \qquad (1)$$

The track angle sign is positive for positive slopes in the yz plane (side view), negative for negative slopes. The $(1/\beta)$ sign is given by the time difference between the last and first TT planes. With this convention, upward-going and downward-going particles have positive and negative speed side, respectively. Given the slope of the CNGS beam and the muon angular resolution, almost all on-time events have a positive speed side (exit angle $\sim 3.3°$). Off-time upgoing events, with a cut $0.75 < (1/\beta) < 1.25$, are selected as atmospheric neutrino events.

For this first analysis, the total livetime is 254 effective days, with the detector running also with the magnetic spectrometer off. The up-going muon sample was selected applying two levels of cuts: the first level cut requires a track unambiguously detected in at least three consecutive TT planes along both projections. The second level cut requires $(1/\beta)$ between 0.75 and 1.25 and 20 TT planes hit. From a full Monte Carlo simulation based on the Honda flux calculations, 6.4 up-going muons were expected for the analysed period. The OPERA detector observed 5 events recognized as muon neutrino induced.

4. Cosmic ray muon charge ratio

The cosmic ray muon charge ratio R_μ is an important observable to shed light on the physics of cosmic ray interactions in atmosphere: its measurement will help to better understand the features of high energy hadronic interactions in the forward region and to improve Monte Carlo models of interactions, constraining the predictions at high energy (above 1 TeV).

Cosmic ray muons are produced when primary cosmic ray nuclei (mainly protons) impinge on the Earth's atmosphere, producing showers of secondary particles. Most of the interaction products are π and K mesons, which decay into muons. Since the cosmic ray primaries are positively charged, there are more positive than negative pions and kaons in the hadronic showers. At high energies, several competing processes can affect the charge ratio. As energy increases, the fraction of muons coming from kaon decays also increases, and since strong interaction production channels lead to a K^+/K^- ratio higher than for π^+/π^-, R_μ is expected to rise. But there are other competing processes involved in the resulting charge ratio value. We also expect a dependence of R_μ on the underground muon multiplicity m_μ, which is related to the energy of the primary cosmic rays and to their chemical composition. For primaries different from protons, the charge excess is reduced and so is the muon charge ratio.

4.1. Data Analysis

The results here presented are based on data recorded during the CNGS Physics Run, from June 18 until November 10, 2008. The total number of events is 403069 corresponding to 113.4 days of livetime.

For this analysis, the basic information required for the charge-momentum measurement is at least one reconstructed $\Delta\phi$ angle in each event (*acceptance cut*). A second cut removes noisy events in HPT stations, potentially dangerous for the muon charge determination (*clean PT cut*). This typically occurs when some drift tubes are fired by secondary particles (δ-rays, showers etc.). Finally, we selected tracks whose deflections are above the experimental resolution: we require $\Delta\phi/\sigma_{\Delta\phi} > 3$ (*deflection cut*). After the application of this cut, the charge-misidentification η (defined as the fraction of tracks reconstructed with wrong charge sign) is reduced from 0.080 ± 0.002 to 0.030 ± 0.001.

The muon charge ratio has been computed separately for single muon events and multiple muon events. Single muon events are selected requiring single tracks in each projected view, well merged in 3D. Multiple muon events are selected by requiring a muon multiplicity ≥ 2 in both views, with tracks identified and merged in 3D.

For this kind of measurement, the main sources of systematic error are due to the HPT alignment accuracy and to the determination of the η value. For both sources, we considered all muon tracks crossing both arms of each spectrometer, thus providing two deflection values $\Delta\phi$ for the same muon track. To evaluate the misalignment contribution, we used the difference $\delta\Delta\phi = \Delta\phi_{arm_1} - \Delta\phi_{arm_2}$, that with perfect alignment should be peaked at zero. Given the peak values for the two spectrometers, we propagated the offset in the charge ratio calculation, obtaining $\delta R_\mu \simeq 0.015$. To evaluate the systematic uncertainty on η, we computed the fraction of tracks with two opposite deflection angle signs, finding η_{real}. The difference between experimental and Monte Carlo misidentification is one-sided (since $\eta_{real} \geq \eta_{MC}$), and corresponds to $\delta R_\mu = 0.007$. The final systematic error is the quadratic sum of its contributions.

In order to provide a result independent from the detector features, we unfolded the charge ratio measured value using the charge-misidentification η, computed with Monte Carlo.

The unfolded single-muon charge ratio is

$$R_{unf}(m_\mu = 1) = \frac{\eta - (1-\eta)R_{meas}}{\eta R_{meas} - (1-\eta)} = 1.377 \pm 0.014\,(\text{stat.})^{+0.017}_{-0.015}\,(\text{syst.}) \tag{2}$$

The unfolded charge ratio for multiple-muon events is

$$R_{unf}(m_\mu > 1) = 1.23 \pm 0.06\,(\text{stat.})^{+0.017}_{-0.015}\,(\text{syst.}) \tag{3}$$

This value is 2.4 σ away from the value for single muon events, consistent with the hypothesis of dilution of R_μ due to the neutron enhancement in the primary nuclei.

5. Conclusions

We presented results on two cosmic ray physics items carried out in parallel to the oscillation physics program of the OPERA experiment. We reported the measured value of the atmospheric muon charge ratio, integrated over all directions, at an average depth of 3800 km w.e. This experimental result will be helpful to constrain phenomenological hadronic interaction models in the very forward region. The first upgoing atmospheric neutrinos were observed; since this item needs a large livetime the analysis is still in a preliminary phase.

References

[1] R. Acquafredda *et al.* [OPERA Collaboration], 2009 JINST 4 P04018.
[2] N. Agafonova *et al.* [OPERA Collaboration], 2009 JINST 4 P06020.

Investigation of $\beta\beta$ decay in ^{150}Nd and ^{148}Nd to the excited states of ^{150}Sm and ^{148}Sm

A S Barabash[1], **Ph Hubert**[2], **A Nachab**[2] **and V Umatov**[1]

[1] Institute of Theoretical and Experimental Physics, B. Cheremushkinskaya 25, 117218 Moscow, Russia

[2] Centre d'Etudes Nucléaires, IN2P3-CNRS et Université de Bordeaux, 33170 Gradignan, France

E-mail: barabash@itep.ru

Abstract. Double beta decay of ^{150}Nd and ^{148}Nd to the excited states of daughter nuclei have been studied using a 400 cm^3 low-background HPGe detector and an external source consisting of 3046 g of natural Nd_2O_3 powder. The half-life for the two-neutrino double beta decay of ^{150}Nd to the excited 0_1^+ state in ^{150}Sm is measured to be $T_{1/2} = [1.33^{+0.36}_{-0.23}(stat)^{+0.27}_{-0.13}(syst)] \cdot 10^{20}$ y. For other $(0\nu + 2\nu)$ transitions to the 2_1^+, 2_2^+, 2_3^+, and 0_2^+ levels in ^{150}Sm, limits are obtained at the level of $\sim (2-8) \cdot 10^{20}$ y. In the case of ^{148}Nd only limits for the $(0\nu + 2\nu)$ transitions to the 2_1^+, 0_1^+, and 2_2^+ excited states in ^{148}Sm were obtained and are at the level of $\sim (4-8) \cdot 10^{20}$ y.

1. Introduction

In this article, results of an experimental investigation of the $\beta\beta$ decay of ^{150}Nd and ^{148}Nd to the excited states in ^{150}Sm and ^{148}Sm are presented (see Ref. [1] for details). The search for $\beta\beta$ transitions of ^{150}Nd and ^{148}Nd to excited states has been carried out using a HPGe detector to look for γ-ray lines corresponding to their decay schemes. A preliminary result for $\beta\beta$ decay of ^{150}Nd to the 0_1^+ excited state of ^{150}Sm was published in Ref. [2].

2. Experimental study

The experimental work was performed in the Modane Underground Laboratory (depth of 4800 m w.e.). A 400 cm^3 low-background HPGe detector was used to measure a 3046 g sample of Nd_2O_3 powder in a special Marinelli delrin box which was placed on the detector endcap. Taking into account the natural abundance there are 153 g of ^{150}Nd (5.64%) and 154 g of ^{148}Nd (5.76%) in the sample. Data were collected for 11320.5 h.

3. Analysis and results

3.1. Search for $\beta\beta$ processes in ^{150}Nd

Double beta decays of ^{150}Nd to 2_1^+ (333.86 keV), 0_1^+ (740.38 keV), 2_2^+ (1046.04 keV), 2_3^+ (1193.73 keV) and 0_2^+ (1255.40 keV) levels in ^{150}Sm have been investigated. The result together with available data on $\beta\beta$ decay of ^{150}Nd from other experimental works are presented in Table 1.

Table 1. Experimental results for $(0\nu + 2\nu)\beta\beta$ decay of ^{150}Nd to the excited states of ^{150}Sm. All limits are given at the 90% C.L.

Excited state	Energy of γ-rays (efficiency)	$(T_{1/2}^{0\nu+2\nu})_{exp}$ (10^{20} y) this work	other works
2_1^+(333.86)	333.9 (2.60%)	> 2.2	> 0.91[3]
			> 24[a] [4]
0_1^+(740.38)	333.9 (2.30%)	$= [1.33^{+0.36}_{-0.23}(stat)^{+0.27}_{-0.13}(syst)]^b$	> 1.5 [5]
	406.5 (2.29%)		> 1 [3]
			> 2.4[a] [4]
2_2^+(1046.04)	712.2 (1.78%)	> 8.0	> 1.4 [6]
2_3^+(1193.73)	1193.7 (0.95%)	> 5.4	> 0.027 [7]
0_2^+(1255.40)	921.5 (1.45%)	> 4.7	> 2 [3]

[a]Only 0ν decay mode
[b]Half-life value for 2ν decay (see Ref. [1] for the details)

Table 2. Experimental results for $(0\nu + 2\nu)\beta\beta$ decay of ^{148}Nd to the excited states of ^{148}Sm. All limits are given at the 90% C.L.

Excited state	Energy of γ-rays (efficiency)	$(T_{1/2}^{0\nu+2\nu})_{exp}$ (10^{20} y) this work	other works
2_1^+(550.26)	550.3 (2.36%)	> 6.6	> 0.03 [7]
0_1^+(1424.46)	550.3 (2.16%)	> 7.9	-
	874.2 (1.83%)		
2_2^+(1454.12)	550.3 (1.11%)	> 3.8	> 0.027[7]
	903.9 (0.87%)		

3.2. Search for $\beta\beta$ processes in ^{148}Nd

A search for the double beta decays of ^{148}Nd to the 2_1^+, 0_1^+, and 2_2^+ excited states of ^{148}Sm was carried out by looking for γ-rays with energies of 550.3, 874.2, and 903.9 keV accompanying these transitions. The result together with available data on $\beta\beta$ decay of ^{148}Nd from other experimental works are presented in Table 2.

4. Conclusion

Double beta decay of ^{150}Nd and ^{148}Nd to the excited states of daughter nuclei was investigated with a high level of sensitivity. The half-life for the $2\nu\beta\beta$ decay of ^{150}Nd to the excited 0_1^+ state in ^{150}Sm is measured to be $T_{1/2} = [1.33^{+0.36}_{-0.23}(stat)^{+0.27}_{-0.13}(syst)] \cdot 10^{20}$ y. The strongest limits for other transitions were established. The sensitivity of this experiment could still be increased by a few times using a pure Nd_2O_3 (or Nd) sample. Also further increases in the sensitivity could be reached using an enriched Nd sample and a multicrystal HPGe installation to study larger masses of Nd samples.

References

[1] Barabash A S, Hubert Ph, Nachab A and Umatov V I 2009 *Phys. Rev.* C **79** 045501
[2] Barabash A S, Hubert F, Hubert Ph and Umatov V I 2004 *JETP Lett.* **79** 10
[3] Arpesella C, Bellotti E, Ferrari N and Zanotti L 1999 *Nucl. Phys. B (Proc. Suppl.)* **70** 249
[4] Argyriades J *et al.* 2008 *Preprint* hep-ex/0810.0248
[5] Klimenko A A, Osetrov S B, Smolnikov A A and Vasiliev S I 2002 *Czech. J. Phys.* **52** 589
[6] Arpesella C *et al.* 1996 *Nucl. Phys. B (Proc. Suppl.)* **48** 247
[7] Bellotti E *et al.* 1982 *Lett. Nuovo Cimento* **33** 273

The BiPo detector

S Cebrian, on behalf of the SuperNEMO Collaboration
Universidad de Zaragoza, Pedro Cerbuna 12, 50009 Zaragoza, Spain
E-mail: scebrian@unizar.es

Abstract. The BiPo detector is intended to measure extremely low radioactive contaminations of ^{208}Tl and ^{214}Bi in very thin layers like the SuperNEMO double beta decay source foils. A modular prototype with 20 capsules (the so-called BiPo1) is running in the Modane Underground Laboratory (France). New approaches based on a more compact geometry (BiPo2) and on the use of the Phoswich technique have been also taken into account and the corresponding prototypes are running at the Modane set-up too. First measurements for samples of interest, with a sensitivity unreachable by Ge detectors, have started. Description of prototypes, results validating the detection techniques and estimates of sensitivity are presented here.

1. Principle and prototypes

The goal of the BiPo detector is to measure ultra-low levels of impurities in double beta source foils reaching activities of A(^{208}Tl)<2 μBq/kg and A(^{214}Bi)<10 μBq/kg, by registering BiPo events in the natural chains as a delayed coincidence between electrons and alpha particles using a sandwich of two plastic scintillators (see Fig. 1). Photomultiplier (PMT) signals are sampled by MATACQ VME digitizing boards (12 bit resolution) at 1 GS/s in 2.5 μs. A new trigger board for measuring also ^{214}BiPo events has been recently installed.

BiPo1 prototype consists of 20 capsules (with total surface of 0.8 m^2) made of polystyrene scintillators (20×20×0.3 cm^3 each) coupled to low radioactive 5-inch PMTs. It is running at Modane Underground Laboratory since February 2008, based on a standard technique with very good proven sensitivity and having a good β/α discrimination capability. One of these capsules has a Phoswich design, with fast BC400 (300 μm) and slow BC444 (10 mm) scintillators read by the same PMT; this approach offers a powerful β/α discrimination and the possibility of running with only one PMT per capsule. A more compact design, using less PMTs and offering spatial resolution has been implemented in the BiPo2 prototype, consisting of two plastic scintillators BC408 (75×75×1 cm^3, total surface of 0.56 m^2) coupled each to 10 low radioactive 3-inch PMTs. Both the Phoswich capsule and the BiPo2 prototype are running at Modane since July 2008.

The next step is the construction of the BiPo3 detector, foreseen for 2010, using 36 capsules (total surface of $\sim$3.5 m^2) with organic plastic scintillators (30×30×0.2 cm^3 each) coupled to low radioactive 5-inch PMTs.

2. Results

The detection technique and efficiency for BiPo1 were validated with an Al foil of known activity and a successful method for β/α discrimination was developed (see details in [1]). Background measurements taken in last months have allowed to identify BiPo events, having the proper time interval between β-α signals (see an example of time interval distribution in Fig. 2), and

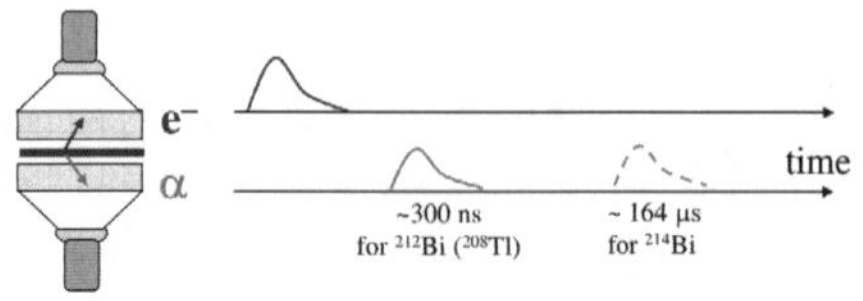

Figure 1. Basic design of the BiPo detector to identify BiPo events.

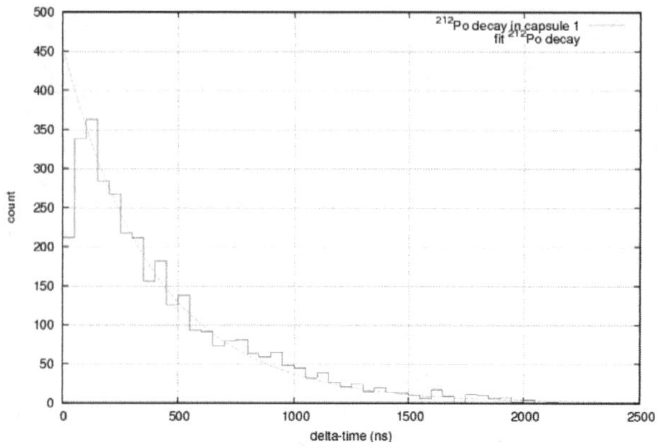

Figure 2. Distribution of time intervals between β-α signals in BiPo1 background measurements and the corresponding exponential fit. ^{212}Bi half-life (299 ns) is well reproduced.

to measure a surface activity on ^{208}Tl of (1.4±0.2) μBq/m^2 (32 events in 258 days × m^2). Assuming this background, a sensitivity of ~3 μBq/kg of ^{208}Tl could be at reach in a 6-month measurement in BiPo3 for a ^{82}Se double beta foil.

Proper distributions of time intervals between β-α signals for identified BiPo events have been also obtained for both the Phoswich capsule and the BiPo2 prototype in measurements with Al foils. The β/α discrimination power is being carefully studied for the Phoswich capsule and a background measurement is running since April 2009 in order to estimate its sensitivity. For BiPo2, a position reconstruction method has been proposed and checked with calibration sources, and will be validated with measurements being taken presently using an Al foil cut like a draughtboard.

3. Summary and outlook

The detection technique of the BiPo detector has been validated in all prototypes for ^{212}BiPo events and measurements for ^{214}BiPo events have already started. Now, the sensitivity goal is at reach in BiPo1 and first sample measurements are underway using this prototype. Construction of a medium BiPo detector (BiPo3) is foreseen in the very next future in order to qualify the first foils for the SuperNEMO experiment [2].

Acknowledgments

We would like to deeply acknowledge Professor Julio Morales, who passed away just when TAUP2009 was being held, for his dedicated work and always kind friendliness.

References

[1] Bongrand M (on behalf of the SuperNEMO Collaboration) 2008 *Journal of Instrumentation* **3** P06006
[2] Vala L (on behalf of the NEMO3 and SuperNEMO Collaborations) 2009 *Nucl. Phys.* B *(Proc. Suppl.)* **188** 62

Journal of Physics: Conference Series **203** (2010) 012132 doi:10.1088/1742-6596/203/1/012132

PAMELA sub-detectors capability in light nuclei identification

Lucia Consiglio, for the PAMELA colaboration
Present address: Department of Physics, Via Cintia M.S.Angelo, 40127, Napoli, Italy
E-mail: lucia.consiglio@na.infn.it

Abstract. The PAMELA (Payload for Antimatter Matter Exploration and Light nuclei Astrophysics) experiment is a satellite-borne apparatus installed on the Resurs DK1 Russian satellite, and launched from the Baikonur cosmodrome on June 15th 2006. It aims to study charged particles in the cosmic radiation focusing on antiparticles and light nuclei. The PAMELA apparatus is made of different sub-detectors: a time-of-flight system, a magnetic spectrometer, a silicon-tungsten electromagnetic calorimeter, an anticoincidence system, a shower tail catcher scintillator and a neutron detector. In this paper the capability of the sub-detectors to identify light nuclei, on the flight data, will be presented.

1. Introduction

One of the aims of the PAMELA experiment [1] is the measurement of the light nuclear component of Galactic cosmic rays in the interval 100 MeV/n -200 GeV/n. The ratios of the spallogenic nuclei (B, Be, and Li) to mostly primary nuclei such as C are particularly important to fix constraints in the cosmic ray propagation models since these ratios are sensitive to the amount of material traversed by GCRs from the source to detection at Earth. In the PAMELA apparatus three different sub-detectors (ToF, tracker and calorimeter) are able to identify light nuclei, with different efficiencies, resolutions and charge ranges. Accurate simulations are in progress to evaluate systematic uncertainties resulting from the various correction factors needed to evaluate fluxes such as uncertainties in the determination of the geometry factor, spallation loss within the instrument, and the tracking efficiency as function of the charge.

2. Nuclei identification by ToF

The charge of an incident particle is derived from six independent ionization energy loss measurements using the six ToF layers. The ToF charge calibration was performed using samples of relativistic helium selected by the tracker to obtain the correction factors for the signal normalization of all PMTs. By correcting the ADC also for the incident angle of the trajectory of the particle it was possible to define the energy release in terms of minimum ionizing particle (mip). The signals were corrected also for non linearities and light attenuation along the paddles. Fig. 1 shows the energy release in mip obtained for a single PMT versus particle velocity. Different coloured bands correspond to the nuclei families from Li to O. Combining measurements from all the 48 PMTs one obtains the mean charge distribution for the overall instrument as shown in fig. 2.

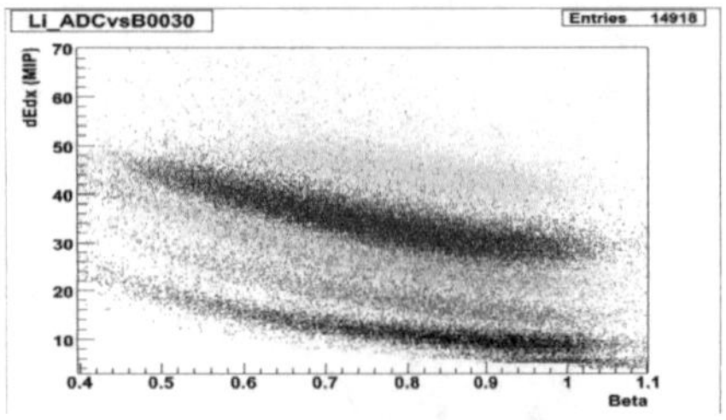

Figure 1. The ionization loss vs. beta for a single channel of the ToF.

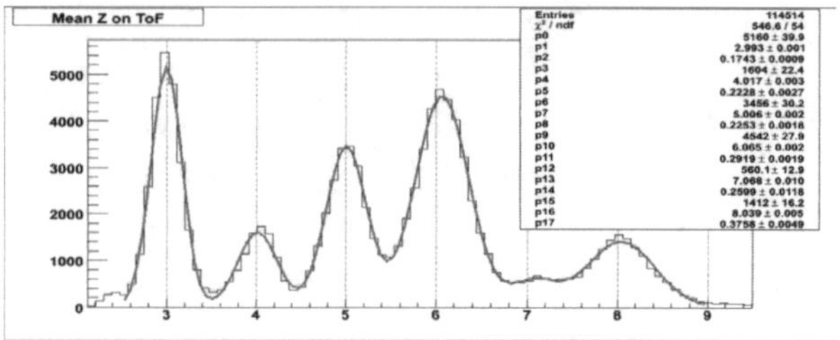

Figure 2. Mean charge distribution for the ToF system

3. Nuclei identification by tracker

Ionising energy loss measurements in the six silicon planes of the magnetic spectrometer allow to determine in an independent way the absolute charge of traversing particles. The signals recorded in the silicon layers of the tracker are grouped in cluster structures, where a cluster is defined as one or more adjacent strips with a signal/noise ratio greater than 4 noise sigmas. In fig. 3 the ionization loss measured in the tracker versus rigidity is shown. The plot is obtained

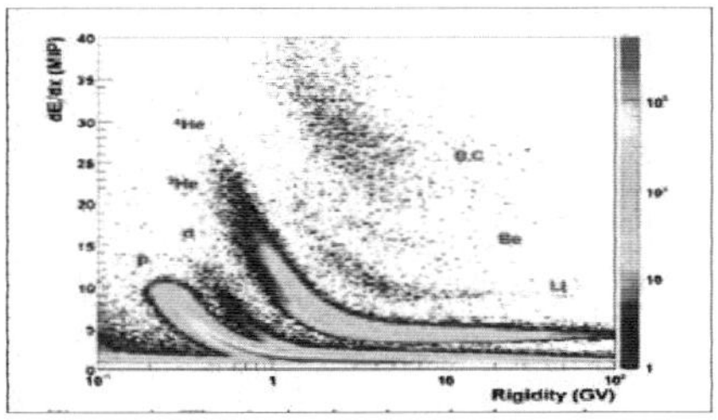

Figure 3. Ionization loss measured by tracker vs. rigidity

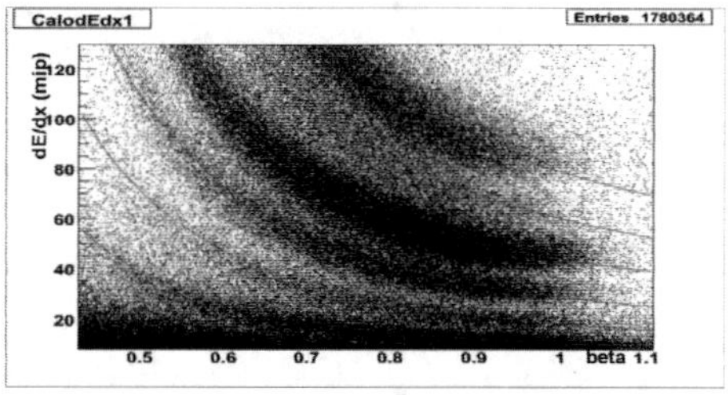

Figure 4. Ionization loss in the calorimeter vs particle velocity

considering the mean of the six measurements. The calibration constant used to convert ADC channels in mip is the same for all the electronic read-out chips. The tracker electronics is going to saturate in correspondence of an energy release greater than about 16 mips. This loss of linearity for high charge values is not surprising since the tracker design was optimized for the detection of relativistic Z=1 particles.

4. Nuclei identification by calorimeter

The charge of the particles can be measured in the calorimeter by considering the energy released in the first plane of the detector which is not covered by tungsten plates. Fig. 4 shows the charge separation obtained with this method for the different nuclei families from Li to O. Obviously the charge separation increases with the number of planes required but the efficiency of the measurement decreases.

References

[1] P. Picozza et al., *PAMELA - A payload for antimatter matter exploration and light-nuclei astrophysics*, Astroparticle Physics 27 (2007) 296-315

Journal of Physics: Conference Series **203** (2010) 012133 doi:10.1088/1742-6596/203/1/012133

Positioning, alignment and absolute pointing of the ANTARES neutrino telescope

F Fehr[1], **C Distefano**[2] **for the ANTARES collaboration**

[1] Erlangen Centre for Astroparticle Physics, Erwin-Rommel-Str. 1, Germany

[2] INFN Laboratori Nazional del Sud, Via S. Sofia 62, 95123 Catania, Italy

E-mail: `fehr@physik.uni-erlangen.de`

Abstract. A precise detector alignment and absolute pointing is crucial for point-source searches. The ANTARES neutrino telescope utilises an array of hydrophones, tiltmeters and compasses for the relative positioning of the optical sensors. The absolute calibration is accomplished by long-baseline low-frequency triangulation of the acoustic reference devices in the deep-sea with a differential GPS system at the sea surface. The absolute pointing can be independently verified by detecting the shadow of the Moon in cosmic rays.

1. Positioning and alignment of the ANTARES telescope

The ANTARES neutrino telescope [1] is operating at a depth of 2.5 km in the Mediterranean Sea, 40 km off-shore from Toulon (France). By detecting Cherenkov light induced by charged particles produced in neutrino reactions, ANTARES aims to reveal cosmic neutrino sources. The detector consists of about 900 photomultipliers arranged on the 12 detection strings, which are anchored on the sea-bed and held upright by freely floating buoys.

Being exposed to the sea current, the ANTARES strings undergo movements of several metres. The essential real-time measurement of the optical sensor positions with the required accuracy of $10 - 20$ cm is performed by an acoustic positioning system [2] which consists of sound emitter-receivers installed at the line anchors, receivers distributed over the lines, and oceanographic probes measuring the sound velocity at the site, in order to translate the propagation time of the acoustic signals between two modules into a distance. An acoustic triangulation is carried out every 2 min. Triggered by the ANTARES clock system, alternately each emitter-receiver module sends signals to the other modules in listening mode. Complementary information on the orientation of the sensors is obtained with a set of compasses and tiltmeters distributed over the detection lines. The full alignment of the detector is based on a fit of the line shape that includes information from the acoustic system, the compass and tiltmeter measurements as well as the cable dimensions, buoyancies and weights of the detector components. The single free parameter in this fit is the sea current which is obtained as a by-product.

The absolute positioning of the detector is accomplished by low-frequency long-baseline acoustic triangulation of the line anchors in the deep-sea with an acoustic reference system on a ship. The coordinates of the reference are measured with a differential GPS antenna. The measurement is repeated for different ship positions. In this way accuracies of about 1 m on the anchored detector components are reached. Systematic studies indicate that the precision of the absolute pointing is better than $0.2°$.

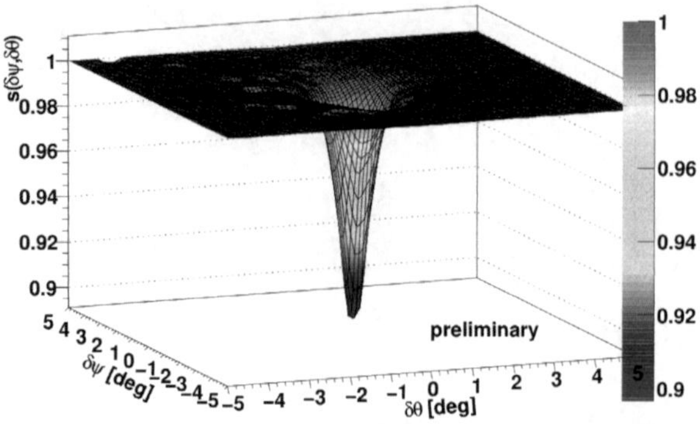

Figure 1. Simulated shape of the Moon shadow: muon flux with shadow effect (H_1) from the direction of the Moon normalised to the full flux (H_0) after analysis cuts.

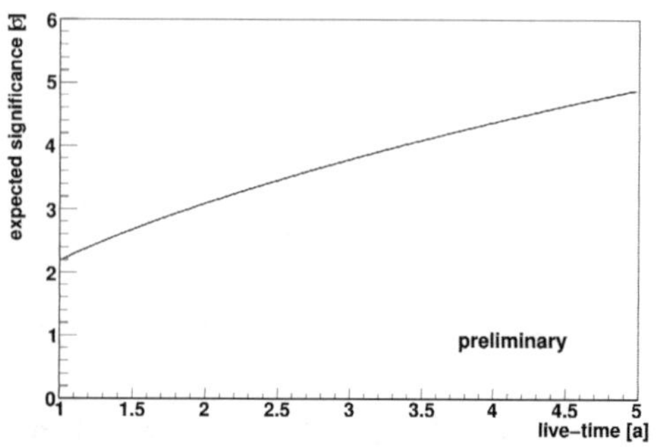

Figure 2. Expected significance (in units of standard deviations) for the observation of the Moon shadow effect with ANTARES as a function of the data-taking time.

2. Detection of the Moon shadow with ANTARES

The Moon casts a shadow in the cosmic ray flux impinging on the Earth [3]. By measuring an event deficit from the direction of the Moon, the absolute orientation of the detector can be checked up to the final analysis level. However, since the Moon (with an average radius of $R_{\text{Moon}} = 0.259°$ and a corresponding solid angle of $\Omega_{\text{Moon}} = \pi R^2_{\text{Moon}}$) covers only a tiny fraction of the sky, good statistics as well as a good angular resolution ($< 1°$) are required.

A detailed simulation of the shadow effect based on a modified version of CORSIKA [4] has been performed. In particular, this simulation includes the geomagnetic field, the angle between muon and cosmic primary, and the zenith angle dependence of the reconstruction quality. From this simulation the expected shape of the Moon shadow $s(\delta\theta, \delta\psi)$ with relative coordinates $\delta\theta = \theta_{\text{evt}} - \theta_{\text{Moon}}$, and $\delta\psi = (\phi_{\text{evt}} - \phi_{\text{Moon}}) \sin\theta_{\text{Moon}}$ is calculated in a region Ω_{View} around the Moon centre position (zenith θ_{Moon}, azimuth ϕ_{Moon}), defined as the muon flux with the shadow effect at $(\delta\theta, \delta\phi)$ normalised to the full flux without the shadow effect. (C.f. figure 1.)

The probability to observe the shadow effect is inferred from a hypothesis test in which the hypotheses *no shadow visible* H_0 and *shadow visible* H_1 are examined [5]. A test statistic $t(x)$ (for a measurement x yielding N events) is defined by the logarithm of the likelihood ratio:

$$t(x) = \log \frac{P(x|\text{H}_1)}{P(x|\text{H}_0)} = \log \prod_{i=1}^{N} \frac{p(\delta\theta_i, \delta\psi_i|\text{H}_1)}{p(\delta\theta_i, \delta\psi_i|\text{H}_0)} = \sum_{i=1}^{N} \log s(\delta\theta_i, \delta\psi_i) - N \log(1 - \frac{\Omega_{\text{Moon}}}{\Omega_{\text{View}}}).$$

A sample of measurements for each of the two hypotheses H_0 and H_1 is simulated to derive the probability distributions $p(t(x)|\text{H}_0, \text{H}_1)$, which are then used to calculate the expected significance of the measurement. The result of the analysis is shown in figure 2. The expected significance of the effect scales with the square root of the live-time, after 2 years the 3σ level is reached. The result is preliminary since systematic effects, in particular the influence of dead channels on the analysis, require further studies. Besides the presented simulation study, the analysis of first ANTARES data is currently ongoing to reveal the shadow effect.

[1] Zornoza J D 2009 *In these proceedings.*
[2] Brown A 2009 *Proc. of the 31th International Cosmic Ray Conference, Lodz, Poland*
[3] Clark G W 1957 *Physical Review* **108** 450
[4] Heck D 1998 *Forschungszentrum Karlsruhe Report FZKA* **6019**
[5] Fehr F 2009 *Ph.D. thesis, Univ. Erlangen*

Journal of Physics: Conference Series **203** (2010) 012134 doi:10.1088/1742-6596/203/1/012134

Background level estimations using Pulse Shape Analysis in new generation Ge experiments

H. Gómez, S. Cebrián, J. Morales, J.A. Villar

Laboratory of Nuclear and Astroparticle Physics. University of Zaragoza. Pedro Cerbuna 12. 50009-Zaragoza (SPAIN)

E-mail: hgomez@unizar.es

Abstract. Some neutrino properties, like its mass, could be determined or at least delimited using germanium detectors enriched in ^{76}Ge and studying the neutrinoless double beta decay of this isotope. In order to have the sensitivity required, the background level should be around 10^{-3} $c/keV/kg/y$ in the region of interest. Some previous studies showed that rejection of background events using Pulse Shape Analysis can be a very powerful tool to reach this level offering high efficiency for neutrinoless double beta decay events detection. To determine the rejection capability of this method and the detection efficiency, a set of simulations considering segmented Ge detectors were developed showing that three-dimensional spatial resolution and not only in radial direction, is necessary in order to reject enough background multi-site events. For this reason an algorithm to generate signals from simulated events and analyze them with three-dimensional sensitivity has been defined and will be presented.

1. Introduction

New generation experiments studying the neutrinoless double beta decay ($0\nu\beta\beta$) of ^{76}Ge using germanium detectors enriched in this isotope, need to reach a background level around 10^{-3} $c/keV/kg/y$ keeping a detection efficiency for $0\nu\beta\beta$ events above 75 % to be able to explore the neutrino effective mass ($< m_\nu >$) region around 50 meV [1]. Preliminary studies [2] concluded that Pulse Shape Analysis (PSA) in segmented detectors having three-dimensional resolution to discriminate signal events from background ones is one of the best options to reach this level. To estimate more accurately the potential of this technique a set of routines to generate electric pulses inside the detector from simulated events and its further analysis have been developed.

2. Simulation features

Pulse generation and analysis routines have been applied to a background model generated using GEANT4 and DECAY0 and composed by the main contributions to the background level existing in previous experiments. $0\nu\beta\beta$ as signal events have been also simulated. The geometry used to carry out the simulations is a single detector composed of a germanium crystal and three layers of copper with different thickness to define the cryostat. Different crystal masses and segmentation schemes have been implemented.

3. Pulse generation and analysis

The PSA with three-dimensional resolution is based on the analysis of the signal produced in the segment where the energy deposit is located (net signal), together with the signals produced in the adjacent segments (induced signals). The generation of all the signals is based on the movement of the charge carriers inside the crystal, that induces a differential charge in the contacts of the segment which depends on the so called *weighting field* [3]. The *weighting field*, that corresponds to the electric field existing in the detector for the net signals, are a technique to treat mathematically the capacitive coupling between the segments based on Ramo's theorem [4]. The representation of the variation of the differential charge along 1000 ns has been considered the electric signal. The shape of these signals is influenced by the *characteristic time* of the transfer function of the electronics associated to the detector as it showed in figure 1.

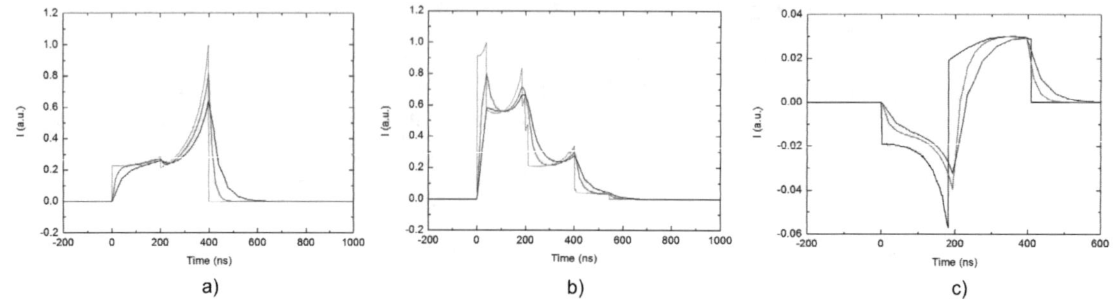

Figure 1. Some examples of generated signals without considering *characteristic time* (black line) and considering 20 ns (red) and 40 ns (blue): net signal of a *monosite* event (a), net signal of a *multisite* event (b), induced signal for a monosite event (c).

Analysis methods to distinguish *Monosite* events (like the signal) from *Multisite* ones (like in many background components) are based on the counting of the singular points for the net signals, since each energy deposit produces two of these points. For the induced signals, the area of lobes is analyzed since it has been found to be different for *Mono* and *Multisite* events of the same total energy. The efficiency of the analysis of the net signals, that only has radial resolution, depends on the *characteristic time*, and the best sensitivity levels are obtained for a value of 40 ns.

4. Results and conclusions

Efficiency to signal and final background level have been evaluated in different scenarios applying the described analysis methods. With them, the sensitivity of the experiment ($< m_\nu >$) can be estimated through the detector and nuclear factors of merit F_D and F_N (see definitions for instance in [5]). The range of sensitivity of the experiment depends on the uncertainty given by F_N. In general the sensitivity level is better for 4-kg detectors and PSA with three-dimensional resolution reach the best sensitivity values comparing with other experimental techniques. For example, the sensitivity level for the most favorable F_N values, are better than 50 meV even with an exposure level below 1000 $kg \times y$, and for this exposure the obtained sensitivity range is 37-53 meV.

References

[1] Avignone III F T 2008 *Rev. Mod. Phys.* **80** 481
[2] Gómez H, Cebrián S, Morales J, Villar J A 2007 *Astrop. Phys.* **28** 435
[3] Kroll Th, Bazzacco D 2001 *Nucl. Instr. and Meth.* A **463** 227
[4] Ettenauer S 2008 *Nucl. Instr. and Meth.* A **588** 380
[5] Morales A 1999 *Nucl. Phys.* B *(Proc. Suppl.)* **77** 335

Journal of Physics: Conference Series **203** (2010) 012135 doi:10.1088/1742-6596/203/1/012135

First operation of an 18-fold segmented n-type HPGe detector in liquid nitrogen

I. Abt, A. Caldwell, D. Lenz, J. Janicsko, J. Liu, X. Liu, B. Majorovits and F. Stelzer

Max-Planck-Institut für Physik in München, Germany

E-mail: ``Jing Liu'' <jingliu@mppmu.mpg.de>

Abstract. For the first time an 18-fold segmented n-type HPGe detector was operated for nearly five months submerged in liquid nitrogen. The performance of the detector was monitored and proven to be stable.

In order to minimize the radioactivity close to the HPGe detectors used in low background experiments, the detectors can be directly submerged into cryogenic liquid [1]. The cryogenic liquid serves as cooling as well as shielding material. Segmented HPGe detectors are used to identify background events. They typically have energy deposits in several different segments while signal events deposit energy in only one segment. A combination of these two techniques is considered to be used in the second phase of the GERDA experiment [2, 3].

Dark matter or neutrinoless double beta decay experiments have to run for a couple of years. If segmented HPGe detectors are to be operated directly in cryogenic liquid for a long period, their stability has to proven. A test stand was built at the Max-Planck-Institut für Physik in München for the simultaneous operation of up to three 18-fold segmented HPGe detectors in cryogenic liquid. As a first test one detector was kept in liquid nitrogen from Apr. 23 to Sep. 15, 2008.

Figure 1 depicts the test stand and two 18-fold segmented n-type HPGe detectors mounted inside. During the relevant period the detector under study was mounted at the top position.

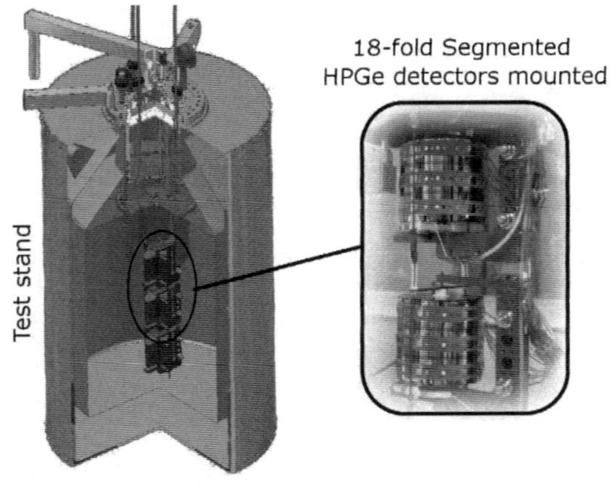

Figure 1. Test stand for the operation of segmented detectors in liquid nitrogen. The 18-fold segmented n-type HPGe detectors mounted inside the test stand are shown in the inset.

The energy resolution and the leakage current of the detector at the operation voltage of 2000 V were monitored during the 146 days of operation. The results are shown in Fig. 2 and 3.

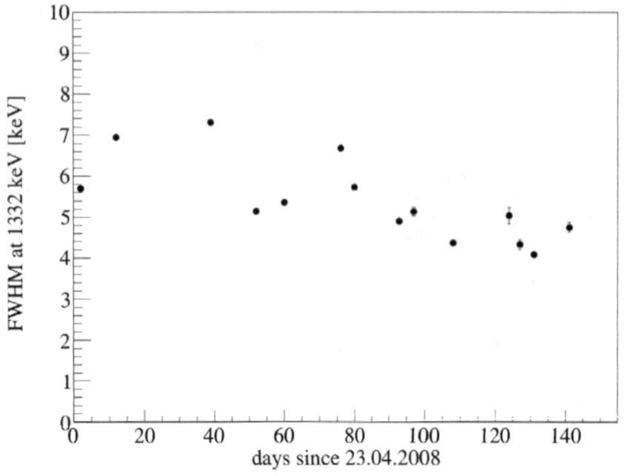

Figure 2. Resolution of the detector as determined from fits to the 1332 keV γ line of ^{60}Co. The uncertainties on most values are hidden by the size of the dots.

Two methods were used to measure the leakage current: (a) a direct measurement using a picoammeter and (b) an indirect measurement comparing the baselines at 0 V and 2000 V. The leakage current stayed constant, around 28 pA. The values given by method (b) was not calibrated. Important is, that the values are constant within errors.

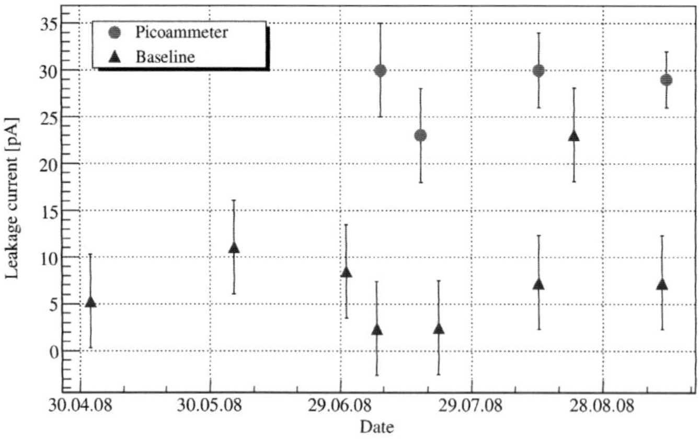

Figure 3. Leakage current of the detector at the operation voltage of 2000 V. The results from measurements using a picoammeter (dots) and uncalibrated baseline shifts (triangles) are given.

To summarize, for the first time, an 18-fold segmented n-type HPGe detector was operated in liquid nitrogen for a long period. No degradation of the performance was observed, indicating that the technique of segmented detectors submerged in cryogenic liquid is mature enough to be used in low background experiments such as dark matter or neutrinoless double beta decay searches.

References

[1] G. Heusser, Annu. Rev. Nucl. Part. Sci. **45** (1995) 543.
[2] I. Abt *et al.*, arXiv:hep-ex/0404039v1.
[3] S. Schönert *et al.*, [GERDA Collab.], Nucl. Phys. Proc. Suppl. **145** (2005) 242.

Journal of Physics: Conference Series **203** (2010) 012136 doi:10.1088/1742-6596/203/1/012136

PAMELA mission: constraints to cosmic rays propagation from nuclei

L Marcelli on behalf of the PAMELA Collaboration
Via della Ricerca Scientifica 1, I-00133, Rome, ITALY
E-mail: laura.marcelli@roma2.infn.it

Abstract. The PAMELA (Payload for Antimatter Matter Exploration and Light nuclei Astrophysics) experiment is a satellite-borne apparatus that performes measurements of the cosmic radiation with a particular focus on antiparticles and light nuclei. Accurate measurements of the elemental composition of cosmic rays are required in order to understand the origin, propagation and lifetime of the cosmic radiation. In particular PAMELA measures the light nuclear component of cosmic rays from Hydrogen up to Oxygen in the interval 200 MeV/n-200 GeV/n. Object of this paper is the presentation of light-charge identification capabilities of PAMELA instrument as evaluated during the first three years of flight.

1. Introduction

The relative abundances of the constituents of galactic cosmic rays (GCRs) provide information about cosmic ray transport within the Galaxy. In particular, cosmic rays of primary origin such as Carbon and Oxygen may interact with the interstellar medium to produce secondary fragments such as Lithium, Beryllium and Boron. The measured ratio of secondary to primary cosmic rays can be used to compute the mean amount of interstellar matter that cosmic rays have encountered before reaching the Earth. One of the most sensitive quantity is the Boron to Carbon ratio (B/C), as B is purely secondary and its main progenitors C and O are primaries. Measuring the energy dependence of the B/C ratio we can infer the diffusion coefficient [1]. Moreover, B/C ratio is the quantity measured with the best accuracy, so that it is ideal to test models.

A better determination of the cosmic ray propagation is fundamental for the search of exotic matter, like dark matter candidates or antimatter produced in exotic processes, since the signature of such processes can be recognized only by knowing with great precision the fluxes due to the conventional production, acceleration and transport models [1, 2, 3].

2. The PAMELA mission

The PAMELA mission is focused on the investigation of antimatter and dark matter, cosmic ray propagation in our Galaxy and Solar System, solar modulation and interaction of cosmic rays with the Earth's magnetosphere.

The PAMELA apparatus comprises a magnetic spectrometer, a Time of Flight system, a silicon-tungsten electromagnetic calorimeter, an anticoincidence system, a shower tail catcher scintillator and a neutron detector. The combination of these devices allows antiparticles to be reliably identified from a large background of other charged particles. For more details see [4].

3. Charge Identification and Intrinsic Detectors Resolution

PAMELA is able to discriminate nuclei events, from Boron up to Oxigen, in the energy range from 200 MeV/n up to 200 GeV/n. In this section the intrinsic detector resolutions in Z reconstruction for nuclei from Z = 3 up to Z = 6 will be presented.

In figure 1, two nuclear events (Boron and Carbon respectively) detected by PAMELA instrument are shown. The hadronic interactions inside the Calorimeter are clearly visible, as well as the activity recorded by the Neutron Detector.

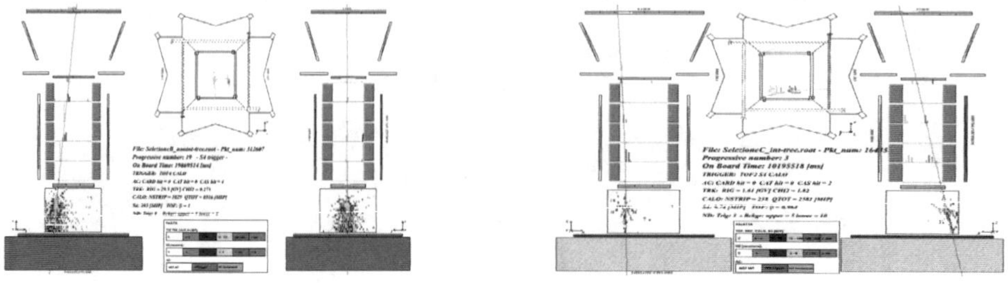

Figure 1. Boron (left) and Carbon (right) as detected by the PAMELA instrument. For each figure, on the left(right) the x, bending view (y, non-bending view) of PAMELA are indicated. A plan view of PAMELA is shown in the centre.

PAMELA can determine the particle charge in a redundant way, by means of the ToF scintillating paddles, the silicon microstrips of the Tracking system and the ministrips of the imaging Calorimeter.

The particle charge can be measured in the Calorimeter by considering the energy released in one, or more than one, plane of the detector. The gaussian fit of the peaks of the charge histogram gives the following peak/sigma: 3.01/0.20; 4.01/0.23; 5.00/0.26; 6.01/0.29.

The ToF system provides the charge of the incident particles by the energy deposit measurements in the scintillators. The six scintillator layers enable six independent charge determinations, thus significantly improving the charge resolution of the whole ToF system. Combining by a weighed average the charges reconstructed with this method it is possible to obtain the following resolution: 3.00/0.11; 4.00/0.14; 4.99/0.18; 5.99/0.21.

Moreover also the Tracking system can be used to determine the absolute value of the charge by multiple measurements of the mean rate of energy loss in the silicon sensors with the following resolution: 2.99/0.16; 4.00/0.23; 4.98/0.28; 6.01/0.34.

4. Conclusions

Preliminary results on PAMELA charge identification capabilities and the consequent charge resolution has been presented in this paper. The ToF system seems to be the most powerful detector.This is only a first analysis of the nuclear components of cosmic rays detected by the PAMELA instruments. This work is in progress. However, PAMELA confirms the possibility to improve our knowledge in the mechanisms involved in the generation, acceleration and propagation of cosmic rays.

References

[1] Castellina A and Donato F 2005 *Astropart. Phys.* **24** 146
[2] Maurin D et al 2002 Galactic Cosmic Ray Nuclei as a tool for Astroparticle Physics *Preprint astro-ph/0212111*
[3] Bottino A, Donato F, Fornengo N and Scopel S 2004 *Phys. Rev.* D **70** 015005
[4] Picozza P et al 2007 *Astropart. Phys.* **27** 296

Journal of Physics: Conference Series **203** (2010) 012137 doi:10.1088/1742-6596/203/1/012137

Hanohano: Hawaiian Antineutrino Observatory

Jelena Maricic for the Hanohano collaboration

Drexel University, 3141 Chestnut St. 12-816, Philadelphia, PA, 19104

E-mail: `jelena@physics.drexel.edu`

Abstract. Design studies are underway for the deep ocean antineutrino observatory Hanohano. The 10 kton monolitic underwater detector will be able to make precision measurement of neutrino mixing parameters (including θ_{13} and neutrino mass hierarchy) if stationed around 60 km offshore, from the nuclear reactor. Hanohano will be a mobile detector and placing it in a mid-Pacific location will provide the first ever flux measurement of geoneutrinos (antineutrinos emitted in the radioactive decay series of uranium and thorium), coming from the Earth's mantle and perform a sensitivity search for a hypothetical natural fission reactor in the Earth's core. Additional deployment at a different mid-ocean location will lead to tests of lateral heterogeneity of uranium and thorium in the Earth's mantle. These measurements would provide an important insight into deep-Earth geophysics, mantle composition and understanding of the Earth's heat flow and sources of energy inside the Earth.

1. Introduction

Hawaiian Antineutrino Observatry (Hanohano) is under development in Hawaii and elsewhere. Hanohano is envisioned as a 10 kton monolitic detector filled with ultra-pure scintillator fluid. Inverse beta decay reaction is used to detect electron anti-neutrinos with energy of 1.8 MeV or higher. The 10 kton target volume is surrounded with an array of 10-inch photomultiplier tubes, placed in 13-inch pressure housings (see Fig. 1 and [1][2]). PMTs will collect scintillating photons from interactions taking place inside the detector. The detector will have energy resolution of $3.5\%/\sqrt{E_{vis}}$, where $E_{vis} = E_{\nu} - 0.8$ MeV is visible energy inside Hanohano. It will be deployed in the deep ocean at depth between 3 and 4 km to obtain sufficient overburden. Low muon incidence rate, combined with thick shielding of the target volume and ultra-pure scintillator will result in a very high detection efficiency. Hanohano will be deployed from the barge, with about 30 minutes needed for descent and ascent. Hanohano's mobility significantly increases its science potential. While a location offshore from the nuclear reactor will allow precision measurement of neutrino mixing parameters including θ_{13} and neutrino mass hierarchy, additional deployment near Hawaii will provide excellent measurement of terrestrial anti-neutrinos.

2. Scientific potential

2.1. Neutrino Mixing Parameters

Understanding of the phenomenon of neutrino oscillations has improved dramatically over the last decade and precision measurements of the most of the neutrino oscillation parameters have been obtained. However, neutrino oscillation angle θ_{13} is still only confined by an upper limit and a question of the neutrino mass hierarchy is still open.

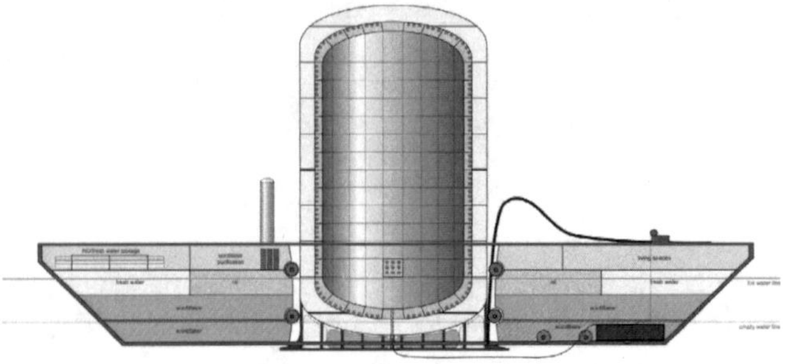

Figure 1. Conceptual design of the Hanohano detector along with the deployment barge.

One year long deployment of Hanohano, 60 km offshore from San Onofre nuclear plant (7 GW power level) in California, or Maanshan reactors (6 GW) power level) in Taiwan could resolve mass hierarchy problem for $\sin^2 2\theta_{13} \geq 0.05$ and measure θ_{13} without the need for the near detector [2].

2.2. Terrestrial Antineutrinos
Measurement of the terrestrial antineutrino flux is of great importance to geophysics. Recent results from KamLAND [3] provided the first ever experimental evidence of terrestrial antineutrino flux. The measurement was rather crude and could only confirm our current understanding of the Earth's composition and energetics. More precise measurement and at different location would shed light and potentially differentiate among different Earth models.

One year long deployment of Hanohano near Hawaii would provide the first ever measurement of geoneutrino flux from the mantle with 25% uncertainty. In addition, remoteness of Hawaii location from man-made nuclear reactors, is ideal for testing the hypothesis for the existence of the natural nuclear fission reactor in the earth's core. This reactor is predicted to be in the energy range from 1-10 TW. Hanohano would set a 99% C.L. limit to the geo-reactor power at 0.3 TW, or a 1 TW reactor (if proven to exist) would be detected at 5σ C.L.

2.3. Other physics opportunities
Ten kton scinitillator detector will detect a strong galactic supernova signal and even search for relic supernovae neutrinos. The signal can be as large as 4 events per year [4]. If the radio purity level is high enough, the flux from sub-MeV solar neutrinos may be measured. Finally, Hanohano can detect proton decay in the following mode: $p \rightarrow \nu + K^+$ and may reach a sensitivity of almost 10^{34} years.

3. Conclusion
Hanohano is a deep ocean antineutrino observatory being developed in Hawaii and other places. One year long deployment, 60 km away from the nuclear reactor can resolve mass hierarchy for $\sin^2 2\theta_{13} \geq 0.05$ without the near detector. Additional one year long deployment in mid-pacific will measure the flux of geoneutrinos from U and Th decay series with 25% uncertainty and either measure or severely limit the power level of the hypothetical geo-reactor. Monte Carlo studies, accompanied by lab and field tests are underway.

References
[1] *http://www.phys.hawaii.edu/ sdye/CEROS_final_report.pdf*
[2] J. Learned *et al* 2008 *submitted to Phys. Rev. D hep-ex/0612022*
[3] T. Araki *et al* 2005 *Nature.* **436** 499
[4] L. Strigari *et al* 2004 *JCAP* **7** 0403

Journal of Physics: Conference Series **203** (2010) 012138 doi:10.1088/1742-6596/203/1/012138

Results of the radio-detection experiment HASRA (Hawaii Askaryan in Salt Radio Array) and prospect for UHE neutrino detection in salt

Radovan Milincic for HASRA collaboration

Drexel University, Philadelphia, PA/ Gloucester County College, Sewell, NJ

E-mail: radovan@physics.drexel.edu

Abstract. Measurements of radio emission, associated with Ultra High Energy Cosmic Ray particles interaction in a dense dielectric material, provide an effective method of their detection and exploration. The Hawaii Askaryan in Salt Radio Array (HASRA) was built to explore possibility of detection of coherent radio Cherenkov emission in rock salt from interaction of cosmic ray protons and air shower particles. Performance of the detector, the implemented detection techniques, and results of 1 year of measurements of Askaryan effect from CR particles with energy above 100 GeV will be reported. Also potential of the future detectors utilizing radio techniques in rock salt will be discussed.

1. Introduction

The Ultra High Energy (UHE) neutrinos are produced at cosmological distances in processes which take place in or around powerful accelerators like AGNs and GRBs, or are remnants from the processes which happened after the inflation period. An additional source of UHE neutrinos are interactions of UHE protons with cosmic microwave radiation. The interaction cross-section of the UHE neutrinos combined with the low expected flux from different sources, makes their detection extremely challenging. A 100 km^3 water equivalent size neutrino telescope can provide a steady monitoring of GZK neutrinos and measurement of UHE neutrino flux. The most promising and cost effective approach to detection of UHE neutrinos and measurement of their flux is via the use of Askaryan effect combined with the radio and acustic detection technique. The Hawaii Askaryan in Salt Radio Array (HASRA) is designed as a test bed detector for radio Cherenkov emission from cosmic ray induced cascade in salt. It was operated at Manoa campus of University of Hawaii, just above Honolulu.

2. HASRA

Fig. 1 shows the apparatus of HASRA experiment. The target dielectric mass is built from 22.68 kg salt blocks, with a total mass of about 20,000 kg. The basic receiving element is a printed circuit board (PCB) broadband dual-linear polarization bowtie antenna. Two arrays of antennas are installed in the horizontal plane, with a good frequency response in range 0.2-2.0 GHz. Together, they formed one trigger system. The energy of the electromagnetic shower above the trigger threshold is estimated using EGS4 Monte Carlo simulation. This results in the threshold value of 14 GeV. Muons, protons and charged pions are particles which significantly

Journal of Physics: Conference Series **203** (2010) 012138 doi:10.1088/1742-6596/203/1/012138

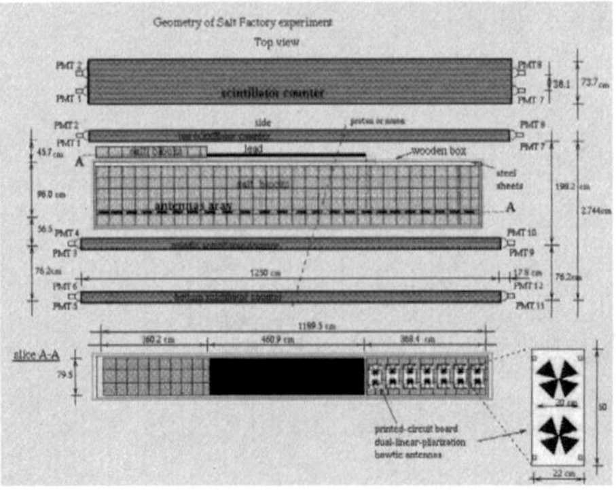

Figure 1. Experimental setup of HASRA.

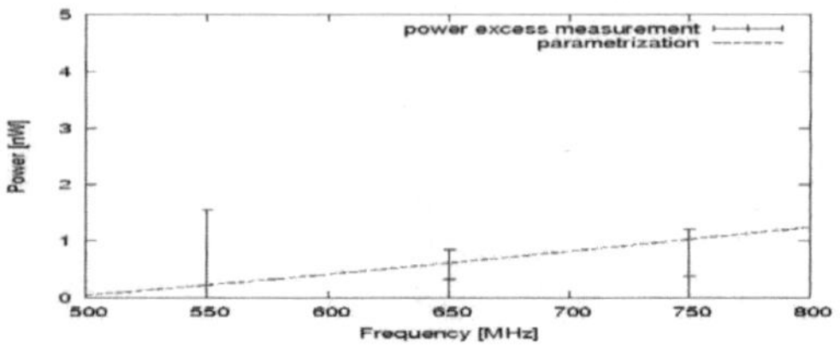

Figure 2. Power excess (500 - 800 MHz).

contribute to the total number of triggered events in the HASRA detector. The sensitivity of the detector is determined under the assumption that the electric field strength from the shower development is similar to the one produced by bremsstrahlung photon bunches in the SLAC experiment. The 1σ threshold based on EGS4 simulation and measured thermal noise is: $E_{1\sigma} = 2.84\ TeV$. We performed several different analysis on the data set. The analysis of power of the recorded signal was performed in the time and frequency domain. We use time domain analytic signal analysis method, and in this way we facilitate coherent addition of signals in the time window of the Askaryan pulse. Matlab Hilbert transform function which returns a complex sequence from real data was used. The results of the time domain analysis with the analytic signal method for the power excess show that the mean value of the power distribution has a small positive offset of (0.0017 ± 0.0115) mW. The analysis of the antenna data to calculate power excess was also done with the fast Fourier transform (FFT). Fig. 2 presents result of the analysis in band 500 - 800 MHz. The spectrum variation in the band influences the magnitude of the standard error. The dashed line is a parametrization based on the previous accelerator measurement.

3. Perspective for UHE neutrino detection and Conclusion

Two calibration measurements:accelerator measurement and HASRA experiment indicate that radio detection seems to be the optimal approach for UHE neutrino detection. The large natural rock salt mushroom shape formations with 3 - 8 km in diameter and 10 - 15 km in depth (dome), found in a Gulf region of USA, are excellent choice to host the detector. The 1 km^3 SalSA detector can be placed within a host dome. The detector consists of 10×10 grid of antenna strings which begin at 750 m and have 2 km in length with 225 m in between string spacing. Full detector sensitivity for estimated GZK ν fluxes from different models (for 3 years of detector livetime) is between 70 and 230 events. The center mass energy ranges from 150 up to 300 TeV providing us with the opportunity to explore phenomena in the regime of the GUT scale.

The HASRA detector was built to study coherent radio Cherenkov emission from muon and proton interactions in the rock salt. Analysis of the data set for the excess of power from the Askaryan pulse shows positive offset in agreement with the expectation. Sensitivity of the detector for the average excess in the collected signal requires a higher average shower energy threshold.

The SalSA detector offers measurement of UHE ν with high statistics, with capability to characterize GZK ν fluxes within one year of operation. The detector has potential for distinguishing between different ν flavors and has an extremely long baseline for neutrino oscillation measurement.

Detection of fast neutrons with LiF and Al_2O_3 scintillating bolometers

N Coron[1], C Cuesta[2], J Domange[2], E García[2], J Gironnet[1], J Leblanc[1], P de Marcillac[1], M Martínez[2], Y Ortigoza[2], A Ortiz de Solórzano[2], C Pobes[2], J Puimedón[2], T Redon[1], T Rolón[2], M L Sarsa[2], L Torres[1] and J A Villar[2]

[1] Institut d'Astrophysique Spatiale, Bât 121, 91405 Orsay, Paris, France
[2] Laboratorio de Física Nuclear y Astropartículas, Universidad de Zaragoza, 50009 Zaragoza, Spain

E-mail: puimedon@unizar.es

Abstract. Scintillating bolometers of LiF and Al_2O_3 can monitor the fast neutrons flux in WIMPs searches. With both materials we merge the traditional fast neutron detection methods of induced reactions and scattering. The ROSEBUD collaboration devoted an underground run in the old Canfranc laboratory to study the response of LiF and Al_2O_3 to fast neutrons from ^{252}Cf. Both bolometers were used simultaneously in a common experimental set-up resembling those of current WIMPs searches, which could give valuable insights into future WIMPs searches with cryogenic detectors as EURECA.

The ROSEBUD collaboration [1] can test several absorbers simultaneously [2] running as a small size multi-target experiment. A time fraction of the last run in the old Laboratorio Subterráneo de Canfranc [3,4] was used to study the nuclear recoil band of LiF, Al_2O_3 and BGO scintillating bolometers as well as the viability of monitoring fast neutrons inside the Pb shielding using the combined data of LiF and Al_2O_3. The irradiation with neutrons from a source of ^{252}Cf located outside the shielding shows their particle discrimination power: nuclear recoils, captures in ^6Li and β/γ events are clearly separated (figures 1 and 2). The Al_2O_3 heat channel was calibrated with a ^{57}Co gamma source, although the irradiation time was not enough to produce high statistics full absorption peaks.

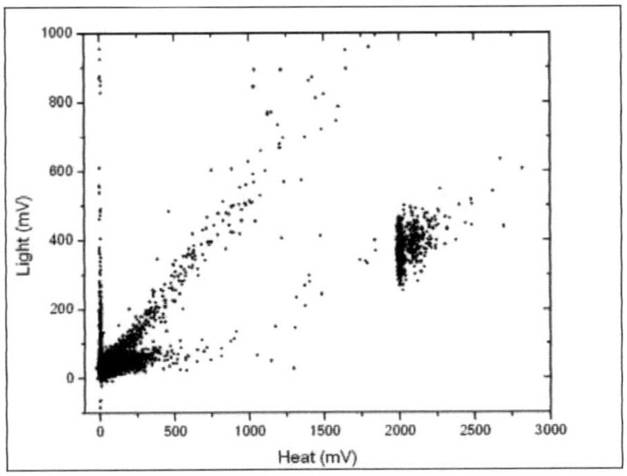

Figure 1. Heat-light plot of LiF. $^6Li(n,\alpha)t$ captures are located around (2000,400) mV.

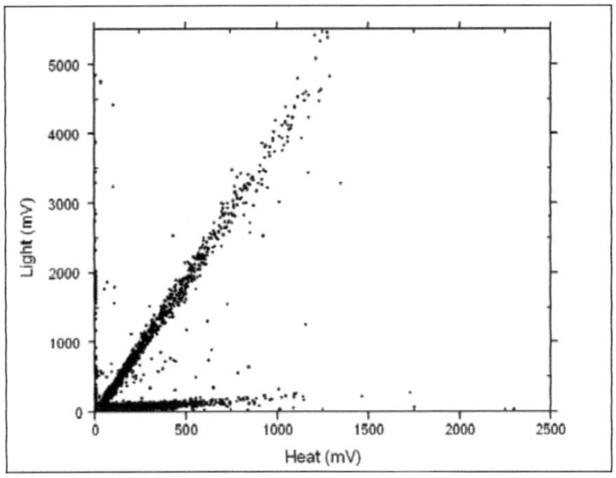

Figure 2. Heat-light plot of Al_2O_3. Nuclear recoil events give lower light output than β/γ ones.

We assumed an isotropic neutron flux around the bolometers and a three-parameter model for the neutron spectrum: $\Phi(E)dE=(\Phi_0/\Gamma(\alpha+1))(E/T)^{\alpha}\exp(-E/T)dE/T$. The isotropy can be justified because 25 cm of Pb flat the anisotropy of the neutrons from the outside point source; the energy dependence corresponds to the shape of the ^{252}Cf neutron spectrum (α=0.5, T=1.29 MeV) and it contains as special cases a translation in lethargy (α=0.5, T<1.29 MeV) and, asymptotically, the equilibrium spectrum ($\alpha \rightarrow$-1, E<<T). It seems valid that Pb produces an intermediate shape with other (α, T) pair.

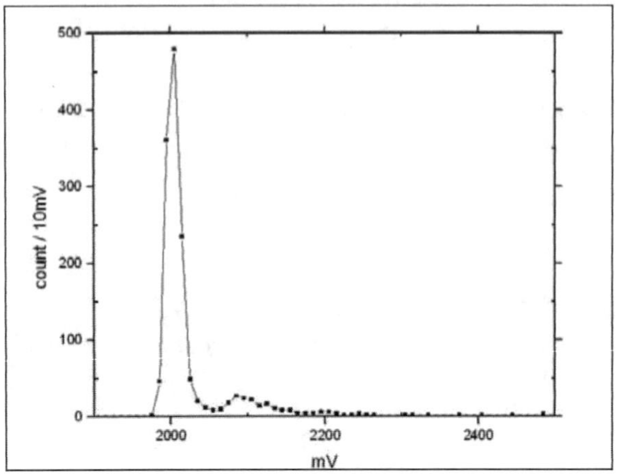

Figure 3. Heat spectrum of the (n,α) events in LiF

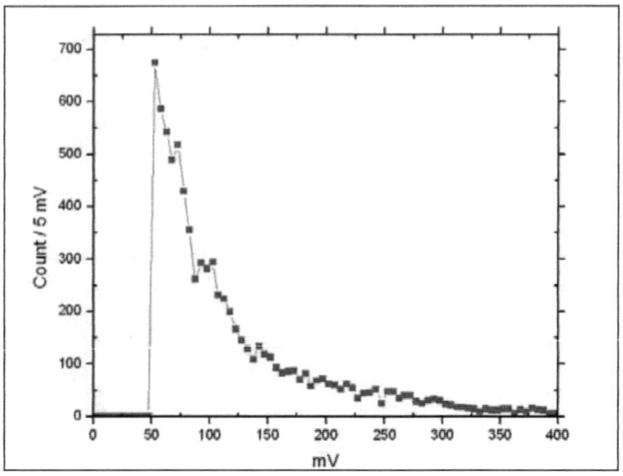

Figure 4. Heat spectrum of recoil events in Al_2O_3

Appropriate cuts on the heat-light plots provided the differential spectra of the heat channel for the ^{6}Li(n,α)t capture (figure 3) and for the elastic scattering in Al_2O_3 (figure 4). The capture by ^{6}Li of thermal neutrons and at the 0.24 MeV resonance can be seen in figure 3 as two peaks at 2000 mV and 2100 mV, respectively. The number (C) of neutron captures at the resonance, the number (A_1) of nuclear recoil events in Al_2O_3 in the interval [50, 100] mV and in the region above 100 mV (A_2) were used to estimate the free parameters α, T and Φ_0. The ^{57}Co calibration gave 37±9 keV at the 50 mV discrimination threshold and 85±4 keV at 100 mV. The choice of 100 mV gave similar values for A_1 and A_2, minimizing their joint contribution to the uncertainties of the three parameters. The efficiencies for each type of event were calculated as a function of the neutron energy with MCNP version 4C [5] and, finally, we obtained a solution with a neutron mean energy of 0.148 MeV, well below the mean energy of the original ^{252}Cf neutrons (1.9 MeV), showing that our model was not incompatible with the integral data C, A_1 and A_2. The parameters α and T were obtained with big uncertainties (-1<α<1 and 0.3 MeV<T<2.5 MeV) which came from the non accurate calibration of Al_2O_3 recoil spectrum and from the statistical uncertainty of the captures at the ^{6}Li resonance. We concluded that the experimental data gave information for neutron energies between 0.1 and 2 MeV and a fast (E > 0.1 MeV) neutron flux of 0.20±0.02 n s^{-1} cm^{-2}. We foresee further improvements exploiting the shape of the differential heat spectrum of recoil events in Al_2O_3.

Acknowledgements

This work has been supported by the Spanish Ministerio de Ciencia e Innovación (FPA2008-03228), the French CNRS/INSU (MANOLIA and BOLERO projects) and the EU project ILIAS (RII3-CT-2004-506222). Y Ortigoza and T Rolón thank the support of UZ/BSCH/Fundación Carolina.

References

[1] Cebrián S, Coron N, Dambier G, García E *et al* 1999 *Astroparticle Physics* **10** 361-8
[2] Cebrián S, Coron N, Dambier G, de Marcillac P *et al* 2004 *Astroparticle Physics* **21** 23-34
[3] Calleja A, Coron N, García E, Gironnet J, Leblanc J *et al* 2008 *J. Low Temp. Phys.* **151** 848-53
[4] Coron N, Gironnet J, Leblanc J *et al* 2008 *Proceedings of Science* PoS(idm2008)007
[5] Briesmeister J F (ed) 2000 MCNP - *A general Monte Carlo N-Particle transport code, version 4C*, LA-13709-M

Technique to measure the cosmic muon flux at the old Canfranc Underground Laboratory

A. Rodríguez, J. M. Carmona, S. Cebrián, T. Dafni, J. Galán, I. G. Irastorza, H. Gómez, F.J. Iguaz, G. Luzón, J. Morales†, J. Ruz, L. Seguí, A. Tomás

Universidad de Zaragoza, Pedro Cerbuna 12, 50009 Zaragoza, Spain

Email: mara@unizar.es

Abstract. A technique to measure the cosmic muon flux distribution has been developed in the old Canfranc Underground Laboratory (2450 m.w.e.).This technique can determine the muon direction and it allows to obtain a three-dimensional distribution of the muon intensity which could be correlated to the mountain profile, as well. The experimental set-up, which has been taking data since November 2008, consists of a Time Projection Chamber (TPC) detector and a veto system composed of two plastic scintillators.

1. Introduction

Muon transport through matter has an important role in many areas of particle and astroparticle physics. Cosmic rays, which consist mostly of protons, interact in the atmosphere producing muons. Owing to their low energy loss, essentially by ionization, they can travel large distances and commonly reach the ground with a mean energy of 3 GeV. The average muon flux at the surface goes as $\cos^2\theta$. The rate in a thin horizontal detector is roughly 1 $cm^{-2}.min^{-1}$ [1]. Muon flux is effectively suppressed at large depths underground, for instance a previous measurement at the old Canfranc Underground Laboratory provided us with a muon flux of (5.04±0.25) x 10^{-3} $\mu.m^{-2}.s^{-1}$[2]. Also, their interaction with the rock or detector materials produces high-energy neutrons becoming an important source of background in experiments looking for rare events in underground sites.

The goal of the technique presented here is to measure the cosmic muon flux and determine the muon direction in the Canfranc Underground Laboratory using a Time Projection Chamber (TPC) detector. The features of this detector allow us to obtain information about several parameters of the muon flux, such as direction, length of its track, interaction angle, and energy deposition. It also permits to discriminate different particles like gammas, electrons and alphas.

2. Experimental set-up

The experimental set-up consists of two different components: a TPC detector and a veto system. The TPC detector used here was installed in the CAST Experiment where it achieved successful results [3]. The gas mixture, Ar (95%) +CH_4 (5%), is located inside the TPC in a volume 10x15x30 $cm^3$with a 10 cm drift.

†deceased

The TPC working concept is simple: muon passing through the detector ionizes the gas liberating electrons, which produce secondary electrons with lower energy while drifting towards the amplification area. Here, the avalanche process takes place, due to the high electric field in the anode wires. Finally electric signals, that contain information about the location and charge production of initial interaction, are generated in the wires and recorded by the acquisition system. The detector has one anode plane and two cathodes planes. Each cathode plane has 96 wires, and the anode plane has 48 wires running perpendicular to the cathodes wires. These planes signals, combined with the trigger time, provide us three dimensional information for each event.

The veto system is located on top of the TPC detector, to obtain fast coincidence signals in time with the TPC trigger. It is made of two plastic scintillators BICRON BC 408 (80x40x5 cm^3). A 6 μs veto time window has been chosen for identification of almost vertical muons.

3. Results

In a first phase, the set-up was installed in a sea level laboratory of the University of Zaragoza. There, it has been possible to check the proper operation of the detector and to measure the vertical muon flux in surface: 80 $\mu.m^{-2}.s^{-1}$. From these data recorded in surface we have also obtained the software cuts necessary to distinguish different particles (see figure 1 left). For instance, to identificate vertical muons and induced events by them. Some of the cuts used are: veto time less than 6 μs, good linear fit to a track (figure 1 right), number of touched anodes 48. Since November 2008, the detector has been recording data in the Canfranc Underground Laboratory.

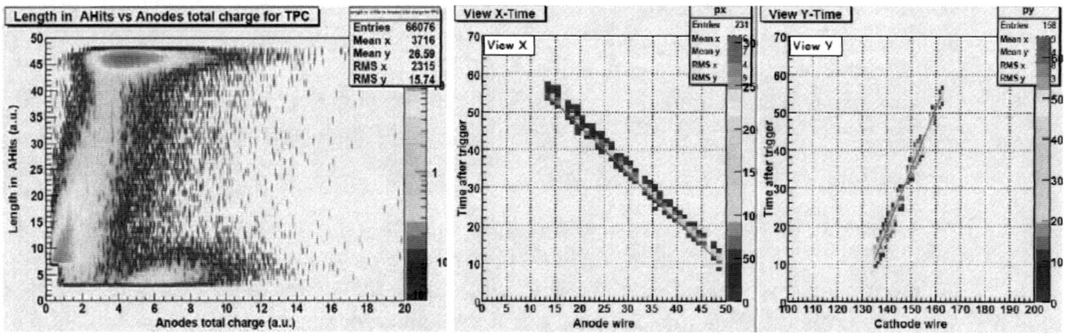

Figure 1: Left: This plot shows clearly distinction of the particles depending on their charge collection in anodes and track length: vertical muons (on top), X-rays (on the left) and alphas (on the bottom). Right: these pictures show the muon track in anodes (left) and cathodes (right).

4. Outlook

This technique to measure muons provides us with the opportunity to know more deeply the features of the energy deposition of muons and to obtain a three dimensional map of their distribution. The measurements carried out in surface have proved the reliability of the detector and its capability to measure muons. The second phase, in the Canfranc Underground Laboratory is undergoing. We expect to have the first results from these measurements in the next future.

References

[1] Amsler C et al 2008 Phys. Lett. B 667, 1.
[2] Luzon G et al 2006 Proceedings of the Sixth International Workshop on the Identification of Dark Matter Word Scientific 514-519.
[3] Autiero D 2007 et al New Journal Physics Vol 9, 6.

R-Process and Mass Varying Neutrinos

Rossi-Torres F, Guzzo M M and de Holanda P C

Instituto de Física "Gleb Wataghin"
Universidade Estadual de Campinas (UNICAMP) - 13083-970
Campinas - São Paulo - Brasil

E-mail: ftorres@ifi.unicamp.br

Abstract. In this brief article we present the relation between neutrino oscillation (in the $\nu_e \to \nu_s$ channel) in the mass varying neutrino scenario, and the r-process nucleosynthesis, trying to put some bounds in the oscillation space parameter from the r-process nucleosynthesis.

Mass Varying Neutrinos (MAVAN) may be a channel for explaining the cosmic coincidence issue and the expansion of the universe [1]. Neutrinos and dark energy densities track each other throughout the lifetime of the universe through the acceleron field, which energy density contributes to the dark sector. The dependence of neutrino mass with neutrino densities will be obtained by the minimization of the potential, which is constructed by the contribution from neutrinos as well as from the scalar potential (V_0). The contribution of neutrinos is obtained from the relic neutrinos of the Big Bang ($n^{C\nu B}$) and neutrino density of the medium ($n_{\nu,medium}$) [2]:

$$V_{C\nu B} = m_\nu n^{C\nu B}, \tag{1}$$

$$V_{\nu,medium}(m_\nu) = \int \frac{d^3k}{(2\pi)^3}\sqrt{(k^2 + m_\nu^2)}f(k). \tag{2}$$

The basic equation one must follow is:

$$n^{C\nu B} + \frac{m_\nu}{E_\nu} n_{\nu,medium} + V_0'(m_\nu) = 0, \tag{3}$$

where m_ν and E_ν are, respectively, the mass and energy of the neutrino. We choose a potential which has a logarithmic form, since we want to satisfy the equation of state $w \approx -1$ (see reference [2] for other kinds of potential that satisfy this condition).

R-process nucleosynthesis produces heavy nuclei beyond iron by fast n-capture compared with beta decay. The most plausible environment yet proposed is the neutrino-heated ejected from the nascent neutron star. Naturally, the r-process can only occur in a neutron-rich medium ($Y_e < 0.5$). The n/p fraction is governed by $\nu_e + n \to p + e^-$ and $\bar{\nu}_e + p \to n + e^+$. If neutrino oscillation occurs [3], even a partial swap to make a exchange of few 10% is enough to shift the medium to a proton-rich state. So it is possible to constraint neutrino oscillation from r-process [4]. Below, Fig.1 and Fig.2 show the survival probabilities of $\bar{\nu}_e \to \bar{\nu}_s$ and the electronic fraction, without the MAVAN scenario. One must remember that the region of $Y_e > 0.5$ must be excluded.

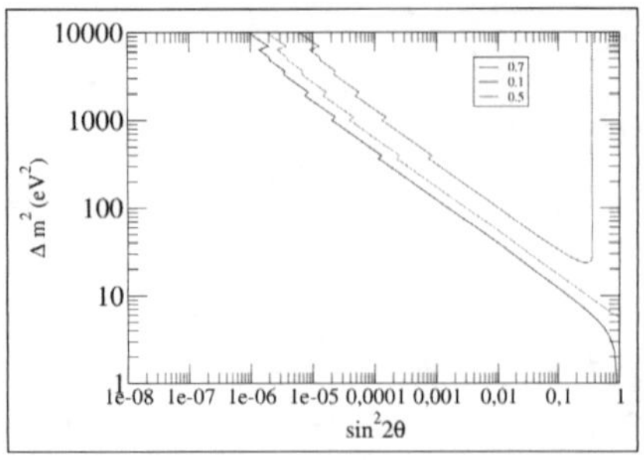

Figure 1. *Survival Probabilities of the process* $\bar{\nu}_e \rightarrow \bar{\nu}_s$.

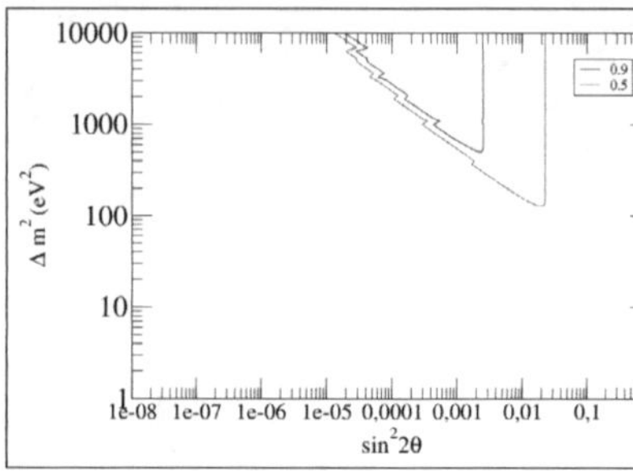

Figure 2. *regions of electronic fraction -* $Y_e > 0.5$ *excluded.*

Applying equation (3), we show Figs.3 and 4, where the vaccum mass squared difference necessary to damage r-process is reached only when neutrinos go out from the supernova. We notice, using this model, that Δm^2 is very low and, solving the evolution equation for neutrinos, $\bar{\nu}_e$ conserves its flavor during all the propagation.

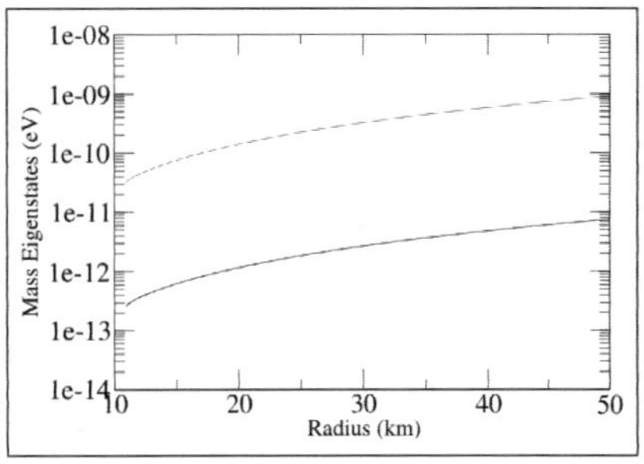

Figure 3. *Evolution of mass eigenstates.*

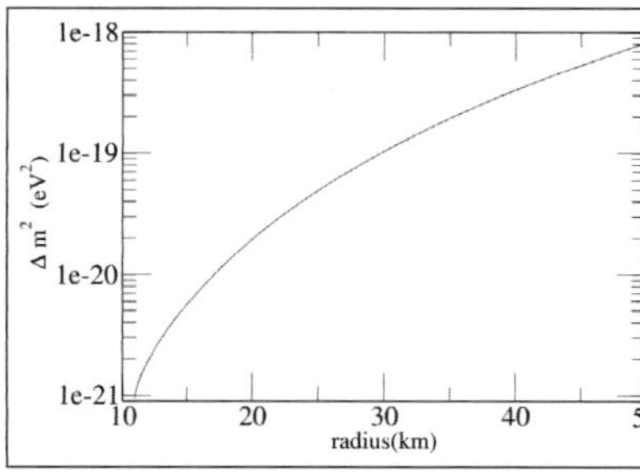

Figure 4. *Evolution of the mass squared difference.*

The inclusion of the MAVAN approach does not modify $Y_e \approx 0.4$, so the r-process is not damaged, since neutrino oscillation does not happen. Also we could say that the condition ($Y_e < 0.5$) does not put any bounds on the oscillation space parameter when we consider MAVAN.

Acknowledgments
The authors would like to thank FAPESP, CAPES and CNPq for several financial supports. F. Rossi-Torres thanks also the TAUP organizing committee for the financial support.

References

[1] Fardon R, Nelson A E and Weiner N 2004 *J. Cosmol. Astropart. Phys.* JCAP 0410(2004)005 (*Preprint* astro-ph/0309800)
[2] Cirelli M, Gonzalez-Garcia M C and Pena-Garay C 2005 *Nucl. Phys.* B **719** 219 (*Preprint* hep-ph/0503028)
[3] Qian Y Z, Fuller G M, Mathews G J, Mayle R, Wilson J R and Woosley S E 1993 *Phys. Rev. Lett.* **71** 1965
[4] Nunokawa H, Peltoniemi J T, Rossi A and Valle J W F 1997 *Phys. Rev.* D **56** 1704 (*Preprint* hep-ph/9702372)

New observation of $2\beta2\nu$ decay of ^{100}Mo to the 0_1^+ level of ^{100}Ru in the ARMONIA[1] experiment

P. Belli[1], R. Bernabei[1], R.S. Boiko[2], F. Cappella[3], R. Cerulli[4], F.A. Danevich[2], S. d'Angelo[1], A. Incicchitti[3], V.V. Kobychev[2], B.N. Kropivyansky[2], M. Laubenstein[4], P.G. Nagornyi[5], S.S. Nagorny[2], S. Nisi[4], F. Nozzoli[1], D.V. Poda[2], D. Prosperi[3], O.G. Polischuk[2], V.I. Tretyak[2], I.M. Vyshnevskyi[2], S.S. Yurchenko[2]

[1]Dipartimento di Fisica, Università di Roma "Tor Vergata" and INFN, Sezione di Roma "Tor Vergata", I-00133, Roma, Italy
[2]Institute for Nuclear Research, MSP 03680 Kyiv, Ukraine
[3]Dipartimento di Fisica, Università di Roma "La Sapienza" and INFN, Sezione di Roma, I-00185, Roma, Italy
[4]INFN, Laboratori Nazionali del Gran Sasso, I-67010 Assergi (Aq), Italy
[5]Chemical Department, Kyiv National Taras Shevchenko University, 01033 Kyiv, Ukraine

E-mail: bernabei@roma2.infn.it

Abstract. Sample of $^{100}MoO_3$ with molybdenum enriched in ^{100}Mo to 99.5% and mass of 1199 g was measured deep underground (3600 m w.e.) in the Laboratori Nazionali del Gran Sasso of INFN (Italy) during 17249 h with a low-background set-up with 4 HP Ge detectors. After $2\beta2\nu$ decay of ^{100}Mo to the 0_1^+ excited level of ^{100}Ru (E_{exc} = 1131 keV), two γ quanta of 540 keV and 591 keV should be emitted in deexcitation process. Both these γ's are observed in the accumulated data as in coincidence spectrum as well in 1-dimensional sum spectrum. Measured half life is $T_{1/2} = (7.0^{+1.1}_{-0.8}) \times 10^{20}$ yr, in agreement with positive results obtained in previous experiments.

Neutrinoless (0ν) double beta (2β) decay is forbidden in the Standard Model (SM) due to violation of the lepton number by two units; however, it is predicted by many SM extensions where neutrinos are naturally expected to be Majorana particles ($\nu = \overline{\nu}$) with small but non-zero mass [1]. This process in not observed to-date, with positive observation reported recently only for ^{76}Ge [2]. In absence of experimental data on $2\beta0\nu$, observation of $2\beta2\nu$ decays is important tool to test theoretical models used for calculations of nuclear matrix elements for 2β processes.

Half lives for $2\beta2\nu$ decay of ^{100}Mo to the 0_1^+ excited level of ^{100}Ru (E_{exc} = 1131 keV) were measured in few experiments [3, 4, 5, 6] in the range of $(5.5-9.3)\times10^{20}$ yr. These results are in some contradiction with earlier paper [7] where only limit $T_{1/2} > 1.2\times10^{21}$ yr was obtained at 90% C.L. Aim of the present experiment was remeasurement of $\simeq$ 1 kg of Mo enriched in ^{100}Mo to 99.5%, used before in [7], to confirm observations [3, 4, 5, 6] or to set more severe $T_{1/2}$ limit.

Measurements were performed in the Laboratori Nazionali del Gran Sasso (LNGS) of INFN (Italy) in underground conditions on the depth of 3600 m w.e. If the 0_1^+ excited level of ^{100}Ru with E_{exc} = 1130.5 keV is populated, two γ quanta with energies of 590.8 keV and 539.6 keV will

[1] ARMONIA: meAsuReMent of twO NeutrIno 2β decAy of ^{100}Mo to the first excited 0_1^+ level of ^{100}Ru.

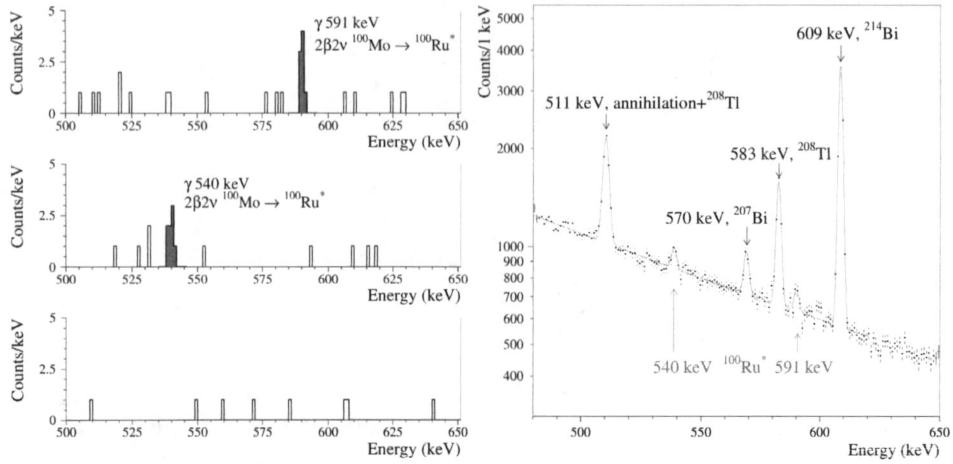

Figure 1. Left: Coincidence spectra accumulated with the $^{100}MoO_3$ sample during 17249 h with the 4 HP Ge set-up, when energy of one detector is fixed as 540 ± 2 keV (top), 591 ± 2 keV (middle) and 545 ± 2 keV (bottom). Right: Sum of spectra of 4 HP Ge detectors in the energy range of 480 – 650 keV.

be emitted in a cascade in subsequent deexcitation. To search for these γ quanta, we use set-up with 4 low-background HP Ge detectors mounted in one cryostat with a well in the center. The HP Ge detectors were of 225.2, 225.0, 225.0, and 220.7 cm^3 volume. Typical energy resolution (FWHM) of the detectors is 2.0 keV at the 1332 keV line of ^{60}Co. The set-up is enclosed in a lead and copper passive shielding and has a nitrogen ventilation system against radon.

Sample of $^{100}MoO_3$ with mass of 1199 g was measured during 17249 h. Data acquisition system gave possibility to take into account coincidence between individual HP Ge detectors. Fixing the energy of one of detectors to the expected energy of γ quanta emitted in $2\beta 2\nu$ decay to ^{100}Ru* (540 or 591 keV; width of window ± 2 keV is in accordance with the energy resolution of HP Ge at these energies), we observe coincidence peak at the corresponding supplemental energy (591 or 540 keV), see Fig. 1 (left). Bottom part of the figure shows background events, when energy window is shifted to the neighbouring value of 545 ± 2 keV. Eight events detected in coincidence correspond to half life $T_{1/2} = (6 \pm 2) \times 10^{20}$ yr for $2\beta 2\nu$ decay of $^{100}\text{Mo} \rightarrow {}^{100}\text{Ru}^*$.

Sum of the spectra of 4 HP Ge detectors is shown in Fig. 1 (right) in the energy range of 480 – 650 keV. Both peaks at 540 keV and 591 keV expected for $^{100}\text{Mo} \rightarrow {}^{100}\text{Ru}^*$ $2\beta 2\nu$ decay are observed in the experimental data. Fit of the spectrum, also shown in Fig. 1 (right), gives number of events in these peaks as 310 ± 54 and 255 ± 51, respectively. Together with efficiencies calculated with the GEANT4 and EGS4, which gave close results, it gives $T_{1/2} = (7.0^{+1.1}_{-0.8}) \times 10^{20}$ yr, consistent with $T_{1/2}$ derived from the coincidence spectrum.

Present observation is in agreement with positive observations [3, 4, 5, 6] and does not confirm negative result [7].

References

[1] Avignone III F T, Elliott S R, Engel J 2008 *Rev. Mod. Phys.* **80** 481
[2] Klapdor-Kleingrothaus H V, Krivosheina I V 2006 *Mod. Phys. Lett.* A **21** 1547
[3] Barabash A S et al. 1995 *Phys. Lett.* B **345** 408
[4] Barabash A S et al. 1999 *Phys. At. Nucl.* **62** 2039
[5] Arnold R et al. 2007 *Nucl. Phys.* A **781** 209
[6] Kidd M F et al. 2009 *Nucl. Phys.* A **821** 251
[7] Blum D et al. 1992 *Phys. Lett.* B **275** 506

AUTHOR INDEX

AUTHOR INDEX

AUTHOR INDEX

AUTHOR INDEX

AUTHOR INDEX

AUTHOR INDEX

AUTHOR INDEX